# METALLURGICAL COATINGS 1989

## VOLUME I

# METALLURGICAL COATINGS 1989

*Proceedings of the 16th International Conference on Metallurgical Coatings, San Diego, CA, U.S.A., April 17–21, 1989*

EDITOR: BRUCE D. SARTWELL

Volume I

Reprinted from the journal
*Surface and Coatings Technology*
Vol. 39/40, Nos. 1–3

ELSEVIER APPLIED SCIENCE
LONDON and NEW YORK

ELSEVIER APPLIED SCIENCE PUBLISHERS LTD
Crown House, Linton Road, Barking, Essex 1G11 8JU, England

*Sole Distributor in the USA and Canada*
ELSEVIER SCIENCE PUBLISHING CO., INC.
655 Avenue of the Americas,
New York, NY 10010, USA

**British Library Cataloguing in Publication Data**
International Conference on Metallurgical
Coatings (*16th: 1989: San Diego, Calif.*)
Metallurgical Coatings 1989: proceedings
of the 16th International Conference on
Metallurgical Coatings, San Diego, CA, USA.
April 17–21, 1989.
1. Metallic coatings
I. Title II. Sartwell, Bruce D.
III. Surface and coatings technology
IV. Thin solid films
667.9

ISBN 1–85166–803–9 (v.1)
ISBN 1–85166–804–7 (v.2)
ISBN 1–85166–805–5 (The set)

**Library of Congress CIP Data Applied For**



Printed in The Netherlands by Krips Repro B.V.

# Preface

The 16th International Conference on Metallurgical Coatings (ICMC) was held in San Diego, California, April 17–21, 1989. This meeting was sponsored by the Vacuum Metallurgy Division of the American Vacuum Society and was co-sponsored by the Thin Film Division and Electronic Materials and Processing Division of the American Vacuum Society and by the Vacuum Society of Japan. The 1989 conference was the largest in the history of ICMC, with three plenary, fifty invited, and two hundred and forty contributed papers being presented.

This Proceedings contains the 131 papers out of 175 submitted that were peer reviewed and accepted on the basis of technical merit and timeliness of submission. As with the 1987 and 1988 conferences, they are being published in one of two Elsevier Sequoia journals, *Surface and Coatings Technology* and *Thin Solid Films*. For 1989, the papers from the symposia on Coatings For Use at High Temperatures, Hard Coatings, and Industrial Equipment and Applications, plus those from the Special Topics sessions on Coatings for Space Applications and Coatings for Corrosive Aqueous Environments, are being published in *Surface and Coatings Technology*. The papers from the symposia on Thin Films for Microelectronic, Superconducting, and Information Storage Applications, Methods of Characterizing Coatings and Modified Surfaces, and Tribological Coatings/Surface Modifications, plus those from the Special Topics session on Optical Coatings: Fabrication and Characterization, are being published in *Thin Solid Films*.

On a subsequent page are listed the winners of the Bunshah Awards, which were established to recognize authors of outstanding papers from each symposium.

The organization of a conference and assembly of a proceedings of this size and magnitude obviously require the efforts of a considerable number of people. The Executive, Program, and International Advisory Committees for ICMC89 are recognized in the following pages, and their efforts are hereby gratefully acknowledged. In addition, the generous support of the firms exhibiting at the conference, also listed separately, is acknowledged.

Specifically related to the Proceedings, the editor would like to thank all those individuals who diligently served as referees of the conference papers and to several of the symposium chairmen who assisted in the review process. These are Robert C. Tucker, Jr., Andrew R. Nicoll, Anthony J. Perry, Angus Rockett, Federico O. Sequeda, Paul H. Holloway, Gregory J. Exarhos, Manfred Kaminsky, and Tali Spalvins. Finally, the editor would like to thank Mary Gray for her excellent work in creating the manuscript database and distributing papers to referees.

BRUCE D. SARTWELL, PROCEEDINGS EDITOR
Washington, D.C., August 1989

# Publication schedule

Volume I

Coatings to resist high temperature corrosion
Coatings to resist wear at high temperature
Thermal barrier coatings
Superhard coatings and emerging technologies
Fundamentals of hard coatings
Hard coating properties
Applications of hard coatings
Hard coating processes
Coating materials, applications and economics
Industrial coating equipment, decorative and foil coil systems
Surface modification techniques and coating enhancements
Coatings for space applications
Coatings for corrosive aqueous environments

Author and subject indexes for Volume I

Volume II

Fundamental studies of surfaces and interfaces in VLSI materials
Materials applications in VLSI technology
Magnetic, magneto-optic and phase-change recording media
Synthesis and characterization of high $T_c$ superconductors
Compound semi-conductor, heterostructure and superlattice thin films
Thin film materials for integrated sensors
Surface and thin film analysis techniques
Microstructural characterization techniques
Mechanical properties of films and coatings
Adhesion measurements of films and coatings
Reduction of friction through coatings/surface modifications
Wear resistance of ceramic, alloy and composite coatings
Analysis and micromechanical testing of tribological coatings
Optical coatings: fabrication and characterization

Author and subject indexes for Volume II

## AWARD WINNING PAPERS

As an incentive for the submission and presentation of papers of the highest quality, the R. F. Bunshah Award (named in honor of the founder of the International Conference on Metallurgical Coatings) is given each year to the authors of the best paper in each of several symposia. The selection is based on the quality of both the oral presentation and the proceedings manuscript. The author(s) of each winning paper receive a medallion and a monetary award.

The following are the winning papers from ICMC89:

1 High-temperature corrosion of alumina-forming coatings for superalloys, G. H. Meier and F. S. Pettit, University of Pittsburgh. Paper appears in Volume I, pages 1–17.

2 Texture and some properties of vapor deposited diamond films, Y. Sato and M. Kamo, National Institute for Research in Inorganic Materials. Paper appears in Volume I, pages 183–198.

3 Surface and interface analysis for the development of VLSI devices, M. Grasserbauer, Technical University of Vienna. Paper appears in Volume II, pages 17–34.

4 Some mechanics for the adhesion of thin films, M. D. Thouless, IBM Research Division, T. J. Watson Research Center. Paper appears in Volume II, pages 397–406.

# Contents

COMMITTEES

## EXECUTIVE COMMITTEE

*General Chairman and Proceedings Editor*
Bruce D. Sartwell
Naval Research Laboratory

*Program Chairman*
Gary E. McGuire
Microelectronics Center of North Carolina

*Local Arrangements Chairman*
Phyllis Greene
Champaign, IL

*Exhibits Chairman*
Paul F. Woerner
GTE Valenite Corporation

*Publicity Chairman*
Gary L. Parr
Research & Development

*Treasurer*
D. C. Carmichael
Battelle Columbus Labs.

*Past Chairman*
William D. Sproul
Northwestern University

*Founder & Past Chairman*
R. F. Bunshah
University of California

PROGRAM COMMITTEES

## SYMPOSIUM A: COATINGS FOR USE AT HIGH TEMPERATURES

*Chairman*
Robert C. Tucker, Jr.
Union Carbide Corporation

*Chairman*
Andrew R. Nicoll
Plasma-Technik AG

*Committee*
William Goward
Turbine Components Corp.

T. Rhys-Jones
Rolls Royce plc

Fred S. Pettit
University of Pittsburgh

Harry E. Eaton
United Technologies Research Center

L. L. Hsu
Solar Turbines Inc.

Dinesh K. Gupta
Pratt & Whitney Group

Thomas A. Taylor
Union Carbide Corporation

G. Johner
Leybold AG

David J. Wortman
General Electric Corp.

## SYMPOSIUM B: HARD COATINGS

*Chairman*
Anthony J. Perry
GTE Valenite Corporation

*Committee*
Aharon Inspektor
N.R.C. Negev

Russell Messier
Penn State University

Sokichi Takatsu
Toshiba Tungaloy Co. Ltd

Helmut W. Holleck
Universität und
Kernforschungszentrum

Allan Matthews
University of Hull

Craig Emsberger
CTS Corporation

David Rickerby
Harwell Laboratory

Hermann A. Jehn
Forschunginstitut für
Edelmetalle
und Metallchemie

Dale McIntyre
University of Illinois

Antti S. Korhonen
Helsinki University of
Technology

H. Randhawa
Vac-Tec Systems Inc.

## SYMPOSIUM C: THIN FILMS FOR MICROELECTRONIC SUPERCONDUCTING AND INFORMATION STORAGE APPLICATIONS

*Chairman*
Angus Rockett
University of Illinois

*Chairman*
Federico O. Sequeda
IBM Almaden Research
Laboratory

*Committee*
Siegfried Hofmann
Max-Plank Institut für
Metallforschung

Frank Bresnock
United Technologies

Wen Y. Lee
IBM Almaden Research
Laboratory

Scott Barnett
Northwestern University

Ken D. Wise
University of Michigan

## SYMPOSIUM D: METHODS OF CHARACTERIZING COATINGS AND MODIFIED SURFACES

*Chairman*
Paul H. Holloway
University of Florida

*Chairman*
Gregory J. Exarhos
Battelle Pacific Northwest Laboratories

*Committee*
Gerald Nelson
Sandia National Laboratories

Michael S. Donley
Air Force Wright Aeronautical Laboratories

Ting Huang
IBM Almaden Research Laboratory

T. T. Sheng
ATT Bell Laboratories

Warren C. Oliver
Oak Ridge National Laboratory

George M. Pharr
Rice University

Siegfried Hofmann
Max-Plank Institut für Metallforschung

Wim van Ooij
Colorado School of Mines

## SYMPOSIUM E: TRIBOLOGICAL COATINGS/SURFACE MODIFICATIONS

*Chairman*
Manfred Kaminsky
Surface Treatment Science, International

*Chairman*
Tali Spalvins
NASA Lewis Research Center

*Committee*
Paul D. Fleischauer
The Aerospace Corporation

D. G. Teer
D. G. Teer Coating Services, Ltd.

Roger N. Johnson
Hanford Engineering Development Lab.

Traugott E. Fischer
Stevens Institute of Technology

Hans Hintermann
Centre Suisse d'Electronique et de Microtechnique S.A.

## SYMPOSIUM F: INDUSTRIAL EQUIPMENT AND APPLICATIONS

*Chairman*
Sundarajan Mutialu
Aultimut International Inc.

*Committee*
R. F. Bunshah
University of California

W. D. Münz
Leybold GmbH

Doug Chambers
XYTORR Corporation

Keith Legg
Ionic Atlanta Incorporated

Piran Sioshansi
Spire Corporation

## SYMPOSIUM S: SPECIAL TOPICS
## (1) COATINGS FOR SPACE APPLICATIONS
## (2) OPTICAL COATINGS: FABRICATION AND CHARACTERIZATION
## (3) COATINGS FOR CORROSIVE AQUEOUS ENVIRONMENTS
## (4) PLASMA PROCESSING

*Committee*
(1) Bruce A. Banks
NASA Lewis Research Center

(1) Donald J. McClure
3M Corporation

(2) Bryant Hichwa
OCLI

(2) Donald M. Friedrich
Battelle, Pacific Northwest Laboratories

(3) David R. Flinn
Tuscaloosa Research Center

(4) Manfred Kaminsky
Surface Treatment Science International

(4) Steve Rossnagel
IBM T. J. Watson Research Center

## INTERNATIONAL ADVISORY BOARD

K. L. Chopra
Indian Institute of Technology

C. Hayashi
ULVAC Corporation

O. Knotek
Technical University Aachen

S. Schiller
Forschungsinstitut Manfred von Ardenne

## EXHIBITORS

Applied Science and Technology, Inc.
Commonwealth Scientific Co.
Corona Vacuum Coaters, Inc.
Denton Vacuum, Inc.
Electro-Plasma
Granville-Phillips Company
Hauzer Techno Coating
Huntington Labs.
ISM Technologies
Implant Sciences Corporation
Innotec Group, Inc.
Interatom GmbH
Ion Tech, Inc.
Kurt J. Lesker Co.
Leybold Inficon, Inc.
Leybold Technologies, Inc.
MKS Instruments, Inc.
Microscience, Inc.
Midwest Tungsten Services
Miller Thermal, Inc.
Perkin Elmer Corporation
Plasmaterials, Inc.
RF Plasma Products, Inc.
Surface Science Laboratories
Techware Systems
Tecvac Ltd.
US, Inc.
Vac-Tech Systems, Inc.
Vacuum, Inc.
Vapor Technologies, Inc.
Wavemat, Inc.

# HIGH-TEMPERATURE CORROSION OF ALUMINA-FORMING COATINGS FOR SUPERALLOYS*

G. H. MEIER and F. S. PETTIT

*Department of Materials Science and Engineering, University of Pittsburgh, Pittsburgh, PA 15261 (U.S.A.)*

(Received March 10, 1989)

## Summary

The results of recent experiments on the cyclic oxidation and hot corrosion of diffusion aluminide and CoCrAlY overlay coatings on nickel-base superalloys are described. It is shown that platinum additions to aluminide coatings produce a substantial increase in resistance to cyclic oxidation and high temperature (Type 1) hot corrosion but have a rather small effect on resistance to low-temperature (Type 2) hot corrosion. The corrosion of aluminide coatings is also shown to be influenced by the composition of the underlying superalloy. Mechanisms for the above behavior are proposed. The effects of yttrium and hafnium on the cyclic oxidation of CoCrAl are illustrated. The results of scanning and transmission electron microscope characterization of the hot corrosion morphologies of CoCrAlY coatings are discussed and a mechanism to account for the microstructures is briefly described.

## 1. Introduction

Gas turbine engines are widely used in propulsion systems and in power generating equipment. Materials used in gas turbines may be subjected to a range of environmental conditions depending upon their location in the engine and the operating characteristics of the unit. Materials used for the high-pressure turbine are exposed to rather severe conditions because temperatures are high and salt deposits can accumulate on the surfaces of hardware. High-pressure turbine materials can be exposed to temperatures in the range 600 - 1100 °C where the upper limit is encountered only intermittently. The formation of deposits on hardware is determined by the composition of the gas and the temperature of the hardware. Depending on the purity of the fuel and ingested air and temperature, deposits such as $Na_2SO_4$ can accumulate on hardware surfaces at metal temperatures in the range from 600 to about 1000 °C. Some fuels can contain vanadium which can lead to the presence of vanadium compounds, such as $NaVO_3$, in the deposits.

*Recipient of the Bunshah Award for the best paper in its Symposium.

0257-8972/89/$3.50 

The severe environments encountered in turbines generally require that the superalloy components be protected by coatings for corrosion protection. The coatings fall into two categories: diffusion coatings and overlay coatings. The most widely used diffusion coatings are the aluminides while typical overlay coatings are physically vapor deposited (PVD) CoCrAlY coatings. The aluminide and overlay coatings are subject to several forms of degradation. In high-temperature applications (*e.g.* aircraft engines) the coatings undergo isothermal and cyclic oxidation and interdiffusion with the superalloy substrate. At lower temperatures the coatings are subject to degradation which is accelerated by the presence of surface deposits such as sulfates (hot corrosion).

An important aspect of minimizing the effects of the above-mentioned form of degradation involves understanding the mechanisms of degradation. The development of this understanding involves the ability to reproduce the corrosion morphologies found in service by conducting well-controlled laboratory experiments. This paper presents the results of experiments used to ascertain the pertinent variables which reproduce the corrosion morphologies observed in service.

## 2. Background

### *2.1. Diffusion aluminide coatings*

Pack cementation is a very common CVD process used to produce diffusion aluminide coatings. In this process a part is immersed in a pack mix and heated in a retort purged with an inert gas. The pack mix includes an aluminum source, an inert filler and an activator to accelerate aluminum transfer from the source to the substrate. The final microstructure depends on the temperature of the pack, the pack mix and any pretreatment of the part to be coated. The coating microstructures that result from the pack process have been divided by Goward and Boone into two types [1, 2], those produced by a high aluminum activity source and those produced by a low activity source. Often these types are referred to as the low-temperature and high-temperature types respectively. This coating classification was established using Jannsen and Rieck's work on diffusion in nickel aluminides [3]. Their work showed that only aluminum atoms are mobile in the aluminum-rich aluminides and nickel atoms are the only mobile species in aluminum-deficient aluminides. Therefore, high aluminum activity coatings grow primarily inward and low aluminum activity coatings grow outward.

Aluminide coatings are degraded via reaction with the environment (corrosion) and by interdiffusion with the substrate. Both processes result in depletion of aluminum from the coating. A significant advance in the technology of diffusion aluminides was made when other elements such as chromium, tantalum, silicon and most notably platinum were incorporated into aluminide coatings [4,5]. The platinum modification to normal diffiusion aluminides results in a much more degradation resistant coating [5 - 11].

Platinum modified aluminide coatings are used commerically in the hot section of turbine engines [11].

Various processing schemes have been developed to add platinum to diffusion aluminide coatings each of which produces coatings with different microstructures. In most of these processes platinum is electroplated as a thin layer (about 3 - 10 $\mu$m) on the surface prior to aluminizing. The platinum aluminides contain either continuous $PtAl_2$, discontinuous $PtAl_2$ or platinum in solution in NiAl in the outer regions of the coating.

### *2.2. CoCrAlY overlay coatings*

CoCrAlY coatings are widely used in the hot section of gas turbines. The coatings can be deposited by electron beam physical vapor deposition (EB-PVD), chemical vapor deposition, or plasma spraying with the EB-PVD technique being the most widely used [12]. The phases present in CoCrAlY are $\alpha$- and $\varepsilon$-Co, $\beta$-CoAl, cobalt yttride and yttrium oxide. The $\alpha$- and $\varepsilon$-phases are face-centered cubic (f.c.c.) and hexagonal close-packed (h.c.p.) solid solutions [13] respectively, $\beta$ is a CsCl-structure intermetallic and the yttride is presumably $Co_{17}Y_2$ [14].

The CoCrAlY overlay coatings are quite resistant to isothermal oxidation. However, they can be degraded during long term cyclic exposures as aluminum is depleted from the coating. The presence of yttrium in the coatings improves the scale adherence and, therefore, the cyclic oxidation resistance. However, the detailed mechanism for this effect has not been totally explained.

Coating degradation can also be severe in the presence of liquid sulfate deposits [15]. While a great variety of conditions can be used to induce deposit-affected corrosion, liquid sulfate deposits at temperatures between 600 and 1000 °C in gases containing oxygen and $SO_3$ provide degradation characteristics representative of operational gas turbines. Two types of hot corrosion attack can be observed, low-temperature (Type 2) hot corrosion at about 600 - 800 °C, and high-temperature (Type 1) hot corrosion at higher temperatures. The low-temperature form of hot corrosion, which produces a characteristics pitting morphology, is particularly severe for CoCrAlY coatings. Several mechanisms [15 - 18] have been proposed for low-temperature hot corrosion but there is no agreement on these mechanisms. Deposition of salts which contain oxides of vanadium or vanadates have also been shown to cause severe corrosion of metallic coatings in the temperature range 600 - 1000 °C [19, 20].

## 3. Experimental procedure

The alloy substrates investigated were Mar M200, IN738, Rene 80, PWA 1480 and CMSX-3. PWA 1480 and CMSX-3 are single crystal superalloys whereas the others are polycrystalline alloys frequently used in the hot section of gas turbines. Two state-of-the-art pack cementation coating processes are used for comparison, namely PWA 73 and CoDep. The platinum-modified coatings were prepared by Turbine Components Inc. Coupons of the alloys (12 mm × 10 mm × 1.5 mm) were grit blasted with alumina particles prior to the deposition of a 3 - 7 $\mu$m thick electroplated platinum layer. In order to deposit the platinum layer a thin platinum wire was welded to each coupon to provide an electrical lead. After electroplating the coupons were exposed to an out-of-the-pack aluminizing process. This process produced an aluminide coating of the discontinuous $PtAl_2$ type. A limited number of platinum aluminide coatings prepared by Howmet were also studied. The CoCrAlY coating studied was an EB-PVD coating of nominal composition Co–24Cr–10.5Al–0.3Y(wt.%) which was deposited on an IN738 substrate by Airco Temescal Corporation. Cast specimens of Co–20Cr–10Al–0.2Y and Co–20Cr–10Al–lHf were also prepared and exposed in some tests to compare with the coated specimens. These alloys were made in a drop caster under an argon atmosphere. The ingots were subsequently annealed in argon for 100 h at 1100 °C to promote homogeneity.

Isothermal oxidation tests were performed in dry air at 1100 °C using a Cahn 2000 microbalance. Cyclic oxidation tests were conducted in dry air at 1100, 1135 and 1200 °C. These tests were conducted using an automatic apparatus which inserted the specimens into the hot zone of a furnace for a period of 45 min then removed them to the cold zone for a period of 15 min.

Acoustic emission tests were used to assess oxide scale damage that occurred on a sample during exposures at high temperature. The specimens were suspended from a platinum waveguide. The acoustic emission signals were analyzed by a Dunegan Endevco distribution analyzer. The tests were run such that the acoustic emission signals were taken during isothermal oxidation for 24 h in air at 1100 °C and during the 15 min cool down of the samples.

Diffusions anneals were carried out on the aluminide-coated samples in order to observe diffusional interactions between the coating and the substrate. The experiments were carried out in argon in a tube furnace at 1200 °C. At periodic intervals the samples were removed and 10 mm × 2.5 mm × 2 mm sections were cut with a low speed saw. Data were obtained for exposures up to 265 h.

High-temperature hot corrosion tests were performed using the cyclic oxidation apparatus described previously but with 1 mg $cm^{-2}$ of $Na_2SO_4$ applied once every 20 h. These specimens were cycled from 1000 °C to room temperature once every hour.

In the case of low-temperature hot corrosion it is necessary to have $SO_3$ present in the gas phase and to apply deposits which are liquid at the test temperature of 700 °C. Low-temperature hot corrosion tests were performed by applying deposits of $Na_2SO_4$–50 mol% $NiSO_4$ to the aluminides or 50% $CoSO_4$ to the CoCrAlY and using gases with an initial room temperature composition of $O_2$ with 100 ppm $SO_2$. These gas mixtures were passed over a platinum catalyst which established a $SO_3$ pressure at 700 °C of about $10^{-4}$ atm. These conditions produced degradation microstructures typical of low-temperature hot corrosion.

Oxidation and corrosion morphologies were evaluated by optical and scanning electron microscope (SEM) analysis and phase identification was performed using X-ray diffraction (XRD) using Cu $K\alpha$ radiation. The hot corrosion morphologies of the CoCrAlY coatings were further evaluated by transmission electron microscope (TEM) analysis.

## 4. Results and Discussion

### *4.1. Diffusion aluminides*

Micrographs showing the microstructures of typical as-processed coatings are presented in Fig. 1. The coatings are a conventional aluminide (PWA 73) and a platinum aluminide on PWA 1480. The light discontinuous phase evident at the surface of the platinum-modified coating is $PtAl_2$ which lies in a matrix of (Pt,Ni)Al. Alpha-refractory metal precipitates can be observed beneath the $PtAl_2$ layer. The PWA 73 coating has a typical three-zone structure.

Isothermal oxidation tests were performed on a number of the coating systems at 1100 °C in air. The platinum-modified coatings usually had smaller

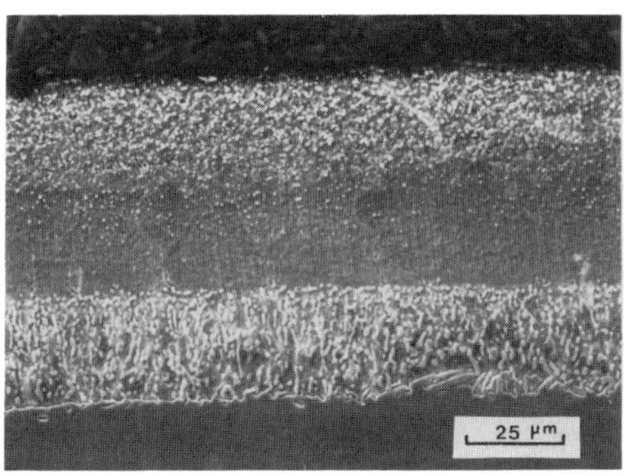


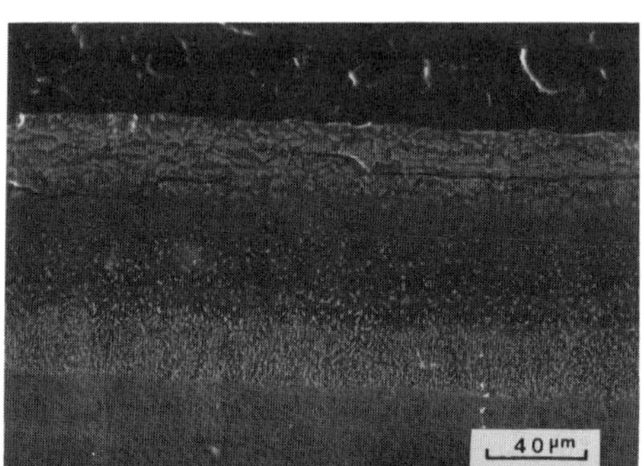


Fig. 1. Cross-sections of as-prepared coatings on a PWA 1480 substrate. The top micrograph is PWA 73 and the bottom one is a platinum aluminide.

weight increases than the conventional aluminide coatings; however, some of the conventional aluminides also exhibited very small weight increases.

Typical cyclic oxidation weight change *vs.* time measurements are presented in Fig. 2 for different coating systems. In all of these tests the platinum-modified coatings performed better than the conventional aluminides. In particular, protective alumina scales were maintained for longer periods of time on the platinum modified coatings compared to the conventional coatings. Some platinum modified coatings exhibited lives greater than others because substrate elements influenced the coating lives. Hafnium-containing substrates such as CMSX-3 generally resulted in longer lives. The surface observations of the scales indicated that a purer alumina grew on the platinum aluminides than the straight aluminides but the amount of scale spalling from both coating types was similar. Therefore, it does not appear that the beneficial effects of platinum are associated with improved scale adherence. Figure 3 shows cross-sections of MAR M200 after cyclic oxidation at 1100 °C. The conventional aluminide shows no $\beta$ phase remaining whereas

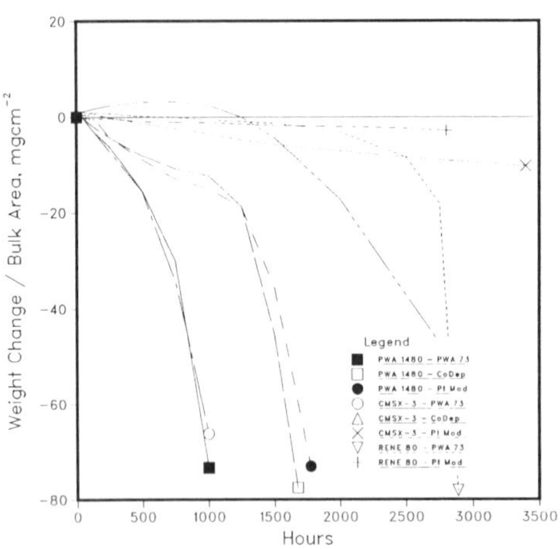


Fig. 2. Cyclic oxidation weight change *vs.* time data at 1100 °C for a number of aluminide coating–substrate combinations.

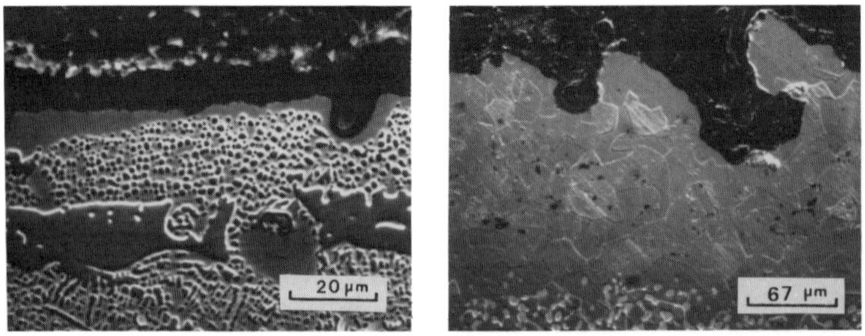


Fig. 3. Cross-sections of aluminide coatings on MAR M200 after cyclic oxidation in air at 1100 °C. The left micrograph is PWA 73 and the right one is a platinum aluminide.

some $\beta$ phase is still evident on the platinum modified coating. An improvement in diffusional stability of the platinum coatings was also apparent during high-temperature annealing. Figure 4 compares the microstructures of a conventional aluminide and a platinum aluminide after annealing for 19 h in argon at 1200 °C. The coatings have lost aluminum by interdiffusion with the substrate and by oxide formation caused by the oxygen impurities in the argon. The platinum coating still consists entirely of the $\beta$ phase while the conventional aluminide has degraded to the lower aluminide $\gamma'$ and even a layer of $\gamma$ solid solution at the surface. Only a small portion of $\beta$ remains in the center of the coating. Thus, the major effect of the platinum is to produce a large initial reservoir of aluminum and to reduce the amount of aluminum lost by diffusional processes.

Results obtained from acoustic emission tests with several coating systems are presented in Fig. 5 and are consistent with the conclusion that

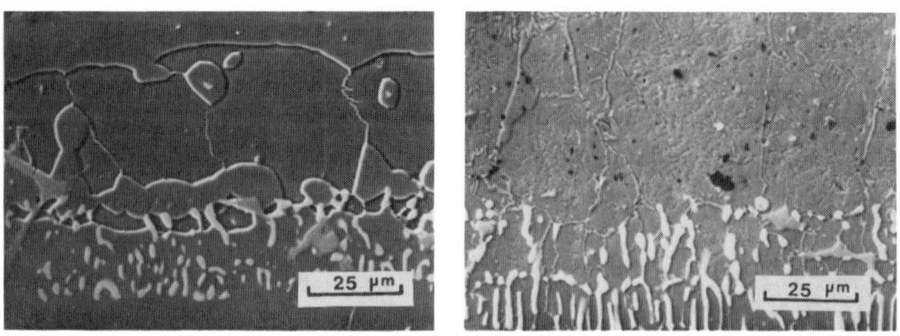


Fig. 4. SEM micrographs of PWA 73 on MAR M200 (left) and a platinum-modified aluminide on MAR M200 (right) after diffusion annealing in argon at 1200 °C for 19 h.

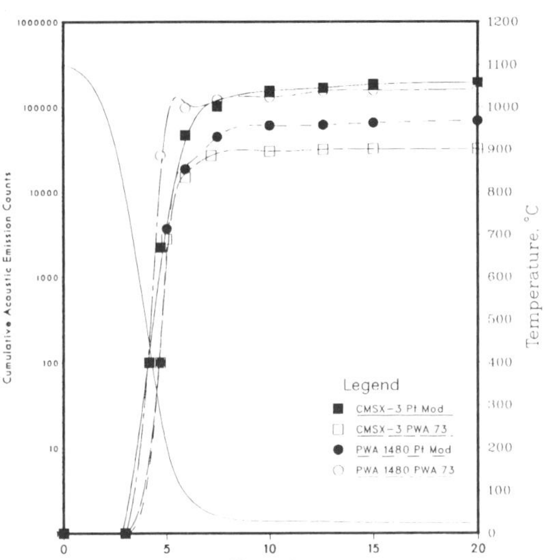


Fig. 5. Cumulative acoustic emission counts taken during cooling of several coating–substrate combinations after oxidation at 1100 °C in air. (Temperature profile is also plotted.)

the platinum modified coatings do not exhibit substantially better oxide scale adherence than the conventional aluminide coatings. Therefore, it is proposed that platinum improves the ability of the coatings to reform alumina scales after spalling by maintaining a large aluminum reservoir.

The various aluminized alloys were exposed to high-temperature hot corrosion conditions which consisted of coating with $Na_2SO_4$ deposits and exposing to air at 1000 °C under thermal cycling. Results obtained from these tests are presented in Fig. 6. The band along the abscissa, labeled "Pt mod. Group" represents data from several platinum modified coatings on PWA 1480, CMSX-3 and Rene 80. These data indicate that the platinum-modified coatings are generally better than coatings that do not contain platinum. Typical microstructures of specimens exposed to cyclic hot corrosion conditions are presented in Fig. 7. The amount of degradation is much more extensive in the coating that does not contain platinum. The data presented in Fig. 6 show that there also is an effect of substrate composition. Coatings generally have shorter lives on Rene 80 compared to CMSX-3 which in turn are not as good as coatings on PWA 1480. The significant difference between these substrates is the amount and type of refractory elements. Rene 80 contains both molybdenum and tungsten, CMSX-3 contains mainly tungsten and PWA 1480 contains predominantly tantalum. In the hot-corrosion literature [15] it is well documented that refractory elements such as molybdenum and tungsten can have a profound effect upon hot corrosion behavior. The

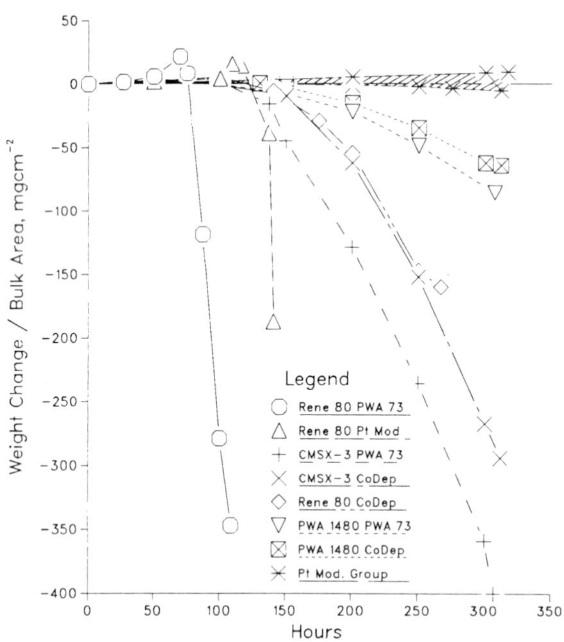


Fig. 6. Cyclic high temperature hot corrosion kinetics for various systems at 1000 °C with 1 mg cm$^{-2}$ $Na_2SO_4$ deposits.

Fig. 7. Cross-sections of coated CMSX-3 after 300 h of cyclic hot corrosion in air at 1000 °C. The top micrograph is platinum modified while the bottom one is PWA 73.

results obtained for the high-temperature hot corrosion tests show that platinum extends the lives of diffusion aluminide coatings, but the lives also depend on the composition of the superalloy substrate. These results are interpreted in terms of the platinum additions excluding refractory metals from the outer regions of the coatings and in improving the ability to reform alumina scales under cyclic conditions.

The weight change data for coating systems exposed to low-temperature hot corrosion conditions are presented in Fig. 8. The weight changes are not as large as those of the high-temperature hot corrosion test but are still very much larger than would have been observed in the absence of a deposit and without any $SO_3$ in the gas phase. Examination of these results show that the weight changes for coatings which contain platinum are usually slightly less than those which do not contain any platinum. This observation is also supported by metallographic examination of the exposed specimens. As can be seen in Fig. 9, less attack of the coating on CMSX-3 with platinum has occurred than of the coating with no platinum. In many cases, however, the differences in morphology between platinum and conventional aluminides was not as marked as that in Fig. 9. No obvious influence of substrate composition was evident in the low-temperature hot corrosion tests.

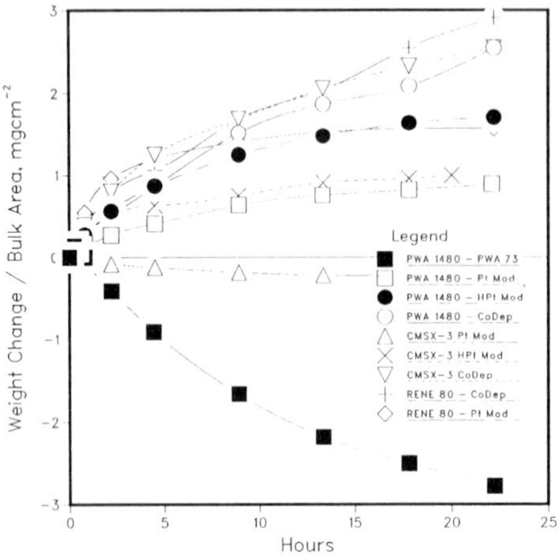


Fig. 8. Low-temperature hot corrosion kinetics at 700 °C and $10^{-4}$ atm $SO_3$ for various systems.

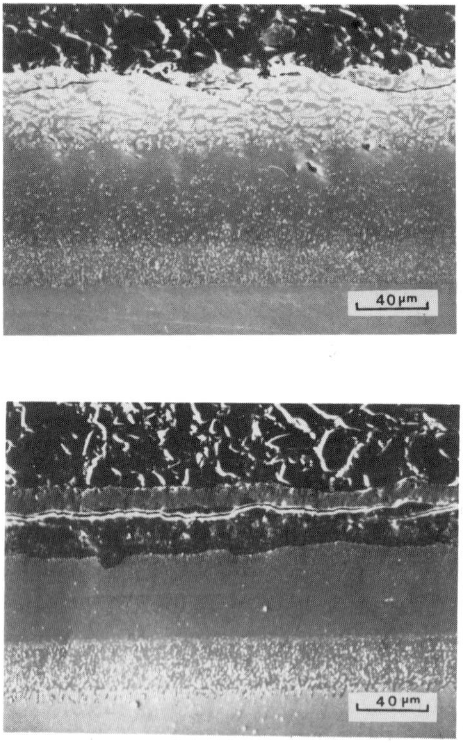


Fig. 9. Cross-sections of CMSX-3 after exposure for 24 h to low-temperature hot corrosion conditions at 700 °C. The top micrograph is platinum modified and the bottom is PWA 73.

*4.2. CoCrAlY overlay coatings*

Figure 10 shows the cyclic oxidation behavior of several cast MCrAl alloys with additions of yttrium or hafnium [21]. There is a substantial difference in behavior from one alloy to the next with CoCrAl showing less cyclic degradation than NiCrAl. All of the alloys show much less degradation than the corresponding alloys without an addition of yttrium or hafnium (data not plotted). The models which propose to explain the effects of reactive element additions for alumina scales are [22]

(1) Oxide pegging — scale is mechanically keyed by oxide intrusions.
(2) Vacancy sink — vacancies are prevented from condensing to form interfacial voids.
(3) Growth stress — scale growth mechanisms altered to reduce growth stresses.
(4) Scale plasticity — scales are able to accommodate stresses by creep.
(5) Chemical bond — oxygen active element directly strengthens the scale–alloy interface or "getters" elements such as sulfur which would otherwise segregate to the interface and weaken it.

Each of the models can explain some aspects of the active-element effect but there are also observations which contradict each, *e.g.* adherence without pegs, spalling in the presence of pegs, spalling without voids, adherent buckled scales etc. The "sulfur gettering" mechanisms, originally proposed by Smeggil and coworkers [23], has been able to explain the behavior of NiCrAl alloys. Additional support for this model comes from experiments where adherent scales are produced for NiCrAl alloys by simply reducing the sulfur levels or preparing ultrapure alloys [24]. Other investigators have concluded that there is no correlation between sulfur segregation and adherence [25]. The acoustic emission data for the CoCrAl alloys on cooling after oxidation are presented in Fig. 11. There is clearly less scale damage for the

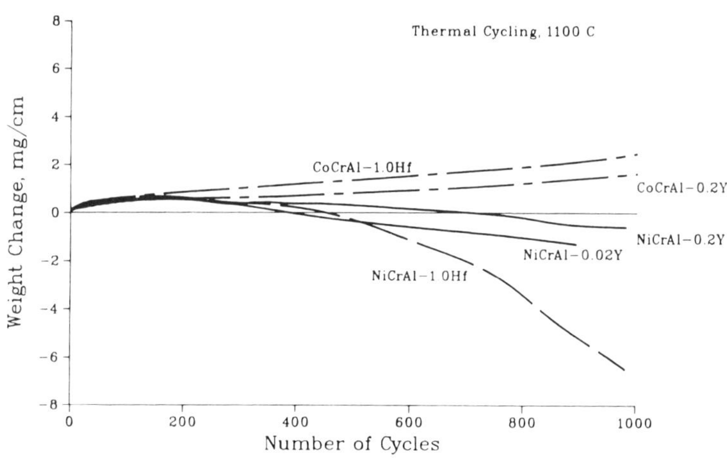


Fig. 10. Weight change *vs.* time for several MCrAl alloys exposed to cyclic conditions in air at 1100 °C.

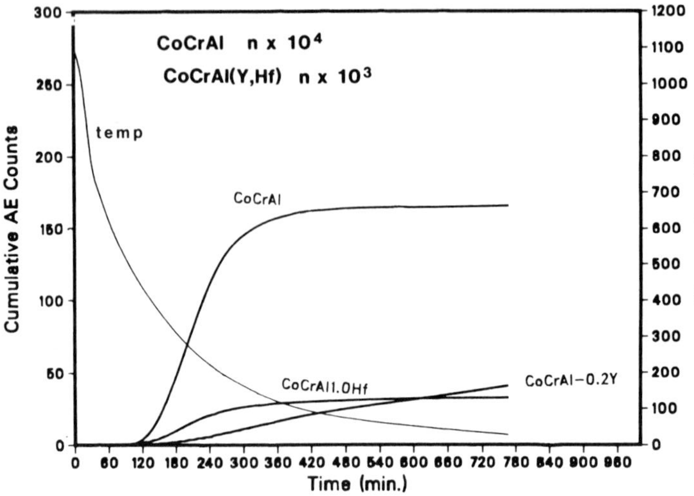


Fig. 11. Cumulative acoustic emission counts *vs.* time for several CoCrAl alloys during cooling after oxidation in air at 1100 °C.

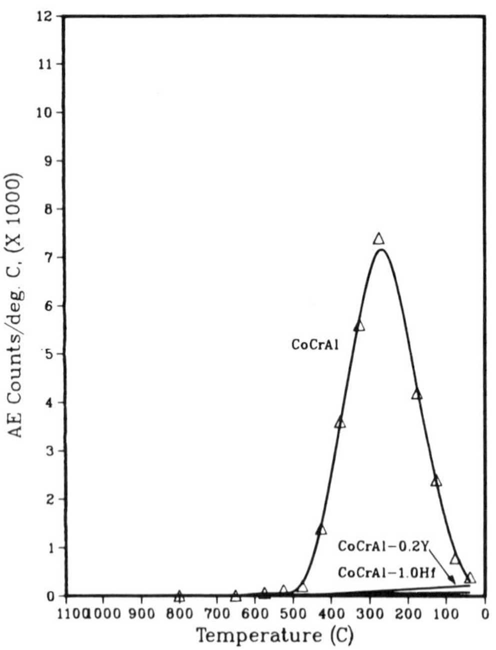


Fig. 12. Acoustic emission data from Fig. 11 plotted as counts $^{\circ}C^{-1}$ during cooling.

alloys with yttrium or hafnium than for the undoped alloys. Figure 12 shows the acoustic emission as a function of temperature as the alloys are cooled. The emission response for CoCrAl starts at a high temperature, goes through a maximum of about 300 °C, and then decreases. The emission from the alloys with yttrium and hafnium starts at about 400 °C and continues to increase until room temperature is reached. Thus, the scale damage for the undoped alloys occurs at higher temperature and is much more extensive than that for the alloys with yttrium and hafnium. This behavior has been interpreted as resulting from a stronger alloy–scale interface in the presence of the yttrium and hafnium [21]. However, these data do not allow a determination as to how the yttrium or hafnium strengthens the interface.

The exposure of CoCrAlY overlay coatings to the low-temperature hot corrosion conditions described above, *e.g.* in marine turbines, is known to cause even more severe degradation than is experienced by the aluminide coatings. The typical morphology is a pitting type as illustrated in Fig. 13

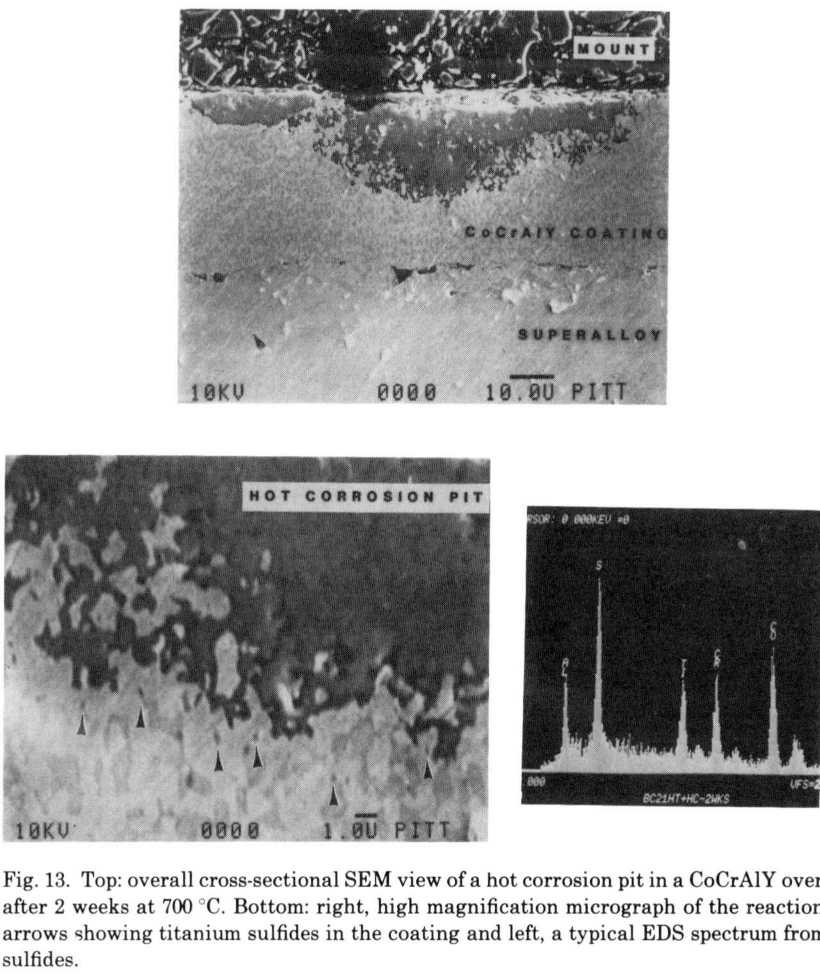


Fig. 13. Top: overall cross-sectional SEM view of a hot corrosion pit in a CoCrAlY overlay coating after 2 weeks at 700 °C. Bottom: right, high magnification micrograph of the reaction front with arrows showing titanium sulfides in the coating and left, a typical EDS spectrum from one of the sulfides.

(top). Higher magnification SEM observation of the corrosion front (Fig. 13 (bottom left)) indicates some preferential attack of the aluminum-rich $\beta$-CoAl phase and the precipitation of sulfides ahead of the corrosion front. EDS analysis of these sulfides shows them to be titanium sulfides. The titanium has diffused into the coating from the IN738 substrate. A combination of SEM, cross-sectional TEM and selected area electron diffraction (Fig. 14) indicates that the bulk of the pit contains $Cr_2O_3$ and a mixed spinel of cobalt, chromium and aluminum. A combination of bright-field and dark-field transmission electron microscopy (Fig. 15) indicated that the metal islands in the lower part of the pit were virtually pure cobalt. The combination of the above morphological observations with thermodynamic and kinetic analyses, which have been eliminated for the sake of brevity, have led to a model for this form of degradation. The aluminum is removed preferentially from the coating and reprecipitates a short distance away from the surface in spinels. This prevents a protective alumina scale from forming and allows the chromium and cobalt to react. Importantly, the cobalt is seen to be the last element to react (*e.g.* Fig. 15). Some of the cobalt is reacted to spinels in the pit while the

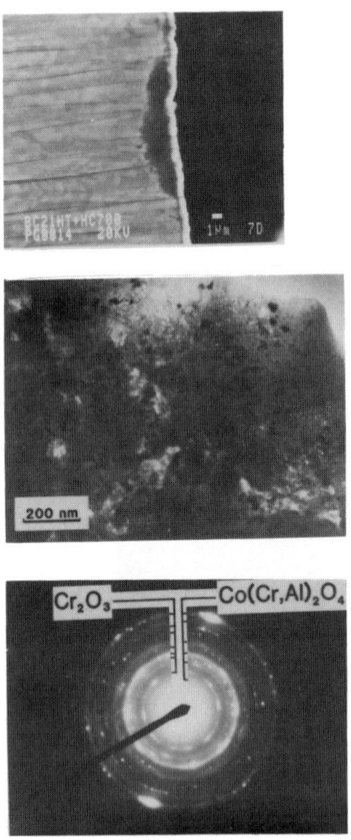


Fig. 14. Top: SEM image of a typical low-temperature hot corrosion pit. Center: TEM image of the corrosion product in the center of the pit. Bottom: selected area electron diffraction from the reaction product in the pit.

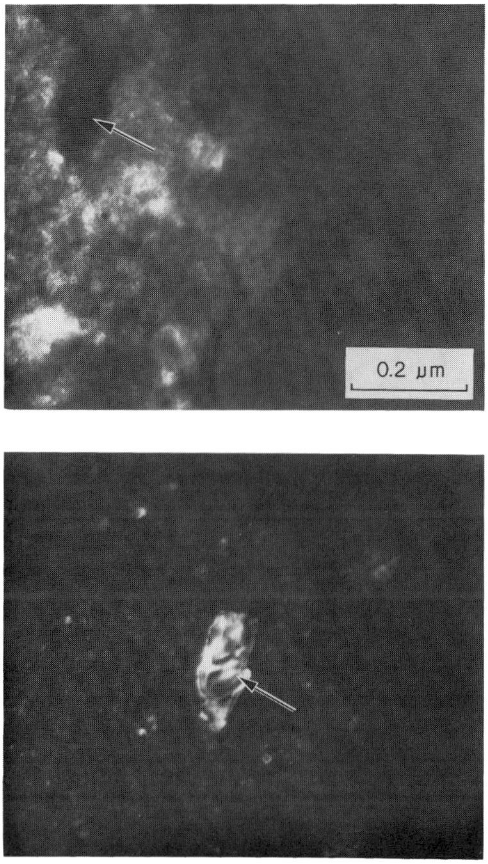


Fig. 15. Top: TEM image of the corrosion product at the base of a low-temperature hot corrosion pit showing a metallic particle at arrow. Bottom: dark-field image of the particle obtained using a f.c.c. cobalt reflection.

remainder is transported to the salt–gas interface, probably by the $Co^{2+}$–$Co^{3+}$ mechanism proposed by Luthra [16], where it is converted to cobalt oxide.

## 5. Summary

(1) Platinum-modified coatings possess improved cyclic oxidation resistance because the platinum-enriched zone of the coating promotes selective oxidation of aluminum with minimized effects of other elements in the coatings. Long-time exposures under cyclic conditions allow substrate elements to affect oxidation behavior. Substrates containing hafnium have longer coating lives.

(2) The high affinity of platinum for aluminum is one of the necessary

characteristics which results in the observed improved coating performances. This characteristic results in platinum aluminum phases that limit transient oxidation, that exclude other elements from the coating surfaces, and that may also impart diffusional stability to the aluminide coatings.

(3) Platinum improved the resistance of diffusion aluminide coatings to attack via high-temperature hot corrosion. This improvement results from the selective oxidation characteristics of these coatings. After long exposures, however, substrate elements affect coating performances and coating lives are generally shorter on the substrates with higher concentrations of molybdenum and tungsten.

(4) Platinum improves the low-temperature hot corrosion resistance of diffusion aluminide coatings but the effect is not as dramatic as in high temperature hot corrosion.

(5) Oxygen-active elements such as yttrium and hafnium improve the adherence of alumina scales on CoCrAl by strengthening the oxide–alloy interface. The data were, however, not sufficient to define the strengthening mechanism.

(6) The mechanism of low-temperature hot corrosion of CoCrAlY coatings was shown to be associated with the preferential dissolution of aluminum and chromium which prevented protective scale formation, followed by dissolution of cobalt.

## Acknowledgments

The authors gratefully acknowledge the financial support of this work by the Office of Naval Research under Contract N00014-81-K-0355/P00007 and the Materials Technology Laboratory (Department of the Army). The experimental contributions of A. Ashary, J. Caola, G. Kim and J. Schaeffer are also acknowledged.

## References

1 G. W. Goward and D. H. Boone, *Oxid. Met.*, *3*, (1971) 475.
2 G. W. Goward and D. H. Boone, *Trans. ASM*, *60* (1967) 228.
3 M. M. P. Jannsen and G. D. Rieck, *Trans. Metall. Soc. AIME*, *239* (1967) 1372.
4 D. A. Joseph, *U.S. Patent No. 3,495,748*, 1970.
5 H. H. Todd, *U.S. Patent No. 33,494,748.*
6 M. J. Fleetwood, *Met. Sci.*, *98* (1970) 503.
7 P. Deb and D. H. Boone, *Microstructural formation and effects on the performance of platinum modified aluminide coatings*, Technical Report, Naval Postgraduate School, 1985.
8 K. Bungardt et al., *U.S. Patent No. 33,677,789*, 1972.
9 G. J. Tatlock, T. J. Hurd and J. S. Punni, *Platinum Met. Rev.*, *31* (1987) 26.
10 G. R. Johnston and P. G. Richards, The relative durabilities of conventional aluminide and platinum-modified aluminide coatings in an operational gas turbine engine, in I. G. Wright (ed.), *Corrosion in Fossil Fuel Systems*, The Electrochemical Society, 1983, p. 456.
11 R. Bauer, K. Schneider and H. W. Grunling, *High Temp. Technol.*, *3* (1985) 59.

12 F. S. Pettit and G. W. Goward, High temperature corrosion and the use of coatings for protection, in *Metallurgical Treatises*, AIME, Warrendale, PA, 1982, p. 603.
13 J. R. Caola, G. H. Meier and F. S. Pettit, *J. Vac. Sci. Technol., A4* (1986) 2915.
14 J. A. Sprague, V. Provenzano and F. A. Smidt, *Thin Solid Films, 95* (1982) 57.
15 F. S. Pettit and C. S. Giggins, Hot corrosion, in C. T. Sims, N. S. Stoloff and W. C. Hagel (eds.), *Superalloys II*, John Wiley, New York, 1987, p. 327.
16 K. L. Luthra, *Metall. Trans., 13A* (1982) 1647, 1843.
17 K. T. Chiang, F. S. Pettit, and G. H. Meier, Low temperature hot corrosion, in R. A. Rapp (ed.), *High Temperature Corrosion*, NACE, 1983, p. 519.
18 R. H. Barkalow and F. S. Pettit, On the mechanisms of hot corrosion of CoCrAlY coatings in marine gas turbines, in *Proc. 14th Conf. on Gas Turbines in a Marine Environment*, Naval Sea Systems Command, Annapolis, MD, 1979, p. 493.
19 R. Jones and C. Williams, *Mater. Sci. Eng., 87*, (1987) 353.
20 M. Seiersten and P. Kofstad, *Mater. Sci. Technol., 3*, (1987) 576.
21 A. Ashary, *Ph.D. Thesis*, University of Pittsburgh, 1985.
22 J. L. Smialek and G. H. Meier, High Temperature Oxidation, in C. T. Sims, N. S. Stoloff and W. C. Hagel (eds.), *Superalloys II*, John Wiley, New York, 1987, Chapter 11.
23 A. W. Funkenbusch, J. G. Smeggil, and N. S. Bornstein, *Metall. Trans., 16A* (1985) 1164.
24 J. L. Smialek, *Metall. Trans., 18A* (1987) 164.
25 C. L. Briant and K. L. Luthra, *Metall. Trans., 19A* (1988) 2099.

*Surface and Coatings Technology, 39/40* (1989) 19 - 27 

# BONDING GLASS CERAMICS TO HIGH TEMPERATURE ALLOYS

F. HONG and D. HOLLAND
*Department of Physics, University of Warwick, Coventry CV4 7AL (U.K.)*
(Received March 10, 1989)

## Summary

Scanning and transmission electron microscopy and energy dispersive X-ray analysis have been used to determine the interactions occurring at the interface during the coating of a Nimonic alloy and titanium metal by lithium silicate glass ceramics. The fast diffusion of chromium from the substrate into the glass is shown to have a major influence on the microstructure and phase composition of the resulting glass ceramic. It is observed that a layer of adhering metal oxide on Nimonic plays an important role in achieving the desired bonding between the alloy and the glass ceramic coating. Lack of such an oxide, in the case of titanium metal, leads to direct contact and reaction of silicates with the metal. This severe reaction between the glass coating and titanium gives rise to a porous structure. It is shown that adherence oxides are needed to control the reaction. Alternatively some form of barrier layer may be necessary.

## 1. Introduction

In many technical areas, where the requirements for bulk material and surface properties are different, coating techniques are often used to overcome incompatibilities. The traditional enameling of metal surfaces has greatly extended their applications (reviewed by Maskall [1]). Glass ceramic coatings show improvements in properties over vitreous coatings, such as increased hardness, improved mechanical strength and heightened refractoriness. The advantages of glass ceramic coatings over ceramics lie in their greater range of thermal expansion coefficient and ease of application. Glass ceramics are, in many situations, the preferred coating materials since their chemical and physical properties can be "tailored" by careful choice of composition and heat treatment to give specific microstructures and crystal phases. Glass ceramic coated metals include copper, tungsten, molybdenum, mild steel, stainless steel, Ni–Cr alloy etc., and their applications range from domestic and medical areas to microelectronics and other industrial fields. [2 - 4]. Oxidation or corrosion protection of refractory metals or alloys in turbine compressor blades is one of many important applications of glass

0257-8972/89/$3.50 

ceramic coatings. Two model coating systems (lithium silicate–Nimonic alloy and lithium silicate–titanium metal) were chosen in our study to investigate the possible bonding characteristics.

It is generally believed that strong glass–metal bonding requires a continuous structural transition from the metallic bonded substrate metal to the essentially ionic-covalent bonded glass coating. Metal oxides act as the transitional layer because they are compatible with both metal and glass. Therefore, a typical glass ceramic coating procedure is to pre-oxidize the substrate metal to produce an adherent oxide layer and then coat the metal with glass. The assembly is usually subject to subsequent heat treatment to convert the glass to glass ceramic. Thus, maintenance of a transitional oxide layer or an oxide saturated glass–metal interface is essential in achieving a successful bond between substrate metal and glass ceramic coating [5].

However, problems remain even if a transitional oxide layer is formed. Two main factors are involved. First, fast dissolution and diffusion of oxides into the bulk glass sometimes occur at the interface. As a result, the chemical bond or equilibrium may not be maintained throughout the whole firing and heat treatment procedure. Additional redox reactions may also take place between substrate metal and glass coating [6]. These interactions bring about various interdiffusions which may change both the metal and glass ceramic compositions or microstructures to a great extent. Second, the preformed oxide layer does not always adhere to the substrate metal, *e.g.* in the case of titanium, $TiO_2$ attaches to the base metal rather poorly. This kind of weakly adhering oxide layer results in poor coating adhesion. Therefore, the interactions between glass ceramic coatings are far from being simple and in the process of coating metals we may encounter very intricate problems. Although glass ceramic coatings have been used widely, little information exists on the interactions occurring during the various stages of coating the assemblies, and especially how these interactions affect the glass ceramic process in terms of microstructure, coating adhesion and other physical and chemical properties. Some results of interfacial reactions of lithium silicate glass ceramic coatings with Nimonic alloy and titanium metal are reported here in order to gain some fundamental understanding about these bonding systems.

## 2. Experimental procedure

The glass ceramic compositions used were based on the lithium silicate binary with various additions, typically of less than 5 wt.% of $Al_2O_3$, $K_2O$, ZnO and $P_2O_5$. The glass composition and the subsequent heat treatment were chosen to give very close thermal expansion coefficient match to the metal to minimize any possible thermal stress occurring at the glass ceramic and metal interface.

The glasses were prepared from Analar grade reagents and melted at 1360 °C for 3 h in a Pt–10%Rh crucible. The melt was quenched into deionized water and remelted to ensure homogeneity. The final product was ball milled to produce a fine powder (approximately 10 $\mu$m) and mixed with an organic binder.

The ink was applied to the metal surface by standard screen printing techniques. After drying, the organic binder was burnt off at 300 °C for 15 min in air and a uniform layer of thickness around 80 - 100 $\mu$m could be produced ready to fire.

Both Nimonic 263 alloy (Henry Wiggin and Company Limited) and titanium metal (Goodfellow Advanced Material) sheets were cut, vapour blasted and degreased. To observe the direct interactions between the metals and glasses, some sheets were coated without being pre-oxidized. Thus, a rough but clean surface is maintained before the assemblies are subject to firing. To determine the roles of metal oxides in bonding, some other sheets were oxidized at the appropriate temperature for a fixed time. The oxidized Nimonic has a well adhering layer of chromium oxide $Cr_2O_3$, while titanium has a layer of loosely attached $TiO_2$. The coated unoxidized metals were fired and heat treated in either $N_2$ or argon to prevent further oxidation. The coated oxidized metals, if the layer is protective, may be fired in air. The whole metal pretreatment and coating conditions are summarized in Table 1.

The fired only coating usually shows a well fused and wetted glass layer provided that there is no severe interaction between the coating and the substrate metal. A two stage heat treatment, consisting of nucleation and crystallization, was used to change the glassy coating into a glass ceramic coating, see also Table 1.

Both fired-only and heat-treated samples were sectioned for scanning electron microscopy (SEM) and transmission electron microscopy (TEM) with energy dispersive X-ray analysis (EDX) to reveal the microstructure and phase compositions. Spot analysis and line scan were used to determine the distribution of the constituent elements across the interface. X-ray diffraction (XRD) analysis was also carried out to identify the crystal phases formed due to either the glass ceramic process or interactions with the substrate.

TABLE 1

Pre-oxidation, firing and heat treatment conditions

| | *Pre-oxidation* | *Firing* | *Heat treatment* |
|---|---|---|---|
| Nimonic | 900 °C air 3 h | 980 °C air<br>5, 10, 20, 40 min | 600 °C air 30 min<br>800 °C air 1 h |
| | None | As above, but in $N_2$ | As above, but in $N_2$ |
| Titanium | 900 °C air 20 min | 1000 °C Ar<br>5, 10, 20, 40 min | 600 °C Ar 30 min<br>900 °C Ar 1 h |
| | None | As above | As above |

## 3. Results and discussion

### *3.1. Coating on Nimonic alloy*

#### *3.1.1. Chromium diffusion*

Rapid diffusion and dissolution of chromium from the substrate alloy into the glass coating were observed in the fired only assemblies, where the alloy was not pre-oxidized. Chromium, a major alloy component, is considered the most active element from the point of view of thermodynamics. The results of spot analysis using EDX shows that chromium can be detected up to 80 $\mu$m away from the interface after only 5 min firing, and with further firing time, the chromium concentration increases.

Pask *et al.* [7] described a two stage process: (1) formation of metal oxide at the interface by oxidation of the substrate metal by the glass and (2) solution of the oxide into the glass. In our specific glass–metal bonding system we propose the following reactions

$$\mathrm{Cr} + \mathrm{ZnO(glass)} = \mathrm{CrO(interface)} + \mathrm{Zn} \tag{1}$$

$$\mathrm{CrO(interface)} = \mathrm{CrO(glass)} \tag{2}$$

The formation of the thin interfacial CrO layer and its disappearance with firing time are confirmed by SEM (Fig. 1). A high concentration of chromium is observed at the interface by X-ray line scanning verifying formation of the interfacial chromium oxide (Fig. 2).

Figure 2 also shows enrichment of zinc at the interface but within the alloy. This indicates that reduction of ZnO in the glass has taken place and, following migration of zinc, leads to its alloying with the substrate to form a zinc-rich zone. This redox reaction proceeds quickly because the CrO layer dissolves.

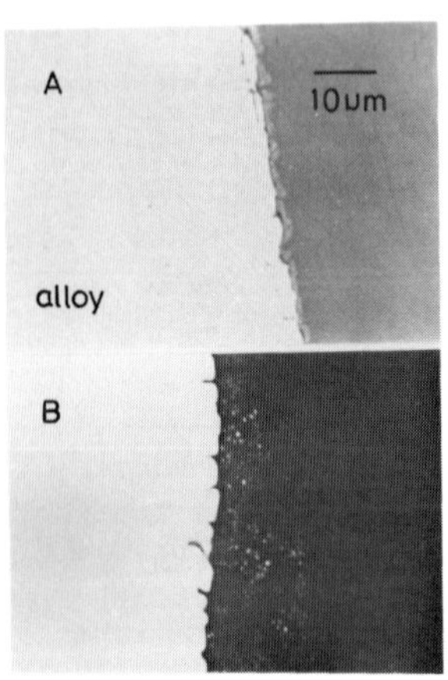


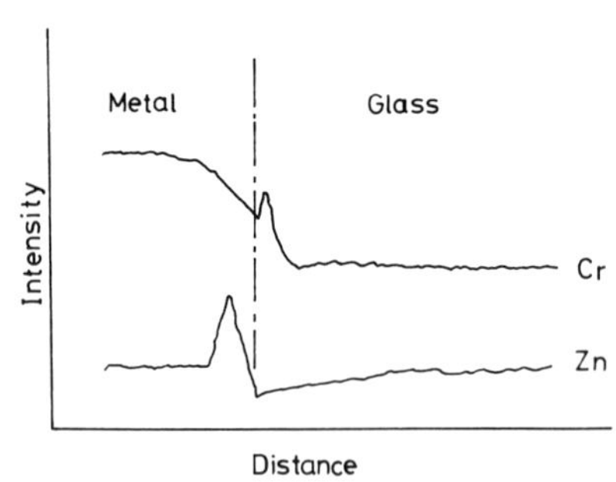


Fig. 1. Existence of an interfacial oxide layer and its disappearance after long firing times.

Fig. 2. Line scanning confirms a high concentration of chromium at the interfacial layer and enrichment of zinc within the alloy.

*3.1.2. Interactions between chromium and phosphorous*

It has been noticed in Fig. 1(b) that a cluster of bright metallic spots are formed extending 5 $\mu$m from the glass–metal interface into the glass after a long time firing. TEM shows the bright particles are actually a needle-like crystal phase with high concentrations of chromium and phosphorous.

$P_2O_5$ is introduced into the glass as a nucleating agent. In lithium silicate glass systems, $P_2O_5$ reacts to form numerous fine $Li_3PO_4$ crystallites which provide nucleation sites on which the desired crystals grow epitaxially [8]. Watkin *et al.* [9] discovered a similar crystal phase in their glass ceramic–Inconel sealing and identified it as $Cr_2P$ or its solid solution with other alloy elements. They proposed that the chromium reacts with phosphorous as follows

$$9Cr(alloy) + 2Li_3PO_4(s) = 2Cr_2P(s) + 3Li_2O(glass) + 5CrO(glass) \quad (3)$$

These reactions consume $Li_3PO_4$ nucleation sites. As a result, the heat-treated glass ceramic coating shows a very coarse structure throughout the reacted zone.

*3.1.3. $LiCr(SiO_3)_2$ on the coating surface*

$LiCr(SiO_3)_2$ crystal phase, confirmed by XRD, develops on the surface of the glass coating after a prolonged firing time. This green phase shows the chromium has diffused all the way through from the substrate alloy. The differential contraction causes the coating to wrinkle.

It is suggested that chromium dissolves into the coating as $Cr^{2+}$ instead of $Cr^{3+}$, as seen in reactions (1), (2), and (3). This can be seen from the lack of the typical green $Cr^{3+}$ colour even when significant amounts of diffused chromium exists before $LiCr(SiO_3)_2$ forms at the surface. Brow [10] reported the existence of $Cr^{2+}$ using XPS analysis in a similar bonding system. In our case, when the $Cr^{2+}$ reaches the coating surface, a further reaction (4) could follow as

$$2CrO(glass) + MO = Cr_2O_3 + M \quad (4)$$

MO could be glass oxides depending on the thermodynamic condition prevailing. The reduced metal may be lost as a vapour or precipitate at the surface. Since a short migration path for the vapour exists at the exterior coating surface the reaction (4) proceeds easily. Some zinc-rich metallic particles have been observed at the coating surface area. Apart from this, because of the limited solubility of $Cr^{3+}$ in the glass [11], $LiCr(SiO_3)_2$ tends to precipitate quickly at the surface.

*3.1.4. The role of the preformed $Cr_2O_3$ layer*

Very different results are obtained if the coated substrates are pre-oxidized. As shown in Table 1, a layer of well adhering $Cr_2O_3$ forms between the glass coating and the alloy. This layer plays a very important role in bonding. Firstly, the layer acts as a barrier to keep glass from reacting directly with the alloy and all the redox reactions discussed above are greatly hindered.

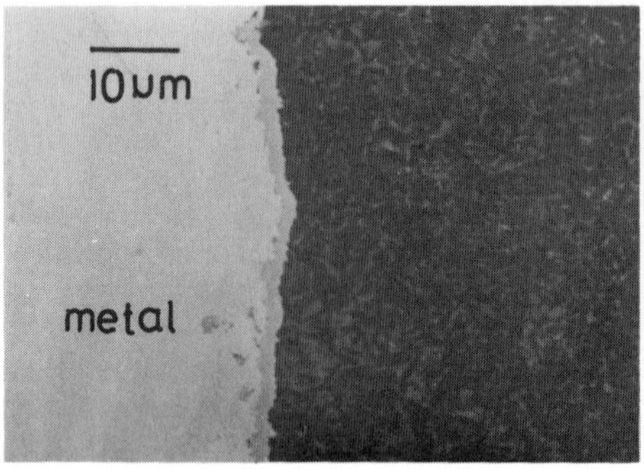

Fig. 3. Fine microstructures can be obtained when the oxide layer hinders the undesired reactions.

Fig. 4. Titanium silicide produced by reaction at the interface.

Secondly, the stability and low solubility of the $Cr_2O_3$ layer provide the saturation of $Cr_2O_3$ in the adjacent glass and the desired bonding conditions are maintained. Thirdly, the $LiCr(SiO_3)_2$ crystals develop slowly along the interface between the glass and oxide after firing for a few hours and this layer proves useful in contributing to the bonding strength [12]. Figure 3 shows the desired microstructures of fired and heat-treated coatings on the pre-oxidized Nimonic alloy.

### *3.2. Coating on titanium*

#### *3.2.1. Interactions between titanium and a glass coating*

Unpre-oxidized titanium was coated with glass and fired under argon, as seen in Table 1. Due to its high reactivity, titanium interacts strongly with the coating to form a layer at the interface (Fig. 4). XRD analysis confirms the layer as $Ti_5Si_3$, after removal of glass by etching in HF. A possible reaction is

$$3SiO_2(\text{glass}) + 5Ti = Ti_5Si_3(\text{interface}) + 3O_2\uparrow \qquad (5)$$

The oxygen released in reaction (5) is trapped in the coating when it fuses and wets the metal. As a result, the coating is very porous. Partial diffusion of oxygen into the titanium metal to form a gas saturated layer is also possible [13]. Reaction (5) reveals the undesired direct contact of titanium with silicates at high temperature. Various reactions have been widely observed [14 - 16].

#### *3.2.2. Role of $TiO_2$ at the coating interface*

Glazing of pre-oxidized titanium sheets showed that reaction (5) can be halted by an intermediate $TiO_2$ layer between the metal and coated glass. A well fused and wetted coating is formed. However, the thin $TiO_2$ layer is not adherent to the metal and, therefore, cracks tend to occur at the interface between the metal and metal oxide. Another problem is the dissolution of $TiO_2$ by the glass. This implies that even if the bond between $TiO_2$ and the base metal were mechanically strong, it is still possible to lose the $TiO_2$ layer eventually and bring the silicate glass in direct contact with titanium metal, which may lead to reaction (5) and further deterioration.

*3.2.3. The importance of adherence oxides*

In the traditional enamelling industry, some oxides, defined as adherence oxides, are used to aid the formation of an oxide saturated interface to achieve sufficient bonding. In dealing with coatings on highly reactive metals, adherence oxides are very necessary. These adherence oxides have a relatively lower formation energy than $SiO_2$; hence, they are less stable and react readily with titanium to form $TiO_2$, which is needed at the bonding interface.

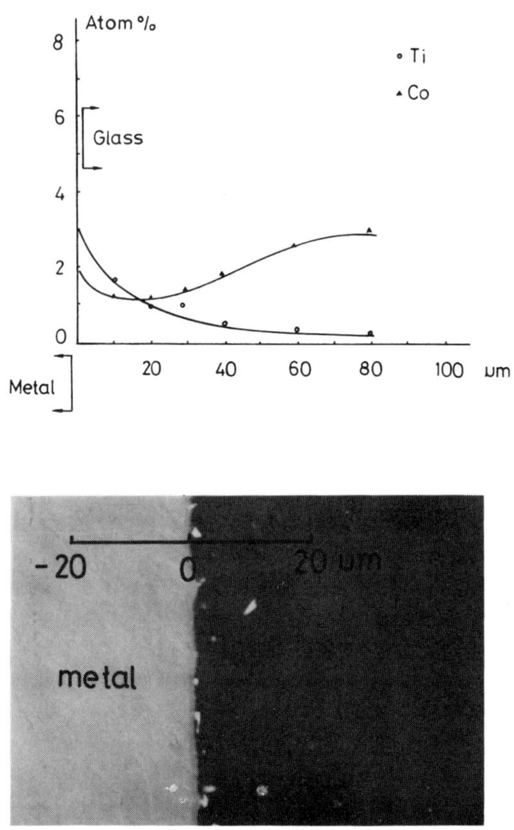


Fig. 5. Reduced CoO forms metallic particles and the interfacial layer is rich in $TiO_2$.

A typical adherence oxide CoO was added to the coating system and the following reaction occurred

$$2CoO + Ti = TiO_2 + 2Co \quad (6)$$

Figure 5 shows the reduced cobalt metallic particles and $TiO_2$-rich zone at the interface. Although reaction (5) still persists to some extent, especially after long firing times, the coated glass ceramic layer is much less porous and wets the titanium metal well.

## 4. Conclusions

Bonding of glass ceramic coatings to high-temperature alloys is a complex process. In the case of reactive substrate metals, severe reactions occur and the interactions and interdiffusions tend to change the microstructures and properties of glass ceramic coatings from those intended.

In the lithium silicate glass–ceramic/Nimonic alloy coating system fast diffusion of chromium from the substrate has been observed. The interactions cause coarsening of the interfacial microstructure and coating surface wrinkling. These interactions can be eliminated by preforming an adherent $Cr_2O_3$ layer.

In the lithium silicate glass–ceramic/titanium metal coating system the non-adhering pre-oxidized $TiO_2$ can not act as a transitional layer due to its mechanical weakness. Direct contact between silicates and titanium forms titanium silicides and oxygen which disrupts the coating. Therefore certain adherence oxides are needed to form a $TiO_2$ saturated interface as well as to hinder the reaction between titanium metal and $SiO_2$.

We are actively seeking a more effective barrier. Thermal cycling of these systems will be conducted to study endurance under service conditions.

## References


1 K. A. Maskall and D. White, *Vitreous Enamelling: A Guide to Modern Enamelling Practice*, Pergamon, Oxford, 1986.
2 A. Ravaglioli and A. Krajewsk, *Ceram. Inf.*, *18* (1983) 583 - 91.
3 E. A. Logan, D. Holland, J. S. Thorp and G. Partridge, *Proc. of the 1988 IEEE Electron. Manuf. Technol. Symp.*
4 W. D. Faust. *Ceram. Eng. Sci. Proc.*, *7* (1986) 702 - 709.
5 J. A. Pask, *Ceram. Bull.*, *66* (1987) 1587 - 1592.
6 J. A. Pask, *Proc. Porcelain Enamel Inst. Tech. Forum*, *33* (1971) 1 - 16.
7 A. P. Tomsia, F. Zhang and J. A. Pask, *J. Am. Ceram. Soc.*, *68* (1985) 20 - 24.
8 T. J. Headley and R. E. Loehman, *J. Am. Ceram. Soc.*, *67* (1984) 620 - 625.
9 R. D. Watkins and R. E. Loehman, *Adv. Ceram. Mat.*, *1* (1986) 77 - 80.
10 R. K. Brown, *J. Am. Ceram. Soc.*, *70* (1987) C129 - C131.
11 M. B. Volf, *Chemical Approach to Glass*, Elsevier, Amsterdam, 1984.

12 A. J. Sturgeon, *Ph.D. Thesis*, Warwick University, 1986.
13 A. Ya. Sitnidova, I. S. Anitov and I. A. Sudomoikin, *J. Appl. Chem. USSR.*, *44* (1971) 1473 - 1479.
14 E. S. Zhmud and A. E. Shmelev, *Inorg. Mat.*, *10* (1974) 1816 - 1820.
15 A. Ya. Sitnidova, A. A. Appen and I. S. Anitov, *J. Appl. Chem. USSR.*, *47* (1974) 1922 - 1926.
16 A. Passerone, G. Valbusa and E. Biagini, *J. Mat. Sci.* *12* (1977) 2465 - 2474.

*Surface and Coatings Technology*, *39/40* (1989) 29 - 42 

# MECHANICAL INTERACTIONS BETWEEN COATINGS AND SUPERALLOYS UNDER CONDITIONS OF FATIGUE

M. I. WOOD

*Engineering Materials Division, ERA Technology, Cleeve Road, Leatherhead, Surrey KT22 7SA (U.K.)*

(Received March 15, 1989)

## Summary

The increasing mechanical demands being placed on high-temperature gas turbine components has generated interest in establishing a sound understanding of the mechanical behaviour of coatings and coated components. This review covers their bulk and *in-situ* properties and behaviour, primarily under cyclic conditions. The close interrelationship between bulk coating properties, the presence and nature of structural defects within the coating and the interaction between coating and substrate are emphasized. The role of deformation modelling and fracture mechanics in advancing our understanding and design capabilities is highlighted.

## 1. Introduction

Coatings have been developed in general to protect gas turbine blading from oxidation and corrosion. Many fulfil this role very well. Apart from a few relatively simple mechanical considerations required because of the presence of coating, *e.g.* loss of load bearing cross-section during the formation of a diffusion coating, the deformation properties of the coatings have seemed to be of secondary importance. An exception has been some military engines where premature cracking under cyclic conditions has resulted in coating developments being guided by considerations of the coating's mechanical properties [1]. However, even there, this has been limited to a modification of the coating's characteristics as measured solely by its ductility [1].

There is now a more general realization that such a single property approach, based on ductility, is inadequate. For example "the relative cracking tendency of two coating systems has been found to completely reverse on two blades of different design and type of operation" [1]. Thermomechanical fatigue (TMF) problems have inevitably become greater because of the continuing increase in gas turbine performance. Even industrial and marine gas turbines are not immune [2, 3]. What is required is an approach based on

0257-8972/89/$3.50 

knowledge of the coating's whole range of mechanical and physical properties, just as if it was a structural alloy.

It is against this background of increased interest that this overview is set. Its objectives are to cover the somewhat more quantitative relationships that have been developed between the basic properties of a coating and its mechanical interaction with the substrate, primarily under cyclic conditions.

## 2. Coating properties

In order to go beyond qualitative interpretations of observed behaviour it is necessary to know the mechanical and physical properties of the coatings. In general little information is available, the only exception being for ductile-to-brittle transition temperature (DBTT) curves. Information on aluminide based coatings is even scarcer than for MCrAlY coatings, primarily because of the experimental difficulties (*e.g.* refs. 4 and 5; the physical properties have usually been measured on simulated coatings, *e.g.* castings of the right composition). Most of the studies of MCrAlY overlay coatings use test pieces cut from thick (1 - 3 mm) plasma-sprayed deposits, although other techniques have been used (Table 1).

Because of the importance of these properties a brief summary is in order.

### *2.1. Expansion coefficients [4, 6 - 9]*

In general, MCrAlY coatings have larger expansion coefficients than most superalloys whilst aluminide based coatings may have values lower than superalloys, but only at the higher temperatures. This property differential manifests itself at low temperatures as a residual stress in the coating, the magnitude of which will also depend on the processing conditions and the substrate in question.

TABLE 1

Coating compositions and fabrication methods for property determination

| *Designation in text* | *Composition* | *Fabrication* | *Reference* |
|---|---|---|---|
| PWA 273 | aluminide coating | *in-situ*, casting | 4, 9 |
| PWA 276 | Ni20Co17Cr13AlY | LPPS[a] | 14, 18, 23 |
| PWA 286 | NiCoCrAlY + Hf + Si | LPPS[a] | 4, 19 |
| CoCrAlY | Co18Cr8AlY | LPPS[a] | 8, 12, 13, 24 |
| NiCoCrAlY | Ni39Co21Cr8AlY | LPPS[a], IGPS[b] | 8, 12, 13, 15, 24, 26 |
| NiCoCrAlYTa | Ni23Co20Cr8.5AlY4Ta | LPPS[b] | 10 |
| NiCoCrAlY | Ni23Co18Cr12.5AlY | EB-PVD[c] | 6 |

[a]Low pressure plasma spray.
[b]Inert gas shielded plasma sprayed.
[c]Electron beam-physical vapour deposition.

*2.2. Elastic constants [6, 9 - 11]*

The Young's modulus of both types of coating is reported to be fairly similar, around 170 GPa at room temperature. This is a little lower than a typical polycrystalline superalloy, around 200 - 220 GPa. However, measurements on single crystal NiAl [11] show that for the binary alloy not only are the elastic constants strong functions of composition, but also that NiAl is elastically anisotropic: *e.g.* for stoichiometric NiAl the Young's modulus varies as

along ⟨100⟩: 100 GPa
along ⟨110⟩: 190 GPa
along ⟨111⟩: 290 GPa

This anisotropy has two consequences for aluminide coatings, one general and one specific to coatings on single crystals. The first is since aluminide coatings often have a grain size which is large relative to the thickness of the coating, the coatings will not be elastically isotropic. The second factor comes from X-ray diffraction work on aluminized single crystals from which it appears that the coatings exhibit a strong texture, *i.e.* preferred orientation [8]. Any analysis of aluminized components must therefore use the appropriate elastic constants, not isotropic values.

*2.3. Tensile properties [6, 10, 12 - 16]*

MCrAlY coatings have large yield stresses at low temperatures (Fig. 1), comparable with many superalloys. However, above 500 - 600 °C this decreases rapidly to a very low value. At around the same temperature there is

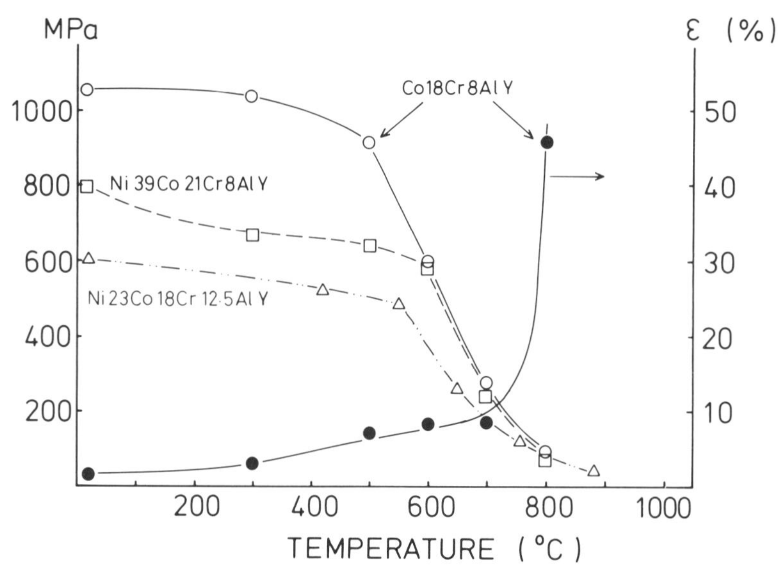


Fig. 1. Yield stress and ductility of freestanding MCrAlY coating [6, 12, 13].

an increase in ductility (cf. DBTT curves [17]), although this seems to occur at a slightly higher temperature (50 - 100 °C) than does the start of the rapid fall in strength [10, 13].

*2.4. Creep [6, 13, 18 - 21]*

The creep strength of MCrAlY coatings falls off very rapidly with temperature, following the temperature dependence of the yield stress [22]. Creep elongations of up to 400% have been reported at the higher temperatures, deformation being superplastic [10, 14]. However, it must be remembered that in common with all superplastic systems, the rupture strain is a strong function of strain rate, hence stress, at any given temperature: increasing the stress reduces the above mentioned ductility to around 10%. Measurements of the creep properties of aluminide coatings have been attempted by using very thin substrates, 0.25 and 0.13 mm before coating [4], but experimental difficulties have hampered the work [5].

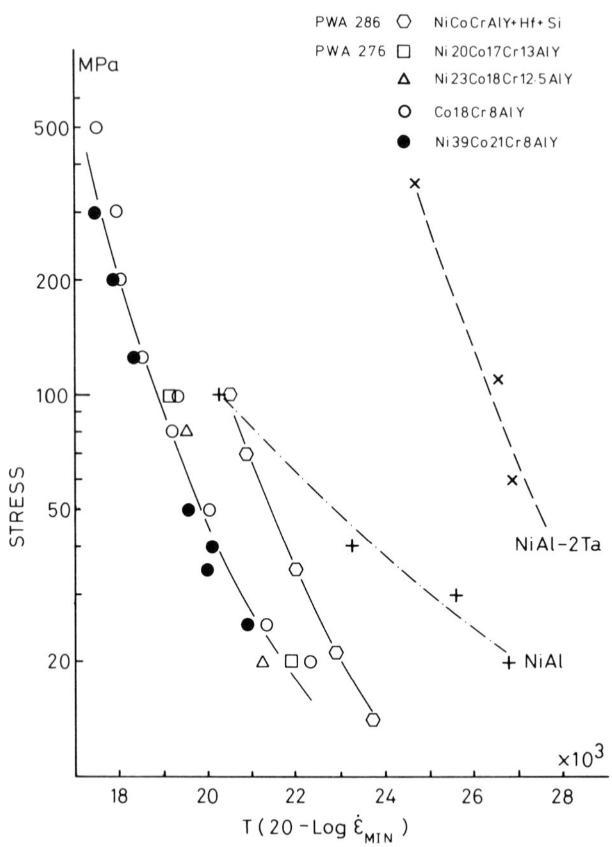


Fig. 2. Stress *vs.* strain rate modified "Larson–Miller" parameter for various coating alloys, ($\dot{\epsilon}_{min}$ in %/h) [6, 13, 18 - 21].

It is instructive to compare MCrAlY coatings and NiAl alloy data on a strain rate basis (Fig. 2) using a "Larson–Miller" type approach (after ref. 6). The creep rate disparity between NiAl and MCrAlY based alloys is clear. Further to this, the creep strength of NiAl is increased markedly by alloying, even with small ternary additions (1 - 3%) of solute such as molybdenum, tantalum, niobium and hafnium [21], elements commonly found in aluminide coatings along with chromium.

*2.5. Fatigue [4, 12, 23, 24]*

Work on free standing coating alloys is restricted to one study on the overlay PWA 276 at 650 and 1050 °C [23], spanning the range when the coating is reasonably strong (490 MPa yield stress) and when it is weak (less than 10 MPa yield stress). The lives at 1050 °C were greater than those at 650 °C on a total strain range basis. Surface observations also revealed that many cracks became visible half way through the 1050 °C test: little or no surface damage was observed during the 650 °C tests. The increase in life was attributed primarily to lower crack growth rates at 1050 °C where multiple branching transgranular crack growth was observed. This was unlike the situation at 650 °C when singular, intergranular cracks were found.

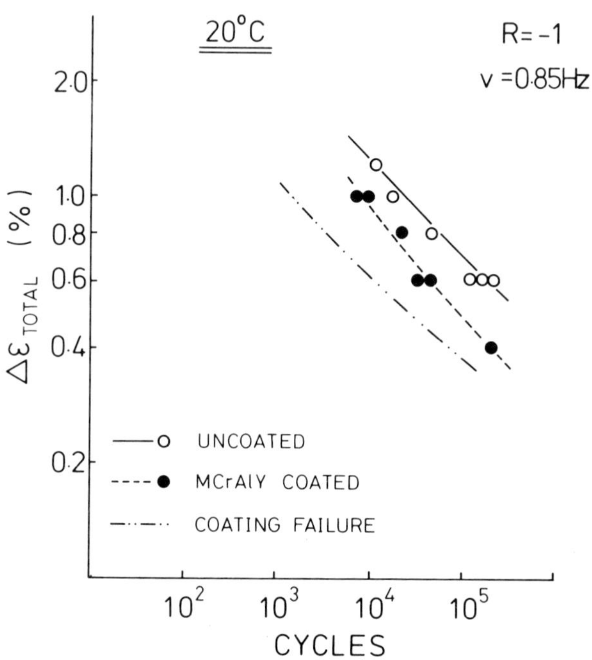


Fig. 3. Fatigue life of *in-situ* MCrAlY coating, the MCrAlY coated and the uncoated single crystal SX60A at 20 °C [24].

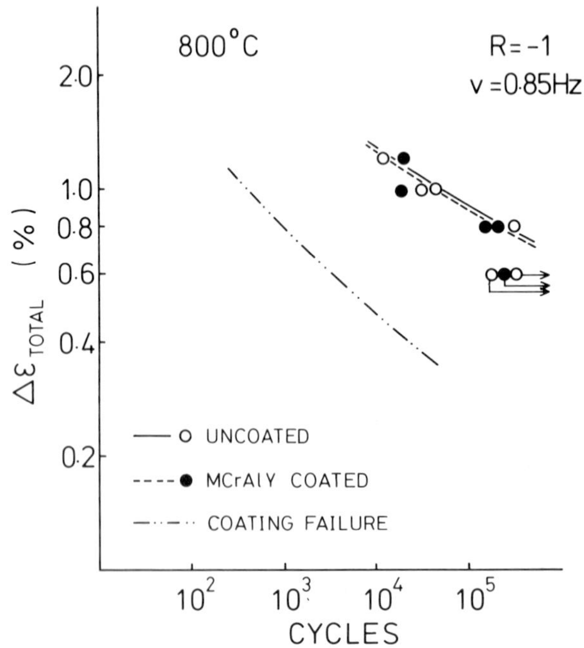


Fig. 4. Fatigue life on *in-situ* MCrAlY coating, the MCrAlY coated and the uncoated single crystal SX60A at 800 °C [24].

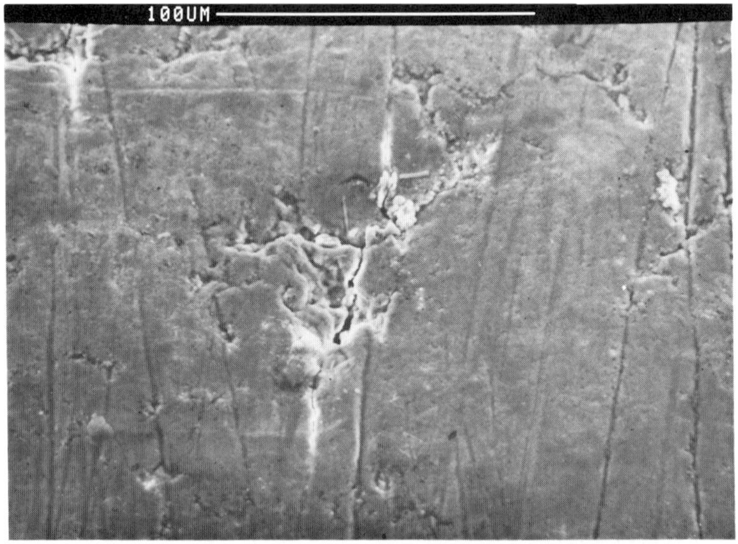


Fig. 5. Crack initiation site during isothermal fatigue test for an *in-situ* NiCoCrAlY coating [8].

There appear to be only two reported studies where attention has been paid to identifying the "life to crack initiation" for an *in-situ* coating during a fatigue test [4, 12], both on coated single crystal substrates. The broader of these two studies (strain controlled LCF) again encompassed the two extremes of the coating's behaviour (20 °C — strong; 800 °C — weak). For both aluminized and NiCoCrAlY/CoCrAlY-coated [001] single crystals the crack initiation life was lower at 800 °C than at 20 °C (cf. Figs. 3 and 4), especially for the aluminide coating [12]. It was possible to identify unambiguously the (surface) site of crack initiation: at both temperatures cracks always formed at unpolished regions of the surface (Fig. 5). (Part of the standard processing cycle for these plasma-sprayed coatings is a mechanical polish which produces a surface finish which is acceptably smooth aerodynamically, but from which it is not necessary to remove all traces of the original sprayed surface.) This requires us to decide whether the properties of an *in-situ* coating are being governed entirely by its macroscopic properties or by 'defects' and processing features.

The apparent difference between the fatigue behaviour of the free standing and *in-situ* coatings with increasing temperature (stronger *vs.* weaker respectively) could be due to this observation, or one of several others, *e.g.* strain rate (affecting ductility and $\Delta\epsilon_{in}$) or the use of a different criterion for failure (a surface crack of 20 $\mu$m or the failure of a test piece of diameter 3 mm).

## 3. Isothermal fatigue

There have been a moderate number of studies of the effects of coatings on fatigue life: a wide range of observed behaviour has been found [17]. However, most explanations have tended to be rather qualitative, mainly because of the lack of coating property data. However, some studies have been more particular in trying to separate coating behaviour from the performance of the coated component, and then relating these observations specifically to coating properties (*e.g.* refs. 4, 12, 23 - 26). As a result of such work it has been possible to draw out three (apparently) separate factors [12] which govern the overall effect of the coating.

(i) The ability of the coating to withstand crack formation and the relationship between this and the coating's structure and mechanical properties.

(ii) The behaviour of cracks in the coating once they have reached the coating-substrate interface. Do they arrest or continue propagating?

(iii) The rate at which (i) and (ii) can operate relative to crack nucleation and growth processes associated with the substrate.

The need to make these distinctions can be seen from the results shown in Figs. 3 and 4 (CoCrAlY or NiCoCrAlY plasma-sprayed [001] oriented single crystals). At room temperature (cf. Fig. 3) cracks could be detected in the coating (Fig. 5) after 10 - 30% of the life of the coated test piece. The presence of the coating reduced the fatigue properties of the single crystal substrate. At 800 °C cracks were detected at less than 2% of the life of the coated test piece whilst the coating had little or no effect on the fatigue properties of the single crystal (Fig. 4). This failure of the coating well before the failure of the coated test piece has been observed in other work, *e.g.* PWA 286 and 273 on the single crystal PWA 1480 [4], where there was no observable difference between the MCrAlY and aluminide coatings in their cracking lives at high temperatures.

As previously discussed, the decrease in coating life with increasing temperature may be linked to an increase in the inelastic strain range consequent on the marked drop in strength. This must then outweigh the increase in ductility, large though this can be (it should be recalled that the ductility at high temperatures is a function of strain rate). However, such an approach does not address the role of "defects" in initiation cracks. Because of the high surface-to-volume ratio for an *in-situ* coating, the importance of surface events will be greatly increased over that found for a conventional test piece. Insufficient modelling and experimental work has been done to reconcile the data currently available, although phenomenological approaches do show some promise [4, 5].

As the *in-situ* coatings failed more rapidly at the higher temperatures the reason for the diminishing effects produced by the coating on the coated test piece's life arises from a decrease in the crack propagation rate, in particular in the vicinity of the coating-substrate interface. At low temperature the few cracks which formed in CoCrAlY/NiCoCrAlY-coated single crystals all propagated into the substrate causing test piece failure [12, 22]. The crack growth planes were also considerably different in the coated specimens [12, 22] especially at the lower temperatures. At 800 °C the large number of coating cracks which formed arrested at the coating-substrate interface. Test piece failure usually occurred by crack growth from porosity. This arrest or retardation at the coating–substrate interface has been observed by others [4, 27] and found to be dependent upon the type of coating (for a given test condition). For example, cracks in a PWA 273 aluminized single crystal (PWA 1480) propagated from the coating into the substrate during an LCF test at 1038 °C whilst cracks in PWA 286 NiCoCrAlY arrested at the interface [4].

Since the stress range associated with a given strain range decreases with increasing temperature, the possibility that the lower crack growth rates are linked to the resultant lower $\Delta K$ needs to be excluded. To demonstrate that this is not the case, the number of cycles required to grow a coating crack from initiation to test piece failure is shown in Fig. 6 as a function of the maximum stress in the substrate. For a crack which has reached the coating–substrate interface the stress intensity value ($K$) at the tip of the crack at different temperatures will be directly related to the maximum stress in the substrate, *everything else being equal.* The slower crack growth rates at 800 °C are clear, especially at the lower stresses, indicating that the physical situation at 20 and 800 °C is different.

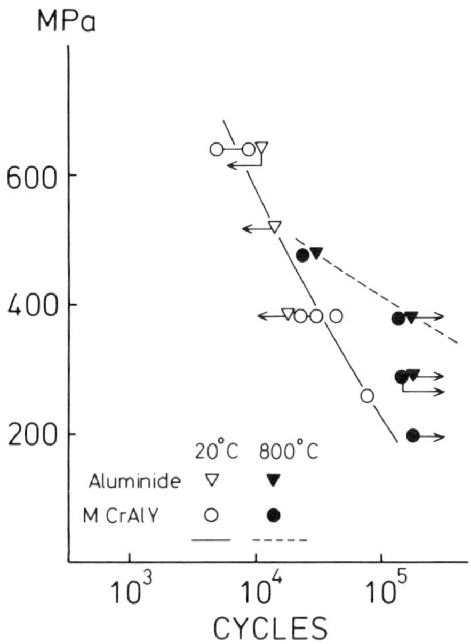


Fig. 6. Maximum stress in the substrate *vs.* number of cycles to grow a coating crack to fail the single crystal test piece [12].

Few explanations have been proffered to resolve the question of propagation *vs.* arrest. Possible explanations have included retardation of short crack growth through a crack closure effect [22, 28] or increased crack tip blunting at high temperatures reducing the stress intensity factor [22, 29]. A third interpretation [12, 24] concentrates on the way load is partitioned between the coating and the substrate, for if the coating cracks, the load it was carrying must be redistributed. At low temperatures the coating is strong and hence any crack will have a large stress intensity at its tip. In contrast, when the coating is weak (high temperature) it carries little load anyway, hence the presence of a crack requires little load to be redistributed. Thus the stress intensity at the tip of a geometrically equivalent crack will be far smaller than when the coating was strong. The crack growth rate beyond the coating will therefore decrease with increasing temperature when the coating softens significantly. Although less explicitly stated, a somewhat similar observation has been made [27] that non-propagating cracks behave as they do because they are "not sufficiently severe from a stress intensity standpoint to reduce the $10^7$ uncoated strength of these superalloys".

This is still a qualitative argument. Verification requires detailed mathematical modelling of the load repartitioning behaviour using constitutive equations to describe the coating's properties. However, it can be seen that of the three factors identified above, crack initiation and propagation are linked through their common dependence on the mechanical properties of the coating.

## 4. Thermomechanical fatigue

Simple experimental data on the TMF properties of coated and uncoated superalloys is fairly readily available, *e.g.* based on tests in hot gas streams or fluidised beds. Various interpretations have been advanced which offer semiquantitative rationales of the observed behaviour, *e.g.* the use of the DBTT curve. However, it is only recently that detailed attempts have been made to analyse and model the deformation and failure of the coating under TMF conditions.

Various forms of TMF cycles can be postulated. However, it is generally agreed that the most damaging cycle is that which imposes a tensile strain on the coating at low temperature and compression at high temperatures (out of phase cycle): *e.g.* the cyclic life [4] for NiCoCrAlY and aluminized PWA 1480 single crystals increased from around 500 - 1500 cycles (respectively) for out of phase TMF to around 10 500 cycles for in phase TMF (the converse of an out of phase cycle). It is this out of phase cycle which is the most relevant test vehicle in attempts to analyse the performance of coatings for gas turbine blading.

One approach to the problem has been phenomenological. This treats the coating as being subject to the same descriptive rules that have been found useful in the analysis of structural alloys, *e.g.* inelastic strain range or hysteretic energy. However, because a coated component is a composite system where failure occurs as the consequence of a series of sequential events, the life of a coated component $N_f$ has to be expressed as [4]

$$N_f = N_c + N_{sc} + N_{sp}$$

where $N_c$ is the number of cycles to initiate a crack through the coating, $N_{sc}$ is the number of cycles for the coating crack to penetrate a small distance into the substrate and $N_{sp}$ is the number of cycles to propagate the substrate crack to failure.

The best descriptive model for $N_c$ was found to be that of tensile hysteretic energy [26]

$$N_c = C(\Delta W_t)^{-b} \nu^{m}$$

where $\Delta W_t$ is the tensile hysteretic energy, $\nu$ is a complex time and temperature dependent frequency term, after ref. 30, and b and m are constants.

The values $\Delta W_t$ are derived from calculated hysteresis loops for the coating. These loops are, in turn, based on constitutive models for the coating. For PWA 286, NiCoCrAlY, the best agreement between an experimental hysteresis loop and a constitutive model was found for the isotropic unified viscoplastic formulation [4, 31]

$$\dot{\epsilon}_{in} = \left(\frac{\sigma - \Omega}{K}\right)^{n}$$

where $\Omega$ (back stress) and $K$ (drag stress) are strain history dependent internal state variables describing kinematic and isotropic cyclic hardening respectively. Using this approach [5] for $N_c$ it is possible to correlate isothermal overlay coating lives and coating lives under TMF conditions to within a factor of around 2.5. Determination of the $N_{sc}$ and $N_{sp}$ terms in the cycle life equation remains to be addressed [4, 5]. Likewise, the aluminide coating model still needs to be developed [5].

One of the objectives of any such work is to predict lives, but it also ought to be capable of indicating why different coatings behave differently. A major failing of an approach based on DBTT curves is its inability to explain why the relative TMF resistance of different coatings should be cycle or blade dependent [1]. A simplistic interpretation of the model would be that the "more ductile" coating would always be best. This is not found in practice, nor in TMF tests, although the data [32, 33] is as yet somewhat limited (Fig. 7). Here there appears to be a switch over in relative performance, with the NiCoCrAlY PWA 286 being best at high strain ranges and the PWA 273 aluminide being (potentially) superior at low strain ranges. As yet no mechanistic reason for this difference appears to have been put forward.

A somewhat different approach to the problem has placed more emphasis on the role of defects within the coating [6, 34] based on work with electron-beam-physical-vapour-deposition (EB-PVD) NiCoCrAlY coatings on a variety of substrates. Once the extra strain induced in the coating by thermal expansion mismatch had been taken into consideration, coating cracking lives could be correlated with the total strain range for out of phase cycling. An important point arising from this, apart from the role of the expansion mismatch strain, is that initiation occurred at flaws. Constitutive models are based on "bulk" property data. A similar difficulty was noted in the isothermal fatigue section between "bulk" and *in-situ* properties. This again raises the question of the importance of the surface condition and nature of structural flaws or weakness in such a high surface to volume "test piece" as *in-situ* coating. Whilst plasma sprayed coatings will not have the

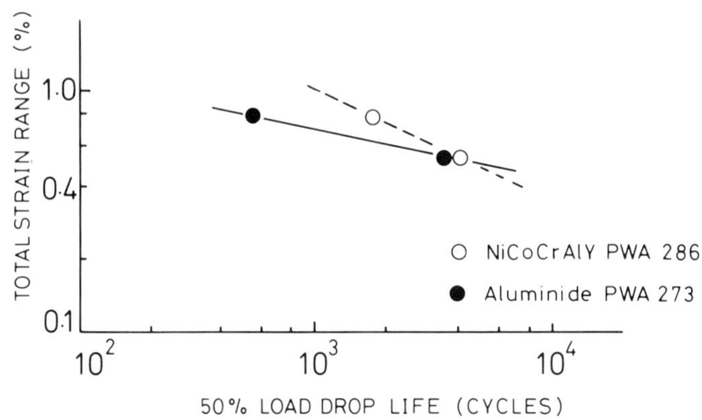


Fig. 7. Total strain range *vs.* life for coated PWA 1480 single crystal for an out of phase TMF cycle: 427 - 1038 °C [32].

same structural features as EB-PVD coatings, they are not metallurgically perfect products. It has been observed in the work on plasma-sprayed NiCoCrAlY PWA 286 that crack initiation occurs at many sites over the coating, with cracks from many sites contributing to produce the failure.

The question of coating crack growth into the substrate has also been analysed [35], the coating crack being regarded as a geometrical extension of any (short) substrate crack, somewhat modified by differences in elastic constants and differential expansion. However, as was the case for the equivalent isothermal case, the coating crack cannot be treated purely as a geometric feature. The load repartitioning will depend upon the temperature dependence of the coating's properties as well as the rate at which the TMF cycle is carried out and will greatly complicate any such analysis, especially for "slow" thermal transients. In addition to all these factors, environmental effects must be included, for TMF lives in vacuum are greater than those in air [26].

## 5. Concluding remarks

This brief survey of the mechanical properties of coatings and coated components has been very selective. Many factors have not been discussed, *e.g.* the effects of oxidation and corrosion, interdiffusion and ageing, or the effects of different coating techniques on the properties of MCrAlY coatings. These are all additional effects which, whilst of importance, further complicate an already complex subject. Although it is clear that advances in understanding and lifing are being made, there are many topics which are unresolved or simply unclear because of the lack of data, *e.g.* the relative merits in a high surface to volume material such as a coating of a phenomenological approach based on "bulk" properties *vs.* an analysis based on crack initiation at surface flaws. However, any fracture mechanics type approach is likely to be complex both because of the size of the defect but also because of the huge range of coating properties encompassed during a single TMF cycle. The answers to other important questions, such as the causes of the differences between MCrAlY and aluminide coatings, is hampered by the lack of suitable data, in this case primarily because of experimental difficulties. Whilst deformation modelling has been actively pursued, the basic metallurgical side of the problem seems to have been somewhat overshadowed, *e.g.* there has been minimal work on crack initiation and growth. Although it is important to be able to life a particular coating or component, there is also the need to know how to modify it metallurgically in order to develop a better coating or improve deposition techniques.

The increased effort and interest shown in this topic over the recent years has brought into clearer focus some of the physical and metallurgical factors governing coating life. However, there is a great deal of work needing to be done before a sound quantitative understanding has been achieved.

## Acknowledgment

The author is grateful to the Directors of ERA Technology Ltd. for permission to publish this article.

## References

1 G. W. Meetham, *Mater. Sci. Technol.*, *2* (1986) 290.
2 M. I. Wood and J. E. Restall, *U.K. Corrosion '87, Brighton, U.K., October 1987.*
3 R. Burgel, *Mater. Sci. Tech.*, *2* (1986) 302.
4 G. A. Swanson, I. Linask, D. M. Nissley, P. P. Norris, T. G. Meyer and K. P. Walker, *2nd Annual Status Report*, NASA-CR-179594, 1987.
5 G. R. Halford, T. G. Meyer, R. S. Nelson and D. M. Nissley, *33rd Int. Gas Turbine and Aeroengine Congress, Amsterdam, The Netherlands, June 1988*, ASME.
6 T. E. Strangman, Thermal fatigue of oxidation resistant overlay coatings for superalloys, *Ph.D Thesis*, University of Connecticut, 1978.
7 L. Pawlowski, D. Lombard, F. Tourenne, F. Kassabji and P. Francis, *High Temperature–High Pressure*, *17* (1985) 611.
8 M. I. Wood, *RAE TR 88-021*, Royal Aircraft Establishment, England, 1988.
9 M. Kaufman *TMS Paper F82-11*, The Metallurgical Society., AIME, Warrendale, PA.
10 J. M. Veys and R. Mevrel, *Mater. Sci. Eng.*, *88* (1987) 253.
11 N. Rusovic and H. Warlimont, *Phy. Status Solidi. (a)*, *44* (1977) 609.
12 M. I. Wood, *ASM Europe Conference, Paris, France, September 1987.*
13 M. I. Wood, RAE Technical Report awaiting publication.*
14 M. G. Hebsur and R. V. Miner, *Mater. Sci. Eng.*, *83* (1986) 239.
15 T. A. Taylor, M. P. Over, B. J. Gill and R. C. Tucker, Jr., *J. Vac. Sci. Tech.*, *A3* (1985) 2526.
16 E. M. Schulson and D. R. Barker, *Scr. Metall.*, *17* (1983) 519.
17 A. Strang and E. Lang, *High Temperature Alloys for Gas Turbines, Liege, October 1982*, Reidel, Dordrecht, 1982, p. 469.
18 M. G. Hebsur and R. V. Miner, *Thin Solid Films*, *147* (1987) 143.
19 G. A. Swanson, I. Linask, D. M. Nissley, P. P. Norris, T. G. Meyer and K. P. Walker, *1st Annual Status Report*, NASA-CR-174952, 1986.
20 J. D. Whittenberger, *J. Mater. Sci.*, *22* (1987) 394.
21 K. Vedula, V. Pathare, I. Aslandis and R. M. Titran, *Mater. Res. Symp.*, *39* (1985) 411.
22 M. I. Wood, J. E. Restall, *High Temperature Alloys for Gas Turbines and Other Applications, Liege, October 1986*, Reidel, Dordrecht, 1986, p. 1215.
23 J. Gayda, T. P. Gadd, R. V. Miner, *Int. J. Fatigue*, *8* (1986) 217.
24 M. I. Wood, *RAE TR 88-057*, Royal Aircraft Establishment, England, 1988.*
25 D. E. Graham and D. A. Woodford, *Metall. Trans.*, *12A* (1980) 329.
26 J. Gayda, T. P. Gabb, R. V. Miner and G. R. Halford, *1987 TMS–AIME Annual Meeting, Denver, Colorado, February 1987*, TMS, AIME, Warrendale, PA, 1987.
27 T. E. Strangman, M. Fuji and X. Ngugen-Dinh, *5th Int. Symp. on Superalloys, Seven Springs, U.S.A., 1984*, TMS, AIME, Warrendale, PA, p. 797.
28 S. Suresh, C. F. Zaminski and R. O. Ritchie, *Metall. Trans.*, *12A* (1981) 1435.
29 J. S. Crompton, S. A. Morley and J. W. Martin, *5th Int. Conf. on Superalloys, Seven Springs, U.S.A., 1984*, TMS, AIME, Warrendale, PA, p. 771.
30 W. J. Ostergren, *J. Testing Evaluation*, *4* (1976) 304.

---

*Available from Defence Research Information Centre, 65 Ground Street, Glasgow G2 8EX, U.K.

31 K. P. Walker, NASA-CR-165533, 1981.
32 G. A. Swanson and R. C. Bill, *21st Joint Propulsion Conf., Monterey, California, July 1985,* AIAA/SAE/ASME/ASEE, Paper AIAA-85-1421.
33 K. R. Bain, *21st Joint Propulsion Conf., Monterey, California, July 1985,* AIAA/SAE/ASME/ASEE, Paper AIAA-85-1366.
34. G. R. Leverant, T. E. Strangman and B. S. Langer, *3rd Int. Conf. on Superalloys 1976, Seven Springs, U.S.A., 1976,* Claitors Pub. Division, 1976, p. 285.
35 T. E. Strangman and S. W. Hopkins, *Ceramics Bull, 55* (1976) 304.

*Surface and Coatings Technology*, *39/40* (1989) 43 - 51 

# FORMATION OF SILICON DIFFUSION COATINGS ON FERROUS ALLOYS FROM THEIR REACTION WITH SILANE

A. L. CABRERA and J. F. KIRNER

*Air Products and Chemicals, Inc., Allentown, PA 18195 (U.S.A.)*

(Received March 8, 1989)

## Summary

A method for surface modification of metals using silane–hydrogen reactive atmospheres was developed which results in the formation of a silicon diffusion coating. The silicon is mainly confined in the near surface of the metal and imparts high-temperature oxidation protection. The silicon coatings were characterized with a wide range of analytical techniques including Auger electron spectroscopy. Kinetics of the surface reaction between silane and the metal substrate, as well as the behavior of the coating in oxidizing environments at high temperatures, were studied by a gravimetric technique. Characterization of these silicon coatings on iron, low carbon steel and stainless steel, as well as their applications for high-temperature oxidation, are discussed.

## 1. Introduction

Addition of suitable amounts of silicon to the bulk of wrought and cast metals and alloys improves resistance to oxidation and carburization at high temperatures [1]. Unfortunately, such high levels of silicon impair the desirable bulk physical properties of metals. However, if the silicon can be confined only to the surface in a coating, improved surface properties are achieved without modifying the bulk properties of the metal.

Processes for siliconizing metals using a silicon hydride at reduced pressures at 900 - 1000 °C were patented in the early 1900s [2, 3]. More recently, Caillet and coworkers [4, 5] produced diffusion coatings consisting of solid solutions and metal silicides on iron and nickel under static conditions at elevated temperatures (1100 °C). Attempts to use $SiH_4$ at lower temperatures have resulted in silicon overlay coatings. In fact, Powell [6] reported that silicon hydrides are advantageous for low-temperature deposition of silicon where diffusion or silicide formation is to be entirely avoided. More recently, the Central Electricity Generating Board of Great Britain patented a process for the formation of elemental silicon coatings using $SiH_4$ on 9%Cr–1%Mo steels [7, 8]. At lower temperatures, only Dubois

0257-8972/89/$3.50

and Nuzzo have formed metal silicides (on nickel but apparently not on iron) using sputter-cleaned metal surfaces under ultrahigh vacuum conditions [9].

The aim of our work was to develop a low-temperature process for producing a silicon diffusion coating using a silicon-containing molecule in a gas mixture, at atmospheric pressure [10]. We chose $SiH_4$ because it is a gas and can easily be blended with other gases. It decomposes at lower temperatures than the halides, and its decomposition products are only silicon and hydrogen. Problems caused by silicon halides such as corrosion of process equipment and volatilization of the substrate are avoided.

In this paper, we report the use of $SiH_4$–$H_2$ gas mixtures at atmospheric pressure at relatively low temperatures to form silicon diffusion coatings on iron- and chromium-containing steels, as well as the application of these coatings for high-temperature oxidation resistance.

## 2. Experimental

### *2.1. Materials*

Iron foil (purity, 99.9975%; thickness, 0.01 cm) was obtained from Alfa; carbon steel (1010) foil (thickness, 0.005 cm) was obtained from Teledyne Rodney; and coupons (2.54 cm × 0.63 cm × 0.16 cm) of 9%Cr–1%Mo steel were obtained from Metal Samples Co.

Gases used were Air Products and Chemicals Research Grade hydrogen, argon, nitrogen, oxygen or blended mixtures of approximately 0.5% $SiH_4$ in hydrogen.

### *2.2. Sample treatment*

Iron and steel foils were siliconized and characterized by Auger electron spectroscopy (AES) in an *in-situ* surface analysis unit which has been described in detail previously [11]. These samples were prepared for X-ray diffraction (XRD) and X-ray fluorescence (XRF) analyses.

Samples with approximate dimensions of 0.76 cm × 1.0 cm were cut from a larger piece of foil and mounted on the manipulator of the deposition–surface analysis system. The samples were heated by a high-current alternating current power supply using two tungsten wires spot welded to the edges of the sample. The temperature of the sample was monitored with a chromel–alumel thermocouple spot welded to one face.

Metal coupons and foils were also treated in a gravimetric system containing a Cahn 2000 microbalance. The sample was suspended by a quartz wire in a quartz flowthrough hangdown tube. The sample was heated with a tube furnace and the temperature was measured using a Type K thermocouple inserted in a quartz thermowell. The dew point was measured using a Panametrics Model 2100 hygrometer and converted to $H_2O$ partial pressure using Goff formulas.

Siliconizing experiments were performed in two steps. The sample was pretreated in flowing dry $H_2$ ($H_2O/H_2 \leqslant 1 \times 10^{-5}$) at 800 °C to remove surface impurities. It was then cooled to the treatment temperature and $SiH_4$ admitted by blending in premixed 0.5% $SiH_4$–$H_2$ in an amount needed to obtain the desired concentration.

Rates of oxidation were determined using *in-situ* microgravimetry. The hangdown tube was purged with inert gas and the furnace temperature was raised to the oxidation temperature. The inert gas was turned off, $O_2$ was admitted and weight was monitored for the desired period of time.

### *2.3. Analyses*

Auger analyses were obtained with a Physical Electronics 560 Spectrometer equipped with a double-pass cylindrical mirror analyzer. Depth profiles were obtained using AES with argon ion sputtering over a 2 mm × 2 mm or 4 mm × 4 mm rastered area. Energy dispersive analysis was obtained with an EDAX PV9500 energy dispersive X-ray fluorescence spectrometer using a microcollimator. XRD patterns were obtained with a Siemens D-500 diffractometer. Samples were scanned at a rapid drive rate for the purpose of phase identification.

## 3. Results and discussion

### *3.1. Treatment of iron and 1010 steel foils with $SiH_4$*

Pure iron samples were pretreated in dry $H_2$ with a dew point of less than −55 °C at 800 °C to prepare a clean surface for deposition. They were then exposed to an atmosphere of 800 ppm by volume $SiH_4$ in $H_2$ and heated to 500, 600 and 700 °C. The samples were immediately analyzed *in-situ* with AES to determine surface elemental composition (depth of 10 Å). The samples exhibit strong lines corresponding to both silicon and iron, indicating that decomposition of $SiH_4$ and diffusion of silicon occur at these low temperatures. Figure 1(a) plots the silicon surface concentration determined by AES as a function of treatment time at 500, 600 and 700 °C. At 500 °C and 600 °C, the silicon surface concentrations are about 10 at.% and 40 at.% respectively, and remain relatively constant as a function of treatment time up to 15 min. At 700 °C, the silicon concentration increases with longer treatment times so that after 15 min the surface is almost entirely silicon.

After removal from the reactor, the siliconized iron samples were analyzed by XRF to determine the elemental composition to a depth of about 3 $\mu$m, which will be referred to as the near-surface region. Figure 1(b) plots the silicon near-surface concentration determined by XRF as a function of treatment time at 500, 600 and 700 °C. For samples treated at 500 °C, the silicon signal is weak and comparable to the signal for a pure iron sample used as a blank, suggesting that the silicon coating observed by AES is very thin. The samples treated at 600 °C show intense silicon signals corresponding to 27 - 29 at.% that remain constant as a function of time. The small difference between the AES and XRF results at 600 °C suggests that the additional silicon deposited at longer times diffuses further into the bulk and

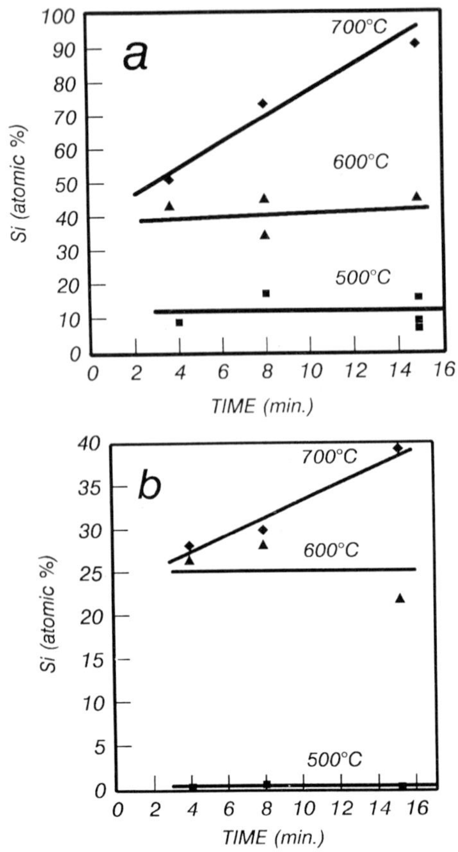


Fig. 1. Silicon concentration of iron siliconized in 800 ppm $SiH_4$–$H_2$: (a) surface silicon by AES; (b) near surface silicon by XRF.

is not detected by the XRF technique. At 700 °C, the silicon near-surface concentration increases with treatment time.

The samples were also analyzed by XRD to determine the phases present after siliconizing. Table 1 lists the phases observed. Two phases, FeSi and $Fe_3Si$, are detected in all cases. The primary phase observed at 500 °C is $\alpha$-Fe; at 600 °C, $Fe_3Si$; at 700 °C, FeSi. This trend to higher silicides at higher temperatures was also observed for silicides formed by interdiffusion between two metal layers [12]. There is also a minor trend to higher silicides with longer treatment times, although this trend is less dramatic. The XRD data agree with the XRF analyses. The depth of penetration for XRD is about 7 $\mu$m; therefore, XRD and XRF characterize the near surface region rather than the surface (top atomic layers).

Rates of reaction of silane with metal surface were determined using an *in-situ* microgravimetry. An iron foil was cleaned in dry hydrogen at 800 °C,

TABLE 1
XRD phases for iron samples siliconized in 800 ppm $SiH_4$–$H_2$

| *Treatment* | | *XRD phase*[a] | | |
|---|---|---|---|---|
| *Temperature* (°C) | *Time* (min) | *α-Fe* | $Fe_3Si$ | *FeSi* |
| 500 | 4 | S | M | — |
| 500 | 8 | S | M | M |
| 500 | 15 | S | M | M |
| 600 | 4 | W | S | M |
| 600 | 8 | — | S | M |
| 600 | 15 | W | S | S |
| 700 | 4 | W | M | S |
| 700 | 8 | W | M | S |
| 700 | 15 | W | M | S |

[a]S: strong diffraction pattern intensity; M: moderate intensity (around 10%); W: weak intensity (less than 10%).

and then treated at 600 °C. Figure 2 displays weight *vs.* time plots for iron and a control quartz coupon with the same geometric surface area.

Iron gains weight rapidly at first and then slowly at a fairly constant rate. After about 10 min, the rate of deposition for iron matches that for the quartz control. This slow silicon deposition rate results from the homogeneous gas-phase decomposition of silane. The initial rapid uptake must then be surface catalyzed. Initially, iron sites on the clean surface, which are continually exposed as silicon diffuses into the bulk, catalyze the silane decomposition. When silicon diffusion slows, the surface iron sites become covered with silicon, and the rate reverts to that observed for the homogeneous reaction.

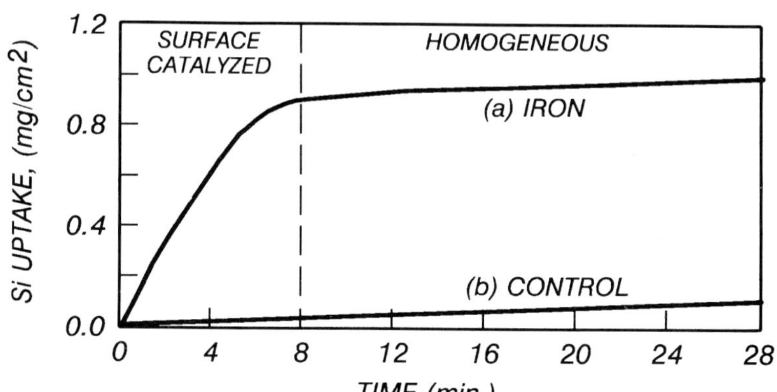


Fig. 2. Weight *vs.* time curve for samples exposed to 0.1% $SiH_4$–$H_2$ at 600 °C: (a) iron foil; (b) quartz control.

*3.2 Control of $H_2O$ level*

The presence of a surface oxide can prevent diffusion of silicon into the metal when it is treated with $SiH_4$. Surface oxides that form on iron and low-alloy steels are relatively easy to reduce to the metal using a $H_2$ treatment. However, when iron is alloyed with significant amounts of chromium, chromium oxide will form on the surface. For these alloys, the levels of oxygen contaminants in the $H_2$ must be kept very low for the atmosphere to be thermodynamically reducing to the oxide.

Several key experiments demonstrated the careful control of the amount of oxygen contaminant in the gas required to produce silicon diffusion coatings on Cr–Fe alloys at low temperatures using $SiH_4$–$H_2$ gas mixtures. Coupons of alloy A182F9 (9%Cr–1%Mo–Fe) were pretreated in $H_2$ and then treated at 500 °C in 500 p.p.m. $SiH_4$–$H_2$ mixtures containing different levels of

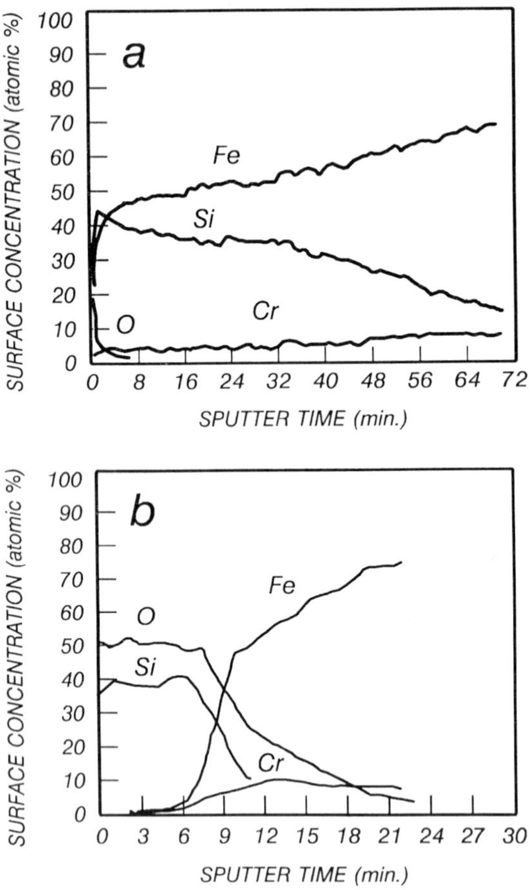


Fig. 3. AES depth profiles of 9% Cr–1% Mo steel siliconized with 500 ppm $SiH_4$ in $H_2$: (a) $H_2O/H_2 \leq 1 \times 10^{-5}$ (sputter rate was 14 nm min$^{-1}$ for first 26 min and 30 nm min$^{-1}$ thereafter); (b) $H_2O/H_2 = 4.5 \times 10^{-3}$ (sputter rate was 14 nm min$^{-1}$).

water vapor in the microbalance system. For one of the samples, the $H_2O/H_2$ ratio was $1 \times 10^{-5}$ for both the pretreatment and treatment steps, whereas for the other the ratio was $4.5 \times 10^{-3}$. Auger depth profiles were obtained and they are displayed in Figs. 3(a) and (b).

The sample that was treated with low concentrations of $H_2O$ ($H_2O/H_2 = 1 \times 10^{-5}$), exhibits a thick siliconized region (about 1.7 $\mu$m), containing iron, chromium and silicon. The silicon concentration decreases from about 40 to about 15 at.% with depth, while the iron and chromium concentrations increase to approximate their values in the bulk (molybdenum is not observed). A small amount of $SiO_2$ formed on top of the silicide when the sample was exposed to air at room temperature during transfer from the treatment system to the surface analysis unit. For the sample that was treated with higher amounts of $H_2O$ ($H_2O/H_2 = 4.5 \times 10^{-3}$), a layer of silicon oxide of thickness of about 0.13 $\mu$m covers the surface. The atomic concentrations suggest a stoichiometry between $SiO_2$ and SiO. There appears to be an oxide region about 50 nm thick below the silicon-containing overlay coating. It is enriched in chromium relative to the concentration of chromium in the bulk. We hypothesize that a chromium rich oxide forms during the pretreatment in wet $H_2$, and that the oxide acts as a barrier to prevent diffusion of silicon into the bulk. It also slows the rate of deposition of silicon by preventing catalysis of $SiH_4$ decomposition by the elements of the substrate.

Comparison of these two treatments demonstrates the critical role of oxygen contaminants. Although the two experiments involve mixtures of the same gases, they yield entirely different results. When the treatment is performed under reducing conditions, a silicon diffusion coating containing silicon and iron is obtained. When higher amounts of $H_2O$ are used, the treatment results in a silicon oxide overlay coating.

### *3.3. High temperature oxidation resistance*

Experiments were performed to demonstrate the utility of this treatment for providing high temperature oxidation resistance. Rates of oxidation of siliconized metal foils were determined in pure $O_2$ at 800 and 1000 °C using microgravimetry.

Foils of 1010 carbon steel were investigated first. The samples were siliconized in a mixture of 0.12% $SiH_4$ in $H_2$ at 600 °C until they gained 0.16, 0.31 or 0.47 mg Si cm$^{-2}$. (A weight gain of 0.16 mg Si cm$^{-2}$ is enough to form an $Fe_3Si$ coating of thickness about 1.3 $\mu$m.) The samples yielded linear oxidation rates over 1 h at both 800 and 1000 °C. Table 2 compares the oxidation rates of siliconized samples with the oxidation rate of an untreated 1010 steel sample. Increasing the silicon level decreases the oxidation rate. For the sample that contained 0.31 mg Si cm$^{-2}$, the oxidation rate at 800 °C was $1.4 \times 10^5$ times lower than untreated 1010 steel. Even at 1000 °C, a sample siliconized to 0.47 mg Si cm$^{-2}$ oxidized at a rate of only 0.60 $\mu$g cm$^{-2}$ min$^{-1}$.

These experiments demonstrate the feasibility of imparting high temperature oxidation resistance by siliconizing the surface of ferrous alloys.

TABLE 2
Oxidation rates of siliconized[a] steel tested at 800 and 1000 °C in 1 atm of $O_2$

| *Steel* | *Silicon level* (mg cm$^{-2}$) | *Oxidation* | | |
|---|---|---|---|---|
| | | *Temperature* (°C) | *Linear rate*[b] ($\mu$g cm$^{-2}$ min$^{-1}$) | *Improvement*[c] |
| 1010 | — | 800 | $2.71 \times 10^4$ | — |
| 1010 | 0.16 | 800 | 1.19 | $2.3 \times 10^4$ |
| 1010 | 0.31 | 800 | 0.20 | $1.4 \times 10^5$ |
| 1010 | 0.31 | 1000 | 1.81 | $1.5 \times 10^4$ |
| 1010 | 0.47 | 1000 | 0.60 | $4.5 \times 10^4$ |
| 430 SS | — | 1000 | 22.6 | — |
| 430 SS | 0.52 | 1000 | 3.0 | 7.5 |
| 304 SS | — | 1000 | 2.8 | — |
| 304 SS | 0.20 | 1000 | 1.9 | 1.5 |

[a]Siliconized in $SiH_4$–$H_2$ at 600 °C.
[b]From 0.5 to 2 h.
[c]Improvement = untreated rate/siliconized rate.

However, carbon steel is not typically used at these temperatures because it does not have the high temperature strength.

Consequently, we also investigated the utility of this treatment for improving the oxidation resistance of alloys used at higher temperatures. Since the chromium that is added to these alloys to give strength at high temperatures also imparts a significant amount of oxidation resistance, the benefits of siliconizing are expected to be lower than the benefits for 1010 steel. Table 2 lists oxidation rates for representative ferritic and austenitic stainless steels. Siliconizing 430 SS improves its oxidation resistance by a factor of 7. The benefits for 304 SS are marginal, in part because the oxidation resistance of the untreated alloy is so good, and the rates are near the detection limits of our microbalance. Additional experiments demonstrate that silizonizing improves the resistance of 304 SS to scaling when exposed to oxidizing atmospheres with thermal cycling.

## 4. Conclusions

These results suggest that a variety of metals and alloys can be siliconized using $SiH_4$–$H_2$ mixtures at relatively low temperatures. However, to obtain diffusion coatings, the level of oxygen contaminant must be carefully controlled to reduce or prevent formation of any oxide layer that may act as a diffusion barrier to silicon. Siliconizing metals in this way increases their high-temperature oxidation resistance.

## Acknowledgments

We would like to thank R. Pierantozzi, R. Miller and J. Armor for technical discussions, and C. Ceccarelli, E. Karwacki and K. Wood for analyses. The technical expertise of J. Zehner and J. Arbogast-Schappell is gratefully acknowledged. Thanks are due to Air Products and Chemicals, Inc., for permission to publish this work.

## References

1 H. W. Grünling and R. Bauer, *Thin Solid Films*, *95* (1982) 3.
2 A. G. Bosnischen Elektrizitats, German Patent 302305 (1915).
3 R. von Boss and E. Kruppa, French Patent 774656 (1932).
4 A. Abba, A. Galerie and M. Caillet, *Mater. Chem.*, *5* (1980) 147 - 164.
5 M. Pons, A. Galerie and M. Caillet, *Mater. Chem. Phys.*, *8* (1983) 153 - 161.
6 C. F. Powell, in C. F. Powell, J. H. Oxley and J. M. Blocher (eds.), *Vapor Deposition*, John Wiley, New York, 1966, p. 403.
7 M. P. Hill and J. F. Mathews, U.K. Patent 2107360 (1985).
8 A. T. Cocks and P. L. Surman, U.K. Patent 1530337 (1978).
9 L. H. Dubois and R. G. Nuzzo, *J. Vac. Sci. Technol.*, *A2* (2) (1984) 441 - 445.
10 A. L. Cabrera, J. F. Kirner, R. A. Miller and R. Pierantozzi, U.S. Patent 4,714,632 (1987).
11 A. L. Cabrera, N. D. Spencer, E. Kozak, P. W. Davis and G. A. Somorjai, *Rev. Sci Instrum.*, *53* (1982) 1888.
12 J. M. Poate, in K. N. Tu and J. W. Mayer (eds.), *Thin Films — Interdiffusion and Reactions*, John Wiley, New York, 1978, p. 359.

*Surface and Coatings Technology, 39/40* (1989) 53 - 64 

# DIFFUSION CHROMIZING OF FERROUS ALLOYS

G. H. MEIER and C. CHENG

*Department of Materials Science and Engineering, University of Pittsburgh, Pittsburgh, PA 15261 (U.S.A.)*

R. A. PERKINS

*Lockheed Palo Alto Research Laboratory, Palo Alto, CA 94304 (U.S.A.)*

W. BAKKER

*Electric Power Research Institute, Palo Alto, CA 94304 (U.S.A.)*

(Received March 7, 1989)

## Summary

The results of an extensive study of the mechanism of formation of chromium diffusion coatings on ferrous alloys are described. The effects of the significant coating variables on the composition, thickness and microstructure of the coatings are discussed. These variables include the pack geometry, substrate composition, type of halide activator, type of "inert filler", time, temperature and the type of the chromium source material. The diffusion of chromium and chromium plus vanadium are considered. The chromium concentration profiles are seen to deviate from those predicted from the Fe–Cr phase diagram. Mechanisms for the coating formation are described.

## 1. Introduction

Pack chromizing has been used for many years as a means to increase the oxidation and corrosion resistance of ferrous alloys [1 - 3]. The coating process involves packing the steel in a powder mixture which contains a chromium source (pure chromium or ferrochromium), an inert filler (usually alumina) and a halide activator ($NH_4Cl$, $NH_4I$, etc.). The pack is heated in dry inert gas or hydrogen for 3 - 4 h at 1000 - 1300 °C to form coatings in the range of 2 - 5 mm thick [1]. A typical commerical coating on a low alloy steel is characterized by an outer chromium carbide layer and an underlying diffusion zone in which the chromium concentration decreases from 30 - 50 wt.% beneath the carbide case to about 10 wt.% at the interface with the steel. Carbides of chromium are present in the grain boundaries, often as massive particles. A decarburized zone is present in the steel just below the coating and is the source of the carbon from which the outer case and carbide

0257-8972/89/$3.50 

particles in the coating are formed. Coatings free of a carbide case and internal carbides can be formed on pure iron, decarburized steels, or steels stabilized with strong carbide formers such as titanium or niobium [1]. Coatings formed on steels with carbon often contain a substantial amount of porosity and random large voids near the surface just below the carbide layer.

Recently, there has been renewed interest in chromizing of low-alloy steels for use in coal conversion atmospheres. The major goal is to improve both the fireside and steamside corrosion resistance of low alloy boiler steels for waterwall and heat exchanger tubing. This application requires the coatings to resist oxidation, sulfidation, and aqueous pitting corrosion. Pitting resistance of Fe–Cr alloys is known to improve with small additions of refractory metals such as molybdenum or vanadium [4]. Therefore, the ability to incorporate these elements into chromized coatings is important.

Despite the long use of the chromizing process a fundamental understanding of the mechanism of coating formation is still lacking. Commercial coatings, while often effective, have been developed by empirical means. In 1985, an exploratory research program was initiated by the Electric Power Research Institute to provide an improved understanding of the mechanisms of coating formation during pack chromizing of ferrous alloys and the mechanisms whereby a second element (*e.g.* molybdenum or vanadium) could be incorporated into the coatings. The goal was to identify how process and material variables influenced coating microstructure and performance in corrosive environments. Some of the results of this project are summarized in this paper.

## 2. Technical background

Early studies of chromizing have been extensively reviewed by Samuel and Lockington [2] and Samuel [3]. A model of the chromizing process as described in these reviews is presented in Fig. 1. The process must be conducted in the absence of air to prevent oxidation of chromium in the pack. Although coatings have been produced in inert atmospheres, dry hydrogen is used in most cases. Hydrogen will reduce chromium oxides and may reduce halides to deposit chromium on the metal surface. Cracked ammonia has also been used and was found to give somewhat more uniform results than hydrogen [2]. The halides have been evaluated as activators with ammonium chloride being used in most cases.

Chromium is transferred to the steel surface by one or more of several possible deposition reactions. Previous investigators [2] recognized three possible reactions:

$$\text{Interchange: } Fe(s) + CrX_2(g) = FeX_2 + Cr(s) \tag{1}$$

$$\text{Reduction: } CrX_2(g) + H_2(g) = 2HX(g) + Cr(s) \tag{2}$$

$$\text{Thermal dissociation: } CrX_2 = X_2(g) + Cr(s) \tag{3}$$

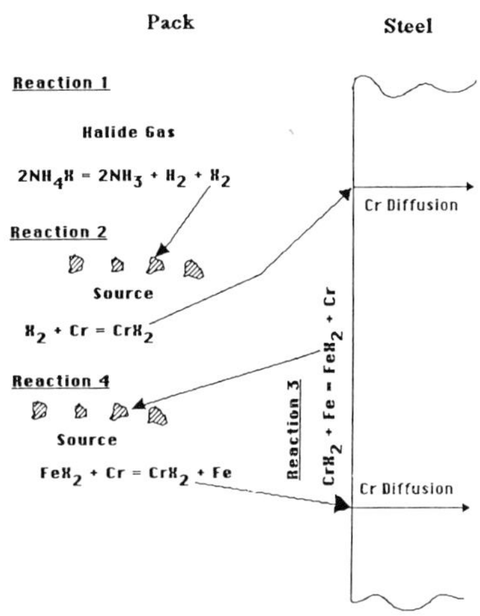


Fig. 1. Schematic representations of the pack chromizing process.

Based on weight change and chemical analysis for total chromium, it was deduced that the deposition reaction using chlorine, bromine or fluorine activators was by hydrogen reduction below 1000 °C and by interchange above 1200 °C. Between 1000 and 1200 °C a mixed reaction occurred with a decreasing amount by reduction on increasing temperature. The reaction was estimated to be 50% by interchange at 1100 °C [2]. Substantial deposition occurred by thermal decomposition for iodide activators.

The deposited chromium diffuses into the iron while the product gases are transported back into the pack and recycle the activator. Coating thickness was observed to increase as $(\text{time})^{0.5}$ after a short initial period (about 2 h) and to be relatively insensitive to the activator used. The rate of chromizing is controlled by the uptake of chromium in the initial period and by the inward diffusion of chromium once a steady state surface concentration is achieved. The steady state concentration is dependent on the rate of the surface reaction and the rate of inward diffusion. The average chromium concentration in a coating was found to decrease as the deposition temperature increased as a result of more rapid inward diffusion relative to the surface reaction. Coating alloys containing $\gamma$-stabilizing elements produced thinner coatings with higher chromium contents while $\alpha$-stabilizing elements produced thick coatings with lower chromium contents. Diffusion is more rapid in the $\alpha$ phase at all temperatures. The tendency is to use the lowest possible temperature compatible with effective surface protection from both a technical and an economical point of view. Processing temperatures normally are in the range of 900 - 1100 °C.

Coating of steels containing carbon has been recognized as a problem area. Substrates with significant amounts of carbon form chromium carbide cases which were considered to shut off chromium diffusion [3]. Avoidance of this problem would be accomplished by the following.

1. Use of low carbon (0.1%) steels for coating.
2. Decarburizing higher carbon steels before coating.
3. Adding stable carbide-forming elements (titanium, niobium, tungsten, molybdenum, vanadium) to the steel [3].
4. Codiffusing Cr and carbide forming elements such as vanadium [2].

Chromizing in a hydrogen atmosphere also could promote surface decarburizing of the steel. Generally, previous investigators believed that the presence of hydrogen was necessary for effective chromizing.

The formation of multielement chromized coatings has been discussed extensively in the Russian literature [5]. Recently Rapp *et al.* [6] have analyzed the problem of codeposition of aluminum and chromium and have successfully prepared Fe–Cr–Al coatings on iron. A major finding in this work was the need to control the pack atmosphere since aluminum halides are much more stable than those of chromium which results in low chromium contents in the coating. This problem was overcome by choice of the activator and by using a Cr–Al masteralloy source in which the activity of aluminum was low.

## 3. Experimental procedure

An experimental program was designed to delineate factors in the process that control the surface concentration of chromium, coating thickness and depth profiles, carbide formation, and formation of voids and porosity. Particular attention was devoted to the role of halide activators, chromium deposition reactions and substrate alloy composition. The chromizing process was analyzed in terms of thermochemistry and diffusion to guide and interpret experiments. Chromizing experiments were conducted and the coatings produced were characterized for all important features.

### *3.1. Thermochemistry and diffusion*

The thermochemistry of metal halide formation and decomposition was evaluated using thermodynamic data obtained from a variety of sources [7 - 9]. The concentrations of the gaseous species in the pack were calculated using a modification of the SOLGAS-MIX program [10]. Chromium diffusion profiles in pure iron were modeled for a range of fixed surface and interface concentrations as established by surface reactions and alloy phase relations. Published diffusion data and alloy phase diagrams [11, 12] were used as the basis for all calculations.

### *3.2. Chromizing experiments*

Pack chromizing was done in a resistance heated tube furnace at 1150 °C under an atmosphere of slowly flowing argon. Two different types of activators were used: (1) $NH_4X$ which releases hydrogen by dissociation on heating and (2) $MCl_2$ which does not contribute hydrogen to the atmosphere. The following activators were used in the tests: $CrCl_2$, $FeCl_2$, $NH_4Cl$ and $NH_4F$.

Experiments were conducted in two different types of pack container designs as illustrated in Fig. 2. The double steel can (Type B) restricts venting of halide vapors to the furnace tube and is designed to maintain a high concentration of halides throughout the run. The alumina thimble (Type A) allows more rapid venting and halide concentrations should drop with time of exposure. (The surface chromium concentration was found to decrease with time for the Type A packs so only data from Type B packs will be discussed in this paper.) The pack mixtures for most chromizing experiments was 30% source (chromium or ferrochromium), 5% halide activator and 55% filler ($Al_2O_3$) by weight. Multielement deposition packs were similar except that additional source materials (vanadium, ferrovanadium) were included.

Four alloys were studied to determine the effect of substrate composition on coating formation as listed below:

1. SA213 Grade T-22 Steel (2.25% Cr–1% Mo)
2. SA213 Grade T-22 Nb Steel (2.25% Cr–1% Mo–1% Nb)
3. Fe–2% Cr
4. Armco iron

The first one is a standard boiler tube steel and contains 0.2% carbon. The second is similar to T-22 but has niobium added to stabilize the carbides. The Fe–2% Cr alloy is a high purity model material prepared by melting electrolytic chromium and iron while the last is a high purity iron deoxidized with titanium. All surfaces were ground on 120 grit SiC paper prior to coating.

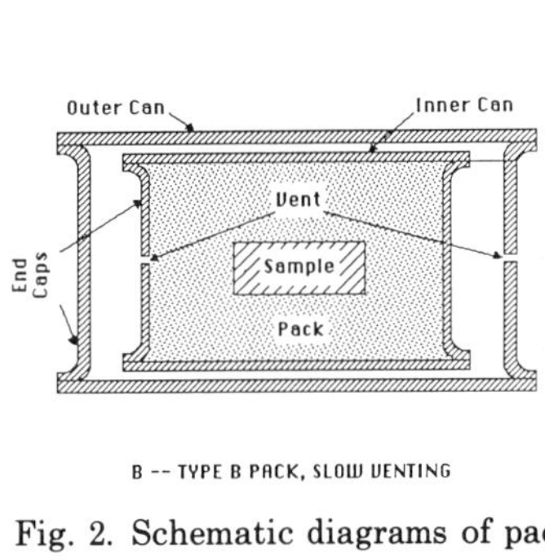


Fig. 2. Schematic diagrams of packs used for chromizing experiments.

In some tests the tube furnace and packs were evacuated and back filled with argon while in others only an argon purge was used. The packs were heated to the process temperature of 1150 °C in 3 h, held at 1150 °C for the prescribed time (0, 1, 3, 5.5, 16 and 28 h) and cooled to 600 °C in 2 h. All packs had a transient exposure of 2 h in the range 950 - 1150 °C during heating and cooling. The process temperature of 1150 °C is that of the samples and was measured during many runs by inserting a thermocouple in the pack adjacent to the specimens.

*3.3. Coating evaluation*

The specific weight changes of each specimen were measured after the samples were wire brushed to remove adhering pack material. The specimens were examined in cross-section by optical metallography and the coating thicknesses were measured. The chromium concentration profile of each sample was determined at intervals of 2 $\mu$m by energy dispersive X-ray spectroscopy (EDS) measurements with the counting time standardized at 90 s. The raw data, after ZAF correction, were correlated to obtain the actual concentrations using a correlation equation which was established from eight Fe–Cr standards in the range of 2 - 30 wt.% Cr. Standards were not used for vanadium analyses.

## 4. Analysis of the chromizing process

*4.1. Thermochemistry*

The results of pack atmosphere calculations for chloride- and fluoride-activated packs are presented in Figs. 3 and 4 respectively. Comparison of the

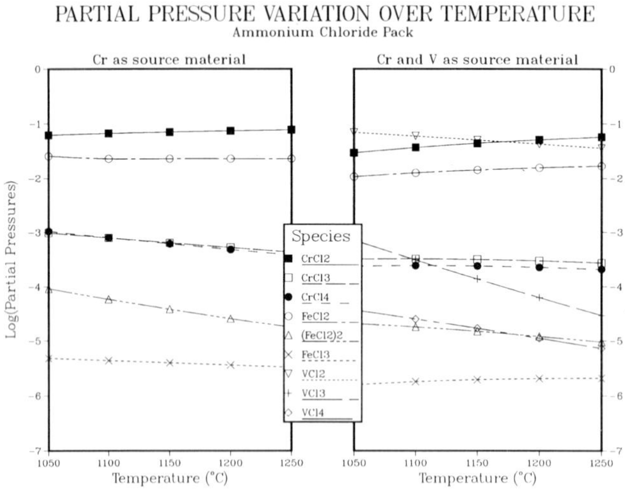


Fig. 3. Atmosphere composition calculations for $NH_4Cl$-activated packs for diffusion coating of iron. Left is straight chromizing. Right is chromvanadizing.

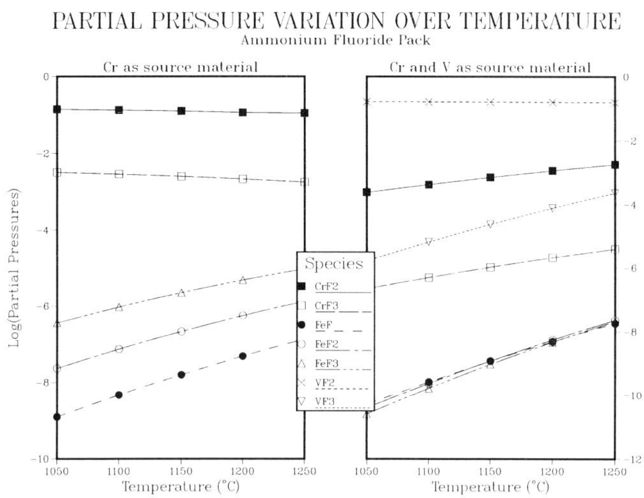


Fig. 4. Atmosphere composition calculations for $NH_4F$-activated packs for diffusion coating of iron. Left is straight chromizing. Right is chromvanadizing.

calculations for straight chromizing indicates that dichloride and difluoride are the major chromium species and have similar pressures. A major difference, however, occurs for the iron species. The dichloride of iron has a high pressure while the difluoride has a negligible pressure. This indicates that deposition by an interchange reaction, such as eqn. (1), can deposit Cr in a chromium chloride pack but is unlikely in a fluoride pack. Comparison of the vanadium-containing packs with those containing only chromium show that $VCl_2$ and $VF_2$ develop high pressures and the presence of vanadium reduces the pressure of the major chromium species, particularly for fluoride packs. Therefore, one would expect lower chromium concentrations in Cr–V codeposition packs.

*4.2. Diffusion*

The development of the chromium diffusion profiles in pure iron was modelled for the case of a fixed chromium concentration being established at the substrate surface by one of the deposition reactions. The Fe–Cr phase diagram indicates that the zone will involve two phases, $\alpha$ and $\gamma$, for temperatures between 912 and 1394 °C. Figure 5 shows the calculated profile for chromizing of pure iron for a fixed surface concentration of 30 wt.%, temperature 1150 °C and diffusion time of 24 h. These calculations allow the following generalizations to be made regarding the characteristics of the diffusion zone:

1. Coating thickness is almost directly proportional to the surface concentration $C_s$ established by the pack.
2. Coating thickness, for a fixed $C_s$, increases substantially with only a small increase in temperature.
3. The diffusion zone thickness in the $\gamma$ field is independent of $C_s$.

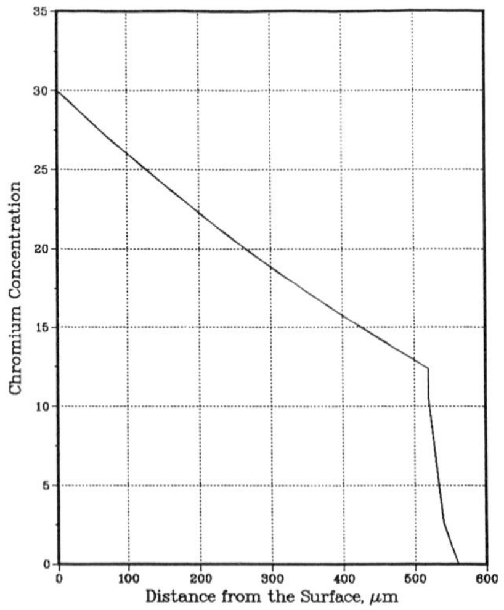


Fig. 5. Calculated chromium concentration profile for chromizing pure iron at 1150 °C for 24 h with a constant surface concentration of 30 wt.% Cr.

This type of calculation also allows evaluation of the effect of solutes in the iron substrate which shift the phase boundaries. Solutes which stabilize $\gamma$ decrease coating thickness while $\alpha$-stabilizers increase it.

## 5. Experimental results and discussion

### *5.1. Surface concentration of chromium*

Figure 6 shows experimental concentration profiles for two coating runs in fluoride activated packs. An alumina filler was used in the packs. Figure 6(A) represents straight chromizing while Fig. 6(B and C) represents chromvanadizing. The surface chromium concentration is seen to be greatly reduced in the presence of vanadium which results from the reduced $CrF_2$ pressure, as described above. The presence of the alumina filler was also found to affect surface concentrations in the fluoride-activated packs. The alumina reacted with the activator to form Al-fluorides and resulted in reduced chromium concentrations compared to alumina-free packs. Similar results were not observed for chloride-activated packs since the chlorides of aluminum are less stable than the fluorides.

### *5.2. Thickness and depth profiles*

The concentration profiles in Fig. 6(A) are similar to the calculated curves except that the experimental coating thickness is always much thinner than the calculated thickness. Figure 7 shows the time dependence of the coating thickness for four substrates coated in $FeCl_2$-activated packs. The time dependence of coating thickness is greater than the value of 0.5 expected for diffusion control. The reason for this and the deviation between the

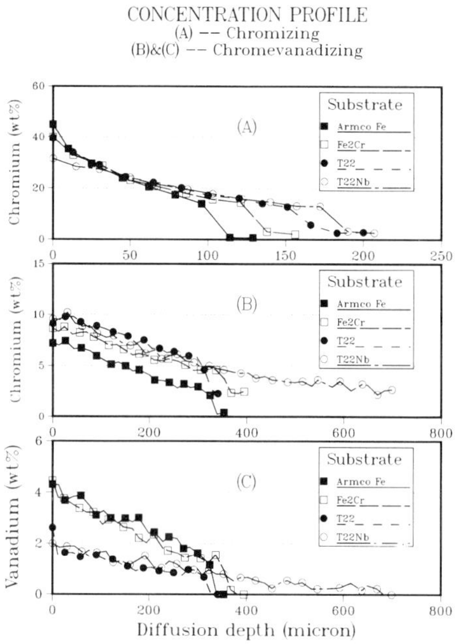


Fig. 6. Experimental concentration profiles for chromizing (A) and chromvanadizing (B and C) of various substrates in $NH_4F$-activated packs for 16 h at 1150 °C.

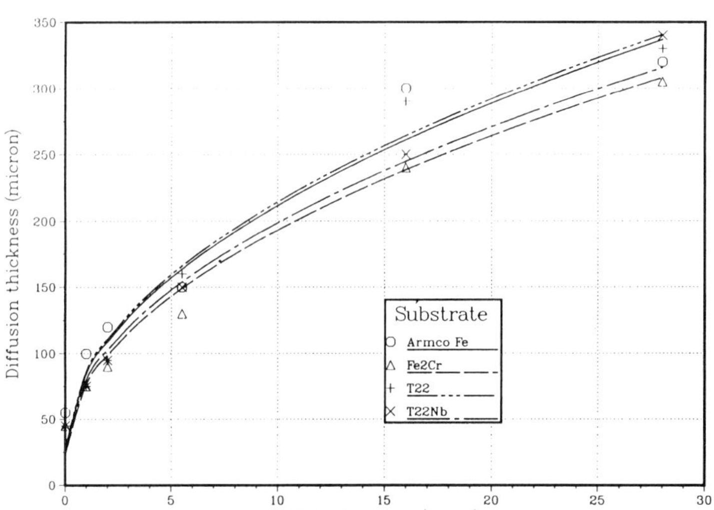


Fig. 7. Coating thickness *vs.* time for various substrates chromized in $FeCl_2$-activated packs at 1150 °C.

experimental and calculated thicknesses is believed to result from non-equilibrium at the $\alpha - \gamma$ interface. Similar behavior has been reported for Fe–Cr diffusion couples [13] where the mobility of the $\alpha - \gamma$ interface was found to be too low to keep up with diffusion. This explains the reduced coating thickness and time dependence greater than 0.5 since a linear and a parabolic process are coupled in this case. This point is being investigated further.

Comparison of Fig. 6(A) with 6(B) indicates the chromium concentration profile penetrates much deeper for the chromvanadized coating even though the surface concentration is lower. This is interpreted in terms of the vanadium stabilizing the $\alpha$ phase in which the chromium diffusivity is two orders of magnitude higher than in the $\gamma$ phase.

The comparison profiles established for each specimen were integrated to obtain the area under the curve. This gives the total chromium uptake which can be compared with the measured specimen weight gain (Table 1). This comparison indicates that for chloride packs an interchange reaction is responsible for virtually 100% of the chromium deposition on carbon-free substrates (iron and Fe–2Cr%). The interchange reaction would result in a weight loss of 7% of the chromium uptake due to the difference in atomic weights of iron and chromium. Similar comparisons cannot be made for carbon-containing steels since part of the weight gain results from formation of an outer chromium-carbide layer. The comparison of weight gain with chromium uptake for fluoride packs indicates good correspondence between the two. This indicates negligible chromium deposition by an interchange reaction. The high stability of $CrF_2$ precludes the possibility of thermal decomposition. Therefore, deposition must occur by disproportionation ($CrF_2$–$CrF_3$) and hydrogen reduction.

### 5.3. Coating microstructures

Figure 8 shows typical cross-sections of a chromium and Cr–V coating on T-22 deposited using a chloride pack. The greater thickness of the Cr–V

TABLE 1
Characteristics of chromized coatings[a]

| *Substrate* | *Activator (source)* | *Coating thickness* ($\mu$m) | *Surface chromium* (wt.%) | *Chromium uptake*[b] (mg cm$^{-2}$) | *Measured* $\Delta m/A$ (mg cm$^{-2}$) |
|---|---|---|---|---|---|
| Armco | $FeCl_2$(Cr) | 378 | 70.5 | 72.0 | −2.7 |
| Armco | $FeCl_2$(FeCr) | 375 | 58.3 | 76.7 | −5.0 |
| Fe–2%Cr | $FeCl_2$(FeCr) | 355 | 50.7 | 57.2 | −10.0 |
| Armco | $NH_4F$(FeCr) | 263 | 22.7 | 29.0 | 24.8 |
| Fe–2%Cr | $NH_4F$(FeCr) | 254 | 20.9 | 27.5 | 25.6 |
| Armco | $NH_4F$(Cr + V) | 353 | 7.4 | 19.4 | 21.8 |
| Fe–2%Cr | $NH_4F$(Cr + V) | 395 | 8.8 | 19.0 | 23.3 |

[a]16 h–1150 °C pack process.
[b]From integrated area under Cr ( + V) concentration profile.

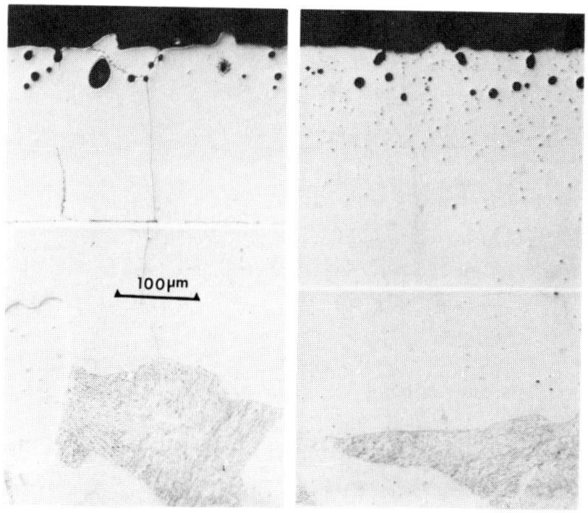


Fig. 8. Cross-sections of T-22 which has been chromized (left) and chromvanadized (right) at 1150 °C in an $NH_4F$-activated pack for 16 h.

coating is evident. The chromium coating is covered with a chromium carbide layer which is absent on the Cr–V coating. The vanadium, which forms much more stable carbides than chromium, has "stabilized" the coating. Substantial amounts of porosity are evident in both coatings. This is believed to result from the counterdiffusion of chromium and iron since coatings in chloride packs are deposited by an interchange reaction. The porosity is generally more extensive in the Cr–V coatings.

## 6. Summary

The formation of chromium and Cr–V coatings on iron from chloride- and fluoride-activated packs has been evaluated. Chloride coatings form by a displacement reaction while fluoride coatings form by a combination of disproportionation reaction and hydrogen reduction. The presence of vanadium results in coatings which are thicker but have lower surface chromium concentrations. Porosity is increased in the Cr–V coatings. The chromizing process has been shown to be partially controlled by migration of the $\alpha$-$\gamma$ interface.

## Acknowledgments

This work was supported by the Electric Power Research Institute under contract RP 2742-1 with the University of Pittsburgh as the prime contractor and Lockheed Missiles and Space Co. as a subcontractor. The authors gratefully acknowledge the support and contributions of D. Marsh and A. Lopeman in conducting all coating experiments and of L. Iosty for preparation and optical metallography of all cross-sectioned specimens. Helpful discussions with Professor W. A. Soffa are also acknowledged.

## References

1 *ASM Metals Handbook*, 1948 Edn, ASM, Cleveland, OH, p. 705.
2 R. L. Samuel and N. A. Lockington, *Metal Treatment and Drop Forging*, *18*, (1951) 354, 407, 440, 495.
3 R. L. Samuel, *Murex Rev.*, *1* (1958) 501.
4 A. J. Sedriks, *Corrosion*, *42* (1986) 376.
5 L. S. Lyakhovich (ed.), *Multicomponent Diffusion Coatings*, translated from Russian, NBS, Washington, D.C., 1987.
6 R. A. Rapp, D. Wang, and T. Weisert, Simultaneous chromizing and aluminizing of iron and iron-base alloys by the pack cementation process," in M. Khobaib and R. Krutenat (eds.), *Metallurgical Coatings*, TMS–AIME, Warrendale, PA, 1987, p. 131.
7 JANAF Thermochemical Tables, 3rd edn., *J. Phys. Chem. Ref. Data*, *14* (1985) Parts 1 & 2.
8 C. F. Wicks and F. E. Block, *Thermodynamic Properties of 65 Elements — Their Oxides, Halides, Carbides, and Nitrides*, Bulletin 605, U.S. Bureau of Mines, 1963.
9 L. B. Pankratz, *Thermodynamic Properties of Halides*, Bulletin 674, U.S. Bureau of Mines, 1984.
10 R. A. Rapp, Ohio State University, private communication.
11 M. A. Kristhal, *Diffusion Process in Iron Alloys*, Israel Program for Scientific Translations Ltd., 1970, p. 177.
12 T. B. Massalski, J. L. Murray, L. H. Bennett, and H. Baker (eds.), *Binary Alloy Phase Diagrams*, ASM, Metals Park, OH, 1986, p. 822.
13 T. Nishizawa and A. Chiba, *Trans. Jpn. Inst. Met.*, *16* (1975) 767.

# THE USE OF TAFALOY 45CT, AN Ni-Cr-Ti ALLOY, AS AN ARC SPRAYED CORROSION BARRIER IN HIGH TEMPERATURE SULFUROUS ENVIRONMENTS

WALTER ZANCHUK

*TAFA Incorporated, 146 Pembroke Road, Concord, NH 03301 (U.S.A.)*

(Received March 12, 1989)

## Summary

A high chromium content nickel alloy, Tafaloy 45CT, was developed to combat corrosion in high temperature sulfurous environments. The alloy is applied by arc thermal spraying at thicknesses of $15 - 20 \times 10^{-3}$ in (381 - 508 $\mu$m) onto surfaces such as water wall tubes in black liquor recovery boilers.

The coating has a bond strength of 6500 lbf in$^{-2}$ (44.82 MPa) and is well matched to the thermal expansion characteristics of steels used for boiler tube construction.

Analysis of the coating indicates the presence of two solid solutions of chromium in nickel and two chromium oxides. The major component of the metallic phases is the high chromium (44%) bulk material. The other components are a small portion of the coating.

Coatings in service for 18 months show surface sulfur concentrations amounting to only 1%. The sulfur is effectively blocked from more penetration by combining with the chromium and the formation of a self healing chromium oxide film on the coating's surface and within the pores with access to the atmosphere. The coating remains well bonded and protective. Coatings have been in service for over four years without failing.

## 1. Introduction

High temperature sulfurous environments, such as those found in black liquor recovery boilers, utility boilers and petrochemical processing create demands on materials which are often unmet. Thermally sprayed coating materials for these conditions must not only be chemically resistant, but must have the bond strength to resist spalling, or catastrophic separation, from the substrate. They must be self sealing to prevent the migration of the corrosion environment through the coating's imperfections to the less corrosion resistant substrate.

0257-8972/89/$3.50 

One coating system which has successfully met these performance criteria is arc sprayed Tafaloy 45CT, a Ni-Cr-Ti alloy original developed as a welding alloy for high sulfur and vanadium environments. The most widespread use now is in black liquor recovery boilers in the pulp and paper industry. Coatings have been in service for over four years with no signs of failure. This service compares favorably to previous thermal spray systems which required replacement in one year or less.

## 2. Tafaloy 45CT development and coating application

The resistance of high chromium content nickel alloys to high temperature sulfidation and oxidation is well documented [1] (see Fig. 1). The difficulty of mechanically working these materials had limited their use to castings and minimally worked wrought forms. A research program at INCO in the late 1970s focused on development of workable welding filler metals to repair these alloys [2]. The research led to chromium composition ranges which had the desired workability and provided the corrosion protection needed for the environments. The addition of titanium to these alloys produced crack free welds. It is felt that the titanium deoxidizes the weld puddle and promotes the solidification of the weld metal as a two phase structure of $\alpha$-Cr in a $\gamma$ matrix, which improves workability. The current composition of the filler metal, known as Filler Metal 72, is 55% nickel, 44% chromium and 1% titanium. It has the highest chromium content of any wire product. It has seen use in applications such as cladding marine diesel engine valves, welding high chromium content alloys and cladding burner tips in high sulfur feedstock gasifiers. Interestingly, it also has excellent resistance to stress corrosion cracking. This property makes it the material of choice in welding repairs in boiling water nuclear reactors where the use of ultrapure water creates severe stress corrosion environments.

During the same time frame in which INCO developed Filler Metal 72, the British Central Electricity Generating Board conducted studies of high chromium content nickel alloys for use with high sulfur content oils and coals [3]. Their results showed chromium contents above 40% are needed to provide protection against sodium sulfate, vanadium pentoxide and other fuel generated corrosive compounds, and to prevent the formation of low melting point nickel sulfides. The results of this study prompted the search for, and selection of, the INCO alloy for arc wire spraying in high temperature sulfurous environments.

Special processing techniques were developed to make the material suitable for the wire feeding and arc formation needs of the wire arc process. In this form the material is known as Tafaloy 45CT. Besides preparing the wire for spraying, great care is taken in spraying at the job site, as most of the applications are done in the field. Spraying parameters are optimized for depositing a $15-20 \times 10^{-3}$ in ($381-508\ \mu$m) coating. A specially designed,

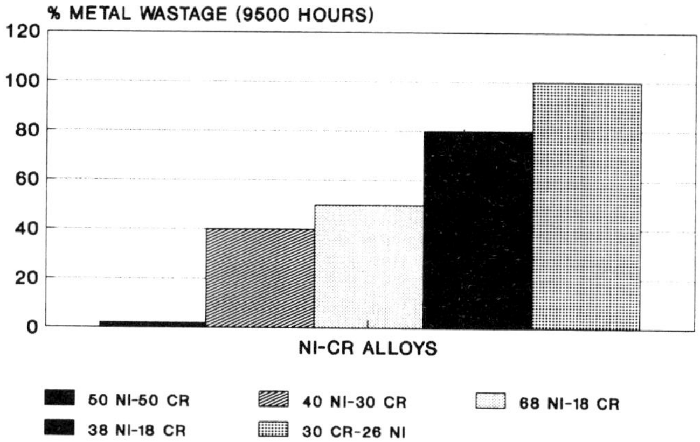


Fig. 1. High chromium content alloys exhibit the best resistance to high sulfur, high vanadium environments. These results show metal wastage as a function of time (9500 h) and temperature (1600 °F; 871.1 °C). Tafaloy 45CT was originally designed to be compatible with 50Ni–50Cr alloys.

boiler tube mounted robot controls the gun's spraying speed, orientation and distance from the tube sheet.

## 3. Tafaloy 45CT coatings — performance and structure

There is a synergy to the marriage of the arc spray process to Tafaloy 45CT. Arc spraying is a high spray rate process whose coatings are very clean (low oxides and porosity) and have high bond strengths. In addition, the efficiency of the process reduces excessive substrate heating, which minimizes thermal stresses and premature separation of the coating from the substrate. Tafaloy 45CT sprays with good bond strengths (6500 lbf in$^{-2}$; 44.82 MPa) and high deposit efficiencies (85%). The thermal expansion characteristics of Tafaloy 45CT are similar to A178 steels, which are often used for boiler tubes. Iron based chromium alloys, which are used for plasma or flame sprayed coatings, are much less like the A178, creating expansion stresses which contribute to spalling.

### *3.1. As-sprayed coatings*

The as-sprayed coating was analyzed to characterize its chemical and structural attributes. Optical metallography shows coating structure similar to many thermally sprayed coatings (Fig. 2). Oxides and pores occasionally interrupt a metallic phase. Closer examination, however, reveals that the metallic phase consists of a highly reflective continuous phase and a subtly contrasted discontinuous phase. This suggests two chemically different metallic phases. The oxides appear as a marbled, interspersed phase and the voids are dark areas. The presence of many small features suggests regions too small for the resolution of light microscopy, about 0.5 $\mu$m.

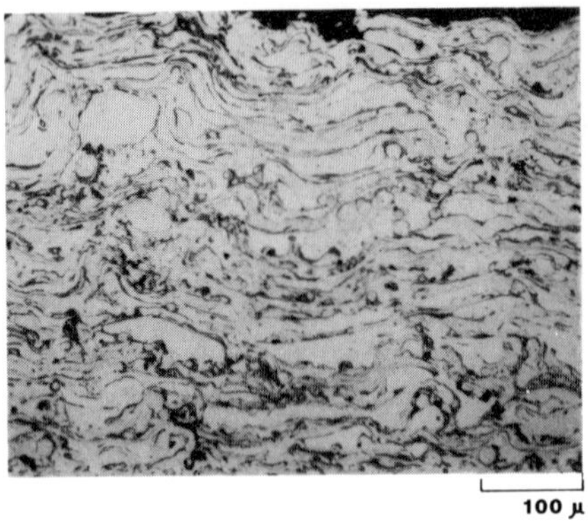


Fig. 2. As-arc-sprayed Tafaloy 45CT shows evidence of good melting, and thinner oxide layers and less porosity than most thermally sprayed coatings. These characteristics add to the alloy's ability to resist sulfur penetration.

Energy dispersive X-ray spectrometry (EDS) was used to analyze the metallic and oxide phases. The continuous phase was 40 - 45% chromium, 54 - 60% nickel and less than 1% titanium. The second metallic phase was a nickel rich phase containing 82 - 96% nickel, 4 - 17% chromium and less than 0.2% titanium. The oxides were 46 - 50% oxygen, 32 - 49% chromium, 1 - 20% nickel and less than 2% titanium.

X-ray diffraction (XRD) found four crystalline phases. Two were f.c.c. solid solutions of chromium in nickel, and the other two were oxides, $Cr_2O_3$ and $Cr_3O_4$. The two solid solutions were easily distinguished since they had very distinct lattice parameters. The oxides were weakly diffracting phases, showing they made up a small fraction of the bulk coating. Elemental compositions of the metallic phases estimated from the XRD data agreed with the EDS analysis. Based on the peak height intensities, the high chromium phase was, by far, the most abundant.

The data suggest that chromium selectively combines with oxygen, diffusing out of the base material, and leaving a high nickel phase. The amounts of oxides and the nickel rich phase are small compared with the bulk alloy.

*3.2. In service coatings*

A section of a black liquor recovery boiler tube wall coated with Tafaloy 45CT was removed after 18 months service for detailed microstructural and chemical analysis.

The cross-section revealed a microstructure similar to the as-sprayed pre-service coating. The oxide and pore structure remained and the bond line still showed excellent adhesion. The coating thickness was $11 - 16 \times 10^{-3}$ in (279 - 406 $\mu$m). EDS analysis showed 80% of the coating unaffected by any sulfur attack. The sulfur affecting the remaining 20% was in areas between deposited layers near the coating surface. Bulk chemical analysis showed less than 1% bulk sulfur content. The overall attack was very uniform.

The coating compared well to the behavior of weld deposited Filler Metal 72. Under the same conditions the bulk sulfur content Filler Metal 72 would be less than 0.01%. However, the sulfur distribution pattern for the coating was similar to that expected for Filler Metal 72. The sulfur was concentrated in areas of high chromium content. These results suggest that the arc sprayed coating behaved much like the bulk material. The porosity of the coating permitted slightly more sulfur to penetrate, but the ability of the chromium to absorb the sulfur and readily form oxides, which seal the porosity, stopped the migration of sulfur to the substrate [4]. The lack of continuous paths through the coating and the absorption of sulfur by the coating appear to reduce the ability of fresh sulfur to penetrate deeper into the coating. Coatings have been in service over four years without signs of spalling or other failures.

## 4. Conclusions

Arc sprayed Tafaloy 45CT has shown the ability to withstand highly sulfurous high temperature environments. Special wire processing, synergy between the process and the material, careful application, and a composition which ideally withstands the environment contribute to this success.

The mechanism of protection appears to be the absorption of sulfur by the chromium rich continuous metallic phase, and the formation of adherent surface chromium oxides which block continuous sulfur migration.

The success of Tafaloy 45CT in recovery boilers opens the way for use in other high temperature sulfurous environments. Utility boiler use is beginning and results at this time are similar to those in the black liquor recovery boilers.

The successful application of 45CT to environments which have eluded other thermally sprayed coatings offers the opportunity for profitable life extension to high temperature structures which are forced to operate on fuels and at demands well beyond their design specifications.

## References

1 M. Fontana and N. Greene, *Corrosion Engineering*, McGraw-Hill, New York, 1967, p. 360.
2 E. Sadowski, Development of a corrosion resistant Ni–Cr–Ti filler metal, *Welding J., June* (1980).
3 M. Thorpe and R. Unger, A new coating for corrosion protection in boilers, *Welding J., February* (1987) 19.
4 P. Kofstad, *High Temperature Oxidation of Metals*, John Wiley, New York, 1966, p. 274.

# EVALUATION OF CERAMIC COATINGS ON SILICON CARBIDE

J. I. FEDERER

*Oak Ridge National Laboratory, Oak Ridge, Tennessee (U.S.A.)*

M. VAN ROODE and J. R. PRICE

*Solar Turbines, Inc., San Diego, California (U.S.A.)*

(Received March 10, 1989)

## Summary

Ten ceramic coatings were plasma sprayed onto a sintered alpha silicon carbide (SiC) tube to determine if the coatings would protect SiC from high-temperature corrosion. Several coatings consisted of mullite or mullite-zircon. Others had base layers of mullite or mullite-zircon, transition layers with increasing amounts of alumina ($Al_2O_3$), zirconia ($ZrO_2$) or yttria ($Y_2O_3$), and outer layers of these last three oxides. Mullite and mullite-zircon coatings, which have coefficients of thermal expansion (CTE) near to that of SiC, survived 60 thermal cycles between 300 and 1300 °C without apparent damage. Coatings having $Al_2O_3$ and $Y_2O_3$ layers cracked, while those with $ZrO_2$ layers cracked and spalled. Survivors were corrosion tested at 1200 °C in an oxidizing atmosphere containing $Na_2CO_3$, which corroded uncoated SiC at a rate of about 21 mm year$^{-1}$. The coatings protected SiC for up to 500 h, but degraded because the atmosphere penetrated and reacted with the coatings and SiC to produce a disruptive silicate reaction product. Denser coatings are needed for extended protection.

## 1. Introduction

Silicon carbide (SiC) ceramics are candidate materials for high-temperature heat exchanger tubes because their combined strength, thermal conductivity, and thermal shock and oxidation resistance at temperatures above 1000 °C are unmatched by most metallic alloys and oxide ceramics. Heat exchangers using the tubes recover heat from industrial furnace flue gases for such purposes as preheating combustion air, preheating materials to be processed, and producing steam. Unfortunately, these materials are corroded by alkali compounds, such as NaCl, NaF, $Na_2SO_4$ and $Na_2CO_3$, which are present in certain industrial furnace atmospheres [1 - 8]. Alkali oxides derived from these compounds react with the normally protective $SiO_2$ film on SiC, which results in enhanced diffusion of $O_2$ and corrosion of the SiC by

0257-8972/89 Elsevier Sequoia/Printed in The Netherlands

oxidation. This type of corrosion is characterized by surface recession beneath a silicate reaction product. Because certain oxide ceramics corrode substantially less than SiC in the same atmospheres, the ability of oxide coatings to protect SiC from high-temperature corrosion has been investigated. The coatings are intended to prevent or limit access of corrosive atmospheres to the SiC. If the atmosphere penetrates the coating through pores, cracks, or other defects, the coating might still protect sacrificially by reacting with and chemically altering the normal silicate reaction product.

## 2. Experimental procedure

Ten coating bands, each of length about 7.5 cm, were applied by conventional plasma spraying to a sintered α-SiC tube (outside diameter, 44.5 mm) which was treated by a proprietary process prior to coating to improve adherence. The development and initial evaluation of the coatings has been described [9-11]. As shown in Table 1, three coatings consisted of mullite ($Al_2O_3$–28wt.%$SiO_2$), one of which had a vitreous $Cr_2O_3$-containing ceramic seal coat. The purpose of the latter was to decrease the permeability of the coating, because plasma sprayed coatings typically contain 5 - 15% porosity. Another coating consisted only of equal parts of mullite and zircon ($ZrO_2$–33wt.%$SiO_2$). The other six coatings, which were multilayered, had mullite or mullite-zircon as the first layer (because their coefficients of thermal expansion (CTE) nearly match that of SiC), intermediate transition layers with increasing amounts of alumina ($Al_2O_3$), stabilized zirconia ($ZrO_2$–8wt.%$Y_2O_3$) or yttria ($Y_2O_3$), and outer layers of one of these last three

TABLE 1

Composition of coatings

| *Designation* | *Layers*[a] (*base–transition–outer*) | *Thickness* | |
|---|---|---|---|
| | | (mm) | (in) |
| ST-1 | M only[b] | 0.58 | 0.023 |
| ST-2 | M only | 0.58 | 0.023 |
| ST-3 | M only | 0.94 | 0.037 |
| ST-4 | M — A[c] | 0.91 | 0.036 |
| ST-5 | M — A — ZY[c] | 1.30 | 0.051 |
| ST-6 | MZ only | 0.53 | 0.021 |
| ST-7 | MZ — A[c] | 0.99 | 0.039 |
| ST-8 | MZ — A — ZY[c] | 1.32 | 0.052 |
| ST-9 | M — Y[c] | 0.61 | 0.024 |
| ST-10 | M — ZY[c] | 0.94 | 0.037 |

[a]M, mullite; A, alumina; ZY, stabilized zirconia; MZ, mullite–50wt.%zircon; Y, yttria.
[b]Very thin layer of vitreous $Cr_2O_3$-containing ceramic on outer surface.
[c]Accomplished in 25 vol.% increments. For example, M — A means 100%M — (75%M–25%A) — (50%M–50%A) — (25%M - 75%A) — 100%A.

compounds. Mullite and zircon, which contain substantial amounts of $SiO_2$, might also react with the normal $SiO_2$ layer on SiC to form a ceramic bond. Coatings ST-5 and -8 each consisted of nine layers, and were the thickest coatings tested. Thicknesses ranged from 0.53 to 1.32 mm (0.021 to 0.052 in).

The coated tube was sectioned to divide each coating into a band of length 2 cm (specimen) for ceramographic examination and a specimen of length 5.5 cm for testing. The specimens were tested for adherence and structural integrity of the coating by thermal cycling to 1300 °C, initially at heating and cooling rates of around 200 °C $h^{-1}$ ("A" cycle), and then at rates of around 500 °C $h^{-1}$ ("B" cycle). Survivors were then corrosion tested at 1200 °C in an atmosphere consisting of $N_2$–21.3wt.%$O_2$–4.2wt.%$H_2O$–0.04wt.%$Na_2CO_3$ in an apparatus previously described [12]. The atmosphere was prepared by mixing 6000 $cm^3$ $min^{-1}$ of air with a vaporized solution (0.25 $cm^3$ $min^{-1}$) of $H_2O$–0.5wt.%$Na_2CO_3$. Each coated specimen was contained in a separate furnace tube (inside diameter, 6 cm) for the corrosion test.

## 3. Results

### *3.1. Thermal cycle tests*

One "A" thermal cycle caused obvious cracks in four multilayer coatings (ST-5, -7, -8 and -10). During 20 "B" cycles the three multilayer coatings with zirconia outer layers (ST-5, -8 and -10) cracked severely or cracked and spalled from the substrates; therefore, these were not suitable for further testing. After a total of 60 "B" cycles single layer coatings (ST-1, -2, -3 and -6) had no obvious cracks, and the other multilayer coatings either had microscopic cracks (ST-4), axial cracks (ST-7) or cracked blisters (ST-9). The range of thermal cycling results in Fig. 1 shows apparently undamaged ST-2, severely cracked ST-8, cracked and spalling ST-5, and completely spalled ST-10.

ORNL-PHOTO 1198-89

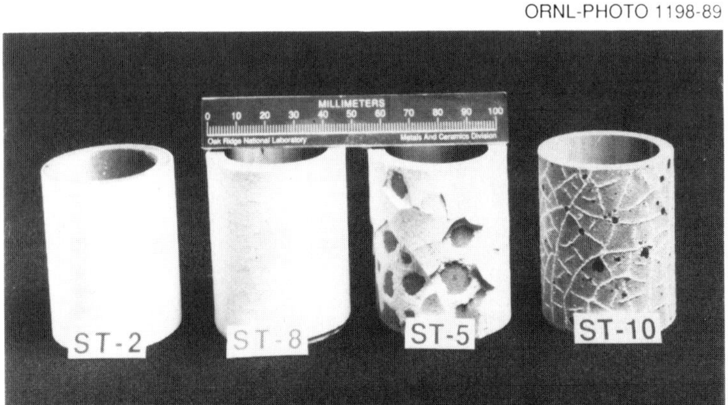


Fig. 1. Thermally cycled coated specimens.

*3.2. Corrosion testing*

Six specimens were corrosion tested: ST-1, -3, -4, -6, -7 and -9. Coating ST-2, which was similar to ST-3 except for thickness, was not corrosion tested. As previously mentioned ST-4 and -6 had a few fine cracks, and ST-7 and -9 had obvious cracks due to thermal cycling. Caps of Kao Tab 95 castable cement (Babcock & Wilcox) were fitted to the specimens to minimize attack on the uncoated ends and inner surface. Specimens were visually inspected during the test after 150, 300 and 500 h. After the first exposure for 150 h an uncoated SiC specimen replaced ST-6, because the latter was obviously damaged. Since the atmosphere was very corrosive to SiC, the behavior of coated specimens could be compared to that of an uncoated specimen. Specimens ST-1 and -7 were removed from the test after an exposure of 300 h. Thus, during corrosion testing, coated specimen ST-6 had one thermal cycle between 25 and 1200 °C, ST-1 and -7 had two cycles, and ST-3, -4 and -9 had three cycles.

Results of corrosion testing are shown in Fig. 2. Coatings ST-1, -3, -6, -7 and -9 degraded by cracking and separating from the SiC. Coatings ST-6 and -7 had full length cracks with a transparent reaction product that wetted the coating adjacent to the cracks, but coating ST-4 had no visual evidence of degradation caused by corrosion. The diameters of the coated specimens, however, increased during the corrosion test. Specimens ST-3 and -9, for example, increased about 3%, indicating that the coatings and/or the SiC corroded, or that the coatings detached. As will be shown, most of the reaction occurred in the coatings.

The coatings were easily separated from the substrates revealing a thin transparent reaction layer. This material was dissolved with concentrated hydrofluoric acid at room temperature for 2 h so that the diameters of the substrates could be measured for comparison with the as-coated diameters. (A

ORNL-PHOTO 1203-89

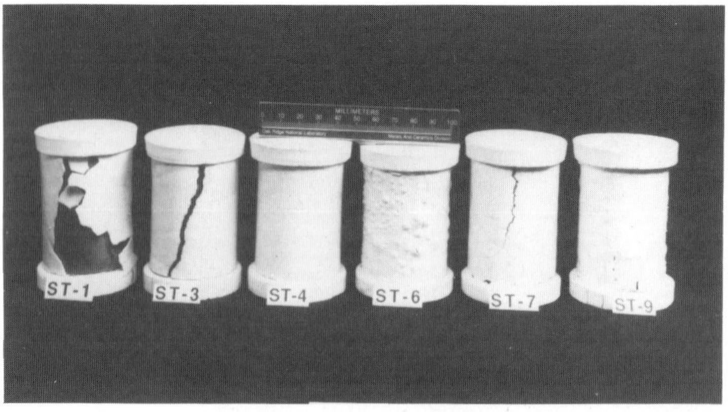


Fig. 2. Coated specimens after corrosion testing at 1200 °C. Specimen ST-6 was tested for 150 h, ST-1 and -7 for 300 h, and ST-3, -4 and -9 for 500 h.

TABLE 2

Change in diameter of coated and uncoated SiC substrates caused by corrosion at 1200 °C

| *Specimen* | *Exposure time* (h) | *Coating diameter*[a] (mm) | *Coating thickness*[a] (mm) | *SiC substrate diameter*[a] | | |
|---|---|---|---|---|---|---|
| | | | | *Before* (mm) | *After* (mm) | *Change* (mm) |
| ST-1 | 300 | 45.39 | 0.58 | 44.22 | 44.32 | +0.10 |
| ST-3 | 500 | 46.10 | 0.94 | 44.22 | 44.15 | −0.07 |
| ST-4 | 500 | 45.90 | 0.91 | 44.07 | 44.03 | −0.04 |
| ST-6 | 150 | 45.21 | 0.53 | 44.15 | 44.20 | +0.05 |
| ST-7 | 300 | 46.13 | 0.99 | 44.15 | 44.15 | 0 |
| ST-9 | 500 | 45.44 | 0.61 | 44.22 | 44.22 | 0 |
| uncoated SiC | 350 | | | 44.27 | 42.57 | −1.70 |

[a]Average of eight equally spaced measurements around the circumference at one axial location.

ORNL-PHOTO 1200-89

Fig. 3. Uncoated SiC after corrosion testing at 1200 °C for 350 h.

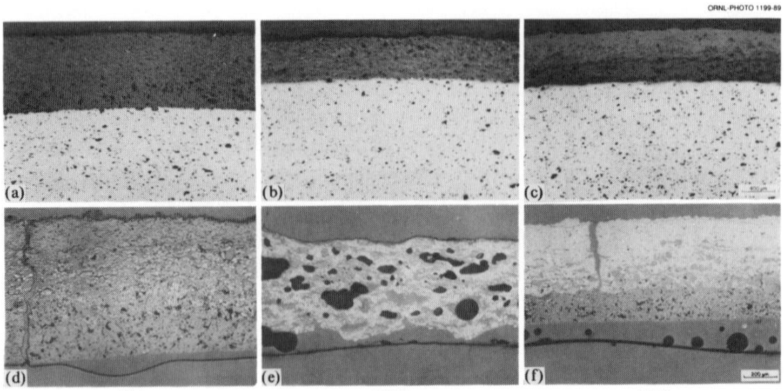


Fig. 4. Microstructures of coatings before (a - c) and after (d - f) corrosion testing.

SiC specimen exposed to the acid in the same manner showed that this treatment did not affect the dimensions of SiC.) Table 2 shows that coated specimens had very small diameter changes, indicating that most of the reaction occurred within the coatings. The diameter changes in Table 2 represent measurement errors of $\pm 0.1$ mm involved in eight separate measurements (equally spaced around the circumference) of coating diameter and thickness required to determine the substrate diameter. The glassy material that formed on the uncoated SiC specimen is shown in Fig. 3. X-ray diffraction showed that this material was amorphous. Neither tridymite nor cristobalite, high-temperature forms of $SiO_2$, were detected. The diameter change for this specimen was much larger in magnitude. The linearly projected annual surface recession rate for the uncoated SiC from the value in Table 2 is 21.3 mm year$^{-1}$ (0.84 in year$^{-1}$).

The microstructures of several typical coatings reveal porosity in the pretest condition (Fig. 4(a) - (c)). The corrosion tested coatings, which detached from the SiC, had a layer of transparent reaction product on the original inner surface, and some of this material had infiltrated or formed within the coatings (Fig. 4(d) - (f)). X-ray diffraction of the reaction product on the inner surface of ST-4 and -6 revealed one or more sodium aluminum silicates (nepheline, $Na_2O{\cdot}Al_2O_3{\cdot}SiO_2$). Sodium silicates might have been present also. This material apparently was molten at 1200 °C, as indicated by its tendency to seep through cracks to the outer surface. Electron microprobe examination showed that the transparent material within the microstructure of ST-4 and -6 (Fig. 4(d) and (e)) consisted mainly of silicon, oxygen, aluminum and sodium. This material probably was similar to the silicate material on the surface.

## 4. Discussion

Coatings consisting of mullite or mullite–50%zircon (ST-1, -2, -3 and -6) had no cracks after 60 B thermal cycles. This behavior is attributed to the close match in CTE between SiC, mullite, and zircon, which have values of 5.5 - 5.6 × $10^{-6}$ °$C^{-1}$. Coatings with $Al_2O_3$, $Y_2O_3$ and stabilized $ZrO_2$ in the transition and outer layers cracked during thermal cycling, the degree of degradation being related to the difference in CTE between SiC and the oxides. Alumina (around 9.3 × $10^{-6}$ °$C^{-1}$) caused either microscopic cracks (ST-4) or several visual cracks (ST-7); $Y_2O_3$ (around 9 × $10^{-6}$ °$C^{-1}$) caused several cracks and cracked blisters (ST-9); and stabilized $ZrO_2$ (around 12 × $10^{-6}$ °$C^{-1}$) caused cracking and spalling (ST-5, -8 and -10).

All corrosion tested coatings consisted of mullite (ST-1 and -3), mullite–50%zircon (ST-6), or had these materials as base layers (ST-4, -7 and -9). All as-prepared coatings contained porosity. The corrosion atmosphere penetrated the coatings and reacted with the normally protective $SiO_2$ layer on the SiC substrates. The atmosphere also reacted with mullite and zircon, which contain 28 wt.% $SiO_2$ and 33 wt.% $SiO_2$, respectively. Corrosion of mullite in alkali-containing industrial furnace atmospheres has been reported [3]. Although all aspects of the corrosion reactions in the coatings have not been determined, a silicate reaction product was evident in every case. Formation of this material within the microstructure and at the interface apparently caused detachment and eventual failure of some coatings.

Uncoated SiC corroded severely. The coatings, however, protected the SiC for the duration of the tests, as shown by essentially unchanged substrate diameters. The silicate reaction product filled the pores and minimized access of $O_2$ to the SiC. These coatings probably would provide substantial protection under isothermal conditions; however, they would probably crack and separate from the substrate during thermal cycling, as in this test.

## 5. Conclusions

Oxide ceramic coatings can protect SiC from corrosion under conditions that severely corrode uncoated SiC. The corrosive atmosphere, however, penetrates through pores in the coatings and reacts with mullite and zircon, which contain substantial amounts of $SiO_2$, and with the $SiO_2$ layer on the SiC to form a new, disruptive silicate reaction product. Protection would be improved by denser, less permeable coatings.

## Acknowledgments

This research was sponsored by the Waste Energy Recovery Program, Office of Industrial Programs, U.S. Department of Energy, under contract DE-AC05-840R21400 with Martin Marietta Energy Systems, Inc., Oak Ridge, Tennessee. The coatings were prepared by Solar Turbines, Inc., San Diego, California, under contract to Gas Research Institute, Chicago, Illinois, contract 5086-232-1233.

## References


1 A. D. Russell, C. E. Smeltzer, and M. E. Ward, Waste heat recuperation for aluminum furnaces, *GRI-81/0160*, Solar Turbines, Inc., San Diego, California, April 1983.

2 J. I. Federer, T. N. Tiegs, D. M. Kotchick and D. Petrak, Analysis of candidate silicon carbide recuperator materials exposed to industrial furnace environments, *ORNL Rep. TM-9677*, Oak Ridge National Laboratory, Oak Ridge, Tennessee, July 1985.

3 J. I. Federer and P. J. Jones, Oxidation/corrosion of metallic and ceramic materials in an aluminum remelt furnace, *ORNL Rep. TM-9741*, Oak Ridge National Laboratory, Oak Ridge, Tennessee, December 1985.

4 N. S. Jacobson and J. L. Smialek, *J. Am. Ceram. Soc.*, *68* (1985)432.

5 J. I.. Federer, J. M. Robbins, P. J. Jones and C. Hamby, Jr., Corrosion of SiC ceramics in synthetic combustion atmospheres containing halides, *ORNL Rep. 6258*, Oak Ridge National Laboratory, Oak Ridge, Tennessee, December 1985.

6 N. S. Jacobson, *J. Am. Ceram. Soc.*, *69* (1986)74.

7 N. S. Jacobson, C. A. Stearns and J. L. Smialek, *Adv. Ceram. Mater. 1* (1986)154.

8 J. I. Federer, *Adv. Ceram. Mater.*, *3* (1988)56.

9 M. van Roode and J. R. Price, The development of protective coatings for ceramic tubular components, *GRI*-88-0005, Solar Turbines, Inc., San Diego, California, December 1987.

10 J. R. Price, R. E. Gildersleeve, M. van Roode and C. E. Smeltzer, A new ceramic recuperation technology for use in corrosive environments, in *Proc. Workshop on Corrosion of Ceramic Materials*, The Pennsylvania State University, 12/13, 1987, pp 81–114.

11 M. van Roode, J. R. Price, R. E. Gildersleeve and C. E. Smeltzer, *Ceram. Eng. Sci. Proc.*, *9* (1988)1245.

12 J. I. Federer, An investigation of ceramic coatings for protection of SiC from high-temperature corrosion, *ORNL Rep. TM-10606*, Oak Ridge National Laboratory, Oak Ridge, Tennessee, February 1988.

*Surface and Coatings Technology*, *39/40* (1989) 79 - 87 

# THIN FILM MOLYBDENUM SILICIDE AS POTENTIAL TEMPERATURE SENSORS FOR TURBINE ENGINES

C. H. HO, S. PRAKASH, C. V. DESHPANDEY, H. J. DOERR and R. F. BUNSHAH

*Department of Materials Science and Engineering, School of Engineering and Applied Science, University of California, Los Angeles, CA 90024-1595 (U.S.A)*

(Received March 3, 1989)

## Summary

Temperature measurements of Mo–Si-based thin-film resistance thermometers were studied. Annealing in an argon ambient at a temperature above 1000 °C for at least 1 h is required to form the stable tetragonal $MoSi_2$ phase. With a crack-free AlN barrier layer of thickness 2 $\mu$m on top, a sensor was tested up to 1200 °C. The resistivity *vs.* temperature characteristic shows the room temperature resistivity and temperature coefficient of resistivity (TCR) of the sensor to be approximately 350 $\mu\Omega$ and $11.95 \times 10^{-3}$ K respectively. No film adhesion problems were observed for at least four testing cycles.

## 1. Introduction

Thin film temperature sensors for turbine applications are still a relatively new area for research [1 - 7], and this is more so with regard to the advanced turbine engines utilizing ceramic–composite blades for higher operating temperature. Sensor materials capable of exhibiting sensitive indications of thermal fluctuations and at the same time functioning in a stable manner at relatively high temperatures in a combustion gas are required.

Among the many possible candidates for temperature measurements up to 1775 K, $MoSi_2$ was selected because of (i) its outstanding oxidation resistance at such elevated temperatures due to the formation of a protective glassy layer and (ii) its optimum value of temperature coefficient of resistivity (TCR) in comparison with the other refractory metal silides [8].

Based on metallographic studies, Berkowitz-Mattuch and Dils [9] concluded that, after being kept at an elevated temperature for a sufficiently long time, the original $MoSi_2$ layer will eventually be converted into $Mo_3Si_2$, $Mo_5Si_3$ and $Mo_3Si$. The lower silicides were reported [9, 10] to be much less oxidation-resistant than $MoSi_2$. Therefore, the performance of the whole resistance-sensor system will deteriorate once the lower silicides start to form.

0257-8972/89 Elsevier Sequoia/Printed in The Netherlands

In this paper, we investigate the evolution of $MoSi_2$ phases throughout the whole processing and testing steps, and the influence of an AlN protective layer on sensor performance. Then, by temperature testing up to 1425 K for five cycles, we obtain the initial results from $MoSi_2$ temperature sensors.

## 2. Principle of resistance temperature sensors

The resistivity of a metal will increase proportionally with temperature. Based on the fundamental theory for a low electrical field, we can write

$$V = IR = I\left(\rho \frac{l}{wt}\right) \tag{1}$$

where $\rho$ is the intrinsic resistivity of material, $l$ is the total length of conducting route of sensor, $w$ is the width of conducting path and $t$ is the thickness of sensor film.

By using a set up similar to that used for four-point probe measurements, we are able to detect the variations of $p$ with temperature by measuring the output voltage with a constant input current, if the other parameters, *i.e.* $l$, $w$ and $t$, are all kept at certain values. The laboratory model of our sensor configuration is shown in Fig. 1.

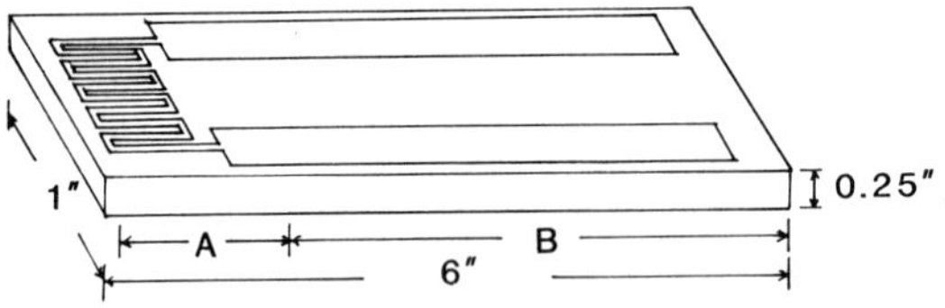


Fig. 1. Schematic diagram of a sensor pattern for resistance thermometry.

In Fig. 1, the portion A is the main pattern of the 3 - 10 $\mu$m thick $MoSi_2$ layer used for temperature sensing. Portion B represent the platinum extention pads to facilitate the attachment of electrical contact.

## 3. Experimental procedure

The $Si_3N_4$ turbine-blade pieces were supplied by NASA. Although polished in the "as-received condition", the surface of this ceramic material still showed peak-to-valley roughness of about 3 - 4 $\mu$m. The surface morphology was studied by a profilometer and scanning electron microscope (SEM) analysis. The results are shown in Figs. 2(a) and 2(b).

The $Si_3N_4$ substrates were then ultrasonically vibrated in acetone for 5 - 10 min and then finally vapor-degreased in isopropyl alcohol.

For each run, two substrates were placed in a direct-current magnetron sputtering system, where the background pressure is normally $5 \times 10^{-6}$ Torr.

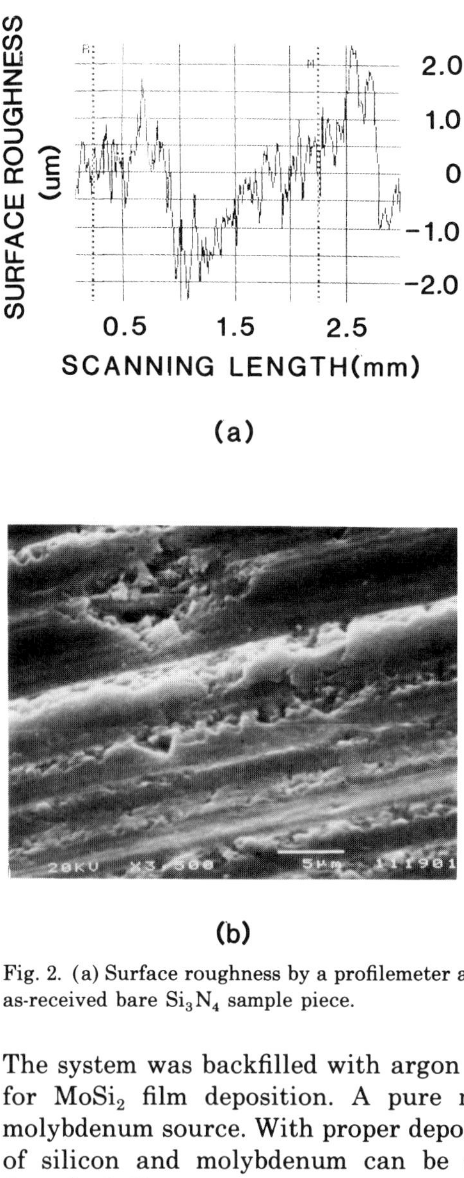


Fig. 2. (a) Surface roughness by a profilemeter and (b) the corresponding SEM micrograph of an as-received bare $Si_3N_4$ sample piece.

The system was backfilled with argon and a mixture of 10% $SiH_4$ in pure $H_2$ for $MoSi_2$ film deposition. A pure molybdenum target was used as the molybdenum source. With proper deposition conditions, the relative contents of silicon and molybdenum can be adjusted to the desired ratio in the deposited film.

Before depositing the protective layer, the samples were annealed in an argon ambient at 1150 °C for 2 h to completely form the stable tetragonal phase of $MoSi_2$. X-ray diffraction was used to identify the phase formation.

The deposition of the protective AlN layer was carried out by plasma-enhanced MOCVD with $NH_3$ and trimethyl-aluminum (carried by $H_2$) used as source gases. The detailed description of the MOCVD system is discussed elsewhere [11]. With the substrate temperature in the range 550 - 700 °C, an AlN layer 2 - 6 $\mu$m thick was deposited.

After the sensor configuration was ready, high temperature testing was carried out by arranging the sample on a 1.5 ft long stainless-steel platform support, wiring four contact leads for constant current input and voltage output measurements, and then connecting the whole set up to a computer system for data acquisition.

## 4. Result and discussion

### *4.1. $MoSi_2$ film*

#### *4.1.1. The effect of silicon content in the film and surface roughness of substrate on the resistivity of film*

The resistivity of $MoSi_2$ films has been a controversial issue with values ranging from 20 to 100 $\mu\Omega$ cm depending on different deposition method used [8]. The trends of resistivity *vs.* silicon content of the $MoSi_2$ films on polished sapphire both before and after the annealing treatment are shown in Fig. 3.

The resistivity of $MoSi_2$ films on the rough $Si_3N_4$ substrates showed higher values than those on smooth surfaces such as polished silicon wafers.

In Fig. 4(a), we can see that the surface morphology of the as-deposited $MoSi_2$ film follows the surface contours of the as-received $Si_3N_4$ pieces with cluster segregation on the ridge part and shadowing effects around the foot area of the ridge.

After annealing at 1150 °C in an argon ambient for 2 h, the fluctuation looked somewhat smoothed out, as shown in Fig. 4(b). However, with a film thickness of 3 - 5 $\mu$m, which is comparable with the surface roughness, the properties, *e.g.* the resistivity, of the annealed film cannot be improved as much as expected.

#### *4.1.2. Compositional profile and the effect of annealing on tetragonal $MoSi_2$ phase formation*

As was discussed above, the Si/Mo ratio is an important factor in determining the performance and lifetime of the sensor. In order to supply silicon to form a better continuous oxide layer, silicon-rich films instead of a nearly stoichiometric disilicide layer were usually deposited. The compositions of our sensor, as analysed by energy dispersive X-ray analysis, normally fell in the $Mo_{0.25}Si_{0.75}$ to $Mo_{0.3}Si_{0.7}$ range.

In Fig. 5(a), which shows the Rutherford backscattering spectroscopy (RBS) profile of an as-deposited $MoSi_2$ film, we can see that a thick silicon-enriched layer exists on the film surface.

Directly heat treating the as-deposited film in air at 1300 °C for 5 h showed thick $SiO_2$ developing on the surface and no $MoSi_2$ forming underneath this thick oxide layer. The compositional profile by RBS of this treatment is shown in Fig. 5(b).

However, a sample after undergoing the heat treatment in an argon ambient gave a totally different result. The evolution of different $MoSi_2$ allotrophs is shown in Fig. 6 by using X-ray diffraction techniques. For the

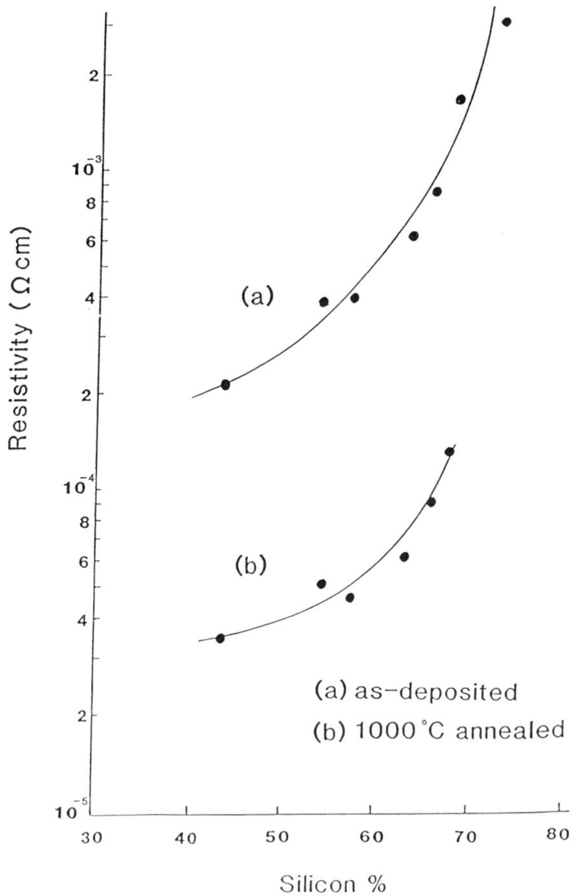


Fig. 3. Resistivity of Mo–Si film *vs.* silicon content. (a) As-deposited. (b) After annealing in an argon ambient at 1000 °C.

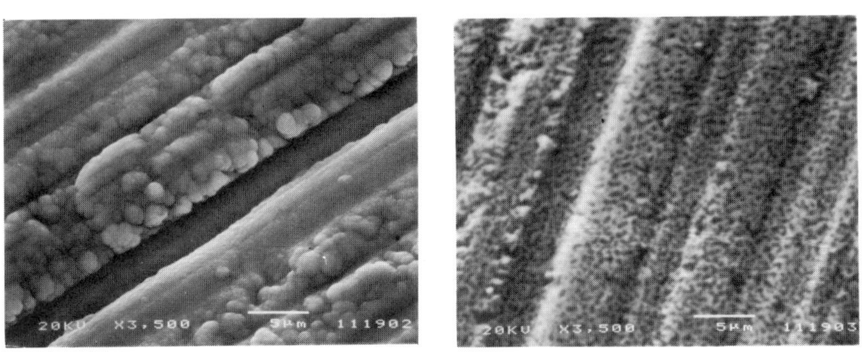

(a) (b)

Fig. 4. SEM micrographs showing surface morphology of $MoSi_2$–$Si_3N_4$ of (a) as-deposited and (b) after annealing in an argon ambient at 1150 °C for 2 h.

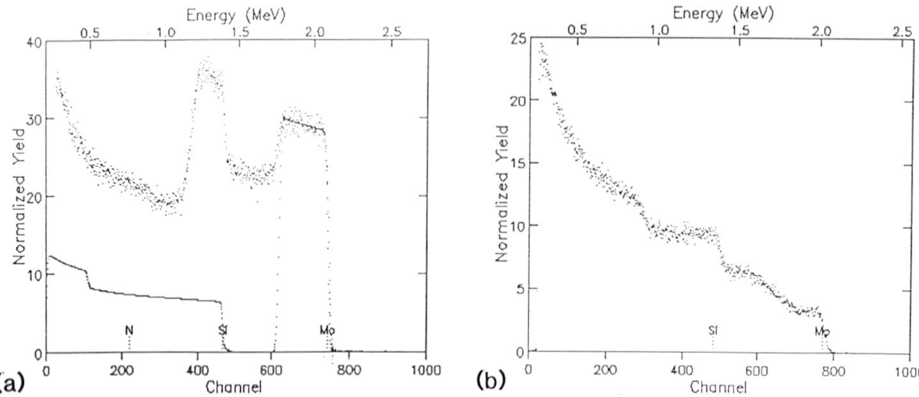


Fig. 5. RBS spectra of Mo–Si film on $Si_3N_4$ substrate. (a) As-deposited. (b) Heat-treated in air at 1200 °C for 5 h.

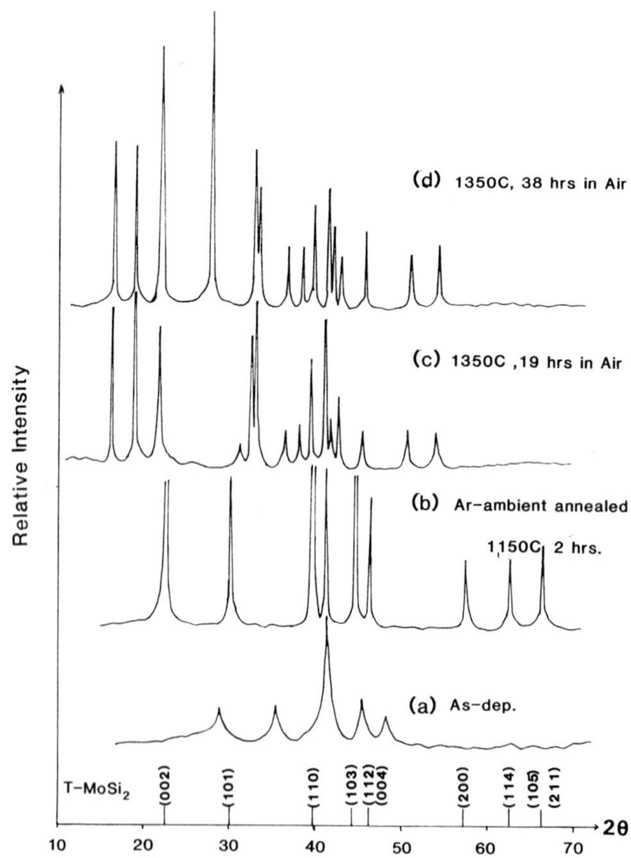


Fig. 6. X-ray diffraction patterns of $MoSi_2$ film. (a) As-deposited. (b) After annealing in an argon ambient at 1150 °C for 2 h. (c) Sample (b) heat-treated in air at 1350 °C for 19 h. (d) Same as sample (c), but for 38 h.

as-deposited film, only the hexagonal (h) phase of $MoSi_2$ is obtained. After annealing at 1150 °C in argon for 2 h, lines from the tetragonal (t) $MoSi_2$ phase become prominent. This result clarified the beneficial effect of argon annealing on the stable t-$MoSi_2$ phase formation.

Also shown in Fig. 6 is the consumption of the t-$MoSi_2$ phase at 1350 °C in air. After heating for 19 h, the amount of t-$MoSi_2$ phase diminished, and only a small amount was left after 38 h, along with other phases. These results imply that the lifetime of the sensor could be much shorter than 19 h without an AlN protective layer.

### *4.2. AlN protective layer*

An AlN protective layer of thickness 2 - 6 μm was deposited to reduce the oxidation rate of the underlying sensor film. The X-ray diffraction pattern of an AlN layer of thickness 5 μm on a silicon substrate showed a crystalline AlN structure at a deposition temperature of 550 °C.

From the cross-sectional SEM micrograph shown in Fig. 7, we can see that the as-deposited MOCVD-grown AlN film exhibits a loosely-packed columnar structure with pillar-axis oriented perpendicular to the substrate surface and observable cracks existing in between the columns.

Therefore, since microcracks exist, a protective layer of thickness 5 μm may not be expected to actually achieve the same protective effect as a fully dense layer of thickness 1 μm. In the next section we will show the test result of a sensor coated with a crack-free AlN protective layer of thickness 2 μm.

### *4.3. High temperature testing of the sensor pattern*

An as-deposited $MoSi_2$ film without argon ambient annealing prior to testing was heated to 1000 °C, since it was presumed that the film would be able to achieve $MoSi_2$ t-phase formation as the high temperature testing proceeded. The result given in Fig. 8(a) shows an unusual semiconducting characteristic up to 900 K, which was completely different from the expected metallic nature of a positive and almost constant TCR. Repeating the cycle to

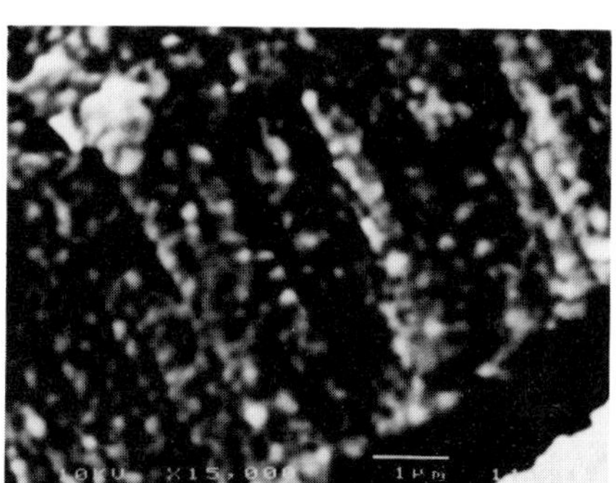


Fig. 7. Cross-sectional SEM micrograph of a AlN film of thickness 5 μm on Si(100) substrate.

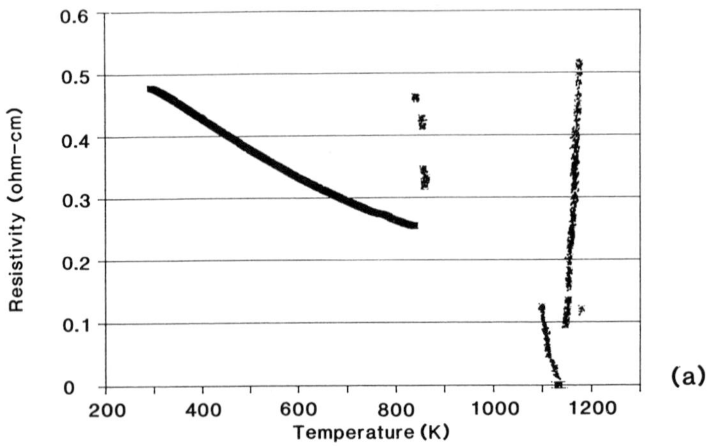


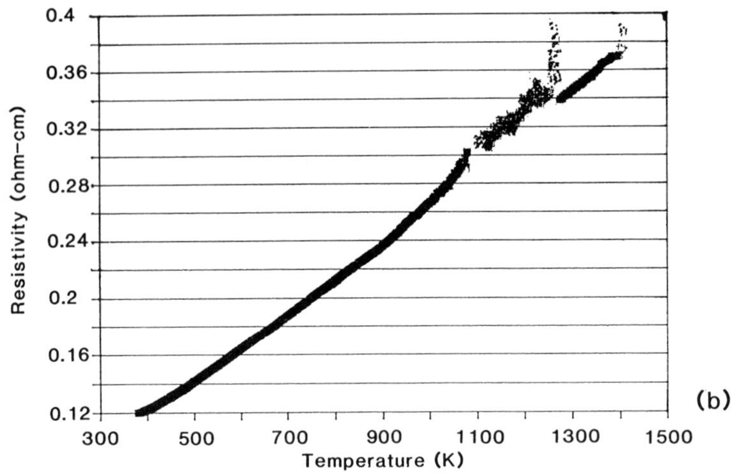


Fig. 8. Test results of a $MoSi_2$ thin-film resistance temperature sensor. (a) As-deposited pattern and no AlN protective layer. (b) Sensor pattern annealed in an argon ambient at 1150 °C for 2 h and coated with a AlN layer of thickness 2 $\mu$m.

1270 K did not cause any improvement. This is consistent with the previous observations of annealing effects.

For comparison, the test results from a patterned sensor with an AlN barrier layer of thickness 2 $\mu$m on top of the argon-annealed $MoSi_2$ film is shown in Fig. 8(b). The sensor operated up to 1425 K. No adhesion problems were observed. The characteristic was linear, and the TCR was approximately $11.95 \times 10^{-3}$ K, which is about 1.87 times that of the reported value [8]. The higher TCR could be attributed partially to the intentionally higher silicon content used to obtain a better protecting effect as described above, and partially to grain size effects, possible contaminations during sputtering, etc.

## 5. Conclusions

From the results obtained, we can draw some conclusions.

(1) In general, the as-deposited $MoSi_2$ film showed an amorphous nature for low $T_{sub}$ and a hexagonal crystal structure for higher $T_{sub}$. Argon-ambient annealing treatment above 1000 °C for at least 1 h is necessary for the formation of a stable t-$MoSi_2$ phase.

(2) The surface roughness of the $Si_3N_4$ substrate proves to be in excess of what is optimal for adherence requirements for a sensor film of thickness 3 - 5 $\mu$m. As a consequence, the $MoSi_2$ film grown on it becomes non-uniform in thickness. If no barrier layer is coated, thin portions of the sensor film get rapidly oxidized and consumed on heating in air, so that the originally conducting path cannot be maintained and the sensor function is lost.

(3) A crack-free barrier layer is able to reduce the oxidation rate of inner layer. An $MoSi_2$ thin-film resistance thermometer with a crack-free barrier layer of thickness 2 $\mu$m was successfully tested up to 1420 K for at least four cycles without failure. We would suggest that if both the silicon content and the thickness of the sensor film can be optimized, the performance of thin film molybdenum silicide temperature sensors may be improved for even higher temperature than 1420 K. We are presently investigating this.

## Acknowledgments

The authors would like to acknowledge the funding of this work under the contract NASA NAG-3-451 and to thank Mr. Ray Holanda for his interest and encouragement and Dr. T. Bardin and her colleagues at Lockheed Missiles and Space Company, Inc. for the RBS analysis.

## References

1 R. R. Dils and P. S. Follansbee, *Proc. of 21st Int. Instrumentation Symp., Vol. 12 in Advances in Test Measurement*, Instrumentation Society of America, Pittsburgh, PA, 1975, pp. 127 - 132.
2 H. P. Grant and J. S. Przybyszewski, Pratt & Whitney Aircraft, Commercial Products Division, *Report No. NASA CR 159782,* 1980.
3 J. C. Godefroy, C. Gageant, D. Francis and M. Portat, *J. Vac. Sci. Technol., A5* (1987) 2917.
4 W. H. Bennethum and L. T. Sherwood, G. E. Aircraft Engines, Advanced Technology Operation Division, *Report No. NASA CR 180900,* 1988.
5 W. H. Atkinson, M. A. Cyr and R. R. Strange, United Technology Corporation, Pratt & Whitney, Commercial Engineering Division, *Report No. NASA CR 182111.*
6 S. Prakash, R. C. Budhani and R. F. Bunshah, Thin-film temperature sensors for high-performance turbojet engines, *Mater. Res. Bull., 23,* (1988), 187 - 195.
7 R. C. Budhani, S. Prakash and R. F. Bunshah, Thin-film temperature sensors for gas turbine-problems and prospects, *J. Vac. Sci. Technol., A4(6),* (1986), 2608.
8 S. P. Murarka, *Silicides for VLSI Applications*, Academic Press, Orlando, FL, 1983, p. 46.
9 J. B. Berkowitz-Mattuck and R. R. Dils, *J. Electrochem. Soc. 112* (1965) 583.
10 R. A. Perkin, *Materials Science and Technology for Advanced, Vol. II, 1964 Golden Gate Metals Conference, San Francisco*, February 1964.
11 S. Prakash, R. C. Budhani and R. F. Bunshah, to be published.
12 *CRC Handbook of Chemistry and Physics*, CRC Press, Inc., Boca Raton, FL, F-120.

# SCANDIA-STABILIZED ZIRCONIA FOR RESISTANCE TO MOLTEN VANADATE–SULFATE CORROSION

R. L. JONES

*Naval Research Laboratory, Washington, DC 20375-5000 (U.S.A.)*

(Received March 10, 1989)

## Summary

Following an acid–base theory of ceramic corrosion developed at the Naval Research Laboratory, we have identified scandia as a stabilizer potentially more resistant to molten sulfate–vanadate attack than yttria, the commonly used zirconia stabilizer. This prediction is supported in the present work where, for example, scandia-stabilized zirconia (SSZ) gave no detectable reaction or destabilization after 160 h exposure to 900 °C molten $NaVO_3$, whereas yttria-stabilized zirconia (YSZ) reacted strongly with the $NaVO_3$ (with $YVO_4$ formed) and was extensively destabilized. Other tests confirmed the superior resistance of $Sc_2O_3$ over $Y_2O_3$ to reaction with both molten sodium vanadate and $Na_2SO_4/SO_3$. The additional cost for scandia, which is expensive, was calculated to be about \$0.15 cm$^{-2}$ of $10 \times 10^{-3}$ in (250 $\mu$m) coating, or less than half the cost, e.g. for platinum in current platinum-augmented aluminide coatings. Development of molten sulfate- and vanadate-resistant SSZ coatings could give large savings by allowing cheaper, low quality sulfur- and vanadium-containing fuels to be burned in gas turbine or diesel engines.

---

## 1. Introduction

Thermal barrier coatings (TBC) of stabilized zirconia allow higher gas temperature and thermal efficiency in gas turbines and diesels, and may provide a 10% or greater increase in fuel economy or power. Yttria-stabilized zirconia (YSZ) is currently the predominant choice for TBC purposes, but yttria is removed from YSZ by reaction with vanadium fuel contaminants, with destabilization and failure of the YSZ resulting. YSZ TBCs are therefore limited to aviation or other "vanadium free" fuel applications. However, low vanadium crude oils amount to only about 25% of the remaining world petroleum reserves, and vanadium-free fuels will become increasingly expensive and scarce in the future. It is thus important to develop vanadium-resistant stabilized zirconia coatings which could reduce national fuel costs by permitting high-efficiency burning of cheaper, and more abundant, vanadium- and sulfur-containing fuels in Navy and industry engines.

0257-8972/89/\$3.50 

In relation to this problem, we have investigated the reactions between the different vanadate compounds and the various ceramic oxides [1, 2], and confirmed: (i) that the critical reaction in the destabilization of YSZ is a *Lewis acid–base reaction* between yttria ($Y_2O_3$) and vanadium pentaoxide ($V_2O_5$), i.e.

$$Y_2O_3(\text{base}) + V_2O_5(\text{acid}) = 2\ YVO_4(\text{salt}) \quad (1)$$

and (ii) that zirconia ($ZrO_2$) itself is resistant to reaction even with pure $V_2O_5$ (activity equals 1), indicating that stabilized zirconia coatings having substantial resistance to vanadium in fuel might be possible if a zirconia stabilizer more acidic than $Y_2O_3$ could be found.

One such candidate stabilizer is ceria ($CeO_2$), which appears to be more acidic than yttria [1]. However, ceria-stabilized zirconia was found to be destabilized as rapidly as YSZ, for reasons still not understood, by molten $NaVO_3$ in laboratory tests [2]. The question therefore remains as to the merit of $CeO_2$ as a vanadate-resistant stabilizer for zirconia.

We have recently identified scandia ($Sc_2O_3$) as a potential corrosion resistant stabilizer for zirconia. Scandia is the most acidic of the rare earth oxides [3], and has a stable $Sc^{3+}$ oxidation state. It is also an effective stabilizer of zirconia, with 5 wt.% $Sc_2O_3$ being sufficient to stabilize tetragonal zirconia [4]. This paper describes experiments which verify that scandia-stabilized zirconia (SSZ) is more resistant to molten vanadate corrosion than YSZ. Our results indicate that SSZ may have substantial superiority over YSZ for use in gas turbines or diesels burning low quality sulfur- and vanadium-containing fuel.

## 2. Experimental procedure

Our experiments fell into three categories: (i) tests of the chemical reactivity of pure $Sc_2O_3$ with sodium metavanadate ($NaVO_3$) and $V_2O_5$, (ii) tests of the sulfation of $Sc_2O_3$ *vs.* $Y_2O_3$ by $SO_3$–$Na_2SO_4$ and (iii) tests of the corrosion of sintered pellets of SSZ and YSZ by molten $NaVO_3$ films. For the vanadate reaction tests, 50 mol% mixtures of $Sc_2O_3$ with $NaVO_3$ or $V_2O_5$ were exposed in platinum boats under quiescent air for periods of 4 - 48 h at 700 and 900 °C. The occurrence of reaction, if any, and the identity of the reaction product were established by X-ray diffraction of the reaction mixture. In the sulfation studies, 50 mol% mixtures of $Sc_2O_3$–$Na_2SO_4$ and $Y_2O_3$–$Na_2SO_4$ were exposed in platinum boats under flowing furnace air (300 ml min$^{-1}$) containing $13 \pm 2$ Pa of equilibrated $SO_2$–$SO_3$ at 700 - 800 °C. The mixtures were exposed until weight gain ceased (usually within 100 h), with the mol% of $Sc_2(SO_4)_3$ or $Y_2(SO_4)_3$ formation being calculated from total weight gain. Reagent chemicals of 99.9% or higher purity were used, except for $NaVO_3$ (98%).

For the ceramic corrosion tests, pellets of diameter 8.2 mm and thickness 2.5 mm of $Sc_2O_3$(5 wt.%)–$ZrO_2$ and $Y_2O_3$(8 wt.%)–$ZrO_2$ were prepared (by J. Wallace, formerly of NRL, now at NIST) using a coprecipitation process, followed by pressing, calcining, repressing at $25 \times 10^3$ lbf in$^{-2}$ and sintering at 1300 °C. Approximately 96% theoretical density and essentially complete tetragonal stabilization were obtained in each case. These pellets were subjected to 700 °C "thin film" and 900 °C "vapor deposition" $NaVO_3$ corrosion tests. In the former, 5 mg of $NaVO_3$ was placed in the center of the pellet surface, and the pellet carefully inserted (in a platinum boat) into the furnace where the $NaVO_3$ melted, producing a molten film of approximately 250 $\mu$m thickness. In the latter, SSZ and YSZ pellets on separate small platinum coupons were placed on a porcelain boat partially filled with $NaVO_3$ so that, at the 900 °C furnace temperature, only vapor phase $NaVO_3$ should contact the pellet surfaces.

After $NaVO_3$ exposure, the pellets were examined by scanning electron microscopy (SEM)–energy dispersive X-ray spectroscopy (EDS) (ISI DS-130 SEM/Kevex 8000 EDS system) and X-ray diffraction to characterize the reaction products and extent of zirconia destabilization.

## 3. Results

### *3.1. Chemical reaction of scandia with vanadium compounds*

Vanadium may form a series of compounds in engines ranging from $Na_3VO_4$ to $V_2O_5$ itself, where the $V_2O_5$ activity and thus Lewis acidity decreases as the $Na_2O$ content of the compound increases. Which compound forms depends upon the relative abundance of the two other major engine contaminants, sodium and sulfur. For marine gas turbines burning low quality sulfur- and vanadium-containing fuel, Luthra and Spacil [5] predict that the predominant vanadium species will be $NaVO_3$ and $V_2O_5$. Sodium metavanadate is formed by the reaction of 1 mole of $Na_2O$ and 1 mole of $V_2O_5$ (*i.e.* $NaVO_3 = 1/2\ Na_2O{\cdot}V_2O_5$), and has a $V_2O_5$ activity substantially lower than that of pure $V_2O_5$ (activity equals 1).

Accordingly, scandia was tested for reaction with $NaVO_3$ and $V_2O_5$ at 700 and 900 °C. X-ray diffraction showed that $NaVO_3$ produced some $ScVO_4$ formation at 700 °C, but none at 900 °C; with $V_2O_5$, formation of $ScVO_4$ occurred at both temperatures. $Sc_2O_3$ is therefore evidently more acidic than $Y_2O_3$, which forms $YVO_4$ readily (cf. reaction (1)) at the $V_2O_5$ activity associated with $NaVO_3$ at both 700 and 900 °C [1]. This finding supports the idea that scandia should be superior to yttria as a vanadate-resistant stabilizer for zirconia.

### *3.2. Resistance of scandia to sulfation by $SO_3$–$Na_2SO_4$*

Yttria has been shown to be leached from YSZ by mixed sulfate formation (6). Mixed sulfate formation is a Lewis acid–base reaction with $SO_3$ as the acid, and can be written as

$$Y_2O_3(\text{in } ZrO_2) + 3\,SO_3 + Na_2SO_4 = Y_2(SO_4)_3\text{–}Na_2SO_4(\text{as mixed sulfate}) \quad (2)$$

Tests where 50 mol% mixes of $Y_2O_3$–$Na_2SO_4$ and $Sc_2O_3$–$Na_2SO_4$ were brought to equilibrium at different temperatures under air containing 13 Pa of equilibrated $SO_2$–$SO_3$ (representative of $SO_2$–$SO_3$ partial pressures in gas turbines burning moderately sulfur-contaminated fuel) gave the results in Fig. 1. Complete sulfation of the oxide corresponds to 50 mol% sulfate formation. As seen in Fig. 1, $Y_2O_3$ is essentially completely sulfated even at 800 °C. Scandia, however, gives only 9.9 mol% sulfate formation at 700 °C, and just 0.3 mol% sulfation at 800 °C. Thus scandia acts as a stronger acid than yttria in reaction with $SO_3$, and SSZ should be significantly more resistant to degradation via sulfation than YSZ, especially above 800 °C.

### *3.3. Resistance of SSZ to destabilization by $NaVO_3$ at 700 and 900 °C*

When the SSZ and YSZ pellets were coated with $NaVO_3$ and heated at 700 °C, the yttria-stabilized pellets reacted strongly with the $NaVO_3$, producing a mass of acicular $YVO_4$ crystals on the ceramic surface (Fig. 2(a)), and giving extensive destabilization (confirmed by X-ray diffraction) of the YSZ tetragonal structure. Conversely, the $NaVO_3$ film on the SSZ pellets remained effectively unreacted (Fig. 2(b)) even after periods of 475 h at 700 °C. Energy dispersive X-ray spectroscopy of the exposed SSZ surface (site C in Fig. 2(b)) gives a $Sc_2O_3$ peak of the same height as the $Sc_2O_3$ peak in virgin SSZ, indicating no depletion or destabilization in these areas. No $ScVO_4$ was detected, even though $NaVO_3$ produced some $ScVO_4$ with *pure* $Sc_2O_3$ at 700 °C, but this is not inconsistent since the activity of $Sc_2O_3$ in solid solution in $ZrO_2$ would be lower than for pure $Sc_2O_3$.

Close examination of the $NaVO_3$ film, however, shows a surface texture at some sites (Site A in Fig. 2(b)) which suggests that a second phase has precipitated within the $NaVO_3$. Metallurgical cross-sectioning of $NaVO_3$-deposited SSZ after 16 h at 700 °C verified that a second phase was being formed

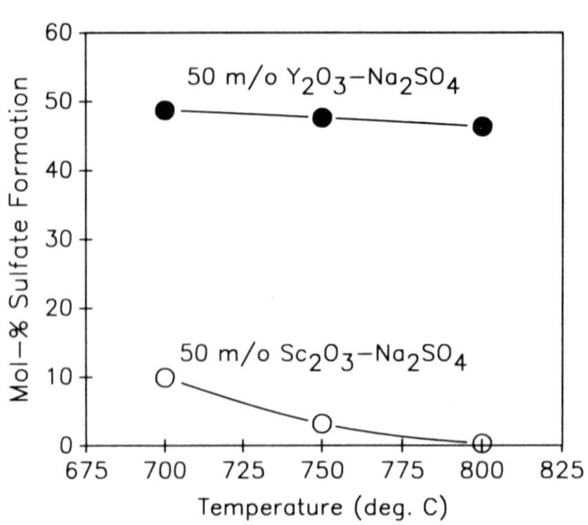


Fig. 1. Sulfation behavior of 50 mol% $Sc_2O_3$–$Na_2SO_4$ *vs.* 50 mol% $Y_2O_3$–$Na_2SO_4$ under 13 Pa of equilibrated $SO_3$–$SO_2$ in air at 700 °C.

(a)

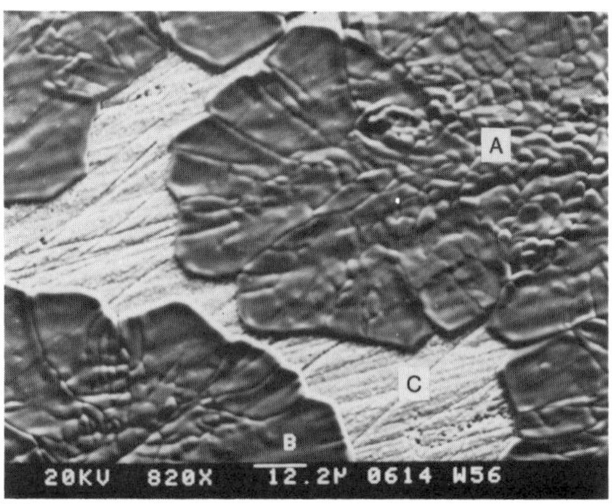


(b)

Fig. 2. (a) Surface of $Y_2O_3$(8 wt.%)–$ZrO_2$YSZ pellet coated with 5 mg of $NaVO_3$ and exposed 475 h at 700 °C. Acicular crystals are $YVO_4$. (b) Surface of $Sc_2O_3$(5 wt.%)–$ZrO_2$ SSZ pellet coated with 5 mg of $NaVO_3$ and exposed 475 h at 700 °C. Compare unreacted $NaVO_3$ film on SSZ in (b) with massive $YVO_4$ formation on YSZ in (a).

within the $NaVO_3$ layer, and that it originated near the ceramic interface. This second phase was shown by EDS to be rich in calcium, scandium and vanadium, and therefore to probably be a calcia–scandia vanadate. When the $NaVO_3$ film in Fig. 2(b) was removed by water washing, X-ray diffraction indicated that the SSZ surface now contained some monoclinic phase, *i.e.* that sufficient $Sc_2O_3$ had been removed by formation of the Ca–Sc–V phase to cause partial destabilization of the SSZ.

Although the destabilization was substantially less than that produced by $NaVO_3$ attack on YSZ, the occurrence of the Ca–Sc–V reaction was worrisome since sea salt has a sizeable content of calcium, and $CaSO_4$ is often found in engine salt deposits. However, in developmental tests made since this original work, no destabilization of SSZ was found when 10 mol%

additions of either CaO or $CaSO_4$ were made to the $NaVO_3$ test salt. This suggests that the calcium in the Ca–Sc–V reaction comes from the ceramic (calcia impurities are common in ceramics), and that the calcia exists in the ceramic in a more reactive state than it would in the molten salt where the calcia basicity would be reduced by acid–base displacement reactions with $NaVO_3$, $Na_2SO_4$ etc. Thus, while calcia impurities may need to be excluded from SSZ, it does not appear that calcium-rich engine salts *per se* will be likely to cause degradation of SSZ.

At 900 °C, "thin film" $NaVO_3$ corrosion tests proved difficult to interpret because, although $NaVO_3$ on YSZ continued to show massive $YVO_4$ formation, the $NaVO_3$ film on SSZ could no longer be easily discerned (as in Fig. 2(b)), possibly because of wicking of the $NaVO_3$ into the SSZ ceramic. To avoid this difficulty, and also to produce thinner films of $NaVO_3$ on the SSZ, a "vapor deposition" technique was adopted for the 900 °C experiments, wherein SSZ and YSZ pellets on individual platinum planchets were placed on top of a porcelain boat partially filled with $NaVO_3$ so that, at temperature, the pellets would be bathed in $NaVO_3$ vapor but not touch the bulk $NaVO_3$ melt. (In actuality, there was reaction between the $NaVO_3$ and porcelain boat, with the melt becoming reddish-brown (indicative of $V_2O_5$) and strongly wetting the porcelain surface. Therefore some deposition may have occurred by thin film creeping, and the pellet deposit may have contained some $V_2O_5$. Neither effect should invalidate the comparative test, however.)

The "vapor deposition" technique was successful in depositing sodium vanadate films on both the SSZ and YSZ pellets. The deposition rates were monitored by weighing the pellets after different times at temperature. The SSZ pellet deposit leveled off at about 5 mg after 160 h total time at temperature, while the YSZ deposit was about 17.5 mg at 160 h and still increasing.

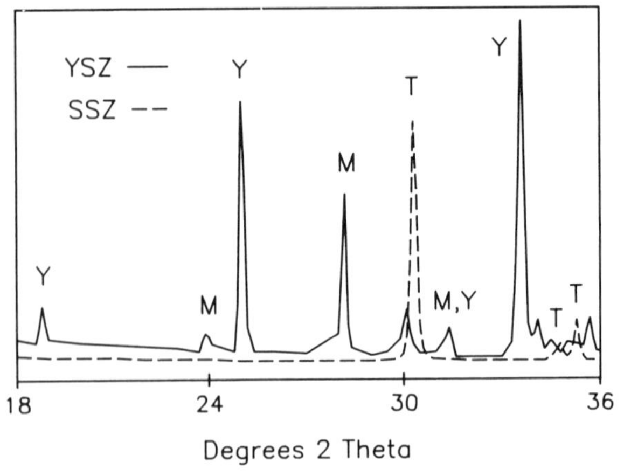


Fig. 3. X-ray diffraction spectra from washed surfaces of SSZ and YSZ pellets after being "vapor" deposited with $NaVO_3$ for 160 h at 900 °C. SSZ has remained stabilized in the tetragonal structure (peaks marked T), while YSZ shows strong $YVO_4$ (marked Y) and monoclinic zirconia (marked M) peaks, indicating extensive reaction and destabilization of YSZ.

Although not understood in detail, these weight gain behaviors would be consistent with there being $NaVO_3$ reaction and $YVO_4$ formation with YSZ, but only deposition of $NaVO_3$ and no $ScVO_4$ formation with SSZ.

Examination of the deposited pellets by SEM–EDS showed continuous salt films on both pellet surfaces, with the EDS patterns from both films exhibiting prominent sodium and vanadium peaks, and approximately the same Na/V peak height ratios, to indicate that they were essentially $NaVO_3$. A strong zirconium peak occurs in the SSZ pattern, but only a weak zirconium peak in the YSZ pattern, which would be consistent with the molten salt film being 3 to 4-times thinner on the SSZ pellet than the YSZ pellet. More importantly, when the salt films were gently washed away, and the SSZ and YSZ pellet surfaces X-rayed, the X-ray diffraction spectra (Fig. 3) showed that, after being coated with $NaVO_3$ for 160 h at 900 °C, the SSZ had remained essentially totally stabilized, whereas the YSZ had reacted to form large quantities of $YVO_4$ and was extensively destabilized.

## 4. Discussion

The prediction, based on our understanding of oxide acid–base reactions in ceramic corrosion [1, 2], that scandia should be superior to yttria as a molten vanadate-resistant stabilizer for zirconia appears confirmed by the present work, where SSZ has been shown to be significantly more resistant to reaction and destabilization by molten $NaVO_3$ at 700 and 900 °C than YSZ. In considering SSZ for further development as a molten vanadate–sulfate resistant coating for engines or other high temperature applications, two points come to mind. First, it remains to be established that SSZ has the thermomechanical properties to withstand the severe thermal shock and thermal–mechanical stresses that occur with engine coatings. Achievement of such properties is not simple, and indeed a large effort has been required to bring YSZ to its present state where it is capable of surviving the thousands of thermal cycles necessary for successful aircraft service. Second, there is the problem of the high cost of scandia.

### *4.1. Cost of scandia for SSZ coatings*

Scandia is expensive and currently costs about the same as platinum ($20 g$^{-1}$). However, platinum itself is employed in gas turbine blade coatings in, for example, the commercially successful, and widely used, platinum-augmented aluminide coatings. In these coatings, a platinum underlayer approximately $0.3 \times 10^{-3}$ in (7.5 $\mu$m) thick is deposited onto the blade prior to aluminization, with the platinum cost being about $0.32 cm$^{-2}$ of coating. On the other hand, the cost for $Sc_2O_3$ in a $Sc_2O_3$(5 wt.%)–$ZrO_2$ coating of $10 \times 10^{-3}$ in (250 $\mu$m) thickness would be about $0.15 cm$^{-2}$, and only about $0.08 cm$^{-2}$ for a $5 \times 10^{-3}$ in (125 $\mu$m) coating, which is of the order of thickness used in many marine gas turbines. Considering that first stage blades for modern, fuel efficient, air-cooled gas turbines may cost up to $1000 each, the slight additional cost for $Sc_2O_3$ would be easily justified if it extends blade life, and allows reliable burning of lower cost and more available fuels containing at least certain levels of sulfur and vanadium.

*4.2. Further research and development*

Since this original work a plasma spray powder of scandia-stabilized zirconia has been prepared, and coatings of SSZ plasma sprayed on test coupons and burner rig test pins. The SSZ powder sprayed well, and produced adherent coatings which were essentially 100% tetragonal. The SSZ coated coupons will be investigated for thermal shock behavior, and resistance to molten vanadate–sulfate mixes under $SO_3$ where the $V_2O_5$ activity will be higher than for pure $NaVO_3$. The SSZ coated pins will be tested in the David Taylor Naval Ship R&D Center burner rig test facility, and compared to other stabilized zirconia coatings, as well as metallic coatings, for resistance to vanadium, sulfur and sea salt contaminants. If these tests are favorable, SSZ will be recommended for engine test, and possibly ultimately commercial development.

## Acknowledgments

Support for this research under the Gas Turbine Component Improvement (Ceramics) Task of the NAVSEA Ship Propulsion Technology Program is gratefully acknowledged. Thanks are also due to Dr. J. Wallace, of NIST, who prepared the YSZ and SSZ ceramic test pellets; to Mr. L. Domingues of Trans-Tech Corp., who prepared the SSZ plasma spray powder; and to Prof. H. Herman of SUNY, Stony Brook, whose group plasma sprayed the SSZ coatings.

## References

1 R. L. Jones, C. E. Williams and S. R. Jones, *J. Electrochem. Soc.*, *133* (1986) 227.
2 R. L. Jones and C. E. Williams, *Surf. Coat. Technol.*, *32* (1987) 349.
3 M. C. Sneed and R. C. Brasted, *Comprehensive Inorganic Chemistry*, Vol. 4, Van Nostrand, New York, 1955, p. 168.
4 R. Ruh, H. J. Garrett, R. F. Domagala and V. A. Patel, *Am. Ceram. Soc.*, *60* (1977) 399.
5 K. L. Luthra and H. S. Spacil, *J. Electrochem. Soc.* *129* 649 (1982).
6 R. H. Barkalow and F. S. Pettit, in J. W. Fairbanks and J. Stringer (eds.), *Proc. 1st Conf. Advanced Materials for Alternate-Fuel Capable Directly Fired Heat Engines*, CONF-790749, NTIS, Springfield, VA, 1979, p. 704.

*Surface and Coatings Technology, 39/40* (1989) 97 - 115



# INVESTIGATION OF COATINGS AT HIGH TEMPERATURE FOR USE IN TURBOMACHINERY

W. TABAKOFF

*Department of Aerospace Engineering and Engineering Mechanics, University of Cincinnati, Cincinnati, OH 45221 (U.S.A.)*

(Received March 26, 1989)

## Summary

Aircraft gas turbines, steam turbines and coal utilization turbines operating under particulate flows are exposed to erosion and performance deterioration. Almost all advanced aircraft turbines today are manufactured with a protective coating on the blades. The particulates in the steam turbines are produced in the boiler from exfoliated oxide scales and carried in the steam to the turbine. In coal utilization turbines, the particulates are by-products of the combustion process. There are many problem areas which need future research to be able to produce better and more durable high-temperature coatings. However, this paper will present only some of the work done at the University of Cincinnati on erosion testing at high temperatures and velocities for different materials and coatings. The testings have been performed with a special high-temperature erosion wind tunnel which simulates the aerodynamic conditions on the blades.

## 1. Introduction

Commercial and military airplanes are often exposed to foreign objects such as birds, hailstones, ice slabs, runway gravel and others which are ingested into engine inlets. The ingestion of particulate matter such as sand and dust leads to the deterioration of these engines both structurally and aerodynamically. Some of the mechanisms that cause foreign object ingestion are (a) vortex from engine inlet-to-ground during high power setting with the aircraft standing or moving on the runway, (b) sand storms transporting sand to several thousand feet altitude and (c) thrust reverser efflux at low airplane speed may blow sand, ice and other particles into the engine inlets. Sand and salt spreading on runways in wintertime contributes to solid particle ingestion by airbreathing engines. Erosive solid particles may also be produced during the combustion process, from the burning of different types of heavy oils or synthetic fuels. Erosion has been pointed out as a problem in other areas such as coal fired boiler systems, fluidized beds and gas turbines

0257-8972/89/$3.50 

burning coal. Research efforts are still needed to provide a thorough knowledge of the various parameters which influence the extent of erosion damage. This knowledge is required to improve the life and the performance of engines operating in an ambient of particulate flow.

Under two-phase flow conditions, the gas and particles experience different degrees of turning through the blade channels. The degree of turning and acceleration or deceleration achieved by the particles depends on the ratio of the viscous forces to the inertial forces experienced by the particles. This leads to variations in particle concentration across the blade channels and causes a change in the properties of the gas flows which alters the engine performance during the period of particle ingestion [1].

If the particles are erosive, impingement of particles on the blade surfaces can cause severe erosion damage, leading to structural failure of the blades. This damage is manifested by pitting and cutting of the blade leading and trailing edges, and a general increase in the blade surface roughness [2]. The overall effect of the above phenomena, from the aerodynamic viewpoint, is an increase in total pressure loss across the blade row.

The erosion problems in military and commercial airplane gas turbine engines are generally reçognized in industry. A study performed on commercial jet engines estimates that a minimum of 2% thrust specific fuel consumption loss is caused by the performance deterioration due to erosion (NASA Conference Publication 2190, May 6 - 7, 1981). The application of inlet vortex dissipating devices has been shown to reduce erosion in engines. However, most of the commercial engine fleets do not have such devices.

The intensity and pattern of the compressor and turbine erosion is dependent on the locations of particle impacts and on the magnitude and direction of their impact velocity relative to the surface. It is very important to be able to calculate accurately the particle trajectories in turbomachines if one is to determine their influence on the engine life and performance.

Three-dimensional particle trajectory calculations through axial flow turbines and compressors were first reported by Hussein and Tabakoff [3, 4]. They studied the effect of particle size, and particle material density on their trajectories and demonstrated their influence on particle impact locations. On the other hand, the three-dimensional particle trajectories through twisted vanes were studied by Hamed [5] who demonstrated that the particle trajectories are strongly influenced by the hub and tip contouring for the different particle sizes.

## 2. Analysis

The equations governing the particle motion in the turbomachinery flow fields are written in a rotating frame of reference using cylindrical coordinates. The three components of the equations of motion in a frame rotating about the engine axis with angular velocity $\omega$ are given by:

$$\frac{d^2r_p}{d\tau^2} = F_r + r\left(\frac{d\theta_p}{d\tau} + \omega\right)^2 \tag{1}$$

$$r_p\frac{d^2\theta_p}{d\tau^2} = F_\theta - \frac{2dr_p}{d\tau}\left(\frac{d\theta_p}{d\tau} + \omega\right) \tag{2}$$

$$\frac{d^2z_p}{d\tau^2} = F_z \tag{3}$$

where $r_p$, $\theta_p$ and $z_p$ define the particle location in cylindrical polar coordinates, $V_r$, $V_\theta$ and $V_z$, represent the relative gas velocities in the radial, circumferential and axial directions respectively, and $\omega$ is the blade angular velocity. The centrifugal force and Coriolis acceleration are represented by the last terms on the right hand side of eqns. (1) and (2). The first term on the right-hand side of eqns. (1) - (3) represents the force of interaction between the two phases, per unit mass of particles. It is dependent on the relative velocity between the particles and the gas flow, as well as the particle size and shape. Under the particulate flow conditions in turbomachines, the effect of the forces due to gravity and to interparticle interactions are negligible compared to those due to the aerodynamic and centrifugal forces. In addition, the force of interaction between the two phases is dominated by the drag due to the difference in velocity between the solid particles and the flow field. The force of interaction per unit mass of solid particles is therefore given by

$$\bar{F} = \frac{3}{4}\frac{\rho}{\rho_p}\frac{C_D}{d}\left\{\left(V_r - \frac{dr_p}{d\tau}\right)^2 + \left(V_\theta - \frac{d\theta_p}{d\tau}\right)^2 + \left(V_z - \frac{dz_p}{d\tau}\right)^2\right\}^{1/2}(\bar{V} - \bar{V}_p) \tag{4}$$

where $\rho$ and $\rho_p$ are the gas and solid particle densities, $d$ is the particle diameter and $C_D$ is the drag coefficient. The drag coefficient is dependent on the Reynolds number, which is based on the relative velocity between the particle and the gas. Empirical relations as shown in ref. 4 are used to fit the drag curve over a wide range of Reynolds numbers.

### *2.1. Trajectory calculations*

The particle trajectory calculations consist of the numerical integration of the eqns. (1) - (4) in the flow field, up to the point of blade, hub or tip impact. The impact location can be determined from the particle trajectory calculations and the local geometry of the impacted surface. The magnitude and direction of the impact velocity relative to the surface, for a large number of particles constitute the essential data needed for the evaluation of the erosion of the various turbomachinery components. The magnitude and direction of particle rebounding velocity after these impacts are dependent on the impacting conditions and the particular particle surface material combination under consideration. Empirical correlations of the restitution parameters are used to calculate the particle rebounding conditions in the calculation of particle trajectories. These correlations [6] are based on the experimental data obtained using high speed photography and laser Doppler velocimetry for particle laden flows over metal samples at various incidence angles and flow velocities in a special tunnel. Two correlation parameters such as the velocity and angle restitution ratios describe the particle rebound characteristics in the two-dimensional tunnel. These ratios were found to be mainly dependent upon the impingement angle and not significantly influenced by the particle size and impact velocity.

The velocities and angles of impact and rebound are considered relative to the rotating blade in the case of rotor blade impacts [7]. Furthermore, in order to use these correlations in the three dimensional particle trajectories, a knowledge of the geometry of the solid surface at the impact location is required. Geometrical data describing the blade, hub and tip configuration should be therefore available during the trajectory calculations to be used in the determination of the impact location and the impingement angle relative to the surface.

### 2.2. *Flow field representation in particle trajectory computation*

Careful consideration must be given to the flow field representation in the particle trajectory calculations [5]. The flow field solution must be chosen to represent the important flow field characteristics in any particular application. Further consideration must also be given to the computer storage required for this data and computer time for interpolating flow properties at each time step in the numerical integration of the particles equations of motion.

The field representation on a number of blade to blade stream surfaces [8, 9] was used in the particle trajectory calculations through the axial flow turbine. This flow modeling in the particle trajectory calculations insured the accurate representation of the three dimensional flow field in the particle trajectory calculations throughout the engine.

### 2.3. *Particle blade interactions*

The particle rebounding conditions after surface impacts are determined from empirical correlations of the restitution parameters. These correlations are based on the experimental data obtained using laser Doppler velocimetry for particle laden flows over metal samples in special tunnels [10]. The measurements at various incidence angles and flow velocities indicate that these ratios are mainly dependent upon the impingement angle for a given particle material combination. The following correlations apply to Kingston fly ash particles impacting 410 stainless steel target [10]

$$\frac{V_{PR}}{V_{PI}} = 1 - 0.90847\,\beta_I + 0.3072\,\beta_I^2 + 0.05695\,\beta_I^3 \tag{5}$$

and

$$\frac{\beta_R}{\beta_I} = 1 - 0.38746\,\beta_I + 0.51442\,\beta_I^2 + 0.45094\,\beta_I^3 \tag{6}$$

while the empirical restitution ratios for sand particles impacting 410 stainless steel are given by [11]

$$\frac{V_{p_R}}{V_{p_I}} = 1 - 2.0586\,\beta_I + 3.0748\,\beta_I^2 - 1.3184\,\beta_I^3 \tag{7}$$

and

$$\frac{\beta_R}{\beta_I} = 1 + 2.60676\,\beta_I - 5.70832\,\beta_I^2 + 2.535\,\beta_I^3 \tag{8}$$

where $V_{p_I}$ and $\beta_I$ represent the particle impact velocity and impingement angle while $V_{p_R}$ and $\beta_R$ are the rebound velocity and rebound angle measured from the impacted surface.

In the case of rotor blade impacts, the velocities and angles of impact and rebound are considered relative to the rotating blade. Furthermore, in order to use the empirical correlations obtained from two-dimensional measurements, in the three-dimensional particle trajectories, a knowledge of the geometry of the solid surface at the impact location is required. The blade surface coordinates at several sections between the hub and tip are used in the trajectory calculations to describe the blade shapes. Interpolation is used to determine the blade surface coordinates and its slope at other locations. The particle rebounding conditions after each blade impact are determined by the restitution correlations of eqns. (5) and (6). The predicted particle dynamics, as well as the resulting particle redistribution pattern are dictated by the blade passage geometry through these impacts [12].

### 2.4. *Blade erosion*

The erosion of a target surface under particle impacts is generally dependent upon the particles impact velocity, impingement angle and on the temperature for a given particle target material combination [11, 13]. Experimental data for targets tested in the erosion tunnel was used to obtain the following empirical equation for the erosion mass parameter $\epsilon$, which is defined as the ratio of the eroded mass of target material to the mass of impinging particles.

$$\epsilon = K_1\left[1 + Ck\left\{K_{12}\sin\left(\frac{90}{\beta_0}\beta_I\right)\right\}\right]^2 V_I^2\cos^2\beta_I(1 - R_t^2) + K_3(V_I\sin\beta_I)^4 \tag{9}$$

where $V_I$ and $\beta_I$ are the impact velocity and impingement angle, respectively. $R_t = 1 - 0.0016\,V_1\sin\beta_1$, $R_t$ is the tangential restitution ratio, $\beta_0$ is the angle of maximum erosion, $CK = 1$ for $\beta_I \leqslant 2\beta_0$, $CK = 0$ for $\beta_I > 2\beta_0$ and $K_1$, $K_{12}$, $K_3$ are constants that depend on the target material and solid particles characteristics. Experimental measurements were obtained in the erosion tunnel for fly ash particles impacting 410 stainless steel at different impacting velocities and impingement angles. These measurements gave the following values for the constants:

$K_1 = 1.505101 \times 10^{-6}$

$K_{12} = 0.296077$

$K_3 = 5.0 \times 10^{-12}$

The computation of blade erosion combines the empirical correlation for the mass erosion parameter (eqn. (9)) with the results of the particle trajectory calculations in the form of impact velocities and impingement angles, for a large number of particle trajectories.

A computer code was especially developed to calculate the blade surface erosion distribution. The program input consists of the blade surface geometrical data and the particle blade impact data as determined by the particle trajectory calculations. A blade surface fitted grid is generated and the impact data is then combined with the empirical equations for the mass erosion parameters to predict the blade erosion intensity and pattern.

## 3. Results and discussion

The particle dynamics numerical computations were conducted for 15 $\mu$m fly ash and 150 $\mu$m sand particle dynamics were carried out through a 0.45 scale model, two stage axial flow oxidizer pump-drive turbine [14]. For the standard inlet air equivalent conditions at 1893 rev min$^{-1}$ and 7.071 lb s$^{-1}$ weight flow, the particle distribution at inlet was statistically specified according to the gas flow. The small ash particles were assumed to be in equilibrium with the gas flow at the turbine inlet, while the larger sand particles had 50% velocity. The presented results show typical sand and ash particle trajectories and blade impact patterns and the blade surface erosion by the ash particles.

Figures 1 and 2 present the projection in the $z$–$\theta$ plane of the computed trajectories for sand and ash particles through the first stage. The main difference between the two sets of trajectories is observed for those particles impacting the blunt leading edge. After rebounding, the larger sand particles deviate more from the flow streamlines and hence tend to impact the opposite surface further upstream (Figs. 1a and 2a). Comparing Figs. 1b and 2b, one can observe a large difference between the ash and sand particle trajectories in the first rotor. The small ash particles mostly interact with the blade pressure side only even though they may impact it more than once. However, many of the larger sand particles impact the blade suction side sometimes twice and they also impact the pressure side.

The main difference between the computed trajectories of the ash and sand particles was found to be caused by their different directions at inlet relative to the blade rows. Comparing Figs. 1a and 2a, no large differences are observed in the direction of the sand and ash particles, in the absolute frame of reference, as they leave the first nozzle. However, the absolute exit

(a) (b)

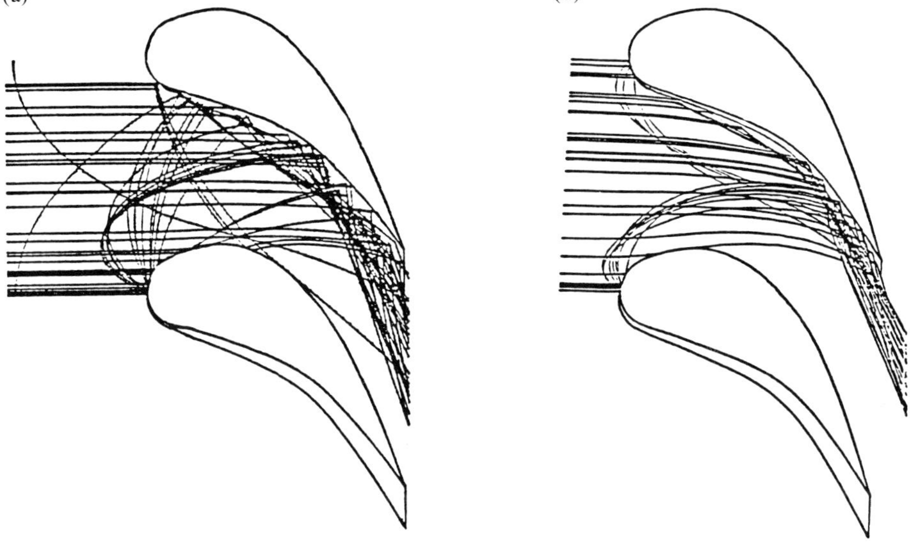

Fig. 1. a, Sand particle trajectories through the first turbine nozzle; b, ash particle trajectories through the first turbine nozzle.

(a) (b)

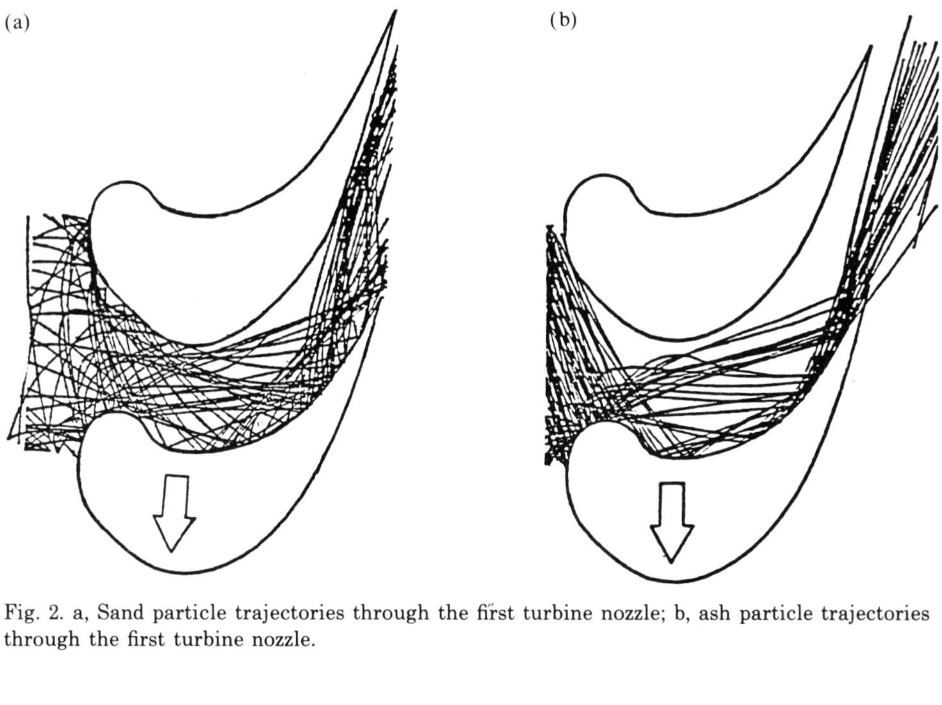

Fig. 2. a, Sand particle trajectories through the first turbine nozzle; b, ash particle trajectories through the first turbine nozzle.

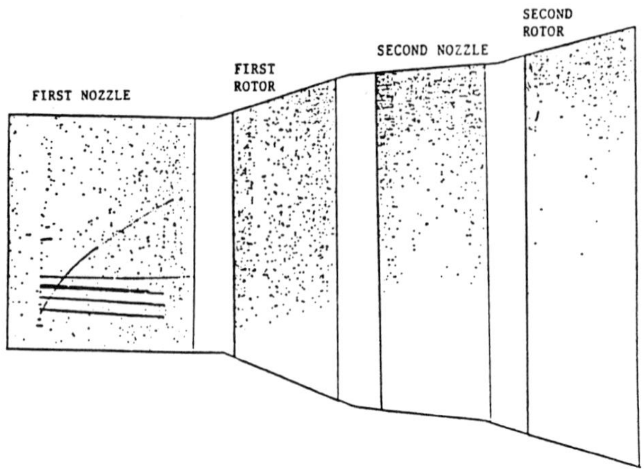


Fig. 3. Blade pressure side impact locations for sand particles.

particle velocities of the larger sand particles are much lower. This results in drastically different directions in the relative frame of reference at inlet to the following rotor. While the small ash particles enter the first rotor with positive incidence angles, the large sand particles enter the same rotor with negative incidence, as can be seen in Figs. 1b and 2b. The high negative incidence angles of the same particles relative to the blades cause the sand particle impacts with the suction surfaces of the first rotor blades. These differences in the relative directions of the sand and ash particles, continue even though they are somewhat reduced in the rest of the blade rows.

Figures 3 and 4 present the locations of blade pressure surface impact by the sand and ash particles respectively. The first observation one can

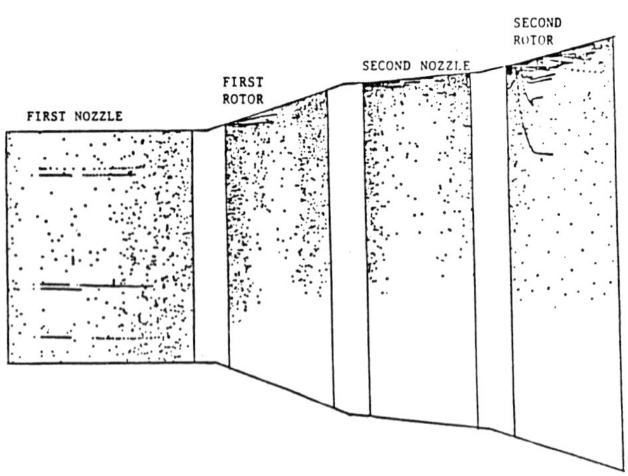


Fig. 4. Blade pressure side impact locations for ash particles.

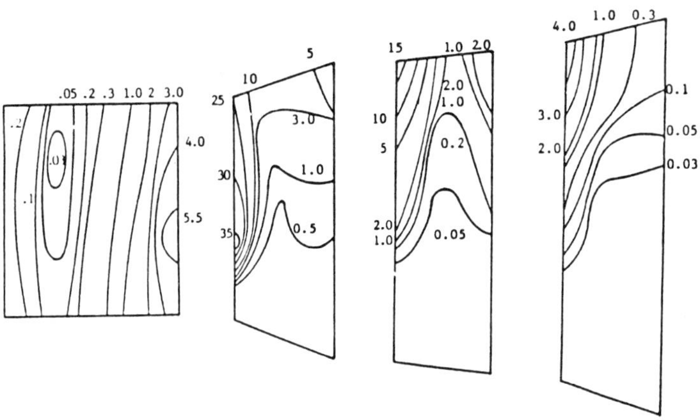

Fig. 5. Blade pressure side erosion distribution for fly ash particles.

conclude from these figures is that the particles start to centrifuge in the radial outward direction after their first rotor impact and this continues through the rest of the turbine. Both particle pressure surface impacts are somewhat similar, but those for the larger sand particles tend to spread more evenly in the chordwise direction. On the other hand, very few of the small ash particle impacts can be observed over the middle third of the blade axial chord, except very near the blade tip.

The greatest difference between the sand and ash impact patterns was observed over the blade suction surfaces. While the number of ash particles with the blade suction surfaces were too insignificant to report, a relatively large number of sand particles impact the blade suction surfaces.

The erosion patterns of the blade pressure surface due to ash particle impacts are presented in Fig. 5. The contours in Fig. 5 represent the blade mass erosion parameter which is defined as the blade mass erosion per unit area of the blade surface per unit mass of ingested particles in $\mathrm{mg\,m^{-2}\,gm^{-1}}$. The maximum local blade surface erosion is observed in the first rotor at the leading edge one-third the blade height from the hub. This is caused by the higher impact velocities in this region, even though the particle impacts are more concentrated towards the tip. The erosion of the first nozzle blade pressure surface is higher near the trailing edge, also due to the higher particle impact velocities, coupled with impingement angles ranging between 25° and 30°. The blade erosion pattern is very similar, however, to the impact frequency pattern over the blade pressure surfaces in the second stage, with the highest erosion rate at the leading edges near the tip. The presented results show the impact patterns with the blade surfaces for the two types of particles. While the pressure surface impact patterns are somewhat different, the suction surfaces are impacted by the sand particles near the leading and trailing edges, but not impacted by the fly ash particles.

*3.1. High-temperature erosion tunnel*

The high-temperature erosion test facility was designed to provide erosion and rebound data in the range of operating temperatures experienced in compressors and turbines. In addition to the high temperatures, the facility properly simulates all erosion parameters which were found to be important from aerodynamics point of view which was previously established at ambient temperatures in another erosion wind tunnel. These parameters include particle velocity, angle of impact, particle size, particle concentration and sample size. Close attention was given to aerodynamic effects to insure that important parameters, such as angle of impact, are not masked or altered.

A schematic of the test apparatus is shown in Fig. 6; it consists of the following components: particle feeder (A), main air supply pipe (B), combustor (C), particle preheater (D), particle injector (E), acceleration tunnel (F), test section (G) and exhaust tank (H).

The equipment functions as follows. A measured amount of abrasive grit of a given constituency is placed into the particle feeder (A). The particles are fed into a secondary air source and blown up to the particle preheater (D), and then to the injector (E), where they mix with the main air supply (B), which is heated by the combustor (C). The particles are then accelerated by the high-velocity air in a constant-area steam-cooled duct (F) and impact

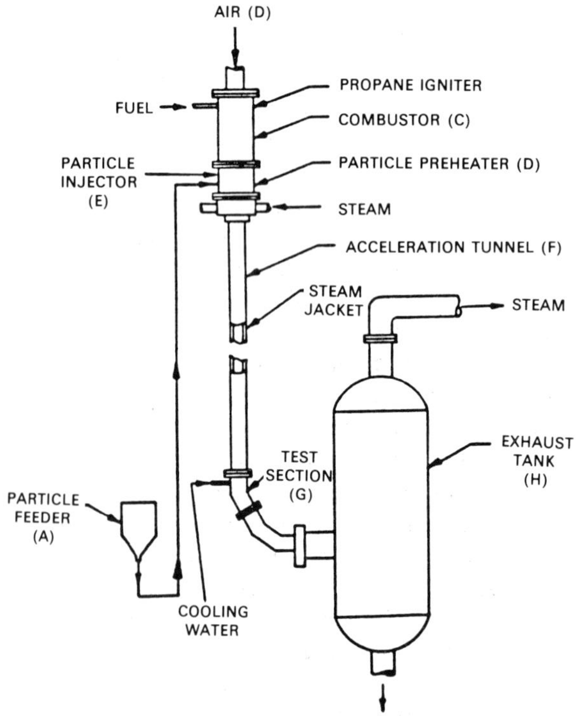


Fig. 6. Schematic of erosion test facility.

the specimen in the test section (G). The particulate flow is then mixed with the coolant and dumped in the exhaust tank. This facility is capable of supplying erosion data at temperatures in the range of ambient to 1093 °C, and particle velocities from 60 to 450 m $s^{-1}$. In this facility, the particle velocity is controlled by adjusting the tunnel air flow, while the impingement angle is set by rotating the sample relative to the flow stream. The sample temperature is controlled through the combustor heating the flow stream which in turn affects the material sample temperature.

### *3.2. Geometry and sizes of specimens and holders used in the experimental study*

Figure 7 shows the schematic diagram of the specimens and the holders with their dimensions. The specimen holders were designed in such a way that the leading edge of the target specimen was protected and the erosion was restricted to the test face in order to avoid the measurement of losses incurred at the complementary angle to the airflow. Most of the holders are fabricated with erosion resistant materials. The specimens used in the present research work were obtained from ductile and nonductile alloys. Care was taken to ensure that the target specimens were used with the same type of surface finish. They were washed in acetone, dried and then weighed carefully before and after erosion testing. The erosion losses are reported in terms of the erosion mass parameter mg $gm^{-1}$ (mg of material eroded per g of abrasive impacted on the specimen surface), and the erosion volume parameter $cm^3\ g^{-1}$ $cm^3$ of material eroded per g of abrasive impacted on the specimen surface). The samples are weighed with electronic analytical balance OHAUS model Galaxy™ with readability in g (0.00001 g).

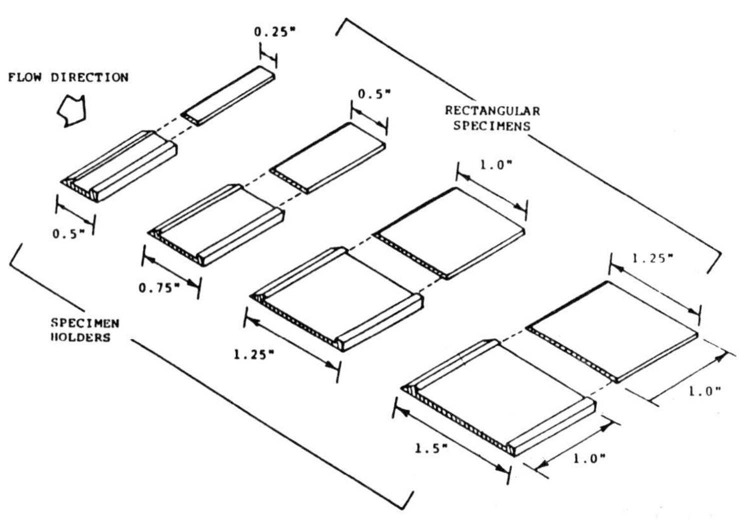


Fig. 7. Schematic diagram of the test specimens and their holders.

### 3.3. Test erosion results

The erosion rates *vs.* the particle impingement angles for different sample temperatures are presented in Figs. 8 and 9 for Ti 6-4 and INCO 718, correspondingly. The erosion rate is maximum for Ti 6-4 at an impingement angle of 30° and for the INCO 718 at 35°. In both cases, the particle size and velocity were 165 $\mu$m and 152 m s$^{-1}$ (500 ft s$^{-1}$) respectively. Similarly, Fig. 10 shows the erosion parameter for 304 stainless steel for three different velocities and two temperatures, 26 and 649 °C (78 and 1200 °F). It may be observed that at the maximum angle of attack (25°), the erosion volume parameter for a particle velocity of 183 m s$^{-1}$ (600 ft s$^{-1}$), from ambient temperature of 26 (78) to 649 °C (1200 °F) increases almost 2.5 times and from a velocity of 305 m s$^{-1}$(1000 ft s$^{-1}$) 6.5 times. These results clearly demonstrate that the equations available for erosion calculations at ambient temperatures cannot be used when high temperature is present.

### 3.4. Thermal barrier coatings of turbomachinery

Thermal barrier coatings in turbomachinery are being used in combustors, compressors and turbines. Presently much testing has been done on turbines and compressor blade coatings. It is already well documented that the newly developed coatings will improve the gas turbine performance. Ceramic coatings can reduce the metal temperature several hundreds of degrees, resulting in significant engine life improvement. The main problem

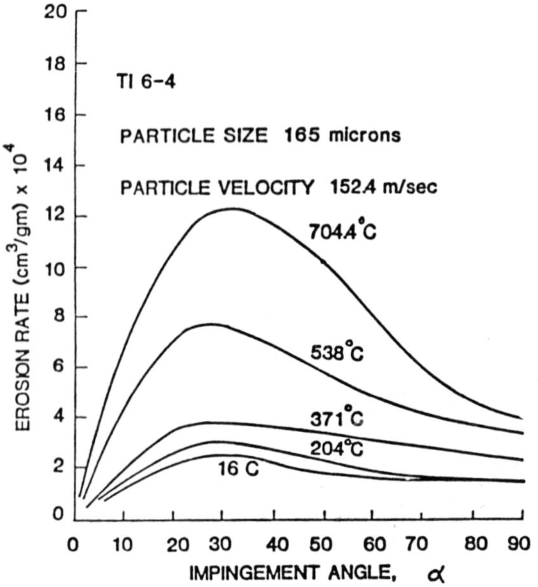


Fig. 8. Ti 6-4 erosion rate *vs.* impingement angle for different temperatures.

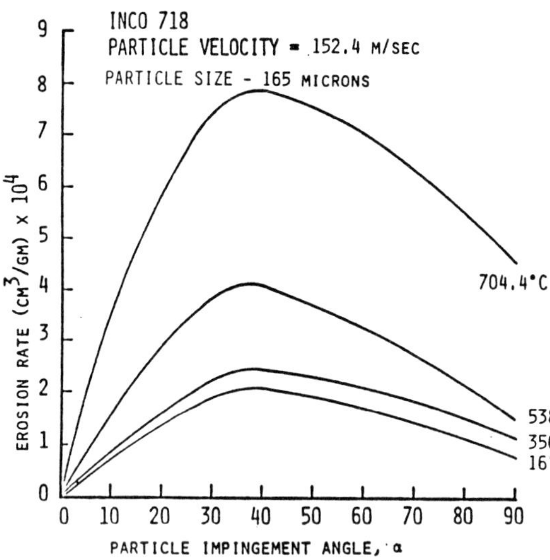


Fig. 9. INCO 718 erosion rate *vs.* impingement angle for different temperatures.

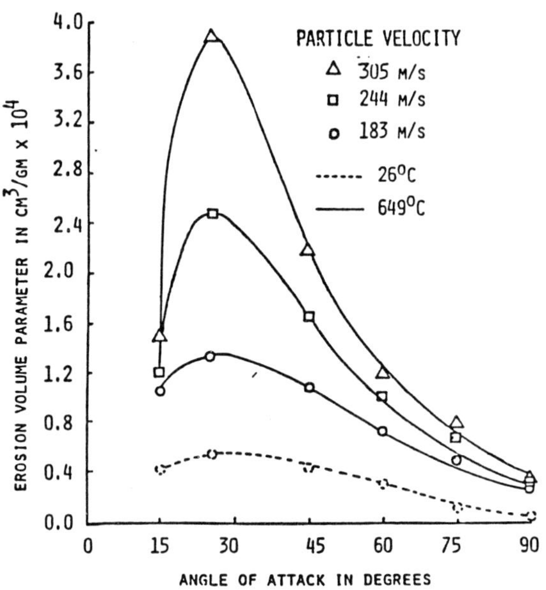


Fig. 10. Effect of angle of attack on 304 steel at 28 and 649 °C for different particle velocities.

presently is how to prevent the coating from failure owing to
(a) thermal cycling of the ceramic layer;
(b) oxidation of the bond coat;
(c) erosion due to gas stream solid particles;
(d) deposition of molten debris.

The major failure modes of thermal blade coating systems are the spallation of the coated material due to thermal cycling. The failures are believed to be caused by thermal stresses and oxidation of the coated bond. However, new improvements have been introduced by going to high density, oxidation resistant bond coats. Generally three methods are available to obtain superior bond coats: shrouded plasma spraying, vacuum plasma spraying and physical vapor deposition.

### *3.5. The effect of particulate erosion*

Many studies were undertaken to measure the particulate erosion rate of thermal barrier coating under similar but accelerated conditions as found in gas turbines. Numerous programs aimed to rank the erosion rates of different thermal barrier coating systems have been carried out at University of Cincinnati. Leading jet engines, steam turbines and industrial gas turbine industries in the U.S. have projects performed at the University of Cincinnati Propulsion Laboratory. Additional work has been performed for other countries such as the U.K., Germany, France, Switzerland, Israel, Japan and others. Unfortunately, the investigated coatings are company properties and therefore cannot be reported. The only published data is part of the work performed for Rolls Royce Inc., supervised by W. Tabakoff and F. C. Toriz. Details of this work are reported in ref. 15.

A short description of the results obtained in ref. 15 is described below. This study was undertaken to measure the particle erosion rate of thermal barrier coating under similar conditions as found in a high-pressure turbine. Nine thermal barrier coating systems were evaluated. Systems evaluated included several variations of yttria stabilized circonia (YSZ), magnesia stabilized zirconia (MSZ), ceria stabilized zirconia (CYZ), Alumina ($Al_2O_3$), calcium titinate ($CaOT_iO_2$) and chromia silica titania (CST). The maximum test temperature was 1600 °F. The test velocities ranged from 400 to 1000 ft/$s^{-1}$. The impingement angles were 20°, 30°, 40°, 60° and 90°. The testings were performed with a mixture of $Al_2O_3$ particles of sizes 8, 40 and 130 $\mu$m.

### *3.6. Erosion test results*

The steady state erosion rate of 90° was compared for different coatings. The ranking is presented in the bar chart in Fig. 11. The standard YSZ was not very erosion resistant and it was used to normalize the erosion rates of all the coatings tested. The best, most erosion resistant thermal barrier coating was the physical vapor deposition (PVD) YSZ. However it did have a higher erosion compared with bare nickel alloy. The two dense coatings also did well. The zirconia titania yttria (ZTY) was second and third. The $Al_2O_3$ was fourth, and about 3 times more erosive that of PVD YSZ. The laser glazed CYZ had a low initial erosion rate. The ($CaOT_iO_2$) rated sixth just ahead of the magnesia MSZ, and the fully stabilized 20° yttria zirconia ranked eighth. The effect of impingement angle on erosion is shown in Fig. 12. Inspection of Fig. 12 shows that the coatings are consistent with the traditional response

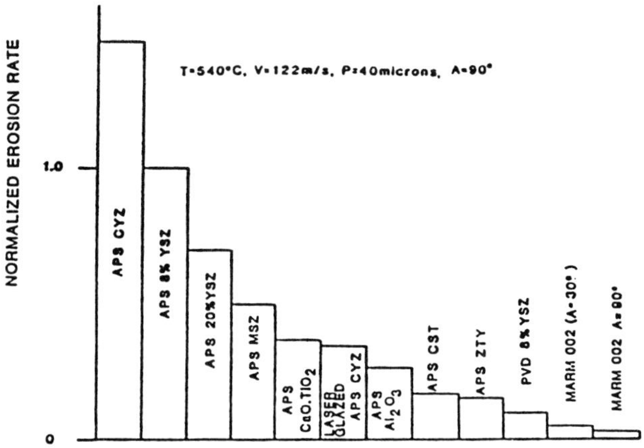


Fig. 11. Normalized thermal barrier coating erosion rate.

of 'brittle materials'. Testings were performed for different impingement angles with different velocities, temperatures and particle sizes. The maximum erosion for all coatings was at 90°. The degree of erosion increased with increasing the impingement velocities, temperatures and particle sizes. For further description see ref. 15. It is generally considered that a 10-fold improvement in erosion resistance are required for any material or coating developed to alleviate the erosion problem.

Basically, three major types of coatings have been used in jet engines: (1) multiphase overlay coatings: tungsten carbide–cobalt; (2) diffusion coatings: chromium–boron and (3) single phase hard compound overlay coatings: titanium diboride. Laboratory and rig erosion testing of plasma deposited and diffusion coatings has demonstrated the potential of 4- to 8-fold reduction in

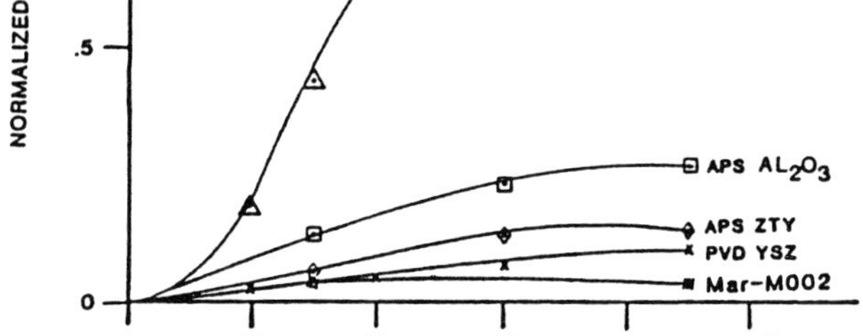


Fig. 12. Effect of impingement angle on erosion.

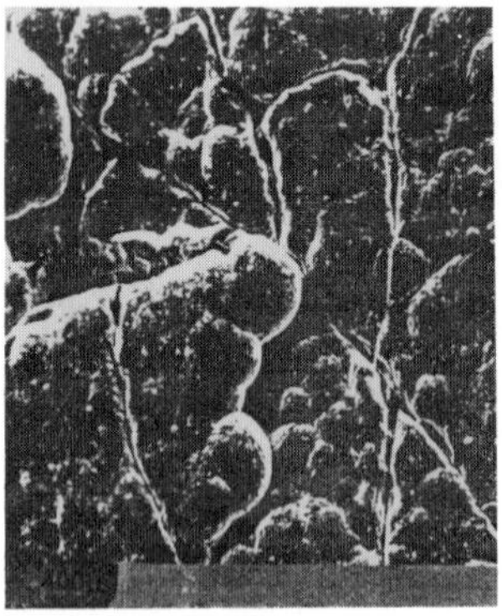

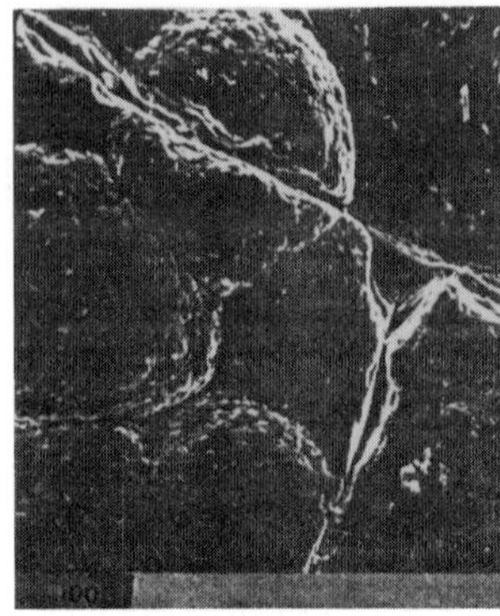

Fig. 13. Scanning electron micrograph of carbide coated surface.

Fig. 14. Enlarged coated surface with cascades.

erosion for the titanium carbide and titanium biboride coatings. Thin ceramic coatings can augment the difference between gas temperature and metal temperature by several hundred degrees. Thus, they are potentially a major step forward by permitting reduced coolant flow, higher turbine inlet temperature, or improving durability by reducing metal temperature and transient thermal stresses. In the last 10 years thermal barrier coatings were developed with a zirconia-based two layer system. From several stabilized zirconia compositions, which included CaO, MgO and $Y_2O_3$ stabilizers, the $ZrO_2$–12 wt.%$Y_2O_3$/MCrAlY system was chosen on the basis of adherence to the MCrAlY bond coat. It offers thermal stress resistance, erosion resistance and stability in the aircraft turbine environment. Figures 13 and 14 are scanning

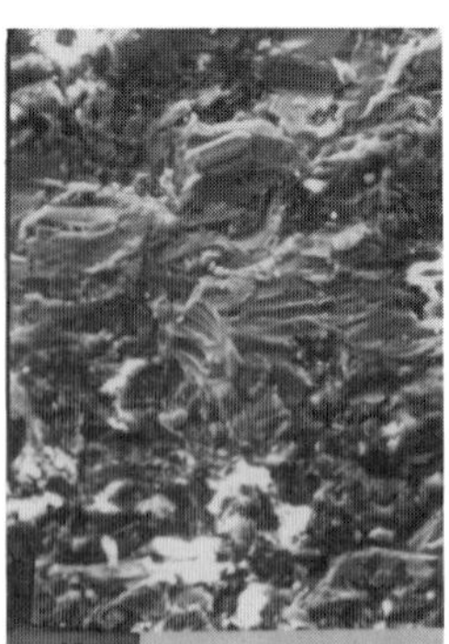

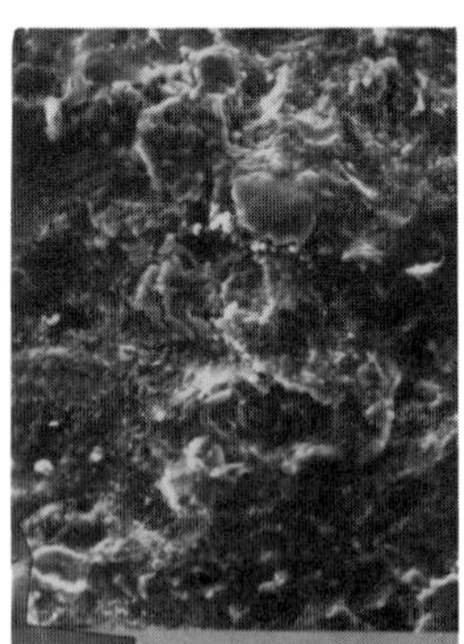

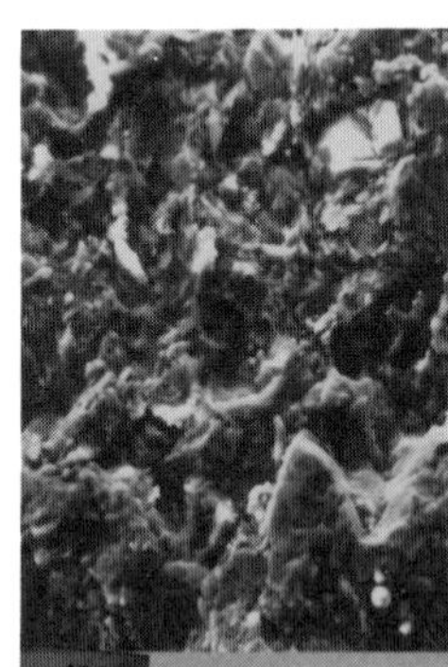

Fig. 15. Scanning electron micrograph of ductile coating at an impingement angle of 20°.

Fig. 16. Scanning electron micrograph of ductile coating at an impingement angle of 45°.

Fig. 17. Scanning electron micrograph of ductile coating at an impingement angle of 90°.

electron micrographs of sand impacted carbide coated surfaces. They clearly show the large cracks produced by sand particle impacts. Additional scanning electron micrographs for ductile coatings are shown in Figs. 15, 16 and 17 for particle impact angles of 20°, 45° and 90° respectively. This type of coating does not have any cracks. Similar problems with regard to erosion and coatings in gas turbines are encountered in steam turbines as well. The erosion of steam turbine components has become a serious technical and economic problem. This erosion damage can contribute to the replacement of components, loss of turbine efficiency and increased maintenance cost. Therefore, Westinghouse has started a coating evaluation program that encompasses erosion resistance evaluation of various coatings, and the effects of various coatings on the substrate tensile, fatigue and creep properties variation of the coated components. At the University of Cincinnati we have studied and tested surface erosion on many Westinghouse coatings and some of the results obtained may be found in ref. 16.

## 4. Conclusions

The results of sand and ash particle trajectory computations are presented in a two-stage turbine. The presented results show the difference between the trajectories and blade impact patterns for the two types of particles. The differences are particularly obvious in the suction surfaces of the blade rows succeeding the first nozzle. The blade suction surfaces are impacted by the sand particles at the leading and trailing edges, but are practically not impacted by the fly ash particles. The critical blade areas with regard to erosion are different from sand and ash particles, since they are strongly influenced by the impact frequency pattern. The erosion of the blade suction surface near the trailing edge by sand particles, in the intermediate blade rows, can have a significant effect on the turbine performance deterioration. The high-temperature erosion wind tunnel used in these investigations has major advantages. Some of them are variable fluid and particle velocities, particles flow rates, and particle sizes can be controlled in a representative aerodynamic environment. In addition, the temperature can be adjusted from ambient to 1600 °F. With regard to the erosion resistance of a given coating material, it was found that it is very much dependent on the process used to apply this coating. High-density bond coats such as those produced by shrouded plasma spray, vacuum plasma spray and physical vapor deposition, resulted in significant erosion improvements.

## Acknowledgments

This research work was sponsored by the U.S. Department of Energy, Contract No. 19X-89628C, Oak Ridge National Laboratory, Oak Ridge, TN. In addition, I would like to thank Rolls Royce, Inc., and F. C. Toriz for permitting me to use some of the published data obtained in our laboratory.

## References


1 W. Tabakoff, Review — turbomachinery performance deterioration exposed to solid particles environment, *J. Fluids Eng.*, *106* (1984) 125 - 134.

2 W. Tabakoff, A study of the surface deterioration due to erosion, *ASME J. Eng. Power, October* (1983) 834 - 839.

3 M. F. Hussein and W. Tabakoff, Computation and plotting of solid particle flow in rotating cascades, *Comput. Fluids, 2* (1974) 1 - 15.

4 M. F. Hussein and W. Tabakoff, Dynamic behavior of solid particles suspended by polluted flow in a turbine stage, *J. Aircraft, 10 (7)* (1973) 434 - 440.

5 A. Hamed, Solid particle dynamic behavior through twisted blade rows, *J. Fluids Eng. 106* (1984) 251 - 256.

6 G. Grant and W. Tabakoff, Erosion prediction in turbomachinery resulting from environmental solid particles, *J. Aircraft 12 (5)* (1975) 471 - 478.

7 B. Beacher, W. Tabakoff and A. Hamed, Improved particle trajectory calculations through turbomachinery affected by coal ash particles, *ASME J. Eng. Power, 104* (1982) 64 - 68.

8 T. Katsanis, Fortran program for calculating transonic velocities on a blade-to-blade stream surface of a turbomachine, *NASA TND 2809*, May 1965.

9 T. Katsanis and W. D. McNally, Revised fortran program for calculating velocities and streamlines on the hub-shroud mid channel stream surface of an axial, radial, or mixed flow turbomachine or annular duct, Vols. 1 and 2, *NASA TND 8430* and *NASA TND 8431, 1977.*

10 W. Tabakoff and M. F. Malak, Laser measurements of fly ash rebound parameters for use in trajectory calculations, *J. Turbomachinery, 109* (1987) 535 - 540.

11 A. Hamed, W. Tabakoff and M. Mansour, Turbine erosion exposed to particulate flow, *ASME Paper GT-86-258*, June, 1986.

12 A. Hamed, Particle dynamics of inlet flow fields with swirling vanes, *J. Aircraft, 19 (9)* (1982) 707 - 712.

13 W. Tabakoff, A. Hamed and J. Ramachandran, Study of metals erosion in high temperature coal gas streams, *ASME J. Eng. Power, 102* (1980) 148 - 152.

14 R. J. Roelke, R. G. Stabe and D. G. Evans, Cold-air performance evaluation of scale model oxidizer pump-drive turbine for the M-1 hydrogen-oxygen rocket engine. II — Overall two-stage performance, *NASA TND-3368*, 1966.

15 F. C. Toriz, A. B. Thakker and S. K. Gupta, Thermal barrier coatings for jet engines, *ASME Paper 88-GT-279*, June 1988.

16 J. I. Qureshi and W. Tabakoff, *Surf. Coat. Technol., 36* (1988) 433.


## Appendix A: Nomenclature

| | |
|---|---|
| APS | air plasma spray |
| $C_D$ | particle drag coefficient |
| CST | chomia silica titania |
| CYZ | ceria yttria stabilized zirconia |
| $d$ | particle diameter (m) |
| $F$ | force of interaction between the gas and the particle |
| MSZ | magnesia stabilized zirconia |
| PVD | physical vapor deposition |
| $r$ | radial distance from the turbomachine axis (m) |
| $V$ | gas relative velocity ($m\,s^{-1}$) |
| $V_p$ | particle relative velocity ($m\,s^{-1}$) |

| | |
|---|---|
| VPS | vacuum plasma spray |
| YSZ | yttria stabilized zirconia |
| $z$ | axial coordinate (m) |
| $\epsilon$ | erosion parameter, mg m$^{-2}$ g or mg g$^{-1}$ |
| $\theta$ | angular coordinate, radians |
| $\rho$ | gas density (kg m$^{-3}$) |
| $\rho_p$ | particle density (kg m$^{-3}$) |
| $\omega$ | rotor speed (radians s$^{-1}$) |
| $\gamma$ | frequency of particle impacts (total number of particle impacts per unit area of the blade surface per unit number of ingested particle) (m$^{-2}$) |

*Subscripts*

| | |
|---|---|
| p | particle |
| r | component in radial direction |
| z | component in axial direction |
| $\theta$ | component in circumferential direction |

*Surface and Coatings Technology*, *39/40* (1989) 117 - 126 

# THE WEAR MECHANISMS OCCURRING IN ABRADABLE SEALS OF GAS TURBINES

M. O. BOREL, A. R. NICOLL, H. W. SCHLÄPFER and R. K. SCHMID

*Corporate R&D and Surface Technology Division, Sulzer Plasma-Technik Inc., Winterthur (Switzerland)*

(Received February 28, 1989)

## Summary

Metallographic investigations of worn abradable coatings have been carried out to identify and characterize the main wear mechanisms occurring in turbomachine seals. The following mechanisms have been found: cutting, smearing, adhesive transfer, crushing, melting and tribo-oxidation. Three other mechanisms occur without blade–seal interaction: erosion, corrosion and high-temperature oxidation. Using the data from these investigations as a reference, rub tests have been conducted on a specially designed high-speed–high-temperature rig. The many correlations obtained between test values and wear mechanisms allow the generation of 'wear mechanism maps' for coatings of the AlSi–plastic family. Mechanism orientated model tests are needed to understand the correlations between structural parameters and coating behaviour during blade–coating interactions. Wear mechanism maps will help modelling coating systems to be able to withstand severe service conditions. Avoiding the onset of adhesive or melting wear and simultaneously favouring cutting wear produces better abradable seals which will help improve both the operational safety and the economy of modern turbomachines.

## 1. Introduction

Modern turbomachines need very small clearances between rotating components (blade tips, labyrinth seals) and the stator case in order to minimize gap losses and so increase efficiency. For this purpose, different types of abradable seals are deposited on the stator case to cope with rotor misalignment, thermal and centrifugal dilatations, and unbalanced parts. The primary requirement of abradable seals is to allow for wear without damage to the blade tip. A secondary requirement, gaining in importance, is that the surface remains smooth after wear in order to reduce aerodynamic losses. Different coating systems have been developed over the past years or are being evaluated at present. In general, the development of abradable systems

0257-8972/89/$3.50 

has taken place empirically and relies on very little knowledge of the actual situation in service. In this work, a more systematic approach to the development of material systems is applied. As a first step, the wear mechanisms occurring in service are identified by examining worn components. The second step consists of laboratory material trials where by widely varying the test parameters, the different wear mechanisms identified in practice are simulated. The conditions leading to mechanism transitions are then analysed and related to the material properties. This step ends with the generation of 'wear mechanism maps' [1]. In the third step, a model is formulated for the relationship between material microstructure and in-service wear mechanisms. Using this model, material systems for improved performances can be designed (owing to its early development stage, this step will not be discussed here).

## 2. Investigation of worn service parts

In order to identify wear mechanisms that the coatings are subjected to, worn abradable seals that had been run in service were investigated. In the preliminary work only abradable seals from non-turbine stages were inspected ($T < 500$ °C). The worn materials (AlSi–plastic and nickel–graphite coatings) were examined non-destructively and destructively where possible. The main characteristics of each identified wear mechanism are described below.

### *2.1. Erosion, corrosion and oxidation*

The erosion, corrosion or oxidation of a coating can be detected in that no contact (rub interaction) has taken place and/or the roughness, surface chemistry and colour has changed. No surface interaction takes place as the abradable surface retreats thereby increasing the gap between blade tip and coating.

The erosion, corrosion and oxidation of a coating are undesirable as they increase the blade gap so reducing engine efficiency.

### *2.2. Cutting*

The high-speed blade tip can act as a cutting tool when it interacts with the coating. The new surface generated by cutting is very smooth and the coating takes on a shiny appearance. A surprising feature is the efficiency of the cutting (material removal) as the coating material at or around the cut shows very little or no deformation (Fig. 1). When cutting occurs, chip separation takes place in a brittle manner even in highly ductile materials such as polymers.

Cutting is a desirable means of abradable seal wear as the blade tip is not damaged, no material transfer takes place and the coating is left with a smooth surface that has favourable aerodynamic properties.

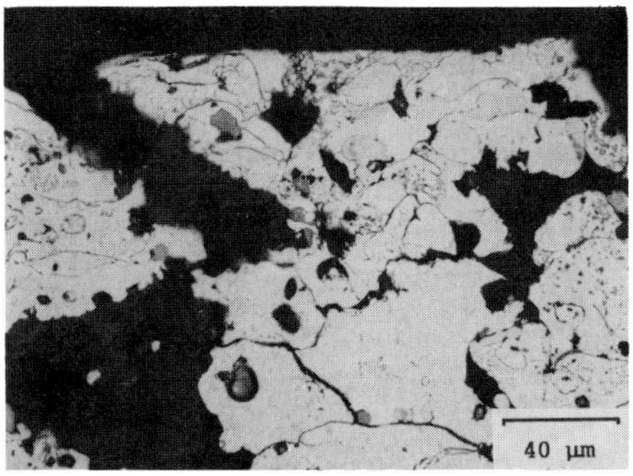


Fig. 1. Microsection through a nickel–graphite coating worn in the pure cutting mode. The subsurface zone is free of deformation.

*2.3. Deformation*

During interaction the coating may be deformed plastically by the blade tip. This deformation can take on many forms even within one coating as these are heterogeneous materials and their individual components deform in different ways.

The surface can become striated on a scale varying from micrometers to millimetres (Fig. 2). Extensive flow of the coating can also take place, leading to large-scale smearing of the metallic phase with the non-metallic components appearing to have fractured in a brittle fashion.

Incursion stresses lead to unwanted densification of the material. When incursion speeds are particularly high and most of the deformation is radial, the term crushing is used [2].

The process of deformation is not favourable as it roughens the surface so leading to aerodynamic losses and compacts the coating. Densification of the coating is unfavourable as this makes future interactions more violent as the coating cannot readily give way to the advancing blade.

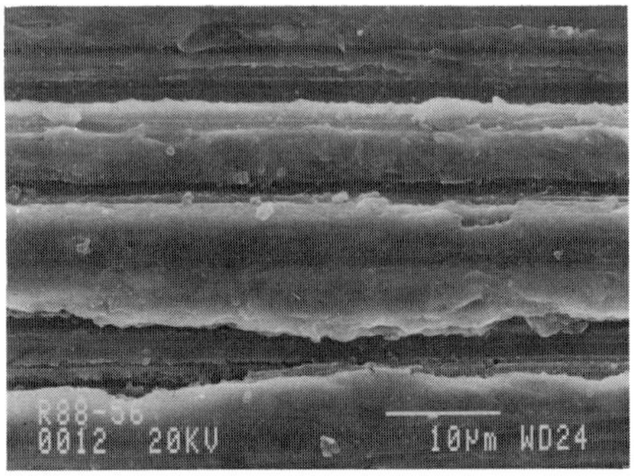


Fig. 2. Scanning electron microscope image of the surface of an AlSi–plastic coating subjected to smearing. Minute surface striations of wave length about 10 $\mu$m are present.

### 2.4. *Adhesive transfer*

This is the transfer of material from the coating onto the blade or vice versa and is often associated with the above process of deformation. The non-metallic particles of the coating may be covered by compacted surface layers that detrimentally affect the behaviour of the abradable seal as they lead to overheating and the formation of thick hard transfer layers. These transfer layers have a horizontal layered substructure due to the successive smearing of the coating and blade tip materials (Fig. 3). Great amounts of frictional heat are released during the metallic transfer so that the substrate may undergo thermal cracking and densification.

Adhesive transfer is detrimental and should be avoided.

### 2.5. *Melting*

The transferred layers mentioned above can have complex chemical compositions. This *in-situ* mechanical alloying may result in a newly formed alloy having a lower melting point than its original constituents. These alloys then melt during the following rubbing interactions forming hard and compact areas upon solidification.

The sectioning of these hard spots shows that local melting followed by rapid solidification takes place. In nickel–graphite coatings, melted layers of up to 125 $\mu$m have been seen to form. These layers had a columnar solidification structure that was extensively cracked (Fig. 4).

The melting of the materials at the rubbing interface is harmful as the transfer of the molten material onto the blade is favoured, and the molten mass forms a hard layer on solidification that can lead to serious blade damage, should a second interaction take place.

### 2.6. *Tribo-oxidation*

This is a minor wear mechanism that does not seem to play a significant role during blade tip coating–interactions but was found to occur at labyrinth seal contacts. At these contacts the labyrinth segments seal against

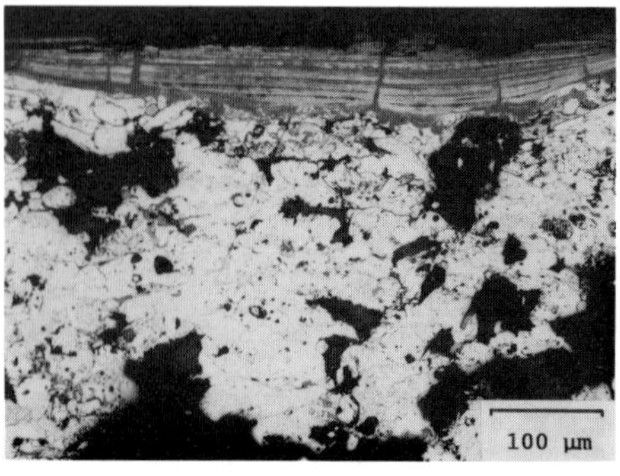


Fig. 3. Microsection through a nickel–graphite coating showing adhesive transfer. The transferred blade material exhibits a layered structure. The coating underwent radial densification (crushing).

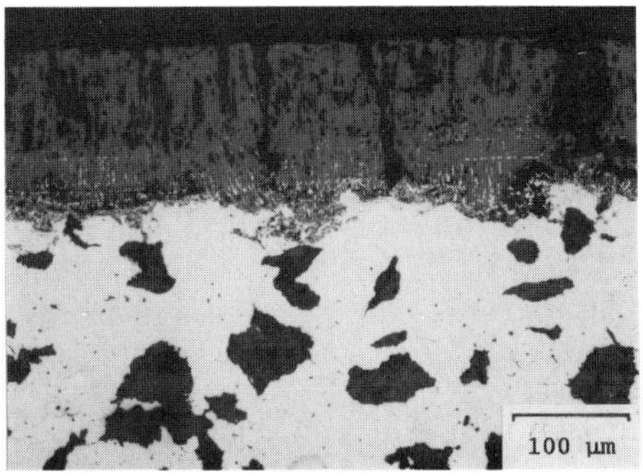


Fig. 4. Microsection through a nickel–graphite coating after melting wear. Molten blade material was transferred to the coating, mixed with the coating surface and quenched by the substrate. Diffusion bonding of the coating particles also occurred.

plasma sprayed coatings. The intermittent rubbing at high speeds and temperatures causes the contacting points to oxidize strongly with the resulting oxide films spalling from time to time [3]. This rubbing induced oxidation is called tribo-oxidation.

Although not a spectacular wear mechanism, it can be responsible for significant wear in short periods of time.

## 3. Rig tests

### *3.1. Test rig description*

To be able to run tests under conditions that produce different wear mechanisms, a rig was designed and constructed that allows for a wide range of all test parameter settings. The following requirements are fulfilled

- Tangential speed range $V_t = 50 - 500$ m s$^{-1}$
- Incursion speed range $V_{inc} = 0 - 3000$ $\mu$m s$^{-1}$
- Coating temperature range $T =$ RT - 800 °C

The incursion speed, temperature and interaction force are controlled via a digital voltmeter. The incursion of the coated sample is actuated by a computer controlled step motor, so that absolutely reproducible tests can be run. Two diagonally opposed blade dummies are fitted in quick change holders. The simple dummy form allows for easy manufacturing of various tip geometries from almost any material. The rotor is driven by an electric motor which has the advantage, over turbine driven systems, that no corrosion affects the wear picture.

Due to the full computer control of the test and rapid exchange of samples a turn around time of 10 - 15 min per test is normally achieved with three to four personnel.

### 3.2. *Test results*

A short test duration is an essential condition when parametric correlations have to be examined. During the first hundred tests the dependence of wear mechanisms on $V_t$, $V_{inc}$, specimen temperature, rub duration, coating heat treatment, blade material and tip geometry were investigated. Certain relationships are discussed in the following.

#### 3.2.1. *Coating temperature dependence on wear mechanism*

Figure 5 shows the temperature measured on the reverse side of the coating, as a function of rub time during two different tests. In test A, the temperature increased by 3 °C from 353 °C initially. Massive metallic transfer of the coating to the blade tip occurred typical of an adhesive wear mechanism. During test B, the temperature increased by about 40 °C without any metallic transfer to either the coating or the blade with a wear mechanism of cutting being observed.

#### 3.2.2. *The correlation between coating roughness, blade weight variation and the wear mechanism*

The data measured for three coatings, AlSi–polyimide, nickel–graphite and AlSi–polyester have been plotted in Fig. 6 using a weighted roughness value ($R = aR_a + bR_z + cR_{max}$) as a function of the mass variation of the blade used in the test ($\Delta G$).

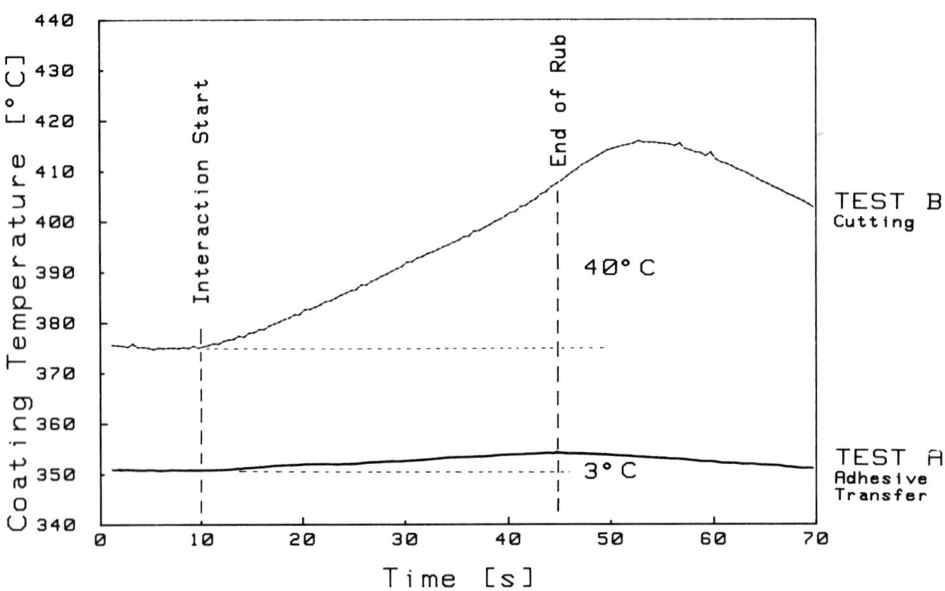


Fig. 5. Specimen temperature *vs.* time for two different rub tests. Test A: adhesive wear with metal transferred from the coating to the blade. The sample temperature remains almost constant. Test B: cutting wear (brittle mode). A temperature increase of more than 40 °C takes place during the rubbing interaction.

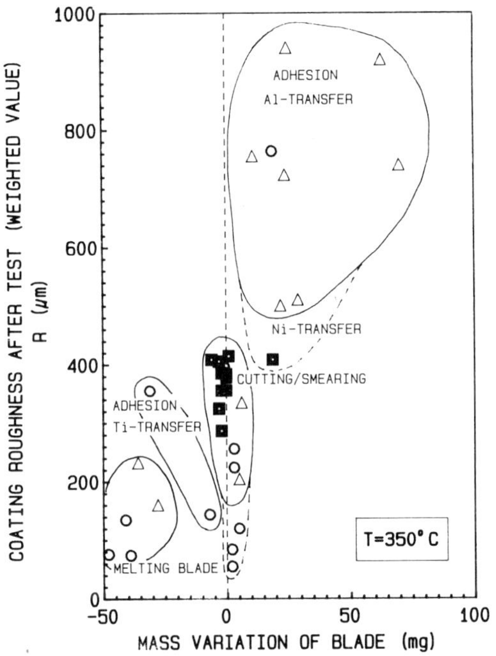


Fig. 6. Plot of the coating roughness, after the test, *vs.* the mass variation of the blade, for the three coating families tested under many different combinations of ($V_t$, $V_{inc}$). Each area on the graph corresponds to a major wear mechanism. ○, AlSi–plastic (solid); Δ, AlSi–plastic (melted); ■, nickel–graphite.

(1) The points in the upper right area of the plot correspond to samples where adhesive transfer from the coating to the blade took place.

(2) Adhesive transfer from the blade to the coating was observed when titanium blades were rubbed at high incursion speeds in the intermediate area at the bottom left.

(3) In the central region, no material transfer takes place ($\Delta G = 0$) and cutting or smearing were identified.

(4) In the lower left area, blade tip melting wear occurred during very high speed tests carried out with comparatively low melting point Nimonic blades.

## 4. Discussion of results

The metallographic investigation of abradables worn in service has shown that different wear mechanisms [4] can be easily identified by either destructive or non-destructive methods. However, the history of in-service interaction conditions cannot be accurately determined as external influences are often superimposed on the wear pattern, thus making any conclusion as to the wear behaviour of the investigated coatings rather uncertain.

Performing model wear tests on a well instrumented test rig is an efficient solution to this problem. It has been shown that systematic correlations exist between the wear mechanisms and the measured test values. From this, 'wear mechanism maps' [1] can be drawn for some of the tested materials (Fig. 7). The tangential speed $V_t$, is plotted on the horizontal axis and the incursion speed $V_{inc}$, on the vertical axis. Mechanisms occurring at a set of conditions are plotted using the initial letter of the relevant mechanism. In wear maps drawn from classical wear experiments, wear rate changes define the mechanism boundaries. In the case of the abradable tests, the wear rate is almost equal to $V_{inc}$, so that other values such as the temperature increase of the coating during the rub, the wear track roughness $R$ or the blade mass variation $\Delta G$ have to be used to determine the wear mechanism.

The results presented in Fig. 6 are summarized in Fig. 7 in the form of two 'wear mechanism maps'. Two AlSi–plastic materials with different plastic filler melting points are compared at a testing temperature of 350 °C. At this temperature, one plastic filler is molten (Fig. 7a) and the other solid (Fig. 7b).

The material containing the low-melting-point plastic underwent

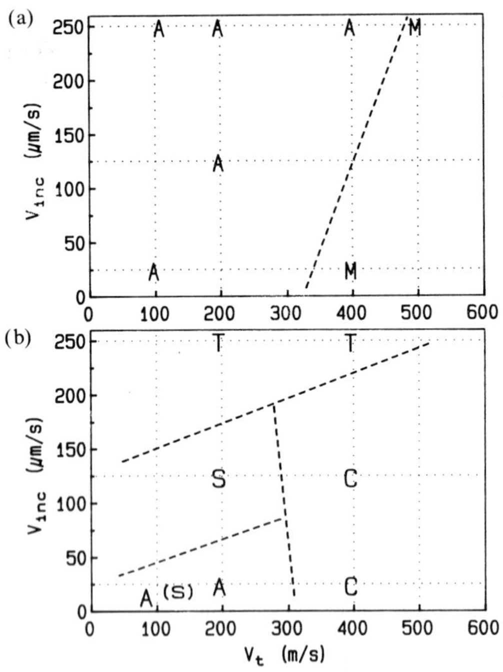


Fig. 7. Wear mechanism maps of two different AlSi–plastic coatings tested at $T = 350$°C. a, Coating with low-melting-plastic (molten at 350 °C); b, coating with higher-melting-plastic (solid at 350 °C). The letters show where tests were run and which mechanism occurred: A, adhesive transfer from the coating to the blades; M, melting wear of blade tip; S, smearing; C, cutting; T, adhesive titanium transfer from the blade to the coating.

adhesive transfer (Fig. 7a, "A") over the major part of the tested speed range. Under adhesive conditions, no temperature rise of the coating occurred during the rub. AlSi was transferred from the coating to the blades, resulting in a high track roughness and a blade mass increase. Although not favourable from a sealing efficiency point of view, this behaviour may be qualified as being safe and unproblematic.

At high blade tip speeds ($V_t \geqslant 400 \text{ m s}^{-1}$), strong frictional heating caused the blade tips to melt ("M") resulting in unwanted blade mass losses and a low track roughness. The liberated particles oxidized and produced heavy sparking. As it results in blade damage, melting wear is harmful and should be avoided in engines.

The material containing the higher-melting-point plastic had a more complex wear behaviour (Fig. 7b). At small $V_t$ and $V_{inc}$ adhesive wear ("A") with coating transfer to the blade tip occurred. These conditions are unlikely to occur under normal operating circumstances. At larger incursion speeds, more material must be removed, in each blade pass, leading to a substantial modification of the stress field under the leading edge of the passing blade. These changing conditions also resulted in a change of wear mechanism. At intermediate incursion speeds, smearing ("S") with a very small amount of coating transfer onto the blade tip took place. As the incursion speeds increased further the mechanism changed to titanium transfer ("T") onto the coating. At very large tangential speeds, ($V_t = 400 \text{ m s}^{-1}$), the coating probably reached its strain rate limit and experienced pure cutting wear ("C"). When this occurred, the coating temperature increased and a very low blade mass variation and track roughness, which are favourable in service, were recorded.

## 5. Conclusions

(1) An investigation of in-service worn abradable seals found the wear mechanisms to comprise the following: cutting, smearing, adhesive transfer, crushing, melting and tribo-oxidation.

(2) The coating temperature during rub testing has been found to depend on the wear mechanism. Adhesive wear induces a small temperature increase while cutting wear causes strong heating of the coating.

(3) Good correlations were observed between wear track roughness, mass variation of the blades and the occurring wear mechanism. The material system (coating–blade) has little effect on these correlations.

(4) The wear mechanisms found on service parts can all be produced on test specimens by varying the velocity ($V_t$, $V_{inc}$) combinations.

(5) Wear mechanism maps have been drawn for several abradable systems. These maps can be used for modelling wear mechanisms and the design of coating systems to provide enhanced performance at elevated temperatures.

## References

1 S. Lim and M. Ashby, *Acta Metall., 35* (1987) 1.
2 R. C. Bill and D. W. Wisander, *Friction and wear of several compressor gas-path seal materials, NASA TP-1128,* 1978.
3 T. F. Quinn, J. L. Sullivan and D. M. Rowson, *Wear, 94* (1984) 175.
4 D. L. Clingman, B. Schechter, K. R. Cross and J. R. Cavanagh, *Lubr. Eng., 39* (1983) 712.

*Surface and Coatings Technology*, *39/40* (1989) 127 - 134 

# APPLICATIONS OF HARD COATING BY LOW-PRESSURE PLASMA SPRAY

HIROSHI TAKIGAWA, MITSUJI HIRATA and MASAMICHI KOGA

*R&D Department, Plant & Machinery Division, Nippon Steel Corporation, 46–59, Nakabaru, Tobata-ku, Kitakyushu 804 (Japan)*

MICHIHISA ITOH and KOICHI TAKEDA

*Future & Frontier Field Research Center, 1st Research Laboratory, Nippon Steel Corporation, 1618, Ida, Nakahara-ku, Kawasaki 211 (Japan)*

(Received March 5, 1989)

## Summary

Low-pressure plasma spray (LPPS) has been recognized as a superior surface modification technology due to the improved coating properties.

In the field of steel-making equipment, hard materials, high temperatures, erosion and corrosion resistant materials are greatly needed.

In this paper, an application of $Cr_3C_2$–NiCr coating by LPPS is mainly discussed. It is observed from X-ray diffraction analysis that the material composition changes very little between prepared powder and coating by LPPS. Application to guiding rollers in a hot strip mill has achieved a life-expectancy more than 10 times that of conventional materials.

## 1. Introduction

In the steel making industry, the cost of maintenance is very high. In the case of Nippon Steel Corporation, it exceeds 1 billion dollars a year, most of which is spent on the repair of surface damage.

Sources of this damage are classified as shown in Fig. 1. This pie chart shows that more than half of the surface problems result from wear. Therefore, wear-resistant surface treatments have been strongly needed.

There are several methods of anti-wear surface treatment (so called "hardfacing"), such as electroplating, arc welding and thermal spray coating (atmospheric). However, they have not achieved sufficient improvement of properties of equipment in steel-making plants. The reasons are

1. Insufficient hardness.
2. Changes of material composition between powder and coating.
3. Insufficient bonding strength.

0257-8972/89/$3.50 

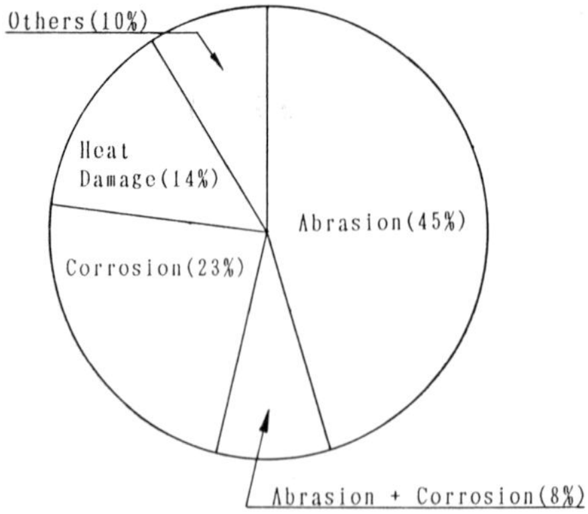


Fig. 1. Sources of damage.

For these reasons, development of advanced hardfacing techniques has been required, and low-pressure plasma spray (LPPS) is one of the answers considered [1].

## 2. Experimental procedure

The principle of LPPS is shown in Fig. 2. Since plasma spray coating is usually carried out in an air atmosphere the metal powder is easily oxidized. Non-oxide ceramics (carbides, borides, etc.) also tend to include oxides. These cause degradation of quality, especially strength. In LPPS, the spray system

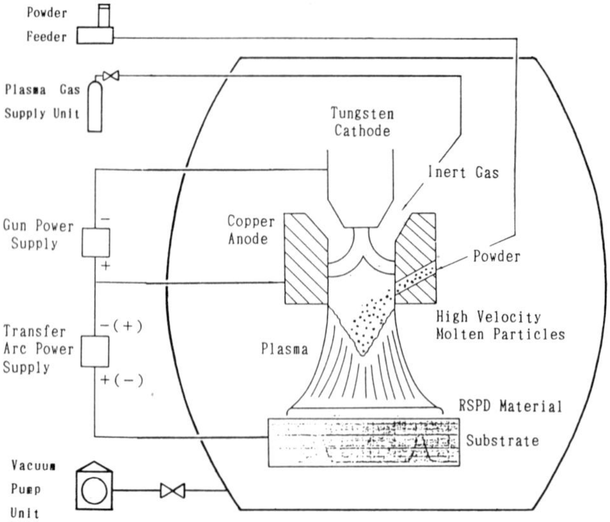


Fig. 2. Principles of LPPS.

is located in a chamber which can be evacuated with a vacuum pump unit. By initial pumping to 0.1 Torr and spraying at a controlled pressure of inert gas, three large advantages appear

1. Little change in material composition (because of the very low oxygen pressure).
2. High tensile strength of coating (because of the very low porosity).
3. High bonding strength (because molten powder impacts at a high velocity on the substrate) [2].

In hardfacing, metal carbide cermets ($Cr_3C_2$–NiCr, WC–Co, etc.) are mostly used because they have both high hardness and ductility. Several tests and evaluations of $Cr_3C_2$–NiCr and WC–Co have been done for coatings produced by LPPS and those by atmospheric pressure plasma spray (APPS).

## 3. Results and discussion

### *3.1. Microstructural observations*

Figure 3 shows a micrograph of the cross-section of the coatings. LPPS coatings have low porosity and little contamination from impurities. All four samples have a bonding layer (NiCr alloy) for relaxation of thermal stress. The $Cr_3C_2$–NiCr APPS coating has peeled off due to insufficient adhesive strength.

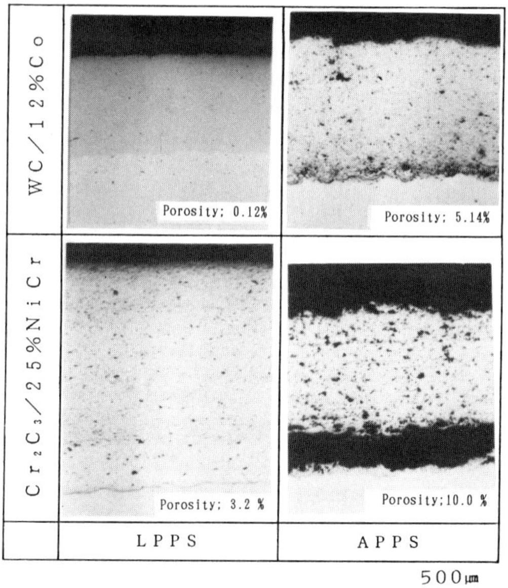


Fig. 3. Optical micrograph of cermet coating by LPPS and APPS.

TABLE 1

Chemical analysis (%)

| | | | *Tungsten* (%) | *Carbon* (%) | *Cobalt* (%) |
|---|---|---|---|---|---|
| WC–17% Co | LPPS | powder | 77.19 | 5.45 | 17.23 |
| | | coatings | 76.87 | 5.16 | 17.12 |
| | APPS | powder | 78.43 | 5.34 | 16.10 |
| | | coatings | 78.32 | 3.54 | 16.86 |
| | | | *Chromium* (%) | *Carbon* (%) | *Nickel* (%) |
| CrC–25% NiCr | LPPS | powder | 69.97 | 12.00 | 15.38 |
| | | coatings | 69.95 | 11.92 | 16.46 |
| | APPS | powder | 67.92 | 11.28 | 20.52 |
| | | coatings | 67.95 | 11.20 | 20.57 |

### *3.2. Chemical analysis*

Table 1 shows the results of chemical analysis of $Cr_3C_2$–25 wt.% NiCr and WC–17 wt.% Co coated by LPPS and APPS. It is obvious that LPPS does not cause a change of material composition. In APPS a considerable decrease in the carbon content has occurred. It is believed that the carbon in WC has oxidized and was emitted as CO or $CO_2$. However, the change of carbon in $Cr_3C_2$–NiCr is small, which suggests that $Cr_3C_2$ is more stable at high temperatures than WC.

### *3.3. X-ray diffraction analysis*

Results of the X-ray diffraction analysis are shown in Fig. 4. The phases in the coating produced by LPPS are the same as the phases in the powder, *i.e.* only the peaks for WC, cobalt, $Cr_3C_2$, nickel and chromium are found. However, analysis of the WC coating by APPS includes peaks of $\alpha$-$W_2C$, $WO_2$ and $WO_3$. From these results, decomposition of carbon is clearly recognized. Result of the analysis of $Cr_3C_2$–NiCr also shows a decrease of carbon which was rather difficult to detect by a chemical analysis.

### *3.4. Hardness measurements*

The temperature dependent of the hardness is shown in Fig. 5. Initially, the hardness was measured during heating from room temperature to 1000 °C, and measured again during cooling to room temperature. According to data shown in this diagram, the $Cr_3C_2$–NiCr coating produced by LPPS is superior in hardness to that produced by APPS in almost the entire temperature range. In the case of WC–Co, the LPPS coating produces a higher hardness than the APPS coating except at high temperatures. This suggests that the APPS WC–Co coating includes $\alpha$-$W_2C$ which has higher hardness than WC but less ductility.

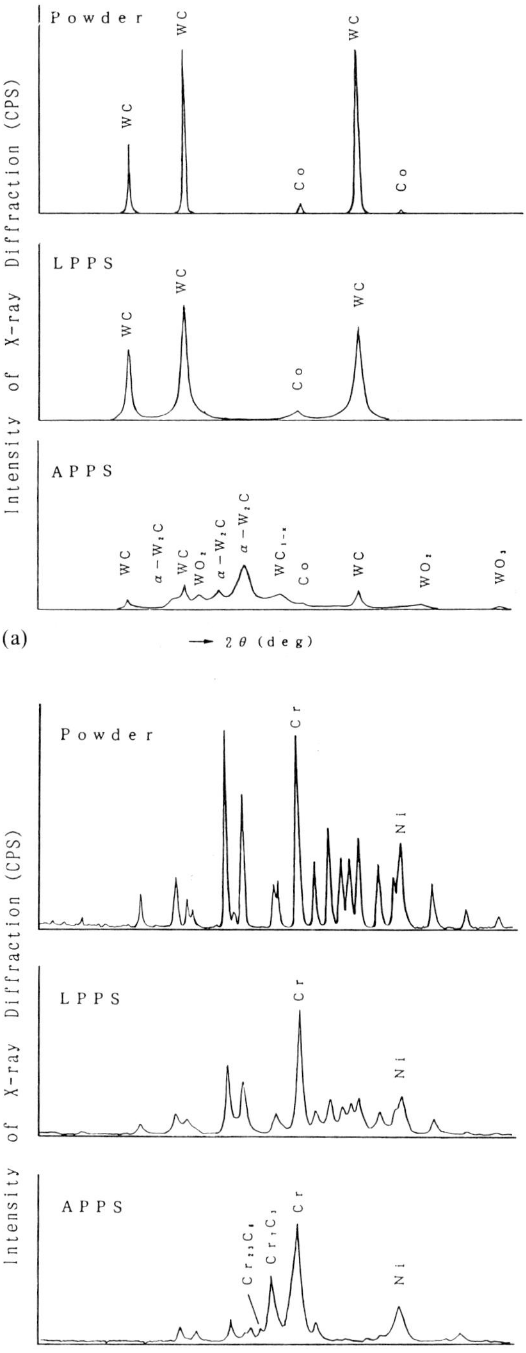


Fig. 4. (a) Results of X-ray diffraction (WC–Co). (b) Results of X-ray diffraction ($Cr_3C_2$–NiCr). Peaks without marking show $Cr_3C_2$.

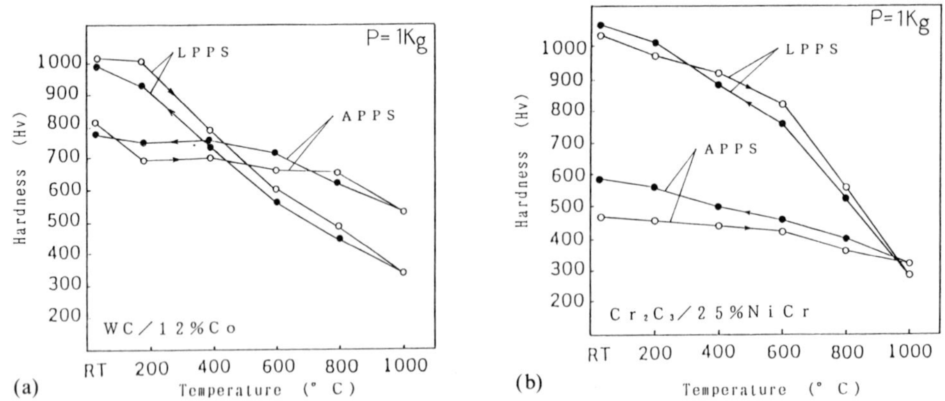


Fig. 5. (a) Hardness as a function of temperature (WC–Co). (b) Hardness as a function of temperature ($Cr_2C_3$–NiCr).

## *3.5. Wear rate measurement*

In general, materials with high hardness have excellent wear-resistant properties. When sprayed coatings are discussed, however, this is not always true due to the effects of porosity and adhesive strength between particles. Therefore, in order to ascertain the wear resistance of LPPS cermet-coating, the wear rate was measured. In this paper, the result of the abrasive wheel wear test is discussed.

The equipment used in this test is shown in Fig. 6. [3]. The specimen is set to touch the abrasive wheel which is wrapped with SiC abrasive paper. This test is done with a reciprocating movement of the specimen. The abrasive wheel is set to rotate a certain degree after each reciprocation. Therefore, the specimen always touches a new position of the abrasive paper.

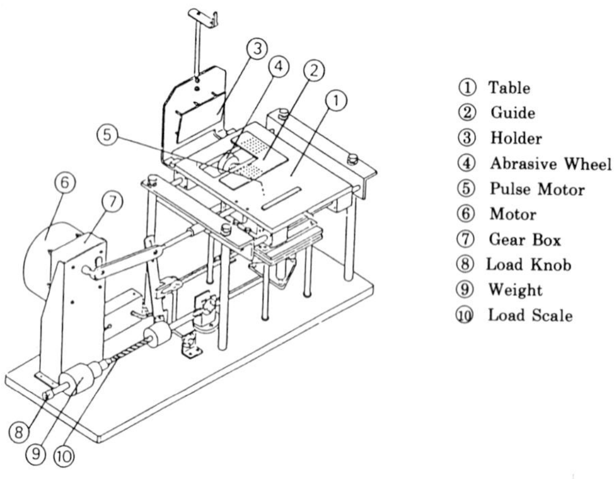


Fig. 6. Schematic diagram of abrasive wear test apparatus.

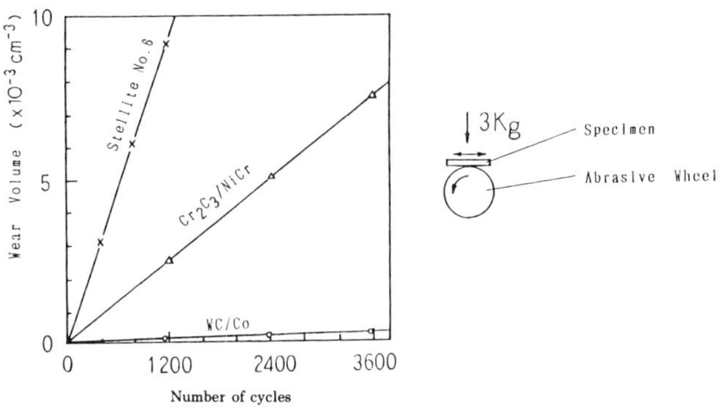


Fig. 7. Result of abrasive wheel wear test.

$Cr_3C_2$–25% NiCr and WC–17% Co coatings produced by LPPS and Stellite No. 6 (as a reference material) were tested. The results are shown in Fig. 7. Both cermets are superior to Stellite No. 6. The results shown for WC–Co indicate good anti-wear properties. This means that the effect of porosity is very small and that particles in the coating have high adhesive strength.

## 4. Applications of LPPS Coating

To confirm the desirable properties of the cermet coatings produced by LPPS, application to guide rollers in a hot strip mill was tried.

The configuration of guide rollers in a hot strip mill is shown in Fig. 8. Guide rollers are used to prevent swaying of stainless steel sheet whose temperature is between 950 and 1000 °C. The major problems of guide rollers

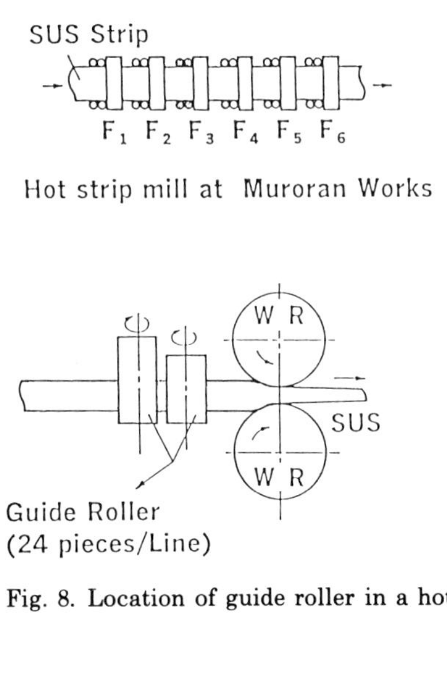


Fig. 8. Location of guide roller in a hot strip mill.

are abrasive wear and adhering of scale. The former shortens the life of the guide rollers. The latter degrades the quality of the stainless steel sheet by rolling in scales which were once adhered to the rollers. Conventionally, guide rollers have been repaired by arc welding after every 4000 tons rolled.

From the results of the property evaluation, $Cr_3C_2$–25% NiCr LPPS-coated guide rollers were manufactured and tested. LPPS-coated WC–Co is better in abrasive wear resistance than LPPS-coated $Cr_3C_2$–NiCr, but WC is easily oxidized at over 500 °C, and the oxides decrease the adhesive strength of particles in the coating. From the condition used, $Cr_3C_2$–NiCr is considered to be more suitable because of its stability at high temperatures and the thickness of the coating is about 1 mm including the NiCr bonding layer.

The coated guide rollers were used initially in July 1986 and are now working without any wear or adhering of scale. In this period no repairs have been needed. This means that the life of the guide rollers are increased by LPPS-cermet-coating by at least a 100 times. Manufacturing cost of LPPS-coated guide rollers are only seven times higher than conventional rollers. That means an excellent economical advantage has been proved.

## 5. Conclusions

(1) Cermet coatings produced by LPPS have much better wear resistant properties than those produced by APPS.

(2) Application of LPPS-cermet coatings to guide rollers in a hot strip mill has increased their life significantly.

## References

1 E. Muelberger, *Proc. 7th Int. Thermal Spray Conf.*, 1973, p. 245.
2 M. F. Smith, *Proc. 1st Plasma-Technik-Symp.* 1988, p. 77.
3 S. Yamashita, S. Yajima and Y. Togawa, *J. Metal Finishing Soc. Jpn.*, *37* (1986) 415.

# DEVELOPMENT OF EROSION RESISTANT REFRACTORY METAL AND ALLOY COATINGS BY MAGNETRON SPUTTERING

D. M. TURLEY

*Materials Research Laboratory, DSTO, PO Box 50, Ascot Vale 3032, Victoria (Australia)*

(Received March 12, 1989)

## Summary

A dual planar magnetron sputtering unit with a fast atom beam (FAB) source has been used to deposit crack-free metal and alloy coatings of chromium, niobium and tantalum on gun steel substrates. Coatings virtually free from nodular defects were obtained by alternately coating and bombarding the surface with excited atoms from the FAB source or by coating at high deposition rates. Scanning electron microscopy was used to characterize the coatings. Flat plates, rifled flat plates, the inside of rifled tubes and a rifled liner were coated. Generally, for the rifled tubes an open columnar type structure was obtained. By bombarding with the FAB source together with depositing a particular alloy combination, however, a dense coherent coating was produced. Firing trials of an alloy coated gun barrel liner have also been carried out.

## 1. Introduction

Erosion of the bore of gun barrels frequently limits the service life of high performance guns. Coating the bore with a refractory metal is a means by which erosion can be reduced and electrodeposited hard chromium plate is often used for this purpose. Hard chromium plate, however, is cracked in the as-deposited condition and these cracks are undesirable as they aid in the formation of deep craze cracks, during firing, which subsequently result in the spalling of chromium plate from the bore [1].

Methods of depositing crack-free less highly stressed chromium from aqueous electrolytes [2, 3], and molten salts [4] are being developed. The FLINAK process, which is a molten salt bath process [4], has advantages in that it can deposit refractory metals such as niobium and tantalum which cannot be deposited from aqueous electrolytes. However, the bath temperature of the process is around 800 °C, which is well above the tempering temperature of the gun steel.

Physical vapour deposition (PVD) processes such as ion plating can deposit refractory metal and alloy coatings below the tempering temperature

0257-8972/89/$3.50 

of gun steel [5 - 7]. In many cases liners, barrel sections and stub barrels have been used as the coating chamber by fitting end plates with a central electrode serving as the source of the coating material. In a recent study [8] refractory metal coatings with good adherence to the substrate were deposited using a planar magnetron. Chromium, niobium and tantalum coatings accommodated the effects of a severe thermal shock (plasma test) and were considered to have potential for gun barrel coatings. The purpose of the present study was to extend this work to study coating morphology on flat and tubular rifled substrates with the aim of producing a coating suitable for a 30 mm liner for test firing. Furthermore, the inside surfaces of the tubes and liner were sputter cleaned by directing a beam of excited atoms from a fast atom beam (FAB) source up the bore. This technique avoids the adherence problems associated with the deposition of sputtered substrate oxide layers onto a central electrode and their subsequent re-deposition back onto the bore surface when coating commences.

## 2. Experimental procedure

The substrates were of gun steel (main alloying elements: carbon 0.36%, molybdenum 0.52%, nickel 2.70% and chromium 0.72%) and were in the form of flat plates (40 mm × 40 mm) some with rifled profiles, 30 mm diameter rifled tubes (38 mm long) and a 30 mm diameter rifled liner (150 mm long).

Coatings were deposited by magnetron sputtering using a MICROVAC 350 system from Ion Tech Ltd. (U.K.). Prior to coating, the substrates were sputter cleaned with the FAB source. For flat substrates the sputter cleaning time was 2 h. It had been previously determined [8] that this removes oxide layers from the substrate resulting in good adhesion. For tubular substrates where the line of sight path of the atoms from the FAB source is tangential to the substrate, a cleaning time of 3 - 4 h was used. The substrates could be rotated about the central axis of the MICROVAC system so that, after FAB cleaning, they could be quickly positioned over either magnetron. Alternatively, the substrates could be oscillated between the magnetrons to produce alloy coatings, or between the magnetrons and the FAB source to produce alternate coating and FAB source bombardment. A schematic sketch showing the position of the substrates with respect to the magnetrons is shown in Fig. 1. Flat substrates, rifled tubes and the 30 mm liner were positioned 80, 38 and 25 mm above the magnetron, respectively. Chromium, niobium and tantalum sputtering targets were used and were in the form of discs which fitted on top of the magnetrons.

FAB cleaning was carried out at a voltage of 6 kV and a current of 5 mA at a pressure of $10^{-3}$ mbar. The magnetron sputtering conditions were: voltage 500 - 600 V, current 0.2 - 0.65 A and pressure $3 - 8 \times 10^{-3}$ mbar. Simultaneous operation of the FAB source and magnetrons was carried out at pressures of $3 - 8 \times 10^{-3}$ mbar.

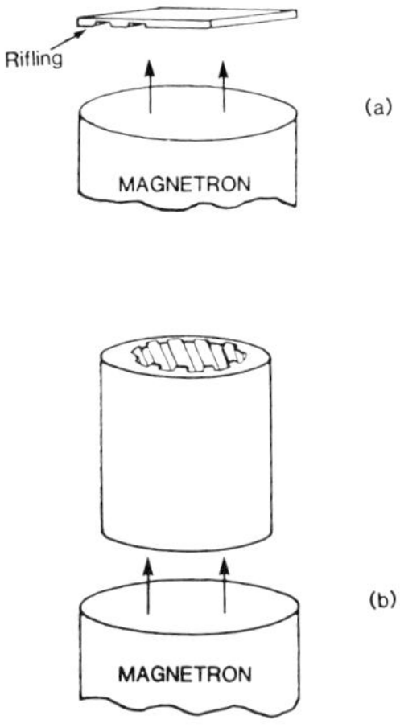


Fig. 1. Schematic sketch showing the position of (a) flat plates and rifled flat plates and (b) rifled tubes and liner, with respect to the magnetrons.

## 3. Results and discussion

### *3.1. Coating of flat plates*

Scanning electron microscope (SEM) examination of coatings showed that coating defects were present, the most prevalent type being nodular growths [9, 10]. These growth defects frequently nucleate at the substrate surface and appear in the coating as inverted cones with their apex at the substrate and their domed bases forming the tops of the nodules protruding above the coating surface (Fig. 2). Detachment of the nodules frequently occurs leaving holes in the coating. Visually, these coatings had a bright, specularly reflecting surface and when examined at high magnification in the SEM the surface between the defects was smooth and featureless.

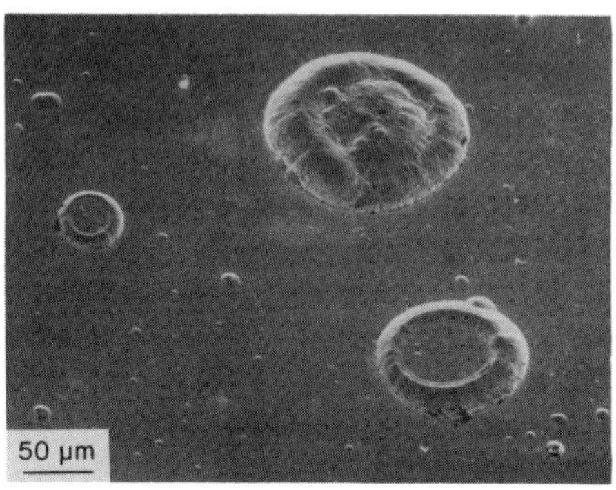


Fig. 2. Scanning electron micrograph of a chromium coating showing the size distribution and frequency of occurrence of nodular defects.

Coating at high deposition rates (10 $\mu$m h$^{-1}$), or oscillating the substrate between the magnetron and the FAB source, resulted in the virtual elimination of nodular growths. Visually, these coatings had a matte-etched appearance and examination in the SEM showed that the surface was faceted. Examination of fracture sections and etched normal sections showed that they were featureless whereas coatings produced at lower deposition rates (2 $\mu$m h$^{-1}$) without FAB source activation had a fine columnar grain structure. A reduction in the incidence of nodules together with grain refinement of the coating has also been reported [10, 11] to occur when a negative bias is applied to the substrate during sputter coating (sputter ion plating).

The absence of nodules at high deposition rates is attributed to the increased energy of the coating atoms striking the substrate which increases the adatom surface mobility which, in turn, increases the likelihood of rearrangement. Bombardment with the FAB source removes material by sputtering and it appears that, in the initial stages of coating deposition, many of the nodule initiation sites are removed. Furthermore, the energy of the argon atom bombardment tends to compact the coating and also enhance surface adatom mobility, which may also reduce nodule nucleation and growth.

According to the model proposed by Thornton [12], the coatings deposited at 2 $\mu$m h$^{-1}$ and $4 \times 10^{-3}$ mbar without FAB bombardment ($T/T_m$ estimated to be 0.1 - 0.2) should have a Zone T structure consisting of tightly packed fibrous grains and this corresponds to the fine columnar grain structure observed. Thornton's model cannot predict the effects of atom bombardment or substrate biasing on coating morphology. Preliminary X-ray diffraction studies, however, have indicated that both the coatings produced by alternate deposition and FAB bombardment and high deposition rates are fine grained.

### *3.2. Coating of flat plates with rifled profiles*

To facilitate coating at the sides of the lands the plates were oscillated backwards and forwards through the plasma.

The coating thickness on the sides of the lands where the line of sight deposition path was parallel to the substrate was only 35% as thick as on the lands and grooves (Fig. 3). This clearly shows the limited throwing power of the system. Attempts to improve the throwing power by operating at increased gas pressure ($10^{-2}$ mbar) were unsuccessful.

The morphology of the coating also varied around the rifling. At the side of the lands, where the line of sight deposition path was parallel to the substrate, the coating had an open columnar structure (refer Fig. 3) whereas on the land tops and grooves it was dense.

### *3.3. Coating of the bores of 30 mm rifled tubes*

The coating flux was directed up the tube so that in this configuration the internal surfaces to be coated were parallel to the line-of-sight deposition

Fig. 3. Optical micrograph of a niobium coating showing the variation in coating thickness between the land top and the side of the land.

path. This results in significantly less deposition than when the coating flux is normal to the substrate (see Fig. 3). Unlike the rifled flat plates, however, there was little variation in coating thickness around the rifling.

The deposition rate decreased along the length of the tube. Typical deposition rates when the substrate was held stationary above the magnetron (pressure $5 \times 10^{-3}$ mbar, current 0.6 A) were 5 $\mu$m h$^{-1}$ at the bottom of the tube decreasing to 2 $\mu$m h$^{-1}$ at the top. The coating on the lands at the end of the tube closest to the magnetron (bottom end) had a very open columnar type structure. The columnar crystals were separated from one another and could be clearly distinguished (Fig. 4). A similar structure was also observed at the sides of the lands and in the grooves. At the top end of the rifled tube a columnar growth structure was also present; however, it was finer.

The coating structure observed in Fig. 4 would not be suitable for gun bores, since the hot erosive gases could penetrate between the crystal growths and attack the substrate. It was therefore considered necessary to densify this microstructure. The effect of alternate coating and argon atom

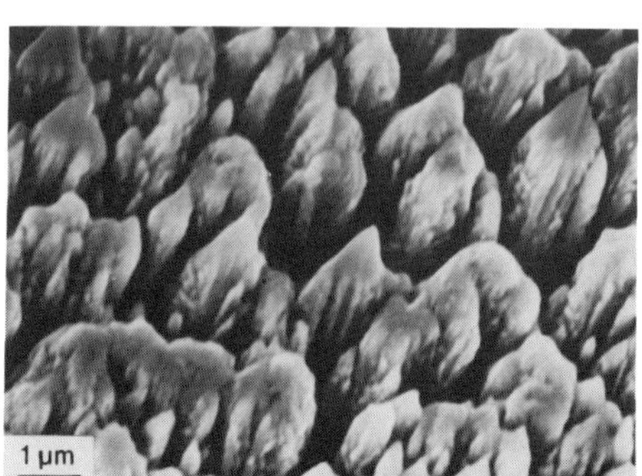


Fig. 4. Scanning electron micrograph of a niobium coating on a land top at the bottom end of a rifled tube showing an open columnar type structure.

Fig. 5. Scanning electron micrograph of a niobium coating on a land top (bottom end) showing that the coating produced by alternate deposition and argon atom bombardment had a less open structure (cf. Fig. 4).

bombardment with the FAB source is shown in Fig. 5 for a niobium coating. FAB bombardment has resulted in a much denser coating (cf. Fig. 4).

In order to further densify the coating FAB bombardment was combined with the alternate deposition of two different metals (alloying) using the two magnetrons. Two types of binary alloys were considered. In the first the two metals had only limited solubility. Since the two metals would be mutually insoluble (50 : 50 composition) it was considered that atoms deposited from the two magnetrons would not build up continuously to form the columnar type growth structure. The Nb–Cr alloy system satisfies this requirement and was chosen for investigation. Conversely, it was considered that when two metals are mutually soluble there would be a greater tendency for the build-up of deposited atoms to be continuous and thus the open type columnar structure would still be formed. Niobium and tantalum are completely soluble in each other and were chosen for comparison.

The Nb–Cr coating was dense, very few nodular growths were present and though some surface structure was discernible the open columnar type growth structure was no longer present (Fig. 6). For the Nb–Ta coating a

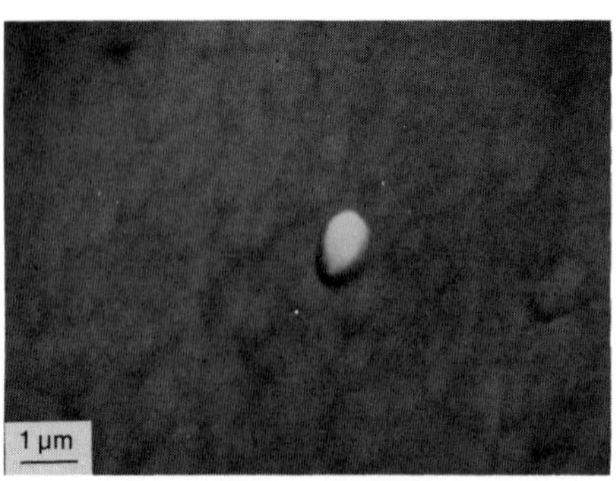


Fig. 6. Scanning electron micrograph of a Nb–Cr coating on a land top (bottom end) produced by alternate deposition and argon atom bombardment. The coating is dense.

columnar growth structure similar to that produced by alternate niobium deposition and FAB bombardment was produced (see Fig. 5). These results support the arguments proposed concerning the formation of dense coatings, however, more alloy systems need to be studied before definite conclusions can be drawn.

### *3.4. Coating a 30 mm liner with Nb–Cr for firing trials*

The promising Nb–Cr coating was deposited in the 30 mm liner.

After FAB cleaning for 16 h the liner was oscillated continually over the FAB source and the two magnetrons for 20.5 h. The liner was subsequently shrink fitted into a 30 mm barrel and 10 rounds fired.

Examination of the coating after firing revealed that much of the coating had been removed from the bore. SEM observations in the commencement of rifling (C of R) region of the bore indicated that pieces of the coating had spalled from the substrate. Optical microscope examination of a transverse section through the C of R region showed that the coating remaining was around 5 $\mu$m thick and this corresponded to the as-deposited coating thickness. Immediately below the coating in the steel substrate was a heat affected zone (HAZ) around 35 $\mu$m deep. The HAZ formed as a result of the substrate being heated into the austenite range ($T > 725$ °C) and subsequently, transforming to martensite on cooling. Both those phase transformations were accompanied by volume changes which would exert stresses at the substrate–coating interface during every firing cycle. During firing, stresses were also generated at the substrate–coating interface because of the difference in the coefficients of thermal expansion between the substrate and the coating. Furthermore, these stresses are expected to increase with increasing substrate–coating interface temperature. In the present case, since the HAZ was around 35 $\mu$m deep the interface temperature would have been appreciably higher than 725 °C. Thus a thicker coating that prevents HAZ formation and lowers the substrate–coating interface temperature should be more resistant to spalling.

## 4. Conclusions

(1) Coating flat substrates at low to moderate deposition rates produced specularly reflecting coatings with a high density of nodular defects and an acicular columnar grain structure.

(2) The nodular growth defects were virtually eliminated either by alternately coating and bombarding the surface with excited atoms from the FAB source or by coating at high deposition rates. These coatings had a matte-etched appearance.

(3) Due to the poor throwing power of the planar magnetrons an uneven coating thickness was obtained around the rifling of flat rifled substrates.

(4) Alternate deposition of chromium and niobium together with excited atom beam bombardment produced dense coherent coatings inside 30 mm rifled tubes. Other alloy systems need to be investigated.

(5) It has been shown that it is feasible to coat liners or the chamber and C of R region of large calibre guns using a planar magnetron source. The coating on the 30 mm liner failed by spalling during test firing primarily because it was too thin to provide an adequate thermal barrier.

## Acknowledgments

This work was largely carried out while the author was attached to the Royal Armament Research and Development Establishment, Fort Halstead, Kent, U.K. and completed at Materials Research Laboratory. The author would like to thank Mr. J. A. Sheward (RARDE) for his support of this work.

## References

1 D. M. Turley, *Wear, 131* (1989) 135.
2 B. A. Wilson and D. M. Turley, *Proc. IMF Conf.*, Brighton, 11 - 14 April, 1989.
3 E. S. Chen, W. Baldauf and R. S. Carter, *Proc. Tri-Service Gun Tube Wear and Erosion Symp.* Dover, New Jersey, 25 - 27 October 1982, p. IV-334.
4 I. Ahmad, G. J. Janz, J. Spiak and E. S. Chen, *Proc. Tri-Service Gun Tube Wear and Erosion Symp.*, Dover, New Jersey, 25 - 27 October 1982, p. IV-299.
5 D. R. Glue, J. A. Sheward, W. J. Young and I. P. Gibson, *RARDE Report 1/81,* June 1981.
6 I. P. Gibson, *Thin Solid Films, 83* (1981) 27.
7 J. F. Cox, Proc, *Tri-Service Gun Tube Wear and Erosion Symp.*, Dover, New Jersey, 25 - 27 October, 1982, p. IV-277.
8 D. M. Turley, *Proc. Int. Tribology Conf.*, Melbourne, Victoria, 2 - 4 December 1987, p. 80.
9 T. Spalvins, *NASA Technical Memorandum 79110,* 1979.
10 C. J. Brooker, *RARDE Memorandum 92/84,* March 1985.
11 D. M. Mattox and G. J. Kominiak, *J. Vac. Sci. Technol., 9* (1972) 528.
12 J. A. Thornton, *J. Vac. Sci. Technol., 11* (1974) 666.

# THERMAL PROPERTIES OF CERIA AND YTTRIA PARTIALLY STABILIZED ZIRCONIA THERMAL BARRIER COATINGS

J. R. BRANDON and R. TAYLOR

*Manchester Materials Science Centre, Grosvenor Street, UMIST, Manchester M1 7HS (U.K.)*

(Received March 9, 1989)

## Summary

Coatings consisting of zirconia alloyed with 12, 15, 20 and 25 wt.% ceria or 4, 6, 8, 10 and 12 wt.% yttria were prepared using the plasma spray process. For each of these coatings measurements were made of thermal expansion parallel to the substrate and of thermal conductivity perpendicular to the substrate. These results are discussed in conjunction with the phase stability of the specimens, as determined by X-ray diffraction. Coatings containing 25 wt.% ceria or greater than 8 wt.% yttria were found to have stable thermal properties. Coatings containing less of these additions exhibited a thermal expansion hysterisis which was most pronounced at the lowest concentrations. The temperature at which the hysterisis occurred was found to reduce with increasing additions of alloying agent. This behaviour was also detected for $ZrO_2$–$CeO_2$ as a change in thermal diffusivity during thermal cycling and is attributed to a reversible tetragonal–monoclinic transformation.

## 1. Introduction

The benefits of thermal barrier coatings applied to critical components in the turbine section of a jet engine have been discussed by a number of authors [1 - 4]. A successful coating should be refractory, chemically inert, possess good mechanical strength and thermal shock resistance, have a low thermal expansion coefficient and exhibit a thermal expansion similar to that of the nickel based superalloy substrate.

Zirconia exhibits a low thermal conductivity and possesses a suitable expansion match with the substrate. However the polymorphism of zirconia necessitates that it be alloyed with other oxides. Several stabilized and partially stabilized $ZrO_2$ compositions have been studied [5 - 7] and the consensus is that coatings exhibiting the best properties comprise a 0.2 - 0.35 mm plasma-sprayed layer of $ZrO_2$ and 6–8 wt.% $Y_2O_3$ deposited over a 0.1 mm thick CoNiCrAlY bond coat [1 - 3, 5]. Within the $ZrO_2$–$Y_2O_3$ system, compositions in the range 6 - 12 wt.% $Y_2O_3$ consist of a non-transformable

0257-8972/89/$3.50 

tetragonal phase t′ [8 - 10]. Amongst alternative systems, the $ZrO_2$–$CeO_2$ system is reported to show promise [11].

The integrity of plasma-sprayed coatings is strongly influenced by the thermal expansion behaviour of the coating with respect to the substrate [12] and the effectiveness of the coating as a thermal barrier depends on the thermal conductivity of the coatings. Both these parameters are in turn influenced by the phases present in the coating and their stability, both long term and short term. In this paper we report thermal expansion and thermal diffusivity data for $ZrO_2$ coatings alloyed with 4, 6, 8, 10 and 12 wt.% $Y_2O_3$ and 12, 15, 20 and 25 wt.% $CeO_2$ and we discuss the results in terms of the phase structure observed.

## 2. Experimental procedure

The $ZrO_2$–$Y_2O_3$ and $ZrO_2$–$CeO_2$ powders were prepared by Magnesium Elektron Ltd. using a coprecipitation process. Particle sizes of these powders lie in the range 10 - 30 $\mu$m. The thermal barrier coatings were plasma sprayed using a Metco 7MB gun, to a set of standard spray parameters prescribed by Rolls Royce. Specimens for thermal diffusivity measurement were sprayed onto nickel based superalloy substrate (C263) discs 1 mm thick × 10 mm in diameter onto which a thin (less than 20 $\mu$m) NiCrAlY bond coat had been applied. The thickness of the coatings were determined by optical measurements to be in the range 150 - 300 $\mu$m. Thermal diffusivity was measured using the laser flash technique [13] within the UMIST apparatus for which full details have previously been published [14]. Measurements were made in a helium atmosphere. The diffusivities of the coatings were measured *in-situ* using an analysis of transient heat flow through two layers developed by Bulmer and Taylor [15]. The temperature range covered is determined by the temperature capability of the substrate, which in this case is approximately 1200 °C.

Specimens for thermal expansion were prepared by spraying a coating approximately 1 mm thick directly onto a nickel substrate which was then dissolved using a chemical etch. Bars approximately 20 mm × 3 mm × 1 mm, were sectioned using a high-speed diamond slitting wheel. Thermal expansion measurements were carried out in air using an alumina pushrod dilatometer at temperatures from 100 to 1000 °C at a constant ramping rate of 2 °C $min^{-1}$. Data from the dilatometer was recorded every 4 min using an IBM PC compatible microcomputer fitted with a 12 bit A/D converter. The stored data was then analysed using a purposed developed spreadsheet program. Values of fractional length change $\Delta l/l$ and expansion coefficient $\alpha$ were calculated.

The phase compositions of the coatings were measured in the as-sprayed state and after testing, using X-ray diffraction. Measurements were made using a Philips horizontal diffractometer PW1380. The fraction of monoclinic phase present in a coating was determined by manually measuring the area of the (111) cubic–tetragonal peak and the ($\bar{1}11$) and (111) monoclinic peaks using the method described by Toraya [16]. The proportions of tetragonal and cubic phases were determined by analysis of the (400) type peaks, using the method described by Miller [9]. The tetragonal phase was identified as t or t′ by surface grinding to induce transformation of tetragonal to monoclinic [17]. Metastable transformable tetragonal zirconia can undergo a stress induced transformation to monoclinic, whereas t′ does not transform.

## 3. Results

The phase structure for all the coatings, both as sprayed and after expansion and diffusivity measurement runs, are shown in Table 1. In the $ZrO_2$–$Y_2O_3$ series of alloys only the 4 wt.% $Y_2O_3$ alloy contains a significant monoclinic content, although a slight amount is present in the 6 wt.% $Y_2O_3$ alloy. The other three alloys consist entirely of the non-transformable tetragonal phase described by Scott [8] and Miller *et al.* [9]. The $ZrO_2$–$CeO_2$ series of alloys show decreasing monoclinic content as the $CeO_2$ content increases. At 25 wt.% $CeO_2$ the coating is wholly tetragonal.

The thermal expansion data for the $ZrO_2$–$Y_2O_3$ coatings are shown in Figs. 1(a) and 1(b). The curves for the expansion behaviour of the 100% t′ coatings containing 8, 10 and 12 wt.% $Y_2O_3$ are identical. During the first heating cycle of the 6 wt.% $Y_2O_3$ coating there was a contraction in length occurring above 250 °C. This is associated with the monoclinic–tetragonal transformation. The tetragonal phase does not transform to monoclinic on cooling and this results in a decrease in length of 0.3% for the 6 wt.% $Y_2O_3$ alloy. The 4 wt.% $Y_2O_3$ alloy exhibits two transformations; retained transformable tetragonal transforms to monoclinic at approximately 250 °C and the reverse

TABLE 1

Phase structure of coatings

| *Alloy composition* | *Coating specimen* | | | | | |
|---|---|---|---|---|---|---|
| | *As-sprayed* | | *Expansion* | | *Diffusivity* | |
| | %m | %t | %m | %t | %m | %t |
| 12 wt.% $CeO_2$ | 44 | 56 | 90 | 10 | — | — |
| 15 wt.% $CeO_2$ | 28 | 72 | 62 | 38 | — | — |
| 20 wt.% $CeO_2$ | 10 | 90 | 43 | 57 | 16 | 84 |
| 25 wt.% $CeO_2$ | — | 100[a] | — | 100[a] | — | 100[a] |
| 4 wt.% $Y_2O_3$ | 42 | 58 | 66 | 34 | — | — |
| 6 wt.% $Y_2O_3$ | $<1$ | $>99$[a] | — | 100[a] | — | 100[a] |
| 8 wt.% $Y_2O_3$ | — | 100[a] | — | 100[a] | — | 100[a] |
| 10 wt.% $Y_2O_3$ | — | 100[a] | — | 100[a] | — | 100[a] |
| 12 wt.% $Y_2O_3$ | — | 100[a] | — | 100[a] | — | 100[a] |

[a]These values refer to the non-transformable t′ phase whereas the other values for tetragonal content refer to possible mixtures of both t and t′ phases.

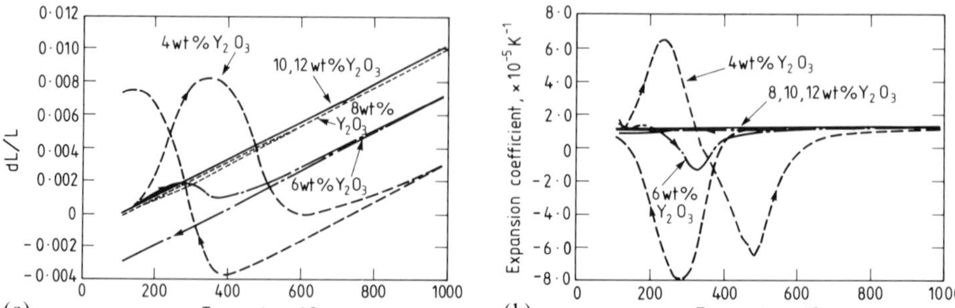


(a) (b)

Fig. 1. (a) Thermal expansion of $ZrO_2$–$Y_2O_3$ coatings. (b) Thermal expansion coefficient of $ZrO_2$–$Y_2O_3$ coatings.

monoclinic–tetragonal transformation occurs above 350 °C. Above 650 °C a linear expansion coefficient of $11 \times 10^{-6}$ is noted. Linear expansion behaviour is noted on cooling down to 400 °C, at which point the tetragonal–monoclinic transformation occurs. This transformation causes a large expansion and a permanent increase in length of 0.8%. This is consistent with the observation (Table 1) of an increased monoclinic content in this alloy after the thermal cycle.

The thermal diffusivity of the 6, 8, 10 and 12 wt.% $Y_2O_3$ coatings is shown in Fig. 2. These show values which range from 2.5 to $4.2 \times 10^{-7}$ $m^2$ $s^{-1}$ and there is no systematic trend with composition. Thermal diffusivity, however, is a structure sensitive property and any compositional changes could be masked by effects such as changes in crack morphology. The microstructure is affected by variables such as particle size and morphology and plasma spray variables [7, 18, 19]. While all the alloys reported here were sprayed under standardized spray conditions, variations in the microstructure of these coatings have been seen [18, 19]. Equally important is the fact that for all these alloys, no change has occurred on cooling. This indicates that no phase changes had occurred.

The thermal expansion of the ceria stabilized coatings are shown in Figs. 3(a) and 3(b). Only the 25 wt.% ceria coating, with an expansion coefficient between 10.5 and $12 \times 10^{-6}$ $K^{-1}$, exhibits linear behaviour. All the other ceria coatings

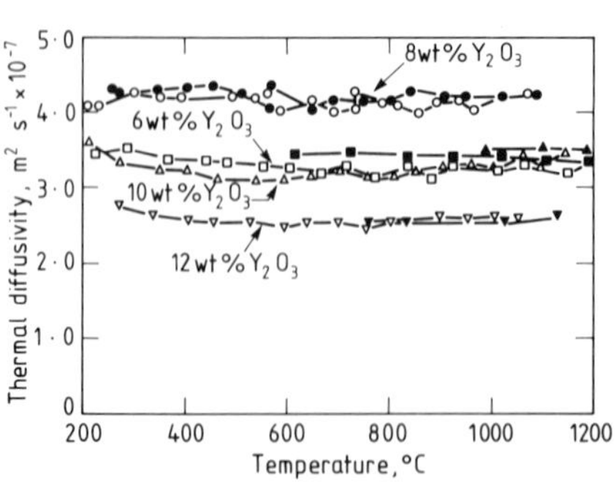


Fig. 2. Thermal diffusivity of $ZrO_2$–$Y_2O_3$ coatings (open symbols heating, filled symbols cooling).

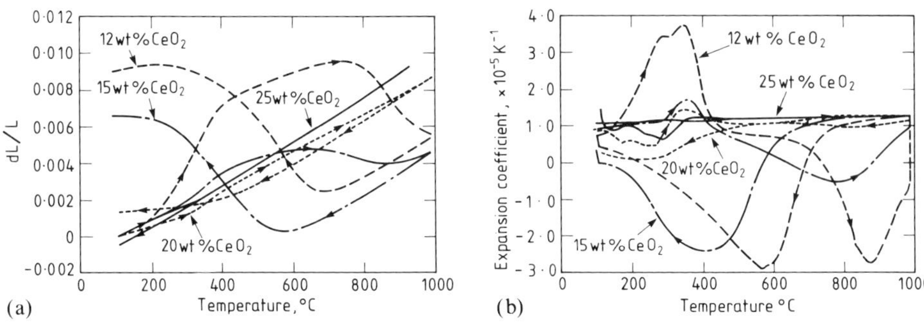


Fig. 3. (a) Thermal expansion of $ZrO_2$–$CeO_2$ coatings. (b) Thermal expansion coefficient of $ZrO_2$–$CeO_2$ coatings.

exhibited non-linear behaviour attributable to the following transformations:

(a) Retained tetragonal → monoclinic (t → m) on heating.
(b) Monoclinic → tetragonal (m → t) on heating.
(c) Tetragonal → monoclinic (t → m) on cooling.

Interestingly, the temperatures at which these transformations occurred varied with $CeO_2$ content. For 12 wt.% $CeO_2$, t → m occurs between 200 and 400 °C and m → t between 650 and 1000 °C on heating, whereas t → m occurs continuously on cooling below 700 °C. In the case of the 15 and 20 wt.% $CeO_2$ alloys, evidence of the t → m transformation occurring at 300 - 400 °C is evident on heating and of the t → m transformation occurring below 600 °C on cooling. However, while the 15 wt.% $CeO_2$ showed the m → t transformation to have occurred above 600 °C, this was not clearly detectable in the 20 wt.% $CeO_2$ alloy. For compositions of 12 - 20 wt.% $CeO_2$ the monoclinic content has increased on heating in air to 1000 °C, although the 25 wt.% $CeO_2$ alloy, originally wholly tetragonal, has remained tetragonal. The overall length change varies with ceria content, decreasing from 0.9% (12 wt.% $CeO_2$) to 0.05% (25 wt.% $CeO_2$).

Thermal diffusivity data from 200 to 1200 °C for coatings of 15 - 25 wt.% $CeO_2$ are shown in Fig. 4. Unlike the data for $ZrO_2$–$Y_2O_3$ alloys, there is a

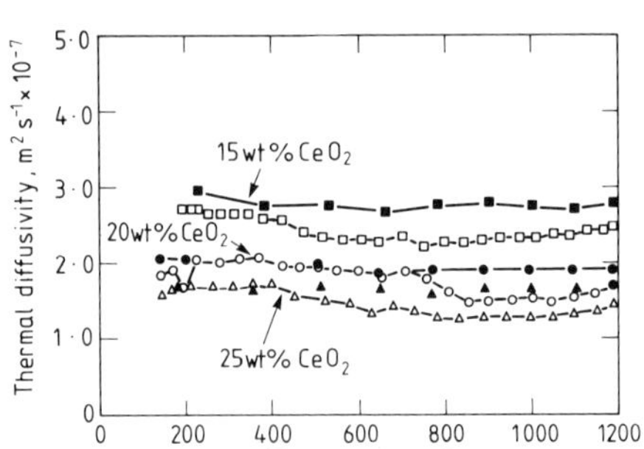


Fig. 4. Thermal diffusivity of $ZrO_2$–$CeO_2$ coatings (open symbols heating, closed symbols cooling).

noticeable decrease in diffusivity with increasing $CeO_2$ content. All coatings exhibited hysterisis effects. Decreases in diffusivity on heating did not, however, correlate with the temperatures at which the phase changes were noted during expansion tests. These temperatures were approximately 400 °C for 15 wt.% $CeO_2$, 700 °C for 20 wt.% $CeO_2$ and 400 °C for 25 wt.% $CeO_2$. Changes in diffusivity can, however, be a result of changes in the crack morphology of the coating, which is a first order effect. Hysterisis was noted in all of the alloys, but only the 15 wt.% ceria coating exhibited a permanent increase in diffusivity on cooling from 1200 °C.

## 4. Discussion

The results presented for the $ZrO_2$–$Y_2O_3$ coatings are consistent with the conclusions of Andersson [10] and other workers, regarding the stability of the t′ phase at compositions less than or equal to 6 wt.% $Y_2O_3$. This is believed to form via an athermal diffusionless martensitic transformation $M_s^{c \to t'}$. This phase is reluctant to undergo the transformation to monoclinic and is slow to destabilize into equilibrium phases even on prolonged heating up to 1300 °C [20]. The small amount of monoclinic present in the 6 wt.% $Y_2O_3$ sprayed coating transforms to tetragonal on heating but does not reappear on cooling. This results in a net contraction of 0.3%. By contrast, the 0.8% permanent change in length noted for the 4 wt.% $Y_2O_3$ coating after the expansion run, is consistent with the observed increase in monoclinic phase content of 24%.

The stabilizing influence of $CeO_2$ is not as great as that of $Y_2O_3$, as shown by the most recent equilibrium diagram due to Tani *et al.* [21] (Fig. 5). The formation of the metastable tetragonal phase during rapid cooling, as encountered during plasma spraying, is believed to be by a martensitic

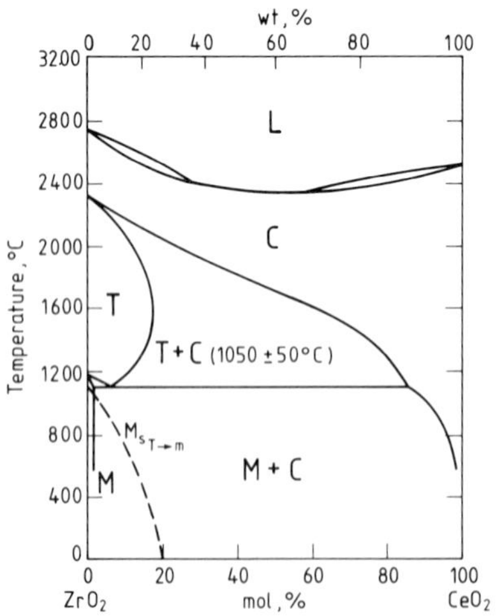


Fig. 5. $ZrO_2$–$CeO_2$ equilibrium diagram after Tani *et al.* [21].

transformation. Based on our thermal expansion data, we have estimated the $M_s^{t \to m}$ transformation temperature as a function of $CeO_2$ content. This is included on the phase diagram and predicts an $M_s^{t \to m}$ temperature below 20 °C for the 25 wt.% $CeO_2$ concentration. The position of this line is in reasonable agreement with an earlier high-temperature X-ray diffraction determination of the $ZrO_2$–$CeO_2$ phase diagram [22].

A general feature of the coatings containing low ceria concentrations is that the transformations m → t and t → m occur over a much wider temperature range than similar transformations in the $ZrO_2$–4 wt.% $Y_2O_3$ coating. This is probably due to compositional variations in the coating. This is possible since coprecipitated particles comprise the second phase precipitated together with $ZrO_2$ on a $ZrO_2$ particle. It is still not clear whether the dwell time in the plasma arc is sufficient to homogenize the particle. Clearly the transformation temperature will be a function of local composition. Also the athermal nature of the transformation is subject to a mechanical constraint factor [23] which prevents the transformation proceeding to completion at a given temperature. This is best shown by the 12 wt.% $CeO_2$ coating, where at 1000 °C the m → t transformation is arrested and the contraction on cooling reverts to a value typical of the tetragonal phase.

The absence of any consistent changes in measured thermal diffusivity for the coatings, combined with the noticeable differences in the monoclinic contents of coatings heated in air or inert atmosphere, indicates that the destabilization of these coatings may be significantly affected by the oxygen partial pressure. This effect is most pronounced in the 15 and 20 wt.% $CeO_2$ coatings which, after heating in the diffusivity apparatus to 1200 °C in helium, show much smaller increases in monoclinic phase fraction than when the same coatings are heated to 1000 °C in air during the expansion run. Due to the reducing conditions in the plasma arc, it is likely that the cerium ion is present in the trivalent ($Ce_2O_3$) state in the as sprayed coating. Oxygen vacancy concentrations will, therefore, be high in these coatings. If the stabilizing effect of trivalent oxide additions is related to the formation of oxygen ion vacancies, as suggested by Heuer *et al.* [24], then it is likely that during heating under an inert atmosphere, and hence low oxygen partial pressures, the cerium oxide remains in the trivalent state and maintains the stability of the coating. To test this hypothesis, the thermal diffusivity of a specimen of 15 wt.% ceria stabilized zirconia was measured in air using a nickel superalloy susceptor. The results shown in Fig. 6 show a decrease in thermal diffusivity above 600 °C consistent with a m → t transformation. During cooling a large rise in diffusivity occurs at a point corresponding to the t → m transformation observed during the thermal expansion tests. These results are similar to those observed for the thermal expansion tests although they occur at different temperatures. It should be noted that the values for thermal diffusivity are higher than those observed for the two layer samples presented in Fig. 4. This increase was caused by laser penetration of the test specimen which could not be mounted on a metal substrate in the usual manner, because the substrate support would oxidize in air. Hence the two sets of data are not directly comparable.

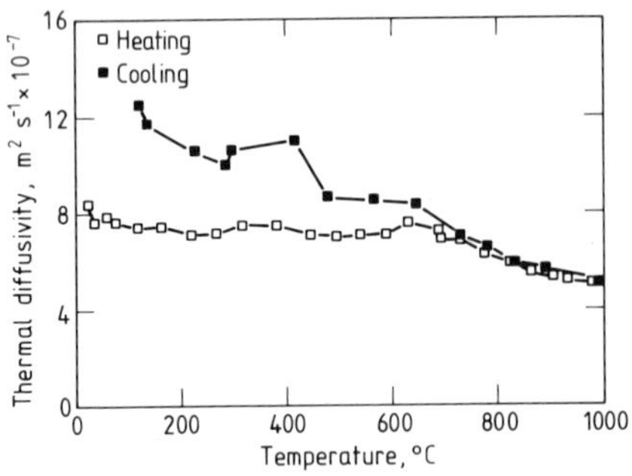


Fig. 6. Thermal diffusivity of $ZrO_2$–15 wt.% $CeO_2$ in air.

## Acknowledgments

The authors would like to thank Rolls Royce plc. for the provision of samples. One of us (J.R.B) would like to thank the Science and Engineering Research Council for financial support.

## References

1 A. Bennett, *Proc. Br. Ceram. Soc., 34* (1984) 207 - 217.
2 A. Bennett, F. C. Toriz and A. B. Thakker, *Surf. Coat. Technol., 32* (1987) 359 - 375.
3 R. A. Miller, *Surf. Coat. Technol., 30* (1987) 1 - 11.
4 T. E. Strangman, *Thin Solid Films, 127* (1985) 93 - 105.
5 S. Stecura, Effects of compositional changes on the performance of a thermal barrier coating system, *NASA Technical Memo. 789761*, 1978.
6 C. H. Liebert, R. E. Jacobs, S. Stecura and C. R. Morse, Durability of zirconia thermal barrier ceramic coatings on air-cooled turbine blades in cyclic jet engine operation, *NASA Technical Memo. X-3410*, 1976.
7 P. Morrell, *Ph.D. thesis*, UMIST, 1985.
8 H. G. Scott, *J. Mater. Sci., 10* (1975) 1527 - 1535.
9 R. A. Miller, J. L. Smialek and R. G. Garlick, in A. H. Heuer and L. W. Hobbs (eds.), *Science and Technology of Zirconia, Advances in Ceramics, Vol. 3*, The American Ceramic Society, 1981, pp. 241 - 253.
10 C. A. Andersson, J. Greggi, Jr. and T. K. Gupta, in N. Claussen, M. Ruhle and A. H. Heuer (eds.), *Science and Technology of Zirconia II, Advances in Ceramics, Vol. 12*, The American Ceramic Society, 1984, pp. 78 - 85.
11 J. W. Holmes and B. H. Pilsner, in D. L. Houck (ed.), *Thermal Spray, Advances in Coatings Technology, Proceedings of the National Thermal Spray Conference, 14 - 17 September 1987, Orlando, Florida, U.S.A.*, ASTM International, pp. 259 - 270.
12 A. S. Grot and J. K. Martyn, *Ceram. Bull., 60* (1981) 807 - 811.
13 W. J. Parker, R. J. Jenkins, C. P. Butler and G. L. Abbot, *J. Appl. Phys., 32* (1961) 1679 - 1684.
14 R. Taylor, *J. Phys. E., Sci. Inst, 13* (1980) 1193 - 1199.
15 R. F. Bulmer and R. Taylor, *High Temp.–High Press., 6* (1974) 491 - 497.
16 H. Toraya, M. Yoshimura and S. Somiya, *J. Am. Ceram. Soc., 67* (1984) C-119 - C-121.
17 R. C. Garvie, R. T. Pacoe and R. H. J. Hannink, *Nature, 258* (1975) 703 - 704.
18 P. Morrell and R. Taylor, *High Temp.–High Press., 17* (1985) 79 - 88.

19 P Morrell and R. Taylor, *Science and Technology of Zirconia III, Advances in Ceramics, Vol. 24,* The American Ceramic Society, in press.

20 J. R. Brandon and R. Taylor, in *Heat Treatment '87,* Institute of Metals, London, 1988, pp. 69 - 77.

21 E. Tani, M. Yoshimura and S. Somiya, *J. Am. Ceram. Soc., 66* (1983) 506 - 510.

22 P. Duwez and F. Odell, *J. Am. Ceram. Soc., 33* (1950) 274 - 283.

23 R. C. Garvie, in A. M. Alper (ed.), *High Temperature Oxides, Part 2,* Academic Press, New York, 1970, pp. 117 - 166.

24 A. H. Heuer and M. Ruhle, in N. Claussen, M. Ruhle and A. H. Heuer (eds.), *Science and Technology of Zirconia II, Advances in Ceramics, Vol. 12,* The American Ceramic Society, 1984, pp. 1 - 13.

*Surface and Coatings Technology, 39/40* (1989) 153 - 159 

# THERMAL CONDUCTIVITY OF METAL–METAL MICROLAMINATE COMPOSITES

M. C. RADHAKRISHNA, H. J. DOERR, C. V. DESHPANDEY and R. F. BUNSHAH

*Department of Materials Science and Engineering, University of California, Los Angeles, CA 90024-1595 (U.S.A.)*

(Received March 3, 1989)

## Summary

Microlaminate composites consisting of alternate layers of metal–metal, metal–ceramic and ceramic–ceramic exhibit anisotropy in thermal conductivity. Thermal conductivity in the direction perpendicular to the laminate plane is significantly lower than in the plane of the laminate. Results of our study on thermal conductivity of Ni–NiCoCrAlY and Ti–CoCrAlY microlaminate composites are presented. A semiquantitative model explaining the thermal conductivity variation in the above systems as a function of number of layers is discussed. An expression correlating the experimental data with the calculated data for the above system is presented.

## 1. Introduction

Microlaminate composites show potential as thermal barrier coatings because in such composites the thermal conductivity in the plane perpendicular to the laminate is markedly low [1]. Bunshah and Movchan [2] had reported that the thermal conductivity of the Fe–Cu microlaminate composites in a direction perpendicular to the plane of the laminates was markedly low. They attributed this reduction in thermal conductivity to interfaces. Francis and Tinkelpaugh [3] have also studied Forsterite–steel laminates and observed that the measured values of thermal conductivities perpendicular to the laminate plane were 7 – 14% lower than the "calculated values". In view of the lack of any detailed investigation into the study of the thermal conductivity of laminates the present study was undertaken. The experimental results showing the effect of number of layers on the thermal conductivity of Ni–NiCoCrAlY and Ti–CoCrAlY microlaminate composites have been published previously [1].

The objective of the present paper is to develop a qualitative model to explain the behavior of the thermal conductivity of microlaminates. A semiquantitative model explaining the observed variation in thermal conductivity

0257-8972/89 Elsevier Sequoia/Printed in The Netherlands

as a function of number of layers for the Ni–NiCoCrAlY and Ti–CoCrAlY microlaminate composite systems is presented in this paper.

## 2. Experimental procedure

The procedure for synthesizing Ni–NiCoCrAlY and Ti–CoCrAlY microlaminate condensates has been described previously [1]. The multilayer condensates were prepared by simultaneous independent evaporation of metal (nickel or titanium) and alloy (NiCoCrAlY or CoCrAlY) from two electron beam sources and alternating the deposition onto a rotating circular stainless substrate heated to about 500 °C.

Thermal conductivity in the direction perpendicular to the plane of the microlaminate was calculated from measured values of thermal diffusivity $\alpha$ as described in ref. (1) and using the relation

$$K = \alpha \rho C_p$$

where $K$, $\alpha$, $\rho$ and $C_p$ are the thermal conductivity, thermal diffusivity, density and specific heat of the microlaminate respectively. The procedures for measuring density and specific heat have been described previously [1].

## 3. Results and Discussion

The thermal conductivity for all the microlaminates as well as for pure components are given in Table 1.

The variation of thermal conductivity with the number of layers for

TABLE 1

Thermal conductivity of metal and laminate specimen

| *Material* | *No. of layers* | *Thermal conductivity* ($W m^{-1} K^{-1}$) | | |
|---|---|---|---|---|
| | | *Measured* | *Model 1* | *Modified model* |
| Nickel | 1 | 82.72 | — | — |
| NiCoCrAlY | 1 | 15.5 | — | — |
| Titanium | 1 | 20.56 | — | — |
| CoCrAlY | 1 | 15.49 | — | — |
| Ni–NiCoCrAlY | 2 | 22.64 | 26.11 | 23.04 |
| Ni–NiCoCrAlY | 24 | 10.57 | 26.11 | 10.31 |
| Ni–NiCoCrAlY | 180 | 8.32 | 26.11 | 7.97 |
| Ni–NiCoCrAlY | 480 | 7.48 | 26.11 | 7.47 |
| Ti–CoCrAlY | 2 | 12.65 | 17.69 | 12.46 |
| Ti–CoCrAlY | 24 | 10.68 | 17.69 | 10.39 |
| Ti–CoCrAlY | 120 | 9.81 | 17.69 | 9.87 |
| Ti–CoCrAlY | 480 | 6.76 | 17.69 | 6.54 |

Ni–NiCoCrAlY and Ti–CoCrAlY microlaminate systems is shown in Figs. 1 and 2 respectively. It can be observed from these plots that:

(1) The thermal conductivity of the laminates decreases as the number of layers are increased for both Ni–NiCoCrAlY and Ti–CoCrAlY systems.
(2) This decrease in thermal conductivity is not linear with the increase in number of layers. It is observed that after a certain number of layers the contribution of interface in reducing the thermal conductivity is not significant.

The thermal conductivity of layered specimen in general has been explained by a model based on an equivalent series or parallel electric circuit depending on the direction of heat flow. This model assumes that there are no interfaces between layers and the layers are in perfect thermal contact. As the above model did not explain the observed thermal conductivity behavior of the laminates, a general model is developed taking into account the existence of finite interfaces between the successive layers of the laminate composites. Both the models are discussed below.

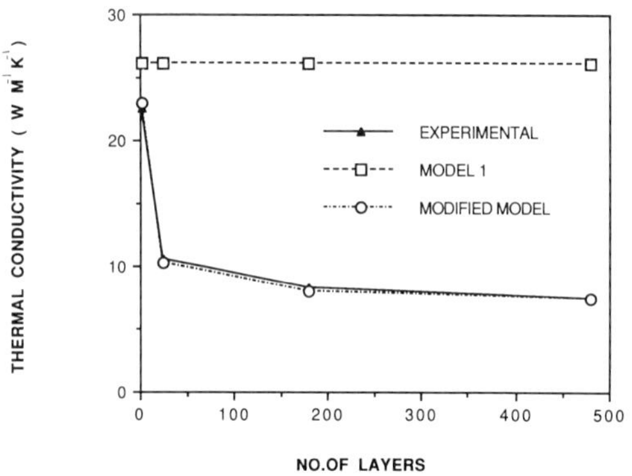


Fig. 1. Thermal conductivity of Ni–NiCoCrAlY microlaminate.

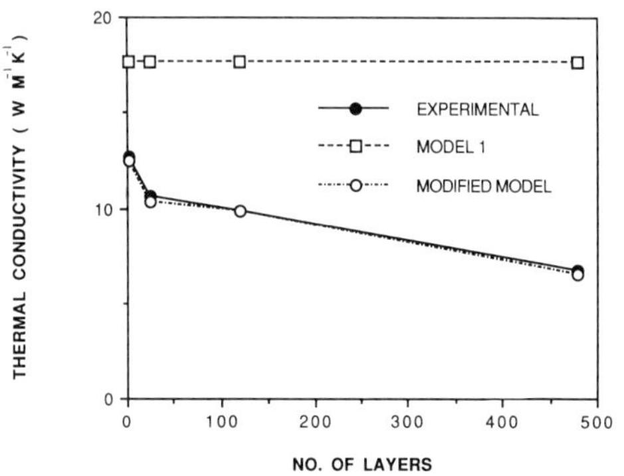


Fig. 2. Thermal conductivity of Ti–CoCrAlY microlaminate.

*3.1. Basic model for thermal conductivity in layered systems*

For a series of alternate layers of two different materials A and B with thermal conductivities $K_A$ and $K_B$ respectively, if the heat flow is perpendicular to the plane of the layers, the effective thermal conductivity $K_{eff}$ of the laminate can be written as

$$\frac{1}{K_{eff}} = \frac{V_A}{K_A} + \frac{V_B}{K_B} \tag{1}$$

where $V_A$ and $V_B$ are the volume fractions of the components A and B respectively. Equation (1) is based upon the equivalent series electric circuit (the heat flow through each layer is equal, but the temperature gradient is different). Moreover the above model assumes that there is no interfacial thermal resistance between the layers. The values of $K_{eff}$ calculated from eqn. (1) are constant, provided the volume fractions are kept constant irrespective of the number of layers as shown in Figs. 1 and 2 for Ni–NiCoCrAlY and Ti–CoCrAlY systems.

However, it is observed that the thermal conductivities of the laminates are not a constant but decrease with increasing numbers of layers for both Ni–NiCoCrAlY and Ti–CoCrAlY microlaminate systems. It is also seen that the observed values of thermal conductivity are much lower than those calculated from the theoretical model (eqn. (1)). The decreases in thermal conductivities are about 60% and 40% for the 24 layer Ni–NiCoCrAlY and Ti–CoCrAlY microlaminate systems respectively. These decreases in thermal conductivities may be due to the interfacial resistance which acts as a barrier to heat flow [2]. However, the decrease in thermal conductivity is not linear with increase in number of interfaces. This indicates that after a certain number of interfaces, the contribution of interfaces in reducing the total thermal conductivity is not significant.

The above observations suggest that the thermal conductivity of laminate composites cannot be explained on the basis of a simple model based on the equivalence of series electric circuit. Hence a modified model is presented which qualitatively explains the variation of thermal conductivity of microlaminates with a number of layers.

*3.2 Modified model*

The previous model assumes that the interfaces are abrupt and there is no interfacial regions. In reality, diffuse interfaces between the two layers do exist. Electron microprobe analysis showed that there is interdiffusion between the layers [1]. Hence the modified model assumes that there is definite thickness associated with the interface due to the interdiffusion of two materials. Figure 3 is the pictorial representation of the microlaminate cross-section. The thermal conductivity varies from $K_A$ to $K_B$ inside the interface.

*3.3 Analysis of the model*

In order to develop a mathematical expression for calculating the effective thermal conductivity of microlaminates, consider a two layer

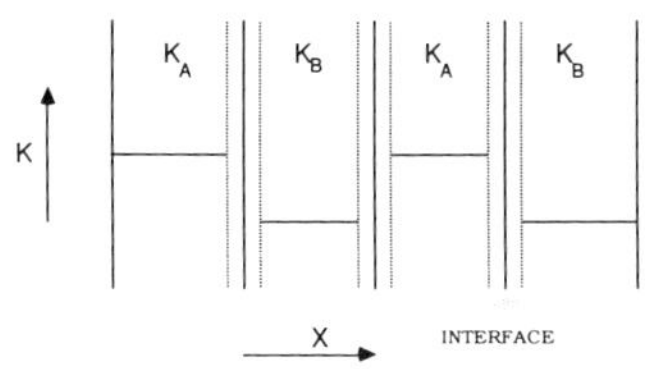


Fig. 3. Pictorial cross-section of the laminate.

microlaminate as shown in Fig. 4. A pure metal A is exposed to an alloying metal B in a furnace maintained at a constant temperature. As the metal B diffuses into the metal A, the thermal conductivity of A is reduced. Empirical data suggests that the thermal conductivity of a pure material is markedly lowered when another metal is added [4]. The effective thermal conductivity $K_{eff}$ can be written as

$$K_{\text{eff}} = K_0[1 - \beta c^n(x,t)]$$

where $K_0$ is the thermal conductivity of the metal A, $c(x,t)$ is the weight fraction of metal B in A normalized to 1, and $n$ and $\beta$ are constants.

If the diffusion length is small compared to $L$, the initial sample thickness, the system can be considered as a "semi-infinite solid". The thermal resistance of such a system can be solved by solving the differential diffusion equation.

$$c = \frac{c'}{2} \operatorname{erfc} \frac{x}{(4Dt)^{1/2}}$$

where $c'$ is the initial weight fraction of metal B in A.

The thermal resistance of the slab is then given by

$$R = \int_0^L \frac{\mathrm{d}x}{K}$$

$$= \int_0^{\delta(t)} \frac{\mathrm{d}x}{K} + \frac{L - \delta(t)}{K_0}$$

If $\delta(t)$ is large enough, $K$ can be assumed to be approximately equal to $K_0$.

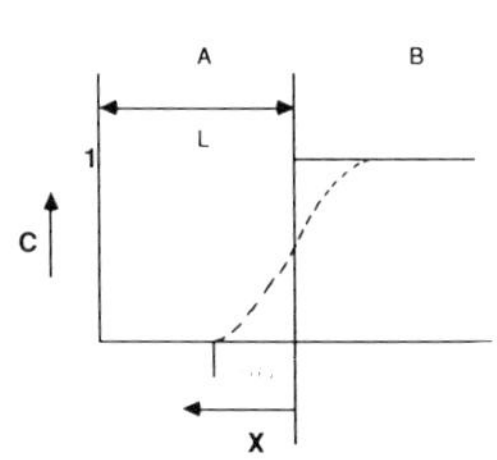


Fig. 4. Pictorial representation of two-layer laminate.

Let $\delta(t) = 4(Dt)^{1/2}$, then $K = K_0[1 - \beta[(c'/2)\ \mathrm{erfc}\{x/2(Dt)^{1/2}\}]^n]$

also

$$R = \sim \int_0^{4(Dt)^{1/2}} \frac{\mathrm{d}x}{K_0[1 - \beta[(c'/2)\ \mathrm{erfc}\{x/2(Dt)^{1/2}\}]^n]} + \frac{L - 4(Dt)^{1/2}}{K_0}$$

$$\frac{R(t)}{R_0} = 1 - \frac{1}{L}\left\{4(Dt)^{1/2} - \int_0^{4(Dt)^{1/2}} \frac{\mathrm{d}x}{1 - \beta[(c'/2)\ \mathrm{erfc}\{x/2(Dt)^{1/2}\}]^n}\right\} \tag{2}$$

where $R(t)$ is the thermal resistance of the laminate as a function of time and $R_0$ is the thermal resistance of metal A.

Even though the Ni–NiCoCrAlY laminate had NiCoCrAlY as one of the layers, it was assumed to be a pure element. As NiCoCrAlY is a cobalt rich alloy, Ni–Co interdiffusion was considered in the Ni–NiCoCrAlY laminate. The interdiffusion coefficient $D$ for Ni–NiCoCrAlY was extrapolated from the literature values [5] and is taken to be $1 \times 10^{-14}\ \mathrm{cm^2\ s^{-1}}$ at 550 °C.

Similarly $D$ for Ti–Co is assumed to be $1 \times 10^{-6}\ \mathrm{cm^2\ s^{-1}}$ for the Ti–CoCrAlY system, as there are no available published values and as very negligible interdiffusion was observed.

Equation (2) is rewritten as shown below for the 480 layer laminate which is considered to be a "symmetrical solid", as the thickness of each layer was very small (2 $\mu$m).

$$\frac{R(t)}{R_0} = 1 - \frac{1}{L}\,4(Dt)^{1/2}$$

$$+\frac{1}{L}\int_0^{4(Dt)^{1/2}} \frac{\mathrm{d}x}{1 - \beta\left[\sum_{n=0}^{\infty} \frac{2(-1)^n}{\pi(n+\frac{1}{2})} \exp\left\{-4(n+\tfrac{1}{2})\frac{\pi^2}{L^2}Dt\right\}\cos^2(n+\tfrac{1}{2})\frac{\pi x}{L}\right]} \tag{3}$$

In order to solve eqns. (2) and (3), the constants and variables are defined in the following way:

$$\beta = \frac{\beta_0}{(c'/2)^n}, t^* = \frac{(Dt)^{1/2}}{L} \text{ and } x^* = \frac{x}{L}$$

A computer program was written to evaluate eqns. (2) and (3) and the thermal conductivity for a 2, 24, 120/180 and 480 layer laminate was calculated. $n$ in the equation is assumed to be 1/3 and $\beta_0$ is varied from $1 - 1 \times 10^{-3}$ to $1 - 1 \times 10^{-6}$.

The values of effective thermal conductivity calculated from eqns. (2) and (3) for the two systems are tabulated in Table 1 and plotted as a function of the number of layers in Figs. 1 and 2.

It is observed that the calculated values of the thermal conductivity from the modified model fits very well with the experimental data.

*4.0 Summarizing remarks and conclusions*

Thick Ni–NiCoCrAlY and Ti–CoCrAlY microlaminate composites with different number of layers were synthesized and their thermal conductivities studied. The results of the above investigations indicate the following.

(1) The above thermal conductivity of the laminate composites perpendicular to the laminate plane decreases with increasing number of interfaces.

(2) The drop in thermal conductivity is associated with the interfaces.

(3) A modified model which takes into account the diffuse interface correlates the calculated value with the observed values.

## Acknowledgments

The authors would like to acknowledge with thanks the financial support given by NASA under the contract NAG 3-701 with Dr. James Nesbitt as Technical Monitor. The modified model was developed under the guidance of Prof. A. Mills of Mechanical Engineering department. The author would like to thank Prof. Fetterman for permitting us to use his $CO_2$ laser set up for thermal diffusivity measurements.

## References

1 M. C. Radhakrishna, H. J. Doerr, C. V. Deshpandey and R. F. Bunshah, *Surf. Coat. Technol.*, *36* (1988) 143 - 150.

2 B. A. Movchan and R. F. Bunshah, *Proc. 7th Int. Conf. on Vacuum Metallurgy, Tokyo, 1982*, Iron and Steel Institute of Japan, 1982.

3 R. K. Francis and J. R. Tinkelpaugh., *J. Am. Ceram. Soc.*, *43* (1960) 560 - 563.

4 Ho et al., *J. Phys. Chem. Ref. Data*, *7* (1978) 979.

5 S. Mrowec, *Defects and Diffusion in Solids*, North-Holland, Amsterdam, 1979.

*Surface and Coatings Technology, 39/40* (1989) 161 - 172 

# FLIGHT SERVICE EVALUATION OF THERMAL BARRIER COATINGS BY PHYSICAL VAPOR DEPOSITION AT 5200 H

F. C. TORIZ, A. B. THAKKER and S. K. GUPTA

*Rolls-Royce Inc., 1895 Phoenix Blvd., Suite 400, Atlanta, GA 30049 (U.S.A)*

(Received March 3, 1989)

## Summary

Thermal barrier coatings (TBC) have the potential to improve the performance of jet engines and have been used in combustors for over 15 years. However, it is only recently that they have been actively used in the harsh turbine environment on nozzle guide vane platforms. It is now intended to use TBCs on turbine vane airfoils and rotating blades where the maximum payoff will be realized. Much work has been done in the last six years towards this goal.

This report will review one such Rolls-Royce program; flight service evaluation of TBCs. It will demonstrate that some TBC systems can survive in the turbine environment for over 16 000 h, but that particulate erosion could be a problem. It will review the service condition of vanes with TBCs applied by physical vapor deposition, and will demonstrate some of the advantages over plasma sprayed coatings such as:

(1) longer thermal cycle lives,
(2) smoother surface finishes,
(3) better surface finish retention,
(4) superior erosion resistance.

---

## 1. Introduction

Rolls-Royce has used ceramic thermal barrier coatings (TBCs) in jet engine combustors for many years, and they are used in most of our engines today. TBCs are being used on high pressure turbine (HPT) nozzle guide vane (NGV) platforms in several engines, including the RB211 535-E4 (40 000 lb thrust class engine) that powers the Boeing 757. Much testing of TBCs has been done on turbine rotor blades, and we look forward to using them in regular service in the near future.

TBCs are of prime interest because of their ability to improve gas turbine performance, at a time when potential improvements from alloy modifications and cooling efficiencies are all but exhausted. Thin ceramic coatings can reduce the component metal temperatures by several hundred

0257-8972/89/$3.50 

degrees resulting in significant life improvement. This life improvement can be traded off for (a) higher inlet temperatures and increased thrust or for (b) reduced cooling air usage and reduced fuel consumption. The use of a 0.025 cm (0.01 in) of ceramic coating can reduce the mass average temperature at cruise by over 125 °C (200 °F). Or the cooling can be reduced, improving the specific fuel consumption.

## 2. Review of work to date

### *2.1. The effect of thermal cycles*

One of the major failure modes of TBC systems is the spallation of ceramic topcoat due to thermal cycling. The failures are believed to be caused by stresses induced by coefficient of expansion ($\alpha$) mismatches and oxidation of the bondcoat. From carousel rig testing [1] we learned several ways to increase the cycle life. For plasma sprayed yttria stabilized zirconia (YSZ), fully stabilized (20% $Y_2O_3$) zirconia was cubic in phase and had a very short cycle life [1]. But, a partially stabilized system (6 - 8% $Y_2O_3$) resulted in a tetragonal phase which was very stable. Further improvements were made using high density, oxidation resistant MCrAlY bondcoats, applied by shrouded plasma spray (SPS), vacuum plasma spray (VPS) or physical vapor deposition (PVD). Applying the ceramic by PVD created a strain tolerant columnar structure, resulting in a 6 times improvement in cycle life.

### *2.2. The effect of oxidation due to time at temperature*

Oxidation of the bond coat is a major mode of failure to TBC systems. The amount of oxidation depends on the time spent at high temperatures. It has been shown [2] that spallation occurs when the oxide at the interface reaches a critical thickness of about 4 $\mu$m. Time at temperature–cycle data can be very costly to acquire, especially when trying to simulate the engine environment and the interaction with thermal cycles. A flight service evaluation (FSE) program was therefore begun to accumulate this data on several TBC systems [1, 3].

### *2.3. FSE program*

After considerable review, the RB211-22B intermediate pressure turbine (IPT) NGV was chosen as the FSE component (see Fig. 1). Prior work on a bench engine had shown that the IPT NGV platform application was considerably more demanding than the combustion chamber environment.

The coatings in phase I were applied only to the inner diameter pressure side platform area. In phase II, the pressure side airfoil of some vanes were coated as shown by the crosshatching in Fig. 1. The nickel base vane is nominally 20 cm long, 7.6 cm wide and 10 cm tall (8 in × 3 in × 4 in). Eight TBC systems were evaluated (Table 1). The coatings were given designations; A was three-layer magnesium zirconate (MSZ), the standard combustor coating. Coating B was two-layer MSZ, without the cermet (a NiCrAl and

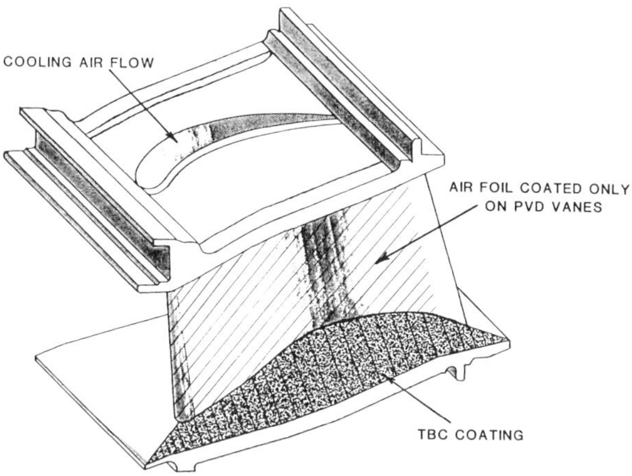


Fig. 1. RB211 IPT NGV showing location of TBCs patches.

MSZ mixture). Coating C used the same NiCrAl bondcoat as A and B, with 6 - 8% YSZ. Coatings A, B, C, G and H were applied by air plasma spray (APS).

Coating E had CoNiCrAlY bondcoat applied by shrouded plasma spray (SPS). Coating F also has a CoNiCrAlY bondcoat applied in a vacuum (VPS) plus a YSZ top coat.

Coatings G, H and I were added in phase II, and therefore have less time on them. Coating G had a CoNiCrAlY bondcoat and YSZ; it had accumulated only 4800 h as of December 1988. Coatings E, F and G should give a direct comparison of the bondcoat performance as applied by three significantly different techniques. Coating H is an APS NiCrAlY and $CaOTiO_2$, which is claimed to be resistant to acid leaching from low grade fuels [4]. It did well in thermal cycle testing. Coating I had a NiCoCrAlY bondcoat plus 8% YSZ, both applied by PVD. (Table 1).

TABLE 1

Coatings used on RB211 IPT NGV FSE

| *Identification letter* | *Bond-coat* | *Ceramic (APS)* | *Bondcoat method* | *On engines* | *Total coated vanes* |
|---|---|---|---|---|---|
| A | NiCrAl + C[a] | MSZ | APS | 1,2,3,4 | 24 |
| B | NiCrAl | MSZ | APS | 1,2,3a | 12 |
| C | NiCrAl | YSZ | APS | 1,2,3,4 | 23 |
| E | CoNiCrAlY | YSZ | SPS | 1,2,3,4 | 23 |
| F | CoNiCrAlY | YSZ | VPS | 1b,2,4a | 28 |
| G | CoNiCrAlY | YSZ | APS | 3a | 16 |
| H | NiCrAlY | $CaOTiO_2$ | APS | 1b,4a | 10 |
| I | NiCoCrAlY | YSZ(PVD) | PVD | 1b | 5 |
| Total | | | | | 141 |

[a] Cermet ≡ NiCrAl + MSZ.

TABLE 2

TBC service evaluation status January 1989

| *Engine build* | *Flight time* (h) | *Time of inspection* (h) | *Time on phase II coatings* (h) | *Original vanes* |
|---|---|---|---|---|
| 1 | 12759 | 4100, 7500, 10100, 12800 | H, I (5200) | 9 |
| 2 | 16322 | 2500, 9800, 16300 | | 15 |
| 3 | 11685 | 6900, 11700 | G (4800) | 14[a] |
| 4 | 8700 | 5500, 8700 | H (3300) | 19[a] |

[a] These sets removed from FSE.

### *2.4. Current status of FSE program*

The flight status for the four engines as of January 1989 is given in Table 2. Engine build 2 was removed in January 1989, with 16 322 h–8120 cycles, and still had 15 of its original 25 coated vanes. Two reports for this engine have been given earlier, at 2539 h [1] and 9800 h [3]. To get the approximate number of cycles, divide the hours by 2 (Table 2).

## 3. Results of plasma sprayed coatings

### *3.1. MSZ*

Coating A lost the most amount of coating (spalling). It was removed from the program at 9800 h with an average area loss of 31% [3]. Twenty data points were used to generate a second order regression analysis curve fit of the data for coating A, which is plotted on semi-log paper in Fig. 2. Bondcoat was missing in the fillet radius of these vanes as early as 2500 h [1].

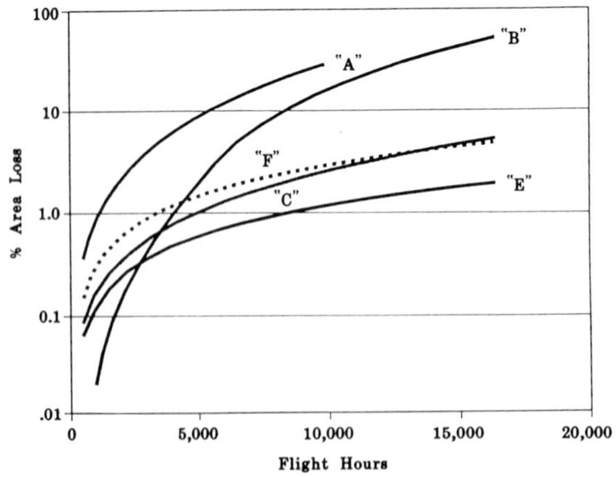


Fig. 2. Loss of ceramic coating (percentage area) on IPT NGVs with time (see Table 1 for coating identification).

Coating B was significantly better than A, but was still not acceptable. The area loss was negligible prior to 3500 h, and only 16% at 10 000 h. By 16 000 h the average loss was close to 50%. Thirty-one points were used to determine the second-order curve as shown in Fig. 2. Bondcoat was missing in the fillet radius at 2500 h [1].

*3.2. Yttria partially stabilized zirconia*

Coating C had significantly less loss, at 10 000 h the average ceramic loss was less than 3% by area. After 10 000 h the loss accelerated such that two vanes had close to 10% loss by 16 000 h. The second-order curve fit used 32 points of data (see Fig. 2). Figure 3(a) shows a vane at 12 800 h with 8% of ceramic loss. Figure 4(a) shows a similar vane at 16 300 h with 10% of the ceramic coating gone. The NiCrAl bondcoat runs cooler than on A or B because of the lower conductivity of the YSZ ceramic, therefore it oxidizes slower. In the fillet radius where the ceramic is minimal, the bondcoat ran hotter and fell off on some vanes early on [1]. References 1 and 3 have more photographs.

Coating E had minimal ceramic loss, the average was less than 1% by area at 10 000 h. Most of those losses were associated with edge effects in the fillet radius; similar to, but less than that shown in Fig. 3(b), where the E-coated vane had 12 748 h. By 16 300 h the average loss was 1%. The worst vane is shown in Fig. 4(b) with 3% area loss. A linear curve fit was adequate with the 60 points of data used to generate the curve fit as also shown in Fig. 2. The bondcoat on all vanes inspected to date was in excellent condition. The coating losses for F were less consistent, with some vanes having very little loss (less than 0.05%), and others as much as 10% at 16 300 h, the average loss (four vanes) was only 3%. The linear curve fit for F in Fig. 2 is for 20 points of data. The bondcoat was in excellent condition, as seen in Fig. 4(c). Because the losses seem to be confined to one or two vanes in each set and in light of the good results other operators [5, 6] have had, we believe some of the coating losses are process related.

*3.3. Calcium titanite*

Coating H has only accumulated 5200 h, but is holding up very well, with only one out of 10 vanes showing significant loss (1.7%), the average

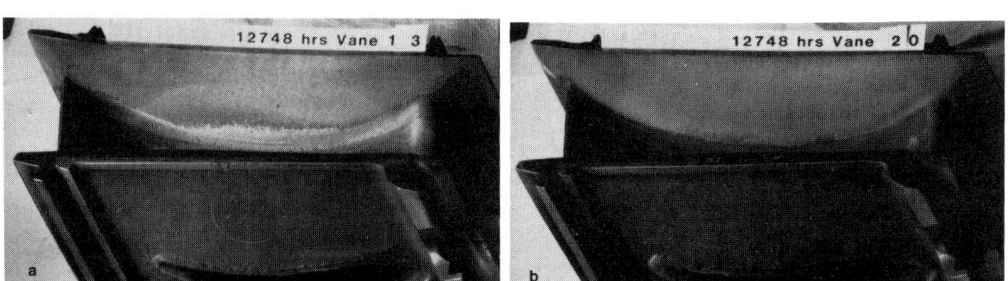


Fig. 3. Ceramic coated IPT NGVs at 1200 h into the FSE: (a) coating C, (b) coating E.

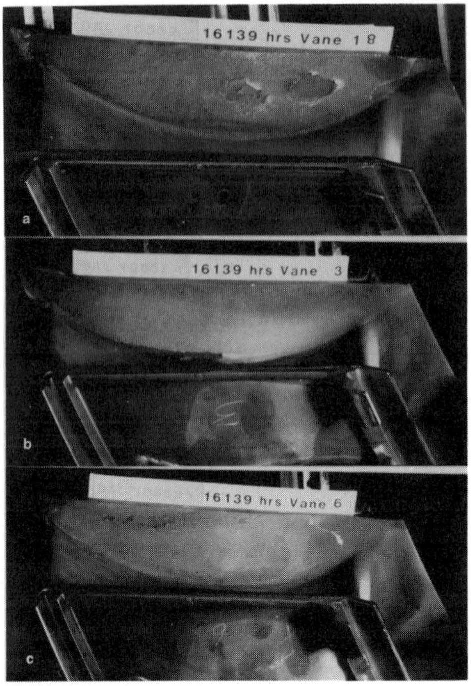


Fig. 4. Ceramic coated IPT NGV at 16 300 h into the FSE: (a) coating C, (b) coating E, (c) coating F.

loss of five vanes at 5200 is less than 0.5%. The curve fit is only for 15 points of data and it is a bit early to come to any conclusions other than it looks very promising.

### *3.4. Ranking of plasma sprayed coatings*

The order of ranking of the TBCs tested is based on the ceramic and bondcoat loss. A - C lost bondcoat on several vanes where not covered by ceramic. Toriz *et al.* [1] has shown the bondcoat loss as early as 2500 h for A and B, and 4100 h for C. This indicates that the bondcoat on C is at or near its oxidation temperature limit when it is on a hot vane. It indicates that C will not be very useful at higher temperatures. The ranking for H is preliminary since it only had 2500 h. The ranking to date (based on ceramic area loss) from this FSE program is as follows.

| Coating system | Hours/data points |
|---|---|
| 1. Coating E SPS CoNiCrAlY + YSZ | 16300/60 |
| 2. Coating F VPS CoNiCrAlY + YSZ | 16300/20 |
| 3. Coating C APS NiCrAl + YSZ | 16300/32 |
| 4. Coating H APS NiCrAlY + CaO.$TiO_2$ | 5200/15 |
| 5. Coating B APS NiCrAl + MSZ | 16300/31 |
| 6. Coating A APS NiCrAl + Cermet + MSZ | 9800/20 |

Many of the vanes had indications of ceramic erosion near the fillet radius at midchord, where it is thinnest because of geometry constraints. This observation prompted a separate program to understand the erosion behavior of TBCs, as reported earlier [9] and is referred to later in this paper.

## 4. PVD TBCs

### *4.1. Benefits of PVD coatings*

Our experience to date indicates there are many benefits to PVD coatings. We have found PVD overlays (and bondcoats) to perform very well; they are of similar density and low oxide content as VPS coatings. PVD ceramic coatings have a unique columnar structure which make them very tolerant.

#### *4.1.1. Improved thermal cycle life*

In carousel rig cycle testing [7], PVD TBCs out performed all the plasma sprayed coatings including SPS and VPS by 50% or more.

#### *4.1.2. Improved initial surface finish*

One of the concerns with the use of TBCs is the loss of aerodynamic performance due to rough surfaces, which can cause laminar-turbulent transition, and increased skin friction. These effects were apparent in numerous cascade, turbine and compressor rig tests involving a range of roughness types [8]. Since APS ceramics are very rough (7.7 $\mu$m, $300 \times 10^{-6}$ in) as sprayed we were reluctant to add them to the airfoils in our FSE program. Even with polishing they were still rougher than desired (3.2 $\mu$m, $125 \times 10^{-6}$ in). However, since PVD ceramics approximately replicate the surface finish of the component applied to, it was added to the pressure side airfoils (2.1 $\mu$m, $80 \times 10^{-6}$ in) of five vanes.

#### *4.1.3. Improved erosion resistance*

An erosion test program run at the University of Cincinnati [7, 9] indicated that PVD YSZ was 10 times more erosion resistant than the standard APS YSZ at a 90° impingment angle. Figure 5 shows the relative erosion rates of several ceramic coatings as a function of impingment angle. At low angles of attack, such as on combustors or vane platforms, PVD is only a 2 times improvement over APS YSZ. But the benefit increases with angle, so that coatings applied to flow path airfoils will see life improvements of up to 1000% (at 90°). By looking at the erosion rate at 90° in Fig. 5, one can obtain the erosion ranking for several different TBC materials, more are found in ref. 7.

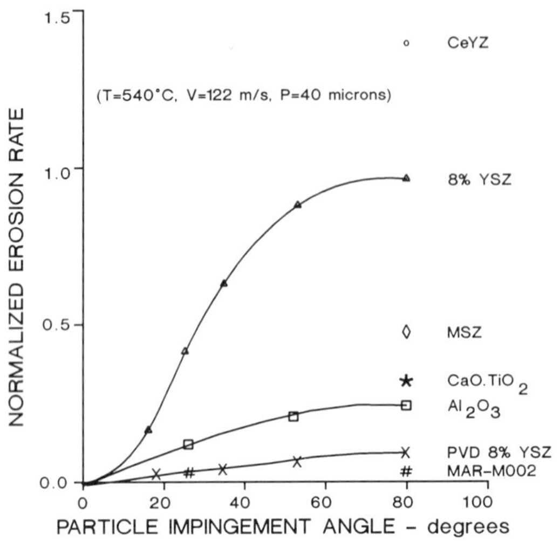


Fig. 5. The effect of impingement angle on the erosion of engine materials. Erosion ranking for six ceramics (five APS) and one metal (substrate) given at 90° impingement angle.

*4.1.4. Improved surface finish retention*

The erosion test program [7, 9] also showed that PVD TBCs have a better surface finish retention with time than the APS TBCs. The as sprayed YSZ was very rough, it measured over 7.7 μm. Most of the erosion test coupons were polished to about 1 μm ($40 \times 10^{-6}$ in) before testing. The PVD YSZ coupons were 1 μm ($40 \times 10^{-6}$ in) as received. When erosion tested, the polished standard YSZ decayed from 1.2 μm ($50 \times 10^{-6}$ in) to over 7.7 μm ($300 \times 10^{-6}$ in) at a 90° impact angle. At a 20° angle the deterioration was much slower, and probably compares to that of the FSE NGV platforms. The PVD coating did not deteriorate much even after 30 g of particles; as can be seen in Fig. 6.

Now to correlate this with the FSE program. Vanes coated with A, B, C and F (Table 1) were polished. Their roughness increased with flight time, as shown in Table 3. None of the E-coated vanes were polished and they stayed at about 7.7 μm up to 12 800 h. From Table 3 we see that it took 9800 h to deteriorate from a 2.8 to 5.1 μm finish at low angles of attack. In Fig. 6 this decay takes approximately 25 g; so we have the rough relationship that 25 g in wind tunnel testing is roughly equal to 10 000 FSE h. Using this relationship, by 8000 h (20 g) the leading edge of an airfoil (at 90°) would have a surface finish of about 7.7 μm. Its surface finish would be 5.1 μm ($200 \times 10^{-6}$ in) in only 1000 h. The PVD YSZ however, will still have a good surface finish 1.5 μm ($60 \times 10^{-6}$ in) at 13 000 h (32 g). The FSE experience data band is overlayed on Fig. 6. Airfoil measurements further indicate that the PVD coating has a better surface finish ($130 \times 10^{-6}$ in) than the bare metal ($185 \times 10^{-6}$ in) at 5200 h (see Table 3).

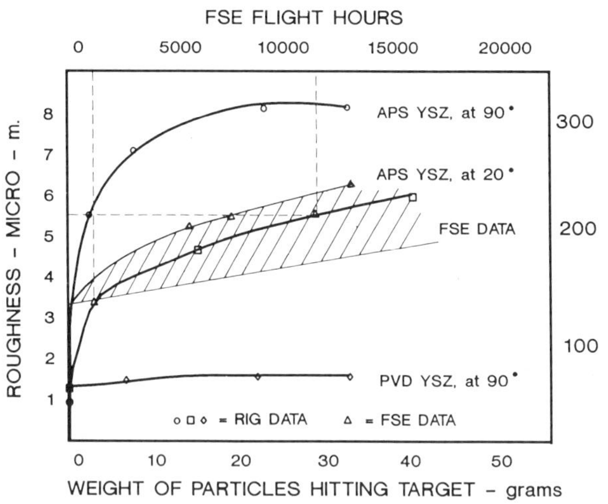


Fig. 6. Ceramics surface roughness variations with erosion–time from rig testing correlated to surface finishes experienced in the FSE program.

TABLE 3

FSE NGV surface roughness (in μm ( $\times 10^{-6}$ in)) with time

| *Coating* | *Time* (h) | | | | | |
|---|---|---|---|---|---|---|
| | 0 | 5500 | 6900 | 7400 | 9800 | 12800 |
| Platform pressure side | | | | | | |
| A | 2.8 (110) | | | 4.3 | 3.5 | |
| B | 2.8 (110) | | 3.3 | 4.3 | 5.1 | 7.7 (300) |
| C | 2.8 (110) | 4.8 | | 5.1 | 5.1 | 5.8 (230) |
| E | 7.7 (300) | | 7.4 | | 7.7 | 7.4 (290) |
| F | 2.8 (110) | 3.4 | | | 3.8 | |
| I | 2.0 (80) | 2.3 | | | | |
| Airfoil pressure side trailing edge | | | | | | |
| I | 2.5 (100) | 3.3 | | | | |
| None | 2.5 (100) | 4.7 | | 5.1 | | 6.2 (245) |

### *4.2. Phase II FSE results*

In phase II of the FSE program five vanes were PVD coated and put into engine build 1b (second rebuild at 7500 h). Both the NiCoCrAlY bondcoat and the 8% YSZ coating were applied by electron beam physical vapor deposition (PVD) by Chromalloy R&T Coating Systems, Orangeburg, NY. The bondcoat was 76 - 127 $\mu$m (0.003 - 0.005 in) thick and the ceramic was 20 - 25 $\mu$m (0.008 - 0.0010 in) thick. The coating was also applied to the pressure side airfoils. The coating tapered off to within 1.27 cm (0.5 in) of the trailing edge, it wrapped around the leading edge by 1.27 cm (0.5 in) (see Fig. 1).

#### 4.2.1. *The 2500 h inspection*

The vanes were inspected at 2500 h when the engine came in for a shop visit, a detailed inspection was not possible as the module containing the IPT vanes was not taken apart at that time. A photograph of a vane prior to service is given in ref. 7. This IPT vane set appeared to have suffered considerably more erosion than any engine inspected to date, even uncoated vanes had visual signs of metal removal on the leading edge (LE) airfoil. Erosion was evident on all of the PVD vanes, but more so on three that had distinct lines outlining the eroded areas, none were eroded through to the bondcoat. Erosion to the platform coating was not noticeable. Two vanes had spalling on the LE, which may have been initiated by small dings which occurred in shipping, one vane (number 2) is shown in Fig. 7. Two vanes had minor spalling on the platforms. The IPT castings are fairly rough, especially on the platforms, and may have reduced the adhesion quality of the PVD coating which performs better on a smooth surface. Some edge chipping was seen, but most of it occurred prior to installation. The coating on the pressure side of all five airfoils was in excellent condition. No mud flat cracking or crazing was evident. The surface finish appeared to be as installed, except on the eroded LEs. The parts did not come out of the engine, so the surface finish was not measured. Table 4 summarizes the area losses.

#### 4.2.2. *The 5200 h inspection*

More recently the engine was removed again and this time the module was taken apart, the PVD coated vanes having 5188 h and 2629 cycles flight experience. There was no further evidence of erosion; however, there was spalling on all the LE coated vanes. The LE spalling was typically 0.38 cm × 3.8 cm along the vertical LE of the vane. Vane number 3 is shown in Fig. 8. The average area loss at the LE is 0.64 $cm^2$. The platform loss increased only slightly to 1.5% by area. The area summary loss is given in Table 4. The pressure side airfoils were still in excellent condition. No mud flat cracking or crazing was seen (Table 4).

Fig. 7. Condition of PVD YSZ coated IPT vane at 2500 h into the FSE program.

TABLE 4

IPT PVD vane area loss at 2500 and 5200 h

| *Platform area loss (%)*[a] | | | *Airfoil area loss* | | | |
|---|---|---|---|---|---|---|
| *Vane* | 2500 h | 5200 h | 2500 h | | 5200 h | |
| | | | $cm^2$ | $in^2$ | $cm^2$ | $in^2$ |
| 1 | 0.5 | 0.9 | 0 | 0 | 2.1 | 0.32 |
| 2 | 0 | 1.1 | 1.3 | 0.20 | 2.2 | 0.34 |
| 3 | 0 | 0.5 | 0 | 0 | 1.0 | 0.15 |
| 4 | 2.1 | 2.1 | 0 | 0 | 2.6 | 0.40 |
| 5 | 2.5 | 2.9 | 0.3 | 0.05 | 0.3 | 0.05 |
| Average | 1.0 | 1.5 | 0.3 | 0.05 | 1.6 | 0.25 |

[a] Platform area is 53 $cm^2$ (8.2 $in^2$).

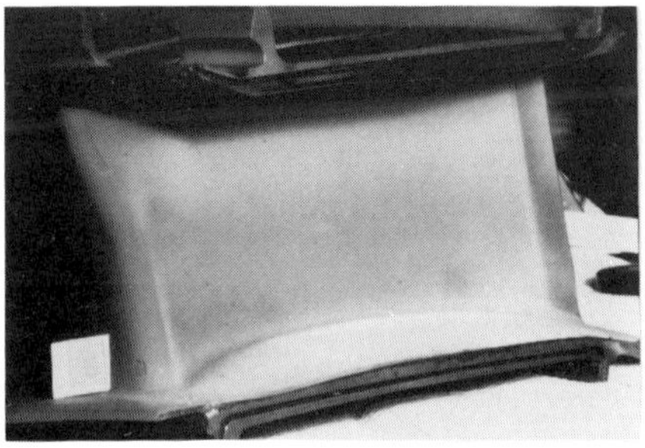

Fig. 8. Condition of PVD YSZ coated IPT vane at 5200 h into FSE program.

## 5. Conclusions

We have demonstrated in this program that TBCs on turbine components can survive for up to 16 000 h in the harsh turbine environment and should therefore be used in earnest.

We have identified the benefits of PVD TBCs, namely they have a superior thermal cycle life, better initial surface finish, better surface finish retention than either APS TBCs, or the bare nickel base alloy, and are more erosion resistant than APS TBCs.

Erosion of TBCs is a potential problem that will have to be dealt with on airfoils or where high angles of impingment occur. Here PVD YSZ or APS erosion resistant top coats can be used with low conductivity ceramic underneath.

$CaOTiO_2$ has done well in thermal cycling and in the FSE program and may be required for industrial, military or mid-eastern operations, where acid leaching is a problem.

It is thought that advanced TBC systems using high density bondcoats such as those produced by SPS, VPS and PVD, and 6 - 8% YSZ ceramics, will be required for future turbine airfoil applications. This is especially true on rotating airfoils.

## References


1 F. C. Toriz, A. Bennett and S. K. Gupta, Rolls-Royce experience with TBCs including high time service evaluation, *ASM 8512-021* (1985).

2 R. V. Hillery, B. H. Pilsner, R. L. McKnight, T. S. Cook and M. S. Hartle, TBC life prediction model development, *NAS3-23493* (1988).

3 F. C. Toriz, A. Bennett, A philosophy for TBCs and its corroboration by 10 000 h service experience on RB211 nozzle guide vanes, *Surf. Coat. Technol., 32* (1987) 359 - 375.

4 J. R. Vogan, A. R. Stetson, Solar Turbines, Inc., Advanced ceramics coating development for industrial/utility gas turbines, *NASA-CR169852* (1982).

5 J. T. DeMasi and K. D. Sheffler, TBC life prediction model development, second annual report, *NASA CR-179508* (1986).

6 R. H. Handschuh, High temperature erosion of plasma sprayed, yttria stabilized zirconia in a simulated turbine environment, *NASA TP2406* (1984).

7 F. C. Toriz, A. B. Thakker and S. K. Gupta, TBCs for jet engines, *ASME 88-GT-279* (1988).

8 R. M. Watt, J. L. Allen, N. C. Bains, J. P. Simons and M. George, A study of the effects of TBC surface roughness of the boundary layer characteristics of gas turbine aerofoils, *ASME 87-GT-202* (1987).

9 T. N. Phys-Jones and F. C. Toriz, Thermal barrier coatings for turbine applications in aero engines, *High Temp. Technol. 7* (2) (1989) 73.

*Surface and Coatings Technology, 39/40* (1989) 173 - 181 

# A REVIEW OF MICROSTRUCTURE AND PROPERTIES OF PLASMA SPRAYED CERAMIC COATINGS

R. McPHERSON

*Department of Materials Engineering, Monash University, Clayton, Victoria 3168 (Australia)*

(Received March 20, 1989)

## Summary

The present understanding of the plasma-sprayed coating formation process is reviewed. The development of porosity, and the very fine voids formed by incomplete inter-splat contact is discussed. The strong relation between porosity and mechanical and thermal properties of ceramic coatings is reviewed. Some of the controlling factors for the grain size and crystal structure within the deposited splats are discussed, particularly in reference to zirconia coatings.

## 1. Introduction

The high temperature (around 10 000 K) achieved in thermal plasmas allow the melting of any material, which does not decompose or vapourize excessively, within a very short time interval. For this reason, plasma spraying of ceramic coatings has become well established as a commercial process over the past 25 years. Plasma spraying basically consists of the injection of powders into a direct current plasma jet, where they are melted and accelerated, and directing the stream of molten particles onto a substrate where they form a coating as they spread and solidify. The simplicity of the process, from a practical point of view, has meant considerable empirical development has taken place with relatively little understanding of the basic principles of coating formation.

Many of the properties of plasma-sprayed coatings depend upon their microstructure, as for other engineering materials; therefore, to properly understand coating properties and to develop new ones, the coating microstructure must be understood. The microstructure itself depends upon the mechanism of coating formation so microstructure is the common factor linking properties to processing details.

0257-8972/89/$3.50 

## 2. The coating formation process

Since coatings are formed by the impact of a stream of particles striking the substrate, the major factors which control the structure of a particular coating are the temperature, velocity and size distribution of the incident particles. The particle size is determined by the feed material but the velocity and temperature distributions depend, in a complex way, on the design and operating conditions of the plasma torch. Direct measurements of particle velocity and temperature distributions in a plasma torch have been related to the coating formation process [1]. Ideally, all of the incident particles which strike the substrate would be completely molten. Unmelted particles bounce off reducing the deposition efficiency and partly melted particles are incorporated within the deposit modifying its microstructure and properties.

When a spherical liquid droplet strikes a flat surface at high velocity, it flattens to a disc but the radially flowing thin sheet of liquid becomes unstable and disintegrates at the edges into small droplets. In the case of plasma spraying the substrate is at a temperature well below the melting point of the droplet and heat transfer to the substrate is rapid so that spreading and break up of the droplet is interrupted by solidification. The individual particles are therefore equivalent to "splat quenched" material with a cooling rate of around $10^6$ K s$^{-1}$. Classical heat transfer treatment of the cooling of a thin plate suggests that the interface structure is determined by a number of small distributed contact points within an effectively non-conducting film arising from entrapped gas [2]. The microstructure of plasma-sprayed coatings is thus related to the mechanism of formation through two predominant factors, the nature of the interfaces between individual lamellae, and the internal structure within lamellae produced by their rapid solidification.

## 3. Interparticle contact

Thermally-sprayed deposits have porosities ranging from a few percent to about 20% depending upon the materials and spraying conditions. Studies of the pore size distribution of alumina coatings, by the mercury intrusion porosimetry (MIP) method, showed a bimodal distribution with coarse pores in the size range 3 - 10 $\mu$m and fine pores around 0.1 $\mu$m [3]. Increasing the power input to the torch, for a given powder particle size, decreased the total porosity and ultimately produced a monodisperse pore size distribution with mean diameter of 0.1 $\mu$m. These results suggest that there are two types of porosity present in a coating, coarse porosity associated with defects in the structure, due to incomplete filling of interstices between previously deposited particles, which is more likely if the impacting particles are not completely molten, and fine pores which are an inherent feature of coatings. The MIP method, however, gives misleading results if the microstructure consists of coarse pores interconnected only by small channels since the pore diameter measured is that of the channels and the fraction porosity is determined by the total connected pore volume. Other techniques must therefore be used to obtain a complete description of pore size distribution.

The porosity of NiAl and nickel plasma-sprayed coatings has been studied using a combination of MIP and quantitative analysis of scanning electron microscope (SEM) images [4]. This approach provides independent estimates of the pore size distribution at larger pore diameters (greater than about 1 $\mu$m). The results were consistent with a two level porosity model and differences in pore microstructure between various coating types could be used to explain differences in coating tensile strength.

The extremely fine porosity observed in thermally-sprayed coatings has been interpreted as incomplete contact between lamellae during the process of coating formation, as referred to earlier [2]. This view is supported by transmission electron microscopy of transverse sections of alumina plasma sprayed coatings [5] which showed narrow pores (around 0.1 $\mu$m) between lamellae suggesting a model for coating structures along the lines illustrated in Fig. 1.

SEM examination of the surfaces of individual lamellae of plasma-sprayed ceramic coatings also shows a network of fine cracks perpendicular to the lamellar plane which are formed by the stresses produced by restraint of the thermal contraction of the lamellae by underlying material during cooling from the solidification temperature [2]. These cracks will not form in metallic coatings because the thermal strain may be accommodated by plastic flow.

Further direct evidence for the micro-porous model for plasma sprayed coatings has recently been provided by a study in which the pore structure of alumina coatings was revealed by a copper plating technique [6]. This delineated the pattern of micro-pores between lamellae together with those within lamellae arising from thermal stresses. Analysis of the data suggested a real area of contact between lamellae of around 20%, close to that postulated in earlier studies [7].

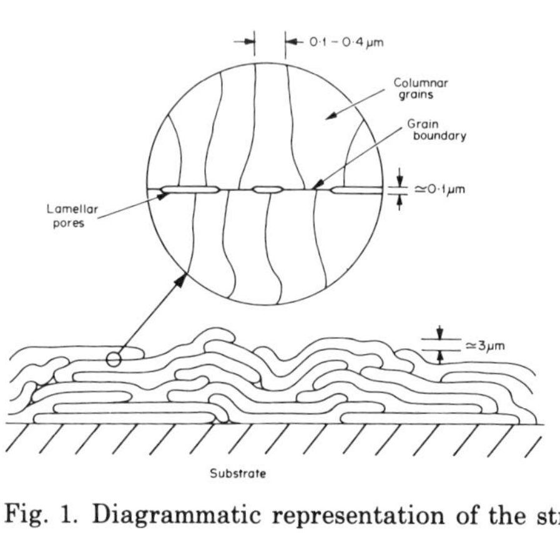


Fig. 1. Diagrammatic representation of the structure of a thermally-sprayed coating.

This model of a hierarchy of porosity is extremely important from the point of view of coating properties since it implies that the fine pores between lamellae are a feature of the structure of coatings sprayed at atmospheric pressure which cannot be eliminated by changing the processing conditions. Spraying in a low pressure atmosphere and/or the use of higher particle velocities may, however, offer some opportunity to reduce this effect.

## 4. Porosity and coating properties

### *4.1. Mechanical behaviour*

Porosity generally degrades the mechanical properties of materials, for example, brittle fracture is extremely sensitive to defect size and the strength is determined by the largest crack-like defect present. Many other properties such as electrical and thermal conductivity and elastic modulus, are reduced by the presence of pores which reduce the effective cross-sectional area of the material. The detailed influence of porosity on these properties depends on the pore size and morphology [8]. However, finely distributed porosity tends to improve the thermal shock resistance of brittle materials and studies of the failure of some $ZrO_2$-based thermal barrier coatings under thermal cycling conditions have shown that increased porosity does in fact increase the number of cycles to failure [9].

The elastic modulus of coatings is much lower than that of the bulk material, for example, the Young's modulus of plasma sprayed alumina is about one-fifth that of a sintered ceramic [10]. This is a much larger reduction than can be explained on the basis of randomly distributed spherical pores but is consistent with the concept of narrow planar pores between lamellae with a relatively small fraction "true contact" area [2]. Heat treatment of the sprayed coating causes sintering of fine pores, which change shape to spherical, the area of true contact increases, and the elastic modulus increases towards that of the bulk polycrystalline material. The Young's modulus of plasma sprayed $ZrO_2$ coatings has been quoted as 48 GPa compared with around 200 for GPa for the sintered ceramic [11], that is around one quarter of the "true" value.

The elastic modulus of Ni–Al plasma-sprayed coatings, measured in tension perpendicular to the coating plane, has been related to the details of pore distribution determined by SEM in conjunction with MIP [12]. The complication of thermal stress micro-cracks within lamellae observed in ceramic coatings is eliminated for these ductile materials which simplifies analysis. The results showed that the fraction real area of contact between lamellae, estimated from the pore size distribution, was similar to that estimated from elastic modulus assuming that the modulus was directly proportional to the real area of contact between lamellae averaged over the specimen. This provides further support for the proposed model for coating structure.

The "strength" of a ceramic coating, as measured for example by the

ASTM tensile adhesive test, will be related to the microstructure through the Griffith equation for the fracture strength of a brittle material:

$$\sigma_f = \left(\frac{2\gamma E}{\pi C}\right)^{1/2} \tag{1}$$

where $\sigma_f$ is the fracture stress, $\gamma$ is the fracture surface energy, $E$ is Young's modulus and $C$ is the length of the fracture initiating defect.

We have seen that the Young's modulus of coatings is much lower than that of the bulk material because of the fine scale porosity and we would also expect the effective fracture surface energy ($\gamma_{eff}$) to be lower, since a crack would pass from one region of good contact to another, so that the effective fracture surface area would be lower. Thus $\gamma_{eff}$ is roughly equal to $a_r\gamma_0$ where $a_r$ is the fraction real area of contact between lamellae and $\gamma_0$ is the fracture surface energy of the dense solid. However, it would be surprising if fracture occurred by propagation of a single crack in a coating. Crack branching would increase the crack surface area by a multiplying factor ($m$) so that $\gamma_{eff} = ma_r\gamma_0$.

Measurements of the critical strain energy release rate ($G_c$) of plasma-sprayed coatings using the double cantilever beam technique [13 - 15] show a wide scatter of values together with variations with crack length which may be related to changes in the degree of crack branching ($G_c = 2\,\gamma_{eff}$).

The range of values of $\gamma_{eff}$ observed from fracture toughness tests, 5 - 20 J m$^{-2}$ for $Al_2O_3$ [13] and 20 - 80 J m$^{-2}$ for $ZrO_2$–6% $Y_2O_3$ [14], are comparable with those of dense polycrystalline ceramics, e.g. 10 - 50 J m$^{-2}$ for $Al_2O_3$ [8], suggesting that if $a_r$ is about 0.25, $m$ is about 4. This multiple crack path effect is additional to other microstructural influences on fracture surface energy observed in polycrystalline ceramics.

The fracture strength of a coating is thus related to its microstructure mainly through the influence of interlamellar porosity on $E$ and $\gamma$. In the absence of larger defects introduced by specimen preparation the largest pores–microcracks produced during spraying would be expected to act as fracture initiating defects and thus influence strength through $C$ in the Griffith equation (Fig. 2).

### *4.2. Thermal conductivity*

The thermal conductivity of plasma-sprayed coatings is also much lower than that of the bulk material. The thermal conductivity of CaO fully-stabilized $ZrO_2$ coatings measured at 600 K in a vacuum is 0.45 W m$^{-1}$ K$^{-1}$ and 0.68 W m$^{-1}$ K$^{-1}$ in argon [16]. Heating at elevated temperatures produces an increase in conductivity (1.8 W m$^{-1}$ K$^{-1}$ after 4 cycles to 2200 K). Similar observations have been made for metallic coatings [17]. All of these results can be explained in terms of the model for fine interlamellar porosity [7]. Conduction occurs only across the true region of contact between lamellae in a vacuum so that the low conductivity can simply be related to the relatively small true contact region in plasma-sprayed coatings. If a gas is present in the pores there is a parallel conduction path and the thermal conductivity of

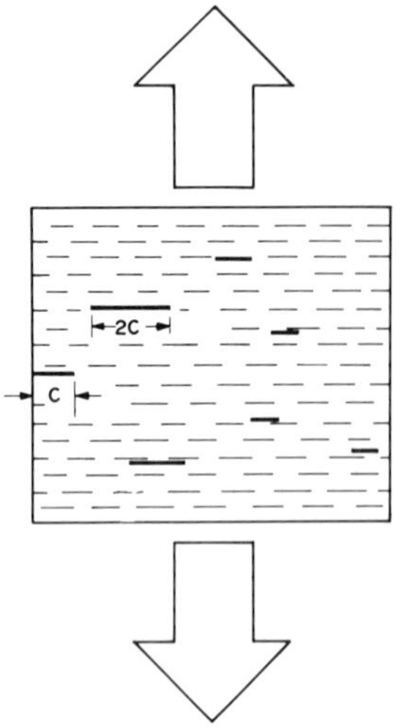


Fig. 2. Diagrammatic representation of a coating under tensile stress perpendicular to the coating plane. The light horizontal dashes represent interlamellar porosity, the bold dashes represent potential fracture initiating defects. *C* is the critical defect size in the Griffith equation.

the coating should increase. If the pores controlling conduction are very narrow, the flow of heat is very greatly reduced because conduction occurs by a different mechanism if the mean free path of the molecules is comparable with the pore width. The good agreement between calculated and experimental values for the thermal conductivity of plasma-sprayed zirconia coatings [7] provides strong support for the interlamellar, microporous model for coatings.

As stated earlier, we might expect improved interlamellar contact in coatings sprayed in a low pressure chamber. One piece of experimental data supporting this view comes from thermal diffusivity measurements of nickel-based alloys sprayed at atmospheric pressure and low pressure [17]. This study showed that the thermal diffusivity of the low pressure sprayed material was approximately twice that of the atmospheric-sprayed deposit. Since the measurements were made in vacuum the results imply a considerably higher real contact area between lamellae in low pressure sprayed coatings. A complete analysis would, however, require an examination of the effect the higher oxide content of air sprayed coatings but this should be a small effect compared with thin planar pores.

## 5. Microstructure of lamellae

Each of the lamellae of which the coating is made up solidifies as a separate entity so that the internal structure of each may be considered individually. The good thermal contact between the impinging droplet and the substrate, together with its small thickness, results in an extremely rapid cooling rate, comparable with that achieved by various "rapid solidification" processes such as splat quenching and melt spinning.

Crystallization of a liquid occurs by nucleation and growth but, if the nucleation rate within the time interval between the melting point and the glass transition temperature is such that no nuclei are formed, the liquid hardens to form an amorphous structure or glass.

Metastable crystalline phases can also form directly from the liquid if it is rapidly cooled. The best known example is thermally sprayed alumina which consists mostly of metastable $\gamma$-$Al_2O_3$ rather than $\alpha$-$Al_2O_3$, the only stable equilibrium phase [18]. The reason for this is that the phase which nucleates from an undercooled melt will be the one with the lowest energy barrier to nucleation [19]. If the phase which nucleates happens to be a metastable phase, it may be retained to ambient temperature if cooled rapidly. Alumina coatings often contain some $\alpha$-$Al_2O_3$ but this can be explained by the presence of $\alpha$-$Al_2O_3$ nuclei introduced into the coating by incompletely fused particles [20] and the fraction of $\alpha$-$Al_2O_3$ is therefore a function of the spraying conditions.

The crystal size and morphology within lamellae depend upon the relative nucleation and growth rates of crystals from the melt. In general, high cooling rates give rise to high nucleation rates and the lamellae are polycrystalline with crystal sizes much smaller than conventionally processed materials. In most cases it seems that rapid nucleation occurs at the cooler surface of the flattened droplet at large undercooling and crystals grow rapidly in the opposite direction to heat flow forming a columnar grain structure. Rapid crystal growth leads to rapid evolution of the heat of fusion, an increase in temperature and suppression of further nucleation but, if the crystal growth rate is relatively low, further nucleation occurs as the liquid undercools and a very fine equiaxed grain structure is formed.

A model for crystal growth in thermally sprayed coatings, commonly referred to in the literature, is based on transmission electron microscope observations by Safai and Herman [21] which showed radial growth of columnar crystals in the outer, thin, regions of aluminium lamellae rather than growth perpendicular to the substrate. They suggested that the growth pattern occurred because the spreading liquid film was not in contact with the substrate and heat flowed radially in towards the centre of the disc as reported for studies of splat quenched metals. This model is, however, based on rather special conditions; the regions examined were within individual particles sprayed onto a polished substrate and were electron transparent, that is, less than 0.1 $\mu$m thick, at least an order of magnitude thinner than the lamellar thickness commonly observed in coatings. The radial growth pattern cannot therefore be regarded as typical of thermally sprayed coatings, as indeed the original authors pointed out.

The small grain size achieved in thermally sprayed coatings has important implications, for example, in $ZrO_2$-based materials. Pure $ZrO_2$ has a cubic crystal structure between its melting point (around 2700 °C) and 2400 °C and transforms on further cooling to a tetragonal structure (t-$ZrO_2$).

At about 900 °C (the martensite start temperature $M_s$) tetragonal $ZrO_2$ usually transforms by a martensitic mechanism to the monoclinic form (m-$ZrO_2$) with a significant volume change. The $M_s$ temperature is reduced for t-$ZrO_2$ containing various stabilizers, *e.g.* MgO, $Y_2O_3$ or $CeO_2$, in solid solution. The reduction in $M_s$ is a function of the type and percentage of stabilizer; however, $M_s$ for a given composition is also reduced as the grain size of the t-$ZrO_2$ solid solution is decreased. If $M_s$ is significantly lower than ambient temperature, t-$ZrO_2$ is retained as a metastable phase. If $M_s$ is just below ambient, the transformation to m-$ZrO_2$ may be triggered by an applied stress and this gives rise to an increase in fracture toughness.

Thermal barrier coatings containing 6 - 8 wt.% $Y_2O_3$ consist almost entirely of t-$ZrO_2$ which does not transform to monoclinic under stress. These coatings are superior to fully stabilized (c-$ZrO_2$) coatings in certain applications, apparently because t-$ZrO_2$, having a higher thermal expansion coefficient and thermal conductivity than c-$ZrO_2$, is more resistant to thermal shock on superalloy substrates.

There could be an advantage in certain applications if transformation toughened zirconia coatings could be prepared. That is, coatings containing a significant proportion of transformable t-$ZrO_2$. Since the grain size of the coating is effectively fixed by the solidification process, such a transformable t-$ZrO_2$ coating can only be achieved by controlling the type and proportion of stabilizer. Recent studies have shown that $ZrO_2$ coatings containing 10 - 15 wt.% $CeO_2$ contain a significant proportion of transformable t-$ZrO_2$ and that they do show evidence of enhanced fracture toughness [14, 15]. It is yet to be demonstrated, however, whether the higher toughness is associated with transformation toughening or arises from some other factor.

## 6. Conclusions

Coatings may be regarded as a composite of rapidly solidified lamellae between which there is only limited real contact. Many of the properties of coatings, particularly mechanical properties, tend to be dominated by the nature of the interlamellar contact. Rapid solidification of the particles within a coating generally results in an extremely fine crystal size, or in certain cases amorphous or partly amorphous structures. Metastable crystalline phases and unusual two-phase structures may be formed. Advantage cannot usually be taken of these structures from a strength point of view, because of the dominating effect of porosity, but in some applications the grain size and crystal structure may have an important influence on properties. For example, the different performance observed between zirconia thermal barrier coatings which are tetragonal phase and those which are cubic. Undoubtedly further advantage can be taken of the unusual structures of thermally-sprayed materials for applications in which intrinsic properties related to crystal structure and crystal size are important.

## References


1 A. Vardelle, M. Vardelle, R. McPherson and P. Fauchais, *Proc. 9th Int. Thermal Spraying Conf.*, Amsterdam, 1980, p. 155.
2 R. McPherson, *Thin Solid Films, 83* (1981) 297.
3 M. Vardelle and J. L. Besson, *Ceram. Int., 7* (1981) 48.
4 R. McPherson and P. Cheang, *12th Int. Thermal Spraying Conf.*, London, June 1989.
5 R. McPherson and B. V. Shafer, *Thin Solid Films, 97* (1982) 201.
6 Y. Arata, A. Ohmori and C. J. Li, *Thin Solid Films, 156* (1988) 315.
7 R. McPherson, *Thin Solid Films, 112* (1984) 723.
8 R. W. Rice, in R. K. MacCrone (ed.), *Treatise on Materials Science and Technology*, Vol. II, Academic Press, New York, 1977, pp. 199 - 381.
9 D. L. Ruckle, *Thin Solid Films, 73* (1980) 455.
10 P. Boch, D. Fargeot, G. Gault and F. P. Platon, *Rev. Int. Hautes Temp. Refract., 18* (1981) 85.
11 W. J. Lackey, D. P. Stinton, G. A. Cerny, L. L.Fehranbracher and A. C. Schaffhauser, *Proc. Int. Symp. Ceramic Components for Heat Engines*, Hakone, 1983, p. 17.
12 P. Cheang and R. McPherson, to be published.
13 P. Ostojic and R. McPherson, *J. Am. Ceram. Soc., 71* (1988) 891.
14 G. N. Heintze and R. McPherson, *Surface Coatings Tech., 34* (1988) 15.
15 G. N. Heintze and R. McPherson, *Surface Coatings Tech., 36* (1988) 125.
16 K. L. Wilkes and J. F. Ladergost, *NASA Contract Report, CR-12114Y* (1973).
17 L. Pawlowski, D. Lombard, F. Tourenne, F. Kassabji and P. Fauchais, *High Temp.–High Press., 16* (1985) 611.
18 N. N. Ault, *J. Am. Ceram. Soc., 40* (1957) 69.
19 R. McPherson, *J. Mater. Sci., 8* (1973) 964.
20 R. McPherson, *J. Mater. Sci., 15* (1980) 3141.
21 S. Safai and H. Herman, *Thin Solid Films, 47* (1971) 295.

*Surface and Coatings Technology, 39/40* (1989) 183 - 198 

# TEXTURE AND SOME PROPERTIES OF VAPOR-DEPOSITED DIAMOND FILMS*

Y. SATO and M. KAMO

*National Institute for Research in Inorganic Materials, 1-1 Namiki, Tsukuba-shi, Ibaraki-ken 305 (Japan)*

(Received April 12, 1989)

## Summary

Polycrystalline diamond films have been prepared by plasma-assisted chemical vapor deposition (CVD) from gaseous mixtures of various methane/hydrogen ratio. The textural and structural changes of the films have been studied by reactive plasma etching, electron microscopy, Raman spectroscopy and density measurement. Changes in elastic constant and thermal conductivity are briefly reviewed. Single crystals of size 100 $\mu$m with good crystalline quality were prepared, while the polycrystalline films grown under the same condition proved to be defective. Finally, it is suggested that the defects characteristic of CVD diamonds may be caused by double bonds of various forms.

## 1. Introduction

Diamond is expected to find many applications because of its unique combination of properties. Practical applications are, however, currently limited because the methods of synthesis has been limited to static and dynamic high pressure techniques. In this context, vapor deposition offers new ways of fabrication which can overcome various constraints of high pressure methods. Capability of coating various materials, freedom from size limitation and higher accessibility to impurity control essential for electronic and optical uses are examples of what may be brought about by CVD.

One of the fundamental problems of vapor-deposited "diamond films" is, however, the possibility of codeposition of the graphitic structure. This is often referred to as the phase purity problem. Even when graphitic components are reduced below detectable amounts, we frequently find double bonds in crystals and films analysed by Raman spectroscopy.

Because of the versatility of potential uses of diamond, the property or properties important for various applications may differ from one application to another. Each property, however, may have structure sensitivity different from the others. Thus, the level of defects or graphitic components allowed

---

*Recipient of the Bunshah Award for the best paper in its Symposium.

0257-8972/89/$3.50 

for specific applications can be variable. It is important to know how the structural features of the CVD deposits depend on growth conditions and how the properties depend on defects and structural variations.

Since earlier works, it has been recognized that the phase purity, defects and growth features critically depend on the substrate temperature and gas composition [1 - 7].

In this paper, changes in structure and properties of films prepared under different methane/hydrogen ratios are discussed. The texture and structures are studied by reactive plasma etching, electron microscopy, density measurement and Raman spectroscopy. The elastic constant and thermal conductivity of polycrystalline films are reviewed as examples of mechanical and thermal properties. A comparison of crystalline quality is made of individual crystals and films grown under the same, carefully controlled conditions. Finally, a brief discussion is given of the defects characteristic of CVD diamonds in terms of double bond formation.

## 2. Texture and structure of films

Changes in surface morphology of polycrystalline films deposited on silicon substrates are shown in Fig. 1. The films were deposited from methane–hydrogen gas mixtures with methane concentrations of 0.5 - 5 vol.%, using a microwave plasma reactor similar to the one used previously [8]. The conditions were: a total pressure of 40 Torr (5.2 kPa), a substrate temperature of 840 - 860 °C and a reaction time of 40 - 80 h. At 0.5%, the surface consists of (111) and (100) faces, with frequent twinning boundaries on the (111) faces. At 2 - 4%, the (100) face predominates and at 5%, the well-defined crystalline morphology disappears.

Surface morphology after oxidative plasma etching is shown on the right hand side column of Fig. 1. The etching was performed by the same apparatus used for the deposition. Air was introduced to a pressure of 10 - 20 Torr (1.3 - 2.6 kPa) and etching was performed under gas flow for 10 - 30 min with a microwave power of 100 - 200 W.

The films formed under lower methane concentrations proved to be more resistant to etching. For the 0.5% sample, it is seen that etch pits are most concentrated at the grain boundaries. Pits are also found along twin boundaries. At 2%, the (100) faces are full of etch pits, while the (100) faces of the 4% sample are smooth and there appears to be facetting. In the case of the 4% film, the intergrain regions are seen to be almost entirely etched away, leaving only crystalline grains. Etch pits are rarely found on the (100) faces, in contrast to the (100) faces of the 2% film. At 5%, much of the film is etched and a texture similar to a spider's web remains predominantly. This texture indicates that the size of the diamond crystallites is fairly small and that they are connected to each other.

Fractured cross-sections of the same samples are shown in Fig. 2. The thickness ranges from 30 to 40 $\mu$m. The fractured sections have been

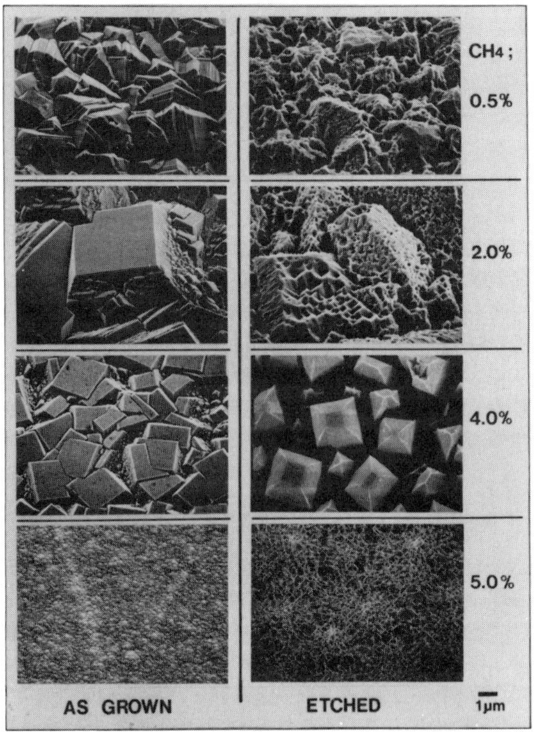


Fig. 1. Scanning electron micrographs of as-grown and etched surfaces of the films prepared under various methane concentrations.

subjected to plasma etching and the results are also shown in Fig. 2. The films formed at 0.5 - 4% are more or less columnar in texture. Interesting features have been found for the 4% sample, in which parallel single crystalline diamond needles or whiskers remain after etching. The axis of the needles was determined to be ⟨111⟩ by X-ray diffraction. From detailed observations, it was concluded that intergrain regions of the 4% film consist of deposits similar to that of the 5% film.

Identification of grain boundaries in fractured sections is often difficult, and etching is useful in revealing the boundaries, as may be seen from the enlarged micrographs shown in Fig. 3. Grain boundaries are apparent after etching. Etch pits are also observed on the fractured surface of crystalline grains, implying that defects are also present within the grains. This is consistent with the surface etch patterns shown in Fig. 1. It is noticed that the texture is essentially columnar and the diameter of the grains increases as they grow vertically. The finer texture near the substrate (bottom) is due to a high nucleation density of nearly $10^9\ cm^{-2}$. Similar textures have been observed for all the samples, except the 5% film which has much finer, isotropic grains.

For reference and confirmation purposes, etching experiments of

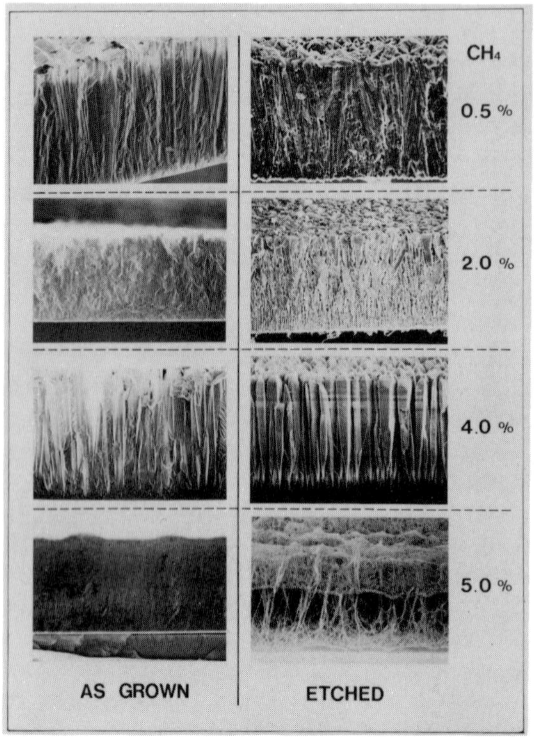


Fig. 2. Cross-sections of the corresponding films shown in Fig. 1. In each of the SEM images, the top is the surface and the bottom is the substrate side. The film thickness ranges from 30 to 40 $\mu$m.

soot-like carbons (highly disordered graphites), natural graphite and diamond have been made. It was observed that highly disordered graphite is etched away much faster than the well crystallized natural graphite, and diamond is by far the most resistant to etching among the three materials. Three low index faces, *i.e.* (111), (110) and (111) faces of synthetic diamond, and the (111) face of natural type Ia diamond have been subjected to etching. It was found that each face has its own etching pattern, but no etch pits have been observed which are similar to those observed with the (111) and (100) faces of the films deposited at 0.5% and 2% methane concentrations (Figs. 1 and 2).

Transmission electron micrographs of the films grown under 0.5% and 5% methane concentrations are shown in Figs. 4 and 5 respectively. For dark field observations, (111) diffraction was employed. Crystallites with well-defined morphology of size 0.2 - 0.3 $\mu$m were observed for the 0.5% sample. Since thin films of submicron thicknesses were used, the grain size is much smaller than that near the surface, where it is about 2 - 3 $\mu$m. In contrast, the grain size of the 5% film is far smaller, the maximum size being 100 nm and most of the grains appear to be 10 - 50 nm. The grain size of the highly disordered graphite was estimated to be no greater than 20 nm.

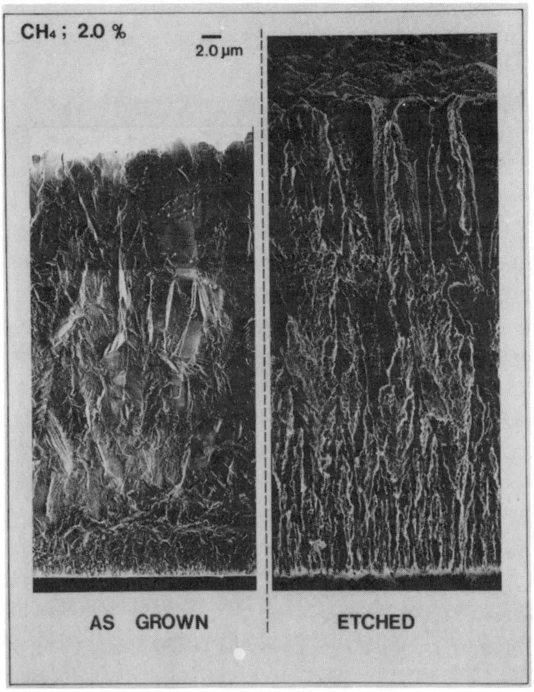


Fig. 3. A close up of the cross sections of the sample prepared under 2% methane shown in Fig. 2.

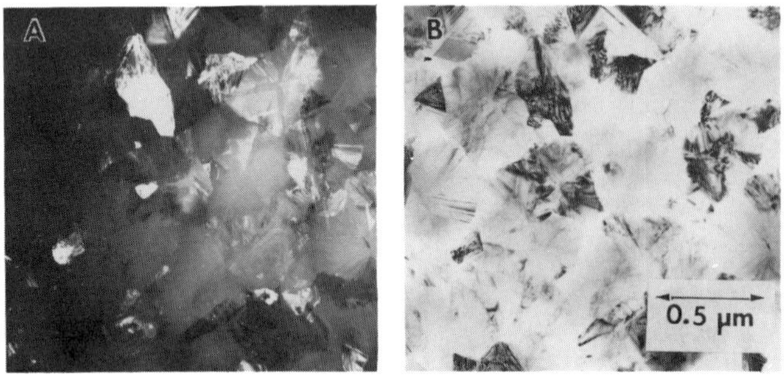


Fig. 4. Transmission electron micrographs of the film prepared under 0.5% methane. Dark field (A) and bright field (B) images.

Raman spectra of the samples are shown in Fig. 6. Drastic changes in the spectra are observed when going from 4% to 5%. The Raman line of diamond centered at about 1333 $cm^{-1}$ is fairly weak for the 5% film. Broad bands which appear simultaneously at around 1360 and 1600 $cm^{-1}$ are due to highly disordered graphite [9], whereas the broad band near 1500 $cm^{-1}$ may be ascribed to other types of double bond structures, presumably linear conjugated double bonds. (It should be noted that a similar band at 1550 $cm^{-1}$

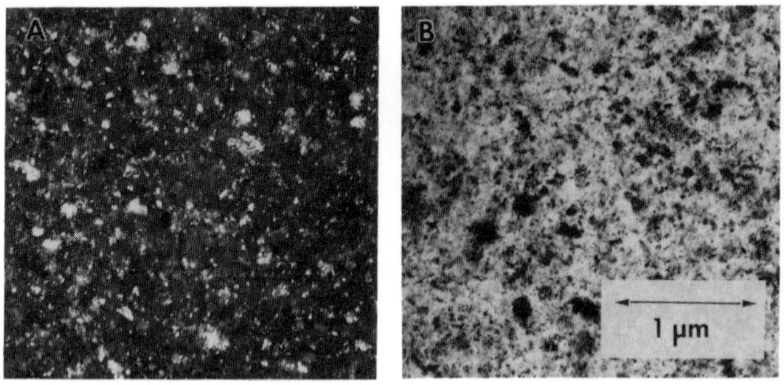


Fig. 5. TEM images of the film prepared under 5% methane. Dark field (A) and bright field (B) images.

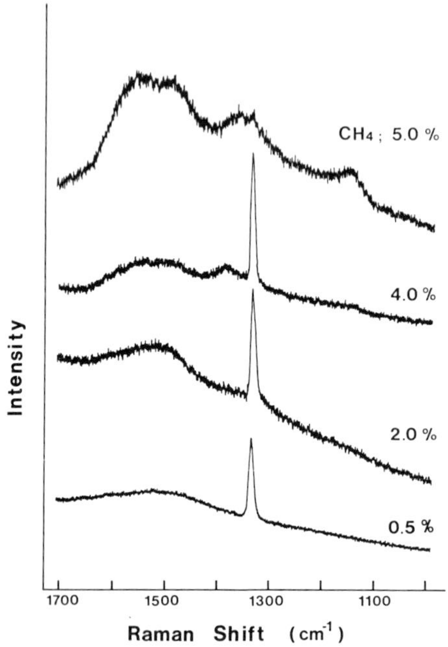


Fig. 6. Raman spectra of the as-grown films shown in Fig. 1.

has been observed with a very thin evaporated film of carbon by Solin and Kobliska [10].) The band near 1500 $cm^{-1}$ is present in all of the spectra shown in Fig. 6, while the signal due to highly disordered graphite is only observable for the 2%, 4% and 5% films.

The density of the films shown in Fig. 1 was measured by the sink–float method and the results are given in Fig. 7. The density of the 0.5% film is in agreement with the density of natural diamond (3.515 g $cm^{-3}$) within 0.3%. The density gradually decreases with increasing methane concentration and rapidly decreases at 5% methane concentration down to 94% of the value of natural diamond.

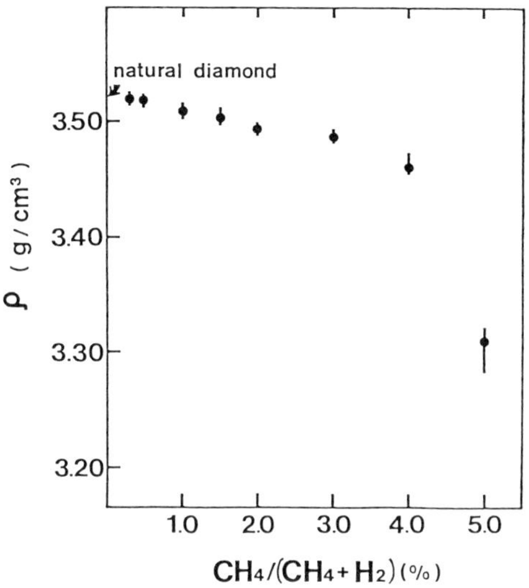


Fig. 7. Density of the films prepared under different methane concentrations. The samples correspond to those in Fig. 1.

The fraction of diamond in this film may be estimated, assuming that the low density is entirely due to graphitic components, the presence of which is evidenced by the Raman spectrum shown in Fig. 6. If the theoretical density of crystalline graphite, 2.25 g $cm^{-3}$, were employed, the weight fraction of diamond in the film is calculated to be 88%, from the observed density of 3.30 g $cm^{-3}$. As seen from the Raman spectrum, the graphitic component in the film is highly disordered, which should be more like carbon blacks or soot rather than well crystallized natural graphite. Therefore, values of 1.6 - 1.8 g $cm^{-3}$ seem to be more practical rather than 2.25 g $cm^{-3}$. If 1.8 g $cm^{-3}$ is employed, the diamond content is calculated to be 93%.

From the above consideration, the diamond content in the 5% film is estimated to be 88 - 93%, with 88% being regarded as the lower limit. It should be pointed out that a well-defined X-ray diffraction pattern of diamond is readily observed with a 5% film of size 10 mm × 10 mm × 0.01 mm, which supports the high content of diamond estimated from the density.

## 3. Elastic constant and thermal conductivity

Properties studied as functions of structural variations are still limited. Here, changes in elastic constant [11] and thermal conductivity as functions of methane concentration are briefly described.

The propagation velocity of surface acoustic waves (SAW) for the samples shown in Fig. 1 was measured by an acoustic microscope equipped with special attachments for this purpose. If we postulate a relation similar to bulk propagation of sound for SAW as

$$V_R = \left(\frac{E}{\rho}\right)^{1/2} \tag{1}$$

where $V_R$ is the propagation velocity of SAW, $\rho$ is the density of film and $E$ is the effective elastic constant for SAW propagation then, using the measured values of velocity and density, the effective elastic constant can be calculated by the derived relation

$$E = V_R^2 \rho \tag{2}$$

and the calculated values are shown in Fig. 8.

The values for natural type Ia and type IIa diamond agree within experimental errors. The calculated elastic constant for 0.5% film is about 90% of that of natural diamond. The value decreases gradually with methane concentration.

Thermal conductivity of polycrystalline films prepared from methane–hydrogen gas mixtures by microwave plasma CVD has been determined by Ono *et al.* [12]. Thermal conductivity along the surface of polycrystalline strips of size 20 mm × 5 mm × 0.01 mm deposited under different methane concentrations is shown in Fig. 9. The absolute values of the methane concentration may not be compared with our results and concentration dependence may be discussed in terms of a relative scale. The Raman spectra of the samples shown in Fig. 10 can also serve as a reference scale for comparison.

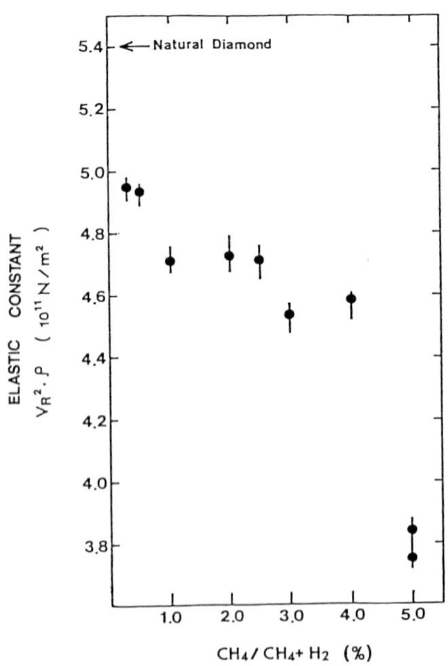


Fig. 8. Elastic constant of the films shown in Fig. 1 and additional samples.

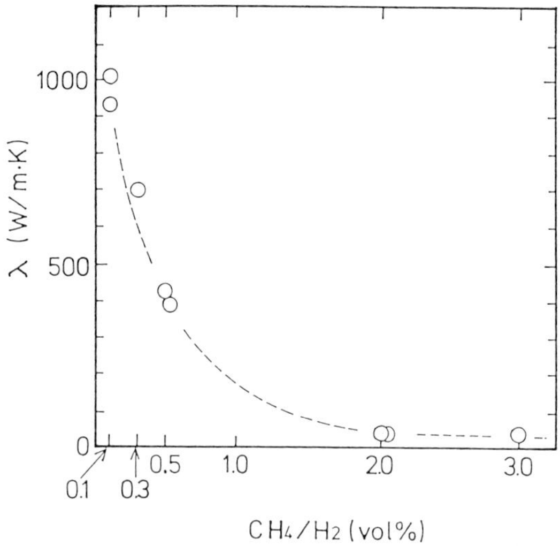


Fig. 9. Thermal conductivity of the films prepared under different methane concentrations and measured at 100 °C [12].

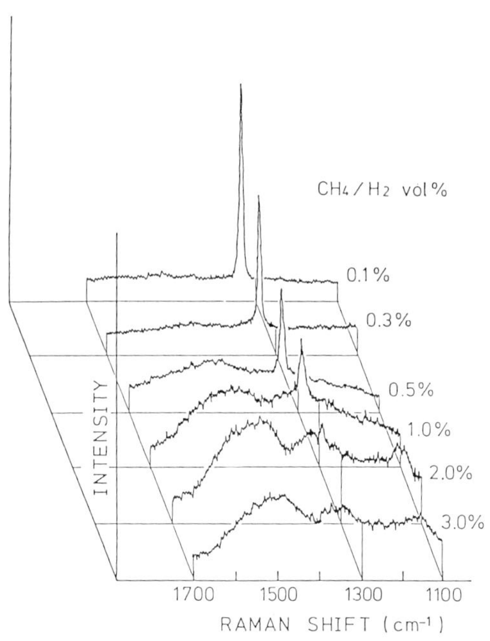


Fig. 10. Raman spectra of the samples used for the thermal conductivity measurement shown in Fig. 9 [12].

It is seen that the thermal conductivity critically depends on the methane concentration. In going from 0.1% to 1%, an increase of 10 times in concentration, the value drops almost by an order of magnitude. Even at 0.3%, where the band near 1500 $cm^{-1}$ appears to be very weak, the thermal conductivity decreases to 70% of the maximum value. In contrast, the dependence of the elastic constant on the methane concentration is far less critical: in going from 0.5% to 5%, the value remains at 78% of the maximum. A higher structure sensitivity of thermal conductivity is also demonstrated by the fact that the value for type Ia is about 1000 W $m^{-1}$ K, whereas that for type IIa is as high as 2500 W $m^{-1}$ K [13], in sharp contrast to the elastic constant, for which practically the same values have been observed with the two types of natural diamond.

It is inferred that the thermal conductivity perpendicular to the film surface for the 2 - 4% film may be higher than that along the surface, because of the columnar texture.

It is seen that a better control of the deposition conditions is required for applications utilizing high thermal conductivity than for those utilizing high elastic constant.

Systematic studies of various properties as functions of growth conditions would be valuable for applications as well as for a better understanding of the structural features of CVD diamond films.

## 4. Single crystals and films

In addition to the gas composition and substrate temperature, evidence has been found that the formation of polycrystalline films gives rise to a marked effect on the crystalline quality.

Comparison of Raman spectrum and luminescence background has been made with individual crystals and polycrystalline films grown under the same conditions on different silicon substrates, one untreated and the other surface-treated to increase nucleation density to facilitate film formation.

An optical micrograph of individual crystals grown on a silicon substrate under 0.5% methane concentration is shown in Fig. 11. Other deposition conditions were the same as those for the samples shown in Fig. 1, except that a small amount of water vapor was added. Single crystals with well-defined morphology, twinned crystals and polycrystals up to 100 $\mu$m in size have been deposited.

Raman spectra similar to the one shown in Fig. 12 was observed with many of the crystals with well-defined crystal habit. The spectrum was recorded using the 514.5 nm line of an argon ion laser with a non-microprobe, conventional set up of illumination and collection optics. The lower spectrum shows that the measured width of the Raman line is around 2.3 $cm^{-1}$ defined as full width at half maximum (FWHM). The value may be compared with the values of around 2.0 $cm^{-1}$ for high pressure synthetic single crystals grown from nickel solution, and 2.0 - 3.0 $cm^{-1}$ observed with the natural type Ia and

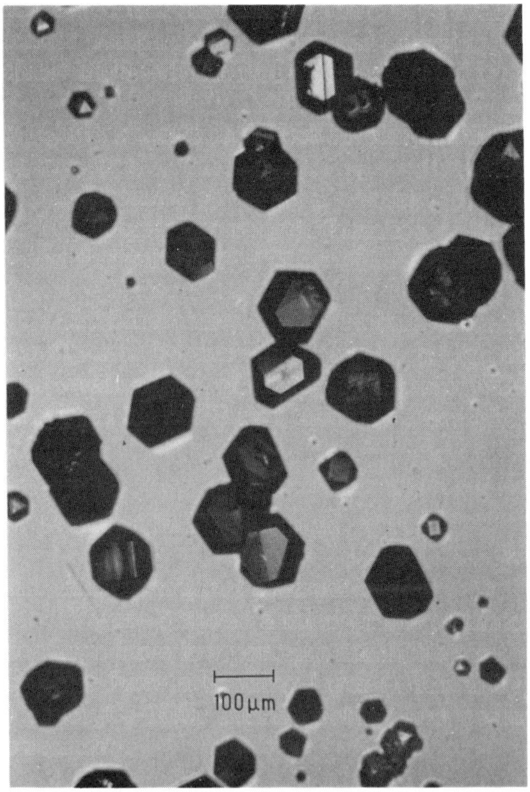


Fig. 11. Optical micrograph (differential interferometry) of individual diamond crystals grown on silicon wafer.

type IIa single crystals that we have used. The peak position was observed to agree within $\pm 0.15\ cm^{-1}$ with those of the high pressure and natural single crystals. The corrected value may tentatively be given as $1332.5 \pm 0.15\ cm^{-1}$, after Solin and Ramdas [14]. The upper spectrum shows that the luminescence background is much lower compared with the Raman signal intensity, and also shows that the broad Raman band centered at 1450 - 1550 $cm^{-1}$ due to double bonds is undetectably weak.

The etch patterns of the (111) and (100) faces of the single crystals or twins have been found to be similar to those of high pressure synthetic diamond and no etch pits similar to those observed with the 0.5% and 2% films (Fig. 1) have been found.

Polycrystalline continuous films of around 25 $\mu m$ thickness were prepared under the same conditions as those for individual crystals, with a shorter reaction time. The Raman spectrum of a film is shown in Fig. 13. The line width is broadened to 7.5 $cm^{-1}$, more than three times that of single crystals, and the peak position is shifted to higher the Raman shift side by around 1.5 $cm^{-1}$, which may be due to strain caused by compressive stress

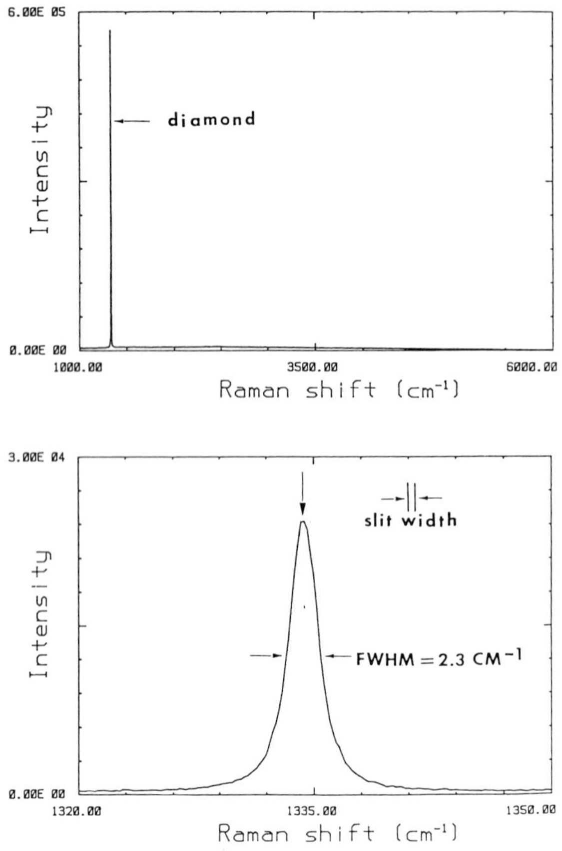


Fig. 12. Raman spectrum of the individual crystal with a diameter of around 100 $\mu$m shown in Fig. 11. The upper spectrum shows luminescence background and the lower spectrum shows line width. The peak position of natural diamond is indicated by the vertical arrow. No correction of the wavenumber has been made.

(*e.g.* see ref. 15). It is also noted that the luminescence is much higher in intensity and that the 1450 - 1550 $cm^{-1}$ band is present, though weak.

Observation of the polished film under a polarizing microscope showed that local birefringence units of a size comparable to the crystallite size are present, which indicates that local strain fluctuation is present in the film.

Broadness and shift of the Raman line of diamond have been pointed out in an earlier report [16], but the anomaly has also been frequently observed with individual crystals. This indicated that the anomaly came not only from film formation but also from structural defects caused by poor control of deposition conditions in earlier works. The effect due to film formation has been first demonstrated unambiguously in the present work.

From the above observations, it may be concluded that optically active defects, double bonds and local strain fluctuation are produced by the formation of the polycrystalline film. The observation suggests that it may

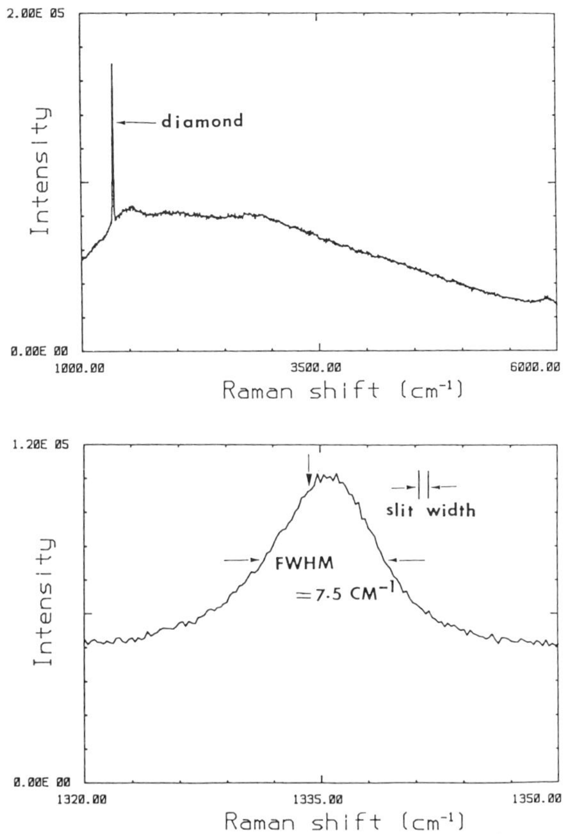


Fig. 13. Raman spectra of the polycrystalline film prepared under the same conditions as the sample shown in Fig. 11. The vertical arrow indicates the peak position of natural diamond. No correction of the wavenumber has been made.

not be practical to reduce the structural imperfections to an unappreciable amount so far as polycrystalline films are concerned. It also supports the tacit understanding that measurements with single crystals are essential for the correct evaluation of the properties of CVD diamond. In the same context, it is noted that examination of individual crystals and autoepitaxial films is necessary for the correct diagnosis of growth conditions.

## 5. Discussion

The film prepared at the lowest concentration of 0.5% contains double bonds but no appreciable amount of graphitic structure. The films prepared at 2% and 4% contain both double bonds and a graphitic phase, the latter being present in the intergrain regions for the 4% film and most probably the same for the 2% film. At 5%, the graphitic component further increases. The growth of diamond is also severely interrupted, resulting in fairly small grains mostly below 50 nm in size, and the grain size of the graphitic components is below 20 nm. For the 5% film, the fraction of diamond has been estimated to be approximately 90%, while the Raman spectrum consists predominantly of the bands due to highly disordered graphite and the signal due to diamond is extremely weak (Fig. 6).

The apparent contradiction between the Raman spectral features and the fraction of diamond estimated from density measurements arises from the fact that diamond and graphite have entirely different optical properties and Raman scattering cross sections: transparent material *vs.* highly absorbing material and normal scattering *vs.* resonance enhanced scattering.

An attempt to correlate Raman spectral features to the fractions of diamond and graphite has been reported by Shroder *et al.* [17]. The present TEM observation of the 5% film supports the fundamental concept of modeling in which a random distribution of spherical grains is assumed. Employment of a much smaller grain size and adjustment of the relative scattering factor may lead to a reasonable correlation. For the films prepared at lower methane concentrations, however, a different model may be necessary because of the oriented texture.

The etch patterns of the 0.5% and 2% films suggest that crystalline grains themselves contain some kind of defects which are found neither in the high pressure synthetic diamonds nor in the natural diamonds used in the present work. Few pits are found with the diamond whiskers formed at 4%, indicating that the whiskers are substantially free from the defects. The Raman spectrum of the whiskers was found to have a very weak band at 1450 - 1550 $cm^{-1}$ and a narrow diamond Raman line comparable to that of high pressure synthetic diamond [18].

It appears that the etch pits are related to double bonds detected by Raman spectroscopy. Further, it seems likely that the double bonds are related to stacking faults which are found in the crystalline grains shown in Fig. 4.

From the observations in the present work and previous works, it seems plausible to postulate that double bonds are always formed in finite probabilities together with single bonds in CVD, and it appears that the presence of double bonds is the origin not only of the graphite phase but also of the defects characteristic of CVD diamonds.

The situation may be described as the ratio $R$ of the numbers of the two types of orbital hybridization, $sp^2$ and $sp^3$, defined as

$$R = \frac{N(sp^2)}{N(sp^3) + N(sp^2)} \tag{3}$$

varies with deposition conditions. (In this discussion, sp hybridization is neglected.)

The value of $R$ is unity for ideal graphite and zero for ideal diamond. For CVD diamond, it is postulated that $R$ takes a finite value, depending on the growth conditions. When $R$ exceeds a certain critical value, phase separation of graphite takes place. Double bonds below the critical limit may be present predominantly at grain boundaries and probably at twin boundaries. A lower concentration of double bonds may be accommodated as defects in the diamond lattice. Stacking faults often found with CVD diamonds may be related to the residual double bonds within the crystal grains.

Apart from various common factors affecting the structural features of films and crystals of a large variety of materials, the control of chemical bonding is an additional factor which should constantly be taken into consideration in the growth of diamond by CVD.

## 6. Summary

It has been shown that the structural features of vapor-deposited diamond films depend critically on gas composition. Oxidative plasma etching, when combined with other methods, proved to be a useful and readily accessible approach to the study of structural features.

The elastic constant of the films was shown to be less structure sensitive compared with the reported behaviour of thermal conductivity.

Single crystals of up to 100 $\mu$m in size with good crystalline quality have been synthesized. In contrast, the polycrystalline films grown under the same conditions have appreciable amounts of double bond and strain. It has been pointed out that examination of individual crystals is essential for the correct evaluation of the property of CVD diamond as well as for the correct diagnosis of the growth conditions.

Finally, it has been briefly discussed that the formation of double bonds may be the cause not only of the graphitic phase but also of defects characteristic of CVD diamonds.

## References


1 B. V. Spitsyn, L. L. Bouilov and B. V. Deryaguin, *J. Cryst. Growth*, *52* (1981) 219.
2 S. Matsumoto, Y. Sato, M. Tsutsumi and N. Setaka, *J. Mater. Sci.*, *17* (1982) 3106.
3 Y. Sato, M. Kamo and N. Setaka, *Proc. 8th Int. Symp. on Plasma Chemistry, Tokyo*, 1987, 2446.
4 H. Tanaka, K. Sato, Y. Saito and H. Miyadera, *Proc. 8th Int. Symp. on Plasma Chemistry, Tokyo*, 1987, p. 2463.
5 R. DeVries, *Ann. Rev. Mat. Sci.*, *17* (1987) 161.
6 A. Badzian and R. DeVries, *Mat. Res. Bull.*, *23* (1988) 385.
7 K. Kobashi, K. Nishimura, Y. Kawate and T. Horiuchi, *Phys. Rev. B*, *38* (1988) 4067.
8 M. Kamo, Y. Sato, S. Matsumoto and N. Setaka, *J. Cryst. Growth*, *62* (1983) 642.
9 F. Tuinstra and J. L. Koenig, *J. Chem. Phys.*, *53* (1970) 1126.
10 S. A. Solin and R. J. Kobliska, *Proc. 5th Int. Conf. on Amorphous and Liquid Semiconductors*, 1974, p. 1251.
11 Y. Sato, C. Hata, I. Ishikawa and M. Kamo, to be published.
12 A. Ono, T. Baba, H. Funamoto and A. Nishikawa, *Jpn. J. Appl. Phys.*, *25* (1986) L808.
13 R. Berman, in R. Berman (ed.), *Physical Properties of Diamond*, Clarendon Press, Oxford, 1965, p. 384.

14 S. A. Solin and A. K. Ramdas, *Phys. Rev. B, 1* (1970) 1687.
15 C. D. Clark, E. W. J. Mitchell and B. J. Parsons, in J. E. Field (ed.), *The Properties of Diamond*, Academic Press, London, 1979, p. 66.
16 S. Matsumoto, Y. Sato, M. Kamo and N. Setaka, *Jpn. J. Appl. Phys., 21* (1982) L183.
17 R. E. Shroder, R. J. Nemanich and J. T. Glass, *Proc. Conf. on Diamond Optics*, San Diego, 1988, p. 79.
18 Y. Sato, C. Hata and M. Kamo, *Proc. 1st Int. Conf. on New Diamond Science and Technology, (Extended Abstracts) Tokyo*, 1988, p. 50.

# ELECTRON MICROSCOPY OF THE GROWTH FEATURES AND CRYSTAL STRUCTURES OF FILAMENT ASSISTED CVD DIAMOND FILMS

R. E. CLAUSING, L. HEATHERLY and K. L. MORE

*Metals and Ceramics Division, Oak Ridge National Laboratory, Oak Ridge, TN 37831-6093 (U.S.A.)*

G. M. BEGUN

*Chemistry Division, Oak Ridge National Laboratory, Oak Ridge, TN 37831-6093 (U.S.A.)*

(Received March 22, 1989)

## Summary

Scanning and transmission electron microscopy (SEM and TEM) have been used to examine the topographic and internal features of diamond films grown by the hot filament assisted chemical vapor deposition method. Films were grown under conditions chosen to provide three distinctly different diamond microstructures. One of these film structures is virtually free of stacking faults and twins within the grains. This paper relates internal and external growth features to each other, to Raman spectra and to the growth conditions. The growth features are discussed in terms of growth mechanisms.

## 1. Introduction

There is much interest and activity in the growth of diamond films by low pressure chemical vapor deposition (CVD) methods. There are many opportunities for the development of unique materials and products which make use of the superlative properties of diamond in the form of extended surfaces, coatings and/or thin layers. Many of the outstanding properties of diamond films are, however, strongly influenced by the perfection of the diamond crystal lattice, impurity concentration, and the microstructure of the films. These factors are in turn all strong functions of the film growth conditions. We have begun to study the influence of growth conditions on the structure and properties of diamond films grown by the hot filament assisted CVD method. This method seems to provide simple, easily characterized and reproducible growth environments.

There is a considerable body of information in the literature on the various methods of growing CVD diamond [1 - 3] and some systematic studies of the effects of changing growth parameters on diamond film morphology

0257-8972/89 Elsevier Sequoia/Printed in The Netherlands

[1, 3 - 5]. There is, however, relatively little work thus far published relating the crystal defects to film morphology and growth conditions [5 - 10]. It is the purpose of this paper to report some of our observations on three types of films grown under controlled and closely related conditions, but each with its own distinctively different morphology. This paper reports the electron microscopy and Raman spectroscopy results which are the first part of an effort to develop the relationship of growth conditions, morphology and crystal perfection to the physical properties of diamond films.

## 2. Experimental procedure

### *2.1. Film growth*

The diamond films investigated in this work were deposited by the hot filament assisted CVD technique. A schematic drawing of the deposition chamber is shown in Fig. 1. The gas used in this study contained 1.0% methane in hydrogen and was supplied in pre-mixed high pressure cylinders. The gas flowed through a pressure regulator and a mass flow controller and finally into the deposition chamber which was a quartz tube of diameter 25 mm. Inside the chamber, the gas was directed to the hot filament through a quartz tube of diameter 6 mm. The products of the exposure of the gases to the hot filament then reached the substrate through a combination of flow

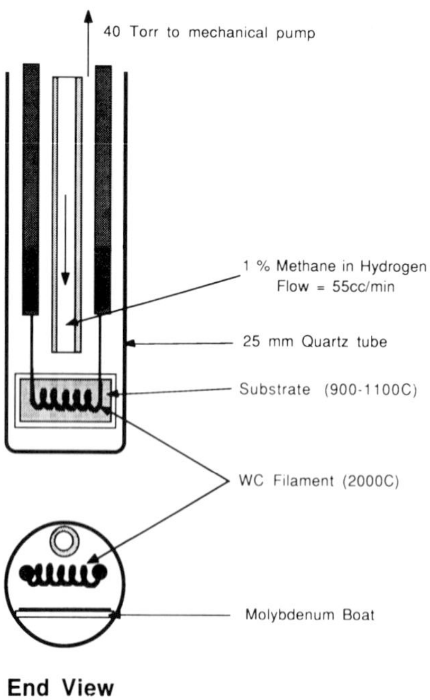


Fig. 1. Diagram of the growth chamber used for growing the diamond films used in this study.

and diffusion processes. The gas was then pumped out through the larger quartz tube with a direct drive mechanical pump. The flow rate for the gas mixture during these studies was 55 std. $cm^3 min^{-1}$, and the chamber operating pressure was 5.3 kPa (40 Torr).

The filament was fabricated by wrapping several turns of tungsten wire (diameter 0.5 mm) around a mandrel (diameter 3 mm). A stable current-controlled direct current power supply was used to heat the filament to the desired temperature. The substrate itself was heated by radiant heat from the filament to the desired temperature (between 900 and 1100 °C for these studies). The substrate was placed below the filament in a molybdenum boat which acted as a radiant heat baffle (reflector). The temperature of the substrate was measured optically by looking at the back of the substrate through two small holes in the boat. A disappearing filament optical pyrometer was used for the measurement. The hot filament temperature was measured through the quartz tube with the same pyrometer. No corrections were made to the pyrometer readings. At the beginning of a new film deposition there was a conditioning period of approximately 1 - 2 h during which the filament carburized and the film began to grow on the substrate. During this time the filament and substrate temperatures had to be adjusted, but then the temperatures of both the filament and substrate stabilized and further adjustments were rarely required over a period of several days. The film deposition rates were from 1 to 2 $\mu m\ h^{-1}$.

The substrates for these films were (100) silicon. The surface was polished just prior to use with 0.5 $\mu$m diamond abrasive and cleaned in ethanol. The purpose of the polishing was to ensure the rapid initiation of a uniform fine grain deposit instead of isolated single crystals.

### *2.2. Electron microscopy*

Specimens for transmission electron microscope (TEM) analysis were prepared in plan view. They were made by mechanical thinning from only the silicon side to a thickness of about 75 $\mu$m, dimpling to a center thickness of about 20 $\mu$m and finally $Ar^+$ ion milling, again only from the silicon side, at 6 kV and 12 ° incidence. The diamond film was protected from redeposition of sputtered contaminants during ion milling by the evaporation of a thin layer of NaCl across the surface. The protective NaCl layer was easily removed by immersion in water at the end of the milling procedure [10]. Transmission electron microscopy was performed in a JEOL 2000FX operated at 200 kV.

The as-grown surfaces of the diamond films, as well as fractured cross-sections, were examined in a Hitachi S800 high resolution scanning electron microscope (SEM) equipped with a field emission gun. The SEM was operated at 5 kV.

### *2.3. Raman spectroscopy*

The Raman spectra were obtained with an HG-2S Ramanor spectrophotometer (Instruments S.A.). This instrument employs a double monochromator with curved holographic gratings, cooled photomultiplier tube and pulse counting electronics. A Nicolet 1170 multichannel analyzer was used to collect the data and allow multiple scans to be accumulated. The spectrometer was coupled to a Nachet NS 400 microscope, and the exciting laser beam was focused on the sample through the microscope objective. The objectives used in this study were 20 × and 40 × . The Raman light was collected through a beam splitting prism at 180 ° to the incident laser beam and focused into the monochromator. Excitation was done with the 514.5 nm line of an argon ion laser with approximately 5 - 10 mW focused on the sample in a spot of 1 - 2 $\mu$m diameter.

## 3. Results

The most common crystal morphology reported in the literature seems to be one which develops with a preponderance of three-fold {111} crystal faces and a few {100} faces such as that shown in Fig. 2(a). The triangular

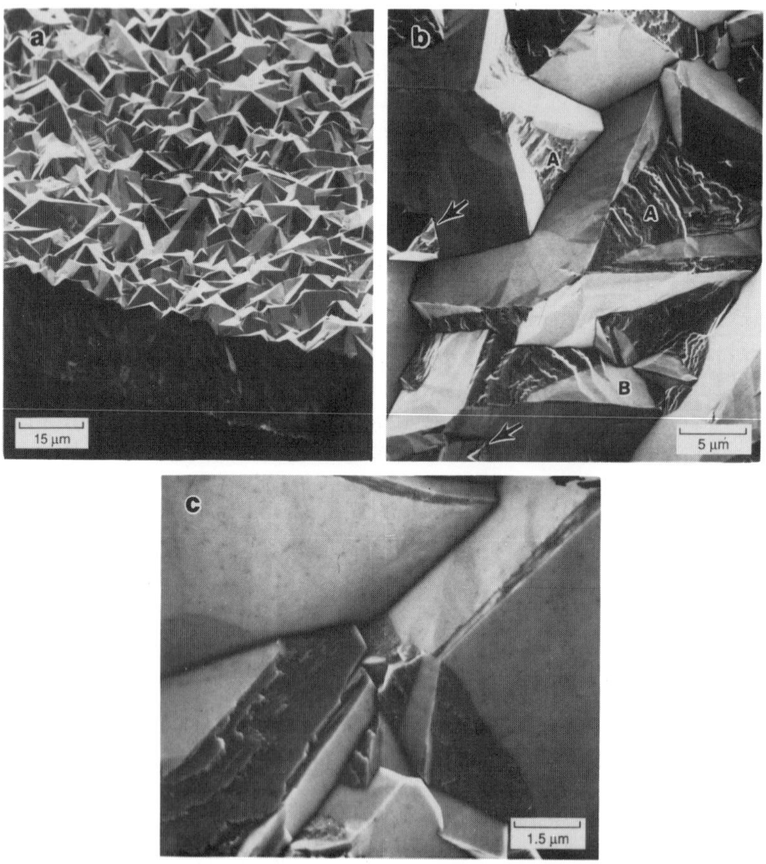


Fig. 2. Scanning electron micrographs of a polycrystalline diamond film grown at 1050 °C showing (a) pyramidal features typical of {111} and {100} faces, (b) the rough growth areas in the saddle regions between relatively smooth {100} faces at higher magnification and (c) growth layers on a {111} surface.

facets on the surface of this film are typical of those obtained in our apparatus, with the use of the conditions described, at substrate temperatures in the range of 1000 - 1100 °C. Close examination of these films reveals many twins. The surfaces of the facets are usually quite flat except where new crystallites nucleate. This nucleation is sometimes in the form of tiny three- or four-sided pyramids, of which the angles and orientations often suggest that they are twins on an underlying {111} face. Figure 2(b) is a higher magnification image of the same film shown in Fig. 2(a). Several twinned nuclei, indicated by arrows, are apparent. Another characteristic of these films is the topographically rough area in the "saddle" where two adjacent facets (labeled "A") meet. One may also see somewhat larger rough areas between neighboring grains. The "saddle" regions often occur where facets on adjacent twins have a re-entrant junction. The material with irregular surface features in Fig. 2(b) appears to be a rapidly growing deposit in such a "saddle" region. These deposits often grow with an irregular layered structure which at very high magnification shows the triangular features of exposed {111} planes. Where this irregular layered structure has advanced substantially, one often observes that deposited material appears to have been transferred to the adjacent flat facets, as can be seen by the extension of the contours of the rough growth over to the adjacent smooth face as in the lower center region (labeled "B") of Fig. 2(b). This probably implies material is being transferred by surface diffusion. It is as though the rough layered deposit is able to capture the reactive species from the gas phase, deposit diamond material and then transfer it to the adjacent relatively smooth faceted surfaces. Figure 2(c) shows a higher magnification image of the edges of several growing {111} layers. The identification of these layers as {111} layers is by crystallographic features revealed at even higher magnification. Curve "a" of Fig. 3 is the Raman spectrum obtained from the film of Fig. 2(a). It shows the relatively strong 1332 $cm^{-1}$ Raman band associated with crystalline diamond. The 1332 $cm^{-1}$ diamond band is 8 $cm^{-1}$ wide at half height. There is also a weak Raman band at 1516 $cm^{-1}$ which is associated with $sp^2$ (graphitic) bonding, which we will discuss below. Curves "b" and "c" will be used to describe other films below. Curve "d" of Fig. 3, the Raman spectrum from a single large diamond crystallite in a film grown at 1050 °C, shows a strong 1332 $cm^{-1}$ crystalline diamond band and a weak 1516 $cm^{-1}$ band. The 1332 $cm^{-1}$ diamond band is 6.5 $cm^{-1}$ wide at half height. The relatively narrow line indicates a well-formed crystal structure.

A TEM image from the same film of Fig. 2 is shown in Fig. 4. Many of the {111} facets evident in the micrographs were also observed via TEM analysis. The grains were often twinned, as indicated by the arrows in Fig. 4. The corresponding selected area diffraction pattern (SADP) recorded with the selected area aperture centered across a twin boundary shows extra reflections due to twinning (A). The boundary (A) has several parallel closely spaced twin planes in it. Twinning occurs on several {111} planes in the micrograph. A high density of twins is evident in some of these boundaries. The larger twin grains are relatively defect free, compared to some other

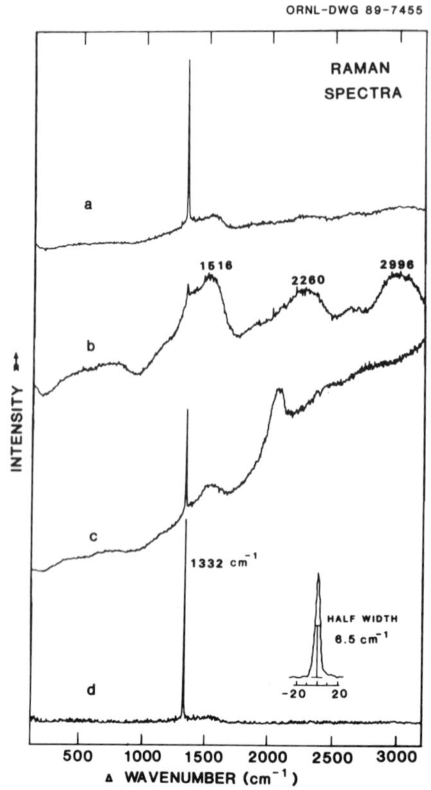


Fig. 3. Raman spectra from polycrystalline diamond films grown at (a) 1050 °C, (b) 920 °C and (c) 950 °C. (d) Raman spectrum from a single large diamond crystallite in a film grown at 1050 °C where the 1332 cm$^{-1}$ diamond band is 6.5 cm$^{-1}$ wide at half height.

areas of the film, the only observed defects being dislocations. Highly defective regions, labeled "B" in Fig. 4, were present between the twins. These regions contained a high concentration of twins or stacking faults as indicated by the streaking normal to the {111} planes in the corresponding SADP (labeled "C"). This diffraction pattern is typical of most of the defective regions. Films which have been grown for long periods of time are thicker and tend to have well developed coarser features and a higher degree of crystalline perfection.

Figure 5 is an SEM image of a microcrystalline diamond film grown at 920 °C which shows the small rounded features typical of films grown under conditions of high hydrocarbon supersaturation. Curve "b" of Fig. 3 shows the Raman spectrum of this film. The 1332 cm$^{-1}$ Raman band is much weaker and broader than that of films grown at higher temperatures. This film also shows a much more prominent broad feature at about 1516 cm$^{-1}$ associated with sp$^2$ (graphitic) bonding and/or less perfect diamond structure. As the grain size is decreased, this latter feature becomes more prominent. Broad Raman peaks at 2260 and 2996 cm$^{-1}$ are probably associated with C–H and

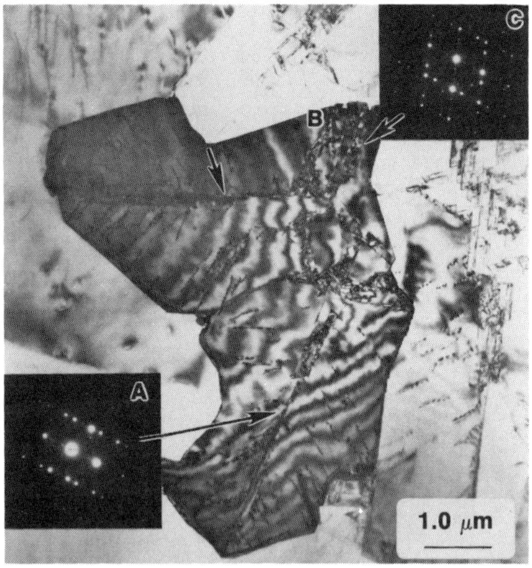


Fig. 4. Transmission electron micrograph of a polycrystalline diamond film grown at 1050 °C.

Fig. 5. Scanning electron micrograph of a microcrystalline diamond film grown at 920 °C showing the small rounded features typical of films grown under conditions of high hydrocarbon supersaturation.

C–$H_2$ groups in the material. When the film is not very thick (*i.e.* the thickness is less than the distance between the initial nucleation sites on the substrate) it may have the characteristic rounded features often referred to as a "cauliflower" microstructure. However, as the film is grown to greater thicknesses, it may, depending on suitable growth conditions, become relatively uniform and flat while still retaining the same microcrystalline character. Figure 6 is a TEM image of a microcrystalline film showing the very small crystallite size and the high density of crystal defects present (mostly twins). The inset selected area diffraction pattern (Fig. 6) was indexed as diamond. The defects, together with the small grain size, no doubt contribute to the broad features seen in the Raman spectrum and will affect the mechanical and physical properties of the films. No evidence of graphite was seen in electron energy loss spectroscopy (EELS) data from films examined, but the experimental sensitivity of EELS for non-diamond components has not been established.

Figure 7(a) is a micrograph of a polycrystalline film grown at 950 °C showing square features typical of {100} faces. For the conditions of this experiment there is a narrow temperature range between those temperatures where microcrystalline films are deposited and where films with pyramidal features dominate. The {100} facets tend to be oriented within about 12 ° of being parallel to the substrate and continue to grow as columnar structures as film deposition continues. The largest of these columnar structures may grow somewhat higher than their surroundings and when they do they have rough nearly vertical {110} sides, as shown by the selected area electron diffraction pattern inset in Fig. 8. Figure 7(b) shows that small pyramidal features with flat facets on their tops often grow epitactically from the sides of the larger columnar structures. Some of these small pyramidal structures, such as those shown in Fig. 7(b), are asymmetrical in that the slopes of the sides of the

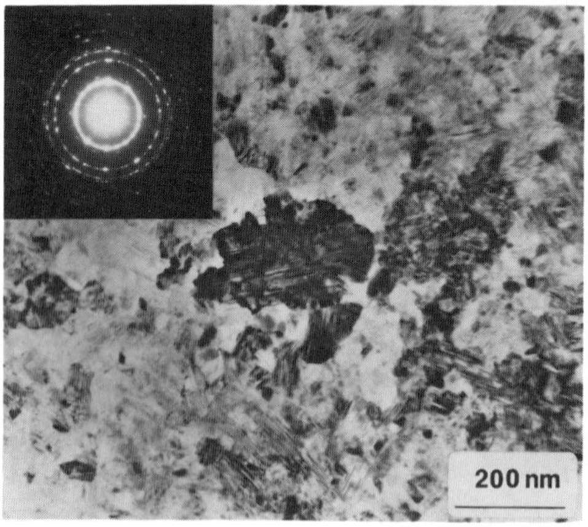


Fig. 6. Transmission electron micrograph of a polycrystalline diamond film grown at 920 °C. Note the very high density of microtwins and other defects.

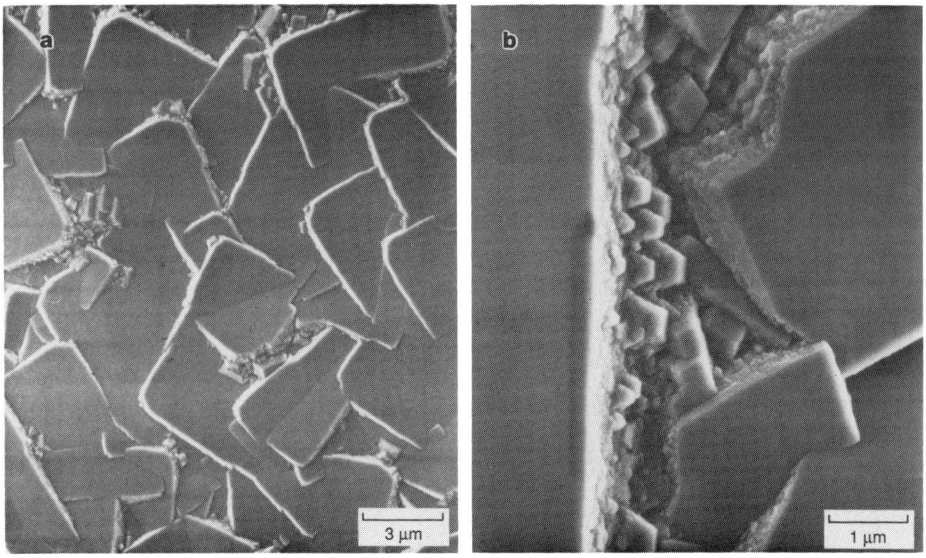


Fig. 7. Scanning electron micrograph of a polycrystalline diamond film grown at 950 °C showing (a) square features typical of {100} faces and (b) the small pyramidal features with flat tops which frequently grow from the sides of the larger columnar {100} structures.

pyramids are different depending on the apparent accessibility of these sides to the gas phase reactants which feed the deposit. Not all of the small structures between the larger grains have well developed surface features. Curve "c" of Fig. 3 is the Raman spectrum of the film grown at 950 °C showing a strong but slightly broadened 1332 $cm^{-1}$ Raman band characteristic of crystalline diamond. The broad characteristic graphitic band near 1516 $cm^{-1}$ is clearly seen. There is also a considerable luminescence apparent in this spectrum. The luminescence was a function of the wavelength of the laser irradiation used for the Raman spectroscopy and both shifted and changed in intensity as the wavelength of the irradiating light was varied. The exact nature of the luminescence was not identified. It is possibly due to crystalline imperfections such as point defects, or organic or other impurities introduced into the film. Some natural diamonds have been found to have a wide variety of luminescent bands [11]. If the nature of the luminescence can be associated with specific imperfections it may become a valuable tool in future investigations.

Figure 8 is a TEM image of the film grown at 950 °C. It was determined from electron diffraction that a majority of the grains were oriented with ⟨100⟩ about 12° from the substrate normal. The [100] diffraction pattern is from the large grain arrowed in Fig. 8. The [100] zone for this particular grain was obtained by tilting about 10 degrees off the film normal. This is consistent with the SEM observation above and X-ray diffraction results which show a [105] texture. Apart from the dislocations, the large grains are relatively defect free. There is a notable lack of twinning within the grains. The regions

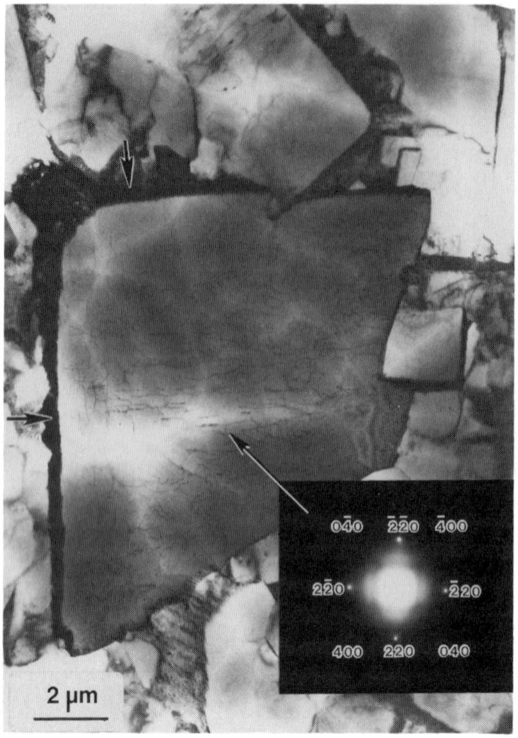


Fig. 8. Transmission electron micrograph of a polycrystalline diamond film grown at 950 °C showing square features typical of {100} faces shown in Fig. 7.

between the large grains, however, were highly defective, as indicated by arrows in Fig. 8. In fact, it was extremely difficult to distinguish individual defects.

## 4. Discussion

The three films described in the previous section were grown under conditions such that the growth mechanism appears to be dominated by different factors. In the case of the films which have pyramidal surfaces, the evidence is that growth often occurs as the result of twin formation on {111} or {100} facets or in layers parallel to {111} planes which originate at the intersection of the plane with another surface. Nucleation of a new layer on a {111} plane is difficult since it requires the addition of two or three carbon atom species [12]. It can be shown that the formation of a new layer (or layers) is much easier (more frequent) under special circumstances such as those found when twins intersect at re-entrant angles. This would explain the apparently rapid growth at saddle points as apparent in Fig. 2(b), since layers could nucleate and propagate from the junction of the re-entrant twin planes by the addition of single carbon atom species. This is readily shown with molecular models of twin intersections and has been discussed by Angus and Hayman [12] and by Frenklach and Spear [13]. The proliferation of microtwins, as shown in the TEM image of Fig. 4, is also a likely result of growth on {111} planes since they are the dominant twin plane and it is easy to build in stacking faults and/or propagate twins as the microstructure grows by step motion along the {111} plane. The height and width of stepped layers such as those shown in Fig. 2(c) can be analyzed to give insights into the relative importance of surface diffusion and the capture rate of feed material. Such an analysis is not, however, the purpose of this paper.

The microcrystalline nature of the deposit formed at 920 °C is probably the result of the high activity of carbon relative to the substrate, since the substrate temperature was the only variable changed. The high carbon activity (supersaturation) and lower surface mobility of surface species are responsible for the continuous nucleation of new crystallites as the deposit grows.

The ability to grow films with {100} faces exposed and nearly parallel to the substrate implies that in this narrow temperature regime the crystallites grow more rapidly in ⟨100⟩ directions than in any other direction. X-ray and selected area electron diffraction both confirm that columns grow in the ⟨100⟩ direction (approximately perpendicular to the substrate) and terminate in (100) facets on the surface. It is commonly thought that the fastest growing faces should grow themselves out of existence. Whatever the explanation, in this case it is apparently not true. Under the growth conditions used in this study the {100} facets remain flat, *i.e.* the SEM images do not reveal growth spirals or steps. Molecular models show that it is possible to add single carbon atom species to flat {100} faces. One would expect that growth in this manner could lead to freedom from growth twins and stacking faults but perhaps to an increase in the point defects present. This appears to be the case, as evidenced by the TEM image (Fig. 8) and the luminescence in the Raman spectra.

In conclusion, diamond films were grown by the hot filament assisted chemical vapor deposition method under conditions chosen to provide three distinctly different diamond microstructures. The topographic and internal structure of these films have been examined by scanning and transmission electron microscopy and characterized by Raman spectroscopy. One of these film structures is virtually free of stacking faults and twins. Growth features were discussed briefly in terms of growth mechanisms. The mechanical and physical properties of these films are now being obtained. We hope to correlate the properties of the films to their structures and thereby to the growth conditions, so as to be able to design and grow better films for specific applications.

## Acknowledgments

This research was jointly sponsored by (1) the U.S. Army Strategic Defense Command under MIPR W31RPD-8-P4030 with the U.S. Department of

Energy and Martin Marietta Energy Systems, Inc., (2) the U.S. Department of Energy, Division of Materials Sciences, Office of Basic Energy Sciences, (3) the U.S. Department of Energy, Assistant Secretary for Conservation and Renewable Energy, Office of Transportation Systems, as part of the High Temperature Materials Laboratory Program and (4) the Exploratory Studies Program of Oak Ridge National Laboratory. The research sponsored by the U.S. Department of Energy was under contract DE-3-ACO5-840R21400 with Martin Marietta Energy Systems, Inc.

## References

1 B. V. Spitsyn *et al., Prog. Crystal Growth Charact., 17* (1988) 79.
2 R. C. DeVries, *Ann. Rev. Mater. Sci, 17* (1987) 161.
3 K. Kobashi *et al., Phys. Rev. B, 38* (1988) 4067.
4 S. Matsumoto and Y. Matsui, *J. Mater. Sci., 17* (1982) 3106.
5 S. Matsumoto and Y. Matsui, *J. Mater. Sci., 18* (1983) 1785.
6 B. E. Williams *et al.*, in G. H. Johnson (ed.), *Diamond and Diamond-Like Materials Synthesis (Extended Abstracts)*, Materials Research Society, Pittsburg, PA, 1988, p. 59.
7 B. E. Williams and J. T. Glass, *J. Mater. Res., 4* (1989) 373.
8 W. Zhu, A. R. Badzian, and R. Messier, *J. Mater. Res., 4* (1989) 659.
9 J. Narayan and A. R. Srivatsa, *Appl Phys. Lett., 53* (1988) 1823.
10 A. T. Fisher and P. Angelini, in G. W. Bailey (ed.), *Proc. 43rd Ann. Meeting of the Electron Microscopy Society of America*, San Francisco Press, San Francisco, CA, 1985, p. 182.
11 D. M. Adams and S. J. Payne, *Faraday Trans., J. Chem. Soc., 70(10)* (1974) 1959.
12 J. C. Angus and C. C. Hayman, *Science, 241* (1988) 913.
13 M. Frenklach and K. E. Spear, *J. Mater. Res., 3* (1988) 133.

*Surface and Coatings Technology*, *39/40* (1989) 211 - 221



# PLASMA CVD DIAMOND DEPOSITION IN C-H-O SYSTEMS

A. INSPEKTOR*, Y. LIOU, T. McKENNA and R. MESSIER

*Materials Research Laboratory, The Pennsylvania State University, University Park, PA 16802 (U.S.A.)*

(Received March 16, 1989)

## Summary

For fixed carbon concentrations (2 vol.%), the O/C ratio has been systematically varied. Measurements of the resulting plasma optical emission were carried out and discussed in relation to the introduction of additional reaction paths to the formation of atomic hydrogen as the result of new active intermediate species containing oxygen. In turn, the optical emission results were related to extensions in the diamond formation region of deposition parameter space. The formation region was determined indirectly from deposition rate and film characterization by scanning electron microscopy, X-ray diffractometry and Raman spectroscopy of each of the samples.

## 1. Introduction

Since early studies of pyrolytic processes in hydrocarbon–hydrogen gas mixtures [1, 2], it has become clear that copious amounts of hydrogen are necessary to accelerate diamond film growth. A high supersaturation of hydrogen atoms (*i.e.* well above the thermal equilibrium value in hydrogen–hydrocarbon mixtures) as well as an elevated surface temperature, typically 700–1000 °C, are considered prerequisites for diamond deposition, regardless of the method employed to activate the deposition process [3]. Further extensions of the formation range of diamond and improvements in the film quality will require: (1) new starting materials to increase the concentration of hydrocarbon radicals and hydrogen atoms in the plasma; (2) additional means to stabilize the $sp^3$ diamond structure of the growing film and (3) additional means for controlling preferential etching of non-diamond carbonic compounds in the coating. This work deals with the first of these three basic requirements.

To date, the use of oxygen-containing additives looks most promising [4 - 7]. Tanaka *et al.* [4] have investigated the effect of water addition to the

*Visiting Research Associate: N.R.C. Negev, P.O. Box 9001, 84190 Beer Sheva, Israel.

0257-8972/89/$3.50 

feed gas on diamond growth in $CH_4 + H_2$ plasmas. They found an increase of the particle size with a parabolic dependence upon the water concentration in the input gas. The relatively large intensities of CH radicals and Hβ atoms were explained by enhanced decomposition of $CH_4$ by atomic hydrogen and OH radicals in the plasma. They also suggested that the removal rate of graphitic by-products in the film is enhanced by its high reactivity with atomic hydrogen and OH radicals from the gas. Similar effect of oxygen on diamond deposition was observed in hot filament chemical vapor deposition (CVD) systems. Hirose and Terasawa [5] found an increased deposition rate when hydrocarbons other than methane, which include oxygen and nitrogen containing reactants, were used. Kawate and Kondo [6] showed that the growth rate of diamond films in $CH_4$–$H_2$–$O_2$ is larger than that without oxygen addition. The suppressed formation of graphitic and amorphous carbon second phases in the film was correlated to a reduced concentration of acetylene in the exhaust gases.

In this paper we examine properties of C–H–O ternary feed gas systems and the effect of the O/C ratio in the feed gas on the graphitic content in the film. The polycrystalline diamond coatings (on a silicon substrate) were produced in a microwave-plasma assisted CVD system from ternary mixtures of carbon, hydrogen and oxygen containing precursors: $CH_4$, $C_2H_5OH$, $CO_2$,$H_2$ and $O_2$. The coatings were characterized by Raman spectroscopy, X-ray diffractometry (XRD) and scanning electron microscopy (SEM). Optical emission spectroscopy (OES) was used to analyze the composition of the plasma gas phase.

## 2. Experimental procedure

### *2.1. The microwave plasma deposition system*

The experimental set up consists of: (1) a bell jar microwave plasma chemical vapor deposition system, manufactured by Applied Science and Technology Co. (ASTeX), with fused silica bell jar of diameter 4 in, 1.5 kW maximum power, 2.45 GHz, (described in detail by Bachmann *et al.* [8]) and (2) emission spectroscopy set up for plasma diagnostics as shown in Fig. 1.

Si(111) wafers of diameter 1 in were used as substrates, which were precleaned and diamond powder (from Raytech spray diamond products) spray seeded on the substrate surface with 0.5 $\mu$m diamond powder. The depositions were carried out with gas mixtures of $H_2 + CH_4$, $H_2 + C_2H_5OH$, $H_2 + CH_4 + CO_2$, $H_2 + CH_4 + C_2H_5OH$, $H_2 + CH_4 + O_2$ and $H_2 + C_2H_5OH + CO_2$, which were introduced into the system with different O/C ratios. The carbon atom concentrations, from $CH_4$, $CO_2$ or $C_2H_5OH$ gases, were kept constant at 2% by volume for all tests in this study, except where noted otherwise. The oxygen atom concentration (from $CO_2$, $C_2H_5OH$ or $O_2$) changed with carbon atom concentration, with the O/C ratio varied from 1/6 to 7/3. The deposition rate was defined as the increased weight (mg) of the substrate divided by the deposition time (h) and the surface area of the

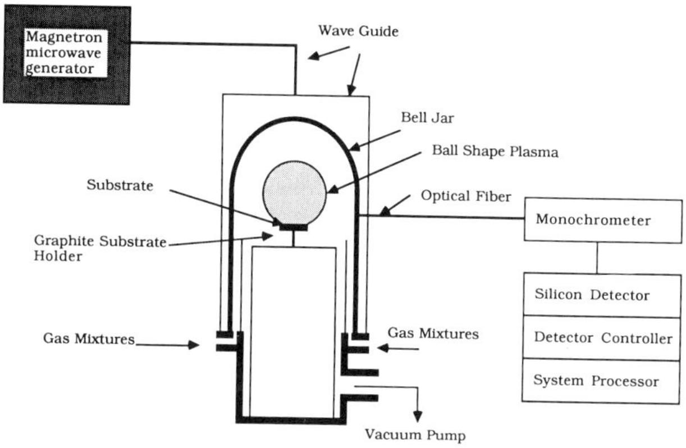


Fig. 1. Schematic diagram of the bell jar microwave plasma CVD system.

substrate ($cm^2$). The other experimental conditions were kept constant throughout all the experiments, *i.e.* substrate temperature of 980 °C (measured with an optical pyrometer; Leeds & Northrup Co.), microwave power of 1030 W (forward power), total pressure of 50 Torr and total flow rate of 300 sccm.

### *2.2. OES*

Identification of the major species present in the plasma was accomplished by the utilization of OES. For these experiments a fused silica optical fiber cable with stainless steel cladding from Oriel Corp. was placed outside the bell jar and in line with the plasma ball. Emissions from the plasma were directed through a slit of 70 $\mu$m by this cable into a HR-320 grating monochrometer from Instruments SA, Inc. The incident light beam was then diffracted by a grating of 300 g $mm^{-1}$ at 500 nm. Upon diffraction the spectra were then analyzed by an OMA 2 emission spectrometer system from EG&G Princeton Applied Research Corp. This system consisted of a Model 1256 intensified silicon detector, Model 1216 multichannel detector controller and Model 1215 system processor, which recorded spectra in the range of 412 - 660 nm.

## 3. Results and discussion

### *3.1. Plasma characteristics*

A major characteristic of the bell jar system is that the plasma ball is located far from the walls of the chamber, minimizing surface effects on the plasma properties and the deposition process. Species which have been identified in this work are the CH radicals (431.4 nm), $C_2$ (Swan band, 563.6, 516.5 and 473.7 nm), atomic hydrogen (Balmer series: H$\alpha$, 656.2 nm; H$\beta$, 486.1 nm; H$\gamma$, 434.0 nm) and $H_2$ molecules. Further references in this work to the $C_2$ and $H_2$ emission lines will pertain to the $C_2$ (516.5 nm) band head and $H_2$ (581 nm) peaks [9, 10].

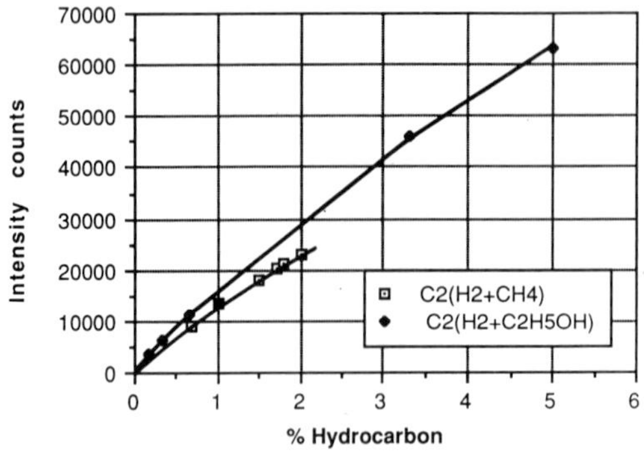


Fig. 2. Variation in intensity counts of $C_2$ *vs.* hydrocarbon concentration in the feed gas of optical emission spectra.

The non-equilibrium nature of plasmas limits the use of OES to only qualitative information on species present in the plasma and their dependency with experimental conditions. As shown in Fig. 2, the intensity of the $C_2$ band head increases linearly with increasing percentage of the carbon-containing monomer ($C_x H_y$) in the feed gas. Variations in the intensity of atomic hydrogen (H$\alpha$ and H$\beta$) with increasing amounts of monomer in the system are shown in Fig. 3. The intensity of both H$\alpha$ and H$\beta$ lines decreased in a polynomial fashion as the percentage of $CH_4$ in the system was increased. This dependency was also present for concentrations of $C_2H_5OH$ greater than 1%. Below 1% $C_2H_5OH$ the intensity of H$\alpha$ and H$\beta$ lines decreases with decreasing pressure, indicating additional processes in the low concentration range. Preliminary measurements of the plasma temperature in the system, based on the ratio of H$\alpha$ and H$\beta$ [11], indicate a corresponding maximum of the plasma temperature at about 1 - 3% $C_2H_5OH$. The reason for this behavior cannot be resolved from this data and a more detailed study is necessary. However, it is possible that small concentrations of atomic oxygen and OH formed in $C_2H_5OH$ discharges may be responsible for the increased formation of atomic hydrogen in this range.

### *3.2. Deposition process*

The deposition rates *vs.* $CH_4$ concentration and *vs.* $C_2H_5OH$ concentrations are shown in Fig. 4. In both cases, a parabolic, asymmetric dependency is observed: the deposition rate reaches a maximum at some point and then decreases sharply with further addition of monomer.

Comparison of the deposition rate measurements and plasma characteristics show that formation of the deposit is controlled by two competing processes. These processes are, first, the dissociation of the starting monomer into various intermediate fragments and by-products (*e.g.* $CH_x$ ($x = 1, 2, 3$) and $C_2$) and, second, the depletion of atomic hydrogen in the plasma. Initially, the concentration of the intermediate species and the formation rate of the deposit increase with increasing monomer concentration in the feed. At

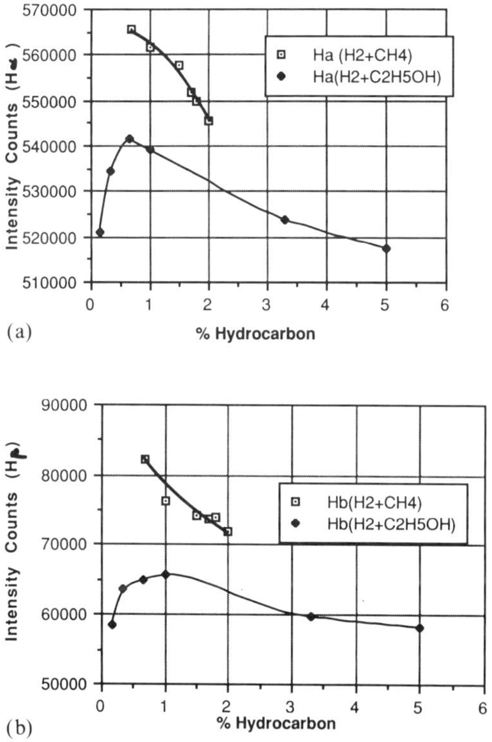


(a)

(b)

Fig. 3. Variation in intensity counts of (a) Hα and (b) Hβ *vs.* hydrocarbon concentration in the feed gas of optical emission spectra.

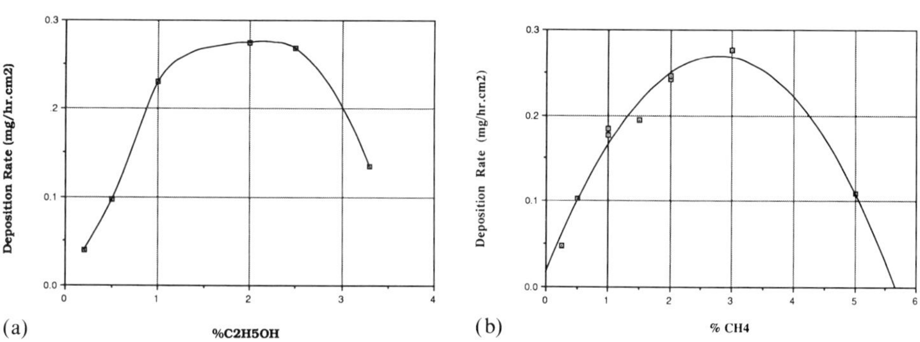


(a)

(b)

Fig. 4. Deposition rates *vs.* (a) $C_2H_5OH$ and (b) $CH_4$ concentrations.

intermediate percentage of monomer, the dependency of the deposition rate on the concentration of atomic hydrogen begins to show. As more monomer is introduced the concentration of atomic hydrogen decreases. This action, in turn, reduces the monomer's decomposition rate and the rate at which deposition occurs.

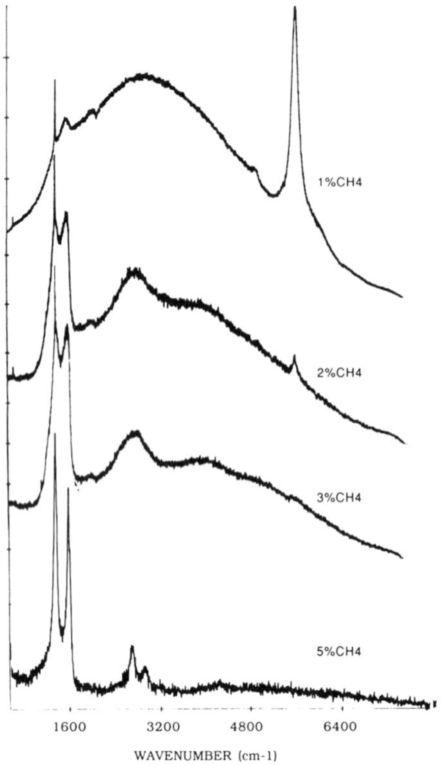


Fig. 5. Raman spectra of the samples prepared from various concentrations of $CH_4$.

The decrease of the deposition rate is accelerated by preferential etching of any non-diamond carbon component in the film. Figure 5 presents Raman spectra of films prepared at increasing concentration of $CH_4$. The non-diamond carbon content of the film is represented by the Raman scattering modes around 1580 $cm^{-1}$ [12]. It is shown that the 1580 $cm^{-1}$ band increases with increasing hydrocarbon concentration in the feed. At low monomer concentrations, the relatively high values of atomic hydrogen concentration have been related to non-diamond carbon re-evaporation. Further increase of the monomer, and reduction of atomic hydrogen, leads to a point where the formation of graphite will be faster than the corresponding etching rate. Then one should expect formation of graphitic coatings and a second maximum in the deposition rate *vs.* monomer's concentration plot.

### *3.3. Intermediate species in the plasma*

The deposition rate *vs.* O/C ratio for different feed gases are shown in Fig. 6. The deposition rates for the mixture of $CH_4 + H_2 + O_2$ at low O/C ratios (from 1/6 to 1/2) are about 20% higher than the deposition rates for the gas mixture of $CH_4 + H_2$. The deposition rates for gas mixtures of $CH_4 + H_2 + CO_2$, $CH_4 + C_2H_5OH + H_2$ and $C_2H_5OH + H_2$ at low O/C ratios

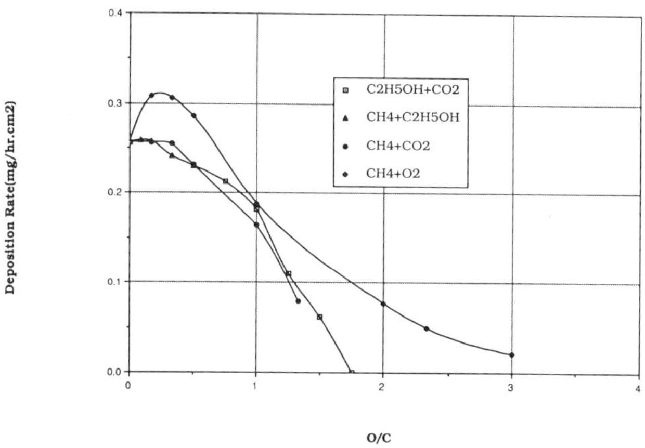


Fig. 6. The deposition rates *vs.* O/C ratio for different feed gases.

are nearly the same as the 2% $CH_4 + H_2$ mixture. At higher O/C gas mixtures (greater than 1), the deposition rates are much lower than the deposition rate of 2% methane in hydrogen, apparently due to the higher etching rate with higher concentrations of the oxygen.

The reasons that the 2% carbon concentration condition was studied can be seen in the Raman spectra of different concentrations of $CH_4 + H_2$ in Fig. 5. The 2% $CH_4 + H_2$ gas mixture gives a higher amount of graphitic carbon in the diamond film. The SEM picture in Fig. 7(a) shows a cauliflower

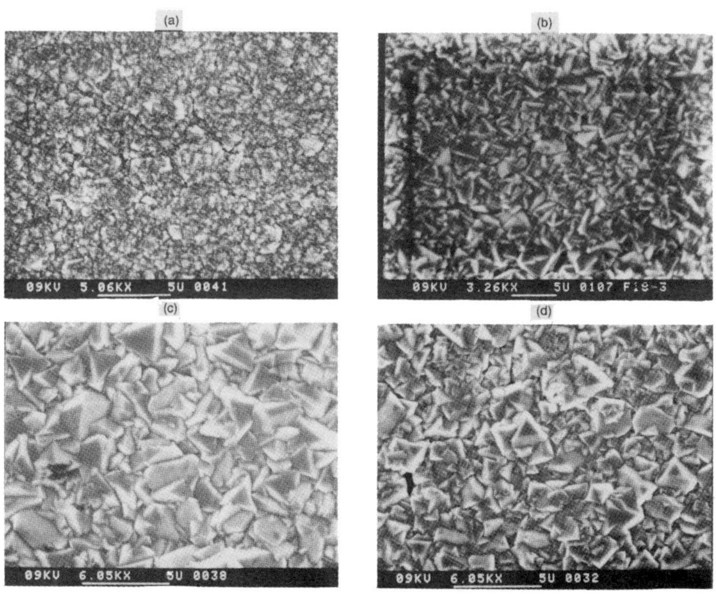


Fig. 7. SEM micrographs of (a) 2% $CH_4$, (b) 1% $C_2H_5OH$, (c) 2% $CH_4 + 1\%$ $O_2$ and (d) 0.7% $CH_4 + 1.3\%$ $CO_2$.

structure and there are no distinct faceted crystals on the film surface. By starting at an off-optimum phase purity condition, improvements could then be seen.

Atomic hydrogen is the active intermediate in the dissociation process and high concentrations of atomic hydrogen are required to accelerate formation of carbonic deposits at the substrate. The nature of the deposit, *i.e.* the ratio between graphitic and diamond content of the film, is determined by the processes at the surface of the growing film [13, 14]. The major feature of C–H–O ternary systems is the presence of additional active intermediates which open new reaction paths in the plasma bulk and at the plasma surface interface. Formation of such intermediate species will be demonstrated in the following discussion of dissociation processes in $CO_2 + H_2$ plasma mixtures.

The decomposition of $CO_2$ in $CO_2 + H_2$ mixture is accomplished by two major processes.

(1) Electron impact dissociation through excited electronic state $CO_2^*$ (with 7 eV threshold energy):

$$CO_2 + e \rightarrow CO_2^* + e \rightarrow CO + O + e \tag{1}$$

and a joint vibroelectronic mechanism through the vibrational system of the $CO_2$ molecules [15]. This reaction is followed by dissociation of CO which can be accelerated by increasing the plasma discharge current density [16]:

$$CO \rightarrow C + O \tag{2}$$

As the decomposition of CO is a surface related process, it can be further enhanced by introducing iron [17] or metallic oxides catalysts [18] into the system.

(2) Hydrogen reduction of the carbon dioxide [19]:

$$CO_2 + H_2 \rightarrow CO + H_2O \tag{3}$$

$$CO + H_2 \rightarrow C + H_2O \tag{4}$$

Both (3) and (4) are second order reactions. The second reaction is the rate determining step. Further dissociation of water leads to formation of H and OH in the plasma:

$$H_2O \rightarrow H + OH \tag{5}$$

Thus, introduction of $CO_2$ to the plasma and its reactions with other constituents of the gas form new active species such as CO, atomic oxygen, OH and $H_2O$. Similar products, although through somewhat different reactions, are present in other C–H–O systems. Each of these open additional reaction paths in the plasma to the formation of atomic hydrogen, atomic oxygen and OH species, to increase the dissociation rate of the starting monomer and to accelerate the preferential etching of graphitic components. The etching effect is demonstrated by the diamond deposition at different O/C ratios in the feed gas.

The SEM microphotographs in Fig. 7 show that the addition of the oxygen improved the microstructures of the diamond film. The faceted crystals in Fig. 7(b), (c) and (d) indicate that they may contain more diamond than Fig. 7(a), which also indicates that the diamond formation range can be extended to even higher $CH_4$ concentrations.

The X-ray diffraction patterns do not show much difference between diamond film deposited at 2% $CH_4 + H_2$ gas mixture (2 vol.% of carbon) and films produced at $H_2 + 1.7\%CH_4 + 0.3\%CO_2$ (2 vol.% of carbon) gas mixture (O/C = 1/3). Both show strong diamond (111) and (220) peaks, weak (311) and (331) diamond peaks and graphite (002) peak. The graphite (002) peak increases with increasing the concentration of $CH_4$ to 5%, the maximum value studied.

The Raman spectra in Figs. 8 and 9 of gas mixtures of $CH_4 + H_2 + CO_2$ and $CH_4 + H_2 + O_2$ show that the Raman peak at 1580 $cm^{-1}$ decreases as the O/C ratio increases. Compared with the Raman spectra of $CH_4 + H_2$ mixtures (in Fig. 2), the graphitic non-diamond modes are lower, showing that the quality of diamond films improves with oxygen additions. Therefore, it is possible to insure better quality diamond films and high deposition rate at the same time just by varying the concentration of the oyxgen in the gas mixture.

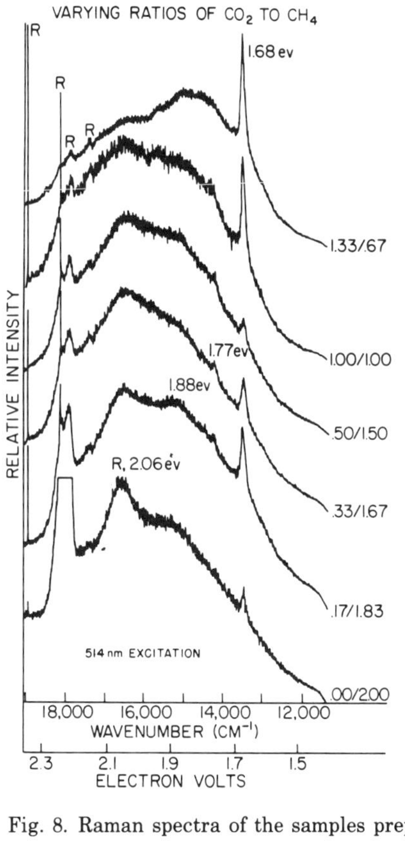


Fig. 8. Raman spectra of the samples prepared from gas mixture $H_2 + CH_4 + CO_2$.

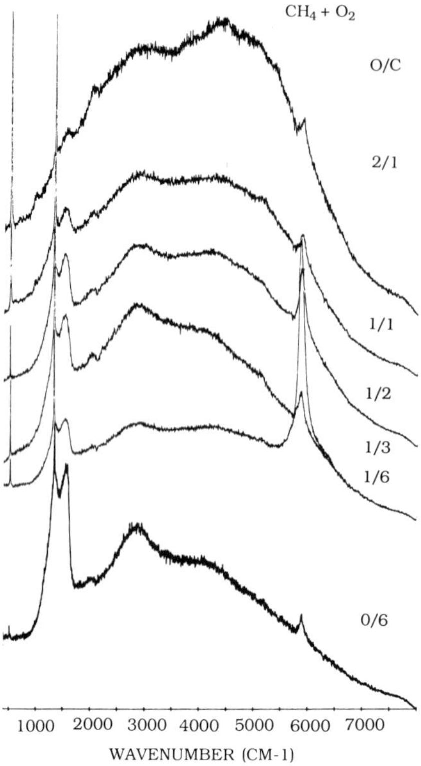


Fig. 9. Raman spectra of the samples prepared from gas mixture $H_2 + CH_4 + O_2$.

## 4. Conclusions

For fixed carbon concentrations (2 vol.%), the O/C ratio has been systematically varied. Measurements of the resulting plasma optical emission were carried out and discussed in relation to the introduction of additional reaction paths to the formation of atomic hydrogen as the result of new active intermediate species containing oxygen. In turn, the optical emission results were related to extensions in the diamond formation region of deposition parameter space. The formation region was determined indirectly from deposition rate and film characterization by SEM, X-ray diffractometry and Raman spectroscopy of each of the samples.

## Acknowledgments

This work was supported in part by the Office of Naval Research (with funding from the Stategic Defence Initiative Organization's Office of Innovative Science and Technology) under Contract No. N00014-86-K-0443 and the Diamond and Related Materials Consortium.

## References


1 B. V. Spitsyn, L. L. Bouilov, and B. Derjaguin, *J. Cryst. Growth, 52* (1981) 219.
2 S. Matsumoto, Y. Sato, M. Kamo and N. Setaka, *Jpn. J. Appl. Phys., 21* (1982) L183.
3 A. Badzian and R. DeVries, *Mater. Res. Bull., 23* (1988) 385.
4 H. Tanaka, K. Sato, Y. Saito and H. Miyadera, in K. Akashi and A. Kinabara (eds.), *Proc. 8th. Int. Symp. on Plasma Chemistry, Vol. 1, Tokyo, 1978,* International Union of Pure and Applied Chemistry, Tokyo, 1987, p. 2463.
5 Y. Hirose and Y. Terasawa, *Jpn. J. Appl. Phys., 25(6)* (1986) L519.
6 T. Kawate and K. Kondo, *Jpn. J. Appl. Phys., 26(9)* (1987) 1429.
7 Y. Saito, K. Sato, H. Tanaka, K. Fujita, S. Matuda, *J. Mater. Sci., 23* (1988) 842.
8 P. Bachmann, W. Drawl, D. Knight, R. Weimer and R. Messier, in G. Johnson, A. Badzian and M. Geis (eds.), *Diamond and Diamond-Like Materials Synthesis (Extended Abstracts)*, Materials Research Society, Pittsburgh, PA, 1988, p. 99.
9 G. Herzberg, *The Spectra and Structures of Simple Free Radicals: An Introduction to Molecular Spectroscopy*, Cornell University Press, Ithaca, NY, 1971.
10 R. W. B. Pearse and A. G. Gaydon, *The Identification of Molecular Spectra*, Chapman & Hall, London, 1963.
11 T. M. Hartnett, *Characterization of diamond deposition in a microwave plasma. M.Sc. Thesis*, Pennsylvania State University, 1988.
12 F. Tuinstra and J. L. Koenig, *J. Chem. Phys., 53* (1970) 1126.
13 W. Zhu, C. A. Randall, A. R. Badzian, and R. Messier, *J. Vac. Sci., Technol., A 7 (3)* (1989) 2315. in press.
14 R. C. DeVries, *Ann. Rev. Mater. Sci., 17* (1987) 161.
15 M. Capitelli and M. Molinari, in S. Veprek and Venugopalaln (eds.), *Plasma Chemistry II (Topics in Current Chemistry)*, Springer, 1986, p. 58.
16 S. Veprek, *J. Cryst. Growth, 17* (1972) 101.
17 E. N. Eremin, A. N. Maltsev and V. L. Invanter, *Russ. J. Phys. Chem., 54* (1980) 80.
18 R. A. Jones and S. S. Barton, *Can. J. Chem., 41* (1963) 1045.
19 E. N. Eremin, A. N. Maltsev and V. L. Invanter, *Russ. J. Phys. Chem., 52* (1978) 204.

*Surface and Coatings Technology, 39/40* (1989) 223 - 233 

# AN AUGER ELECTRON SPECTROSCOPY STUDY OF THE SURFACE OF Ni–Mo CEMENTED Ti(CN) SUBSTRATES USED FOR DEPOSITION OF CRYSTALLINE DIAMOND PARTICLES

R. C. McCUNE and R. E. CHASE

*Research Staff, Ford Motor Company, Dearborn, MI 48121 (U.S.A.)*

W. R. DRAWL

*Materials Research Laboratory, The Pennsylvania State University, University Park, PA 16802 (U.S.A.)*

(Received March 22, 1989)

## Summary

The surface composition of Ni–Mo cemented Ti(CN) tool inserts was studied by Auger electron spectroscopy under various conditions associated with diamond particle deposition in a microwave plasma using $CH_4$–$H_2$ mixtures as the feedstock. The presence of a nickel-rich phase at the surface did not appear to prohibit diamond particle growth, but rather reduced the growth rate relative to a pure TiC surface. Surfaces deliberately enriched in the nickel binder phase did, however, appear to show poor adherence for continuous diamond films grown on them.

## 1. Introduction

The deposition of crystalline diamond particles and films from the vapor phase has become a relatively widespread technology [1, 2]. Among the envisaged applications of diamond films are low-cost tool coatings. While existing grades of compacted or polycrystalline diamond materials are robust and cost effective in many machining applications [3, 4], it is expected that vapor-deposited diamond films may offer new opportunities in applications which were unapproachable with polycrystalline diamond.

The performance of early diamond-coated inserts [5] in non-ferrous metal cutting, while surpassing that of their uncoated counterparts, falls short of the expected performance from compacted polycrystalline diamond in the same application. The difficulties in producing a robust, diamond-coated insert or tool are mainly those of film adherence [6] and of intrinsic mechanical properties relative to the underlying substrate. Thus, while it is possible to arrive at diamond films by several deposition processes, the films so produced are often highly stressed and non-adherent.

0257-8972/89 Elsevier Sequoia/Printed in The Netherlands

The initial approach of simply coating the unmodified insert has met with only limited success for the family of cobalt-cemented tungsten carbides. Both the high degree of thermal expansion mismatch and apparently adverse effects of the cobalt binder [7] have led to poorly adherent coatings. Removal of the binder from the surface layer through chemical reactions occurring in conjunction with the plasma process [7] or by chemical pre-etching of the surface [8] appeared to mitigate these effects.

Since cemented carbide substrates are an attractive choice for diamond film deposition based on a number of practical considerations, the behavior of the binder under the conditions of film growth is of importance both in designing composite tools and in understanding the nucleation and growth processes of the coatings. To date, most of the reported results have focused on the Co–WC family of cemented carbides. In this work, we report the results of several observations of surface composition of Ni–Mo-cemented TiC inserts with differing pre-deposition conditions, both before and after exposure to a microwave plasma used for diamond deposition. In general, exposure of a cemented carbide surface to these conditions produces distinguishable compositional changes, which may play a role in subsequent processes of diamond growth and particle or film adhesion.

## 2. Experimental procedure

### 2.1. Materials

Cemented carbide inserts were selected from the commercially-available, steel finishing grade of the Ni–Mo–Ti(CN) family. These materials are microstructurally complex, due to the addition of molybdenum to the binder, or addition of $Mo_2C$ to the compact in order to promote "wetting" of the TiC grains [9]. Figure 1(A) illustrates a typical microstructure for this material, where the nominally TiC grains are surrounded by an alloy-phase shell containing titanium and $Mo_2C$. A stereological analysis of the binder-phase content of the diamond-polished surface indicated a surface area fraction of approximately 8.5%.

Various surface conditions of the Ni–Mo–Ti(CN) were assessed in conjunction with diamond particle growth. In the "as-sintered" condition, the insert surface was enriched in the binder phase (primarily nickel) as a consequence of the liquid-phase sintering process. In the "as-finished" condition, the insert surface finish was produced by grinding or polishing to the final tool configuration following the sinter operation. The "as-sintered" and "as-finished" surfaces of the Ni–Mo cemented Ti(CN) material are shown in Figs. 1(B) and 1(C) respectively. The "as-finished" surface of the tool insert is irregular; showing a high degree of porosity and finishing scratches. For this reason, the polished and sputter-cleaned surface of Fig. 1(A) was utilized as a reference surface for later Auger spectroscopy comparisons of surface composition.

In an attempt to study diamond particle growth on a comparatively

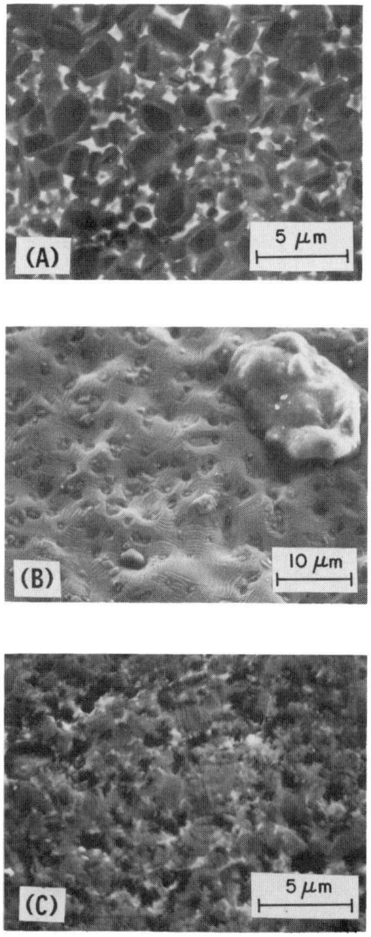


Fig. 1. Ni–Mo–Ti(CN) surface morphologies. (A) diamond-polished, (B) as-sintered and (C) as-finished.

"pure" TiC surface, a commercial tool formulated with a chemical vapor deposition (CVD) TiC layer over a base Co–WC insert was used. This tool was studied in the "as-received" condition; no attempt being made to alter the CVD surface by diamond polishing. Auger spectroscopy of the CVD TiC-coated material later revealed residual cobalt on the surface of the as-received inserts, presumably an artifact of the CVD process. Table 1 summarizes the materials used in this study and their various surface conditions.

*2.2. Diamond deposition*

Diamond deposition and hydrogen plasma exposures were conducted at the Materials Research Laboratory at Pennsylvania State University, utilizing a Toshiba microwave system. This system operates at 2.45 GHz, with a maximum power output of 1500 W. The microwave cavity surrounds a quartz

TABLE 1

TiC materials and surface preparations

| *Material* | *Surface condition* | *Remarks* |
|---|---|---|
| Ni–Mo–Ti(CN) | as-sintered | 2% $CH_4$, 1000 °C, 5 h, seeded |
| | as-finished | 2% $CH_4$, 1000 °C, 5 h, seeded |
| | as-finished | 1% $CH_4$, 1000 °C, 5 h |
| | diamond polished | 1% $CH_4$, 1000 °C, 5 h |
| | diamond polished | $H_2$ plasma, 1000 °C, 1 h |
| CVD TiC-coated Co–WC | as-received | 1% $CH_4$, 1000 °C, 5 h |

tube with a nominal diameter of 1 in, which contains the specimen platform of either BN or graphite supported on a quartz rod. A viewing port allows observation of the specimen surface during the deposition process and measurement of approximate surface temperature by optical pyrometry. Typical operating conditions for diamond deposition were 90 Torr total system pressure; 1 - 2% methane in hydrogen at a total flow of 100 s.c.c.m., with power adjusted to yield an indicated substrate temperature of approximately 1000 °C. For pure hydrogen plasma treatments, similar operating conditions were selected, with the total pressure maintained only through hydrogen flow.

In the first of several experiments with Ni–Mo–Ti(CN) materials, a seeding technique was utilized to promote diamond film growth. In this technique, the surface was "polished" using a cotton swab with 0.25 μm diamond paste, the residues of which were removed by rinsing with methanol. In later experiments, unseeded growths were preferred so as to understand the specific effects of the binder metals. There was no attempt to account for diamond residues from either metallographic polishing or routine finishing steps with the insert materials. In this regard, the only surface not in contact with diamond in any form prior to the deposition process was the CVD TiC-coated Co–WC.

### *2.3. Auger electron spectroscopy (AES)*

AES was conducted using a Perkin–Elmer, Physical Electronics Model 660 scanning Auger microprobe (SAM). Auger spectra were typically acquired with a beam energy of 10 keV, which was also used for imaging the surface in both the SEM and SAM modes of operation. The instrument incorporates a duoplasmatron ion source for sputter ion etching of the surface with xenon gas at energies of typically 2000 eV, at an angle of incidence less than 30°. Exposure of the surfaces to ambient environments between Auger analysis and the plasma processing was unavoidable.

## 3. Results

### *3.1. Seeded depositions*

Figures 2(A) and (B) compare the Auger spectra for the Ni–Mo–Ti(CN) in the reference condition (diamond polished and sputtered surface) and for

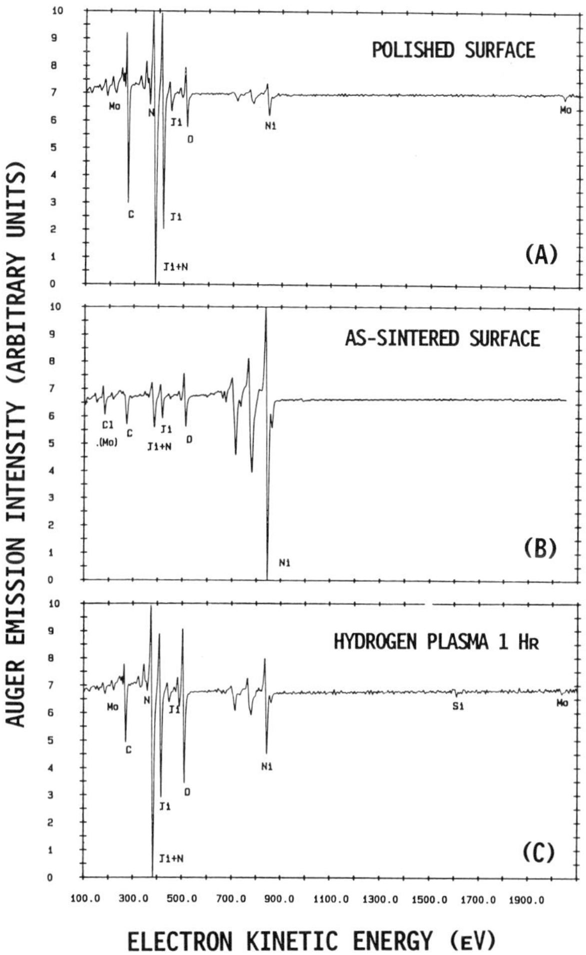


Fig. 2. Broad-area Auger electron spectra for Ni–Mo–Ti(CN): (A) diamond-polished surface, (B) as-sintered and (C) hydrogen plasma treated, 1 h.

the as-sintered surface (exhibiting the binder phase enrichment), respectively. In each case, an attempt was made to remove the atmospheric contaminants by sputtering prior to acquisition of the spectrum. The Auger spectrum for the sputter-cleaned, diamond-polished surface (Fig. 2(A)) indicates the carbon to be present as a bound carbide (presumably to titanium and molybdenum).

Seeded diamond growths on the as-sintered and the finished tool surfaces produced somewhat differing results. Figures 3(A) and (B) respectively show micrographs taken in the centers of the as-sintered and as-finished specimens exposed to the same growth conditions (2% methane, 1000 °C, 90 Torr). In general, the as-sintered, nickel-enriched surface showed a larger number of discrete particles of somewhat smaller size than those encountered

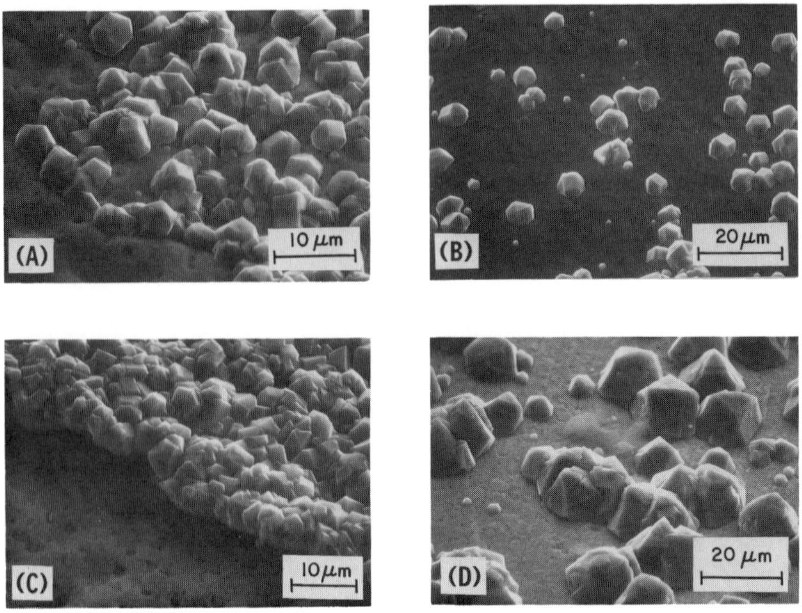


Fig. 3. Seeded diamond depositions on (A) as-sintered surface, near center, (B) as-finished surface, near center, (C) as-sintered surface, near edge and (D) as-finished surface, near edge.

on the finished tool surface. Near the edges of the insert pieces (Fig. 3(C) and (D)), the as-sintered surface showed regions of continuous, but non-adhering film growth, while the finished tool surface showed yet larger, but discrete particle growth. In the case of the as-sintered surface, the underlying topography, reminiscent of Fig. 1(B) can be observed. Remnants of non-adherent, but continuous films were more likely to be observed on the as-sintered surfaces than on the as-finished insert surface, which showed a tendency for discrete particle growth.

*3.2. Unseeded depositions*

In a second experiment, portions of the TiC-coated WC–Co insert and an as-finished Ni–Mo–Ti(CN) insert were coated simultaneously in a 1% methane plasma at 1000 °C for a period of 5 h. The growths on these two substrates are compared in Figs. 4(A) and (B). In general, the TiC-coated insert was found to show a larger number and greater range of particle sizes than that observed on the as-finished Ni–Mo cemented Ti(CN) material.

Two additional treatments were conducted on specimens of a polished Ni–Mo–Ti(CN) specimen similar to that shown in Fig. 1(A). In the first, the specimen was exposed only to a hydrogen plasma for a period of 1 h at 1000 °C. This was believed to be effectively a "decarburizing" condition for the material. In the second case, an unseeded deposition of diamond particles was performed at 1000 °C, with a 1% methane atmosphere. For the case of exposure to the hydrogen plasma only, the surface developed a nickel

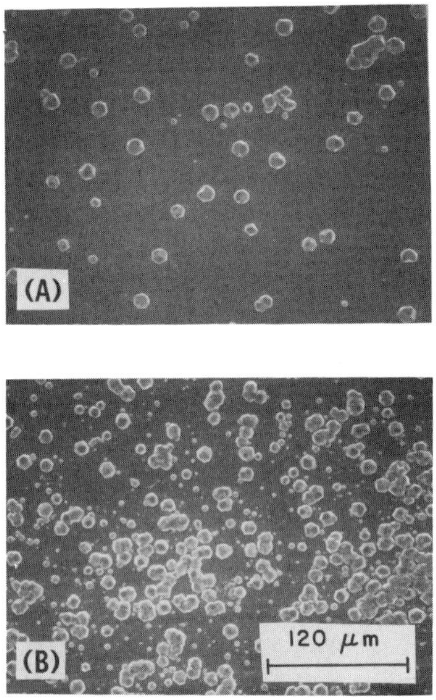


Fig. 4. Comparison of unseeded diamond growths in the same plasma environment on (A) as-finished Ni–Mo–Ti(CN) and (B) TiC-coated Co–WC.

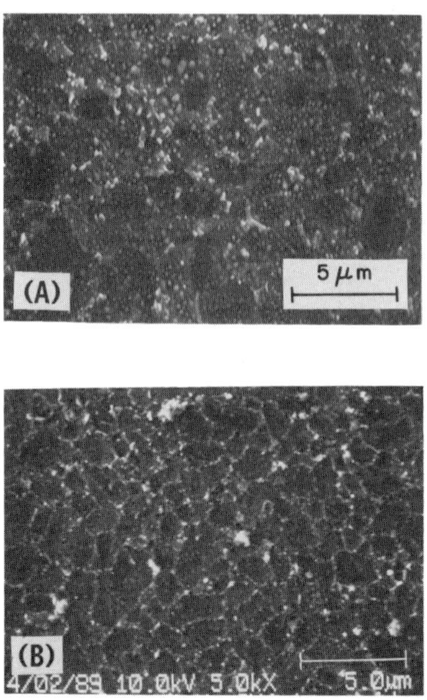


Fig. 5. Microwave plasma treatment of diamond-polished surface of Ni–Mo–Ti(CN): (A) hydrogen plasma, 1000 °C, 1 h, (B) 1% $CH_4$ in $H_2$, 1000 °C, 5 h.

enrichment suggested by the Auger spectra of Fig. 2(C), relative to that of the reference (polished) surface. Figure 5(A) shows an SEM image of the hydrogen treated surface and apparent exudation of nickel-enriched particles. The surface of the methane treated specimen (Fig. 5(B)) is completely different in its morphology and composition. In the region chosen for analysis, no diamond particles were observed and the nickel-rich binder regions observed in Fig. 1(A) appear to have diminished in size, with new particulate precipitation at the surface. The brightest particles observed on the surface contained primarily titanium, carbon and smaller amounts of oxygen. Further investigation is required to elucidate the composition of the smaller particles observed.

## 4. Discussion

### *4.1. Binder composition effects on nucleation and growth of diamond films*

A distinction should be maintained between the behavior of pure metals with regard to diamond deposition from plasma or hot filament sources, and the behavior of the binder phase in cemented carbides. As shown by Kikuchi [5], cobalt and stainless steel are poor substrates for diamond growth compared to carbide-forming elements tungsten, tantalum, niobium and molybdenum. Joffreau *et al.* [10] developed a model for diamond particle growth based on the surface and bulk diffusion of carbon in refractory metals. They proposed that the propensity for nucleation and growth of diamond on the surfaces of a refractory metal is determined by the rate at which carbon is transported from the surface to the bulk, either by dissolution or by growth of a carbide phase at the metal surface. The latter situation is a metastable condition for some transition metals, since a supersaturation with respect to carbon may be reached at the outermost surface in advance of carbon diffusion into the bulk. It is on this metastable carbide layer that diamond growth occurs; the rates for such processes being determined by the relative surface and bulk diffusivities of carbon and propensity for carbide layer formation in the substrate.

The situation for elemental cobalt, nickel and iron is in strong contrast to the refractory metals considered above, since these metals have considerable solubility (several atomic per cent) for carbon at the temperatures of optimum diamond deposition (*e.g.* 1000 °C), as well as rapid diffusivities of carbon (as much as 4 orders of magnitude higher in some cases) relative to the refractory metals considered by Joffreau *et al.* [10]. One expects that pure solvent metal substrates would not sustain diamond nucleation at elevated temperatures, as long as the overall carbon content of the solid solution remains below the solubility limit. Graphite precipitation from nickel [11] and cobalt [12] solid solutions is indeed observed only when the solid solubility for carbon is exceeded at elevated temperatures. Such works suggest that the diamond phase could exist in metastable form in contact with saturated solvent metals in the presence of atomic hydrogen. In low-temperature processes, both the solubility and diffusivity of carbon are greatly reduced, and substrates such as nickel can be considered for diamond growth [13].

In a cemented carbide, the carbon activity in the binder phase can extend over a range of values, depending on temperature and the ambient atmosphere, leading either to graphite precipitation at high carbon saturation, or alternatively to formation of eta phases at substoichiometric carbon content [14]. In the case of the as-sintered Ni–Mo–Ti(CN) material, surface enrichment in the binder phase did not prohibit diamond nucleation and film growth, and in fact, actually appeared to promote more extensive diamond growth compared to that occurring on the surface of an as-finished tool. This suggests that the excess binder phase present at the surface of this material was indeed saturated with respect to carbon.

In contrasting the diamond particle growth on unseeded surfaces of an as-finished Ni–Mo-cemented Ti(CN) insert, and that of a TiC-coated Co–WC insert exposed to the identical plasma environment, an overall lower nucleation density was observed on the nickel-cemented insert, suggesting, in this comparison, that a carbon excess was reached more readily on the TiC-coated surface, leading to more extensive particle growth. This behavior suggests that the apparent negative influence of the binder metal in the diamond growth process is related to the kinetics of forming a carbon excess at the surface of the cemented-carbide substrate.

### *4.2. Binder metal effects on film adherence*

The continuous diamond films deposited directly on cemented carbide substrates have been characterized by non-adhesion and spalling of the films in service. Isolated diamond particles grown on carbide surfaces produced by polishing, etching or otherwise removing the binder phase prior to deposition, however, appear to be adherent.

The poorly adherent nature of the "solvent" metal–diamond interface may perhaps be understood by extrapolation of earlier experiences with the wetting of diamond and graphite surfaces by solvent metals with varying carbon contents [15 - 18]. In these experiments, it was generally found that increasing the carbon content of solvent–metal sessile droplets resulted in higher contact angles with graphite. The implication of such studies to the problem of diamond film adherence on cemented-carbide insert surfaces would be that "wetting" in the sense of solid–solid interfacial strength is reduced with the increasing carbon content of the binder metal. An optimum condition for adherence would therefore be one where the carbon content of the binder metal is a minimum. In general, this condition will not be realized for cemented carbide substrates at elevated temperatures due to dissolution of carbon from the atmosphere and surrounding carbide particles to the solubility limit prior to precipitation of diamond.

### *4.3. Surface active nature of the binder phase*

The sintering process for cemented carbides derives from the "wetting" of the carbide grain surfaces by the binder metal [19]. It was felt that fresh carbide surfaces could be prone to such wetting by the binder metal at the elevated temperatures associated with the diamond deposition process, leading to both inhibition of diamond nucleation, as well as poorly adherent films. In the extreme case of binder-phase wetting observed with the as-sintered materials, only film adherence was affected; the binder phase having a high enough carbon activity to permit diamond growth. On the other hand, in comparing the pure TiC surface of the CVD-coated insert with the as-finished Ni–Mo–Ti(CN), the presence of the binder phase indeed appeared to limit the rate at which nucleation and growth occurred, possibly due to the kinetics of carbon saturation. "Wetting" of the carbide grain surfaces by nickel from the binder phase appeared only to occur for the reducing hydrogen plasma atmosphere; surface precipitation of carbon as graphite or diamond, and TiC precipitation occurring preferably in the carburizing plasma. The effect of surface layer enrichment in the nickel binder metal in this material may thus not be related so much to the inability to produce diamond films, but rather to producing resultant films having poor adherence. This effect, in tandem with the expected high thermal stress states for the continuous diamond films in contact with cemented carbides, suggests a situation wherein the films are prone to failure. Elimination of the binder layer at the outermost surface, either by etching or by addition of a barrier layer (such as TiC on Co–WC) may consequently permit more adherent diamond films to be produced by restricting the occurrence of a poorly bonded interface.

## Acknowledgments

The authors wish to thank Tom Whalen of the Ford Motor Company and Dave Moskowitz of Iscar Ceramics for review of the manuscript and valuable discussions.

## References

1 J. C. Angus and C. C. Hayman, *Science*, *241* (1988) 913.
2 K. E. Spear, *J. Am. Ceram. Soc.*, *72* (2) (1989) 171.
3 G. F. Wilson and D. K. Bruschek, in F. W. Gorsler (ed.), *Cutting Tool Materials*, American Society for Metals, Metals Park, OH, 1981, p. 297.
4 J. C. Miller, *Paper No. G-T81-035*, Society of Die Casting Engineers, Cleveland, OH, 1981.
5 N. Kikuchi and H. Yoshimura, in *New Diamond*, Japan New Diamond Forum, 1988, p. 42.
6 R. C. McCune, D. W. Hoffman, T. J. Whalen and C. O. McHugh, in J. C. Bravman, W. D. Nix, D. M. Barnett and D. A. Smith (eds.), *Thin Films: Stress and Mechanical Properties*, Proc. Vol. 130, Materials Research Society, Pittsburgh, PA, 1989, p. 261.
7 M. Murakawa, S. Takeuchi, H. Miyazawa and Y. Hirose, *Surf. Coat. Technol.*, *36* (1988) 303.
8 K. Shibuki, M. Yagi, K. Saijo and S. Takatsu, *Surf. Coat. Technol.*, *36* (1988) 295.
9 H. Doi, in Inst. *Phys. Conf. Ser. No 75, Int. Conf. Science of Hard Materials, Rhodes, Greece*, Hilger, London, 1986, p. 489.
10 P.-O. Joffreau, R. Haubner and B. Lux, *Int. J. Refract. Hard Met.*, *7* (1988) 186.
11 F. J. Derbyshire, A. E. B. Presland and D. L. Trimm, *Carbon*, *10* (1972) 114.

12 F. J. Derbyshire, A. E. B. Presland and D. L. Trimm, *Carbon*, *13* (1975) 111.
13 R. J. Markunas, R. A. Rudder, D. J. Vitkavage, R. Burns and J. Posthill, *Paper TP-7, SDIO/IST-ONR, Diamond Technology Initiative Symp.*, Arlington, VA, 1988.
14 H. E. Exner, *Int. Met. Rev.*, *243* (24) (1979) 149.
15 M. Humenik and T. J. Whalen, in J. R. Tinklepaugh and W. B. Crandall (eds.), *Cermets*, Reinhold, New York, 1960, p. 6.
16 Yu. V. Naidich and G. A. Kolesnichenko, *Poroshkovaya Metallurgiya*, *1* (1961) 55.
17 Yu. V. Naidich, V. M. Perevertailo and G. M. Nevodnik, *Soviet Powder Metallurgy*, *10* (1971) 45.
18 M. Humenik, D. W. Hall and R. L. van Alsten, *Met. Prog.*, *81* (1962) 101.
19 M. Humenik and N. M. Parikh, *J. Am. Ceram. Soc.*, *39* (1956) 60.

*Surface and Coatings Technology, 39/40* (1989) 235 - 240 

# DIAMOND COATING OF A LONG SUBSTRATE BY USE OF A COMBUSTION FLAME

M. MURAKAWA, S. TAKEUCHI and Y. HIROSE

*Nippon Institute of Technology, 4 – 1 Gakuendai, Miyashiro-Machi, Saitama-Prefecture 345 (Japan)*

(Received March 9, 1989)


## Summary

Synthesis of diamond films has hitherto been attained by chemical vapour deposition (CVD) methods as well as other gas phase methods. These methods are characterized by the use of vacuum chambers; the degree of vacuum being 0.5 - 40 kPa. This is a disadvantage when a long substrate or one having a large area is to be coated. This paper proposes a new diamond coating technique using oxyacetylene combustion whereby one can coat an elongated tungsten carbide substrate (in principle, there is no limit to the length of the substrate) with a diamond film; such coatings have previously been practically impossible because of the limited space of the vacuum chamber. The adhesion of the coating has been measured quantitatively using the scraper-type adhesion tester developed by the authors.


## 1. Introduction

Various techniques have been used in the synthesis of diamond films by chemical vapour deposition (CVD) methods, *e.g.* microwave plasmas, thermal filaments, r.f. plasmas, plasma jets, etc. These methods, however, depend on reduced pressures for the synthesis and, therefore, require a reaction chamber. This imposes a limitation in that when an article of large size or length is to be coated with diamond film the size of the article in question will be limited by the size of the reaction chamber. In contrast, a new diamond synthesis method (the Hirose method) developed by one of the authors [1] and studied by Hanssen *et al.* [2] uses combustion to produce the plasma, and the use of a reaction chamber becomes unnecessary. This is because the combustion flame itself acts as a kind of reaction chamber. The Hirose method also has an advantage in that it has a higher rate of deposition of diamond, *i.e.* 100 - 200 $\mu$mh$^{-1}$, as compared with conventional CVD methods. It will allow an article with a relatively large area or length to be coated in a relatively short period of time.

When using the Hirose method, a coating operation based on moving a

0257-8972/89/$3.50 

relatively small sized torch is desirable because the use of a sufficiently large torch would be expensive and technically difficult. However, one of the biggest problems of the Hirose method yet to be checked is the graphitization of the already-deposited diamond which is subjected to the trailing outer flame portion. This diamond is in an oxidizing atmosphere (because of its contact with the surrounding air) and at a high temperature (assumed to be more than 2000 °C). This leads to the degradation or disappearance of the diamond film.

Accordingly, examination has been made of the possibility of carrying out the Hirose method by coating a long substrate using a linearly moved torch, in an effort to develop a technique whereby articles with a large area can be coated with diamond film.

## 2. Experimental procedure

Figure 1 shows a schematic illustration of the diamond coating apparatus used in the experiment. Oxygen gas ($O_2$) and acetylene gas ($C_2H_2$) were used to produce a combustion flame, using a torch commercially available as a middle pressure cutting tube for use as an oxyacetylene gas tube. The torch was mounted in position on a movable table operated with an electric motor. The substrate used was made of WC–Co alloy (including 6.5 wt.% Co) and had dimensions of 5 mm (h) × 10 mm (w) × 50 mm (l). The substrate was mounted on a water-cooled copper plate; the longitudinal direction of the substrate being coincident with the direction of movement of the torch. In the coating operation using the combustion flame, a copper dummy substrate was provided in longitudinal abutment with the WC–Co alloy substrate to help in obtaining a uniform heating–cooling cycle of the whole WC–Co alloy substrate even after the substrate had been coated. The WC–Co alloy substrate was subjected to a light scratch treatment with diamond powder paste before coating.

Table 1 shows the conditions for diamond synthesis. The flame consists

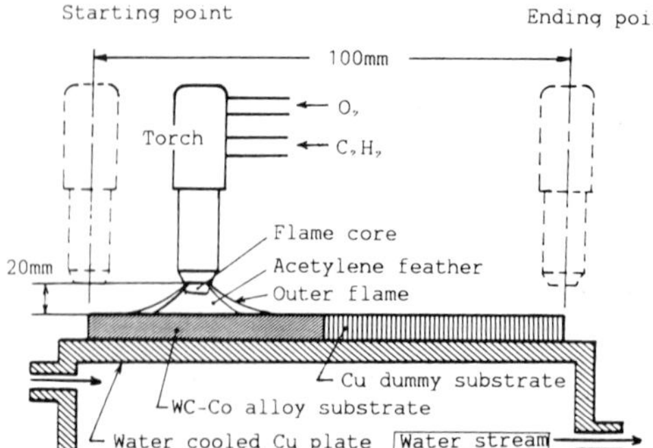


Fig. 1. Schematic illustration of the combusion gas flame method (Hirose method) for diamond coating.

TABLE 1

Conditions for synthesizing diamond

| | |
|---|---|
| Flow rate of oxygen ($l\,min^{-1}$) | 10.0 |
| Flow rate of acetylene ($l\,min^{-1}$) | 11.2 |
| Acetylene feather length (mm) | 30 |
| Substrate temperature (°C) | 1000 - 1100 |
| Torch speed ($mm\,min^{-1}$) | 3.3 |
| Distance between the torch and the substrate (mm) | 20 |

of an oxyacetylene flame with an acetylene excess of about 10%; this can be confirmed by the fact that an acetylene feather of length 30 mm may be observed if the flame is released in the air without inserting the substrate. When the substrate is put into the acetylene feather, if the temperature of the substrate (measured with a pyrometer) is controlled within a range of from 900 °C to 1100 °C, diamond will be deposited onto the WC–Co alloy substrate. In the case of a stationary flame the diamond deposition speed will be 80 - 100 $\mu m h^{-1}$.

## 3. Experimental results and consideration

### 3.1. Quality of the diamond film

Figure 2 shows scanning electron microscope (SEM) micrographs of the diamond film obtained with the Hirose method in its torch moving mode.

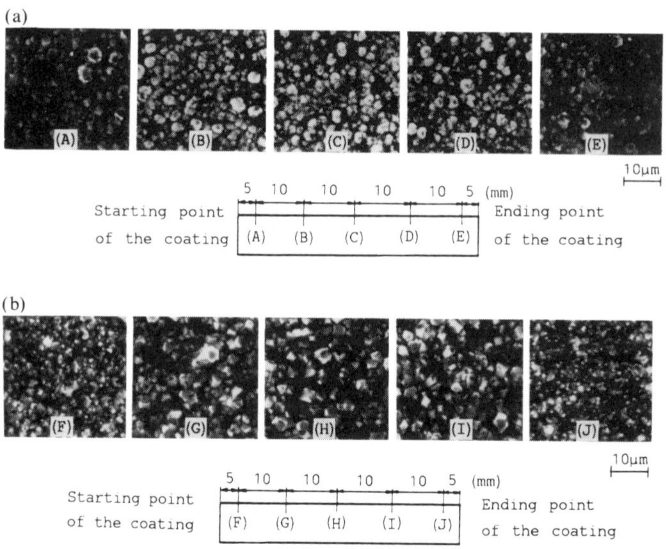


Fig. 2. SEM micrographs of the diamond films obtained: a, ball-like diamond film coated onto a long WC–Co alloy substrate; b, well-crystallized diamond film coated onto a long WC–Co alloy substrate.

From this figure, it can be seen that diamond coating by moving the torch is quite possible, and that either a coating consisting of ball-like diamond (type a) or a coating consisting of well-crystallized diamond (type b) has been obtained over the entire surface of the WC–Co alloy substrate, although each end portion (5 mm) of the substrate has a slightly different appearance to the other portions. It is noted here, however, that a type b coating was only obtained in three cases out of 30; the reason has yet to be elucidated.

With regard to the speed of the movement of the torch, a speed higher than 6 mm min$^{-1}$ has been found to produce an incomplete or mottled coating with a portion of the substrate being exposed. Figure 3 shows a thin film X-ray diffraction pattern for the diamond coatings. For both type a and b coatings, a peak for tungsten carbide in the substrate and peak lines (111), (220) and (311) for diamond crystal can be observed, *i.e.* proof that the coating obtained consists of diamond. Moreover, it is seen that the line peaks corresponding to the diamond with a well-defined morphology are stronger than those of ball-like diamond, indicating the better quality of the former crystal. Figure 4 shows a Raman spectrum of a diamond film with a well-defined morphology; an explicit diamond peak being observed at

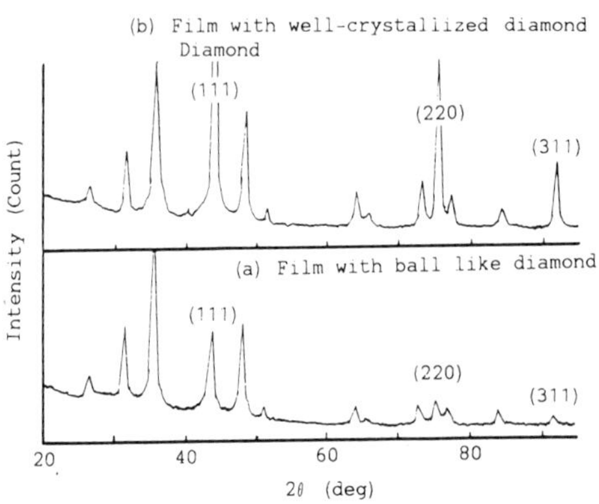


Fig. 3. Results of X-ray diffraction analysis of the diamond films.

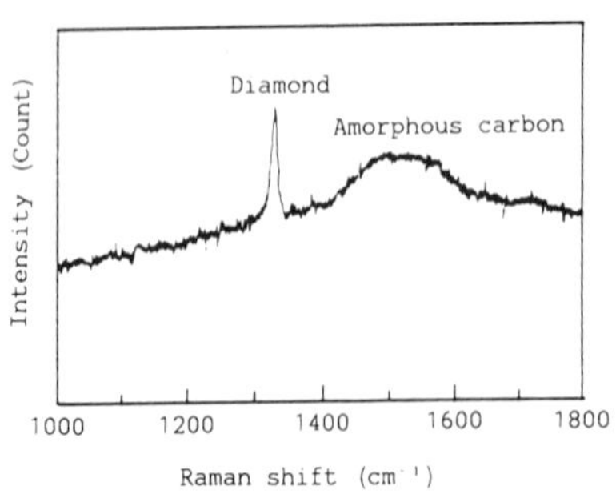


Fig. 4. Raman spectrum of a diamond film with well-defined morphology.

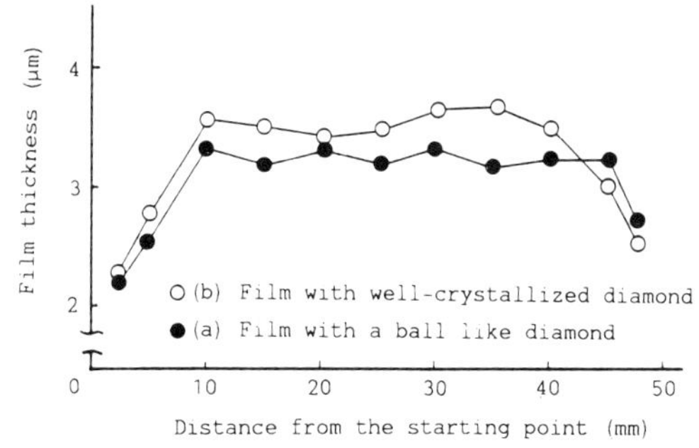


Fig. 5. Film thickness along the substrate length.

1333 $cm^{-1}$. A broad peak at around 1550 $cm^{-1}$, showing the inclusion of amorphous carbon, can also be seen. Figure 5 shows two examples of the film thickness distribution along the substrate length; one for the film with the ball-like diamond and the other for the well-crystallized diamond. Both films have a thickness of about 3.5 $\mu$m in the middle portion of the substrate; however, they have a reduced thickness within 5 mm of the start and end points of the coating. This may be a result of the rise in temperature of the substrate in these portions during coating.

An explanation will now be given on whether or not the diamond coating was degraded when in contact with the outer edge of the combustion flame. Coatings of both type a and b do not seem to be affected by contact with the outer edge of the flame, apart from the end portions of the substrate as mentioned above. Judging from the colour of the substrate and the temperature measurements (using a pyrometer), it is assumed that the temperature of the portion of the substrate where diamond had been deposited and subsequently subjected to the sweeping action of the outer edge of the flame is less than 800 °C. At this temperature the diamond film will not decompose even if the flame atmosphere is reductive or oxidative.

### *3.2. Adherence of the diamond film to the substrate*

Figure 6 shows the results of the scraping-force measurements obtained with the scraper-type adhesion tester [3] designed to evaluate quantitatively the adhesive strength of diamond coatings.

Since the tester has proved able to give quantitative comparisons only when the thickness and quality of the coatings to be compared are the same, direct comparison of adherence between the ball-like diamond film and the well-crystallized diamond film of different film quality is not possible. However, the difference in the adherence at various parts of the substrate can be measured and compared quantitatively for each type of film since the thickness and quality of the film portions in question are the same. Except for the end portions of the substrate, where a somewhat reduced adherence prevails, both types of film have a substantially uniform adherence distribution along the majority (80%) of the length of the substrate. This may be a result of

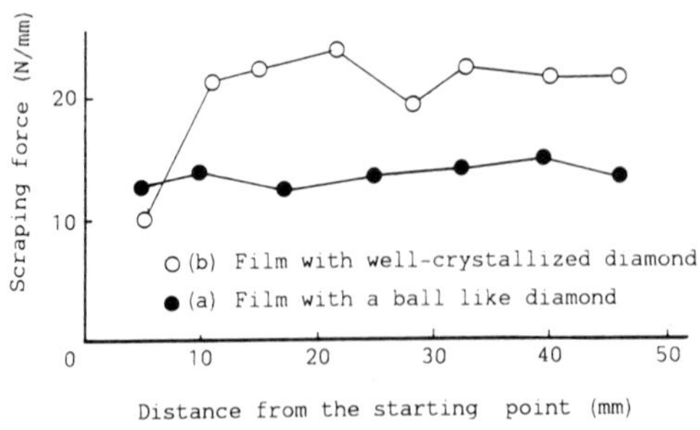


Fig. 6. Scraping force along the substrate length.

mounting the dummy substrate adjacent to the coated substrate so that a uniform cooling effect is obtained throughout the coated substrate.

## 4. Conclusion

The possibility of depositing diamond films onto a long substrate has been checked with respect to the Hirose method (using a combustion flame), wherein a torch of conventional size is moved linearly over the substrate. The results show that substrates of WC–Co alloy having a width of 10 mm and a length of 50 mm can be coated with films of either ball-like diamond or well-crystallized diamond along the whole length of the substrate. In principle it may be possible to coat a longer substrate if the substrate is kept at a constant temperature appropriate for deposition, *i.e.* by employing appropriate cooling means for the substrate.

## Acknowledgments

This work was supported in part by the Amada Foundation for Metalwork Technology and in part by Showa Denko K. K. We extend our thanks to Mr. Katsuro Kishi and Mr. Yuji Murakami, both senior students of our Institute, for their contribution in carrying out the experiment.

## References


1 Y. Hirose, Synthesis of diamond using combustion flame in the atmosphere, *Progr. and Abstr. 1st Int. Conf. on the New Diamond Science and Technology*, Tokyo, 1988, p. 38.

2 L. M. Hanssen, W. A. Carrington, J. E. Butler and K. A. Snail, Diamond synthesis using an oxygen–acetylene torch, *Mat. Lett.*, *7* (1988) 289.

3 M. Murakawa and S. Takeuchi, *Thin Solid Films*, *181* (1989) 443.

*Surface and Coatings Technology, 39/40* (1989) 241 - 252 

# COMBUSTION SYNTHESIS OF DIAMOND

W. A. YARBROUGH, M. A. STEWART and J. A. COOPER, Jr.

*Materials Research Laboratory, The Pennsylvania State University, University Park, PA 16802 (U.S.A.)*

(Received March 6, 1989)

## Summary

Diamond has been grown using an unmodified oxyacetylene brazing/ welding torch in air. The typical experiment uses a water cooled substrate holder with a torch nozzle to substrate distance which varies from around 1 mm to 3 - 4 cm depending upon total gas flow, nozzle size, $O_2$:$C_2H_2$ ratio etc. The deposits obtained to date have not been uniform in morphology, growth rate or Raman signature. Although polishing of silicon substrates clearly enhances nucleation density in most such experiments, the very high (greater than $10^8$ cm$^{-2}$) nucleation densities seen with either microwave or hot filament syntheses have not been seen with the "torch" method. The high gas temperatures employed (around 3000 °C) necessitates the cooling of substrates which introduces large and difficult to reproduce thermal gradients. Much work remains to be done if smooth, thin and/or uniform coatings are to be obtained this way. Diamond can be obtained by acetylene combustion using a flat flame burner at pressures of 25 and 50 Torr by adjusting the $O_2$:$C_2H_2$ ratio to around 1.16. This development offers a means for the study of such flames and the eventual development of a satisfactory model for diamond formation from them.

## 1. Introduction

The synthesis of diamond by chemical vapor deposition is now well established with the critical experiments reproduced in numerous laboratories by many different workers [1 - 3]. Common to most of these experiments is the production of atomic hydrogen, which is believed by many, for various reasons, to be responsible for the growth of diamond rather than graphite [4, 5]. Diamond synthesis by the combustion of acetylene was originally reported by Hirose and Kondo [6] early in 1988 and confirmed by workers at the Naval Research Laboratory [7, 8] and at General Electric [9] as well as in this laboratory. Oxyacetylene flames are expected to produce gas temperatures of around 3000 °C [10], depending upon the proportions of acetylene and oxygen, and whether or not other gases are present. At these temperatures

0257-8972/89/$3.50 

much of the hydrogen liberated in the combustion process will be present as the monoatomic species and hence the flame deposition of diamond is not necessarily inconsistent with many current ideas on the gas phase synthesis of diamond. The use of combustion to grow diamond represents a potentially important development from both technological and scientific points of view. The technological importance is related not only to the relatively low initial capital investment required, but also to the observation that diamond can be grown in the ambient atmosphere using relatively portable equipment. The science of vapor phase diamond growth remains relatively immature, in spite of recent developments, and combustion chemistry offers a new set of experimental possibilities against which various theories can be tested.

## 2. Experimental procedure

### *2.1. Atmospheric pressure "torch" experiments*

These experiments used a commercially available brazing or welding torch with the flame directed against a water cooled substrate surface. As the gas temperatures and velocities are high, the substrate must be cooled efficiently to avoid either the destruction of the substrate or the rapid transformation of any diamond that might be deposited to graphite. This requirement led to experimental restrictions on the shape and size of the substrate to those which can be efficiently cooled and which will tolerate the large thermal gradients involved.

The experimental apparatus consisted of a water cooled sample holder, or stage, made of copper, the torch, and mass flow controllers for regulation of the gas flows. Small fragments, 0.5 - 1 cm in diameter, of single crystal silicon wafers (arsenic, boron and phosphorous doped as well as undoped) were used as substrates and most of these were polished with submicron diamond powder prior to deposition. Substrate temperatures were estimated using an optical pyrometer, without emissivity correction, and found to vary over the range 800 - 1100 °C. The torch (single nozzle) was used with interchangeable nozzle sizes of 0.66 and 0.97 mm as measured with a stereomicroscope.

The $O_2$:$C_2H_2$ ratio was varied from 0.85 to around 1.1 with the most consistent results being obtained at ratios close to unity. Total gas flow rates varied with nozzle size and were in the range of 1.8 - 3.0 sl $min^{-1}$ (standard liter per minute). The torch flame consists (at around 1.0 or lower $O_2$:$C_2H_2$ ratios) of three distinguishable zones. The first (closest to the nozzle) is colorless and transparent and is believed to be a region in which the premixed gas emerging from the nozzle is heated to a high enough temperature to ignite and burn, producing a sharply conical, stable flame front, 1 - 5 mm from the torch tip depending on flow rate and nozzle size. The second zone is a bright white or incandescent combustion zone in which the primary combustion of acetylene with premixed oxygen occurs. The extent of this second zone depended very sensitively on the $O_2$:$C_2H_2$ ratio used and somewhat less sensitively on the total flow rate and nozzle size. Diamond deposition was observed with the substrate placed in this region of the flame. The third zone (farthest from the nozzle and the largest volume of the flame) is a blue transparent region and deposition was not observed in this portion of the flame. The second zone of a premixed oxygen deficient oxyacetylene flame is where the principal reaction products of atomic hydrogen, $H_2$ and CO are formed. The third zone is essentially a diffusion flame in which the products burn with atmospheric oxygen [11].

### *2.2. Low pressure, flat flame deposition*

A commercially available, water cooled, flat flame burner (McKenna Products Inc., Pittsburg, CA) was installed in a vacuum bell jar for the experiments at low pressure. Gas flow rates were controlled using MKS Inc. mass flow controllers (Model 1259B controller with 247C readout and power supply) and the chamber pressure controlled and monitored using an MKS Inc. pressure control system (PDR C-2C with 252A readout and 122AA Baratron$^{TM}$). The burner surface consists of a porous, sintered stainless steel disc (6.0 cm diameter) containing a cooling coil through which water flows to cool the burner surface and prevent ignition of the gas mixture inside the mixing chamber upstream of the burner surface. The substrates used were 1 in diameter, ⟨111⟩ boron doped silicon wafers. Substrates were mounted both horizontally, facing the burner surface, and vertically with their broad surfaces normal to the burner surface and were held in place by a graphite clamp.

## 3. Experimental observations and discussion

### *3.1. Atmospheric pressure "torch" experiments*

Although diamond could be readily grown over a narrow range of gas compositions, significant morphological differences were seen to exist between these growths and those typical of low pressure plasma and hot wire experiments. Figure 1 compares the deposits obtained on polished and unpolished ⟨111⟩ boron doped silicon wafer fragments after 15 - 20 min of deposition. The substrate of Fig. 1(a) was polished with diamond powder prior to deposition and any loose diamond powder removed by wiping with a clean lint free tissue. The substrate of Fig. 1(b) was unpolished. Although polishing with diamond clearly enhanced nucleation, it was not sufficient to ensure the rapid development of a continuous coating. Polishing with diamond powder or paste, as opposed to other abrasives, is known to greatly enhance the nucleation density of diamond on silicon and possibly other substrates [12, 13], but its efficacy in flame growth of diamond at atmospheric pressure appeared limited or non-existent. Indeed the high nucleation densities (greater than $10^8$ cm$^{-2}$) normally seen on diamond polished silicon, which are desired for the rapid development of a continuous coating, were not obtained

(a)

(b)

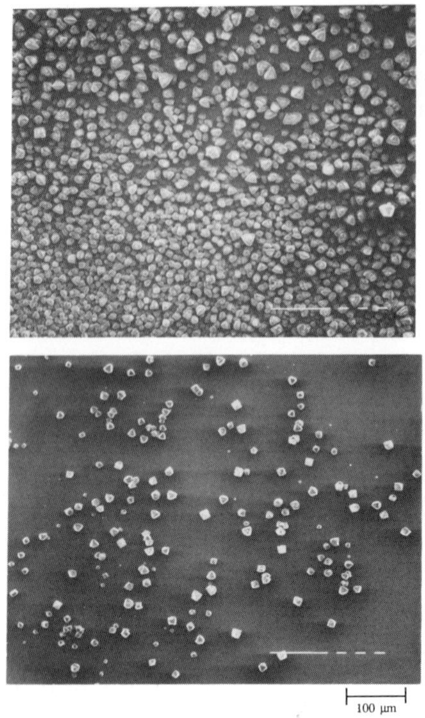


Fig. 1. Diamond deposited onto ⟨111⟩ boron-doped silicon wafer fragments in 15 and 20 min respectively using 0.66 mm diamond nozzle at $O_2$ and $C_2H_2$ flow rates of 1.2 slm. (a) Wafer polished with submicron diamond prior to deposition and, (b) unpolished. Nucleation density is higher in the diamond polished sample, but is non-uniform and relatively low.

in any of the torch depositions which showed well crystallized diamond. Nucleation densities in both cases were low and variable, relative to either hot wire or low pressure plasma methods. Nucleation densities of around $10^4$ to $10^5$ cm$^{-2}$ were found for untreated silicon surfaces and around $10^6$ cm$^{-2}$ on silicon previously polished with submicron diamond, whereas those seen on diamond polished silicon using either low pressure plasma or hot wire techniques are around $10^8$ cm$^{-2}$ or greater. Relatively isolated growth is not necessarily inherent to the technique as we have found that a continuous coating is readily achieved if high nucleation density and significant diamond growth is first obtained by another method and then continued using atmospheric combustion. In earlier work, one of us (W.A.Y.) obtained smooth continuous coatings using a thermal deposition technique [14]. These coatings are optically smooth and consist of diamond, as well as non-crystalline carbon, with a nanometer scale microstructure. A fragment of one of these coatings (1.5 $\mu$m thick) on silicon was used as the substrate and diamond growth continued for 30 min with the torch at an $O_2$:$C_2H_2$ ratio of 1.00. Figure 2 shows the surface morphology at two different locations on this sample. Although continuous diamond growth across the surface is readily

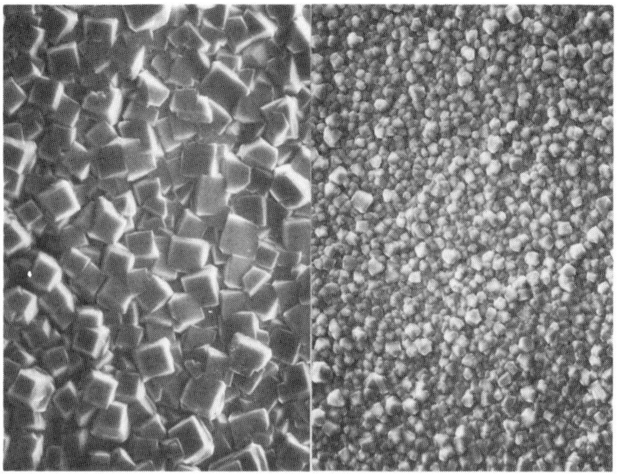


Fig. 2. Diamond deposition, using the same conditions as in Fig. 1, for 30 min onto a nanocrystalline diamond coating. Left-hand and right-hand sides of the figure are of two different regions of the same deposit. Continuous diamond growth across the entire surface is obtained; however, the morphology varies with location.

obtained, the morphology varies with position on the surface. The previously deposited diamond was removed from the surrounding area, leaving bare silicon around the region of the deposit. Presumably the surrounding diamond was oxidized by the oxygen rich periphery of the flame. Removal of the surrounding diamond might be avoided through the use of a non-oxidizing shield gas (*e.g.* hydrogen or helium); however, this has yet to be demonstrated.

The goal of obtaining a uniform and reproducible morphology proved to be elusive. Figure 3 shows a deposit obtained after 40 min using a 0.66 mm torch tip with flow rates of 1.20 sl min$^{-1}$ for both $O_2$ and $C_2H_2$. Although diamond is clearly obtained (as can be seen from the Raman pattern for this material in Fig. 5(a)), the morphology varies across the deposit and the coating is extremely rough. If longer deposition times (about 1 h or longer) were used then a coarse columnar or particulate structure develops, an example of which is seen in Figure 4. Although improvements might be made through the use of modified torch designs, several limitations appear inherent to the method. The first of these is that linear gas velocities and temperatures are highly non-uniform. Consequently there are presumably large variations in the composition of the gas both radially to and along the axis of the flame. This coupled with the very large thermal gradients involved (in both substrate and flame) render it exceedingly difficult to obtain uniform and reproducible morphologies or rates of growth. As deposition continues and the morphology coarsens, thermal and compositional gradients would become larger in the vicinity of deposits which have grown further from the substrate, leading to the development of an even less uniform morphology.

500 μm

Fig. 3. Diamond deposition, as in Figs. 1 and 2, continued for 40 min onto a diamond polished ⟨111⟩ boron-doped silicon wafer. The Raman pattern for this deposit is seen in Fig. 5.

100 μm

Fig. 4. 66 min diamond deposition using the same conditions as in the previous figures. The morphology coarsens rapidly on continued deposition.

Absolute growth rates, estimated from the sizes attained by individual crystals or crystal masses, were relatively large and of the order of 1 $\mu$m min$^{-1}$. However this was only over a small area (1 - 10 mm$^2$) using a high flow rate of $C_2H_2$ (1 - 2 sl min$^{-1}$). Hence if one assumes a continuous film growth of around 1 $\mu$m min$^{-1}$ over about 1 mm$^2$ of surface a volumetric growth rate of $10^{-6}$ cm$^3$ min$^{-1}$ or $3 \times 10^{-4}$ mol min$^{-1}$ can be estimated. A flow rate of 1 sl min$^{-1}$ $C_2H_2$ implies that less than $10^{-4}$ mol% of the acetylene is converted to diamond.

*3.2. Low pressure, flat flame deposition*

In order to establish a system that might yield greater reproducibility, and in which the flame chemistry might be more readily studied, work was undertaken to establish if diamond could be grown by combustion at low pressure. At gas pressures of 25 - 50 Torr a stable and uniform flame front about 1 mm above the burner surface could be established using total flow rates of 3 - 8 sl min$^{-1}$. The position of the flame above the burner was sensitive to both the gas composition and pressure, and relatively insensitive to total flow rate. Substrates were mounted from 5 to 35 mm above the surface of the burner, as measured from the bottom edge of vertically mounted substrates or the polished (bottom) side of horizontally mounted substrates. Upon reducing the gas pressure from atmospheric to 50 Torr the third zone of the flame seen with the atmospheric torch (the transparent blue region) disappeared entirely so that all visible combustion in a closed system at low pressure is essentially "zone 2" combustion. Visible combustion at 100 Torr or less extended 10 - 15 cm above the burner surface. Figure 5 shows Raman patterns for deposits obtained at low pressure. It was found that more consistent, though not necessarily more uniform, coatings were obtained with vertically mounted, as opposed to horizontally mounted, substrates. Auxiliary cooling of the substrate was found to be unnecessary as heat transfer to the substrate decreased with the reduction in gas density. Substrate temperature varied with the position of the substrate in the flame, with the gas composition, and across the surface of the substrate. The edges of the substrates were the hottest and varied from 850 to 1050 °C. In many experiments horizontally mounted substrates showed evidence of deposition only at the very edge of the wafer. It is suspected that this resulted, in part, from the formation of a more or less stagnant gas layer adjacent to the wafer surface facing the flame. More surprisingly it was found that the equimolar mixtures of $O_2$ and $C_2H_2$ which readily produced diamond in the torch experiments, produced only graphite or graphitic deposits at low pressure. The Raman pattern can be seen in Fig. 5(d). However, by shifting to $O_2$ rich $O_2$:$C_2H_2$ ratios it was possible to obtain diamond as well as some non-diamond carbon. Figure 6 shows the deposit obtained on the bottom edge of a vertically mounted wafer (60 min with an $O_2$:$C_2H_2$ ratio of 1.16 at 50 Torr). Diamond was obtained most consistently at $O_2$:$C_2H_2$ ratios of 1.16 - 1.18, well above the $O_2$:$C_2H_2$ ratios used with the torch at atmospheric pressure. Additions of $H_2$ to the flame at low pressure appeared to have little or no effect. Absolute growth rates in the

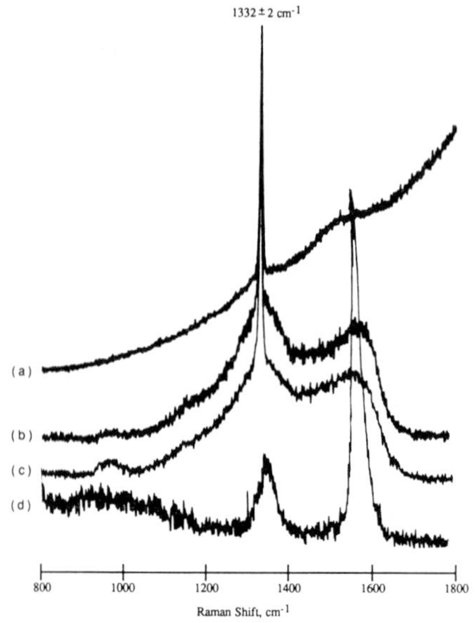


Fig. 5. Raman patterns for four different deposits: (a) diamond grown at atmospheric pressure using 0.66 mm torch nozzle ($O_2$:$C_2H_2$ ratio of unity); (b) diamond grown at 50 Torr using an $O_2$:$C_2H_2$ ratio of 1.16; (c) diamond grown at 50 Torr using $O_2$:$H_2$:$C_2H_2$ proportions of 1.16:0.26:1.00; and (d) graphitic deposit obtained at 50 Torr using an $O_2$:$C_2H_2$ ratio of unity.

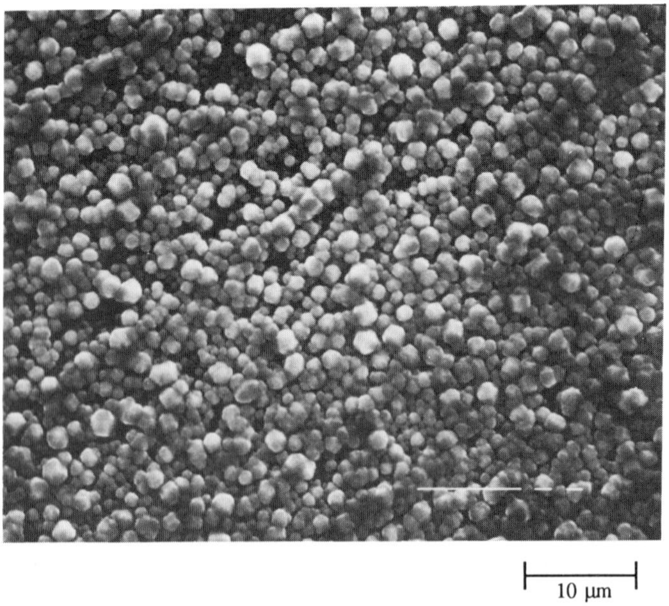


Fig. 6. Diamond deposit obtained at 50 Torr using an $O_2$:$C_2H_2$ ratio of 1.16. The Raman pattern is Fig. 5(b).

low pressure experiments were much lower than those achieved at atmospheric pressure and typical of those seen with low pressure plasma and hot wire methods, *i.e.* 0.5 - 2.0 $\mu m\, h^{-1}$. Much reduced growth rates would be expected as only a small fraction of the flame was intersected by the substrate.

### *3.3. Some thermodynamic considerations in the combustion growth of diamond*

A free energy minimization routine (SOLGASMIX), originally introduced by Eriksson [15], was used to calculate the mole fraction of solid carbon (as graphite) which should be deposited, assuming equilibrium between the gas phase and the solid. Enthalpy of formation and absolute entropy data for 46 possible gas phase species and the one invariant solid (graphite) at 1273 K were obtained by calculation from the NASA Thermodynamic Database [16]. SOLGASMIX uses the formation reactions to establish the stoichiometric relationships between each of the possible species (including the elements) and the Gibbs free energy equation ($\Delta G = \Delta H + T\Delta S$) to determine the total free energy of the system as a function of composition. This function is minimized, at the temperature and pressure of interest, by the Lagrangian method of undetermined multipliers using the mass balance constraints established by the formation reactions and the specified initial composition of the system. Curve (a) in Fig. 7 shows the predicted yield of solid carbon (graphite) for each mole of $C_2H_2$ consumed as a function of temperature at atmospheric pressure. Hence the experimental yield of solid carbon is 3 - 4 orders of magnitude less than that predicted if one assumes that the gas mixture is reversibly quenched to the substrate temperature.

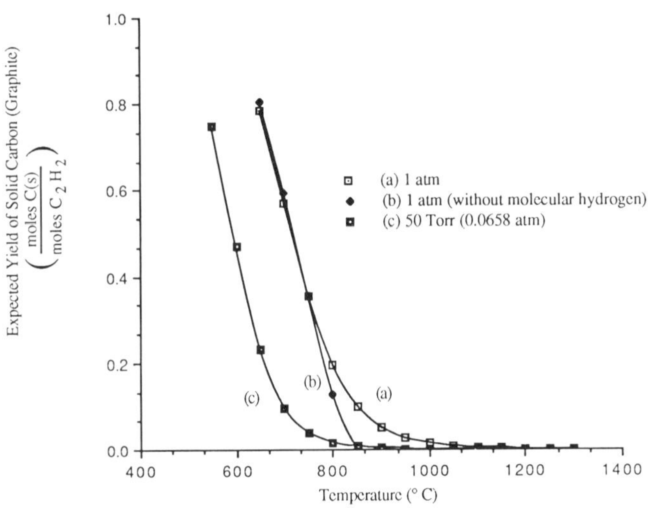


Fig. 7. Expected yield of solid carbon, as graphite, as a function of temperature in an equimolar mixture of acetylene and oxygen: (a) at 1 atm; (b) at 1 atm, if formation of molecular hydrogen is prohibited; and (c) at 50 Torr (0.0658 atm).

An interesting result is obtained if one lifts this assumption with respect to hydrogen and examines an alternative steady state. The extreme case is to remove molecular hydrogen from the database as a possible species and require that all free hydrogen in the system exist as atomic hydrogen. This is equivalent to two physical assumptions. The first is that all hydrogen liberated in the combustion process is liberated as atomic hydrogen. The second is that the formation of molecular hydrogen is sufficiently slow so as to be negligible. The effect of these assumptions on the calculation is to significantly suppress the deposition of solid carbon (curve (b) in Fig. 7) at temperatures of around 750 °C or greater. At lower temperatures there is, interestingly, little or no effect except to slightly increase the yield of solid carbon, primarily through the equilibrium; $CO + 2H^\circ = C(s) + H_2O$. This implies that a superequilibrium concentration of atomic hydrogen, which has been suggested as important to the formation of diamond instead of graphite [4, 5], will also reduce the molar efficiency of deposition at elevated temperature. At atmospheric pressure the predicted atomic hydrogen partial pressure for a gas temperature of 3000 °C (using calculated 3000 °C data for each specie) in an equimolar mixture is intermediate between these extremes at 0.161 atm (0.226 atm for $H_2$) or a mole ratio of atomic to molecular hydrogen of 0.71 (53% dissociation).

These same calculations were performed at reduced pressure with the result that the expected efficiency of carbon deposition decreases as the pressure is reduced. Curve (c) in Fig. 7 shows the predicted molar efficiency of carbon deposition at 50 Torr. The predicted reduction in molar efficiency at reduced pressure is opposite to the trend expected for the thermal decomposition of a hydrocarbon (*e.g.* methane) and results primarily from a shift of the equilibrium; $C(s) + CO_2 = 2CO$, to the right as the pressure is reduced. The reduced rate of deposition observed at 25 - 50 Torr (over that seen at 760 Torr) is consistent with this prediction, but since only a small portion of the flame was sampled no conclusion can be drawn as to its experimental relevance. The low pressure depositions were not sufficiently uniform to allow for an estimate of relative growth rate. The observation that at low pressure diamond could be deposited at $O_2$:$C_2H_2$ ratios as high as 1.18, and that to obtain diamond it was necessary to increase the oxygen content of the flame, appears to defy simple rationalization and supports the argument that deposition, at least in these experiments, is controlled by relative kinetics and that equilibrium thermodynamics is of little value.

## 6. Conclusions and future work

A reasonable model for growth of diamond from oxyacetylene flames at atmospheric pressure can be constructed using essentially the same ideas which have been proposed to rationalize the CVD of diamond by other methods, *e.g.* the production of atomic hydrogen and reactive carbon species at pressures and temperatures where solid carbon is expected to be stable.

However the morphological development of these deposits is qualitatively different, and, although substrate pretreatment with diamond powder can be seen to influence nucleation, the effect is not nearly as large as has been observed with some of the other methods. The finding that diamond could not be obtained at low pressure using the $O_2$:$C_2H_2$ ratios successful at atmospheric pressure could be rationalized as a result of the reduction of atomic hydrogen partial pressure. However this does not readily explain why any form of solid carbon (least of all diamond) should be deposited at $O_2$:$C_2H_2$ ratios as high as approximately 1.2, at low pressure and at substrate temperatures of 9000 - 1000 °C. It appears that a successful model for the growth of diamond under these conditions is more likely to be obtained by kinetic rather than by thermodynamic analyses. Flat flame burners have been extensively used for this purpose and represent a way of establishing reproducible control in the study of flame chemistry [17, 18]. Consequently diamond deposition at low pressure, using a flat flame burner, is one step towards a more thorough understanding of the reaction chemistry.

## Acknowledgments

This work was supported in part by the Office of Naval Research (with funding from the Strategic Defense Initiative Organization's Office of Innovative Science and Technology) under contract N00014-86-K-0443 and the Diamond and Related Materials Consortium of The Pennsylvania State University. The Raman spectra were obtained by Mr. Eric Plesko and Mrs. Diane Knight and their assistance is gratefully acknowledged.

## References

1 G. H. Johnson, A. R. Badzian and M. W. Geis (eds)., *Diamond and Diamond-Like Materials Synthesis, Extended abstracts of the Materials Research Society Symposium*, April 5 - 9, 1988, Reno, Nevada.

2 A. R. Badzian, P. K. Bachmann, T.Harnett, T. Badzian and R. Messier, *Diamond Thin Films Prepared by Plasma Chemical Vapor Deposition Processes, Proceedings of the European Materials Research Society Meeting*, June, 1987, Les Editions de Physique, Paris, 1987, pp. 63 - 77.

3 R. Messier, A. R. Badzian, T. Badzian, K. E. Spear, P. Bachmann, and R. Roy, From diamond-like carbon to diamond coatings, *Thin Solid Films, 153* (1987) 1 - 9.

4 B. V. Spitsyn, L. L. Bouilov and B. V. Derjaguin, *J. Cryst. Growth, 52* (1981) 219.

5 M. Frenklach and K. E. Spear, Growth mechanism of vapor-deposited diamond, *J. Mater. Res.*, *3* (1988) 133 - 140.

6 Y. Hirose and N. Kondo, *Japan Applied Physics Society Meeting*, March 29, 1988.

7 L. M. Hanssen, W. A. Carrington, J. E. Butler and K. A. Snail, Diamond Synthesis using an oxygen–acetylene torch, *Mater. Lett.*, 7 (1988) 289.

8 W. A. Carrington, L. M. Hanssen, K. A. Snail, D. B. Oakes and J. E. Butler, "Diamond growth in $O_2 + C_2H_4$ and $O_2 + C_2H_2$ flames, *Metal. Trans. A*, in press.

9 P. G. Kosky, Private Communication, Feb. 16, 1989. (Publication Pending).

10 R. J. Steffensen, J. T. Agnew and R. A. Olsen, *Tables for Adiabatic Gas Temperature and Equilibrium Composition of Six Hydrocarbons, Engineering Extension Series No. 122*, Purdue University, Lafayette, Indiana, 1966.
11 R. L. Pecsok, L. D. Shields, T. Cairns and I. G. McNilliam, *Modern Methods of Chemical Analysis*, 2nd edn., John Wiley, New York (1976), pp. 243–251.
12 W. A. Yarbrough, A. Kumar and R. Roy, Effect of substrate preparation in the vapor phase deposition of diamond, Paper No. N3.8, *Fall Meeting, Materials Research Society* 1987, Boston, MA.
13 T. Obata and M. Tajiri, *JP 86,163,276* (C1.C23C16/26) 23 July 1986.
14 W. A. Yarbrough and R. Roy, Nanocomposite diamond films by hot filament assisted CVD, in G. H. Johnson, A. R. Badzian and M. W. Geis (eds.), *Diamond and Diamond-Like Materials Synthesis, Extended Abstracts of the Materials Research Society Symposium*, April 5 - 9, 1988, Reno, Nevada, pp. 33 - 38.
15 G. Eriksson, Thermodynamic studies of high temperature equilibria, *Chemica Scr., 8* (1975) 100 - 103.
16 A. Burcat, in W. C. Gardiner, Jr., (ed.), *Combustion Chemistry*, Chapter 8, Springer-Verlag, New York, 1984.
17 D. W. Senser, J. S. Morse and V. A. Cundy, Construction and novel application of a flat flame burner facility to study hazardous waste combustion, *Rev. Sci. Instrum., 56* (1985) 1279 - 1284.
18 J. A. Miller and G. A. Fisk, Combustion chemistry, *Chem. Eng. News, 65* (1987) 22 - 46.

*Surface and Coatings Technology, 39/40* (1989) 253 - 264 

# STRUCTURE AND PROPERTIES OF c-BN FILM DEPOSITED BY ACTIVATED REACTIVE EVAPORATION WITH A GAS ACTIVATION NOZZLE

K. INAGAWA, K. WATANABE, K. SAITOH, Y. YUCHI and A. ITOH

*Institute for Super Materials (Tsukuba), ULVAC JAPAN Ltd., Tōkōdai, Tsukuba, Ibaraki 300-26 (Japan)*

(Received March 9, 1989)

## Summary

This paper describes the structure and properties of cubic boron nitride (c-BN) film deposited using activated reactive evaporation with a gas activation nozzle. c-BN films deposited directly on various substrates showed very poor adhesion. It was found that a hexagonal phase of a few hundred ångstroms thick was formed before the c-BN phase appeared and this resulted in poor adhesion. The adhesion characteristics of the c-BN film to the substrate were significantly improved by inserting silicon nitride interlayer. The micro Vickers hardness of a c-BN film of thickness 1.5 $\mu$m was 5500 - 6300 kg mm$^{-2}$. As verified by a scratch test, a high substrate temperature effected good adhesion. The coefficient of friction for the c-BN film in contact with stainless steel was about 0.35 in air at ambient temperature, whereas at $1.7 \times 10^{-3}$ Pa at 400 °C it was extremely low, 0.02.

## 1. Introduction

Cubic boron nitride (c-BN) is a very promising material which, after diamond, has the highest hardness and thermal conductivity, along with important characteristic properties such as high electrical insulation, chemical and thermal stability. At present, c-BN is widely used in manfacturing industries in cutting tools and grinding wheels on iron alloys. c-BN has also shown promise in electronic devices designed to operate at high temperatures [1].

c-BN is obtained either by direct transformation at temperatures higher than 2000 K and pressures above $115 \times 10^2$ Pa [2], or in the presence of catalysts at temperatures higher than 1700 K and pressures above $45 \times 10^2$ MPa [3].

Recently, several vapor deposition processes, which are simpler than the high pressure techniques, have been tried. In 1979 Sokolowski reported the deposition of a dense phase of a wurtzite-type BN using reactive pulse

0257-8972/89/$3.50 

plasma crystallization [4]. In 1980, Weissmantel *et al.* [5] prepared a BN film containing the cubic phase by a reactive ion beam plating technique. Satou and Fujimoto [6] were the first to show a clear electron diffraction pattern of the c-BN phase contained in an amorphous BN film deposited by ion vapor deposition. After that, many processes for the preparation of c-BN films by physical vapor deposition (PVD) and chemical vapor deposition (CVD) were investigated. For example, activated reactive evaporation (ARE) [7], ARE with a gas activation nozzle [8], laser evaporation (with an ion beam) [9], ionized evaporation [10], arc-like plasma enhanced ion plating [11], sputtering (dual beam) [12], reactive r.f. sputtering [13], plasma chemical transport [14], ion beam [15], r.f. plasma CVD thermally activated with tungsten filament [16] and ECR plasma CVD [17] have all been reported.

The authors investigated c-BN films formed by modified ARE accompanied with a gas activation nozzle using two electron beam evaporation sources, *i.e.* a hollow-cathode discharge (HCD) gun and a conventional electron beam (EB) gun [8, 18]. The results obtained can be summarized as follows. (1) Stoichiometric BN was obtained using the gas activation nozzle. (2) It was necessary to apply a r.f. bias potential to the substrate to accelerate the formation of the c-BN phase. (3) It was found that adding argon gas to the reactant gas enhanced the formation of the cubic phase. The adhesion of the c-BN film, however, was very poor.

Next, the adhesion characteristics of the c-BN films were investigated [19]. This paper reports the structure of the BN films near the interface with the substrate and the improvement of the adhesion by inserting an interlayer. Some properties on the hardness, the evaluation of the adhesion by scratch tests and the tribological characteristics of the c-BN films are also presented.

## 2. Experimental procedure

The experimental conditions for preparing the cubic phase have been previously described [18]. In this experiment, a c-BN film was deposited on the various substrates using a conventional EB gun. The experimental conditions are shown in Table 1. A nozzle current of 7 A and a peak-to-peak r.f. bias voltage of 600 V, causing a self-bias voltage of about $-300$ V, were included in the conditions of c-BN formation indicated in the previous work. Other experimental procedures were the same as described in ref. 18.

When an interlayer was inserted between the substrate and c-BN film, the co-evaporation system shown in Fig. 1 was used. The evaporation ratio of boron and of the additive element was controlled separately and monitored with quartz oscillator rate meters. The conditions for c-BN film formation are shown in Table 1. Silicon, iron, cemented carbide, etc., were used as substrates.

The crystal structure of the film was determined by infrared (IR) spectroscopy. The depth profiles of the elements on the deposited film was investigated by Auger electron spectroscopy (AES). The film hardness was

TABLE 1

Deposition conditions for c-BN films prepared by modified ARE

| | |
|---|---|
| Substrate temperature (°C) | 400 |
| Argon flow rate (sccm) | 13 |
| Argon pressure (Pa) | 0.11 |
| $N_2$ flow rate (sccm) | 7 |
| $N_2$ pressure (Pa) | 0.053 |
| Activation nozzle current (A) | 7 |
| Activation nozzle voltage (V) | 77 |
| Tungsten filament current (A) | 37 |
| p–p r.f. bias voltage (V) | 600 |

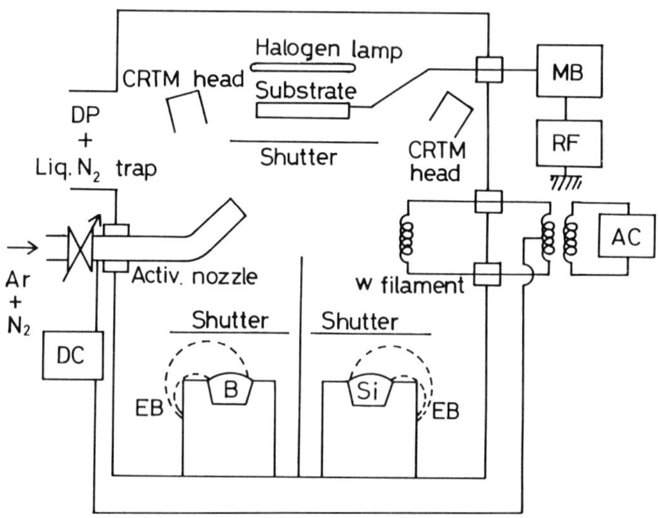


Fig. 1. Schematic diagram of the modified ARE equipment with a co-evaporation system.

measured by a micro Vickers hardness meter. The adhesive or cohesive strength [20] of the c-BN films was estimated with a scratch tester. The coefficient of friction was measured by a ring-on-disk type tribotester, which could be used in vacuum and at high temperatures.

## 3. Results and discussion

### *3.1. Direct deposition of c-BN films*

c-BN films about 3000 Å thick were deposited directly on various substrates, *i.e.* silicon, titanium, iron, nickel, copper, stainless steel, titanium carbide, tungsten carbide (cemented carbide), aluminum nitride, titanium nitride, silicon nitride, alumina and silicon oxide. The deposited c-BN film peeled off from all the substrates after a period of time in normal air, but in dry air or in a vacuum the film did not peel. c-BN film having a thickness of greater than about 5000 Å peeled off even during deposition. These facts suggest that c-BN films have a weak moisture resistance and large internal stress. To understand the cause of this poor adhesion, the structure of the film near the interface with the substrate was investigated. This was performed using two methods. In one, the c-BN film thickness was reduced using a xenon ion beam. In the second, a very thin BN film was achieved by varying the deposition time.

Figure 2 shows the result of the thinning method. As shown in Fig. 2b, a c-BN film was etched by 2 keV xenon ions, then the IR spectra corresponding to the various BN film thicknesses was measured. The observed IR spectra are shown in Fig. 2a. The IR spectrum of step 1 (indicating the film surface) shows the cubic phase only, but the step 2 spectrum (of the slightly etched film) shows both cubic and hexagonal phases. Steps 3 and 4 show a complete hexagonal phase. Thus, the film structure close to the substrate was found to be of a hexagonal phase in spite of the fact that conditions for cubic phase formation had been used. The depth profile of the B/N ratio did not change towards the substrate, and the stoichiometric composition was maintained.

Figure 3 shows the intensity of the IR absorption band of both the cubic and hexagonal phases of the BN films prepared by varying the deposition times under the previous conditions of c-BN formation. It is clear that the hexagonal phase is formed in the initial stage only up to 1 min. Its thickness is a few hundred ångstroms. After 1 min, the cubic phase begins to grow and

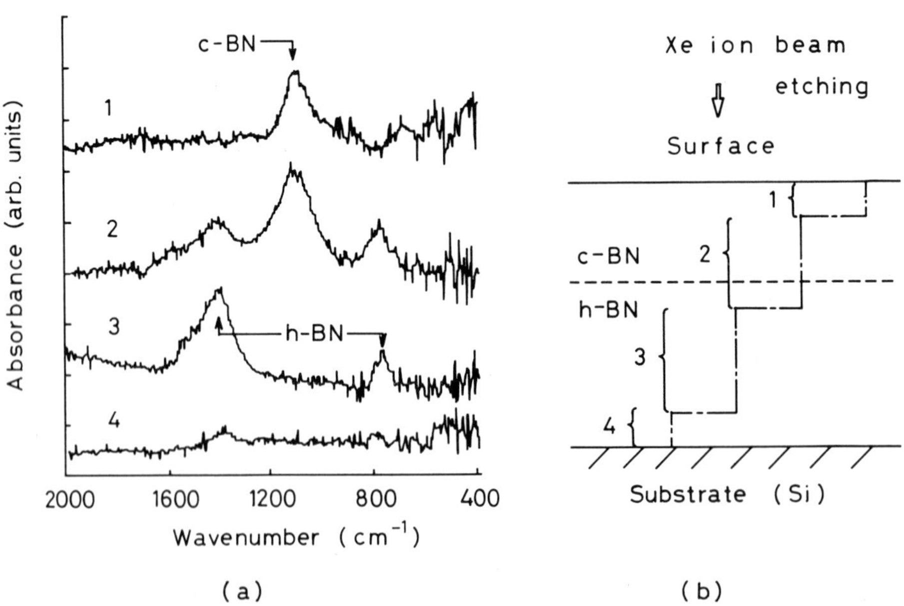


Fig. 2. a, IR spectra of BN film near the interface with the substrate. The film was deposited under conditions for cubic phase formation. Notations 1, 2, 3 and 4 show the steps of the thinned BN film. b, Schematic cross-sectional view of BN film.

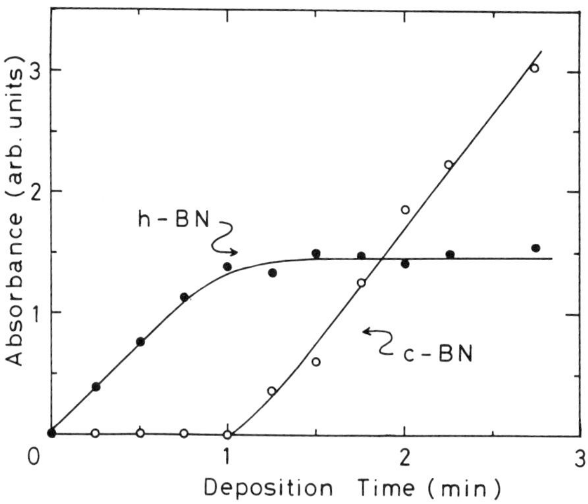


Fig. 3. Intensity of the IR absorption band of the BN film with varying deposition times. c-BN, 1050 cm$^{-1}$; h-BN, 1400 cm$^{-1}$.

the formation of the hexagonal phase stops. The cubic phase increases with the deposition time. This result agrees with the former results based on the BN film reduced by xenon ion beam etching. From both results, it can be considered that the thin hexagonal boron nitride (h-BN) layer formed prior to the cubic phase has essentially poor adhesion and causes peeling-off of the c-BN film over-growth.

*3.2. Inserting an interlayer between the substrate and c-BN film*

In order to improve the adhesion, the effect of an interlayer between the substrate and c-BN film was investigated. The interlayer, which consisted of a pure additive element layer and mixed compound layer having a graded composition of the additive compound, and finally pure c-BN on top, was formed by evaporating the additive element simultaneously with boron in a nitrogen atmosphere using a co-evaporation system equipped with two conventional EB guns. Elements such as carbon, silicon, germanium, aluminum, iron, titanium and chromium were examined as the additive.

Table 2 shows the experimental results of the effect of the additive elements on the adhesion of the c-BN film in air.

Only silicon significantly improved the adhesion of the c-BN film. With aluminum, the adhesion of the c-BN film was slightly improved in the case of a film 0.3 $\mu$m thick, but for a film 1.5 $\mu$m thick the film peeled off. In specimens with poor adhesion, the c-BN film peeled off at the boundary near the zero content of the additive element.

Next, the effect of gradual changes in the silicon content of the interlayer on the adhesion of the c-BN film was examined. Figure 4 shows several schematic diagrams of the gradual change of the evaporation rate to prepare interlayers with decreasing silicon contents. The c-BN film deposited

TABLE 2

Effect of additive elements constituting interlayer on adhesion of c-BN film in air[a]

| *Additive element* | *Adhesion* | |
|---|---|---|
| | *Thickness of c-BN* 0.3 μm | 1.5 μm |
| Carbon | − | − |
| Silicon | ++ | ++ |
| Germanium | − | − |
| Aluminum | + | − |
| Iron | − | − |
| Titanium | − | − |
| Chromium | − | − |

[a] ++, Good adhesion; +, adhesion; −, exfoliation.

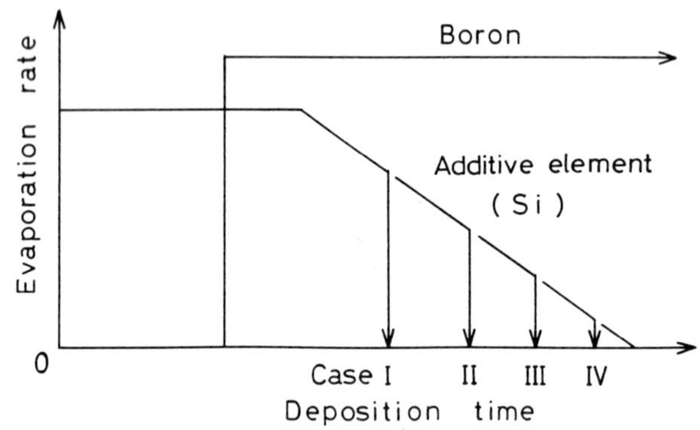


Fig. 4. Schematic diagram of the evaporation rate used to prepare the interlayer with gradually decreasing silicon content.

after cases I and II peeled off at the boundary near the silicon element zero in air, while the film deposited after cases III and IV did not peel off and showed excellent adhesion. Thus, an abrupt change of silicon content to zero, such as cases I and II, hardly improved the adhesion of c-BN film. Figure 5 shows the AES depth profiles of boron, silicon and nitrogen on the thin film prepared in the same way as case IV, resulting in excellent adhesion. It is clear that the silicon content gradually decreases towards the c-BN film of the left side of Fig. 5.

Table 3 shows the film sturcture of a mixed layer prepared with different compositions corresponding to those before the change of silicon content to zero in Fig. 4. In films deposited under conditions corresponding to cases I and II (poor adhesion), B–N bonds of the h-BN type (1380 and 780 $cm^{-1}$) and a Si–N bond of the $Si_3N_4$ type (840 $cm^{-1}$) were observed. Whereas in films corresponding to cases III and IV (excellent adhesion), a B–N bond of the c-BN type (1050 $cm^{-1}$) was observed, the presence of a Si–N bond could

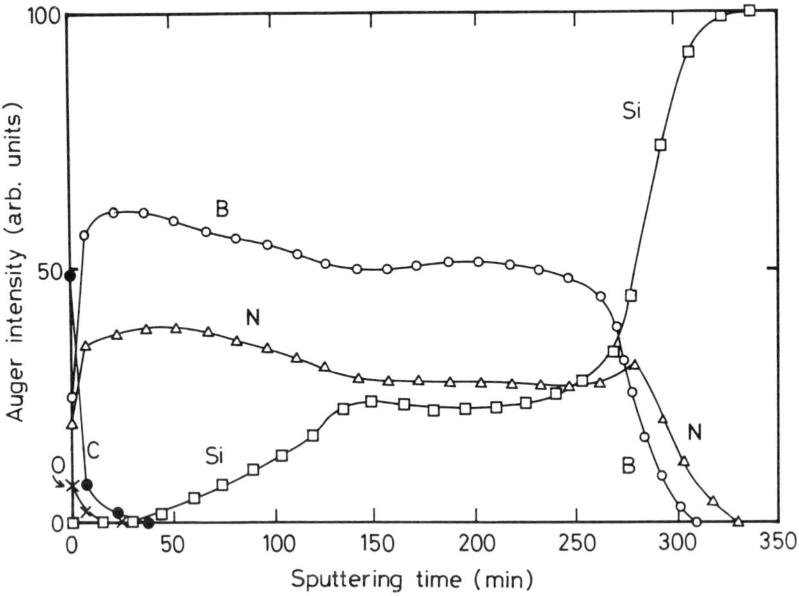


Fig. 5. AES depth profile of the interlayer designed to increase the adhesion of the c-BN film.

TABLE 3

Film structure of mixed layer with different silicon contents

| *Silicon content* | *Bonding type* | |
|---|---|---|
| | *Boron nitride* | *Silicon nitride* |
| Case I | Hexagonal | $Si_3N_4$ |
| Case II | Hexagonal | $Si_3N_4$ |
| Case III | Cubic | — |
| Case IV | Cubic | — |

not be ascertained because the IR absorption intensity corresponding to $Si_3N_4$ was too close to the noise level. Thus, the final silicon content should be this small in order to obtain excellent adhesion.

As mentioned above, a c-BN film with excellent adhesion was deposited by inserting a silicon nitride interlayer. The silicon content had to be gradually decreased to zero rather than suddenly decreased. It is supposed that silicon nitride in the boundary region near the substrate increases the adhesion of the c-BN film.

### *3.3. Properties of the c-BN films*

#### *3.3.1. Hardness*

The micro Vickers hardness of BN films deposited on cemented carbide inserts (JIS K10 for cast iron cutting) was measured under a load of 10 gf.

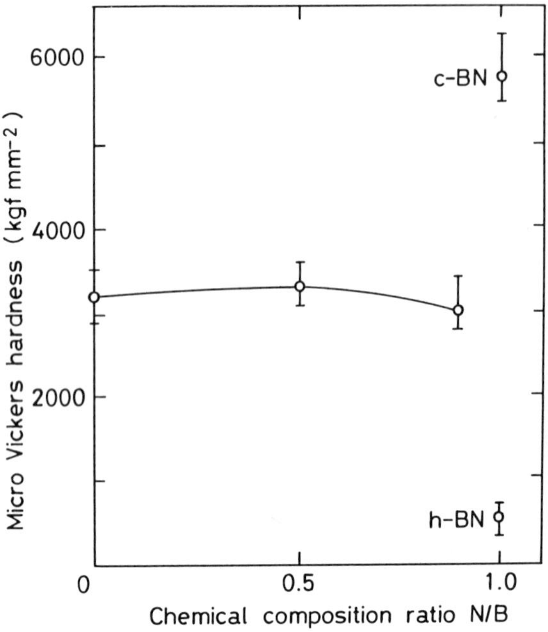


Fig. 6. Micro Vickers hardness of BN films deposited by modified ARE.

Figure 6 shows the relation between the hardness of the BN film and the chemical composition ratio N/B, as determined with an electron microprobe X-ray analyzer. The hardness of a c-BN film about 1.5 $\mu$m thick was very high, 5500 - 6300 kgf mm$^{-2}$, *i.e.* as high as converted compact c-BN (Knoop hardness 4600 - 6600 kgf mm$^{-2}$ with an indenter load of 3500 gf) [21]. The film thickness of the specimens without the c-BN film was about 2 $\mu$m. The hardness of the h-BN film was very low. The film hardness of the specimens with N/B ratios of 0, 0.5 and 0.9 was considerably higher in comparison with the h-BN, although broad IR absorption bands corresponding to the h-BN phase were observed in the films with N/B ratios of 0.5 and 0.9. It is possible that both films were hard because they were the mixtures composed of h-BN and pure boron.

*3.3.2. Adhesion*

Scratch tests were applied to estimate the adhesive or cohesive strength of the c-BN film deposited on cemented carbide inserts at various substrate temperatures. A Rockwell diamond cone (200 $\mu$m radius) was used to scratch the surface of the c-BN-coated specimen at a constant speed of 10 mm min$^{-1}$ and under a constant load.

Figure 7 shows photographs of the scratched specimen surfaces. The c-BN film of specimen a was deposited at a temperature of 330 °C and its thickness including the interlayer was 1 $\mu$m. Photographs b, c and d in Fig. 7 are the results at different positions of the cemented carbide insert, on

(a) Load 7 kgf
( Tsubs. 330 °C )

(b) 7 kgf
( 600 °C )

(c) 9 kgf
( 600 °C )

(d) 10 kgf
( 600 °C )

Fig. 7. Scratched surface of the c-BN film coated on cemented carbide insert by modified ARE.

which the film of 0.8 $\mu$m thick was deposited at a temperature of 600 °C. In the case of a 7 kgf load, adhesive failure at some points arose in specimen a, whereas there was no failure of the film in specimen b coated at a temperature higher than specimen a. It is clear that the high substrate temperature increased the adhesion of the c-BN film with the substrate. In specimen c, a scratch of 9 kgf load caused cohesive failure due to the brittleness of the film. Many cohesive failures were observed in specimen d tested under a load of 10 kgf.

*3.3.3. Coefficient of friction*

The outside and inside diameters of the ring (material: JIS SUS 304 stainless steel) used in the ring-on-disk-type tribotester were 25 mm and 20 mm respectively. A c-BN film of 0.4 $\mu$m thick was deposited at a temperature of 330 °C on a silicon wafer for application to the disk.

Figure 8 shows the relation between the coefficient of friction and the applied load in air at ambient temperature. The coefficient of friction was nearly constant (0.35) within the region of 600 - 1000 gf. Figure 9 shows the relation between the coefficient of friction and the sliding time in a vacuum

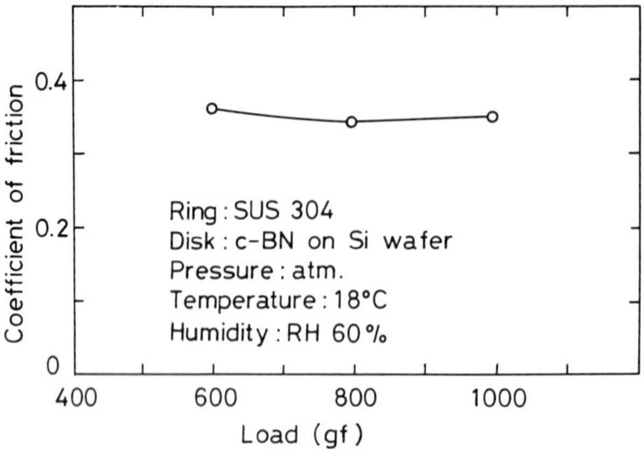


Fig. 8. Relation between the coefficient of friction and the applied load on c-BN films deposited by modified ARE.

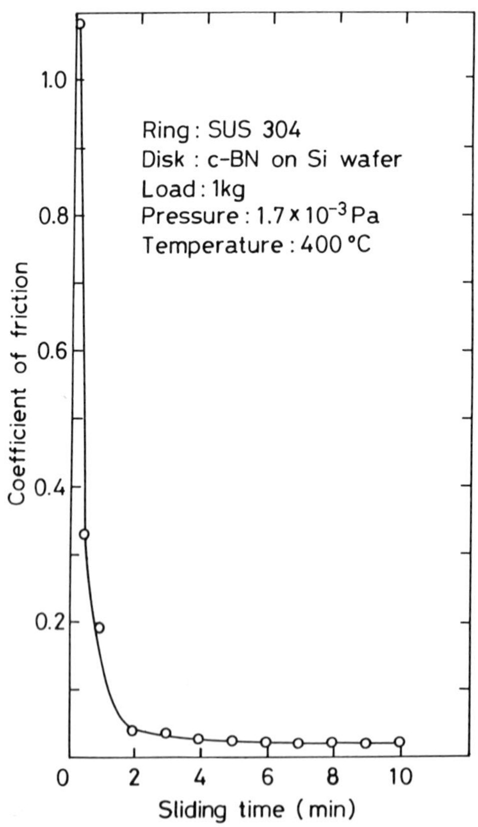


Fig. 9. Relation between coefficient of friction and sliding time in vacuum at high temperature on c-BN films deposited by modified ARE.

of pressure $1.7 \times 10^{-3}$ Pa at a temperature of 400 °C. The coefficient of friction was high at the initial stage, but then reduced to a very small value of 0.02 after rapidly decreasing with the sliding time. This value is lower than the lowest one (0.05 - 0.2) for the case of TiC–TiN among various metal–metal, metal–ceramic (film) and ceramic (film)–ceramic (film) combinations at ambient temperature without lubrication [22]. Further experiments are needed to clarify the reason for this.

## 4. Conclusion

The structure and properties of c-BN films prepared by a modified ARE process were investigated. In spite of the conditions for c-BN phase formation, a h-BN phase of a few hundred ångstroms in thickness was deposited before c-BN film deposition. The c-BN film deposited directly on various substrates displayed very poor adhesion with the substrates. A c-BN film with excellent adhesion was obtained by inserting a silicon nitride interlayer between the substrate and c-BN film.

The micro Vickers hardness of the c-BN film was very high, 5500 - 6300 kgf $mm^{-2}$. Scratch tests showed that a high substrate temperature provided good adhesion of the c-BN film with the substrate. The coefficient of friction was about 0.35 in air at ambient temperature, whereas it was very low, 0.02, in a vacuum of pressure of $1.7 \times 10^{-3}$ Pa at a temperature of 400 °C.

## Acknowledgments

The authors would like to thank Dr. C. Hayashi, the president of ULVAC JAPAN Ltd., for this encouragement throughout this work. They also would like to acknowledge Mr. K. Sekino for his valuable cooperation in the AES observations.

## References

1 O. Mishima, J. Tanaka, S. Yamaoka and O. Fukunaga, *Science, 238* (1987) 181.
2 F. P. Bundy and R. H. Wentorf, Jr., *J. Chem. Phys., 38* (1963) 1144.
3 R. H. Wentorf, Jr., *J. Chem. Phys., 26* (1957) 956.
4 M. Sokolowski, *J. Cryst. Growth, 46* (1979) 136.
5 C. Weissmantel, K. Bewilogua, D. Dietrich, H.-J. Erler, H.-J. Hinnerberg, S. Klose, W. Nowick and G. Reisse, *Thin Solid Films, 72* (1980) 19.
6 M. Satou and F. Fujimoto, *Jpn. J. Appl. Phys., 22* (1983) L171.
7 K. Chopra, V. Agarval, V. D. Vankar, C. V. Deshpandey and R. F. Bunshah, *Thin Solid Films, 126* (1985) 307.
8 K. Inagawa, K. Watanabe, I. Tanaka, K. Saitoh and A. Itoh, *Proc. 9th Symp. on ISIAT, Tokyo,* The Research Group of Ion Engineering, Kyoto, 1985, p. 299.
9 N. Yasunaga, S. Mineta, N. Tarumi, A. Obara, M. Ikeda, Y. Furuta and M. Kohata, *Proc. 5th Int. Conf. on Production Engineering, Tokyo, 1984,* p. 478.

10 T. Mohri, O. Iijima and Y. Namba, *Shinku (J. Vac. Soc. Jpn.), 29* (1985) 307.
11 T. Ikeda, Y. Kawate and Y. Hirai, *Shinku (J. Vac. Soc. Jpn.), 31* (1988) 968.
12 F. Shimokawa, H. Kuwano and K. Nagai, *Proc. 9th Symp. on ISIAT, Tokyo*, The Research Group of Ion Engineering, Kyoto, 1985, p. 467.
13 M. Mieno, T. Yoshida and K. Akashi, *Nihon Kinzoku Gakkaishi (J. Jpn. Inst. Met.), 52* (1988) 199.
14 S. Komatsu, T. Yoshida and K. Akashi, *Proc. 9th Symp. on ISIAT, Tokyo*, The Research Group of Ion Engineering, Kyoto, 1985, p. 421.
15 W. Halverson and D. T. Quinto, *J. Vac. Sci. Technol., A3* (1985) 2141.
16 Y. Ichinose, H. Saitoh and T. Ichiguro, *Proc. 9th Symp. on ISIAT, Tokyo*, The Research Group of Ion Engineering, Kyoto, 1987, p. 469.
17 A. Chayahara, H. Yokoyama, T. Imura and Y. Osaka, *Jpn. J. Appl. Phys., 26* (1987) L1435.
18 K. Inagawa, K. Watanabe, H. Ohsone, K. Saitoh and A. Itoh, *J. Vac. Sci. Technol., A5* (1987) 2696.
19 K. Watanabe, K. Saitoh, Y. Yuchi, K. Sekino, K. Inagawa and A. Itoh, *Proc. 11th Symp. on ISIAT, Tokyo*, The Research Group of Ion Engineering, Kyoto, 1987, p. 347.
20 P. Laeng, P. A. Steinmann and H. E. Hintermann, *Proc. ISTFA*, 1986, p. 67.
21 F. R. Corrigan and F. P. Bundy, *J. Chem. Phys., 63* (1975) 3812.
22 R. F. Bunshah, *Thin Solid Films, 107* (1983) 21.

*Surface and Coatings Technology, 39/40* (1989) 265 - 273 

# SYNTHESIS OF BN FILMS BY THE PLASMA CVD WITH VARIOUS SOLIDS: $BH_3NH_3$, $H_3BO_3$ AND $NaBH_4$

H. SAITOH, T. HIROSE, H. MATSUI, Y. HIROTSU and Y. ICHINOSE

*Nagaoka University of Technology, 1603–1, Kamitomioka, Nagaoka, Niigata 940-21 (Japan)*

(Received April 1, 1989)

## Summary

We have studied the formation conditions of cubic boron nitride (c-BN) formed by the dissociation of $BH_3NH_3$, $H_3BO_3$ and $NaBH_4$ in the reactive gases $H_2$ or $NH_3$, and also by the plasma excitement of the dissociated gases using r.f. induction and tungsten filament heating. It was found in these deposition processes that the tungsten filament heating greatly contributes to the formation of c-BN as compared with the contribution of r.f. induction.

The deposition ratio c-BN/amorphous BN (a-BN) in the films deposited with $BH_3NH_3$–$H_2$ changes with the change of the emission intensity ratio $I(H\alpha)/I(H_2^*)$ in the gas plasma, where the wavelengths of the $H\alpha$ and $H_2^*$ photoemissions are 656 nm and 603 nm respectively. In particular, when the ratio $I(H\alpha)/I(H_2^*)$ becomes maximum, the deposition rate of c-BN also becomes maximum. In the $H_3BO_3$–$NH_3$ and $NaBH_4$–$NH_3$ systems, it was found that the deposition ratio c-BN/a-BN varies with the emission intensity ratio $I(N_2^+)/I(NH_3^*)$ and $I(N_2^*)/I(NH_3^*)$, where the wavelengths of the $N_2^+$, $N_2^*$ and $NH_3^*$ photoemissions are 391 nm, 406 nm and 564 nm respectively.

## 1. Introduction

The synthesis of cubic boron nitride (c-BN) from hexagonal BN (h-BN) catalyst systems was first announced by Wentorf [1] in 1957. Subsequently, c-BN was obtained by the direct transformation from h-BN by Bundy and Wentorf [2]. BN films have been synthesized by various processes such as boron evaporation with $N_2^+$ ion-beam bombardment [3], activated reactive evaporation using the hollow cathode discharge [4] and ECR plasma chemical vapor deposition (CVD) [5]. In the plasma CVD method, various raw materials and reactive gases such as $B + N_2$ [3, 6], $B + N_2 + Ar$ [4, 7], $B_2H_6 + NH_3 + H_2$ [8] and $B_3N_3H_6$ [9] have been used for the synthesis. The c-BN structure, however, used to be formed partially in the amorphous or turbostratic BN (a-BN) structure.

We have investigated the synthesis of c-BN and found a new method

0257-8972/89/$3.50 

using r.f. plasma CVD thermally assisted by a tungsten filament [10]. In the case of synthesis with $B_2H_6 + NH_3 + H_2$, it was found that the deposition ratio c-BN/a-BN depends on the conditions of the r.f. power, filament temperature and gas pressure. The c-BN structure was especially formed under conditions when the tungsten filament temperature was higher than 1600 °C and the r.f. power was higher than 50 W. Thus, it is necessary to the formation of c-BN that hydrogen radicals are excited into higher activated energy levels in the gas plasma [10].

In the present report, relations between formation conditions and crystal structures of BN with respect to various solids, *i.e.* $BH_3NH_3$, $H_3BO_3$ and $NaBH_4$, as raw materials will be described and discussed.

## 2. Experimental procedure

Figure 1 shows a schematic diagram of the apparatus for thermally assisted r.f. plasma CVD. The system is composed of two main parts, *i.e.* a plasma chamber and a substrate heating chamber with a tungsten filament situated just above the substrate. The solid raw material (0.1 g) is placed in a ceramic boat in the plasma chamber. In a typical run, the metal chamber is evacuated to a pressure of 1.3 mPa using the diffusion pump system. After heating the substrate to a temperature of 800 °C, the solid raw material is evaporated and a reactive gas is induced. The evaporated solid raw material and the induced gas then make a gas mixture. The mixed gas is excited into the plasma state by the r.f. induction of 13.56 MHz. The excited plasma is further thermally activated by the heating of the tungsten filament. The reactive gases used here were selected corresponding to the raw solid materials: hydrogen ($H_2$) for $BH_3NH_3$, and $NH_3$ for $H_3BO_3$ and $NaBH_4$. The BN films were deposited on silicon wafers. The deposition conditions are summarized in Table 1.

The crystal structure analyses of as-deposited films were made by transmission electron microscope (TEM) observation and infrared (IR)

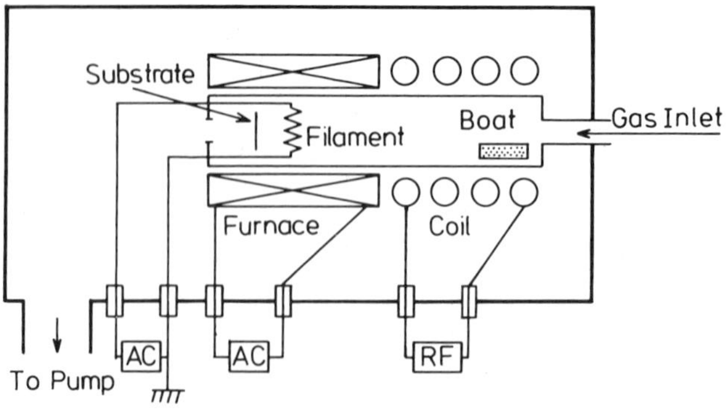


Fig. 1. Schematic diagram of the r.f. plasma CVD apparatus.

TABLE 1

Deposition conditions

| | |
|---|---|
| filament temperature | 1100 - 1700 °C |
| r.f. power | 0 - 200 W |
| gas pressure | 20 - 500 Pa |
| substrate temperature | 800 °C |
| substrate | silicon |
| deposition time | 0.5 h |

absorption measurement. The c-BN and a-BN have strong absorption bands at 1080 $cm^{-1}$ and 1380 $cm^{-1}$ (c plane) respectively. The deposition ratio of c-BN/a-BN can be obtained by the intensity ratio of the IR absorption bands, since the absorption intensity is a function of the film thickness. In this work, the deposition ratio was obtained by this method. The photoemission spectra of the plasma with and without tungsten filament heating were measured in the visible range by optical emission spectroscopy.

## 3. Results and discussion

### 3.1. $BH_3NH_3$–$H_2$ system

In this system, c-BN was always synthesized under the conditions of filament temperature greater than 1600 °C and r.f. power greater than 60 W. The reactive gas pressure was found to affect the c-BN deposition rate. Figure 2 shows the IR absorption spectra measured for films deposited at filament temperatures from 1100 to 1700 °C. The r.f. power and the gas pressure were kept constant at 100 W and 70 Pa respectively. The films deposited below 1500 °C have no absorption peak; however, those deposited at

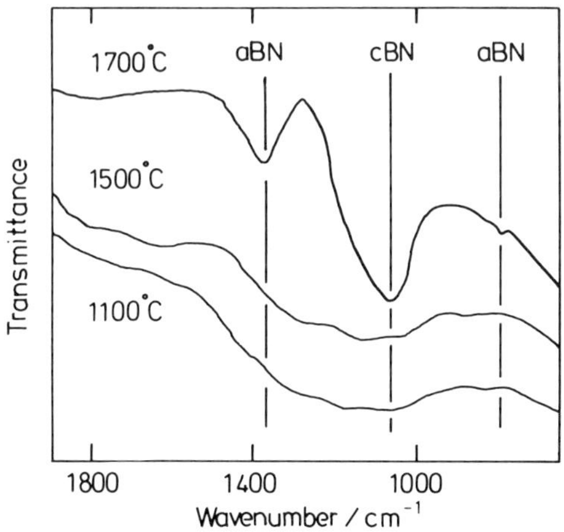


Fig. 2. IR absorption spectra of the films deposited with $BH_3NH_3$–$H_2$.

1700 °C show several absorption peaks. A strong peak near 1080 cm$^{-1}$ corresponds to the $F_2$ vibrational mode of c-BN, and a medium absorption peak at 1380 cm$^{-1}$ and a weak one at 800 cm$^{-1}$ are assigned to the $E_{1u}$($c$ plane) and $A_{2u}$(out of $c$ plane) vibrational modes of a-BN respectively. According to the TEM observation of the films deposited at a filament temperature of 1700 °C, many aggregates of crystalline particles with a particle size of about 30 nm were embedded in the amorphous structure. The mean size of the aggregates and their average distance were about 500 nm and 700 nm respectively. Electron diffraction patterns from the amorphous region showed halo-rings corresponding to those of a-BN or turbostratic BN. However, diffraction patterns of the zinc-blende-type structure having a lattice constant of $a_0 = 0.362$ nm were observed in the crystalline region. Thus, the a-BN and c-BN structures were recognized in the films deposited at a filament temperature of 1700 °C.

The IR absorption spectra were measured for the films deposited at r.f. powers from 0 to 200 W. The filament temperature and the gas pressure were kept constant at 1700 °C and 70 Pa. Intensity ratios $A_{\text{c-BN}}(1080\ \text{cm}^{-1})/A_{\text{a-BN}}$ (1380 cm$^{-1}$) for the IR absorption spectra are shown in Fig. 3. Both the a-BN and c-BN structures were recognized in the films deposited at r.f. powers above 80 W, but the c-BN was not formed in the films deposited at 0 W. The ratio $A_{\text{c-BN}}/A_{\text{a-BN}}$ increases with increasing r.f. power since the deposition rate of c-BN increases rapidly, contrary to that of a-BN which decreases above 100 W.

When the reactive gas pressure was changed, the IR absorption intensity ratio changed as shown in Fig. 4, where the r.f. power and filament temperature were kept constant at 200 W and 1700 °C respectively. The ratio increases with increasing gas pressure up to 70 Pa and then decreases rapidly with a further increase of gas pressure. The increase of the ratio is caused by the increase of the deposition rate of c-BN as compared with that of a-BN. The decrease of the ratio resulted from the rapid increase of the deposition rate of a-BN above 70 Pa.

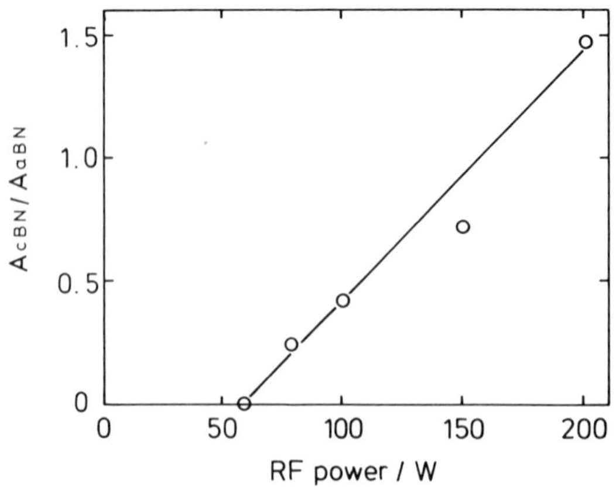


Fig. 3. IR absorption intensity ratio as a function of r.f. power.

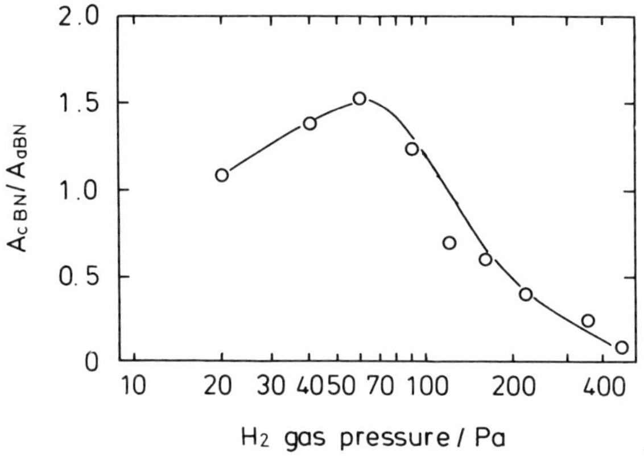


Fig. 4. IR absorption intensity ratio as a function of gas pressure.

According to these results, it was found that the deposition ratio c-BN/a-BN depends upon deposition conditions such as the filament temperature, r.f. power and gas pressure.

According to the plasma photoemission spectrophotometry, photoemissions from several radicals such as $H_2$* (hydrogen molecule radical) and H (atomic hydrogen) were observed in the reactive gas plasma throughout the whole visible region. In particular, photoemissions with wavelengths of 603 nm and 656 nm, corresponding to the radicals $H_2$* and H$\alpha$ respectively, are very strong. Figure 5 shows the emission intensity ratio $I(H\alpha)/I(H_2^*)$ and $I(H\alpha)/I(H\alpha_0)$ as a function of filament temperature, where $I(H\alpha_0)$ is the constant emission intensity of atomic hydrogen excited by r.f. induction only (without filament heating). The ratio $I(H\alpha)/I(H_2^*)$ is nearly constant below 1500 °C, but becomes much stronger above that temperature. This is because

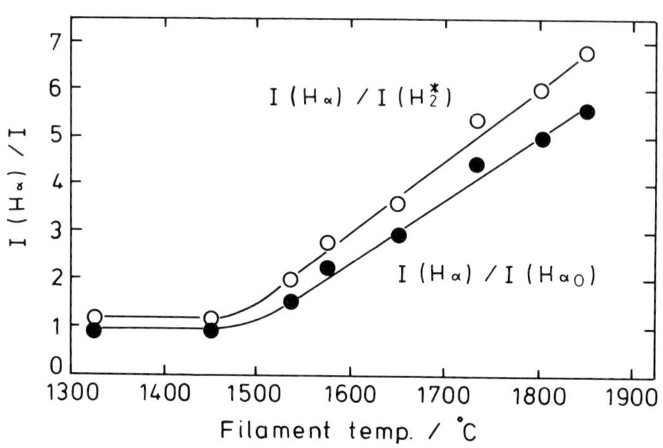


Fig. 5. Photoemission intensity ratios $I(H\alpha)/I(H_2^*)$ and $I(H\alpha)/I(H_2^*)$ as a function of filament temperature: r.f. power; 100 W; gas pressure; 70 Pa.

the degree of dissociation of hydrogen increases rapidly with increasing temperature above 1500 °C. In the higher temperature region (greater than 1500 °C), dissociation by filament heating is found to be predominant.

The emission intensity ratio $I(H\alpha)/I(H_2^*)$ as a function of r.f. power is shown in Fig. 6. The emission intensities begin to stabilize above 30 W. The hydrogen molecule is thought to be a little dissociated to the hydrogen atom by r.f. power higher than 30 W. The ratio $I(H\alpha)/I(H_2^*)$ increases slightly with increasing r.f. power.

The ratio $I(H\alpha)/I(H_2^*)$ as a function of gas pressure is shown in Fig. 7. The ratio shows the gradual decrease with increasing gas pressure to 600 Pa. The decrease of the ratio results from the decline of the dissociation rate, that is to say, the decrease in the amount of radicals.

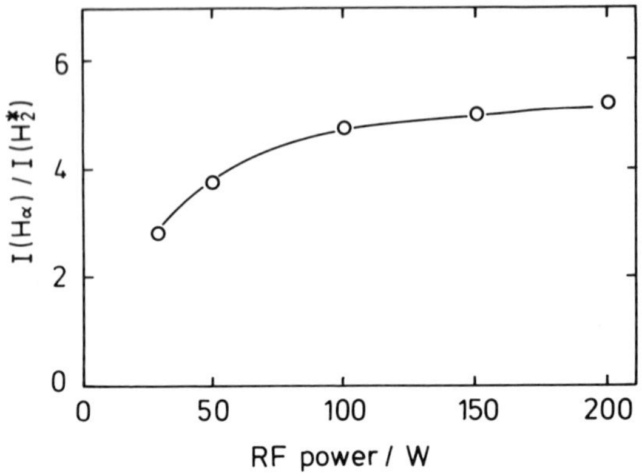


Fig. 6. Photoemission intensity ratios $I(H\alpha)/I(H_2^*)$ as a function of r.f. power: filament temperature; 1700 °C; gas pressure; 70 Pa.

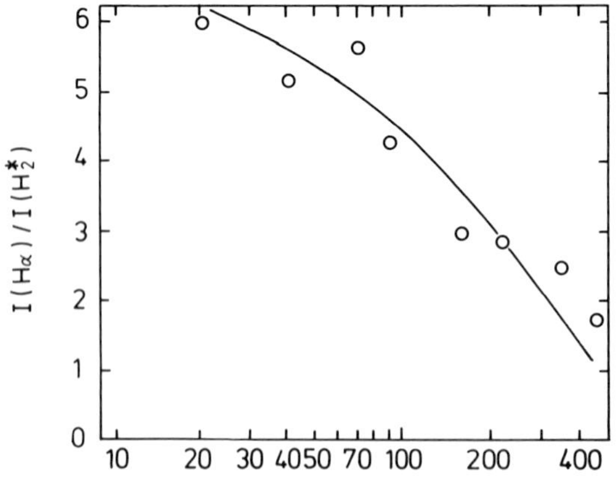


Fig. 7. Photoemission intensity ratios $I(H\alpha)/I(H_2^*)$ as a function of gas pressure: filament temperature; 1700 °C; r.f. power; 100 W.

According to the above results, it was found that the deposition ratio c-BN/a-BN is closely related to the emission intensity ratio $I(H\alpha)/I(H_2^*)$. In particular, when the ratio $I(H\alpha)/I(H_2^*)$ becomes maximum, the deposition rate of c-BN also becomes maximum.

*3.2. $H_3BO_3$–$NH_3$ and $NaBH_4$–$NH_3$ systems*

The IR absorption intensity ratios for these systems change almost in the same manner as shown in the case of the $BH_3NH_3$–$H_2$ system with increasing r.f. power and gas pressure. The ratios in these systems $A_{c\text{-}BN}/A_{a\text{-}BN}$ are shown in Fig. 8 as a function of r.f. power. Both the a-BN and c-BN structures were characterized in the films deposited at various conditions including a condition without r.f. induction. For gas pressure, the intensity ratios are shown in Fig. 9(a). The ratios have broad maxima in the gas pressure range between 20 and 50 Pa. The result of plasma spectrophotometry of $NH_3$ excited by both r.f. induction and filament heating is shown in Fig. 9(b). The photoemission intensity ratios $I(N_2^+)/I(NH_3^*)$ and $I(N_2^*)/I(NH_3^*)$ have broad maxima in the gas pressure range between 10 and 100 Pa. By comparison, it is found that the deposition ratio c-BN/a-BN depends upon the emission intensity ratios $I(N_2^+)/I(NH_3^*)$ and $I(N_2^*)/I(NH_3^*)$.

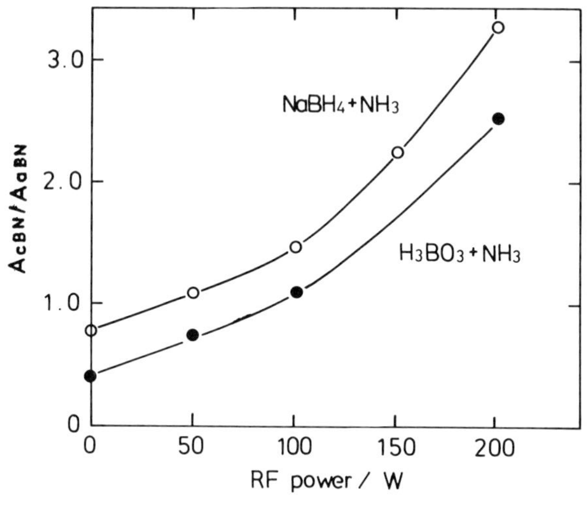


Fig. 8. IR absorption intensity ratios of the films deposited at a filament temperature of 1700 °C and a gas pressure of 40 Pa as a function of r.f. power.

## 4. Conclusion

We have studied the formation conditions of c-BN formed by the dissociation of $BH_3NH_3$, $H_3BO_3$ and $NaBH_4$ in the reactive gases $H_2$ or $NH_3$, and also by plasma excitement of the dissociated gases using r.f. induction and tungsten filament heating.

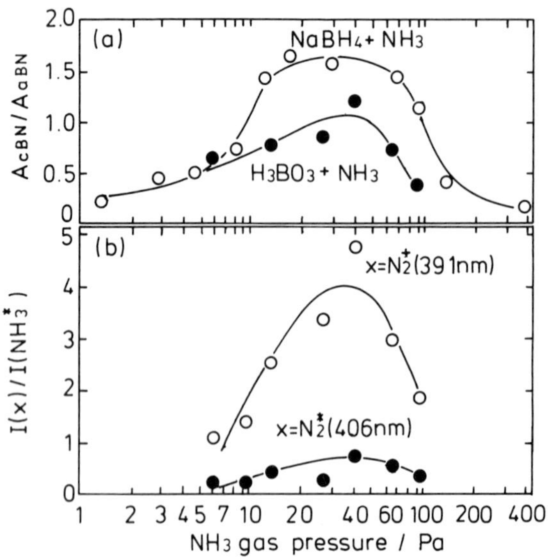


Fig. 9. (a) IR absorption intensity ratios of the films deposited at a filament temperature of 1700 °C and a r.f. power of 100 W as a function of gas pressure. (b) Photoemission intensity ratios of the reactive gas plasma as a function of gas pressure.

It was found in these deposition processes that the tungsten filament heating greatly contributes to the formation of c-BN as compared with the contribution of r.f. induction.

The deposition ratio c-BN/a-BN in the films deposited with $BH_3NH_3$–$H_2$ changes with the change of the emission intensity ratio $I(H\alpha)/I(H_2^*)$ in the gas plasma. In particular, when the ratio $I(H\alpha)/I(H_2^*)$ becomes maximum, the deposition rate of c-BN also becomes maximum. In the $H_3BO_3$–$NH_3$ and $NaBH_4$–$NH_3$ systems, it was found that the deposition ratio c-BN/a-BN varies with the emission intensity ratio $I(N_2^+)/I(NH_3^*)$ and $I(N_2^*)/I(NH_3^*)$.

## Acknowledgment

This study was supported partially by a grant-in-aid for scientific research from the Ministry of Education, Science and Culture.

## References

1 R. H. Wentorf, Jr., *J. Chem. Phys.*, *26* (1957) 956.
2 F. P. Bundy and R. H. Wentorf, Jr., *J. Chem. Phys.*, *38* (1963) 1144.
3 M. Satou and F. Fujimoto, *Jpn. J. Appl. Phys.*, *22* (1983) L171.
4 K. Inagawa, K. Watanabe, H. Ohsano, K. Saitoh and A. Itoh, *Proc. 10th Symp. on ISIAT '86*, Tokyo, 1986, p. 381.

5 A. Chayahara, H. Yokohama, T. Imura, Y. Osaka and M. Fujisawa, *Proc. ISPC-8*, Tokyo, 1987, p. 2440.
6 P. F. Zhou, T. Mori and Y. Namba, *Sinku*, *28* (1985) 581.
7 F. Shimokawa, H. Kuwano and K. Nagai, *Proc. 9th Symp. on ISIAT '85*, Tokyo, 1985 p. 467.
8 H. Miyamoto, M. Hirose and Y. Osaka, *Jpn. J. Appl. Phys.*, *22* (1983) L216.
9 W. Halverson and D. T. Qointo, *J. Vac. Sci. Technol.*, *A3* (1985) 2141.
10 Y. Ichinose, H. Saitoh and T. Ishiguro, *Proc. 11th Symp. on ISIAT '87*, Tokyo, 1987, p. 2440.

# A MICROSTRUCTURAL INVESTIGATION OF Au–a-C:H FILMS

HERMANN KOEBERLE, MARTIN GRISCHKE, FRITZ THIEME and CARSTEN BENNDORF

*Department of Physical Chemistry, University of Hamburg, Bundesstrasse 45, D-2000 Hamburg 13 (F.R.G)*

RÜDIGER MEMMING

*Institute for Solar Energy Research, Sokelantstrasse 5, D-3000 Hannover (F.R.G)*

(Received March 8, 1989)

## Summary

Gold-containing a-C:H films, deposited in a combined CVD and PVD process, were investigated by electric conductivity measurements, X-ray diffraction (XRD) and X-ray induced photoelectron spectroscopy (XPS). The gold concentration was varied between 3 and 91 at.%. For high metal concentration, 40 vol.% Au, or greater the Au–a-C:H films revealed a metallic behavior: the temperature coefficient was nearly constant between 4 and 300 K. For concentrations of 40 vol.% Au or less a drastic decrease of the conductivity was observed, interpreted as percolation barrier, with a change from 'metallic' to 'semiconducting' behavior. In this region, the conductivity strongly decreased with decreasing temperature. With XRD gold clusters within the amorphous a-C:H matrix down to concentrations of 6 at.% Au were detected; the cluster size decreased with decreasing gold content. XPS shows a broadening and a shift of the gold 4f doublet to lower binding energy (B.E) with decreasing gold content. This is related to the decrease of the average Au–Au coordination number with decreasing gold cluster size.

## 1. Introduction

The growing interest in the technical application of a-C:H coatings (amorphous hydrogen-containing carbon films) is based on the outstanding properties of these films. They are well suited as protective coatings against corrosion due to their dense structure; they show a good tribological behavior with a high hardness and low wear and friction values. Pure a-C:H films are insulating (conductivity $\sigma \approx 10^{-11}\,\Omega^{-1}\,cm^{-1}$). However, the unsatisfactory adhesion of pure a-C:H coatings to metals (especially steel) complicated their technical application. Dimigen and coworkers [1] demonstrated that the adhesion problem could be overcome when, during the deposition process of a-C:H, metals (like tantalum, ruthenium, tungsten or iron) are introduced into the a-C:H matrix. For metal-containing a-C:H films the adhesion to metals is significantly increased; the hardness and the tribological properties are conserved. The electrical conductivity even for very small metal concentrations drastically increases. For example, a-C:H with a tantalum concentration of 2.1 at.% shows a conductivity of $10^{-2}$ to $10^{-1}\,\Omega^{-1}\,cm^{-1}$, compared to $10^{-6}\,\Omega^{-1}\,cm^{-1}$ or less for pure a-C:H. The good conductivity combined with the excellent tribological properties of metal-containing a-C:H films opens new applications, *e.g.* for electrical gliding contacts.

In order to shed light into the microstructure of the metal containing a-C:H films, X-ray induced photoelectron spectroscopy (XPS) and X-ray diffraction (XRD) studies combined with conductivity measurements (4 - 300 K) have been performed mainly with Ta–a-C:H and Ti–aC:H films [2 - 5]. An important conclusion was the formation of carbides with a metallic conductivity behavior within the a-C:H matrix. The percolation barrier (transition from metal-like to semiconducting or insulating behavior) is found at around 20 vol.% Ta, which is surprisingly low. The relatively high conductivity even for 2.1 at.% Ta can only be understood assuming a change in the conductivity of the a-C:H matrix itself (within a certain radius of the TaC clusters).

In the present study gold was chosen for the deposition into a-C:H matrix because it does not form any carbide. From these experiments we expected further insights into the microstructure of metal-containing a-C:H films and the conductivity mechanisms.

## 2. Experimental procedure

The apparatus used for the deposition of Au–a-C:H films is shown schematically in Fig. 1.

Two circular electrodes with diameters of 15 cm were arranged vertically at a distance of 6 cm. The top electrode was used for the sputtering of gold; this is referred to as the cathode. The r.f. power (13.56 MHz) was coupled to this electrode via a matchbox. The bottom electrode, referred to as the anode, served as a support for the substrates; this electrode was

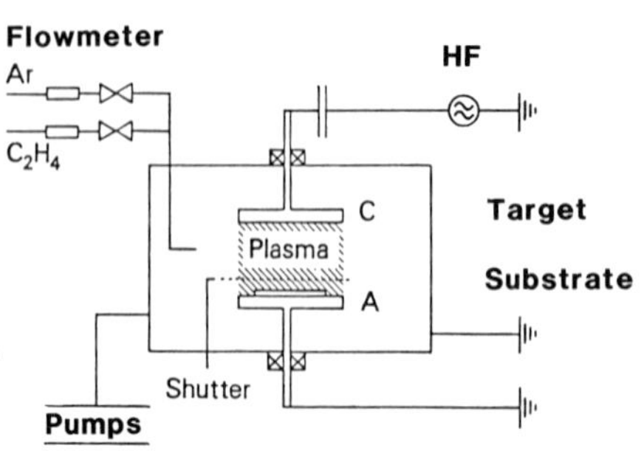


Fig. 1. Schematic drawing of the experimental set-up. C, cathode (gold target); A, anode (silicon or quartz glass substrate); HF, high frequency 13.56 MHz power supply.

grounded. During deposition, the temperature of the substrate could be controlled via external heating in the range 20 - 270 °C. The plasma for a-C:H deposition and gold sputtering was fed with a mixture of argon and $C_2H_4$ with separate flow control.

Prior to the deposition of the a-C:H films the silicon or quartz glass substrates were cleaned *in-situ* by argon ion bombardment.

For the preparation of the gold-containing a-C:H films, a constant argon flow into the deposition chamber leading to a pressure of $3 \times 10^{-2}$ mbar ($30 \text{ l s}^{-1}$ pumping speed) was adjusted; a direct current bias of $-1.5$ kV at the gold cathode was established with an r.f. power of 500 - 800 W. With this configuration, gold from the cathode was sputtered at a deposition rate of around $15 \text{ Å s}^{-1}$ onto the substrate. The deposition of a-C:H onto the substrate (and target) occurs on introducing $C_2H_4$ into the plasma. The carbon:metal ratio in the films could be controlled via the $C_2H_4$ flow. With increasing $C_2H_4$ flow the metal content of the a-C:H films decreased. We noted that the simultaneous deposition of a-C:H and metal on the anode (substrate) was a complicated process and one which is still not completely understood. a-C:H deposition occurred on the cathode as well as on the anode at the same time. From the cathode, a-C:H (or graphite or carbides in the case of reactive metals like tantalum) and metal was sputtered onto the substrate. Further, a direct deposition of a-C:H from the plasma onto the substrate also took place. The deposition of a-C:H onto the cathode proved to be disadvantageous for the preparation of films with low metal concentrations (2 at.% metal or less). The sputter efficiency on the cathode was inhomogeneous; for higher $C_2H_4$ flows some parts of the cathode were covered with a-C:H, preventing the sputtering of gold. Finally, for even higher $C_2H_4$ flows, the whole target (cathode) was covered with a-C:H, poisoning the sputtering of metal.

For the conductivity measurements ($\sigma$) the a-C:H (gold) films were deposited onto quartz glass. To allow measurements based on the standard 4-point method [7] the a-C:H (gold) films had to be deposited with a specific geometrical outline [2]. This was achieved using Ni–Be–Ni sandwich masks mounted on the quartz substrates. The compositions of the Au–a-C:H films were determined by electron probe microanalysis (EPMA) [8]. With this method the carbon and gold content, and also the argon and oxygen impurities, could be determined within $\pm 1$ at.%. However, the hydrogen content of our films was unknown. Despite this uncertainty the elemental concentrations determined with EPMA are summed up to 100% for better comparison.

XRD spectra from Au–a-C:H samples were measured with a Philips PW 1700 spectrometer. The spectra are compared to standard powder spectra from gold.

XPS measurements were performed in an ESCALAB 5 system (Vacuum Generators) equipped with an X-ray source allowing Mg $K\alpha$ and Al $K\alpha$ excitation. The spectra reported here were measured with Al $K\alpha$ excitation (1486.6 eV) and a spectrometer resolution of $\delta E = 0.6$ eV.

Prior to XPS measurements, the Au–a-C:H samples were sputter-cleaned to remove surface contamination. Care was taken to separate surface effects (removal of contamination) from effects due to argon ion bombardment (preferential sputtering) from the bulk effects of the Au–a-C:H film.

## 3. Results and discussion

### *3.1. Conductivity measurements*

Figure 2 summarizes conductivity data obtained for Au–a-C:H films with different metal concentrations. These data are compared with the conductivity of Ta–a-C:H deposited at 20 and 270 °C and to data from Perrin *et al.* [9] for gold in CF (fluorocarbon) films. For comparison with these latter data and with percolation theory, the gold concentrations were converted from atom percent to volume percent, assuming a density of 1.4 g cm$^{-3}$ for the a-C:H matrix and the bulk gold density for the gold particles.

The Au–c-C:H data show that the conductivity decreases continuously with decreasing gold content. There is a steep decrease of conductivity for films approaching the 40 vol.% percolation barrier. This point for Au–c-C:H films coincides with the percolation barrier observed by Perrin *et al.* [9] for Au–CF films.

In the latter case, this barrier describes the transition from metallic conductivity (where the gold clusters are in contact) to insulating behavior (where the gold clusters are isolated). From the percolation theory it can be

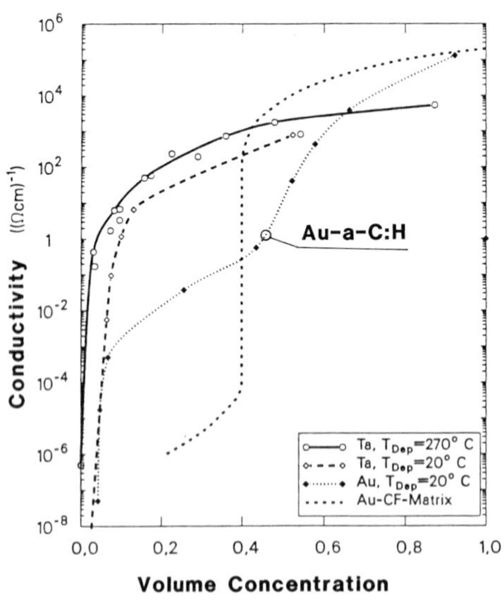


Fig. 2. Conductivity of Au–a-C:H and Ta–a-C:H films as a function of metal concentration (vol.%). For comparison, data from Perrin *et al.* [9] for Au–CF are included. For Ta–a-C:H two series of films are shown, deposited at $T_{\mathrm{dep}} = 270$ °C or $T_{\mathrm{dep}} = 20$ °C respectively.

deduced that in ideal cases this transition barrier (40 vol.%) is not strongly dependent on the metal cluster geometry itself [10]. Below 40 vol.% Au, however, the conductivity remains higher for Au–a-C:H as compared with the Au–CF system.

As we will show below, the interpretation of the percolation barrier for Au–a-C:H is supported by the temperature dependency of the conductivity; above 40 vol.% Au we find metallic behavior, below 40 vol.% Au semiconducting behavior.

Figure 2 shows two experimental series obtained for Ta–a-C:H deposited at 20 and 270 °C respectively. We note that there is no sharp percolation barrier for 40 vol.%. Instead, the conductivity stays at relatively high values up to 10 vol.% metal or higher. Temperature dependent measurements show that the transition from "metallic" to "semi conducting" behavior does indeed occur for Ta–a-C:H in the range 10 - 20 vol.% TaC [2].

Figure 3 shows the temperature dependency of $\sigma$ for four different Au–a-C:H films. Note the double logarithmic scale. The samples with 55 or 48 at.% Au show virtually no change in $\sigma$ during the cooling cycle from 300 to 4 K. For example, the film with 55 at.% Au shows a conductivity of 430 $\Omega^{-1}$ $cm^{-1}$; at 300 K this value increases to 555 $\Omega^{-1}$ $cm^{-1}$ at 4 K.

The conductivity behavior changes drastically for gold concentrations below 40 at.%. Examples are shown in Fig. 3 for two films with 39 and 22 at.% Au. The conductivity decreases rapidly by nearly a factor of $10^4$ during cooling from 300 to 4 K; below $\sigma \approx 10^{-5}\ \Omega^{-1}\ cm^{-1}$ the conductivity measurements proved to be difficult and were terminated.

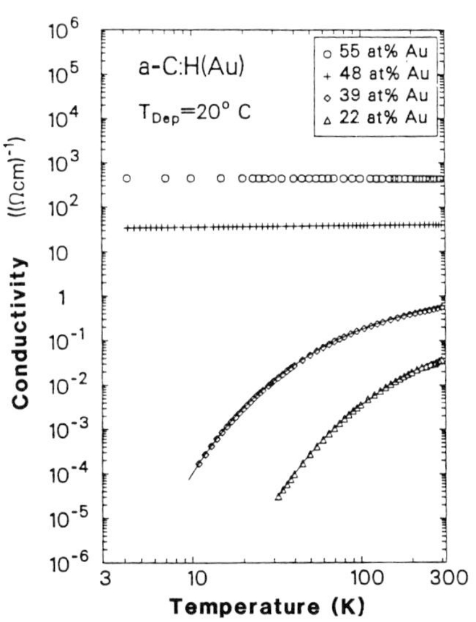


Fig. 3. Conductivity of Au–a-C:H films as a function of temperature. Note the nearly temperature-independent $\sigma$ for 48 and 55 at.% Au and the strong dependency for 39 and 22 at.% Au.

The temperature dependency of $\sigma$ in the range of metal concentration below 40 at.% can be described by a thermally activated process, similar to the conductivity of semiconductors. Following Hill [11], we can describe the temperature dependent conductivity with an Arrhenius-like equation

$$\sigma = \sigma_0 \exp\{-(T_p/T)^p\} \tag{1}$$

This equation gives an excellent fit to the experimental data for $p = 0.61$ and $p = 0.55$ for the films with 39 or 22 at.% Au respectively.

We conclude that there is a definite transition from metal-like to semiconducting behavior for Au–a-C:H films; this transition occurs between 48 (metal-like behavior) and 39 at.% Au (semiconducting behavior); this corresponds in Fig. 2 to concentrations of 44 and 52 vol.% Au. The transition observed in the temperature dependency of $\sigma$ coincides with a drastic decrease of $\sigma$ as a function of metal concentration (Fig. 2). As we will demonstrate below, XRD measurements show the existence of gold clusters within the a-C:H matrix. The diameter of these clusters increases with gold content. Therefore, in the region of metallic conductivity we assume that the gold clusters are in contact with each other. Due to the structural disorder of gold clusters and the complicated current paths through the films the electron scattering is not dominated by temperature dependent electron–phonon processes but rather by temperature independent elastic scattering from grain boundaries and crystal imperfections. This is the reason for the nearly temperature independent conductivity for gold contents of 39 at.% or more.

For 39 at.% Au and below, we still observe gold clusters by XRD. However, the cluster size and the gold content has decreased. There is no longer a continuous path available through the film from joining gold clusters. The conductivity is now dominated by electron tunneling or electron hopping from one cluster to the next. This process is thermally activated and leads to the observed temperature dependency of $\sigma$. The conductivity found for Au–a-C:H for metal concentrations below the percolation barrier, *i.e.* 0 - 40 vol.% Au, is nearly a factor of $10^2$ higher than for the Au–CF system described by Perrin *et al.* [9]. Further, we note a plateau of the conductivity in the range of 39 - 5 at.% Au; below 5 at.% Au $\sigma$ decreases from around $5 \times 10^{-4}\,\Omega^{-1}\,\mathrm{cm}^{-1}$ to below $10^{-7}\,\Omega^{-1}\,\mathrm{cm}^{-1}$ (2.1 at.% Au) at 20 °C. The relative drastic increase of $\sigma$ for very small gold concentrations is believed to be due to a modification of the a-C:H matrix through the presence of the metal. One possible explanation is that the $sp^2$ (C=C double bond) to $sp^3$ (C—C single bond) ratio of the a-C:H matrix is changed by the metal and that the C=C bonds created support the electron mobility over shorter distances (of the order of 3 - 5 Å).

In the case of Ta–a-C:H, the transition from metal to semiconducting behavior (from the temperature dependency of $\sigma$) occurs for 12 - 14 at.% (around 20 vol.%) TaC. This transition does not coincide with the drastic decrease of $\sigma$ for low tantalum concentrations (Fig. 2). Rather, the conductivity is relatively high, even for tantalum concentrations of 5 at.%. This behavior deviates from a simple percolation theory. We believe that, similarly to Au–a-C:H in the concentration range of 5 - 40 at.%, tantalum or TaC modifies the a-C:H matrix which now contributes to the conductivity.

We note, finally, that in the range of technical applications, *i.e.* 10 - 20 at.% metal, the conductivity of Ta–a-C:H a factor of $10^4$ higher than that of Au–a-C:H. In spite of the higher bulk conductivity of gold, the more reactive tantalum modifies the a-C:H conductivity more strongly, resulting in relatively high $\sigma$ values.

*3.2. X-ray diffraction measurements*

The XRD patterns obtained for Au–a-C:H for different gold concentrations are shown in Fig. 4. In the lower part of Fig. 4 the diffraction positions and relative intensities for polycrystalline f.c.c. gold are included.

The gold film sputtered in pure argon (91 at.% Au) shows the expected f.c.c. structure with a (111) texture. With decreasing gold content in the Au–a-C:H films the gold diffraction signal continuously decreases. We note, however, that the diffraction position clearly indicates the f.c.c. structure of gold; the half width of the diffraction peaks, however, has considerably increased. Further, the intensity ratio for different planes is the same as for the polycrystalline gold reference spectrum, indicating an isotropic orientation of the gold clusters. Even for the film with 3.6 at.% Au, weak peaks from the (111) and (200) gold f.c.c. planes seem to be present when compared with the diffraction spectrum from the quartz substrate. Evaluation of the clusters size from the half width of the diffraction peaks reveal 150 Å or greater for the 91 at.% Au film, and 50 Å and 40 Å for the 48 at.% or 39 at.% Au–a-C:H films respectively.

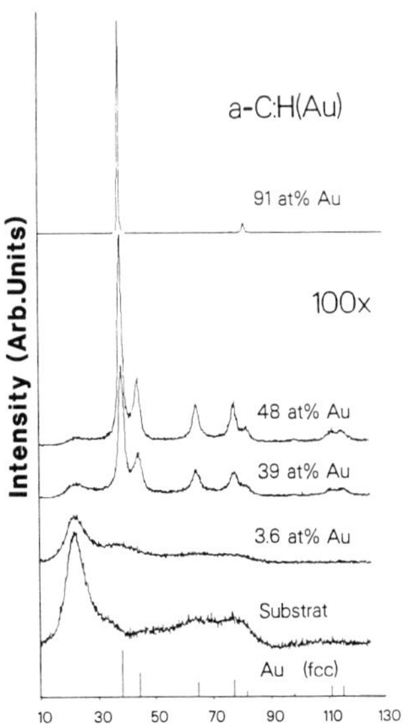


Fig. 4. XRD patterns from Au–a-C:H films with different gold contents. The lowest spectrum (vertical bars) gives the line position and relative intensities for polycristalline f.c.c. gold.

We conclude that XRD studies clearly show the presence of gold clusters inside the a-C:H matrix. The orientation of these clusters is isotropic; the structure is f.c.c. with a nearly unchanged unit cell distance. We note, however, that the XRD technique is not capable of detecting dispersed gold atoms within the a-C:H matrix; such atoms could influence the conductivity of the a-C:H matrix and could be of importance for the electrical conductivity at very low gold concentrations.

### *3.3. X-ray induced photoelectron spectroscopy measurements*

Figure 5 shows the gold 4f XPS spectra obtained for three different Au–a-C:H films with 4, 22 and 45 at.% Au content. Spectra from pure gold film (91 at.%) and from a bulk polycristalline gold reference sample are also shown. Owing to a spin–orbit interaction, the gold 4f core level is split into a doublet with $4f_{7/2}$ and $4f_{5/2}$ peaks. In agreement with literature data, the $4f_{7/2}$ gold peak is found at a B.E. of 83.6 eV. For better comparison of the peak positions and the peak half width, the gold 4f spectrum for each different sample is always scaled to the same height, although the absolute intensity decreases with decreasing gold content in the a-C:H films.

It should be noted that both peaks of the gold 4f doublet shift to higher B.E. and broaden considerably with decreasing gold content. The B.E. of the Au $4f_{7/2}$ peak for Au–a-C:H with 4 at.%, for example, is shifted by $\delta E = 0.4$ eV to higher energy; the half width is increased from $\Gamma = 1.6$ eV (bulk reference) to $\Gamma = 2.0$ eV. Examination of the carbon 1s peak reveals no change in comparison to metal free a-C:H films. The latter behavior is different to that of Ta–a-C:H [6]. In Ta–a-C:H two different carbon 1s states with B.E. values of 284.5 and 286.5 eV were detected, which could be related to carbon from the a-C:H matrix (284.5 eV) and to carbon reacted with tantalum, forming TaC (B.E. 286.5 eV). For Au–a-C:H only one carbon 1s state with a B.E. of 284.5 eV was found. This finding is consistent with the non-existency of AuC.

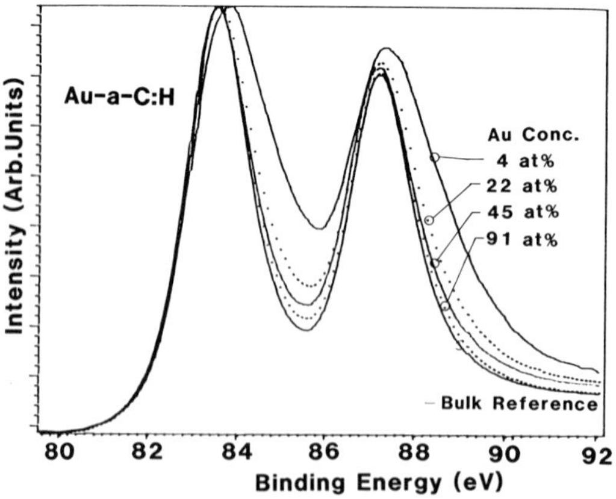


Fig. 5. X-ray excited gold 4f photoelectron spectra from Au–a-C:H films with different gold contents. For better comparison, the intensity of the gold 4f doublet is always scaled to the same vertical height.

In the case of Ta–a-C:H, the tantalum core levels were influenced by the chemical reaction with carbon [5, 6]. However, in the case of Au–a-C:H films, a reaction between gold and carbon can be excluded and the observed shifts and peak broadening of gold 4f have to be interpreted in a different way. We believe that the observed effects in the gold 4f spectra are due to the changing size of the gold clusters within the a-C:H matrix. It is well documented in the literature, that the binding energy of a surface atom is to some extent different from that of the bulk. Further, in some cases the core level binding energy shift $\delta E$ of a metal atom can be directly related to the coordination number of the atom.

For gold clusters deposited on carbon support, Mason [12] determined the dependence of the gold 4f shift as a function of the average gold coordination number within the gold clusters; he found a linear increase of $\delta E = 1.0$ eV when the average coordination number was decreased from $n = 12$ (bulk gold to $n = 0$ (single gold atom).

With this linear dependency derived by Mason [12], the largest shift of $\delta E = 0.4$ eV (Fig. 5), corresponds to an average coordination number of around 8. Because the gold cluster size distribution within the a-C:H matrix is unknown, we do not have a separate method to prove this value. However, the general trend observed with the XRD analysis, showing a decrease of gold cluster size with decreasing gold content, is confirmed by the XPS analysis. Further, the relatively low coordination number of $\bar{n} = 8$ could be a hint for the presence of dispersed gold atoms ($\bar{n} = 0$) within the a-C:H matrix (apart from larger clusters).

In conclusion, XPS analysis shows that for Au–a-C:H, in contrast to Ta–a-C:H, no metal carbides are formed. The observed shifts of the gold 4f B.E. is related to a decrease of the gold cluster size with decreasing gold content.

## 4. Conclusions

We briefly summarize the main conclusions which can be drawn from our present investigation of the Au–a-C:H system.

(1) The existence of metallic gold clusters within the a-C:H matrix is shown by XRD and XPS studies.

(2) Conductivity measurements reveal a metal–semiconducting transition near a percolation barrier of around 40 vol.% Au. Above 40 vol.% Au the Au–a-C:H films show a nearly temperature independent conductivity $\sigma$. Below 40 vol.% Au the strong temperature dependency of $\sigma$ can be described by a thermal activated process with an Arrhenius-like behavior.

(3) Below the percolation barrier, in the range 5 - 40 vol.% Au, the plateau in the conductivity data (around $10^{-3}$ to $10^{-2}$ $\Omega^{-1}$cm$^{-1}$) is suggested to be due to a modification of the a-C:H matrix in the neighborhood of gold; due to an increase of the sp$^2$ to sp$^3$ ratio, the a-C:H matrix may contribute to the conductivity.

(4) The conductivity of Au–a-C:H compared with Ta–a-C:H is low. In particular, in the range of technical applications, 10 - 20 at.% metal, the conductivity of Ta–a-C:H is higher by a factor of $10^4$. This is due to the higher reactivity of tantalum leading to the formation of TaC (metal-like conductivity) and to a higher contribution of a-C:H conductivity (through the modification by tantalum).

## Acknowledgments

We thank Dr. Dimigen (Philips Research Laboratories, Hamburg) who stimulated the present investigations. We are indebted further to Dr. P. Willich (Philips Research Laboratories, Hamburg) for EPMA measurements. Finally, we acknowledge the financial support by the Bundesministerium für Forschung und Technologie (project number 13 N 5374-7).

## References

1 H. Dimigen, H. Hübsch and R. Memming, *Appl. Phys. Lett.*, *50* (1987) 1056.
2 H. Koeberle, *Ph.D. Thesis*, University of Hamburg, 1989.
3 H. Koeberle and R. Memming, *Proc. E-MRS Symp. 17* (1987) 485 (Les Editions de Physiques, Strasbourg).
4 M. Grischke, A. Brauer, C. Benndorf, R. Memming and F. Thieme, *Proc. E-MRS Symp. 17* (1987) 491 (Les Editions de Physiques, Strasbourg).
5 C. Benndorf, M. Grischke, A. Brauer and F. Thieme, *Surf. Coat. Technol.*, *36* (1988) 171.
6 C. Benndorf, M. Grischke, M. Brauer, H. Koeberle, R. Memming and F. Thieme, Proc. Int. Symp. on Trends and Appl. in Thin Films, Strasbourg, 1987, *Suppl. to Vide, Couches Minces*, *235* (1987) 485.
7 L. J. van der Pauw, *Philips Res. Rep. 13* (1958) 1.
8 P. Willich, in J. D. Brown and H. Pachwood (eds.), *Proc. 11th Congr. X-ray Optics and Microanalysis*, University of Western Ontario, London, Canada, 1986, p. 238.
9 J. Perrin, B. Despax, V. Hanchett and E. Kay, *J. Vac. Sci. Technol.*, *A4* (1986) 46.
10 E. Kay, in F. Träger and G. Zu Putlitz, (eds.), *Metal clusters*, Springer-Verlag, Berlin, 1986, p. 151.
11 R. M. Hill, *Phys. Status Solidi, Sect. A*, *35* (1976) k29.
12 M. G. Mason, *Phys. Rev. B*, *27* (1983), 748.

*Surface and Coatings Technology, 39/40* (1989) 285 - 292 

# TARGET EFFECTS DURING THE DEPOSITION OF Ti–a-C:H FILMS

J. T. HARNACK, F. THIEME and C. BENNDORF

*Department of Physical Chemistry, University of Hamburg, Bundesstrasse 45, D-2000 Hamburg 13 (F.R.G.)*

(Received March 9, 1989)

## Summary

The deposition of Ti–a-C:H films from a combined physical vapour deposition (PVD; sputtering of titanium) and chemical vapour deposition (CVD; deposition of $C_xH_y$ radicals) process is described. Erosion measurements and position sensitive X-ray photoelectron spectroscopy (XPS) measurements on the target are correlated with the hydrocarbon content in the plasma and with the composition of the deposited Ti–a-C:H films. These measurements allow the separation of the different processes ($C_xH_y$ radical deposition onto the anode and cathode, sputtering of titanium and $C_xH_y$ from the target) involved in Ti–a-C:H film formation and in the poisoning of the target.

## 1. Introduction

The technical applications of hard coatings have been increased considerably in the recent past. In this field, amorphous carbon films, a-C:H, (often referred to as diamond-like) are of increasing interest. This is due to their extraordinary properties such as high hardness, low wear and friction values combined with an optical transparency even in the infra-red and red region of visible light.

The principal disadvantage of the application of a-C:H films as a hard protective coating has proven to be the high internal stress within the films, leading to adhesion failures especially on metal substrates [1 - 3]. As demonstrated by Dimigen and coworkers [1 - 3], through the simultaneous incorporation of metals into the a-C:H matrix during film deposition, the stress and adhesion problems of pure a-C:H films can be overcome, and the favorable properties of metal free a-C:H coatings are conserved. These films show an electrical conductivity which can be modified over a large range, opening the way for new applications, such as electrical gliding contacts.

0257-8972/89/$3.50 

## 2. Experimental procedure

The typical technique to deposit metal free a-C:H films is by PECVD from an r.f. excited hydrocarbon plasma. The r.f. power is capacitively coupled to the cathode, which also serves as a mounting for the substrates. The properties of the *cathodic* deposited a-C:H films are severely dependent on the self bias of this electrode.

However, it is also possible to deposit a-C:H films onto the anode, which in this case is unpowered. The change from a cathodic to an anodic process is possible because the polymerization of a-C:H occurs from hydrocarbon radicals in the gas phase [4]; in this case the dense a-C:H structure is due to the bombardment with high energy reflected neutrals from the cathode. For the *anodic deposition* of a-C:H, metal can be incorporated into the growing a-C:H layer via sputtering from the cathode.

The simultaneous deposition of a-C:H and metal onto the anode (substrate) is achieved from an r.f. plasma fed with a mixture of hydrocarbon and argon; the sputtering of the metal occurs mainly via $Ar^+$ bombardment of the negatively biased cathode (sputtering target). The metal content of the a-C:H films can be controlled by the ratio of argon/hydrocarbon gas fed into the plasma. It depends further on the nature of the hydrocarbon used, on the target metal, on the pressure inside the vacuum system and finally on the r.f. power coupled to the target.

We note that the anodic deposition of metal containing a-C:H layers is a rather complicated process and not yet completely understood. Plasma polymerization of a-C:H occurs at the same time onto the anode and onto the cathode. On the cathode, sputtering of metal by high energy (1.5 keV) $Ar^+$ ions is a competitive process. The equilibrium between both cathodic processes determines the efficiency of metal sputtering. This efficiency of metal sputtering from the cathode and the deposition of $C_xH_y$ radicals onto the anode is responsible for the metal content in the a-C:H layers.

To shed more light onto the deposition process of a-C:H layers we report on measurements where we analyzed not only the deposited films but also the sputter targets. These experiments were performed in a small physical vapour deposition (PVD)–chemical vapour deposition (CVD) system developed in our own laboratory. The characteristic data are summarized in Table 1. A schematic line drawing of the sputtering system is shown in Fig. 1.

The diameters of our electrodes (38 mm) are rather small. This diameter was chosen to allow the introduction of the complete titanium target into the X-ray photoelectron spectroscopy (XPS) measuring system via a fast entry air lock. The plasma pressure of 1.5 Pa is, however, in a range which is commonly used in standard equipment with larger target diameters. To achieve a steady discharge, a separation between cathode (titanium target) and anode (substrate) of 21 mm was chosen. Under these conditions, the light emission from the plasma is inhomogeneous, indicating hollow-cathode glow effects. We believe this could be due to an expanding of the discharge against the shielding of the anode.

TABLE 1

Characteristics of the PVD–CVD deposition system with two vertically aligned electrodes

| | |
|---|---|
| electrode diameters: | 38 mm |
| electrode distance: | 21 mm |
| pressure during deposition: | 1.5 Pa |
| r.f. power: | 70 W |
| self bias voltage of cathode: | −1500 V |
| target material (cathode): | titanium |
| plasma gas: | $C_2H_4$–Ar (0 - 5 vol.% $C_2H_4$) |

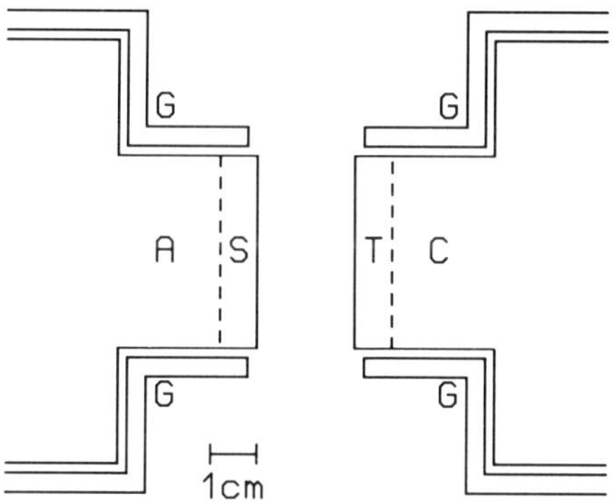


Fig. 1. Schematic line drawing of the sputtering system used. (A) anode, in this case grounded; (S) substrate; (C) cathode, r.f. powered; (T) titanium target; (G) grounded dark room shielding of electrodes.

## 3. Results and discussion

### *3.1. Erosion of the titanium target*

To get an insight into the competitive process of sputtering and a-C:H deposition onto the target we first measured the erosion process of the titanium target in a pure argon plasma. Figure 2 shows the erosion of the target as a function of the radial distance from the center. We note that there is obviously a quite inhomogeneous erosion with a very high sputter efficiency near the periphery of the circular target; this is believed to be due to hollow-cathode glow effects as mentioned above.

In Fig. 2 we can distinguish two different regions on the target. In the centre of the target (0 - 9 mm) there is a region with a relatively moderate erosion; this is referred to as region 1. The second region, referred to as region 2, is observed near the periphery of the target (9 - 18 mm); it shows a drastic increased erosion (up to a factor of five). This two-region process exists also in $C_2H_4$–Ar plasmas. The two regions with inhomogeneous sputter efficiency allow us to investigate the dependence of the target processes from different $Ar^+$ ion fluxes (regions 1 and 2). We believe that the inhomogeneous

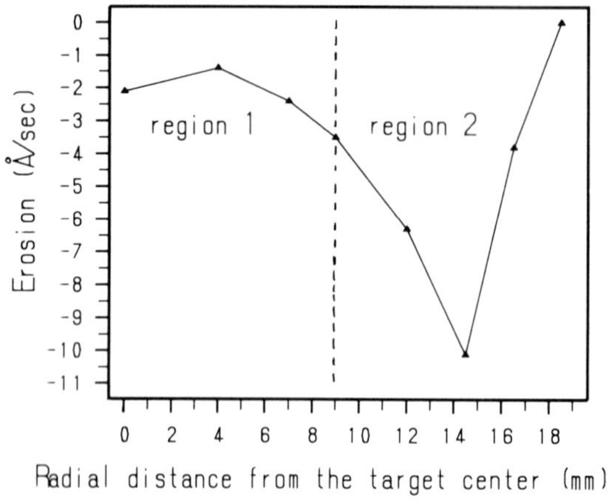


Fig. 2. Erosion of the titanium target as a function of the circular distance from the target center. These data were obtained with a Sloan Dektak II scanner.

$Ar^+$ ion flux has only a quantitative influence on the occurring processes; the qualitative behavior described below, namely the observation of different species (titanium, TiC and a-C:H) on the target, is not influenced by the inhomogeneous erosion of the target.

*3.2. Position sensitive XPS analysis from the titanium targets*

After six different experimental runs with different $Ar/C_2H_4$ ratios the titanium targets were analyzed with XPS as a function of the radial distance from the target center. We note that for our spectrometer, the position sensitive detection is limited to a distance of $\pm 4$ mm. Surface contamination on the targets, which originates from post-process deposition of $C_xH_y$ and from oxidation in air, had to be removed by $Ar^+$ ion bombardment in the spectrometer immediately before the measurement. This is described elsewhere [5]. For a quantitative evaluation of the target composition, atomic sensitivity factors determined by Wagner *et al.* were used [6] and converted to the ESCALAB 5 system used by us [5]. Hydrogen cannot be detected with XPS. Despite this uncertainty, the elemental concentrations determined with XPS were, for better comparison, summed up to 100%.

We believe that the high surface sensitivity of the XPS methods is well suited to our problem, because the processes of sputter emission occur in the topmost surface layers. However, we cannot exclude that the argon sputter cleaning inside the XPS measuring system also removes parts of the target surface, which may contain "active" species for the Ti–a-C:H film deposition.

Figure 3 summarizes the XPS results from the titanium targets after six different experimental runs; depicted are always the concentrations of titanium and carbon (from a-C:H deposition) measured in regions 1 and 2. For region 1 (Fig. 3(a)) we observe a sharp transition from the pure target metal (90 at.% Ti for a hydrocarbon content below 0.2 vol.% of the plasma gas) to

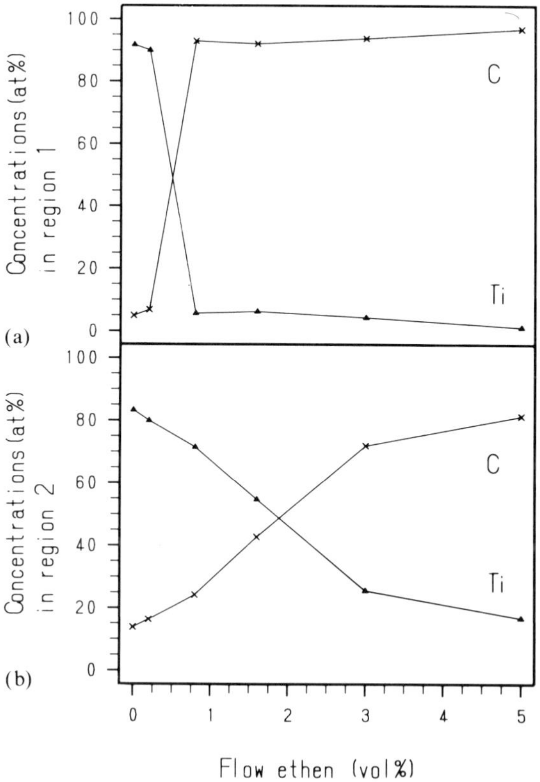


Fig. 3. (a) Titanium and carbon concentration in region 1 of the target determined as a function of increasing $C_2H_4$ content in the plasma; the data were obtained for six different experimental runs. (b) Same measurements as (a), for region 2.

a carbon covered target (less than 10 at.% Ti for hydrocarbon contents greater than 0.8 vol.% of the plasma gas). This behavior is quite different in region 2. Here we observe a rather smooth nearly linear transition from a carbon free to a carbon covered titanium target (Fig. 3(b)).

The behavior found in region 1 is believed to be due to the low erosion intensity, already observed for the sputtering in pure argon (Fig. 2). We suggest that the build-up of a continuous a-C:H layer in region 1 cannot be removed under these conditions. We note that the rapid transition in region 1 to a carbon covered surface is not only reflected in the total carbon concentration but also in the chemical nature of the carbon. For low carbon concentrations, 10 at.%C, XPS analysis shows that all of the carbon is in a carbidic state, *i.e.* directly interacting with titanium. This situation changes when the titanium target is (in region 1) nearly completely covered with carbon (90 at.%C). Now the XPS analysis shows that the polymeric state of the carbon (due to a-C:H or graphite) is dominant.

We conclude that due to the low sputter efficiency, region 1 has only a minor influence on the composition of the anodic deposited Ti–a-C:H film.

This behavior is quite different for region 2. Here we observe an equilibrium between the titanium metal and carbon concentration on the target for a wide regime of hydrocarbon gas concentrations in the plasma. With increasing hydrocarbon concentration (0 - 5 vol.%) the polymeric part of carbon increases in the same way as the total carbon concentration. At the same time, we observe a decrease of the carbidic carbon from 11 at.% C (0.2 vol.% $C_2H_4$) to 3.5 at.% C (5 vol.% $C_2H_4$), proportional to the decrease of titanium in this region.

We can conclude, however, that in region 2 a stoichiometric titanium carbide is not built up on the target even for higher carbon concentrations; the metal character always dominates. For example, for 5 vol.% $C_2H_4$ in the gas mixture, we observe a titanium concentration in region 2 of 17.8 at.% Ti; the carbidic part of the carbon under these conditions is only 3.5 at.% C (4.3% of the total carbon concentration). During the PVD–CVD process the $Ar^+$ ion bombardment of the titanium target leads to an ion-mixing between titanium and the deposited $C_xH_y$ radicals. In region 2 this mixing could produce a reaction layer of TiC with a thickness in the range of 10 atomic layers. This layer, however, would not be detected in our XPS experiments because of its removal by the $Ar^+$ ion sputter cleaning procedure.

Consistent with the erosion measurements in pure argon, we conclude from the XPS measurements discussed above that the sputter efficiency even for Ar–$C_2H_4$ is much higher in region 2 compared with region 1; consequently the compositions of the anodic deposited Ti–a-C:H films are mainly governed by the processes occurring in region 2 on the cathode.

### *3.3. Erosion within an argon–hydrocarbon gas mixture*

The XPS results discussed above for region 2 demonstrated that for the maximum hydrocarbon concentration, 5 vol.% ethylene, used in this investigation, metallic titanium still can be detected with XPS on the cathode. The question which we will address in this section concerns the reduction of metal sputtering due to the cathodic deposition of a-C:H.

For this purpose the maximum erosion in region 2 was measured as a function of the $C_2H_4$ concentration in the process gas. To directly show the relation between carbon coverage and reduction of the erosion rate, the erosion rate determined with a Sloan Dektak II profile scanner as a function of XPS determined carbon coverage is plotted in Fig. 4. We note that the carbon concentration measured after sputtering in pure argon is due to surface contamination of the titanium target.

We observe a drastic decrease of the sputter efficiency (erosion) with increasing $C_2H_4$ content. The erosion rate decreased to below 10% of the rate in pure argon for a carbon concentration of 40 at.% C on the titanium target; this corresponds to a hydrocarbon content of 1.6 vol.% in the plasma gas. Although the titanium concentration in region 2 is, under these conditions, still relatively high (60 at.% Ti), the Ti sputtering seems to be reduced due to simultaneous deposition of $C_xH_y$ species onto the cathode from the gas phase and its subsequent sputtering. As a consequence, the sputtering of carbon prevails.

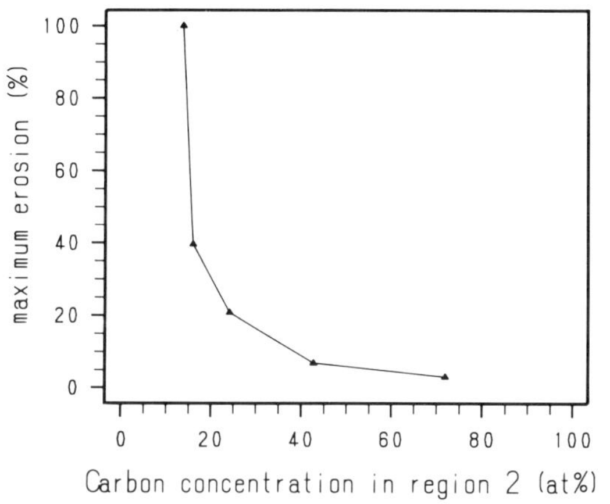


Fig. 4. The maximum erosion of the titanium target (at a distance of about 14 mm from the target center) in an Ar–$C_2H_4$ mixture compared to the erosion in pure argon. The ratio of erosion in Ar–$C_2H_4$ to erosion in pure argon is given as a function of carbon concentration in region 2.

### *3.4. Quantitative XPS analysis of the deposited Ti–a-C:H films*

To correlate the process observed on the sputter target (cathode) with the properties of the anodic deposited Ti–a-C:H films, the composition of these films was analyzed by XPS. These measurements are summarized in Fig. 5. The total carbon and titanium concentrations measured in the deposited Ti-a-C:H films are depicted as a function of the carbon concentration in region 2 of the target.

The drastic decrease of the titanium content in the Ti–a-C:H films reflects the drastic decrease of titanium sputtering with increasing carbon deposition onto the target. We note, however, the decrease of titanium

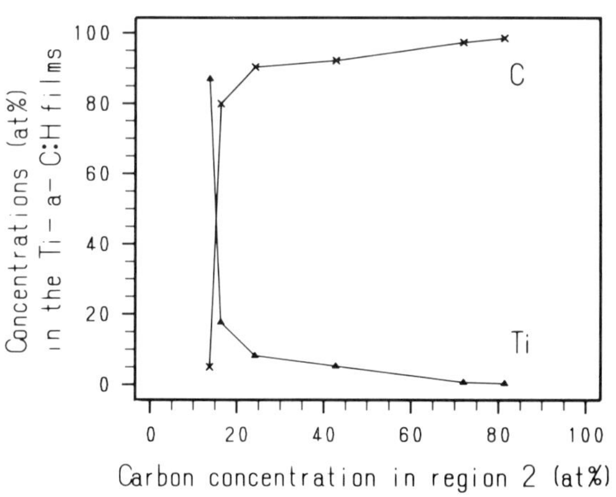


Fig. 5. The titanium and carbon concentrations in the deposited Ti–a-C:H films were determined with XPS and these values are given as a function of carbon concentration on the titanium target in region 2.

concentration in the Ti–a-C:H films is much steeper than the decrease of erosion rate. This is certainly due to the fact that carbon deposition onto the substrate film occurs via two processes, sputtering of carbon (and titanium) from the target and direct deposition from the gas phase. Both processes contribute to the carbon content in the Ti–a-C:H films and increase the C/Ti concentration ratio (by lowering the titanium concentration).

## 4. Conclusions

In the present study, the deposition and erosion processes occurring on a titanium target during the deposition of Ti–a-C:H films from an Ar–$C_2H_4$ gas mixture were studied with XPS and erosion measurements. The data from the target were correlated with the composition of the deposited Ti–a-C:H films. We briefly summarize the main conclusions which can be drawn from our present investigations.

(1) The erosion measurements on the target show the existence of two different regions. The region (referred to as region 2) near the periphery of the sputter target shows a sputter efficiency which is increased up to a factor of 5.

(2) The C/Ti ratio in region 2 determines the composition of the deposited Ti–a-C:H films. The metal sputtering from this region significantly decreases due to carbon deposition for increasing hydrogen content in the plasma.

(3) A relatively high sputter efficiency on the target (as in region 2) is advantagous for the control of metal content in the deposited Ti–a-C:H films.

## Acknowledgments

The authors express their gratitude to Dr. Dimigen of the Philips Research Laboratory, Hamburg for having the opportunity to use the profile scanner. Financial support of this investigation by the Bundesministerium für Forschung und Technologie (project number BMFT 13 N 5374/7) is gratefully acknowledged.

## References

1 H. Dimigen, H. Hübsch and R. Memming, *Appl. Phys. Lett., 50(16)* (1987) 1056 - 1058.
2 K. Enke, H. Dimigen and H. Hübsch, *Appl. Phys. Lett., 34(4)* (1980) 291 - 292.
3 H. Dimigen and H. Hübsch, *Philips Tech. Rev., 41* (1983/84) 186 - 197.
4 Y. Catherine and A. Pastol; *Les Editions de Physique 17, E-MRS Conference, Strasbourg*, 1987, pp. 145 - 151.
5 M. Grischke, A. Brauer, C. Benndorf, F. Thieme and P. Willich, *Fresenius Z. Anal. Chem., 333* (1989) 299 - 303.
6 C. D. Wagner, L. E. Davis, M. V. Zeller, J. A. Taylor, R. M. Raymond, L. H. Gale; *Surf. Interface Anal., 3* (1981) 211.

*Surface and Coatings Technology, 39/40* (1989) 293 - 300 

# PREPARATION OF $TiB_2$ AND $TiB_xN_y$ COATINGS BY PACVD

H. KARNER, J. LAIMER and H. STÖRI

*Institut für Allgemeine Physik, Technische Universität Wien, Wiedner Hauptstraße 8 - 10, 1040 Wien (Austria)*

P. RÖDHAMMER

*Metallwerk Plansee, 6600 Reutte (Austria)*

(Received March 25, 1989)

## Summary

Thin films of $TiB_2$ and $TiB_xN_y$ have been deposited by plasma-assisted chemical vapour deposition (PACVD) in a gas mixture of Ar, $H_2$, $N_2$, $TiCl_4$ and $BCl_3$, using both r.f. and pulsed direct current plasmas. The influence of the $BCl_3/TiCl_4$ ratio and of the $N_2$ flow rate on the chemical composition and the mechanical properties of the coatings as well as on the formation of different phases has been studied in detail. The $TiB_2$ coatings, deposited at rates of approximately 2 $\mu m\ h^{-1}$ reach Vickers hardness values up to 3000 HV0.02 and exhibit a fracture amorphous (r.f. plasma) or fibrous (direct current plasma) structure. They contain less than 1 at.% Cl. The $TiB_xN_y$ coatings consisting of the face-centered cubic $TiB_xN_y$, the hexagonal $TiB_2$ and the orthorhombic TiB phase, reach a Vickers hardness up to 3000 HV0.02. At a deposition rate of 2 $\mu m\ h^{-1}$ they contain approximately 1 at.% Cl. The coverage of the edges of cemented carbide inserts is excellent.

## 1. Introduction

Coating with a hard material is now common practice for the improvement of the performance of many kinds of tools. Even if the conventional chemical vapour deposition (CVD) process is most commonly used today [1 - 5], plasma-based so called physical vapour deposition (PVD) techniques [6 - 8] are continuously gaining importance. CVD offers an excellent uniformity of the deposited film, but the high deposition temperature of approximately 1000 °C is a major drawback. With PVD techniques like sputtering or ion plating the deposition can be performed at temperatures below 500 °C, but to achieve a good film uniformity it is necessary to rotate the work pieces in the vacuum chamber during the deposition process. Plasma-assisted chemical vapour deposition (PACVD) offers a favourable combination of features of both types of process, that is deposition at moderate temperature with good film uniformity without movement of the samples.

0257-8972/89/$3.50 

We developed a PACVD reactor for the deposition of cemented carbide inserts with hard coatings and studied the formation of $TiB_2$ and TiBN films under various experimental conditions.

## 2. Experimental procedure

### *2.1. Deposition apparatus*

The PACVD apparatus is shown in Fig. 1. The reaction chamber consisted of a steel vessel with a diameter of 320 mm and a height of 300 mm. The samples were placed in a substrate holder of high transparency in the middle of two grounded outer electrodes, through which the feed gases were introduced into the reaction volume over a special gas distribution system. The gases were exhausted through a pumping port in the bottom part of the reactor by a roots blower and a rotary vane pump. The pressure, which could be varied from 0.5 to 9 mbar, was controlled by a throttle valve and measured by a diaphragm-type gauge.

As feed gases argon, $H_2$, $N_2$, $TiCl_4$ and $BCl_3$ have been used. The flow rates of all gases were controlled by mass flow controllers. The main components of the gas mixture were argon and $H_2$ with small amounts of $N_2$ (less than 5%), $BCl_3$ and $TiCl_4$ (less than 1%).

The r.f. discharge power (13.56 MHz) was coupled into the plasma via a matching network connected to the substrate holder, where a self bias developed. Alternatively a pulse plasma generator was used, which produced direct current pulses of variable duration and repetition rate covering a frequency range from 25 kHz to direct current. The temperature of the substrate electrode was measured by a thermocouple installed in its interior. Because of the strong r.f. interference during deposition, the temperature measurements in the r.f. mode were carried out only at the end of the deposition process, whereas in the pulsed direct current mode the temperature could be monitored during the deposition process. In the r.f. mode a

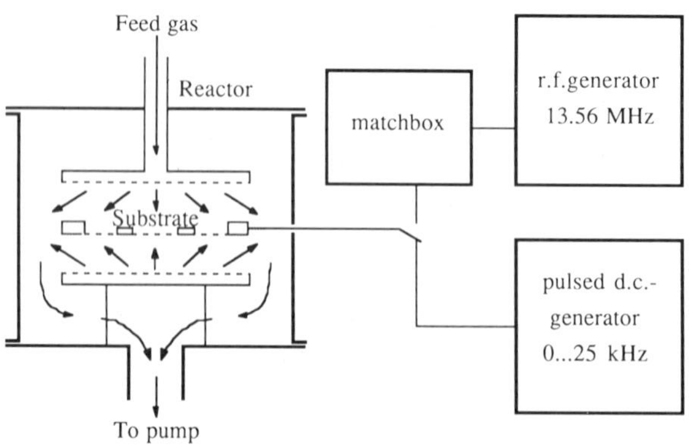


Fig. 1. Schematic of the PACVD deposition apparatus.

power up to 2500 W, in the pulsed direct current mode up to 1200 W was used. The deposition temperatures commonly were in the range between 400 and 600 °C. Because of the lack of an additional heating, the deposition temperature was primarily determined by the discharge power.

### *2.2. Film characterization*

The chemical composition of the coatings was determined by Auger electron spectroscopy (AES) using a commercial Auger system (PHI 10-260) with a double-pass cylindrical mirror analyzer (CMA). TiN, $TiB_2$ and titanium have been used as standards. In the case of the TiBN films, the nitrogen KLL peak coincides with the titanium LMM peak at 387 eV. For the purpose of evaluation the two signals were separated by the method published by Laimer *et al.* [9]. The chlorine LMM peak at 181 eV nearly coincides with the boron KLL peak at 179 eV, so that both peaks could not be separated. Therefore the chlorine content in the $TiB_xN_y$ and $TiB_x$ coatings was determined by X-ray fluorescence analysis (XRF). The atomic concentration ratio Cl/Ti was determined by comparison of the XRF signal ratios Cl/Ti to signal ratios obtained from $TiN_x$ coatings with known chlorine content. The chlorine content of these $TiN_x$ coatings has been determined by AES.

Phase identification and preferred orientation of the coatings were determined by X-ray diffraction (Philips PW 1050/80) using Cu $K\alpha$ radiation. The morphology and the thickness of the coatings were determined by investigation of the fracture cross-sections using scanning electron microscopy (SEM type Jeol JSM 35CF).

Molybdenum or niobium plates were used as substrates for the above mentioned analysis.

The mechanical properties of the coatings, such as hardness and adhesion, were measured using a Vickers hardness tester and a scratch tester (CSEM Revetest). Smoothly polished cemented carbide inserts were used as substrates for the determination of the mechanical properties.

## 3. Results and discussion

### *3.1. $TiB_x$ coatings*

Figure 2(A) shows the results of the AES measurements on $TiB_x$ coatings as a function of the $TiCl_4$ and of the $BCl_3$ flow respectively, while the sum of the $BCl_3$ and $TiCl_4$ flows remained constant at 10 std. $cm^3\ min^{-1}$. These coatings have been deposited in the r.f. plasma at a temperature of approximately 500 °C and a self bias of $-300$ V. The film, showing an atomic ratio of B/Ti = 1.6 exhibits in the XRD pattern a peak at approximately 43.5° with a large halfwidth of 3°. This reflection can be attributed to the hexagonal $TiB_2$ (101) phase with a lattice enlargement in *c* direction of 7%, or can be explained by an overlap of the hexagonal $TiB_2$ (101) and the orthorhombic TiB (102) reflections, corresponding to diffraction angles of 44.6° and 42.2° respectively. Figure 3(A) shows the XRD pattern of a $TiB_x$ film, deposited at

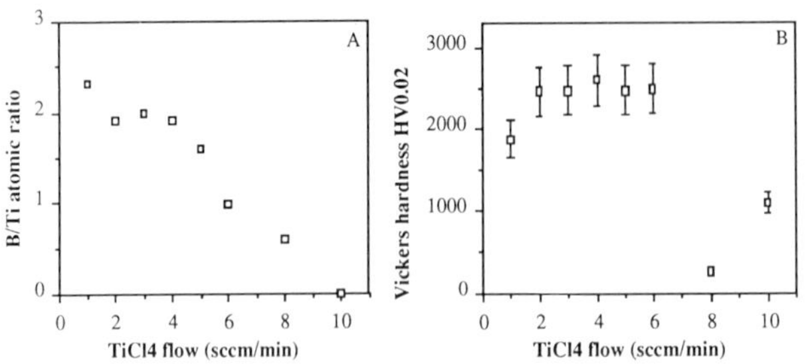


Fig. 2. B/Ti atomic ratio (A) and Vickers hardness HV0.02 (B) of $TiB_x$ films deposited in the r.f. plasma as a function of $TiCl_4$ flow ($TiCl_4 + BCl_3 = 10$ std. $cm^3\ min^{-1}$).

a ratio of $BCl_3/TiCl_4 = 7/3$, which is typical for the coatings deposited in the range $6/4 \leqslant BCl_3/TiCl_4 \leqslant 8/2$. The XRD peaks clearly correspond to the (101) and (100) reflections of the hexagonal $TiB_2$ phase, showing a lattice enlargement in *c* direction up to 1%. At a higher ratio of $BCl_3/TiCl_4 = 9$ the preferred orientation changes to (001) and the intensity of the reflections is strongly increased, as can be seen from Fig. 3(B). A similar influence of gas relative partial pressures on the preferred orientation of the hexagonal $TiB_2$ crystallites

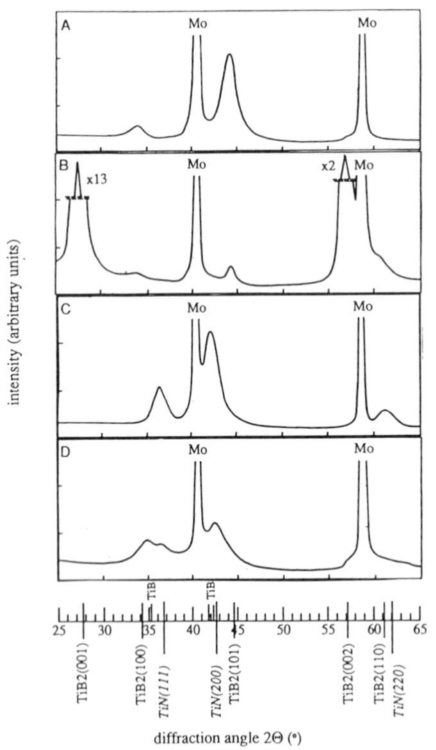


Fig. 3. XRD patterns of films deposited in the r.f. plasma: (A) $TiB_2$ film, (B) $TiB_{2.3}$ film, (C) $TiB_{0.3}N_{1.0}$ film, (D) $TiB_{0.9}N_{0.6}$ film.

in CVD-deposited films has been reported by Zeman *et al.* [4]. The halfwidth of the XRD reflections decreases from 2° for the coating deposited at $BCl_3/TiCl_4 = 6/4$ to 0.5° for the coating deposited at $BCl_3/TiCl_4 = 9$. The atomic ratio of B/Ti amounts to approximately 2 in the range $5/5 \leqslant BCl_3/TiCl_4 \leqslant 8/2$ and the chlorine content is lower than 1 at.% Cl, whereas the coating deposited at $BCl_3/TiCl_4 = 9$ exhibits a ratio of B/Ti = 2.3 and a higher chlorine content of approximately 3 at.% Cl. The hardness of 2500 HV0.02 for the stoichiometric $TiB_2$ coatings is lower than reported by other authors, who have prepared $TiB_2$ coatings by CVD [1 - 3] and by sputtering [6 - 8], where hardness values ranging from 3000 to 4500 have been measured. The lower hardness of the PACVD-coated $TiB_2$ films may be attributed to a highly imperfect crystallization of the nearly stoichiometric films, which corresponds to the large halfwidth of the XRD reflections of these coatings. In spite of the narrow XRD reflections of the coating deposited at $BCl_3/TiCl_4 = 9$, the hardness decreases to 1900 HV0.02, which may be due to the excess of boron and the high chlorine content.

To study the influence of the deposition temperature, we have deposited $TiB_2$ coatings in the pulse direct current plasma at different temperatures by varying the duty cycle of the pulsed source. Figure 4 shows the hardness as a function of the deposition temperature of coatings, deposited at a ratio of $BCl_3/TiCl_4 = 7/3$. There seems to be a slight increase of the hardness with increasing deposition temperature. The decreased hardness at lower deposition temperature could as well be an artefact, reflecting the influence of the cemented carbide substrate with a hardness of 1700 HV0.02, as the film thickness at low deposition temperatures is only 0.5 $\mu$m compared to 1.1 $\mu$m at higher temperature. The coating exhibit a preferred orientation in the (100) and (101) directions with a smaller halfwidth of 1.1°, compared to the halfwidth of 2° for the coatings deposited in the r.f. plasma. Therefore it may be assumed, that the higher hardness of 3000 HV0.02 is caused by a higher degree of crystalline order produced in the pulsed direct current plasma compared to the r.f. plasma. An increase in the hardness with decreasing halfwidth of the XRD reflections has also been reported by Pierson *et al.* [2].

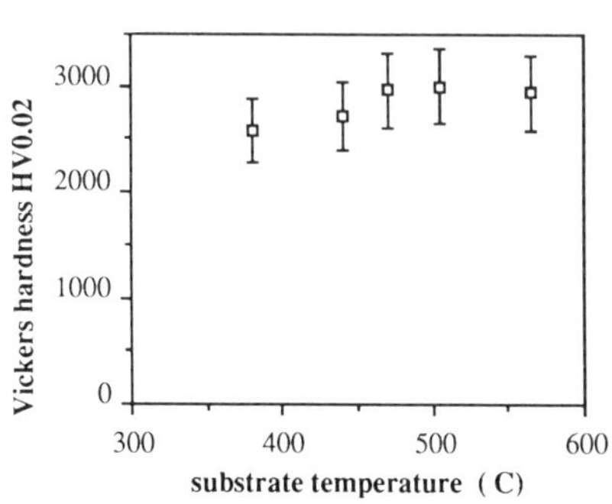


Fig. 4. Vickers hardness HV0.02 as a function of substrate temperature of films deposited in the pulsed direct current plasma at $BCl_3/TiCl_4 = 7/3$.

The $TiB_x$ coatings with $x = 2$ deposited in the r.f. plasma show a fracture amorphous structure, whereas the coatings deposited in the pulsed direct current plasma show in many cases a fibrous structure (Fig. 5). A critical load of 40 N for adhesion has been achieved.

### 3.2. $TiB_xN_y$ coatings

Starting in the r.f. mode with the parameters for the deposition of a stoichiometric TiN film and substituting gradually $BCl_3$ and $TiCl_4$ at a constant total flow of the chlorides, the XRD patterns show only peaks corresponding to the face-centered cubic (f.c.c.) TiN phase up to a ratio of $BCl_3/TiCl_4 = 1$. Figure 3(C) shows the XRD pattern of the film deposited at $BCl_3/TiCl_4 = 1$. Figure 6 clearly shows an increase of the lattice parameter of the f.c.c. TiN phase up to 1.5% with increasing $BCl_3$ flow rate. Additionally, a broadening of the halfwidth of the XRD reflections from 1.0° to 2.2° is observed. AES measurements show a continuous increase of the boron content up to 13 at.% B, corresponding to a composition of $TiB_{0.3}N_{1.0}$. It can not be excluded that there is some amount of boron or BN formed. From the

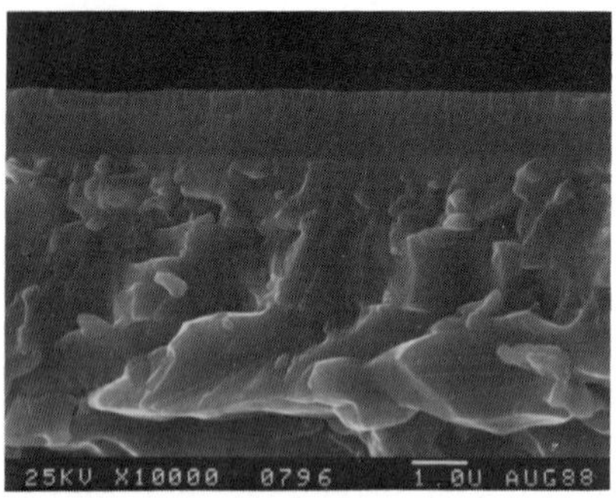


Fig. 5. Fracture cross section of a $TiB_2$ coating deposited in the pulsed direct current plasma.

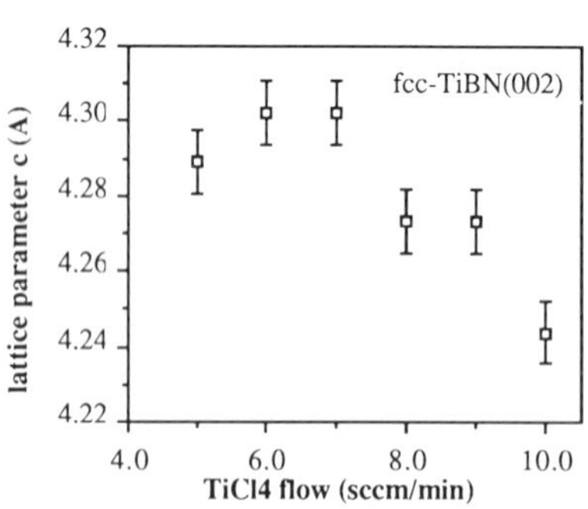


Fig. 6. Lattice parameter $c$ of the f.c.c. $TiB_xN_y$ phase as a function of $TiCl_4$ flow ($TiCl_4 + BCl_3 = 10$ sccm min$^{-1}$, $N_2 = 200$ sccm min$^{-1}$), derived from the (002) reflection. The films have been deposited in the r.f. plasma.

increasing lattice parameter one might conclude that a f.c.c. TiBN phase is formed by incorporation of boron into the TiN lattice. This has also been reported by other authors [5, 8]. The coatings show a fibrous to grained structure in the fracture cross-section. At higher $BCl_3$ flow ($BCl_3/TiCl_4 \geqslant 6/4$) a dramatic change of the film properties is observed. The hardness decreases from 2500 HV0.02 for f.c.c. TiBN to approximately 300 HV0.02, a lack of reflections in the XRD pattern indicates an amorphous structure and the films become electrically insulating. The solubility of boron in TiN at the deposition temperature of 500 °C seems to be exceeded under these conditions and a new type of XRD amorphous coatings is formed.

If the nitrogen flow at a ratio of $BCl_3/TiCl_4 = 1$ is decreased, the XRD patterns exhibit, beside the reflections corresponding to the f.c.c. TiBN phase, weak reflections corresponding to the hexagonal $TiB_2$ phase in (100) orientation (Fig. 3(D)). Because of the large halfwidth of the $TiB_2$ peak at 34.2° and the TiN peak at 36.8°, a weak reflection of a TiB peak at 35.3° would not be seen. Also the most intense TiB peak at 42.2° overlaps with the TiN peak at 42.65°. Therefore it cannot be excluded that also the orthorhombic TiB phase is present in these coatings. From the AES measurements (Fig. 7) it can be seen that the decrease of the $N_2$ flow results in a continuous increase of the boron content and a decrease of the nitrogen content. This is reflected in an increasing amount of the $TiB_x$ phases and a decreasing amount of the f.c.c. TiBN phase. The assumption of the presence of the orthorhombic TiB phase is supported by the XRD pattern of the coating deposited without $N_2$ (see Section 3.1.).

A continuous transition in the fracture morphology from a fibrous structure of the coatings consisting only of the f.c.c. TiBN phase, to an amorphous structure of the coatings consisting of a mixture of the f.c.c. TiBN and the $TiB_x$ phases can be observed.

In particular, the coatings consisting of the mixture of the f.c.c. TiBN phase and the $TiB_x$ phases show high hardness values of 3000 HV0.02. The chlorine content is approximately 1 at.% Cl. The adhesion of these coatings is still not satisfactory, as the critical load is only 25 N.

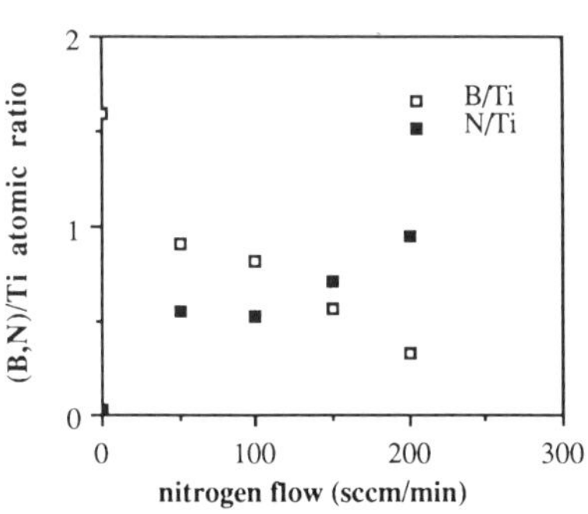


Fig. 7. B/Ti and N/Ti atomic ratio of $TiB_xN_y$ coatings, deposited in the r.f. plasma at $BCl_3/TiCl_4 = 1$, as a function of $N_2$ flow.

## 4. Conclusions

We have developed a new type of reactor for the plasma-assisted chemical vapour deposition (PACVD) of thin films, which enables a deposition on all surfaces of objects of complex geometry. The great advantage of this technique, compared to conventional methods is, that no movement of the samples is necessary. Deposition rates have been approximately 2 $\mu$m h$^{-1}$ at a moderate deposition temperature of 500 °C for both, $TiB_2$ and TiBN coatings.

$TiB_2$ coatings show hardness values up to 3000 HV0.02 and a satisfactory adhesion with a critical load of 40 N. The chlorine content is lower than 1 at.% Cl. The lower hardness as compared to the hardness reported by other authors is attributed to a low degree of crystalline order.

The TiBN coatings, showing a hardness of 3000 HV0.02, consist of a mixture of the f.c.c. TiBN, the hexagonal $TiB_2$ and most likely the orthorhombic TiB phase. These coatings adhere poorly to the substrate, as only a critical load of 25 N is obtained. Their chlorine content amounts to approximately 1 at.% Cl.

## Acknowledgment

The authors wish to thank the "Forschungsförderungsfonds für die gewerbliche Wirtschaft" for the financial support of this work.

## References

1 H. O. Pierson and E. Randich, *Thin Solid Films, 54* (1978) 119.
2 H. O. Pierson and A. W. Mullendore, *Thin Solid Films, 72* (1980) 511.
3 A. J. Caputo, W. J. Lackey, I. G. Wright and P. Angelini, *J. Electrochem. Soc., 132* (1985) 2274.
4 F. Zeman, J. Stamberger, J. Mayerhofer and A. Kulmburg, *Proc. 8th Int. Conf. on Chemical Vapour Deposition*, 1987, p. 87.
5 J. L. Peytavy, A. Lebugle, G. Montel and H. Pastor, *High Temp.–High Press., 10* (1978) 341.
6 J. T. Prater, *Surf. Coat. Technol., 29* (1986) 241.
7 T. Shikama, Y. Sakai, M. Fukutomi and M. Okada, *Thin Solid Films, 156* (1988) 287.
8 Ch. Mitterer, M. Rauter and P. Rödhammer, *Proc. 12th Plansee Seminar, Reutte*, 1989, in press.
9 J. Laimer, P. Braun, H. Störi, F. P. Viehböck, P. Rödhammer and K. Kailer, *Microchim. Acta Suppl., 10* (1983) 177.

*Surface and Coatings Technology, 39/40* (1989) 301 - 314 

# REACTIVE DEPOSITION OF HARD COATINGS

J. MUSIL, S. KADLEC, J. VYSKOČIL and V. POULEK

*Institute of Physics, Czechoslovak Academy of Sciences, Na Slovance 2, 180 40 Prague 8 (Czechoslovakia)*

(Received March 16, 1989)


## Summary

The reactive deposition of thin films is analyzed by means of a microphysical model. This model shows the way in which to (i) eliminate the hysteresis effect and (ii) control the deposition process. Special attention is devoted to physical parameters, such as fluxes of ions $\nu_i$, metallic particles $\nu_m$ and reactive gas particles $\nu_r$, the energy of particles bombarding the growing film and the substrate temperature which play a fundamental role in the deposition process, and to deposition systems with enhanced ionization. A correlation between (i) process parameters, (ii) microstructure, texture, phase and chemical composition of films, and (iii) resulting film properties is discussed in detail for Ti–N films. A comparison of arc evaporation and magnetron sputtering is also given. The need for equipment providing independent control of the deposition rate and ion bombardment of films is emphasized.


## 1. Introduction

Hard coatings embody a large range of materials, from metals and alloys through refractory compounds to diamond. Up to now, mainly transition metal nitrides and carbides, especially TiN, TiC, Ti(C, N), have achieved widespread application in surface engineering. This is due to the fact that these materials exhibit many extraordinary properties such as a high melting temperature, hardness, wear resistance, electrical and thermal conductivity. These properties are, however, determined by the chemical and phase composition, crystalline structure and microstructure of the film. It means that properties of films created depend on the kind of deposition process and processing conditions. Therefore, the relationships between resulting film properties and their microstructural features are being intensively investigated in many laboratories [1 - 13].

Hard coatings can be deposited by both chemical vapour deposition (CVD) and physical vapour deposition (PVD) processes. This paper reviews the reactive deposition of hard coatings by PVD, and mainly by plasm assisted PVD (PAPVD) processes. PAPVD processes are non-equilibrium

0257-8972/89/$3.50 

ones and the plasma plays a very important role in deposition. In general, it can be said that (1) electrons control the plasma chemistry by excitation, dissociation and ionization of a working atmosphere, and (2) ions control the microstructure of films.

Therefore, classical high-temperature processes, used in CVD, can be replaced by the low-temperature plasma (electron temperature $T_e \approx 1$ - $5$ eV) assisted PVD and ion bombardment of growing films with energies from 1 eV to 1 keV (1 eV = 11600 K). Here, it is, however, necessary to note that the ion bombardment is not fully equivalent to a high deposition temperature. It is due to the fact that the interaction of an ion with the film surface is localized and followed by a rapid quenching ($10^{12}$ - $10^{14}$ K s$^{-1}$). Metastable structures which are created in the presence of energy released by bombarding particles are thus frozen in. The ion bombardment of growing films can increase the surface mobility of adatoms, displace surface and bulk atoms, and supply sufficient energy necessary to activate chemical reactions.

The kind and degree of film modifications due to the ion bombardment of growing films depend on (1) the kinetic energy of incident ions and (2) the ratio of fluxes of bombarding ions $\nu_i$ and condensing atoms $\nu_m$, *i.e.* $\nu_i/\nu_m \propto i_s/a_D$, where $i_s$ is the substrate ion current density and $a_D$ is the deposition rate.

Experiments show that changes in film properties occur for energies of about 1 eV per atom and higher [7].

The stoichiometry $x$ of reactively deposited films is determined by the ratio of fluxes of reactive gas particles $\nu_r$ and condensing atoms $\nu_m$. The distribution of $\nu_i$, $\nu_m$ and sometimes also $\nu_r$ inside the deposition chamber is not homogeneous. This fact strongly complicates the creation of films with the same properties on substrates placed in different positions. Thus the scaling of the deposition process can be difficult.

## 2. Deposition systems with enhanced ionization and activation

A fundamental problem in PAPVD processes is to ensure sufficient ionization of the working atmosphere, *i.e.* a mixture of the vapour of deposited metal, reactive and inert gases, a homogeneous distribution of individual species inside the chamber, and to extract sufficient ion currents to the substrates. The basic principles of the most widely used ion-assisted deposition (IAD) techniques, including those with enhanced plasma ionization, photon stimulation and reactive gas activation, are illustrated in Fig. 1.

Figures 1(a) and 1(b) show ion-plating (IP) systems with a substrate bias $U_s$, based on evaporation by high- and low-voltage electron beams (HVEB and LVEB) respectively. While the HVEB system, due to very low collision cross-sections, needs additional ionization, the LVEB system ensures very effective ionization itself. A hollow cathode (Fig. 1(b)) can also be used for ionization in the system with HVEB evaporation. Figures 1(c)–1(e) show magnetron IP sputtering systems with enhanced ionization produced by electron emission (Fig. 1(c)) [14], a hollow cathode (Fig. 1(d)) [15] and a

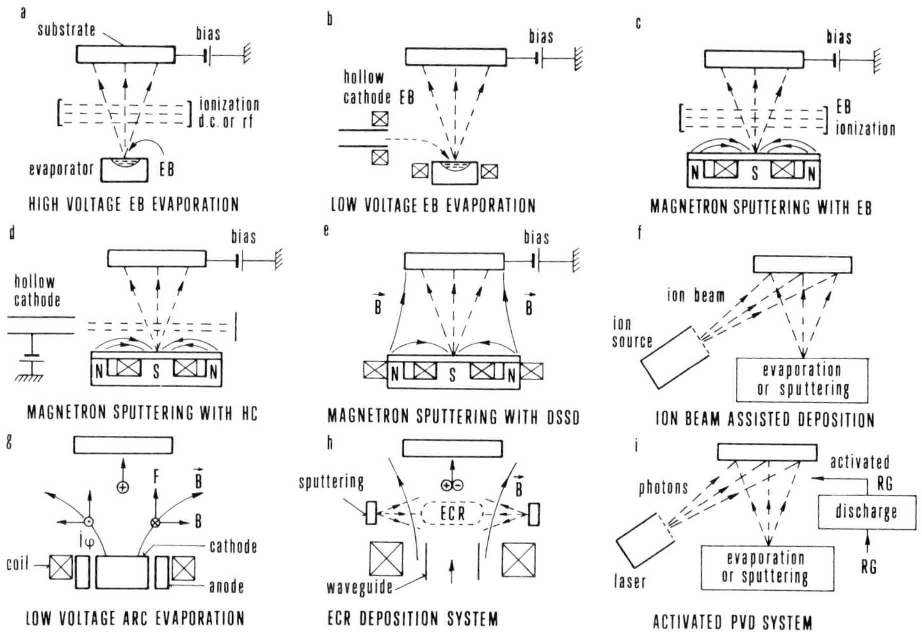


Fig. 1. Basic principles of devices for plasma, ion and photon assisted PVD techniques.

double-site-sustained discharge (DSSD) (Fig. 1(e)) [16]. Figures 1(f) - 1(i) show deposition systems which can be operated without bias because the energy necessary for controlled growth of films is delivered by means other than the acceleration of ions by a bias on the substrate, *e.g.* by an independent ion beam (Fig. 1(f)) [7], acceleration of ions in a magnetic field of a low-voltage arc evaporator (Fig. 1(g)), plasma acceleration in an inhomogeneous magnetic field (Fig. 1(h)) and by photons (Fig. 1(i)). An efficient approach for reactive deposition may be discharge activation of the reactive gas before its introduction into the deposition chamber, as shown in Fig. 1(i).

### *2.1. Sputtering systems with enhanced ionization*

A considerable disadvantage of magnetron sputtering is the strong decrease of the ion current to substrates with increasing distance of substrates from the magnetron target $d_{S\text{–}T}$. This is due to magnetic confinement of the plasma in the vicinity of the magnetron target.

In principle there are two possible ways to increase the ion current density $i_s$ on substrates in magnetron sputtering, *i.e.* by (1) additional gas ionization, *e.g.* using a hot cathode electron emitter (Fig. 1(c)), a hollow cathode electron source (Fig. 1(d)), or a direct current, r.f. or microwave additional discharge or (2) a change of distribution and strength of the magnetron magnetic field, *e.g.* using the unbalanced magnetron (UM) and DSSD (Fig. 1(e)).

Using UM and DSSD, high $i_s$ can be transported over large source to substrate distances $d_{S\text{–}T}$ [16]. Ion current densities of several mA cm$^{-2}$ can be easily obtained on substrates placed in $d_{S\text{–}T}$ up to about 200 mm. Typical

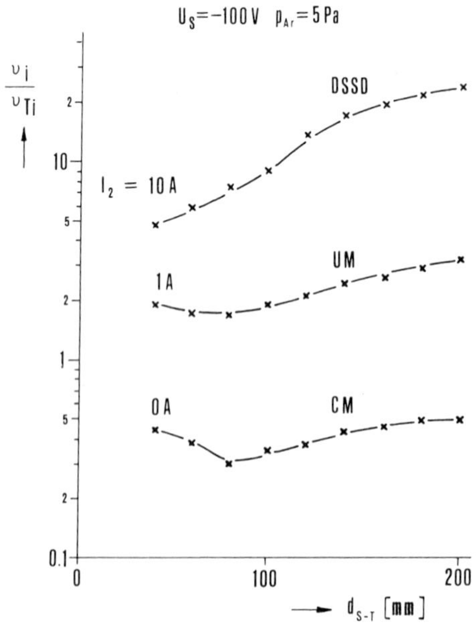


Fig. 2. Typical ratios of $\nu_i/\nu_{Ti}$ as functions of source to substrate distance $d_{S-T}$ for CM, UM and DSSD at the magnetron discharge power $P_d = 500$ W, $I_d = 1$ A. $I_2$ is the current in the external electromagnet of magnetron [16].

ratios of $\nu_i/\nu_{Ti}$ for ion and titanium atom fluxes impinging on substrates as a function of $d_{S-T}$ for the conventional magnetron (CM), the UM and DSSD are given in Fig. 2.

## 3. Sorption of reactive gas on surfaces and compound formation

During a condensation of metallic particles on surfaces of substrates and deposition chamber walls the compound is formed by the chemisorption of the reactive gas. The dominant parameter defining the composition of created film is the ratio $\nu_r/\nu_m$. If we assume that every new monolayer of the metal reacts with reactive gas without diffusion, the surface coverage is equal to the film stoichiometry $x$. Then, $x$ can be expressed, from the simplified kinetic model [17], in the form

$$x = \frac{1}{1 + (1/n_r\sigma_{rs})(\nu_m/\nu_r)} \tag{1}$$

where $n_r$ is the number of atoms in a reactive gas molecule and $\sigma_{rs}$ is the sticking coefficient of the reactive gas. While $\nu_r$ is often distributed almost uniformly over the whole deposition chamber (small pressure gradient), $\nu_m$ strongly depends on geometrical factors (substrates and target position and orientation) and process parameters (total and partial pressures). The coefficient $\sigma_{rs}$ depends on the substrate temperature $T_s$ and the energy and activation of impinging particles.

The total consumption ${}^s\phi_r$ of reactive gas in the growing film on substrates and walls (in the units pressure × volume/time) is [13]

$$^s\phi_r = \frac{k_B T_0}{n_r} \int_{A_s} x\, v_m \mathrm{d}A_s \tag{2}$$

where $A_s$ is the total area of substrates and walls, $k_B$ is the Boltzman constant and $T_0$ is the input gas temperature. The dependence ${}^s\phi_r = f(p_r)$, which can be measured by mass flow meters, enables us to determine the boundary between the metallic, transition and reactive modes and the critical pumping speed [18], see eqn. (5).

The total number of metallic atoms coming onto substrate surfaces per unit time is equal to the number of emitted particles from the target per unit time $\dot{M}$, *i.e.*

$$\dot{M} = \int_{A_s} v_m \mathrm{d}A_s \tag{3}$$

### *3.1. Processes on target*

Two basic processes are being run on the target surface during the reactive deposition of films [17, 19, 20]: (1) emission of particles from the target by sputtering or evaporation and (2) chemisorption of reactive gas on the target surface and formation of compounds with lower sputtering or evaporation rates.

As a result of these competing processes, an equilibrium state on the target surface is reached with a steady state of relative surface coverage by the reaction product. The results of calculations are given in Fig. 3(a) (magnetron sputtering) and Fig. 3(b) (arc evaporation) where $\dot{M}$, $x$ and ${}^s\phi_r$ are given as a function of normalized parameter $p_r/P$ (partial pressure of reactive gas divided by input power). The normalized parameters make it possible to scale process parameters and to prepare films with prescribed stoichiometry [13].

### *3.2. Reactive gas balance in reactive deposition process*

During the deposition the reactive gas flowing into the deposition chamber is pumped by two mechanisms: (i) by the pumping system and (ii) by a gettering in the growing film. An equilibrium state exists between these processes and the reactive gas inlet rate $\phi_r$ can be expressed in the following simplified form

$$\phi_r = p_r S_r + {}^s\phi_r \tag{4}$$

where $S_r$ is the pumping speed of the pumping system for the reactive gas.

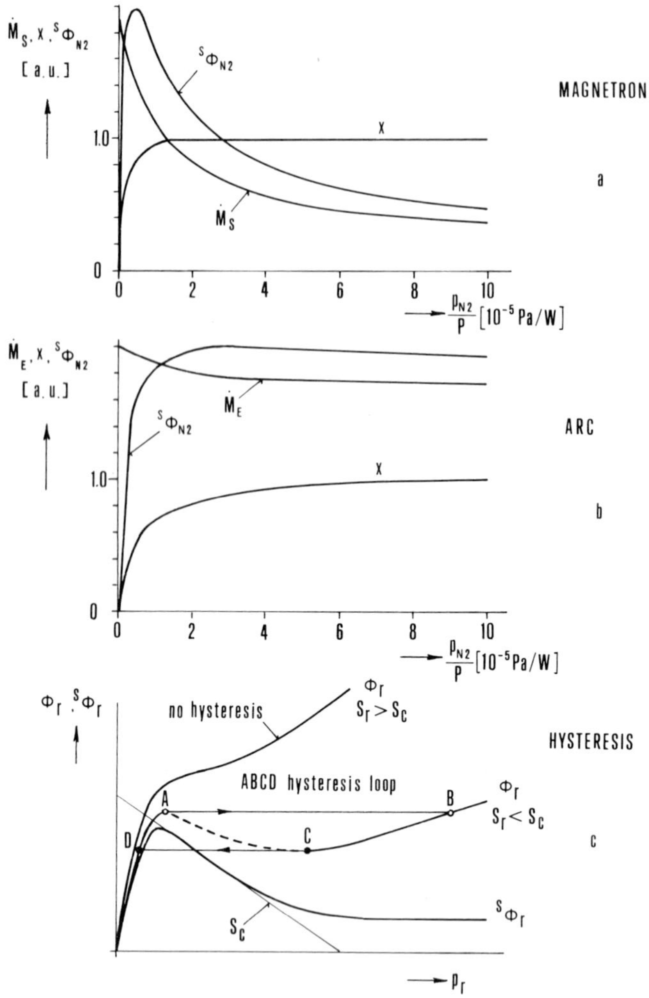


Fig. 3. $\dot{M}$, $x$ and ${}^s\phi_{N_2}$ as functions of $p_{N_2}/P$ calculated for $P = 3$ kW and a target area of 20 cm$^2$; (a) for magnetron sputtering and (b) for arc evaporation. (c) $\phi_r$ and ${}^s\phi_r$ as functions of $p_r$ for different pumping speeds $S_c$.

### *3.3. Hysteresis effect, critical pumping speed*

Hysteresis is a well known undesirable effect in reactive deposition [9, 11]. It occurs in cases when the dependence $\phi_r = f(p_r)$ decreases with increasing $p_r$ in a certain interval of $p_r$ due to formation of the compound on the target surface and due to the decrease of $\dot{M}$. A simple analysis of stability of eqn. (4) shows that the solution is unstable and the hysteresis occurs when the derivative $(d/dp_r)\,\phi_r = S_r + (d/dp_r)^s\phi_r < 0$. The hysteresis effect can be, however, fully eliminated when a sufficiently high $S_r$, satisfying the condition $S_r > S_c$, is used. The $S_c$ is called the critical pumping speed and it is defined by the formula

$$S_c = \left( -\frac{d^s\phi_r}{dp_r} \right)_{max} \tag{5}$$

Figure 3(c) illustrates the elimination of the hysteresis effect by increasing $S_r$.

$S_c$ does not depend on the target power, $U_s$ and $d_{S-T}$, but depends on $T_s$, the total pressure $p_T$ [18] and especially on the effective target area [17].

*3.4. Control of reactive deposition process*

Control of the reactive deposition process is strongly complicated by the hysteresis effect. Two closed loop systems were developed for control of the reactive process: (1) an optical control system [10], based on the measurement of intensity of metal emission lines, connected with a reactive gas valve or a power supply unit and (2) a partial pressure control system which keeps a selected partial pressure $p_r$ by a loop with the mass analyser and a regulation gas valve [11].

Both these systems need a very fast closed loop including special fast regulation gas valves and have a low accuracy of control for deposition processes in the metallic mode where the partial pressure is low (see Fig. 3) and intensities of spectral lines of the metal are almost constant [10].

Good reproducibility of film production can be achieved by full elimination of hysteresis using a vacuum system with the pumping speed $S_r > S_c$, see eqn. (5).

## 4. Reactive sputtering of $TiN_x$ films

*4.1. Dominant parameters*

The properties of created films depend especially on the ratios $\nu_r/\nu_m$ and $\nu_i/\nu_m$ and on the energy of impinging particles and $T_s$. The ratio $\nu_r/\nu_m$ determines mainly the film stoichiometry; the ratio $\nu_i/\nu_m$ and the energy of ions influence especially the film microstructure [3] and preferred orientation.

$T_s$ strongly influences not only the film microstructure and preferred orientation, but also the film composition and adhesion. Substrates are usually heated by conventional heaters, accelerated particles and/or radiation from the source of metallic particles. Many deposition systems use accelerated particles for the substrate heating and for keeping $T_s$ constant during deposition. The substrate heating rate and steady state $T_s$ depend on the power density $P_s$ on the substrate surface. Figure 4 clearly illustrates this, where the calculated $T_s$ *vs.* time parameter $t\,(1/\rho c)(S/V)$ is displayed for different $P_s$($\rho$, $c$, $S$ and $V$ are the density, the specific heat, the surface and the volume of substrate respectively) [21, 22].

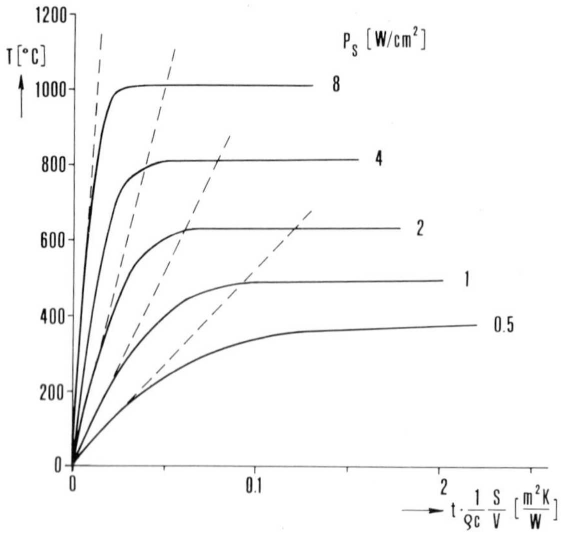


Fig. 4. Substrate temperature $T_s$ as a function of the time parameter for different substrate power densities $P_s$.

### 4.2. *Effect of process parameters on chemical and phase composition and texture of films*

Typical curves of $p_{N_2}$, $a_D$, Vickers hardness HV and stoichiometry $x = N/Ti$ as functions of $\phi_{N_2}$ are given in Fig. 5. Figure 5 compares experimental data with predictions of theoretical calculations [17]. The microhardness HV has a typical maximum $HV_{max}$ around $x \approx 0.7$. The hardest films are $\delta$-TiN and are characterized by maximum internal stresses [13]. Stoichiometric TiN films are a little harder than those prepared by thermal equilibrium processes (*e.g.* by CVD) and practically do not change with $\phi_{N_2}$.

The preferred orientation of films strongly depends not only on their chemical and phase composition but also on their microstructure which develops under the control of $T_s$, energy, number and kind of bombarding particles, and $p_T$. Also, it depends on the substrate surface and material, and the film thickness [13]. For an illustration, Fig. 6 displays intensities of X-ray reflections as functions of the ratio $\phi_{N_2}/I_d$ for two sets of Ti–N films prepared at (a) $T_s = 150$ °C, $U_s = 0$, $i_s = 0.66$ mA cm$^{-2}$ and (b) $T_s = 500$ °C, $U_s = -1.9$ kV, $i_s = 0.97$ mA cm$^{-2}$. Figure 6 shows that while films prepared at low $T_s$ are single-phase $\alpha$-Ti(N) or $\delta$-TiN$_x$ films, films prepared at high $T_s$ and high $|U_s|$ contain the $\varepsilon$-$Ti_2N$ phase. The development of film texture with increasing $\phi_{N_2}/I_d$ (scaling parameter) is different for both sets of films. At present there is, however, little data available to formulate general rules governing the film texture.

### 4.3. *Titanium nitride films containing $\varepsilon$-$Ti_2N$ phase*

An equilibrium phase diagram of a Ti–N compound shows that the $\varepsilon$-$Ti_2N$ phase arises in a narrow interval around 33 at.% N. Production of

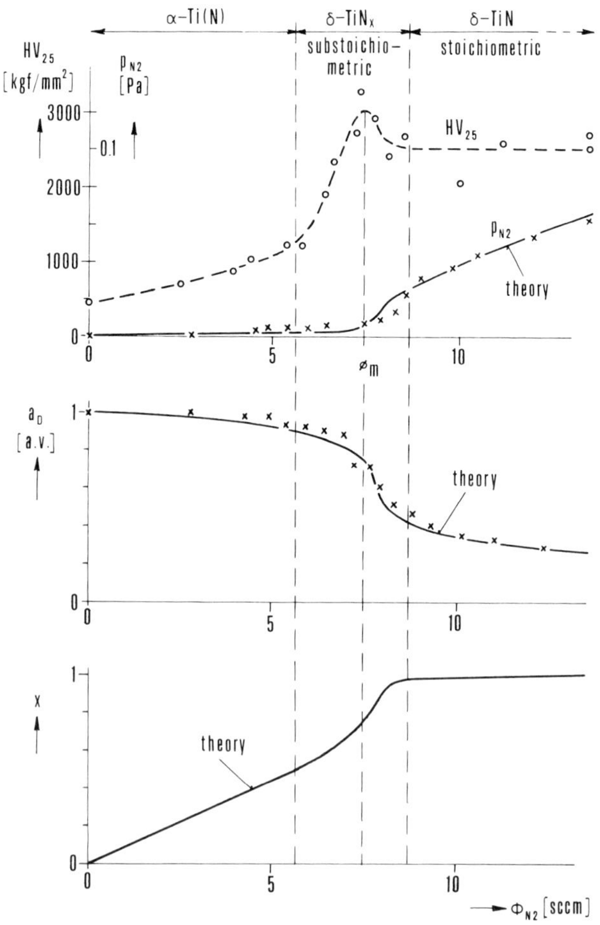


Fig. 5. Typical dependences $p_{N_2}$, HV, $a_D$ and $x$ *vs.* $\phi_{N_2}$ for magnetron sputtering at $P_d = 1.5$ kW, $I_d = 3$ A, $T_s = 150$ °C, $p_T = 0.3$ Pa, $d_{S\text{–}T} = 50$ mm.

Ti–N films containing the $\varepsilon$-$Ti_2N$ phase by PVD methods is, however, very difficult. It is due to the fact that production of the $\varepsilon$-$Ti_2N$ phase depends simultaneously on many process parameters, mainly on $U_s$, $i_s$, $T_s$, $a_D$ and of course on the film stoichiometry [23].

It was found that there is a combined parameter $S_E$ defined as

$$S_E(\mathrm{J\,kg^{-1}}) = \left(\frac{i_s U_s}{a_D}\right)_{T_s} \tag{6}$$

which makes it possible to control the creation of the $\varepsilon$-$Ti_2N$ phase. This parameter represents the energy delivered to a weight unit of growing film. It arises in the case when $S_E \gtrsim S_{thr}$, where $S_{thr}$ is the threshold value of $S_E$ [23].

According to some papers [24], Ti–N films containing the $\varepsilon$-$Ti_2N$ phase should be the hardest composition. Therefore, the microhardness of Ti–N

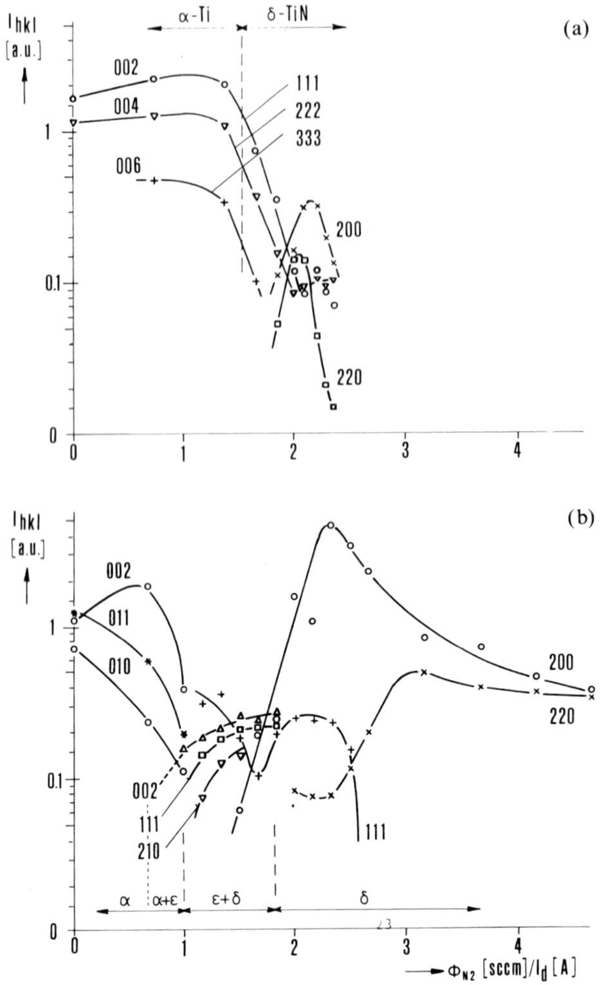


Fig. 6. Intensities of X-ray reflections from Ti–N films sputtered at (a) $U_s = 0$ and (b) $U_s = -1.9$ kV.

films was investigated as a function of phase composition (see Fig. 7). It was found that the hardest sputtered films contain no $\varepsilon$-$Ti_2N$ phase but only the $\delta$-$TiN_x$ phase.

### *4.4. Effect of ion bombardment on properties of TiN films*

The ion bombardment of growing TiN films has a crucial effect on their microstructure and crystalline orientation. These properties are affected not only by $U_s$, but also by $i_s$. It seems that the substrate power density $P_s = U_s i_s$ controls at least the TiN film texture. Figure 8 shows the corrected intensities $I_{hkl}$ of X-ray reflections as functions of $i_s$ and $U_s$. In Figs. 8(a) - (c) it can be seen that the development of the texture as a function of $P_s$ is very similar

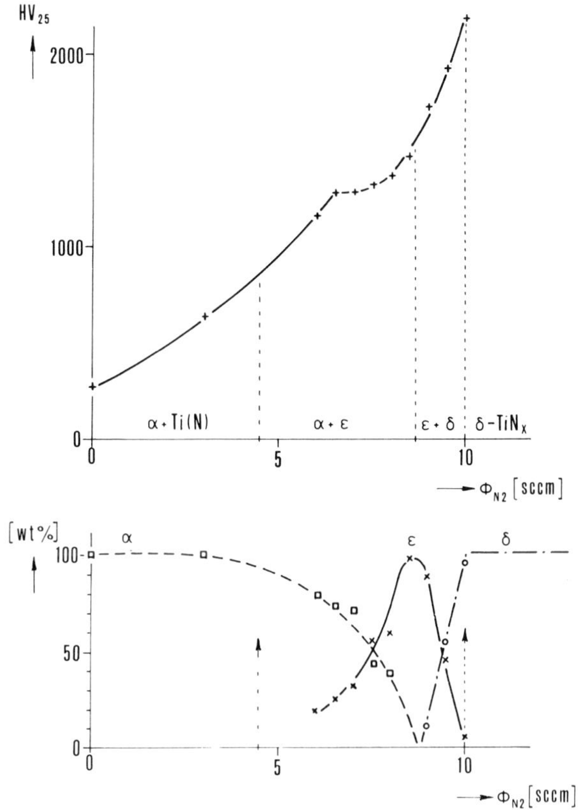


Fig. 7. HV and phase composition of Ti–N films sputtered at $P_d = 3$ kW, $I_d = 6$ A, $U_s = -1$ kV, $i_s = 2.2$ mA cm$^{-2}$, $T_s = 550$ °C, $p_T = 0.5$ Pa, $d_{S\text{-}T} = 50$ mm as functions of $\phi_{N_2}$.

for the three sets of films prepared under different conditions and by different methods.

The film microstructure is affected by both $i_s$ and $U_s$ and also by $p_T$. Experiments showed that TiN films magnetron sputtered at $U_s = -100$ V, $T_s = 500$ °C, $p_T = 5$ Pa, $d_{S\text{-}T} = 150$ mm and $i_s < 2$ mA cm$^{-2}$ are matt black, have a porous microstructure, low microhardness and contain a high amount of carbon impurities. On the contrary, films deposited under the same conditions at $i_s \gtrsim 2$ mA cm$^{-2}$ are compact, dense, bright golden films with no detectable carbon content. In spite of these differences the stoichiometry of all TiN films deposited at very different values of $i_s$ between 0.2 and 7.5 mA cm$^{-2}$ is almost the same and equals to $x \simeq 1$ [16].

Very recent sputtering experiments carried out at low pressures below 0.1 Pa and high $i_s$ indicate that compact and dense TiN films with $HV_{10} = 2500$ kg mm$^{-2}$ can also be prepared at low $U_s$ starting with the floating potential $U_{fl}$ (under conditions of Fig. 8(b)). These experiments indicate that it is possible to deposit low stressed compact films with different crystallographic orientations [25].

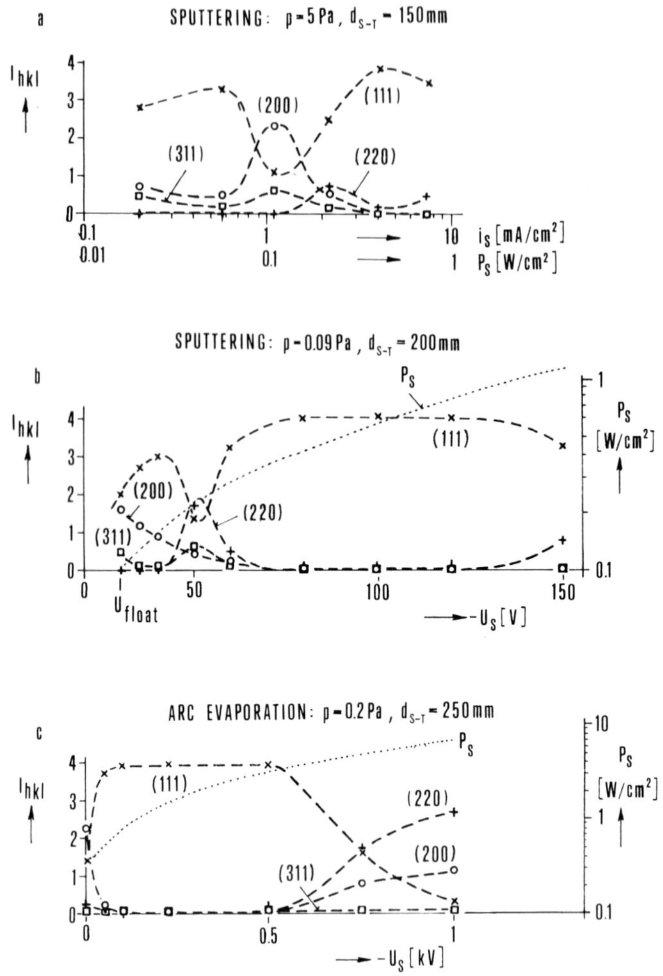


Fig. 8. $I_{hkl}$ of X-ray reflections as functions of $i_s$ and $U_s$.

## 5. Comparison of reactive magnetron sputtering and reactive arc evaporation of TiN films

Over the last decade coating technologies based on magnetron sputtering and arc evaporation have found widespread industrial applications. In spite of this fact a comparison of these techniques, their possibilities, advantages and limitations including the properties of the films created still remains an open physical problem and a question widely discussed by users.

Differences in both methods are caused by differences in the process of particle emission from the material of the cathode. This results in differences in (i) energies of particles (10 eV per atom and 1 - 500 eV per reflected neutral in sputtering, about 50 eV per ion of evaporated material), (ii) the ionization

TABLE 1

Typical parameters of reactive arc evaporation (AE) and sputtering using the CM and UM

| | AE | CM | UM |
|---|---|---|---|
| Energy for material emission (keV atom$^{-1}$) | 0.25 | 1 | 1 |
| Ionization of metal particles (%) | 80 | 10 | 10 |
| $\dot{M}_E/\dot{M}_s$ at $p_{N_2}=0$ | 4 | 1 | 1 |
| $\dot{M}_E/\dot{M}_s$ at $p_{N_2}=0.4$ Pa, $P=3$ kW | 20 | 1 | 1 |
| Power loss in cathode (%) | 30 | 90 | 90 |
| $d_{S-T}$ (mm) | 200 - 400 | 40 - 80 | 50 - 200 |
| $i_s$ (mA cm$^{-2}$) at $d_{S-T}=50$ mm[a] | 150 | 0.5 | 0.5 - 10 |
| $i_s$ (mA cm$^{-2}$) at $d_{S-T}=200$ mm[a] | 10 | 0.1 | 0.1 - 5 |
| $v_i/v_m$ | 1.5 | 0.4 | 0.3 - 15 |
| at $d_{S-T}$ (mm)[a] | 300 | 50 | 200 |
| Critical pumping speed $S_c$(l s$^{-1}$) | 50 | 250 | 250 |
| Macroparticles | yes | no | no |
| Coating of complex shapes | easy | difficult | yes |

[a] $P=2$ kW, $U_s=-100$ V, $p_{N_2}=0$.

(1 - 10% in the sputtering process and about 80% in the case of titanium arc evaporation) and (iii) the reactive gas sorption (see Figs. 3(a) and 3(b)). These differences influence the process of film growth in very different ways.

A detailed analysis [26] shows that possibilities of both evaporation and sputtering techniques are well characterized by their physical parameters which are summarized in Table 1.

## 6. Concluding remarks

The main problem of thin film PAPVD processes is a strong interrelation between $\nu_i$, $\nu_m$, $\nu_r$ and energies of condensing particles $E_c$ and bombarding ions $E_i$. This means that it is not possible to control $a_D$ and the bombardment of films independently. Therefore, the number of combinations $\nu_i$, $\nu_m$, $\nu_r$, $E_c$, $E_i$ is limited and very probably the production of all possible film microstructures cannot be realized. Moreover, certain differences can be expected when growing films are bombarded by host or guest ions. For example, certain microstructures will be produced by the sputtering process and others by the evaporation process.

Drawbacks of PAPVD processes stimulate the development of new deposition devices based on combined processes [6, 7], see Figs. 1(f) and 1(i). The main advantage of these devices is an independent control of $\nu_m$ emanating from the evaporating or sputtering source $\nu_i$, $E_i$, kind and angle of incidence of ions bombarding the growing film in IAD systems (Fig. 1(f)) or of the flux of photons $\nu_{ph}$ illuminating the growing film in laser assisted PVD systems (Fig. 1(i)).

The independent control of the deposition rate $a_D$ and the modification of created films makes it possible to investigate basic relations between the film microstructure and process parameters. This is of fundamental importance for both the basic research and industrial production of new materials with prescribed properties.

## References


1 J. A. Thornton, *J. Vac. Sci. Technol. A4 (6)* (1986) 3059.
2 J. E. Sundgren, *Thin Solid Films, 128* (1985) 21.
3 K. H. Müller, *J. Vac. Sci. Technol. A4* (1986) 184.
4 G. I. Grigorov and I. N. Martev, *Thin Solid Films, 143* (1986) 177.
5 W. D. Münz, *J. Vac. Sci. Technol., A4 (6)* (1986) 2717.
6 N. Savvides, *Thin Solid Films, 163* (1988) 13.
7 S. M. Rossnagel and J. J. Cuomo, *Vacuum, 38 (2)* (1988) 73.
8 D. S. Rickerby and R. B. Newbery, *Vacuum, 38 (3)* (1988) 166.
9 A. G. Spencer, R. P. Howson and R. W. Lewin, *Thin Solid Films, 158* (1988) 141.
10 S. Schiller, U. Heisig, K. Steinfelder, J. Strümphel, R. Voight and G. Teschner, *Thin Solid Films, 96* (1982) 235.
11 W. D. Sproul, *Surf. Coat. Technol. 33* (1987) 73.
12 A. J. Perry, *J. Vac. Sci. Technol., A6 (3)* (1988) 2140.
13 J. Musil, S. Kadlec and J. Vyskočil, *Thin Solid Films, 167* (1988) 107.
14 R. Adachi and K. Takeshita, *J. Vac. Sci. Technol., 20 (1)* (1982) 98.
15 J. J. Cuomo and S. M. Rossnagel, *J. Vac. Sci. Technol., A4 (3)* (1986) 393.
16 S. Kadlec, J. Musil, W. D. Münz, G. Håkanson and J.-E. Sundgren, *Surf. Coat. Technol., 39/40* (1989) 67.
17 S. Kadlec, J. Vyskočil and J. Musil, in A. Golanski, V. T. Nguyen and E. F. Krimmetl (eds.), *Proc. Symp. on Photon, Beam and Plasma Enhanced Processing*, Strasbourg, 1987, p. 161.
18 S. Kadlec, J. Musil and J. Vyskočil, *Vacuum, 37* (1987) 729.
19 A. S. Penfold, *ASVC Ann. Techn. Conf. Proc., New Orleans, LA, 1986,* ASVC, New Orleans, LA, 1986, p. 381.
20 S. Berg, H.-O. Blom, T. Larsson and C. Nender, *J. Vac. Sci. Technol., A5 (2)* (1987) 202.
21 A. Matthews and D. T. Gethin, *Thin Solid Films, 117* (1984) 261.
22 J. Vyskočil and J. Musil, *Proc. 7th Int. Conf. on Ion and Plasma Assisted Techniques,* 31 May - 2 June 1989, Geneva, p. 106.
23 V. Poulek, J. Musil, R. Černý and Kužel, *Thin Solid Films, 170* (1989) L55.
24 B. E. Jacobson, R. Nimmagadda and R. F. Bunshah, *Thin Solid Films, 63* (1979), 333.
25 S. Kadlec, J. Musil, W. D. Münz and V. Valvoda, *Proc. 7th Int. Conf. on Ion and Plasma Assisted Techniques*, 31 May - 2 June 1989, Geneva, p. 100.
26 J. Vyskočil, J. Musil, S. Kadlec and W.-D. Münz, in E. Broszeit *et al.* (eds.), *Plasma Surface Engineering*, DGM Informationsgesellschaft, Oberursel, 1989, pp. 661 - 668.

*Surface and Coatings Technology, 39/40* (1989) 315 - 328 

# ENGINEERING WITH SURFACE COATINGS: THE ROLE OF COATING MICROSTRUCTURE

D. S. RICKERBY and S. J. BULL

*Materials Engineering Centre, Harwell Laboratory, UKAEA, Didcot, Oxfordshire OX11 0RA (U.K.).*

(Received March 23, 1989)

## Summary

There is now great interest in the tailoring of the surface properties of a bulk material by the application of surface coatings, with a view to imparting to that surface specific characteristics which cannot be attained in the bulk material. If the maximum benefits are to be realised for a particular coating–substrate system, it is of fundamental importance that the interrelationship between the coating microstructure and its properties are understood. By reference to work on physical vapour deposited coatings such interrelationships are explored, and the importance of the coating microstructure in dictating many of the physical properties of the coating, for example hardness, adhesion, levels of internal stress, composition etc., are highlighted.

## 1. Introduction

Materials science is concerned with the relation between structure and macroscopic properties and this is true for both bulk and coated materials. According to Hornbogen [1] there exist six structural levels of matter down to the final elementary particles (Fig. 1), and it is the interactions which occur between these different levels of structure which dictate the properties of engineering materials. These interactions may start just above one atomic spacing (*e.g.* in the thickness of a grain boundary) and extend over many grain diameters when considering contact phenomena at a free surface in bulk ceramics [2] or the optical properties of thin films [3]. The structure–property relationships derived for thin films reflect this complex situation and are further complicated by the realization that, in general, coatings are not deposited at equilibrium and contain high concentrations of lattice vacancies, dislocations etc., which can vary from grain to grain [4].

One feature common to vapour deposited coatings is that solid material is distributed in an array of fairly closely packed columns that run perpendicular to the substrate, and it is the columnar microstructure that has considerable influence over the physical behaviour of the film; the microstructure is

0257-8972/89 Elsevier Sequoia/Printed in The Netherlands

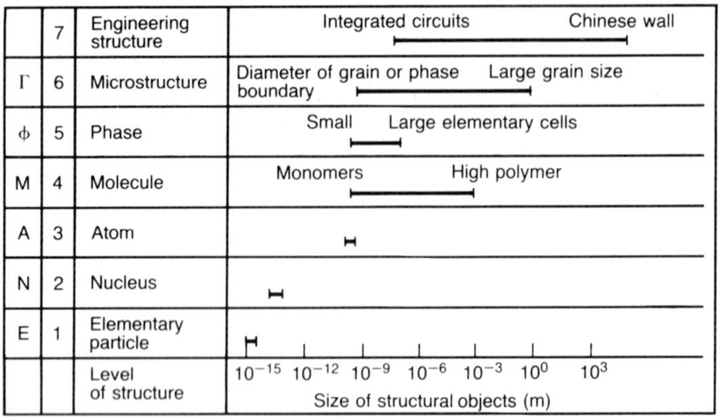


Fig. 1. The seven levels of structure (after Hornbogen [1]).

anisotropic and so are most thin film properties [5]. Many attempts have been made to model film morphology, which range from a consideration of adatom mobility on the growing surface [6 - 8] to computer simulation [9 - 11] and evolutionary growth models [12 - 14], but as yet it is impossible to predict the properties of vapour deposited coatings from a knowledge of their deposition conditions [14].

Our review focuses on the role of coating microstructure in dictating many of the engineering properties of physical vapour deposited (PVD) coatings, and how these are correlated with process and systems parameters. The various structure zone models (SZMs) are assessed and again related to changes in the systems parameters in an attempt to describe the way in which the various PVD microstructures produced by sputter, arc and evaporation sources relate to the physical properties of the coating–substrate system.

## 2. Microstructure of PVD coatings

### *2.1. SZMs*

To understand and control the physical structure of thin films is an important aspect in PVD technology, and an essential feature of thin films is that they are formed from a flux of atoms that approaches the substrate from a limited set of directions. Consequently, the microstructure of the coatings is columnar in nature and this anistropy in structure is translated across into many of the physical properties of thin films. Some undesirable consequences of such a microstructure are the provision of short-circuit diffusion paths along the channels formed between closely packed columns, and the ability of the microstructure to act like a sponge, capable of absorbing moisture from the atmosphere and desorbing part of it when the relative humidity decreases [15, 16]. On the other hand, films with a well-defined uniaxial ansiotropy are employed for magnetoresistive transducers, while non-magnetic films are used for resistors, elastoresistive devices and textured surfaces of silicon and germanium for solar absorption and optical data storage [17, 18].

Movchan and Demchishin [6] were the first to classify thin film microstructures, and identified three distinct structure zones as a function of the homologous temperature $T/T_m$ (where $T$ represents the substrate surface temperature and $T_m$ is the melting point). The low-temperature zone 1 structure ($T/T_m < 0.25$ - $0.3$) corresponds to low adatom mobility and consists of tapered columns with domed tops. In zone 2 ($0.3 < T/T_m < 0.5$), surface diffusion becomes important and the structure is of a straight columnar region and has a smooth surface topography. With increasing temperature ($T/T_m > 0.5$) bulk diffusion becomes the dominant process and zone 3 structures are characterized as being equiaxed in nature akin to recrystallized metals. Later work by Thornton [7] demonstrated that the presence of a sputtering gas could modify the model suggested by Movchan and Demchishin and a further region was identified called zone T which consisted of poorly defined fibrous grains. Computer simulation has contributed significantly to a recognition of the fundamental importance of atomic level shadowing and has shown that shadowing combined with limited surface mobility can explain the basic zone 1 structure [9, 19]. The breakdown of the zone 1 structures with increasing $T/T_m$ has also been simulated by Müller, where the adatoms attain sufficient surface diffusion to remove the effects of atomic shadowing [20].

*2.2. The effect of substrate bias on microstructure*

As an alternative to increasing deposition temperature, zone 1 structures can be overcome by subjecting the growing films to bombardment by particles having sufficient energy so that the resultant momentum exchange causes coating atoms to fill the voided boundaries. Messier *et al.* [18] have suggested improvements to the SZMs which account for the evolution of morphology with increasing film thickness and the effect of both thermal and bombardment induced mobility. In this idealized model all the distinct levels of physical structure column–void sizes are considered and assigned sub groups 1A, 1B, 1C, 1D and 1E as indicated in Fig. 2. This model draws

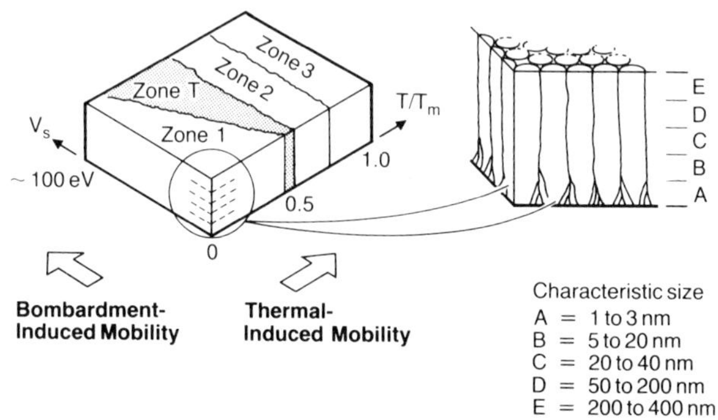


Fig. 2. Revised structure zone model for zone 1–zone T conditions showing the effect of energetic particle bombardment (after ref. 18).

attention to the fact that ion bombardment promotes a dense structure of the zone T type, but also that these atomic movements can have a thermal induced or bombardment induced origin. During ion bombardment Müller has shown by a two-dimensional computer simulation model that the void filling process involves a combination of forward sputtering of coating atoms into the void region, impact induced surface diffusion, and local heating such that recrystallization takes place during the growth and the underlying crystal structure is adopted [20, 21].

Clearly the application of a substrate bias during the deposition of a coating has a profound effect upon the growth and resultant microstructure of PVD films [22 - 25]. The unbiased film shows an open columnar microstructure (zone 1 type, Fig. 3(a)) whilst the biased film appears more dense, the individual columns being much less well defined (zone T type, Fig. 3(b)), in agreement with the SZM of Meissier *et al.* [18]. The microstructures shown in Fig. 3, taken from the work of Rickerby *et al.* [26], will have quite different physical properties due in the main to changes in the packing density of the columnar grains which make up the coating [5, 27]. These changes in properties which can be expected in going from zone 1 to zone T can be seen by reference to Fig. 4, which shows the variation in internal stress, hardness and coating–substrate adhesion as a function of substrate bias voltage for sputter deposited TiN coatings, although similar trends are expected to hold for other PVD technologies. With zone 1 microstructures Rickerby and coworkers [5, 26 - 29] suggest that the open columnar boundaries ensure that any residual stresses in the deposit, either growth or derived from thermal mismatch, are kept to a minimum and by virtue of the open microstructure the reduced load bearing capacity of the surface means that the hardness of coating is dramatically reduced over that of the more dense zone T microstructures. A direct consequence of the denser microstructures associated with the zone T region of the SZMs is that substantial stresses can be generated within the coating and this leads to an increase in stored elastic energy which is accomplished by a reduction in the level of coating adhesion

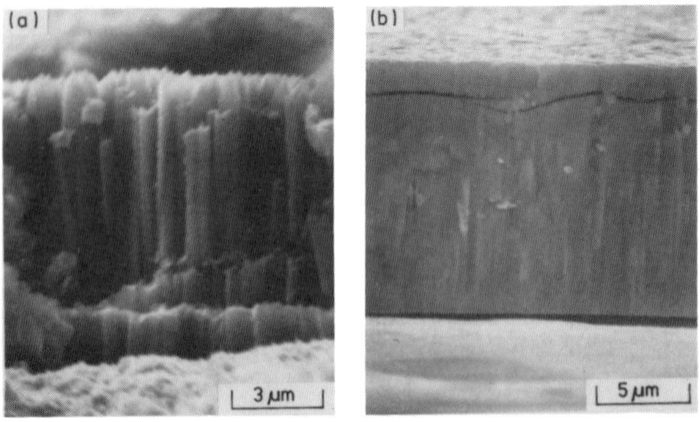


Fig. 3. Scanning electron fractographs (secondary electron image) of (a) unbiased and (b) biased TiN coatings deposited onto a low-alloy steel by sputter ion plating (after ref. 26).

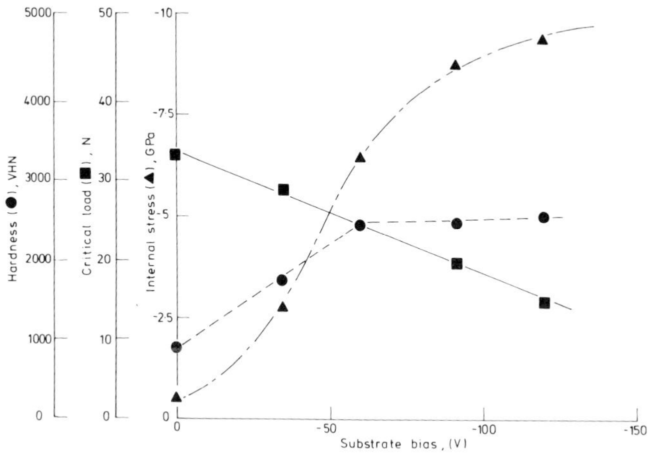


Fig. 4. Variation in coating hardness, internal stress and critical load for coating detachment with substrate bias voltage for TiN coatings deposited onto austenitic stainless steel by the sputter ion plating process (after ref. 28).

($L_c$). Similar observations of a reduction in critical load with increasing residual stress have been reported for TiC coatings deposited by chemical vapour deposition (CVD) [30]. The apparent increase in density of a PVD coating when moving from zone 1 to zone T can also be correlated to the levels of internal stress present in the coatings [27].

*2.3. The effect of internal stress on coating microstructure*

Although the structure zone models can explain the various microstructures which occur in PVD coatings as a function of both thermal and ion induced atom movements, no account is taken of the substrate onto which the deposit is formed. The influence of the substrate on coating microstructure is two-fold; the substrate crystal structure defines the growth habit of the deposit through changes in the distribution, orientation and size of the crystallites nucleated on the growing surface [5, 31 - 33]; the nature of the substrate, in terms of its thermal expansivity, affects the levels of internal stress in a film and this stress is compressive when $\alpha_f < \alpha_s$ and tensile when $\alpha_f > \alpha_s$, where $\alpha_f$ and $\alpha_s$ are, respectively, the thermal expansion coefficients of film and substrate [5]. This second point regarding the degree of mismatch between the expansion coefficients of coating and substrate is very important since the role of internal stress in determining the nature of the final coating microstructure has received little attention. For instance, Aubert *et al.* [34] have correlated the compressive internal stress with the density of magnetron-sputtered molybdenum, and Rickerby [5, 27] has suggested similarly that compressive internal stresses may be of paramount importance in determining the porosity of TiN and other PVD coatings. Thus, internal stress, through variations in thermal expansivity of coating and substrate, constitutes another previously unconsidered degree of freedom in the construction of PVD SZMs.

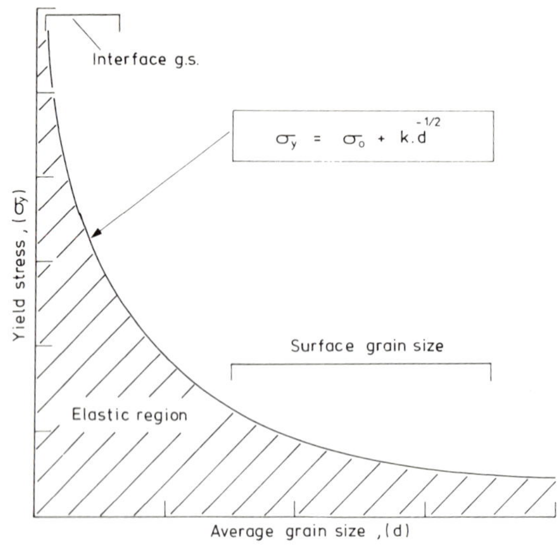


Fig. 6. Schematic variation in yield strength with average grain size for a PVD coating. It should be noted that the interface region can support higher elastic stresses (the hatched region) owing to its smaller grain size (after ref. 39).

From the above discussion, the increased strength of grain boundaries which are formed by an increase in deposition temperature, application of a substrate bias voltage to promote ion bombardment of the film, or the correct choice of substrate material ($\alpha_s > \alpha_f$) has an important bearing on the physical properties of PVD films. Figure 7 shows that the film hardness of many ceramic materials increases with deposition temperature which constrasts strongly with the behaviour of pure metal films [43]. Although the commonly used argument to explain such effects is that the strength of a ceramic material decreases with increasing defect concentration, and as a consequence of the

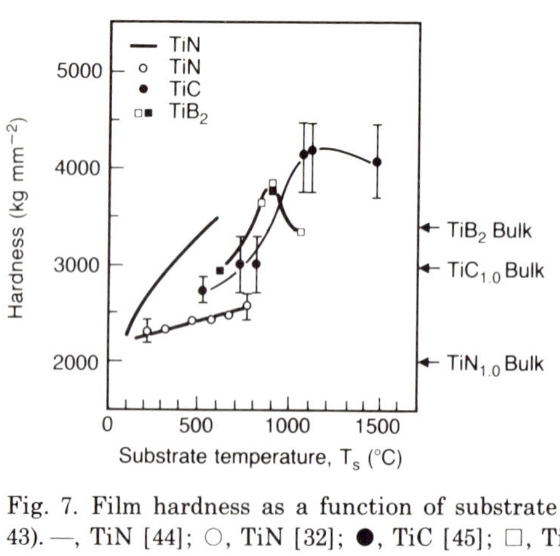


Fig. 7. Film hardness as a function of substrate temperature for ceramic materials (after ref. 43). —, TiN [44]; ○, TiN [32]; ●, TiC [45]; □, $TiB_2$ [46, 47].

higher deposition temperature there is a reduction in defect concentrations and therefore hardness increases as shown in Fig. 7. However, it is now clear that a more plausible explanation based on the strength of grain boundaries is more in keeping with the SZMs proposed for PVD films [43]. For types 1, T and 2 films the microstructural-induced anisotropy which develops during growth can result in physical properties which are unique to PVD films. For example, it has been reported that PVD coatings are significantly harder when tested parallel to their growth direction compared with values derived from the other two orthogonal directions [41, 48]. In contrast, coatings that exhibit a fine equiaxed grain structure (zone 3 as produced by CVD) do not show a hardness anisotropy effect [48]. In addition many of the X-ray observations on PVD films stem from the nature of this oriented growth, and the interactions which occur between the individual columnar units as a function of ion bombardment intensity and deposition and/or service temperature are described by Rickerby and coworkers [26, 27, 49, 50].

In this section, attempts have been made to indicate the interdependence of ion bombardment, microstructure, deposition temperature, mechanical properties and internal stress. Although the structure zone models can explain the microstructural observations in terms of thermally induced and ion bombardment induced adatom mobility the role of substrate has not been taken into account, and this is particularly important for those coating–substrate combinations which result in high thermal compressive stresses acting to densify the coating. Structure-induced anisotropy arises from a competitive growth process, which in turn leads to the development of properties which are unique to PVD deposits and which are a sensitive function of the strength of the grain boundaries which form between the growing columns.

## 3. Applications for PVD coatings

In this section we shall explore the uses for PVD thin films and illustrate the advantages–disadvantages which arise as a direct consequence of the microstructure-induced anisotropy.

### *3.1. Electrical and optical properties*

Recently, there has been considerable interest in TiN as a diffusion barrier in multilevel metallization schemes for integrated circuits due to its chemical stability and electrical properties [51, 52]. Both type 1 and type T microstructures have been assessed, with a view to controlling film microstructures to obtain reliable device performance. For the zone 1 structures, the fibrous microstructure leads to low stresses and a high resistivity due to oxygen contamination on the surface of the columnar structure which results from its high surface to volume ratio [53]. In contrast the much denser zone T microstructures have low oxygen contents, but suffer mechanical failure due to the very large compressive stresses set up in the film by the application of a substrate bias [54]. The gradual removal of oxygen from titanium nitride films in moving from a zone 1 to a zone T structure is illustrated by the work of Poiterin *et al.* [55]. The reduction in film resistivity with increasing ion bombardment (substrate bias voltage) may be explained by a number of factors and these are discussed fully in the literature [33, 56, 57].

The preferential resputtering of oxygen from thin films of TiN is not unique and application of a substrate bias during film growth can remove elements such as sodium, potassium and calcium which are often associated with greases and dust [58]. What is interesting in the case of TiN developed for microelectronic applications is that the zone T films (low resistivity and oxygen contents) were found to stop silicon diffusion, but the formation of intermetallics at the Al–TiN interface resulted in decomposition of the TiN films; a problem which was solved by introducing a sacrifical layer of titanium between the aluminium and the TiN, which leads to the formation of $Al_3Ti$ without destroying the TiN diffusion barrier [59]. However, zone 1 films (high resistivity and oxygen contents), though they were permeable to silicon due to their porous microstructure, were found to stop this damaging interaction and a new metallization concept based on a combination of zone 1 and zone T layers is under development [51].

The optical properties (refractive index) of thin films are related to the packing density $p$ and the shape of the individual columns which comprise the microstructure. Pulker and Jung [60] and Ogura [61] consider the columns to have closely packed circular bases with cross-sectional areas that expand, contract or remain constant in the direction of growth. These simple geometrical models have found practical applications in predicting the optical behaviour of thin films and are fully discussed by Hodgkinson and Wilson [15].

### *3.2. Tribological properties*

The use of surface coatings to improve the wear resistance of contacting surfaces is now well established [62], and hard, wear-resistant coatings such as TiN are finding an increasing number of applications. However, the interrelationship between the microstructure and properties of the coating and its wear resistance in many instances still needs to be established although some progress has been made in this area [36, 63, 64]. For example Matthews and Sundquist [36], using a pin-on-disc wear test, has shown that TiN coatings which have a strong (200) orientation are more wear resistant than a corresponding film with a (111) orientation and the same hardness. This improvement may in part be due to the increase in coating densification which occurred under the higher ion bombardment conditions which gave the {200} texture, *i.e.* a move towards the denser zone T microstructures. The improvements in abrasive wear resistance of titanium nitride coatings in going from a zone 1 to a zone T microstructure is clearly illustrated by the work of Bull *et al.* [64]. For TiN films deposited under zone 1 conditions a rapid initial increase in abrasive wear is observed as the porous TiN coating is removed by a ploughing mechanism. With increasing bias, the load support offered by the denser zone T microstructure increases and the wear rate decreases. However, with high degrees of ion bombardment the levels of internal stress in zone T microstructures are seen to increase, and due to a reduction in coating–substrate adhesion there is a loss in abrasive wear resistance (see Section 2.3). Clearly the abrasive wear resistance of thin TiN coatings is a sensitive function of the degree of ion bombardment which takes place during deposition. Zone T structures which have low levels of internal stress offer the most wear-resistant coatings, but the work of Bull *et al.* [64] illustrates that the coating microstructure, and its effect on hardness, internal stress and adhesion, is an important factor in determining the dominant wear mechanism. Under conditions of adhesive wear the transfer of iron from the sphere to the disc (during a sphere-on-disc test) is increased in the case of the open zone 1 microstructure since iron becomes trapped in the open inter-columnar regions, and this leads to an increase in wear. Akin to the abrasive wear results, best adhesive wear properties are observed for hard, dense coatings as produced under zone T conditions which are promoted by the use of a substrate bias voltage [65].

### *3.3. PVD vs. CVD coatings in metal cutting applications*

The application of hard coatings to cemented carbide metal cutting tools is now accepted practice [66, 67] and the main deposition technology used is CVD. Recently, however, PVD has been demonstrated to be competitive against CVD, and the PVD coatings have certain microstructural advantages over their CVD counterparts [68]. Quinto and coworkers [67 - 69] have compared the structure–property relationships for CVD and PVD coatings in metal cutting applications as a function of coating process parameters. For CVD TiN films, which are formed under thermal equilibrium, well-defined large-angle grain boundaries are produced with typical grain sizes around 0.5 $\mu$m and the structures can be represented as zone 2 in the SZMs. In contrast, PVD films are extremely fine grained and individual crystallites contain a large concentration of point defects [5, 69] due to non-equilibrium deposition. Consequently PVD coatings are consistently harder than CVD coatings and this has an important bearing on the abrasive wear resistance of coated tools. Another important factor in determining the performance of TiN coated cemented carbides in metal cutting applications is the levels of residual stress in the coating, and whilst CVD coatings are under a relatively low tensile stress those formed by PVD show high compressive stresses. The work of Quinto *et al.* [68] endorses the idea that this compressive residual stress in the plane of a PVD coating improves the surface fracture strength or toughness of coated tools which leads to better performance in milling tests. In addition the high processing temperature used in the CVD process anneals out any beneficial compressive stresses which exist in uncoated tools, whilst PVD TiN coating leaves the stress state of the substrate unaffected, which again improves the resistance to cracking of PVD-coated carbide tools. The decrease in flank wear of PVD TiN over both uncoated and CVD-coated materials is correlated with increased surface hardness since the wear mechanism is mostly abrasive. There is also an increase in crater wear resistance for PVD over CVD coatings, which is somewhat surprising since it is generally believed that crater wear is governed by chemical interaction or solubility considerations at the tool–chip interface [70, 71] and that has led Quinto *et al.* [68] to postulate that there must be an abrasive wear component in the crater wear process.

The work of Quinto *et al.* [68] agrees in broad terms with that of Kamachi *et al.* [72], and both investigators indicate that a better understanding of the structure–property relationships for coated materials can lead to significant improvements in coating performance and the correct choice of deposition process for a given application.

## 4. Conclusions

The SZMs for PVD coatings can account for the microstructural changes which arise due to an increase in deposition temperature, and the application of a substrate bias voltage during deposition to promote coating densification by low energy ion-bombardment. These microstructural changes have also been simulated by computer models. However, in all cases no account has been taken of the role of internal stress, and thereby substrate, on the densification of coating microstructure and theoretical work is needed in this area to support the experimental observations.

Structural anisotropy arises in PVD films due to the competitive growth processes which occur during deposition, and as a consequence many of the physical properties of PVD deposits are related to the packing density (for example, zone 1 *vs.* zone T) of the individual columns which make up the coating. In many application areas correct control of PVD microstructure, through an understanding of the structure–property relationships which occur, allows optimum coatings to be chosen for a diverse range of uses, from thin films in microelectronic circuits to thicker films in tribology.

## Acknowledgment

Copyright of this paper belongs to the United Kingdom Atomic Energy Authority.

## References

1 E. Hornbogen, *Acta Metall.*, *32* (1984) 615.
2 T. F. Page, in D. Taylor (ed.), *Science of Ceramics 14*, The Institute of Ceramics, 1988, p. 41.
3 H. Angus Macleod, *J. Vac. Sci. Technol.*, *A4* (1986) 418.
4 A. J. Perry, *J. Vac. Sci. Technol.*, *A6* (1988) 2140.
5 D. S. Rickerby and P. J. Burnett, *Thin Solid Films*, *157* (1988) 195.
6 B. A. Movchan and A. V. Demchishin, *Fiz. Met. Metalloved.*, *28* (1969) 653.
7 J. A. Thornton, *Ann. Rev. Mater. Sci.*, *7* (1977) 239.
8 J. A. Thornton, *J. Vac. Sci. Technol.*, *A4* (1986) 3059.
9 A. G. Dirks and H. J. Leamy, *Thin Solid Films*, *47* (1977) 219.
10 P. Ramanlal and L. M. Sander, *Phys. Lett.*, *54* (1985) 1828.

11 K. H. Müller, *J. Appl. Phys.*, *58* (1985) 2573.
12 R. Messier and J. E. Yehoda, *J. Appl. Phys.*, *58* (1985) 3739.
13 J. E. Yehoda and R. Messier, *Appl. Surf. Sci.*, *22/23* (1985) 590.
14 R. Messier, *J. Vac. Sci. Technol.*, *A4* (1986) 490.
15 I. J. Hodgkinson and P. W. Wilson, *CRC Crit. Rev. Solid State Mat. Sci.*, *15* (1988) 27.
16 N. Kumar, M. G. Fissel, K. Pourrezaei, B. Lee and E. C. Douglas, *Thin Solid Films*, *153* (1987) 287.
17 A. Belu-Marian, R. Manaila, L. C. Nistor, V. Teodorescu, G. Dinca and D. Serbanescu, *Thin Solid Films*, *167* (1988) 1.
18 R. Messier, A. P. Giri and R. A. Roy, *J. Vac. Sci. Technol.*, *A2* (1984) 500.
19 D. Henderson, M. H. Brodsky and P. Chaudhari, *Appl. Phys. Lett.*, *25* (1974) 641.
20 K. H. Müller, *Phys. Rev. B*, *35* (1987) 7906.
21 K. H. Müller, *Surf. Sci.*, *184* (1987) L375.
22 D. M. Mattox and G. J. Kominiak, *J. Vac. Sci. Technol.*, *9* (1972) 528.
23 D. M. Mattox, *J. Vac. Sci. Technol.*, *10* (1973) 47.
24 R. D. Bland, G. J. Kominiak and D. M. Mattox, *J. Vac. Sci. Technol.*, *11* (1974) 671.
25 R. W. Knoll and E. R. Bradley, *Thin Solid Films*, *117* (1984) 201.
26 D. S. Rickerby, A. M. Jones and B. A. Bellamy, *Surf. Coat. Technol.*, *37* (1989) 111.
27 D. S. Rickerby, *J. Vac. Sci. Technol.*, *A4* (1986) 2809.
28 D. S. Rickerby and A. Matthews, *Rev. Powder Met. Phys. Ceram.*, in press.
29 P. J. Burnett and D. S. Rickerby, *Surf. Eng.*, *3* (1987) 195.
30 L. Chollet, H. Boving and H. E. Hintermann, *J. Mater. Energy Syst.*, *6* (1985) 293.
31 M. K. Hibbs, J.-E. Sundgren, B. E. Jacobsson and B.-O. Jacobsson, *Thin Solid Films*, *107* (1983) 149.
32 M. K. Hibbs, B.-O. Johansson, J.-E. Sundgren and U. Helmersson, *Thin Solid Films*, *122* (1984) 115.
33 J.-E. Sundgren, *Thin Solid Films*, *128* (1985) 21.
34 A. Aubert, J. Danroc, A. Gaucher and J. P. Terrat, *Thin Solid Films*, *126* (1985) 61.
35 U. Helmersson, J.-E. Sundgren and J. E. Greene, *J. Vac. Sci. Technol.*, *4* (1986) 500.
36 A. Matthews and H. D. Sundquist, in T. Takagi (ed.), *Proc. Int. Ion Engineering Congress, ISIDT '83, Kyoto, Japan, September 1983*, IEE (Japan), Tokyo, 1983, p. 1325.
37 H. M. Gabriel, in T. Takagi (ed.), *Proc. Int. Ion Engineering Congress, ISIDT '83, Kyoto, Japan, September 1983*, IEE (Japan), Tokyo, 1983, p. 1311.
38 M. Kobayashi and Y. Doi, *Thin Solid Films*, *54* (1978) 67.
39 D. S. Rickerby, G. Eckold, K. T. Scott and I. M. Buckley-Golder, *Thin Solid Films*, *154* (1987) 125.
40 P. J. Burnett and D. S. Rickerby, *Thin Solid Films*, *148* (1987) 41.
41 P. J. Burnett and D. S. Rickerby, *Thin Solid Films*, *148* (1987) 51.
42 V. Valvoda, R. Cerny, R. Kuzel, J. Musil and V. Poulek, *Thin Solid Films*, *158* (1988) 225.
43 J.-E. Sundgren and H. T. G. Hentzell, *J. Vac. Sci. Technol.*, *A4* (1986) 2259.
44 W. D. Münz and G. Hessberger, *Vak. Tech.*, *30* (1981) 78.
45 A. C. Raghuram and R. F. Bunshah, *J. Vac. Sci. Technol.*, *9* (1972) 1289.
46 H. O. Piersson and A. W. Mullendore, *Thin Solid Films*, *72* (1980) 511.
47 H. O. Piersson and A. W. Mullendore, *Thin Solid Films*, *95* (1982) 99.
48 Y. Enomoto, in K. Kamada and M. Fukutomi (eds.), *Proc. Workshop on First Wall Coatings, Institute of Plasma Physics, Nagaya University, November 28-29, 1980*, Publication IPPJ-551, 1982, p. 202.
49 D. S. Rickerby, S. J. Bull, A. M. Jones, F. L. Cullen and B. A. Bellamy, *Surf. Coat. Technol.*, *39/40* (1989) 397.
50 D. S. Rickerby, A. M. Jones and A. J. Perry, *Surf. Coat. Technol.*, *36* (1988) 631.
51 N. Kumar, K. Pourrezaei, B. Lee and E. C. Douglas, *Thin Solid Films*, *164* (1988) 417.
52 M. Wittmer, in W. D. Sproul, J. E. Greene and J. A. Thornton (eds.), *Physics and Chemistry of Protective Coatings*, American Institute of Physics, Conference Proceeding No. 149, New York, 1986, p. 116.

53 D. S. Williams, F. A. Baiocchi, R. C. Beairsto, J. M. Brown, R. V. Knoell and S. P. Murarka, *J. Vac. Sci. Technol.*, *B5* (1987) 1723.
54 I. Suni, M. Maenpaa, M.-A. Nicolet and M. Luamajarvi, *J. Electrochem. Soc.*, *130* (1983) 1215.
55 J. M. Poiterin, G. Lemperrere and J. Tardy, *Thin Solid Films*, *97* (1982) 69.
56 Y. Igaraski and H. Mitsuhashi, *Thin Solid Films*, *70* (1980) 17.
57 B.-O. Johansson, J.-E. Sundgren, J. E. Greene, A. Rockett and S. A. Barnett, *J. Vac. Sci. Technol.*, *3* (1985) 303.
58 D. S. Rickerby and R. B. Newbery, *Vacuum*, *38* (1988) 161.
59 J. Wittmer, *Appl. Phys. Lett.*, *37* (1980) 540.
60 H. K. Pulker and E. Jung, *Thin Solid Films*, *9* (1971) 57.
61 S. Ogura, *Ph.D. Thesis*, Newcastle upon Tyne Polytechnic, 1975.
62 D. Dawson, *Wear*, *103* (1985) 189.
63 D. S. Rickerby and P. J. Burnett, *Surf. Coat. Technol.*, *33* (1987) 191.
64 S. J. Bull, D. S. Rickerby, T. Robertson and A. Hendry, *Surf. Coat. Technol.*, *36* (1988) 743.
65 S. J. Bull, D. S. Rickerby and A. Jain, in press.
66 W. Schintlmeister and O. Pacher, *Metall.*, *28* (1974) 690.
67 D. T. Quinto, *J. Vac. Sci. Technol.*, *A6* (1988) 2149.
68 D. T. Quinto, A. T. Santhanam and P. C. Jindal, *Mater. Sci. Eng.*, *A105/106* (1988) 443.
69 D. T. Quinto, G. J. Wolfe and P. C. Jindal, *Thin Solid Films*, *153* (1987) 19.
70 B. M. Kramer, *J. Vac. Sci. Technol.*, *A4* (1986) 2870.
71 J. A. Yeomans, *Ph.D. Thesis*, Cambridge University, (1987).
72 K. Kamachi, T. Ito and T. Yamamoto, *Surf. J. Int.*, *1* (1986) 82.

*Surface and Coatings Technology, 39/40* (1989) 329 - 337 

# SPATIAL POTENTIAL AND ELECTRON ENERGY DISTRIBUTION MEASUREMENT AROUND PLASMA REACTION ZONE IN TRIODE BARE OF (Ti, Al)N

JUNG HOON JOO and BONG HEE HAHN

*Department of Metallurgical Engineering, Seoul National University, 151-742 Seoul (Korea)*

(Received March 8, 1989)

## Summary

By inserting electric probes around the metal–metal–gas reaction zone a few centimeters above an electron beam heated molten pool, it was found that the potential distribution could be measured. Nitrogen pressure, substrate bias and electrode geometry were varied, and plasma characteristics of the process were measured. By inserting an electron emitter between the activated reactive evaporation probe and molten pool, electron temperature was increased from 5.4 to 15.8 eV and a double peaked potential profile was obtained. The electron energy distribution departed from Maxwellian at its high energy tail. As the nitrogen pressure was increased, the first potential peak at 4 cm above the molten pool increased and electron emitting heating power was increased. Increasing the substrate bias from −200 to −2000 V decreased the potential at 4 cm above the molten pool at nitrogen pressures up to 4 mTorr. However, above 4 mTorr the tendency was reversed for electron emitting heater power in the range 2 - 26 A.

## 1. Introduction

Usually, ion-plating processes produce high energy ions and neutrals around a substrate biased up to a few kilovolts. Various ionization efficiencies have been reported by many authors [1 - 3]. In electron-beam-heated molten pool-type evaporation sources, a glow discharge can be easily initiated by application of +100 V. Our ion-plating system has electron-beam gun of 4 kV and the plasma was maintained only with nitrogen gas. To enhance the ionization around the glow region above the molten pool, we inserted a thermionic emitter and an anode above the aluminum evaporation source. The plasma probe technique with an electrostatic single probe was used to obtain information on electron density and temperature with sufficient spatial and time resolution.

The energy distribution of the incoming flux onto the substrate is the main interest of ion-plating engineers. The plasma generating region has

0257-8972/89/$3.50 

three species, titanium, aluminum and nitrogen. Some of them are mainly ionized by low-energy electrons of 50 - 100 eV. The proper position of the thermal electron emitting filament and the anode should be determined in order to give the greatest substrate current density with a moderate deposition rate. In this experiment, the effect of thermal electron emission was studied by using four electrostatic probes in the parallel axis direction. Electron temperature, density and spatial potential were used to optimize triode-type biased activated reactive evaporation (BARE).

## 2. Experimental procedure

Figure 1 shows our deposition system. A 4 kV electron-beam gun was used for titanium evaporation and a thermal evaporation source for aluminum. A ring-shaped ARE discharge electrode was situated 1 cm above the molten titanium pool. When +100 V was applied to it, a discharge current of 1.5 - 2 A flowed through the load resistor.

The thermal electron emitting filament was made of a tungsten wire of diameter 100 $\mu$m. The heating power was controlled over the range 2 - 26 A and a negative bias of up to −80 V could be applied. The source - substrate distance was about 10 cm. The substrate holder was a circular stainless disc

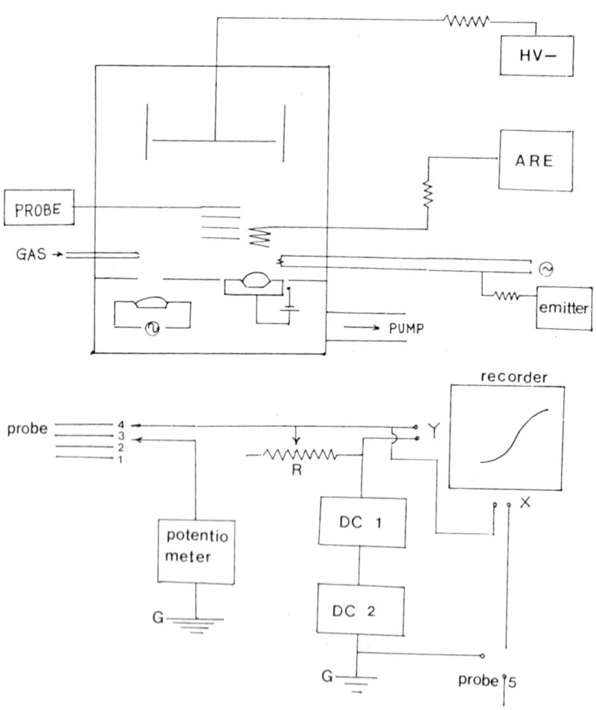


Fig. 1. Experimental setup.

with a surface area of 100 $cm^2$. Electrical probes were parallel to each other and 10 mm apart. Probe characteristics were obtained through a simple circuit with two direct current power supplies and a variable resistor for current conversion into a voltage signal to drive an $X$–$Y$ recorder.

The stainless steel chamber was earthed and could play the role of an anode in the discharge system. A commercial mechanical diffusion pump system was used to evacuate the system down to $2.2 \times 10^{-5}$ Torr.

The titanium evaporation rate was kept constant using a heating power of $3.5\ kV \times 140\ mA = 490\ W$. Electric probes are very susceptible to surface contamination from organics and metals deposited during previous runs [4]. Therefore, before measuring the $I$–$V$ characteristic curve, each probe was sputter cleaned by biasing at $-125$ V for 1 min. At this point the tungsten probe glowed white, which was sufficient to clean the surface of the probe.

The four probes used for potential measurement were connected to a voltmeter with an internal resistance of 10 MΩ and the $I$–$V$ characteristic of each probe was obtained by biasing from $-50$ to $+100$ V. The effect of electron density variation and spatial potential distribution was observed in X-ray determined crystallite size changes and hardness.

## 3. Results and discussion

### *3.1. Substrate current measurement*

The deposition was carried out at a total pressure of a few millitorr, which was measured by a Schultz ionization gauge. Figure 2 shows that substrate current increased with an emitter bias up to 60 V and saturated over 60 V. When the anode was biased at $+50$ V and placed in 1 cm away, the emitter heating current was 10 A. For most inert gases (argon, neon, xenon, krypton) electron impact ionization is highest at an energy of 100 eV [2]. The

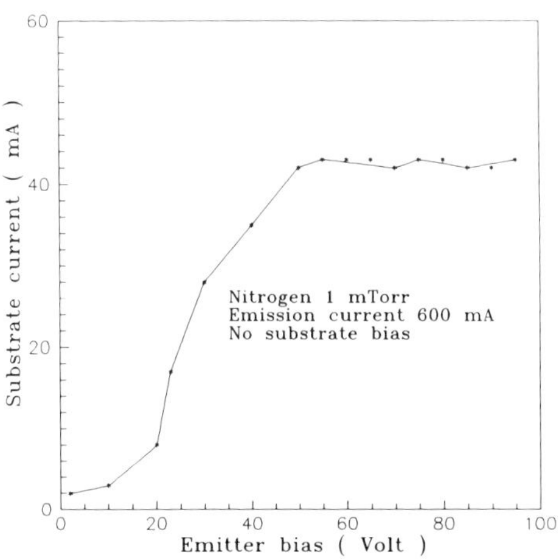


Fig. 2. Substrate current *vs.* cathode bias for the electron emitting filament.

cathode bias at which the substrate current started to saturate moved to a higher value at lower gas pressures.

In most diode ion-plating systems, substrate current saturated immediately at the threshold substrate bias voltage. Figure 3 shows a linear increase of the substrate current with the substrate bias at constant electron emission (nitrogen). When the nitrogen pressure increased from 1 to 4 mTorr (Fig. 4) the substrate current increased to 80 mA. The area of the circular disc specimen holder was about 100 cm$^2$, so that the substrate current density was 0.8 mA cm$^{-2}$ maximum with only a nitrogen gas discharge.

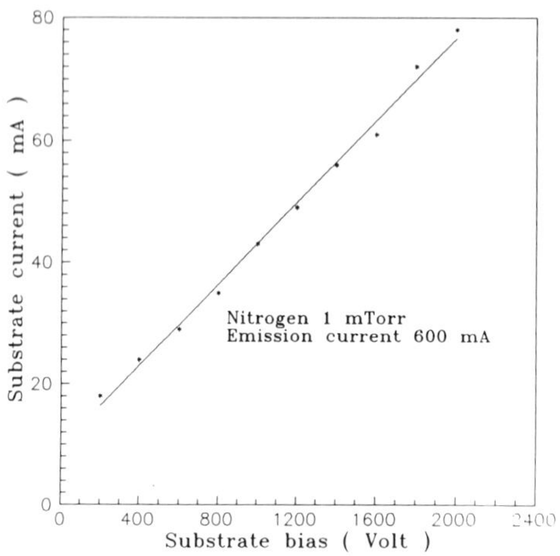


Fig. 3. Substrate current *vs.* substrate bias at an argon discharge of 1 mTorr and an electron emission current of 80 mA.

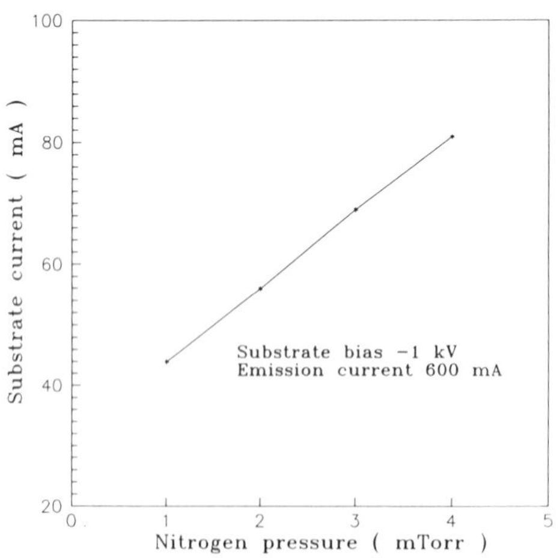


Fig. 4. Substrate current *vs.* nitrogen pressure at a substrate bias of −1 kV and an electron emission current of 80 mA.

During deposition of (Ti, Al)N, the substrate current density was 0.45 mA $cm^{-2}$ at $1 \times 10^{-3}$ Torr nitrogen pressure and $-1$ kV substrate bias in the diode BARE system. With the aid of a thermal electron emitting filament, the substrate current density was 1.2 mA $cm^{-2}$ maximum under the same deposition conditions.

*3.2. Electrostatic plasma probe measurement*

In the determination of electron temperature and densities from *I–V* characteristics during (Ti, Al)N deposition, the sheath area determination is one of primary concerns. We used the Laframboise theory [5] on the interpretation of probe data for reproducible results.

At the diode-type BARE, authors reported electron temperature up to 10 eV at a substrate bias voltage of 1 - 1.8 kV and an electron density of $10^9$ $cm^{-1}$ [3, 6]. In this experiment we observed electron temperatures from 16.9 to 20.2 eV with a substrate bias of $-1.2$ to $-1.6$ kV. The electron temperature around the metal–metal–gas reaction zone plays a role in the ionization of the reaction species. Most of the compound forming reactions are believed to occur at the heated substrate, but the role of the plasma zone is to enhance collision frequency in the glow region. Collision frequency calculations based on gas kinetics without charged particle collision report that the gas phase reaction is 10% of the total compound formation [7].

Figure 5 shows the typical potential energy distribution measured in our system. All of the curves for different nitrogen gas pressures and electron emitting powers showed a similar structure — doubly peaked in the *z*-direction. The peak point in this curve was 1 cm away from the electron emitting filament. The potential can be interpreted using the first Maxwell equation, *i.e.* Poisson's Law. The more positive the probe potential, the more ions present. The highest point was the place where low energy electrons first ionized the gas molecule. For increased gas pressures from 1.0 to 2.0 mTorr,

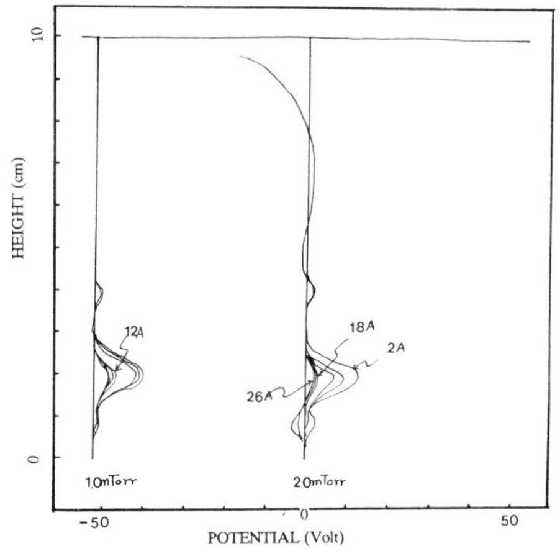


Fig. 5 Potential distribution *vs.* heating power of the electron emitting filament at two different pressures.

the positive potential was increased by a few volts. As the electron emitting heater current increased, its potential decreased due to excessive electrons. The mean breadth of this potential peak was about 1 cm at full width at half maximum. The second peak was the point where the second ionizing collision took place. The gas pressure was 1.0 - 3.0 mTorr and its mean free path was about 1 cm.

In Fig. 6, potential variation by substrate bias during deposition was shown. As the substrate's negative potential was increased, the positive potential around 4 cm above the molten pool also increased. The source–substrate distance was 10 cm, but cathode dark space thickness could not be observed due to the white glow of the molten pool. Therefore, these curves could be interpreted as the increase of ionization density with increasing substrate bias, as previously reported [5].

Figure 7 shows an anomaly in the potential distribution when nitrogen

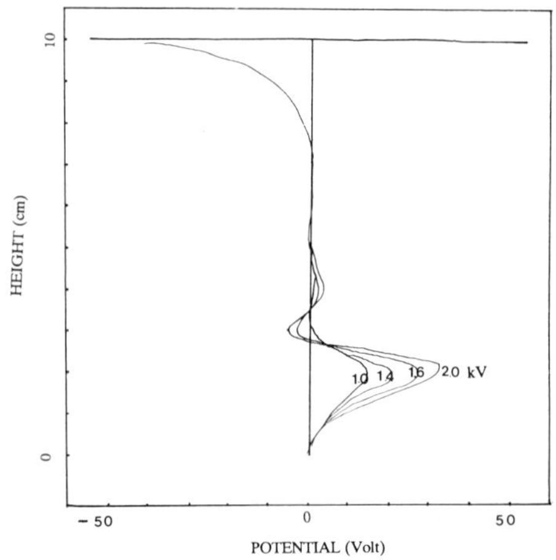


Fig. 6. Potential distribution *vs.* substrate bias during deposition at 1 mTorr.

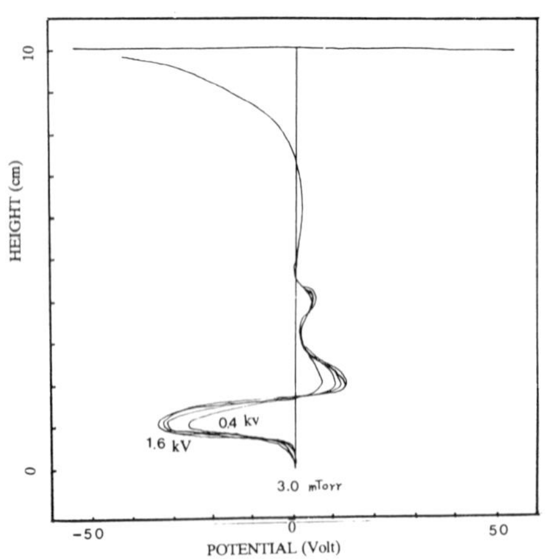


Fig. 7. Potential distribution *vs.* substrate bias at a nitrogen pressure of 3.0 mTorr.

was introduced into the reaction zone up to 3.0 mTorr. This inversion of potential resulted in the formation of negative ions ($N_2^-$ or $N^-$) by electron attachment at this pressure. And the other is nitrogen molecule's high secondary electron yielding coefficient. In the deposition of TiN, the critical pressure was 1 mTorr of nitrogen gas. In the case of (Ti, Al)N, we used the same evaporation rate of titanium, so that excessive nitrogen was needed to obtain the same transient behaviour. Further plasma-physics-oriented study of (Ti, Al)N deposition is required in order to explain the potential inversion.

In Fig. 8, the variation of electron density by the change of thermionic electron emitter current is shown at three different nitrogen pressures. As the nitrogen pressure increased from 1 to 3 mTorr, the corresponding electron density increased to $4 \times 10^9\,cm^{-3}$, where the substrate bias was $-600$ V. Under this electrode geometry, the substrate current increases linearly as the substrate bias voltage from $-200$ to $-1200$ V (Fig. 9). Therefore, it is not in

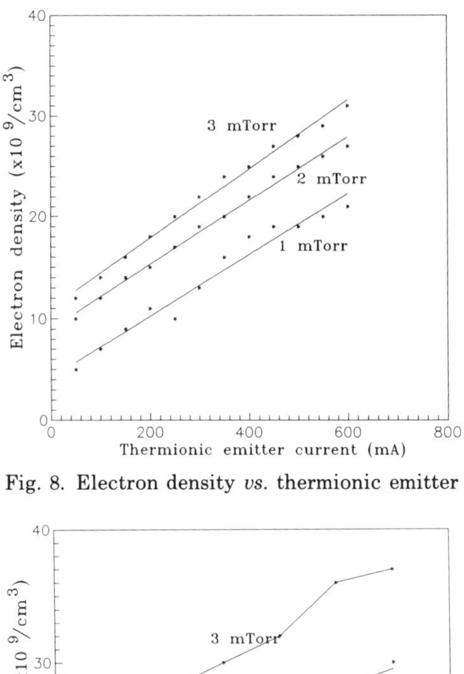


Fig. 8. Electron density *vs.* thermionic emitter current at three different nitrogen pressures.

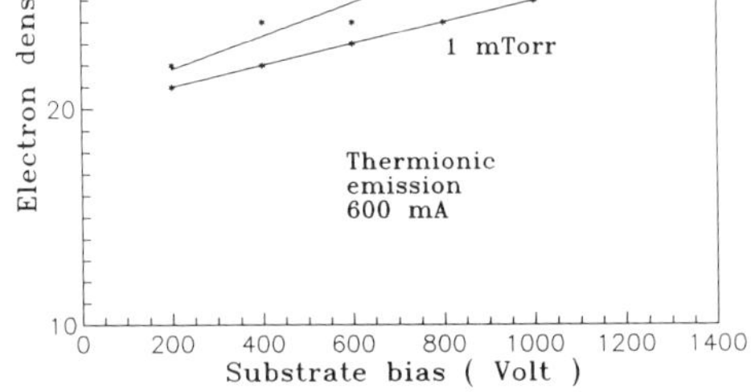


Fig. 9. Electron density *vs.* substrate bias at three different nitrogen pressures.

the space charge limit region, where the substrate current should saturate over the critical bias voltage.

In order to study the effect of electron density on the crystallographic properties of the film, the grain size of the film was measured by peak broadening analysis (Fig. 10). As the nitrogen pressure increases from 1 to 4 mTorr, the grain size increases to 380 Å. However, in Fig. 11 the Knoop hardness number shows a minimum value at 3 mTorr nitrogen pressure. The microhardness is closely related to the concentration of aluminum in the film. The average aluminum concentration is from 17 to 23 at.% Al as measured by the inductive coupled plasma method. In the range of 1 to 3 mTorr nitrogen pressure, the aluminum concentration is nearly the same, therefore the

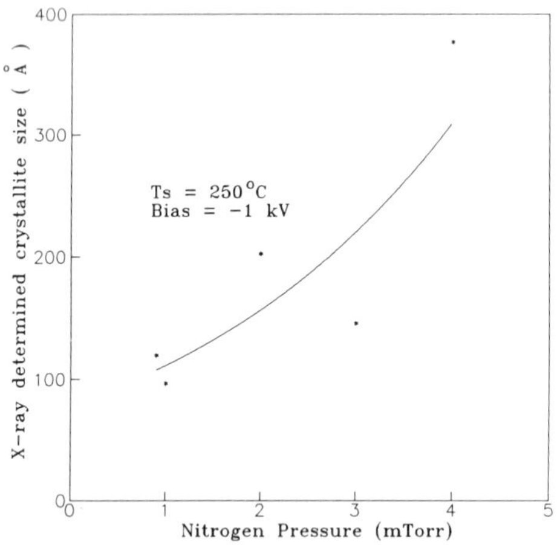


Fig. 10. X-ray determined crystallite size *vs.* nitrogen pressure.

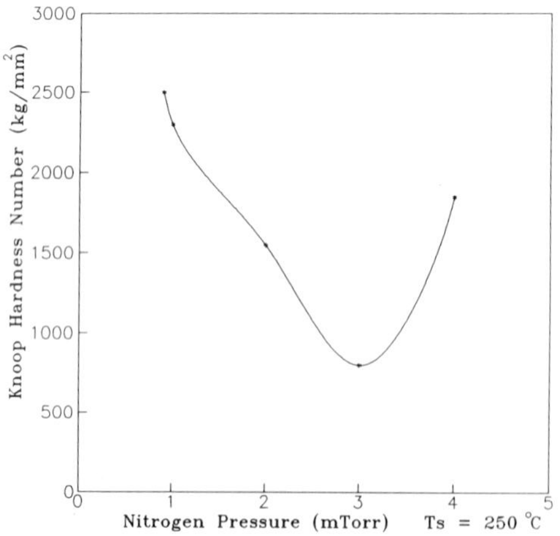


Fig. 11. Knoop hardness *vs.* nitrogen pressure.

change of the microhardness values can be considered as the effect of the incoming ion flux during the deposition. As the nitrogen pressure increases, the electron denisty increases linearly in this range.

## 4. Conclusions

(1) A high substrate current density (1.2 mA $cm^{-2}$) was obtained in the (Ti, Al)N system using triode-type BARE.

(2) The electron temperature in the triode system was increased to 15 - 20 eV from lower than 10 eV in diode BARE.

(3) Potential distribution around the metal–metal–gas reaction zone showed a doubly peaked structure for most gas pressures and substrate biases.

(4) Abnormal potential inversion occurred at 3.0 mTorr nitrogen pressure with a substrate bias from 0.4 to 1-6 kV and was related to a rapid increase of negative ion formation.

## Acknowledgment

We wish to express our gratitude to colleagues at the X-ray Laboratory of Seoul National University for helpful discussions.

## References

1 K. S. Fancey and A. Matthews, *Surf. Coat Technol., 33* (1987) 17.
2 H. Freller and H. Haessler, *Surf. Coat. Technol., 36* (1988) 219.
3 P. Saulnier, A. Debhi and J. Macher, *Vacuum, 34* (1984) 765.
4 P. M. Chung, *Electric Probes in Stationary and Flowing Plasmas: Theory and Applications*, Springer, Berlin, 1975.
5 Laframboise, *University of Toronto, Institute for Aerospace Studies Report No. 100 (1966).*
6 B. H. Hahn, J. H. Jun and J. H. Joo, *Thin Solid Films, 153* (1987) 115.
7 Y. Matsumura, H. Uchida and Yen C. Huang, *Proc. First Int. Conf. on Plasma Surface Engineering*, F.R.G., 1988 (in press).

*Surface and Coatings Technology*, *39/40* (1989) 339 - 354 

# FACTORS AFFECTING THE TRIBOLOGICAL BEHAVIOUR OF THIN HARD TiN AND TiC COATINGS

T. F. PAGE and J. C. KNIGHT

*Department of Mechanical, Materials and Manufacturing Engineering, Materials Division, The University, Newcastle upon Tyne, NE1 7RU (U.K.)*

(Received March 16, 1989)

## Summary

The use of thin hard coatings, often of ceramic materials, appears to offer the possibility of controlled improvement of the tribological behaviour, surface damage resistance and corrosion protection of many engineering components. However, it is only recently that the full complexities of the behaviour of coated systems are being appreciated. The mechanical response of a coated surface to scratching and asperity penetration is complex, involving some properties intrinsic to the coating (*e.g.* coating material and thickness), some controlled by processing (*e.g.* coating microstructure and residual stress state), some determined by the substrate (*e.g.* flexure relative to the coating and deformation in response to stresses transmitted through the coating) and some controlled by interfacial adhesion. Also, raising the temperature of coated systems, for example either in genuine high temperature applications or by frictional heating during ambient temperature service, can be detrimental to coating performance and integrity in that it may alter the microstructure, create pores which allow environmental ingress, relax the residual stress state of the coating and affect interfacial bonding. Based on the authors' own work with TiN and TiC coated steels, this paper discusses these factors and highlights the problems of tailoring coatings to particular applications, predicting coating performance and assuring quality and reliability.

## 1. Introduction

Thin, hard, ceramic coatings are used in surface engineering as a means of attempting to confer the attractive tribological and chemically-protective properties of ceramics onto the surfaces of components made of other materials, *e.g.* metals. Conversely, it is expected that the generally higher toughness of most substrates will prevent the catastrophic brittle fracture often associated with bulk ceramics.

However, from the outset, it is clear that there are a number of important and unavoidable consequences of using thin coatings. First, the

0257-8972/89/$3.50 

overall mechanical and tribological response of the surface will be some *composite* function of the coating and the substrate; the coating essentially modifying the stresses transmitted to the substrate which itself eventually supports the contact load. Second, there will be *scale effects* in this composite response arising not only from deformation scale-effects intrinsic to both the coating and substrate themselves but also from the thickness of the coating. Third, it is an implicit — but often ignored — requirement that the coating should be able, without failure, to deform to follow the flexure and deformation of the substrate. This requirement of *conformation* not only demands a high fracture stress and significant ductility of the coating but also requires an assessment to be made of substrate behaviour (*i.e. mechanical compatibility* of the coating and substrate combination). Finally, there is the problem of ensuring sufficient adhesion of the coating to the substrate such that transmitted loads, flexure, thermal expansion mismatch effects and shear during surface contact do not result in spalling. All these factors will be sensitive to coating identity, microstructure and stress state, and thus can partly be determined during coating deposition. Others, *e.g. chemical compatibility* of the coating with the substrate, will be sensitive to the exact identities of the materials chosen.

Another area of concern is that of maintaining microstructural and property integrity during service, especially if either cumulative damage occurs to the coating or if significant frictional or environmental heating results in detrimental annealing or thermal cycling of the coating. Few workers have yet explored the variation of coating behaviour with temperature and environment though this emerges as a critical issue in assessing the performance of a coated sytem.

Based on the authors' work [1 - 4] this paper addresses these issues from the viewpoint of the deformation response of coatings observed in indentation and scratch test experiments.

## 2. The tribological properties of bulk ceramics

Before discussing the advantages and limitations of ceramics used as coatings, it is helpful to outline their attractive tribological properties in order to establish how successfully this behaviour can be achieved in thin films.

A primary attraction of many oxides, carbides and nitrides is their high stiffness and high hardness. However, hardness is a complex property and, for ceramics, shows very marked dependence on both contact scale and temperature (*e.g.* refs. 5 - 8). A pronounced scale effect (the indentation size effect (ISE)) usually occurs such that penetration hardness rises rapidly as the indentation size is reduced and plastic flow becomes more difficult (*e.g.* refs. 5, 6, 8). This is of critical importance in tribology where contact scales are usually far smaller than typical hardness indentation sizes. Thus, not only are ceramics harder than expected at the scales of wear processes but the rank order of a group of materials may also be different to that expected (*e.g.* ref. 9). However, most ceramics exhibit very marked decreases in hardness over modest increases in temperature: for example, the hardness of SiC materials typically falls from around 3000 VHN at room temperature to less than 1000 VHN at 800 °C even though this temperature increase is small compared to the melting temperature (*e.g.* refs. 7, 8). This again can lead to changes in the expected rank order of a group of materials. Recent work has shown how these effects may be computed using a hardness information data base as a precursor to constructing wear maps [9].

Indentation tests have clearly established that plastic flow does occur in response to the surface deformation of most ceramics even at room temperature (*e.g.* refs. 10 - 12) and thus contributes to the ploughing component of friction. Given the limited slip system in most ceramics, plasticity effects tend to show marked directional anisotropy (*e.g.* refs. 10, 12). Also, ceramics generally exhibit poor adhesion to other surfaces though this is often caused by the presence of surface absorbates (especially water) and is thus also thermally sensitive (*e.g.* ref. 13).

Brittleness is often seen as the price paid for high hardness, but indentation experiments have clearly established another scale effect such that fracture-dominated wear only occurs above some threshold indentation load ($p^*$) (*e.g.* ref. 7). Indeed the transition from very mild plasticity-dominated wear to severe fracture-dominated wear with increasing contact severity is well-documented (*e.g.* refs. 14, 15). Also well established are the characteristic radial and lateral crack arrays often formed around contact sites in ceramics (*e.g.* ref. 7). Crack paths are usually controlled by the sample microstructure with interfaces playing an important role — an effect now exploited to increase the toughness of ceramic-matrix composites (*e.g.* ref. 16). One immediate effect of using thin coatings is obviously to limit the depth penetration of these cracks into the composite surface.

Ceramics also show a marked environmental dependence of behaviour with both plasticity and fracture responses being dependent on absorbates such as water (the so-called chemomechanical effects) (*e.g.* refs. 17, 18). Chemical-reaction-controlled mild wear by either the hydration of surface layers or the oxidation of non-oxide ceramics is well documented (*e.g.* ref. 19) as are the beneficial effects on friction of such adhesion-inhibiting absorbates (*e.g.* ref. 13). At elevated temperatures, chemical reactions can also occur between ceramics and hot metals (for example in metal cutting (*e.g.* ref. (20)) and a data base for predicting such reactions as part of coating selection has been described [21].

In summary, therefore, ceramics potentially offer low friction, very wear-resistant surfaces of a type found with no other materials but whose behaviour is a sensitive function of contact scale, contact severity and environment. However, for many coating materials, little is known about their bulk properties at least in the microstructural states encountered when used as thin films. Thus, while the following sections discuss the behaviour of ceramics when used as coatings, it is sometimes experimentally difficult to separate genuine bulk properties from those conferred by the state and thickness of the coating itself.

## 3. The behaviour of ceramics as thin hard coatings

### *3.1. The effective hardness of a coated (composite) surface*

Indentation hardness testing is a widely used technique for the study of near-surface deformation behaviour. While the penetration resistance of a surface is of obvious interest with respect to ploughing by hard asperities during wear, indentation is also one of the only tests available for probing the small-scale, local, mechanical properties of surfaces.

The indentation hardness response of a coating–substrate composite system is more complex than that of a single-component system. Depending on coating thickness and indentation load, both components will contribute to the overall hardness response in an amount somehow related to the deforming volumes of coating and substrate materials. However, the shape and extent of these volumes are difficult either to determine experimentally or to predict. Further, the individual ISE behaviour of the two components complicates theoretical modelling. However, the quest for reliable models is driven by the need both to predict the scale-sensitivity of hardness for tribological purposes and also to allow extraction of the properties of the coating alone for quality control.

Two models are often used. The first is due to Johnson and Hogmark [22] and is based on a law of mixtures concept which considers projected contact areas. The original version excluded ISE effects and proposed that the effective hardness of the composite should be related to the relative contributions from the coating and the substrate by the expression:

$$H_c = H_s + \left\{2c\left(\frac{t}{d}\right) - c^2\left(\frac{t}{d}\right)^2\right\}\{H_f - H_s\} \tag{1}$$

where $H_c$, $H_f$ and $H_s$ refer to the hardness of the composite, the film and the substrate respectively, $t$ is the coating thickness, $d$ the indentation depth and $c$ a constant depending on the indenter geometry. Incorporation of ISE behaviour into this model has been found to yield good agreement with experiment over a wide range of indentation sizes (*e.g.* refs. 2, 23).

A second approach, based on "spherical cavity" deformation volumes in the coating and substrate (see Fig. 1(a)) has been proposed by Burnett and Rickerby [23]. This extends the ISE-incorporating model of Burnett and Page [24] (based on a volume law of mixtures originally proposed by Sargent [25]) to give the expression:

$$H_c = \frac{V_f}{V} H_f + \frac{V_s}{V} \chi^3 H_s; \qquad H_f > H_s \tag{2}$$

where $H_c$, $H_f$ and $H_s$ are as defined in eqn. (1), $V_f$ and $V_s$ are the deforming volumes of coating and substrate respectively. $V = (V_f + V_s)$ the total

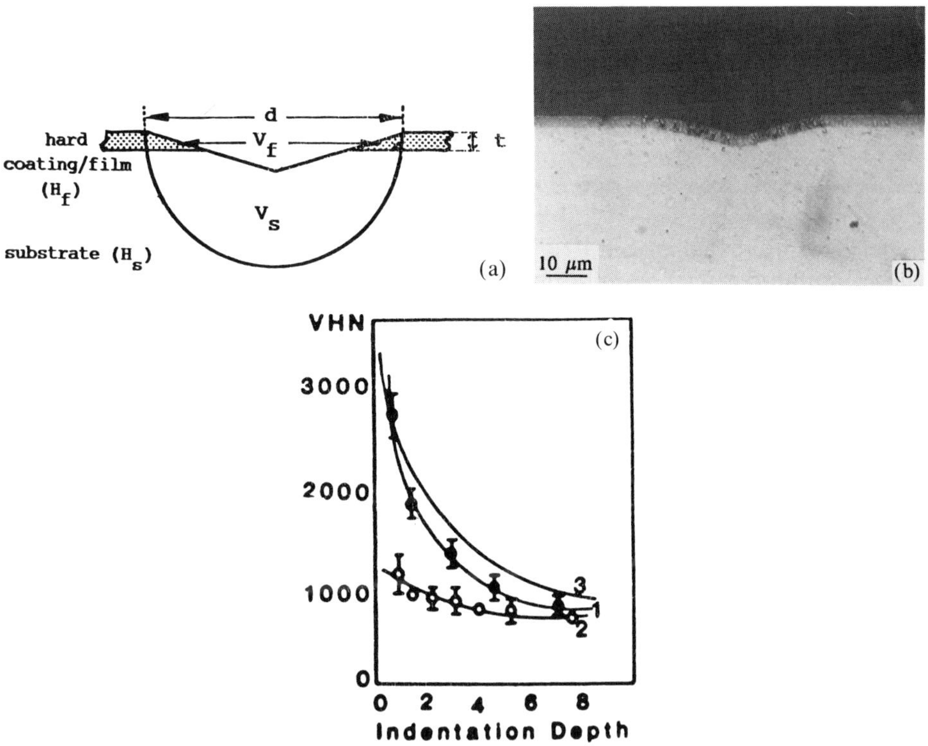


Fig. 1. (a) The typical deformation geometry assumed in the majority of models for calculating the composite hardness of a coated surface. $V_s$ and $V_f$ are the deformation volumes in substrate and coating film respectively. (b) The geometry of the usual real deformation situation. The coating is not cut through but bent to follow the indentation contour. (Courtesy of P. M. Ramsey.) (c) Typical hardness *vs.* indentation depth results: 1, coated composite system; 2, substrate alone; 3, predictions of the Burnett and Rickerby model.

deforming volume and $\chi$ is an "interface parameter" expected to be a function of the mismatch between the plastic zone radii in the coating and substrate. Like the Johnson and Hogmark model with ISE, this model also yields good agreement with experimentally determined hardness behaviour over a wide range of indentation sizes (see Fig. 1(c)). All such models agree that the coating properties dominate the composite response at shallow indentation penetrations while the substrate dominates at very large contact scales (see Section 3.7). It is concerning the behavioural fit in between these extremes that the models differ in detail.

Based on experimental observations, one common assumption of these models has to be dispelled and that is the cutting through of the coating by the indenter (see Fig. 1(a)). A more realistic — and commonly observed — response is that shown in Fig. 1(b) where the coating (supported by the substrate) is bent and stretched by the indenter to conform to the indentation profile. This often occurs without cracking of the coating (but see Section 3.5) and thus implies significant ductility of the coating to an (in-plane) elongation of around 8% for a Vickers indenter. This behaviour should be incorporated into future refinements of hardness models.

As will be shown in Section 3.5, this indentation hardness response is a very significant function of temperature.

### *3.2. The effect of coating microstructure on deformation response*

The microstructure of many hard coatings is far from homogeneous. Depending on coating conditions, either equiaxed or crystallographically aligned columner grains (or some mixture of the two) can occur with varying levels of porosity both in the coating and along the substrate–coating interface. Often, significant variations of all the effects are found in any one production sample (*e.g.* see Fig. 2 and Section 4). All these factors are expected to influence the hardness of the coating, though the small scales involved have usually prevented detailed investigation being made. However, columnar structures have been found to exhibit pronounced hardness anisotropy between plan and profile sections (*e.g.* refs. 26, 27). Such microstructural effects are not only important in determining the effective hardness of the coating but also need to be incorporated into hardness models.

What is also clear, however, is that the microstructural genesis of the coating has a very significant controlling effect on the residual stress state of the coating (*e.g.* ref. 28). In turn, these stresses can exert a considerable influence over both hardness and fracture behaviour (see Sections 3.3 and 3.5).

### *3.3. Contact-induced fracture behaviour*

Figure 3 shows typical contact-fracture responses occurring in TiN coatings. Generally, the crack patterns associated with residual-stress-free bulk ceramics are not observed. Rather, many coatings show no extensive fracture (*e.g.* Fig. 3(c)) until contact loads (around 1000 gf) which are high compared to typical bulk ceramic $p^*$ values (5 - 200 gf). When observed, crack arrays tend either to emanate from the indentation corners and edges (presumably with a limited-depth radial style of geometry (*e.g.* Fig. 3(a)), or to form nested arrays within the indentation profile presumed to be created

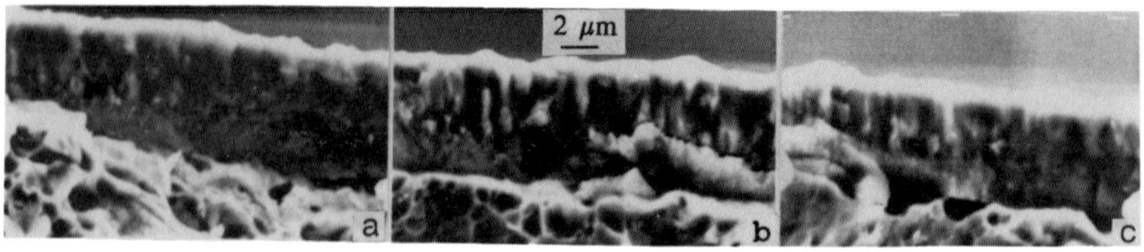


Fig. 2. SEM micrographs (secondary electron images) of a (fractured) cross-section showing the microstructure of a CVD-TiN coating: (a), (b) and (c) were taken from different regions of a single specimen. Note the growth of fine equiaxed grains next to the substrate followed by a porous, facetted columnar structure in the latter stages of deposition. Large, isolated voids are also evident at the coating–substrate interface.

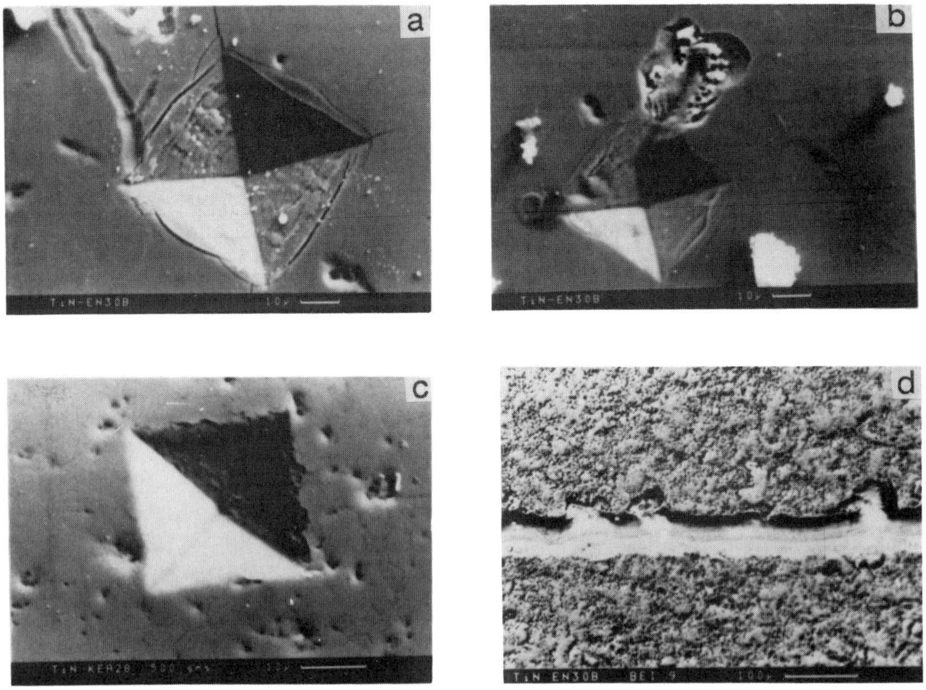


Fig. 3. SEM micrographs showing typical examples of the room temperature response of a CVD TiN-steel composite system to indentation and scratch tests: (a), (b) and (c) TiN-EN30B indented with a Vickers indenter at loads of 1000, 300 and 500 gf respectively. (a) Both nested cracks within the indentation profile and radial cracks emerging across the surface from the indentation corners. (b) Spalling of a portion of coating outside the indentation profile (presumably) from a weakly-bonded interfacial region. (c) No extensive fracture. (d) Removal of the coating by a hemispherical-tipped scratch diamond of radius of curvature 200 $\mu$m under a normal load of 1000 gf. Limited cracking of the coating is evident to the sides of the track.

by tensile fracture as the coating is stretched into the indentation [2] (see Figs. 3(a) and 6). Occasional spalling of the coating from the substrate is a further feature (Fig. 3(b)) but is often associated with interfacial pores or poor adhesion. Figure 3(d) shows the limited cracking associated with a scratch track even when this has penetrated through to the substrate.

The general inhibition of cracking can be attributed partly to the limited thickness of the coating but mostly to the beneficial effects of the high residual compressive stresses retained in the coatings after fabrication (*e.g.* ref. 2). This situation changes dramatically if this stress is relaxed by annealing (see Section 3.5).

Crack genesis is often associated with pores and other flaws both in the coating and along the coating-matrix interface (*e.g.* ref. 2). Fully dense coatings are thus beneficial and the incidence of porosity either by poor processing or by annealing (see section 3.6) has deleterious consequences for the onset of fracture-dominated wear.

### 3.4. *The influence of substrate deformation behaviour*

The deformation characteristics and flow geometry of the substrate beneath the coating both have significant influence on coating behaviour (*e.g.* ref. 1). At low contact stresses, the substrate response to the forces transmitted by the coating is to remain elastic. However, as the substrate yields under the influence of higher transmitted stresses, the situation shown in Fig. 4(a) can occur. For metallic substrates, plastic deformation from such a contact can result in displaced material moving upwards and away from the contact area to form pile-ups. This creates tensile flexure of the coating and may lead to localised through-thickness fracture outside the contact zone [1]. However, the significance of this pile-up depends on the modulus-to-yield stress ratio ($E/Y$) for the substrate with radial displacements and less pile-up becoming more dominant for substrate materials of lower $E/Y$ ratios [25].

The pile-up effect is clearly shown in Fig. 4(b) which is a cross section through a wear track on an evaporated TiN-coated tool steel. Figure 4(c) shows a similar occurrence for a TiN-coated stainless steel with copious interfacial debonding cracks possibly resulting from stresses created by

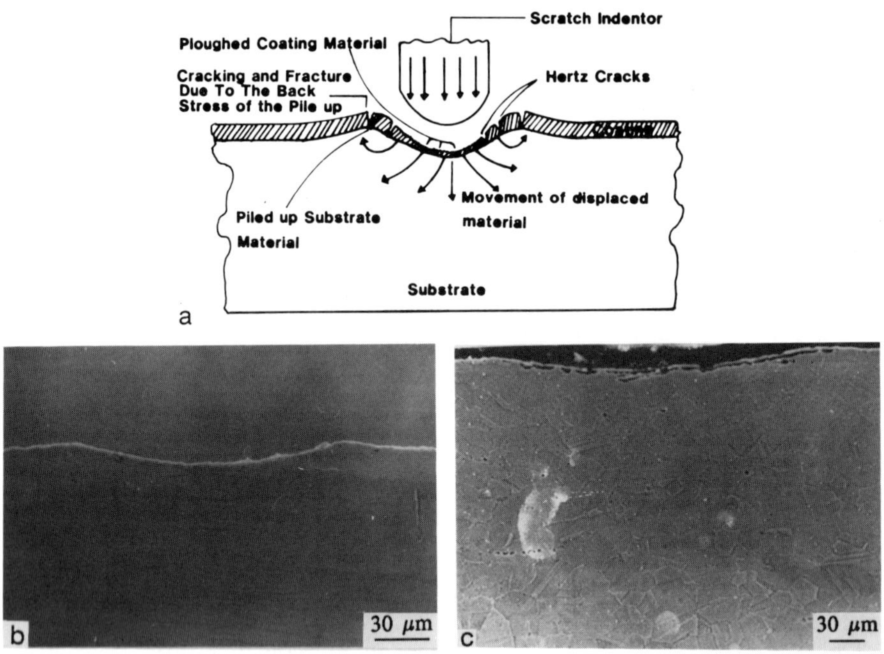


Fig. 4. (a) A schematic representation of the possible effects controlled by yielding and flow of the substrate beneath the coating during scratch testing. The most important feature is pile-up of the matrix to the sides of the track creating local tension (and potential cracking) in the coating. (b) A cross-section through a scratch track made at a sub-critical load of 1000 gf with a diamond hemisphere on a TiN-coated BO1 low-alloy tool steel sample. Note the pile-ups of matrix material. (c) As for (b) but at a higher load of 1500 gf on an austenitic stainless steel sample. Interfacial debonding cracks are evident under the track possibly resulting from differential elastic recovery of the coating and substrate after deformation.

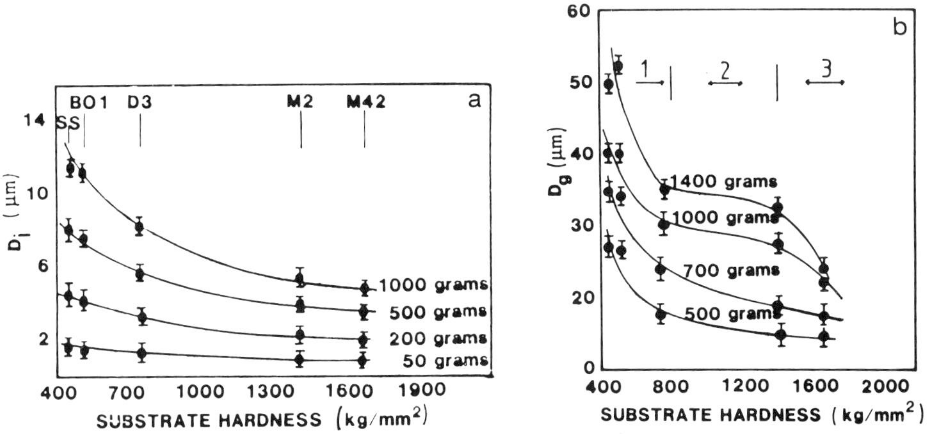


Fig. 5. (a) Vickers indenter penetration ($D_i$) plotted as a function of load and substrate hardness. At low loads, the penetration is essentially independent of the substrate and is dominated by the coating. (b) Scratch indenter penetration ($D_g$) into steel substrates of various hardnesses plotted as a function of load. Three distinct regimes of behaviour occur and possibly reflect the various energy dissipative processes occurring during scratching [1]: (1) plastic flow of the substrate; (2) plastic flow and limited microcracking of the substrate; (3) extensive substrate microcracking at high substrate hardnesses.

differential elastic recovery of the coating and substrate upon unloading [1]. Thus, compatibility between the coating and the elastic and plastic properties of the substrate emerge as being of considerable significance in determining the performance of composite systems.

As expected from the hardness models, the substrate also contributes towards indenter–asperity penetration effects. For example, Fig. 5(a) shows the penetration of a static Vickers indenter plotted as a function of load and substrate hardness [1]. As expected, the large load penetrations are less for harder substrates while the very low load penetration is independent of substrate hardness and essentially only reflects the properties of the coating itself. Figure 5(b) shows a similar plot for the scratch depth produced by a moving hemispherical stylus [1]. In this more complex stress transfer case associated with a moving indenter, three distinct regimes of behaviour seem to occur as a function of substrate hardness and load and have tentatively been assigned to the differing plasticity–microfracture responses of the substrate [1].

Energy absorption may also occur in the substrate and appear as a contribution to the overall friction behaviour (see Section 3.7 and ref. 3).

This important area of how the substrate influences the surface contact response of the coating is one clearly requiring further study.

### *3.5. The variation of hardness and fracture response with increasing temperature*

The temperature-variant response of coating contact behaviour is of obvious importance in any tribological application either at elevated temperature or when significant localised frictional heating is expected.

Figure 6(a) shows the hardness–temperature response of a TiC-coated steel tested in vacuum [2]. Two features of the response are of concern. First, the composite hardness falls dramatically over a temperature rise of around 500 °C and may even be inferior to the substrate alone at above about 300 °C [2]. This implies that considerable detrimental changes are occurring in the coating. Second, a very significant hysteresis in response is observed such that the hardness values recorded on cooling are considerably less than those on heating. This effect is most pronounced on the first heating–cooling cycle [2, 4] and again points to irreversible changes occurring in the coating. A simple analysis suggests that the decrement in heating–cooling hardness $\Delta H$ at any temperature is of the order of the stress relaxed within the coating

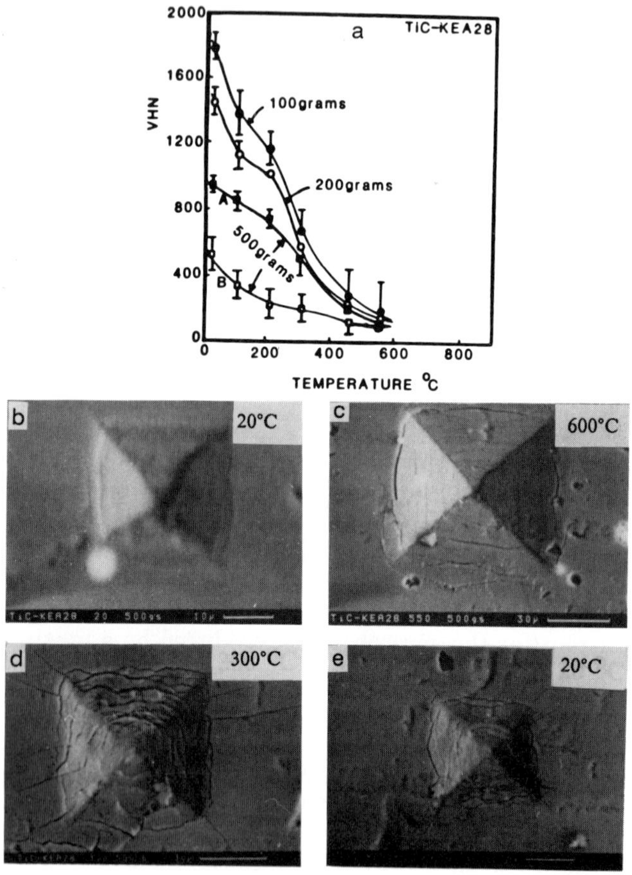


Fig. 6. (a) The temperature dependence of the Vickers Hardness of TiC-coated KEA28 tool steel at various loads. At all three loads, the composite hardness drops rapidly with modest increases in temperature. The two curves for the 500 gf load show the hardness hysteresis experienced between heating and cooling cycles. The micrographs show the change in indentation fracture behaviour exhibited by a TiC-coated steel sample indented during heating ((b) and (c)) and cooling ((d) and (e)). Essentially, progressive relaxation of the residual stress in the coating leads to more pronounced fracture. Especially dominant are the nested arrays of cracks within the indentation profile.

and such measurements have been made [2]. This demonstrates that temperature-variant microhardness testing may be a valuable means of quantitatively examining the stresses within coatings on a point-to-point basis [2].

The relaxation of the residual stress leads to significant and deleterious changes occurring in the contact-induced fracture response of the coatings (see Figs. 6(b) - (d)). As the stresses relax, pronounced nested and radial crack arrays become evident at contact loads at which no fracture occurred prior to thermal cycling [2, 4].

Not only does this behaviour have serious implications for the tribological uses of coatings at even modest service temperatures, but such extensively cracked coatings will also cease to offer protection against chemical attack of the substrate when used in aggressive environments.

### *3.6. The effects of annealing in air*

Figure 7 shows the microstructural effect of annealing a TiN-coating in air for 30 min at 700 °C. The coating has become porous and X-ray diffraction analysis, together with colour changes, suggest that an oxidation reaction of the type

$$2\,TiN + 2O_2 \rightarrow 2TiO_2 + N_2$$

is occurring with oxygen diffusion into the coating leading to the formation of $TiO_2$ and nitrogen bubbles [4]. From indentation experiments stress relief is also believed to have occurred [4]. The porosity now leads to compaction rather than fracture of the coating accompanying indentation (see Fig. 8(a)) with markedly easy debonding of the coating matrix interface (Fig. 8(b)).

Again, exposure of the coating to elevated temperatures produces effects which are deleterious to continued good performance. A further disadvantage is that oxygen diffusion through the coating (either through cracks or connected porosity) can lead to oxidation of the substrate with oxide growth beneath the coating itself [4].

The conclusion is that TiN coatings are sensitive to both temperature and combinations of temperature and environment, and that their attractive properties may not be maintained even after relatively modest exposures.

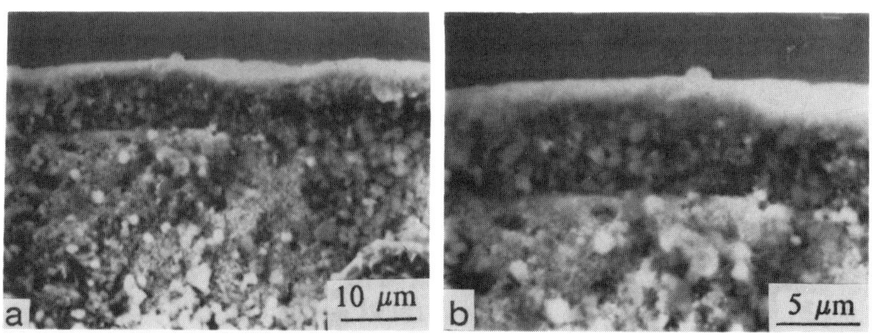


Fig. 7. SEM micrographs of a sectioned PVD-TiN coating (on M2 tool steel) annealed in air for 30 min at 700 °C. Formation of porosity in the coating is evident.

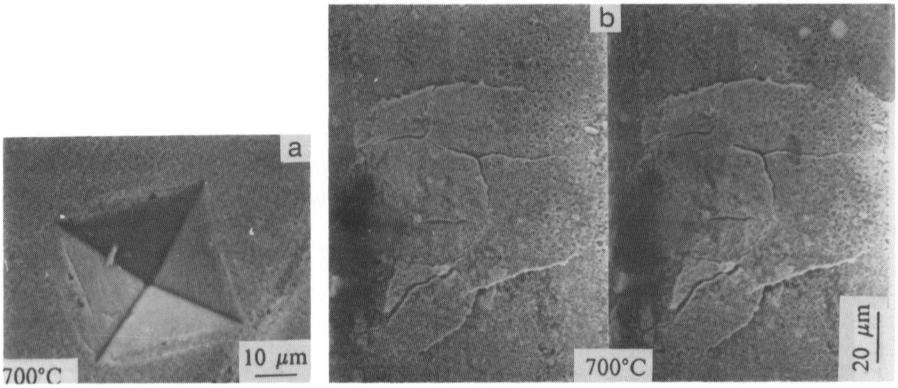


Fig. 8. Changes in indentation fracture behaviour (500 gf, Vickers indenter) after annealing a PVD-TiN-coated M2 tool steel in air at 700 °C. (a) The now porous coating compacts rather than cracks. (b) A stereo pair revealing extensive debonding and lifting of the coating from the substrate. (The coating has lifted from the substrate over virtually the whole area bounded by the cracks.)

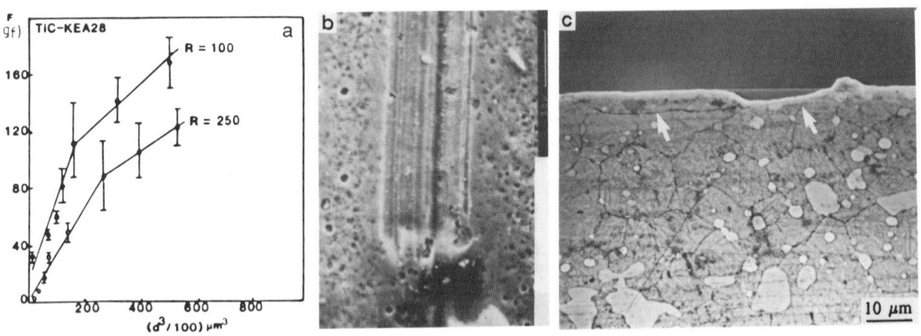


Fig. 9. (a) Plots of frictional force *vs.* $d^3$ (where $d$ is the scratch track width) for a CVD-TiC coating on a KEA28 steel scratched with hemispherical indenters of radii 100 and 250 μm. The change in slope of these plots indicates a change in the ploughing friction term as the deformation zone extends through the coating to the substrate. (b) A typical ploughed track on a PVD-TiN coating. (c) A polished and etched cross section through a scratch track on a PVD-TiN–M42 composite showing the coating having been bent to conform to a track deeper than the coating thickness. Cracking is also evident (arrowed) in the substrate in the vicinity of the scratch.

### *3.7. Friction behaviour*

Our previous work [3] has established that the coating deformation mechanisms responsible for friction is predominantly ploughing and deformation of the coating itself (Fig. 9(b)) with adhesion playing little role in producing frictional forces. As with hardness indentations, the coating can be pushed down into the substrate without spalling (Fig. 9(c)) and thus continue to provide protection to the substrate even when wear scars are deeper than the coating thickness. Again, substrate deformation behaviour is also important. For ploughing, plots of frictional force against $d^3$ (where $d$ is the scratch track width) should be linear with a slope related to the geometry of the stylus and the deformation characteristics of the material. Such a plot is shown in Fig. 9(a) for TiC on KEA28. There are clearly two straight line segments of the plots and it is believed that the behaviour at low values of $d^3$ is dominated by the coating properties with the substrate modifying this response by absorbing energy (*e.g.* by plastic flow or cracking) at larger and/or deeper deformation [1, 3].

### *3.8. Coating–substrate adhesion*

It is clear that the adhesion of the coating to the substrate is of critical importance in preventing spalling of the coating. However, not only is adhesion difficult to control during processing (it will be a sensitive function of interfacial bonding, cleanliness and topography, etc) but it is even more difficult to measure. A further complexity is the way in which interfacial adhesion may control interfacial shear strength and thus the way in which stresses are transferred from the substrate to the coating. While strong adhesion seems desirable, the optimum interfacial shear strength is less easy to define and probably varies from one application to another.

Scratch testing has become the current universal method for the assessment of coating adhesion (*e.g.* ref. 29) with a "critical load" ($L_c$) being sought (for a given stylus geometry) above which the coating is removed from the substrate. Since the substrate can be exposed by combinations of spalling, ploughing and through-thickness fracture, it is not clear how meaningful this assessment is. New methods of properly appraising (and separating) the effects of coating adhesion, interfacial strength and coating strength are urgently needed.

However, even with the critical load approach, it is evident that both substrate deformation and temperature play a critical role in determining coating integrity. For example, Fig. 10 shows $L_c$ plots for a constant thickness

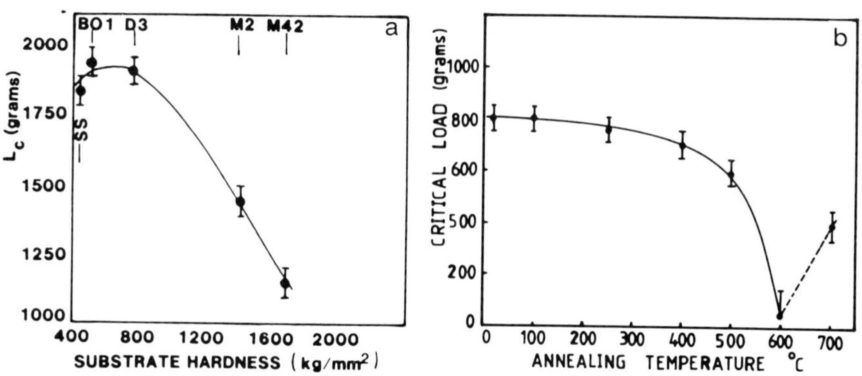


Fig. 10. (a) A plot of $L_c$ (defined as the normal load at which some part of the substrate is first exposed during scratching) as a function of substrate hardness for PVD-TiN coatings scratched at room temperature. (b) $L_c$ plotted as a function of annealing temperature in air for a PVD TiN–M2 sample. Note the dramatic decrease observed on annealing at temperatures greater than around 400 °C and associated with the development of porosity.

TiN coating as functions of both substrate hardness and annealing temperature. In Fig. 10(b), the rapid decrease in $L_c$ with annealing above around 400 °C is attributed to interfacial degradation by $N_2$-porosity [4], while the role of substrate hardness (and hence possibly varying differential plastic flow) in promoting debonding is evident in Fig. 10(a) but less easy to fully explain [1].

## 4. The variability and reliability of coating performance

While various coatings may show good properties on small-scale, carefully prepared, laboratory samples, such behaviour is rarely uniformly apparent over large areas of components prepared under real production conditions. Our experience is that the grain structure, porosity, interfacial adhesion and residual stresses in commercial coatings all show considerable point-to-point variability with consequent lack of reproducibility in mechanical properties (*e.g.* Fig. 2). This has to be an area of concern and improvement if thin hard coatings are to be fully exploited as a surface engineering technique.

## 5. Summary and conclusions

This paper has drawn several themes from the wide variations in the behaviour of composite, coated, surfaces observed under different experimental conditions. The following points emerge as being of importance in determining the behaviour of thin hard coatings:

1. The scale of the surface deformation zone relative to the coating thickness.
2. The mechanical properties of both the coating and the substrate in the microstructural states encountered in the composite system.
3. The ductility and tensile failure strain of the coating.
4. The yield, deformation and flow properties of the substrate.
5. The elastic compatibility of the coating and substrate.
6. The residual internal stress in the coating.
7. The thermal stability of the coating/substrate properties and microstructure when used at elevated temperatures or when exposed to any local thermal excursions.
8. The environment in terms of oxidation of both the coating and the substrate.
9. The adhesion of the coating–substrate interface.
10. The shear strength of the coating substrate interface.

In addition, there may be upper limits on coating thickness — in terms of integrated residual stresses, flaw sizes, ductility etc. — to ensure acceptable or even optimum performance. However, to date, our range of samples has not enabled us to explore such an effect.

Thus, identifying a coating for a particular application not only depends on the expected bulk properties of the coating material, but also its properties in the as-processed condition and its chemical and mechanical compatibility with the substrate. Once all the effects controlling coating properties are fully understood, it may be that coating selection and optimisation could be performed through calculations on a data base.

However, despite the identification of the parameters listed here, it is also clear that considerable progress has to be made in other areas such as controlling and measuring interfacial properties before all the potential benefits of coatings can be fully utilized.

## Acknowledgments

The authors are grateful to Professor R. N. Parkins for the provision of laboratory facilities at the University of Newcastle upon Tyne. The project has been supported by Shell, Thornton Research Centre (U.K.). It is a pleasure to thank other members of the Ceramic and Tribology Research Group at Newcastle (and formerly at Cambridge University — particularly Dr. I. M. Hutchings) for helpful discussions. Ms. Gail Hepple kindly typed the manuscript.

## References

1 J. C. Knight, T. F. Page and I. M. Hutchings, *Thin Solid Films*, *177* (1989) 117.
2 J. C. Knight, T. F. Page and I. M. Hutchings, *Surf. Eng.*, *5* (1989) 213.
3 J. C. Knight, T. F. Page and I. M. Hutchings, *Surf. Eng.*, *6* (1989) in press.
4 J. C. Knight, in K. C. Ludema (ed.), *Wear of Materials 1989*, ASME, New York, 1989, p. 575.
5 P. M. Sargent and T. F. Page, *Proc. Br. Ceram. Soc.*, *26* (1978) 209.
6 S. J. Bull, E. Yoffe and T. F. Page, *Philos. Mag. Lett.*, *59* (1989) 281.
7 M. G. S. Naylor and T. F. Page, *J. Microsc.*, *130* (1985) 345.
8 J. T. Czernuszka and T. F. Page, *J. Mater. Sci.*, *22* (1987) 3907.
9 P. M. Ramsey and T. F. Page, *Br. Ceram. Trans. J.*, *87* (1988) 74.
10 G. R. Sawyer, P. M. Sargent and T. F. Page, *J. Mater. Sci.*, *15* (1980) 1001.
11 A. Kelly, *Strong Solids*, Clarendon Press, Oxford, 1966.
12 C. A. Brookes and A. R. Parry, *Mater. Sci. Eng.*, *A105/106* (1988) 143.
13 D. H. Buckley, *NASA Technical Memorandum, NASA-TM-X-68046* (1972).
14 A. G. Evans, in D. A. Rigney (ed.), *Fundamentals of Friction and Wear, ASM Materials Science Seminar*, Pittsburgh, PA, 1980, ASM, Metals Parks, OH, 1981.
15 M. A. Moore and F. S. King, *Wear*, *60* (1980) 123.
16 K. T. Faber and A. G. Evans, *Acta Metall.*, *31* (1983) 565.
17 N. H. MacMillan, in R. M. Latanision and J. T. Fourie (eds.), *Surface Effects in Crystal Plasticity*, Noordhof, Leyden, 1977, p. 629.
18 J. T. Czernuszka and T. F. Page, *J. Mater. Sci.*, *22* (1987) 3917.
19 P. A. Dearnaley and E. M. Trent, *Met. Technol.*, *9* (1982) 60.
20 R. Barrett and T. F. Page, in K. C. Ludema (ed.), *Wear of Materials, 1989,* ASME, New York, 1989, p. 441.
21 B. M. Kramer and P. K. Judd, *J. Vac. Sci. Technol.*, *A3* (1985) 2439.
22 B. Jonsson and S. Hogmark, *Thin Solid Films*, *114* (1984) 257.
23 P. J. Burnett and D. S. Rickerby, *Surf. Eng.*, *3* (1987) 69.

24 P. J. Burnett and T. F. Page, *J. Mater. Sci.*, *19* (1984) 845.
25 P. M. Sargent, *Ph.D. Thesis*, Cambridge University (1979).
26 K. Stevens and A. Douglas, *Tribology — Friction, Lubrication and Wear, 50 years on*, Institute of Mechanical Engineers, London, 1987, p. 629.
27 D. S. Rickerby and P. J Burnett, *Thin Solid Films*, *157* (1988) 195.
28 D. S. Rickerby, S. J. Bull, A. M. Jones, F. Cullen and B. A. Bellamy, *Surf. Coat. Technol.*, *39/40* (1989) 397.
29 P. A. Steinmann, P. Laeng and H. E. Hintermann, *Mater. Technol.*, *13* (1985) 85.

*Surface and Coatings Technology, 39/40* (1989) 355 - 363



# THE EFFECT OF $N_2$ PARTIAL PRESSURE, DEPOSITION RATE AND SUBSTRATE BIAS POTENTIAL ON THE HARDNESS AND TEXTURE OF REACTIVELY SPUTTERED TiN COATINGS

W. D. SPROUL, P. J. RUDNIK and M. E. GRAHAM

*Basic Industry Research Laboratory, Northwestern University, 1801 Maple Avenue, Evanston, IL 60201-3135 (U.S.A.)*

(Received March 16, 1989)

## Summary

The $N_2$ partial pressure, target power and substrate bias potential all have a significant effect on the hardness and crystallographic orientation of TiN coatings prepared by high-rate reactive sputtering on cemented carbide substrates. Variations in both the substrate bias potential and the target power had a strong affect on the hardness and texture of the TiN coatings. Vickers microhardness values (50 gf) of the TiN coatings varied between 970 and 3290 kgf mm$^{-2}$ as the bias was stepped between 0 and $-200$ V and the target power between 2.3 and 10.0 kW. The texture coefficient of the TiN coatings with a constant $-100$ V bias was strongly (111) at the low power level of 2.3 kW, but the texture became random as the power was increased to 4.0 kW and above with the same $-100$ V bias. At a fixed power level of 10 kW, the orientation of the TiN coatings changed from (111) at zero bias to (200) between $-20$ and $-100$ V, to strongly (220) at $-150$ and $-200$ V. The hardness of TiN coatings with N/Ti ratios of 0.70 - 1.01 produced by varying the $N_2$ partial pressure during deposition had a narrow range between 3140 and 3400 kgf mm$^{-2}$, but the texture coefficient varied from random to strongly (220). The adhesion of the TiN coatings was measured with the scratch test, and a critical load of $8.0 \pm 0.5$ kgf was the same for all samples except for the ones at the lowest target power (2.3 kW), low biases (0 and $-20$ V) and the lowest N/Ti ratio (0.70).

## 1. Introduction

The ability to reactively sputter TiN, ZrN and HfN at high deposition rates using automatic feedback control of the nitrogen reactive gas has been available for several years now [1], and selected properties of these coatings produced at high deposition rates have been reported [2 - 5]. Typically, TiN, ZrN and HfN have ben reactively sputtered at a target power level of 10 kW and a substrate bias of $-100$ V direct current, while occasionally, as for ZrN,

0257-8972/89/$3.50 

results were also reported for a power level of 9 kW [6]. These high power levels were dictated by the desire to use these hard, wear resistant coatings in industrial tribological applications where cost is a major factor.

Not all sputtering applications require very high deposition rates, but until recently [7 - 9], not much had been disclosed about the properties of coatings such as TiN produced under reactive gas automatic feedback control at lower target power levels. It has now been shown [9] that the sputtering target power, which is proportional to the deposition rate, has a major effect on the hardness and crystallographic orientation of TiN coatings. Other sputtering deposition parameters, such as substrate bias voltage, reactive gas partial pressure, substrate to target distance, substrate temperature and total system pressure influence the properties of reactively sputtered coatings. In this current work, two of these parameters, substrate bias voltage and reactive gas partial pressure, were systematically varied during high-rate reactive sputtering of TiN, and the effect of these changes on coating hardness, adhesion and crystallographic orientation were studied. This paper reports on the findings from this work.

## 2. Experimental procedure

All TiN high-rate reactive sputtering experiments were carried out in a Materials Research Corporation (MRC) 902M In-line sputtering system. The argon (purity, 99.999%) was introduced into the system at one end of the chamber and the $N_2$ (purity, 99.999%) was introduced through a manifold located around the planar direct current magnetron sputtering cathode. The MRC Mu Inset sputtering target was VP grade titanium. For all of this current work, the cathode power level was set at a constant 10.0 kW. In the bias series, the negative direct current substrate bias potential was varied between 0 and 200 V in eight steps set at 0, 20, 40, 60, 80, 100, 150 and 200 V; the $N_2$ partial pressure was 0.13 mTorr and the deposition rate was 0.41 $\mu$m min$^{-1}$. In the composition series, the bias was fixed at $-100$ V, and the $N_2$ partial pressure, as measured by a Leybold Inficon optical gas controller (OGC) system, was set at 0.025, 0.05, 0.10, 0.13, 0.16 and 0.20 mTorr.

The partial pressures of the $N_2$ and argon gases were measured during sputtering with both a differentially pumped mass spectrometer system and with the OGC system. The mass spectrometer measured the partial pressures of the gases at the end of the sputtering chamber. The OGC sensor head, which does not have to be differentially pumped, was connected directly to the sputtering atmosphere through a stainless steel tube (length 8 cm, diameter 6 mm) that pierced the sputtering target at a point along the long axis of the rectangular sputtering target, 12.5 cm from the center of the target. Since the OGC sensor measures the density of the gas atoms and since the density of the atoms is reduced in front of the sputtering target during deposition due to gas heating and rarefaction [11, 12], the OGC measures an apparent partial pressure at the face of the target [9]. The OGC can measure the partial pressure of two gases simultaneously [13] and, for this work, it was configured to measure argon and $N_2$.

Partial pressure readings on both the OGC and the mass spectrometer were calibrated against pressure gauges on the sputtering system. The $N_2$ partial pressure was referenced to the ion gauge, and the argon partial pressure was referenced to an MKS Baratron capacitance manometer. The $N_2$ partial pressure during reactive sputtering was controlled with either the OGC through its control loop or the mass spectrometer feedback control system [1], which consists of a modified UTI 100C mass spectrometer and an MKS 260 mass flow controller. Both systems provided good control of the reactive gas. Overall system pressure was maintained at 8 mTorr using the MKS Baratron capacitance manometer and MKS 260 mass flow controller for the argon. Although not part of these experiments, it is also possible to control the argon partial pressure with the OGC.

Substrates were GTE Valenite cemented carbide cutting tool inserts, grade VC-3, style SPE-43-E and glass slides. The Vickers hardness of these carbide substrates (1000 gf) was 2010 kgf $mm^{-2}$. All substrates were cleaned in acetone and methanol and were given a 10 min 1.5 kW r.f. sputter etch before being coated. A coating thickness of 5.0 $\mu$m was applied to the cemented carbide inserts and glass slides.

The thickness of steps in the coating on the glass slides was measured with a Dektak surface profilometer. Vickers hardness of each coating was measured with a Leco DM-400 FT microhardness tester with flat field optics and a maximum 1000X magnification. All microhardness indentations in the coatings were done at a load of 50 gf for 15 s, and at least seven indentations were made on each sample. The X-ray diffraction experiments were carried out in an Scintag XDS 2000 PAD V diffractometer with Cu $K\alpha$ radiation. The scratch adhesion tests were done with a CSEM manual scratch tester at intervals of 0.5 kgf.

## 3. Results and discussion

### 3.1. Coating hardness

The Vickers microhardness values and the critical loads [14] from the CSEM scratch test are given in Table 1 for the bias series samples. At biases between 0 and −60 V, the hardness of the TiN is less than the bulk TiN value [15] of 2000 kg $mm^{-2}$, but between −80 and −200 V, the hardness is greater than that of the bulk material. This ramping up and leveling off of the hardness of the reactively sputtered TiN is similar to what Håkansson *et al.* [16] reported for reactively sputtered $TiAlN_2$. The change in hardness is due to the ion bombardment of the coating as it is formed. The energy imparted to the growing surface by the ion bombardment helps to anneal out imperfections in the coating, but above a certain energy level there is more damage induced from the bombardment than is annealed out [16, 17]. It appears that this effect is observed in comparing the hardnesses of the samples produced at biases of −150 and −200 V.

TABLE 1

Bias series: the Vickers microhardness values (50 gf) and adhesion scratch test critical loads of TiN coatings reactively sputtered with different substrate biases

| *Substrate bias* (V) | *Microhardness* (kgf $mm^{-2}$) | *Scratch adhesion critical load* (kgf) |
|---|---|---|
| 0 | $1440 \pm 63$ | 5.5 |
| −20 | $1520 \pm 86$ | 6.5 |
| −40 | $1630 \pm 46$ | 7.5 |
| −60 | $2140 \pm 97$ | 8.0 |
| −80 | $2660 \pm 99$ | 7.5 |
| −100 | $3070 \pm 69$ | 8.0 |
| −150 | $3290 \pm 104$ | 8.0 |
| −200 | $3110 \pm 106$ | 7.5 |

The N/Ti ratio, under the conditions studied here, had a very small effect on the hardness of the TiN coatings as can be seen from the data in Table 2 for the composition series. This result is surprising because Sundgren *et al.* [18] and Schiller *et al.* [19] have reported that the hardness of TiN coatings made by reactive sputtering is a function of composition, and Musil *et al.* [20] gave the hardness of $TiN_x$ as a function of the $N_2$ partial pressure which was set prior to sputtering. Schiller *et al.* reported that the hardness of $TiN_x$ reached a maximum value of 3000 kgf $mm^{-2}$ at a N/Ti ratio of 0.6 and decreased to 2000 kgf $mm^{-2}$ at the stoichiometric composition. Sundgren *et al.* reported that the hardness of $TiN_x$ increased rapidly as the nitrogen content in the coating increased and reached a maximum value of 2200 kgf $mm^{-2}$ slightly before the stoichiometric composition. Nitrogen rich films showed a rapid decrease in coating hardness. Musil *et al.* found a peak in the hardness of $TiN_x$ at a $N_2$ partial pressure of 0.32 mTorr.

The above three groups of researchers all prepared their $TiN_x$ films under different reactive sputtering conditions than were used in this study.

TABLE 2

Composition series: the deposition rate, N/Ti ratio, Vickers microhardness values (50 gf) and adhesion scratch test critical loads of TiN coatings reactively sputtered at different $N_2$ partial pressures

| *$N_2$ partial pressure* (mTorr) | | *Deposition rate* ($\mu$m $min^{-1}$) | *N/Ti ratio* | *Hardness* (kgf $mm^{-2}$) | *Critical load* (kgf) |
|---|---|---|---|---|---|
| *OGC* | *Mass spectrometer* | | | | |
| 0.025 | 0.075 | 0.48 | $0.70 \pm 0.03$ | $3240 \pm 104$ | 3.0 |
| 0.050 | 0.10 | 0.45 | $0.86 \pm 0.03$ | $3260 \pm 143$ | 7.5 |
| 0.10 | 0.15 | 0.42 | $0.97 \pm 0.01$ | $3220 \pm 62$ | 8.0 |
| 0.13 | 0.18 | 0.38 | $0.97 \pm 0.02$ | $3250 \pm 90$ | 8.0 |
| 0.16 | 0.20 | 0.36 | $0.96 \pm 0.02$ | $3170 \pm 44$ | 8.5 |
| 0.20 | 0.25 | 0.34 | $1.01 \pm 0.03$ | $3400 \pm 167$ | 7.5 |

All three used flow control of the reactive gas, but Sundgren *et al.* used an r.f. diode sputtering system with the substrate grounded whereas Schiller *et al.* used a Plasmatron (direct current magnetron) source with a −20 V bias on the substrate. Musil *et al.* used a direct current planar magnetron with the substrate grounded. It is apparent from the differing results of these authors and our own that there is a large matrix of interrelated parameters that affect the characteristics of sputtered coatings. Owing to the differences in the bias and power levels, deposition rate and mode of reactive gas control, the properties of the coatings may not respond in the same way for a change in a given parameter. For example, the dependence of hardness on composition that was observed at low power and low deposition rate may be subordinated at the high deposition rate achieved with partial pressure control, *i.e.* the contribution to hardness due to the N/Ti ratio may be masked by a stronger structural contribution from the nature of the film growth. It is believed that the use of $N_2$ partial pressure control is responsible for the difference in hardness dependence reported here and that observed in previous work.

Within the region of the parameter matrix used in this work, the substrate bias produces a large change in coating hardness as the bias is varied from 0 to −200 V, and this change in hardness is slightly less than what was observed for the series of coatings prepared at different power levels [9] as is shown in Table 3. Even though the bias series does not have as large a difference in hardness as the power series does, the change in hardness for both sets is very large. Both of these parameters have a major influence on the hardness of the TiN coatings.

### 3.2. *Crystallographic orientation*

The X-ray diffraction analysis of the coatings shows a wide range in crystallographic orientation as a function of bias, power, and composition as is shown in the texture coefficient [21] plot in Fig. 1. For the power series [10] shown in Fig. 1a, the TiN coatings made at the low power level of 2.3 kW are very strongly (111) oriented; but at power levels of 4.0 kW and above, the orientation is basically random [9].

TABLE 3

Power series: the deposition rate, Vickers microhardness values [9] (50 gf) and adhesion scratch test critical loads of TiN coatings reactively sputtered at different target power levels

| *Power* (kW) | *Deposition rate* ($\mu$m min$^{-1}$) | *Hardness* (kgf mm$^{-2}$) | *Critical load* (kgf) |
|---|---|---|---|
| 2.3 | 0.12 | $970 \pm 15$ | 2.0 |
| 4.0 | 0.20 | $2410 \pm 74$ | 7.5 |
| 6.0 | 0.30 | $2740 \pm 48$ | 7.0 |
| 8.0 | 0.39 | $3110 \pm 84$ | 8.0 |
| 9.0 | 0.45 | $3230 \pm 75$ | 8.5 |
| 10.0 | 0.48 | $3160 \pm 137$ | 8.0 |

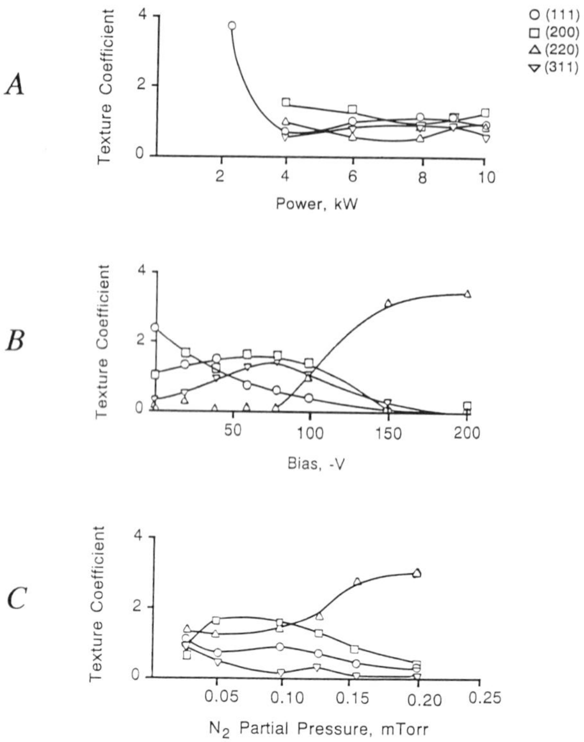


Fig. 1. Texture coefficients of the reactively sputtered TiN coatings as a function of (A) target power, (B) substrate bias, and (C) $N_2$ partial pressure.

In the bias series shown in Fig. 1b, the coatings made at low bias have a strong (111) orientation; however, as the bias was increased negatively to $-80$ V, the (200) and (311) texture coefficient increased as the (111) texture coefficient decreased. Above this point, the (111), (200) and (311) peaks all decreased while the (220) texture coefficient, which had been very small until this point, increased rapidly. At $-150$ and $-200$ V, the coatings have a very strong (220) texture.

Kobayashi and Doi [22] also saw, for reactive ion plated TiN coatings, large changes in crystallographic orientation with respect to substrate bias, but they used biases from 0 to $-950$ V. At zero bias they had a (200) orientation, at intermediate voltages a (111) orientation, and at high voltages a (220) orientation. Although they observed these changes, the specific relationships are different from what was found for the reactively sputtered TiN coatings produced for this study.

The composition series of samples, which were made at different nitrogen partial pressures, also showed large changes in orientation, but the differences in orientation are related more to changes in partial pressure than to the actual N/Ti ratio. The N/Ti ratio increased rapidly as the $N_2$ partial pressure, as measured by the OGC, increased from 0.025 to 0.10 mTorr;

but as the partial pressure was increased further to 0.20 mTorr, the N/Ti ratio stayed essentially constant at 0.97. As these partial pressure changes were occurring, the orientation of the coatings, as shown in Fig. 1c, changed from a mild (220) orientation at a partial pressure of 0.025 mTorr, to a slightly stronger (220) orientation at a partial pressure of 0.13 mTorr and finally to a strong (220) orientation for a partial pressures of 0.16 and 0.20 mTorr. This leveling out of N/Ti ratio has been reported earlier [23], but the actual $N_2$ partial pressures at the face of the target were not known at that time.

Sundgren *et al.* [24] showed that the orientation of their reactively sputtered TiN coatings always had a (200) orientation over the N/Ti range of 0.35 - 0.50. The normalized intensities changed with composition, but the films were always (200) oriented. Musil *et al.* [20] showed the relative intensities for the (111) and (200) peaks in their reactively sputtered TiN coatings for the partial pressure studied, and they found that the films had an orientation that was basically random. Since both of these groups used flow control of the reactive gas and since the substrate was grounded, it is difficult to make a direct comparison to our results. Again, it is felt that these differences are primarily due to the use of partial pressure control in this work.

### *3.3. Coating adhesion*

Although the scratch test does not give an absolute value of coating adhesion, it is a good test to show the relative values of adhesion for the same type of coating on the same kind of substrate. In general, the adhesion of these TiN coatings on the carbide substrate is excellent, and the critical scratch test loads are all approximately $8.0 \pm 0.5$ kgf, but there are a few exceptions. At the lowest target power of 2.3 kW and the lowest N/Ti ratio of 0.70, the critical load is substantially less than the average of 8.0 kgf. Low bias (0 and $-20$ V) also produced a low critical load, but not to the extent that was found for the low power and low N/Ti ratio situations.

## 4. Conclusions

(1) Substrate bias potential strongly influences the hardness and crystallographic orientation of reactively sputtered TiN coatings. As the bias is increased negatively from 0 to $-200$ V, the hardness of the coatings goes from 1440 kgf mm$^{-2}$ to a peak value of 3290 kgf mm$^{-2}$ at $-150$ V before it drops off slightly to 3110 kgf mm$^{-2}$ at $-200$ V. The (111) orientation at zero volts bias changes to random between $-20$ and $-100$ V, and goes very strongly (220) at $-150$ and $-200$ V. At the low bias levels of 0 and $-20$ V, the scratch test critical load is less than the $8.0 \pm 0.5$ kgf value found for all other bias levels.

(2) Target power also strongly influences the hardness and orientation of the TiN coatings. At the low target power of 2.3 kW, the coating is relatively soft with a hardness of 970 kgf mm$^{-2}$, but the hardness increases rapidly as the power increases and reaches a value of 3160 kgf mm$^{-2}$ at 10.0 kW. Orientation of the coatings changes from a very strong (111)

orientation at 2.3 kW to a random orientation at 4.0 kW and above. Above 4.0 kW target power, the scratch test critical load is $8.0 \pm 0.5$ kgf and is not a function of target power.

(3) The $N_2$ partial pressure during deposition had little effect on the hardness and adhesion of the TiN coatings, but it did affect the composition and crystallographic orientation strongly. For all compositions reported in this work, the hardness was in the range of 3170 - 3400 kgf mm$^{-2}$, and the critical load was $8.0 \pm 0.5$ kgf except for the lowest N/Ti ratio of 0.70 where it was 3.0 kgf. Crystallographic orientation of the this group of coatings depended more on the $N_2$ partial pressure during sputtering than it did on the actual composition: orientations ranged from random to (200) to strongly (220) as the $N_2$ partial pressure increased.

## Acknowledgments

The authors wish to thank Mr. Carl Strandberg of the Materials Research and Development Laboratory of GTE Valenite Corporation, Troy, Michigan for the Auger chemical analysis of the $TiN_x$ coatings and Dr. Anthony Perry and Mr. Mark Jagner of the same laboratory for their help with the analysis of the X-ray diffraction work.

## References

1 W. D. Sproul and J. A. Tomashek, *U. S. Patent 4,428,811* (1984).
2 W. D. Sproul, *Thin Solid Films, 107* (1983) 141.
3 W. D. Sproul, *Thin Solid Films, 118* (1984) 279.
4 W. D. Sproul, *J. Vac. Sci. Technol., A4 (6)* (1986) 2874.
5 W. D. Sproul, in W. D. Sproul, J. E. Greene and J. A. Thornton (eds.), *Physics and Chemistry of Protective Coatings*, American Institute of Physics Conference Proceedings, No. 149, New York, 1986, pp. 157 - 172.
6 A. J. Perry, M. Georgson and W. D. Sproul, *Thin Solid Films, 157* (1987) 255.
7 C. Ernsberger, J. Nickerson, T. Smith, A. E. Miller and D. Banks, *J. Vac. Sci. Technol., A4* (1986) 2784.
8 C. Ernsberger, A. J. Perry, L. P. Lehman, A. E. Miller, A. R. Pelton and B. W. Dabrowski, *Surf. Coat. Technol., 36* (1988) 605.
9 W. D. Sproul, P. J. Rudnik and C. A. Gogol, *Thin Solid Films, 171* (1989) 171.
10 A. J. Perry, M. Jagner, W. D. Sproul and P. J. Rudnik, *Surf. Coat. Technol., 39/40* (1989) 387.
11 D. W. Hoffman, *J. Vac. Sci. Technol., A3* (1985) 561.
12 S. M. Rossnagel, *J. Vac. Sci. Technol., A6* (1988) 19.
13 C. A. Gogol and R. Mueller, *Surf. Coat. Technol., 36* (1988) 939.
14 A. J. Perry, *Thin Solid Films, 81* (1981) 357.
15 L. E. Toth, *Transition Metal Nitrides and Carbides*, Academic Press, New York, 1971, p. 7.
16 G. Håkansson, J.-E. Sundgren, D. McIntyre, J. E. Greene and W.-D. Münz, *Thin Solid Films, 153* (1987) 55.
17 I. Petrov, L. Hultman, U. Helmersson, J.-E. Sundgren and J. E. Greene, *Thin Solid Films, 169* (1989) 299.

18 J.-E. Sundgren, B.-O. Johansson, S.-E. Karlsson and H. T. G. Hentzell, *Thin Solid Films*, *105* (1983) 385.
19 S. Schiller, U. Heisig, G. Beister, K. Steinfelder, J. Strümpfel, Chr. Korndörfer and W. Sieber, *Thin Solid Films*, *118* (1984) 255.
20 J. Musil, L. Bardos, A. Rajsky, J. Vyskocil, B. Dolezal, G. Loncar, K. Dadourek and V. Kubicek, *Thin Solid Films*, *136* (1986) 229.
21 D. S. Rickerby, A. M. Jones and B. A. Bellamy, *Surf. Coat. Technol.*, *37* (1989) 111.
22 M. Kobayashi and Y. Doi, *Thin Solid Films*, *54* (1978) 67.
23 A. J. Perry, M. Georgson and W. D. Sproul, *Thin Solid Films*, *157* (1988) 255.
24 J.-E. Sundgren, B.-O. Johansson, S.-E. Karlsson and H. T. G. Hentzell, *Thin Solid Films*, *105* (1983) 367.

*Surface and Coatings Technology, 39/40* (1989) 365 - 376 

# CORRELATION OF INTERFACE STRUCTURE WITH ADHESIVE STRENGTH OF ION-PLATED TiN HARD COATINGS

C. C. CHENG, A. ERDEMIR and G. R. FENSKE

*Materials and Components Technology Division, Argonne National Laboratory, Argonne, IL 60439 (U.S.A.)*

(Received March 9, 1989)

## Summary

Interlayers of titanium have been used to improve the adhesion of ion-plated TiN coatings. However, the mechanisms of the improvement of adhesion remain unclear. In this study, we have applied analytical electron microscopy on cross-section specimens and Auger electron spectroscopy to investigate the corrleation of the interface structure with the scratch adhesion of ion-plated TiN coatings with and without titanium interlayers on M50 steel substrates. Our results have shown that the titanium interlayer can reduce the internal stresses at the interface region. Furthermore, this titanium interlayer can transform into TiC phase at high substrate temperature both during and after ion plating. These types of treatment can in turn increase the adhesion of the TiN coatings via a better interface contact and stronger chemical bonds.

## 1. Introduction

It is known [1, 2] that performance (*e.g.* hardness, adhesion) of a hard coating can be significantly influenced by process parameters (*e.g.* substrate bias, temperature). However, there is still lack of a fundamental knowledge to clarify how these process parameters affect the formation or changes of the microstructure of coatings, and the correlation of these microstructures with the resulting coating performance. Hence, microstructural studies of coating are of great importance because they provide a better understanding of the relationship between the process parameters and the resulting performance [3].

Good adhesion is considered one of the most important prerequisites of engineering coatings for use in tribological environments. If the interface adhesion is poor, coating detachment may occur and lead to premature failure. The adhesion strength of a coating depends on the extent of both chemical and physical interactions between the coating and substrate materials and on the microstructure of the interface region. Poor adhesion may be

0257-8972/89/$3.50 

attributed to: low degree of chemical bonding, poor interfacial contact, low fracture toughness and high internal stresses [4]. Accordingly, interface structure plays a very important role in determining the adhesive strength of coatings, and consequently, it is of some urgency to have a better understanding of the correlation between the interface structure and its related film-substrate adhesion.

Reactive ion-plated TiN hard coating has become popular in the tool industry because of its ease of application and reliable performance [5]. However, the adhesive strength of ion-plated TiN coatings to tool substrate materials is still a concern. In recent years, several reports [1, 6] have revealed that a thin layer of titanium at the interface results in a substantial improvement in adhesion of the deposited TiN coatings. It is postulated that the titanium interlayer acts a a graded interface [7] that avoids the compositionally abrupt change at the sharp interface between a coating and metal substrate (TiN–steel). Nevertheless, the effects of this titanium on the enhancement of the coating adhesion still remain unclear, primarily due to the lack of systematic microstructural studies at the interface region.

In this study, the interface regions of several reactive ion-plated TiN coatings deposited with and without titanium interlayer were structurally and compositionally studied by analytical electron microscopy on cross-section specimens and Auger spectroscopy. Scratch adhesion tests were performed on these TiN coatings to evaluate their adhesion. The objective of this study is to clarify the role of the titanium interlayer and to determine how structural transformations due to high substrate temperature or postannealing treatment affect the adhesive strength of TiN coatings.

## 2. Experimental procedure

### *2.1. Preparation of the substrate and coating materials*

The substrates for TiN coating were prepared by cutting an AISI M50 steel rod into discs of thickness 4 mm. The discs were polished progressively with diamond paste down to 0.25 $\mu$m. Following a series of ultrasonic cleanings in distilled water, acetone, and methanol, the specimens were loaded into the substrate holder of an ion-plating system that was equipped with a hollow-cathode discharge (HCD) electron beam gun for the evaporation of pure titanium. Specimens were sputter cleaned in the ion plating chamber in an argon plasma for 10 min. The specimens were heated with a radiant heater. When the specimens were at the desired deposition temperatures (100, 300 and 500 °C), the HCD gun was energized to deposit titanium into the existing plasma which was sustained by a bias voltage of 100 V negative direct current with a chamber pressure of 0.1 Pa. The flow of nitrogen gas was started simultaneously either at the start of deposition of titanium or after a layer of titanium approximately 100 nm thick was deposited. Several specimens that were ion-plated at 100 °C were post-annealed at 500 °C for 45 min in a vacuum of better than $10^{-4}$ Pa.

*2.2. Cross-sectional specimen preparation for analytical electron microscopy*

The samples (M50 steel with TiN coatings) were cut into 2 mm × 1 mm × 10 mm slabs. Two slabs were then embedded coating-to-coating in epoxy in a stainless steel tube of diameter 3 mm. Discs with a thickness of 0.5 mm were cut from the tube and thinned down by grinding to a thickness of about 80 μm. Further, the center of the specimen was reduced to about 20 μm by a dimpler. Ion-beam milling was used to thin the specimens to electron transparency. During the milling, tantalum masks covered slightly more than one-half of the specimen to avoid the maximum preferential etch when the ion beams were parallel to the interface. In this work, the ion sources were operated at a voltage of 6 kV and an ion current of 2 mA. The specimens were first etched at an angle of 20° to the surface. In the final etching stage, the angle was reduced to 15°.

The transmission electron microscope (TEM) examination was conducted with a Philips EM420 analytical electron microscope operated at 120 kV. The microscope interfaces with a Gatan electron energy loss spectrometer with an EDAX9900 data acquisition system.

*2.3. Auger electron spectroscopy*

The chemical analyses of the TiN film and the film-substrate interface were performed with a Physical Electronics model 600 scanning Auger multiprobe. Depth concentration profiling of the prominent species (*i.e.* carbon, oxygen, titanium, nitrogen and iron) was carried out on radially taper-sectioned areas of the TiN coatings near the film-substrate interface. An argon ion gun with an acceleration energy of around 4 keV and a current density of around 49 $\mu$A cm$^{-2}$ was used to sputter section the TiN at preset time intervals. The Auger data were automatically collected at these intervals. $Ta_2O_5$ was selected as the calibration standard; under these conditions, it showed a sputtering rate of approximately 13 nm min$^{-1}$.

*2.4. Adhesion test*

An LSRH (Laboratoire Suisse de Recherches Horlogeres) Revetest scratch adhesion tester was used to evaluate the adhesion of TiN films to the substrate steels. The test consisted in applying a continuously increasing load (100 N min$^{-1}$) on the coating surfaces by a Rockwell C scratching point while the sample was displaced at a constant speed (13.3 mm min$^{-1}$). An optical microscope was used to observe the surface damage caused by the scratching point. The distance between the start of the scratch and the disbonding site was measured, and from this distance the critical load was assessed.

## 3. Results

*3.1. The TiN coating ($T_{sub}$ = 500 °C) without a titanium interlayer*

TiN coatings were deposited on M50 steel and 440C steel substrates at both 100 and 500 °C. Following the deposition, these TiN coatings tended to buckle up and delaminate from the substrates except the one deposited at 500 °C on 440C steel substrate. This suggests that direct deposition of a TiN coating on steel substrates may accumulate large compressive stresses in the coating. In order to qualitatively ensure that there were high compressive stresses at the interface, we reduced the 440C steel substrate (with a 2.2 $\mu$m TiN coating and deposited at 500 °C) down to around 200 $\mu$m thick by means of grinding and polishing from the uncoated side. It could be visually observed that the thin flat piece was forced to assume a convex dome shape which substantiated that there were indeed high compressive stresses in the film.

During the preparation of cross section specimens for TEM study, the similar problem of internal stress was encountered. When the thickness of cross section specimens were reduced to a certain value by either polishing or dimpling, the TiN coatings tended to detach and break away from the substrates. Figure 1a shows the TiN film well attached to the substrate when the cross section specimen was about 60 $\mu$m thick. However, when the same specimen was thinned down to around 30 $\mu$m, the TiN coating became detached and broken, as shown in Fig. 1b. Accordingly, the internal stresses seem to play a predominant role in the adhesive strength of this sample.

Figure 2 shows a bright-field electron micrograph of the TiN film. The structure was columnar and the size of the columns, near the interface region, was found to be on the order of 25 nm, which was about the same size as that in the sample deposited at 500 °C with a titanium interlayer (see Section 3.5).

### 3.2. *The TiN coating ($T_{sub}$ = 100 °C) with a titanium interlayer*

As can be seen in Fig. 3, the interface region consisted of a thin interlayer around 150 nm thick. The Auger depth profile and electron diffraction pattern, shown in Fig. 3, identify this layer as pure titanium. Furthermore, as shown in Fig. 3, the boundaries between the substrate–titanium interlayer as well as the titanium interlayer–TiN coating were very distinct and straight. As far as the microstructure morphologies were concerned, the titanium interlayer consisted of equiaxed grains whose sizes were about 30 - 50 nm in diameter. The TiN film grown on the top of the titanium

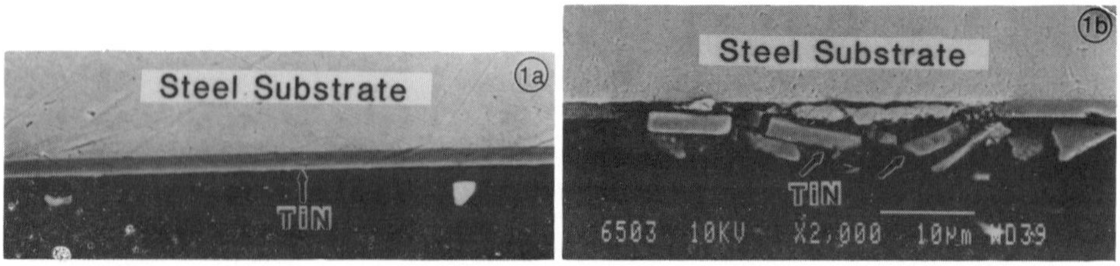


Fig. 1. The TiN film was well attached to the steel substrate when the cross section specimen was about 60 μm thick (a) and the same TiN film was found detached and broken when its thickness was reduced to about 30 μm (b).

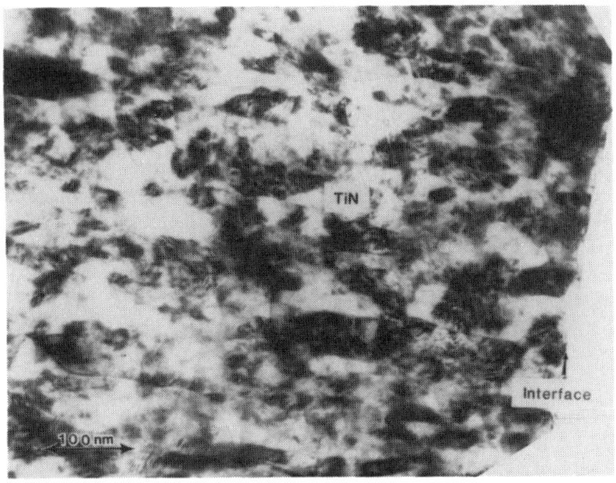


Fig. 2. A bright-field electron micrograph of the TiN film deposited at 500 °C without a titanium interlayer.

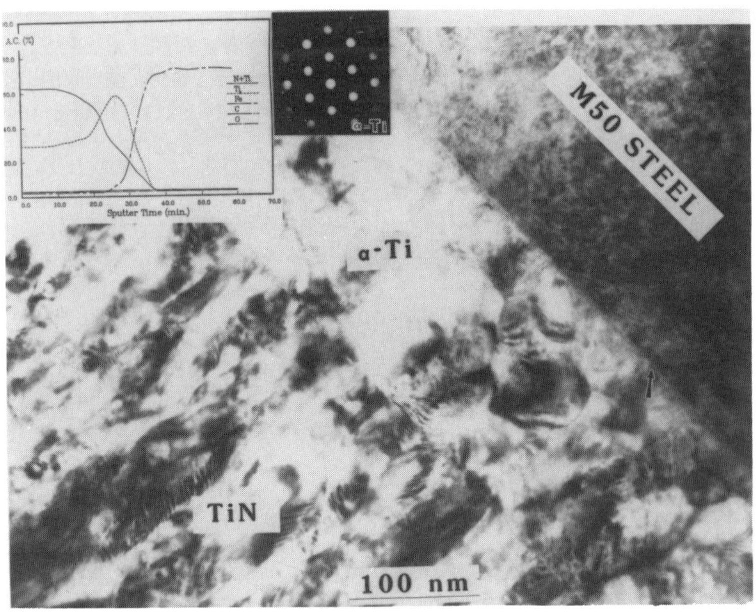


Fig. 3. A bright-field electron micrograph of the interface region of TiN film deposited at 100 °C with a titanium interlayer. The arrow shows a gap between the substrate–titanium interface.

interlayer was composed of columnar grains about 15 nm in size. The equiaxed grain structure of the titanium interlayer is not surprising. It has been reported [8] that, during the evaporation deposition, a substrate bias can result in a change in morphology from columnar to fine, equiaxed grains.

A very thin (about 2 - 3 nm) interfacial layer between the substrate–titanium interlayer was observed and arrowed in Fig. 3. A through-focal series, as shown in Fig. 4, indicated [9] that this interfacial layer was made up of a

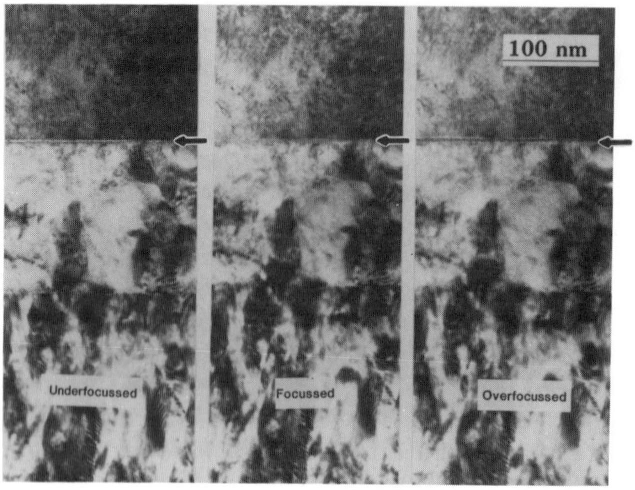


Fig. 4. A through-focal series of the substrate–titanium interface for the TiN film deposited at 100 °C.

low-atomic-density phase or vacancies. It is believed that this poor interfacial contact at the interface between the substrate–titanium interlayer may have weakened the adhesion of this sample.

### 3.3. *The TiN coating ($T_{sub}$ = 300 °C) with a titanium interlayer*

The microstructure of the interlayer in this sample was found to be different from the one deposited at 100 °C; the interlayer consisted of a mixture of pure titanium and a phase possessing the NaCl crystal structure, as shown in Fig. 5. Electron diffraction patterns indicated that this phase may be either TiN or TiC, because both have the NaCl crystal structure and also have fairly similar lattice parameters (0.4239 nm for TiN *vs.* 0.4327 nm for TiC). In this study, we also used the combination of electron energy loss spectroscopy (EELS) and Auger electron spectroscopy, shown in Fig. 5, to conclusively identify it as a TiC phase. Evidently, carbon had diffused out of the M50 steel substrate to form the TiC phase [10].

According to the Auger depth profile in Fig. 5, the TiC phase was limited to regions near the substrate. However, TEM studies showed that many TiC grains extended from the substrate–titanium interface all the way to the titanium–TiN interface. In addition, many of these TiC grains were rectangular or triangular in shape which suggested that the diffusion of carbon may take place along several favourable crystallographic paths or planes.

The size of the TiN columns, near the titanium–TiN interface of this sample, was about 15 nm which is about the same as that of the one deposited at 100 °C. Interfaces between substrate–titanium as well as titanium–TiN were, again, distinct and straight. The small gap of the low-atomic-density phase or vacancy between the substrate–titanium interface was also

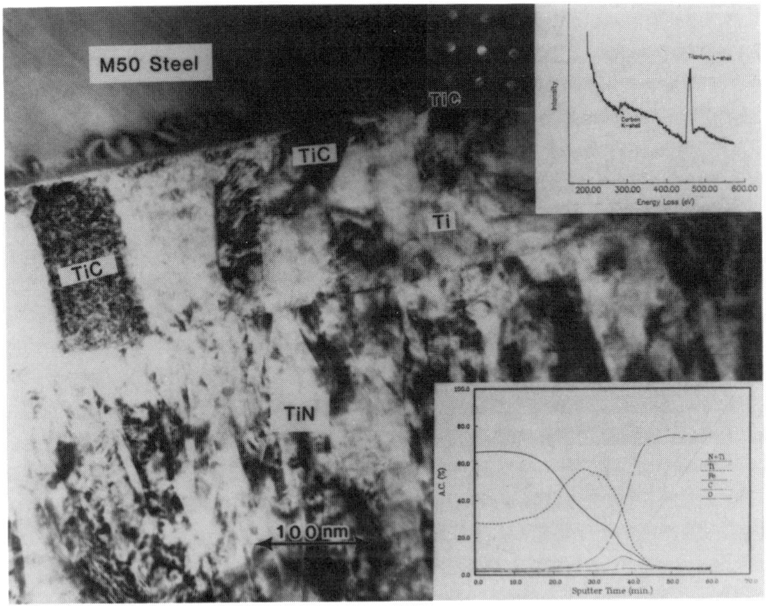


Fig. 5. A bright-field electron micrograph of the TiN film deposited at 300 °C with a titanium interlayer. The combination of electron diffraction, electron energy loss spectrometer and Auger depth profile identifies the TiC phase at the interlayer region.

observed in this sample. However, it was not as wide as that in the one deposited at 100 °C.

### 3.4. *The annealed TiN coating ($T_{sub} = 100\ °C + T_{ann} = 500\ °C$) with a titanium interlayer*

The 500 °C annealing resulted in a phase transformation of the titanium interlayer. The titanium interlayer of this sample was found to be fully transformed into a TiC phase. Again, the electron diffraction pattern shows only the NaCl crystal structure. However, the Auger and EELS resulted positively identify it as a TiC phase. No trace of α-Ti phase was discovered in the interlayer region of this sample.

The interlayer of this sample consisted of two grain morphologies as shown in Fig. 6: small equiaxed grains (about 10 nm) were seen in regions near the substrate–interlayer interface, whereas relatively large grains (about 100 nm) were found for the rest of the interlayer regions. The interface between substrate–interlayer was seen to be sharp and straight. Moreover, the gap found in the samples deposited at 100 and 300 °C was not seen in this sample. The interlayer–TiN interface of this sample become distorted and become less pronounced. Intuitively, the adhesive strength of this sample should be stronger than that of previous ones because of the better interface contact at the substrate–interlayer as well as the irregular shape of the interlayer–TiN interface [11].

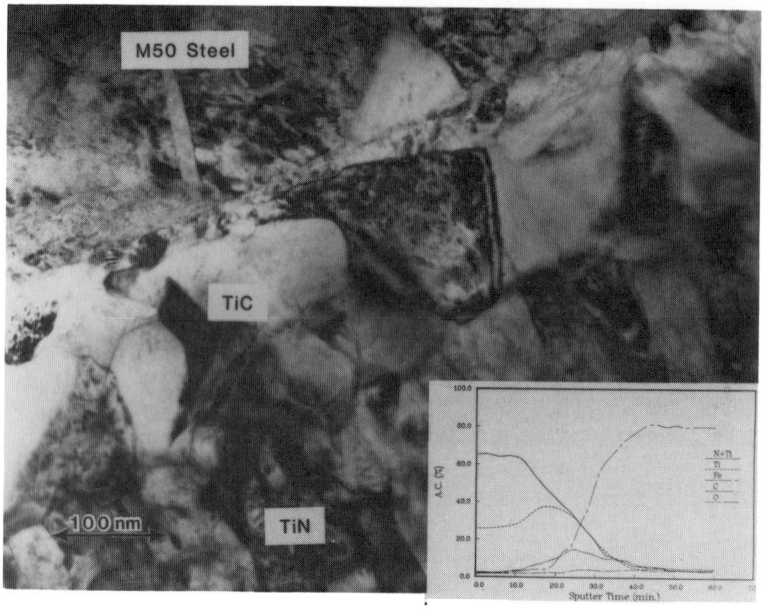


Fig. 6. A bright-field electron micrograph of the TiN film deposited at 100 °C with a titanium interlayer and a post-annealing at 500 °C. The titanium interlayer was fully transformed to the TiC phase.

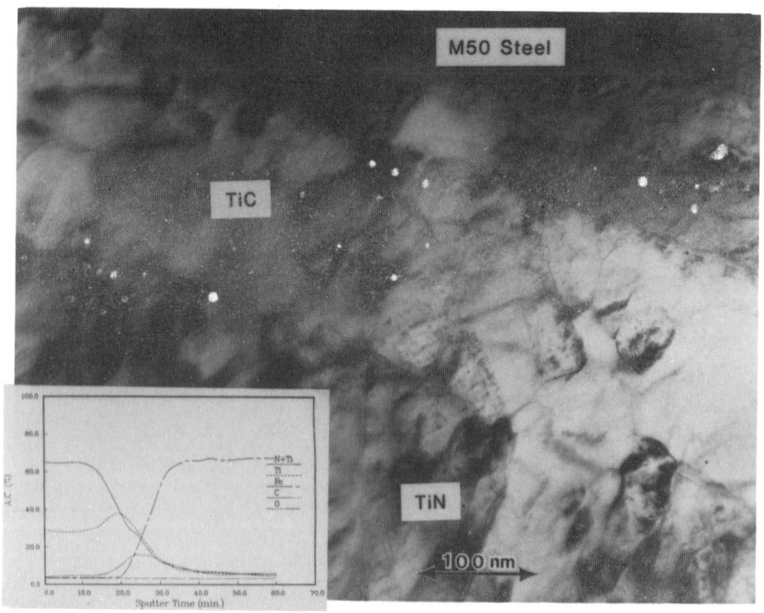


Fig. 7. A bright-field electron micrograph of the TiN film deposited at 500 °C with a titanium interlayer.

### 3.5. The TiN coating ($T_{sub}$ = 500 °C) with a titanium interlayer

The coating deposited at 500 °C exhibited chemical and microstructural characteristics similar to those of the 500 °C annealed sample, as illustrated in Fig. 7. However, in the interlayer region of this sample, the population of small TiC grains (around 10 nm) near the substrate was higher compared to that of the 500 °C annealed sample. It suggests that the nucleation density of the titanium or TiC grains of this sample is higher compared to that of the annealed sample.

It is interesting to note that the size of the TiN columns in this sample was similar to that of the coating deposited at the same temperature without a titanium interlayer. Undoubtedly, the existence of the titanium interlayer did not increase the nucleation density of the TiN film, which may enhance the adhesive strength, as people previously believed [6].

### 3.6. Results of the adhesion test

Table 1 gives a summary of the results of the scratch adhesion test on those samples presented above. Because the tests were performed under similar conditions, it is reasonable to use the critical loads as direct indicators of the adhesive strength of the samples. However, the critical load for the decohesion of coating normally has a linear relationship with the thickness of tested coatings [12] and also is highly influenced by the hardness of substrate. It appears that the scratch test can only provide comparative information about the adhesion at the substrate–coating interface. Thus, in this study, the trend of the adhesive strength, shown in the column 8 of Table 1 for those samples is more meaningful than the values of critical load.

TABLE 1

Results of the scratch adhesion test

| *Sample* | *Titanium interlayer* | $T_{sub}$ (°C) | $T_{ann}$ (°C) | *Substrate hardness* (kg mm$^{-2}$) | *Coating thickness* (μm) | *Critical load* (N) | *Adhesive strength* |
|---|---|---|---|---|---|---|---|
| 1 | No | 500 | — | 430 ± 20[a] | 1.57 | 5 ± 1 | Weak |
| 2 | Yes | 100 | — | 810 ± 18 | 1.92 | 36 ± 3 | ↓ |
| 3 | Yes | 300 | — | 808 ± 26 | 2.39 | 54 ± 4 | |
| 4 | Yes | 100 | 500 | 802 ± 13 | 1.92 | 50 ± 10 | |
| 5 | Yes | 500 | — | 800 ± 20 | 1.81 | 61 ± 3 | Strong |

[a]The substrate material was 440C steel.

## 4. Discussion

During the deposition of a coating, the nature and condition of substrate strongly influence the nucleation and formation of the TiN film. These parameters, thereafter, control the interface structure and its performance.

Better chemical bonding, more interface contact, and less internal stress can markedly improve the adhesion of the coating. To achieve these goals, elevated substrate temperature [13], high-energy bombardment through the interface [14] and post-annealing [11] are perhaps the most effective approaches.

As far as the internal stress is concerned, it is now well established that many deposition processes for hard coatings may leave high compressive stresses in the film side of the interface. As explained in Section 3.1 of this study, the TiN coating without a titanium interlayer experienced a serious internal stress problem. It is interesting to note that the hardness of 440C steel substrate was significantly reduced after the 500 °C deposition, while the rest of three substrates, *i.e.* 440C (100 °C) and M50 (100 and 500 °C) still possess the hardnesses similar to their originals (around 800 kg $mm^{-2}$). It is suspected that the reduction of substrate hardness may help to relax the stresses at the interface and further to retain the TiN coating of the 440C steel substrate from delamination.

Generally, the compression in the film side and the tension in the substrate side results in a tensile component normal to the film–substrate interface [15]. Figure 1b illustrates that there is indeed a tensile stress normal to the interface which causes the delamination of the coating. On the contrary, the internal stress was not a problem for samples deposited at the same substrate temperature, 500 °C, with a titanium interlayer. Consequently, one of the significant benefits of the titanium interlayer was that it dramatically reduced the internal stresses.

The existence of the titanium interlayer eliminates the direct contact of TiN and the steel substrate, but on the other hand two interfaces are created (*i.e.* substrate–titanium and titanium–TiN). Obviously, the interface with a weaker adhesive strength determines the interfacial adhesion of those samples. Considering the case of the as-deposited TiN ($T_{sub} = 100$ °C) coating, the interface between substrate–titanium interlayer apparently has weaker strength due to the presence of the small gap shown in Figs. 3 and 4. This gap causes poor interface contact and may act as a crack initiator. The cause of this gap remains unclear. However, the entrapment of argon atoms at the outset of the ion plating process and low mobility of arriving adatoms may be the possible causes.

Post-annealing generally has significant effects in reducing thermal stresses which are caused by the large differences in the thermal expansion coefficients between the film and substrate. In addition, it is beneficial for promoting the interfacial adhesion if atoms of substrate and film may interdiffuse through the interface during the annealing treatment. In this study, the annealing temperature (500 °C) perhaps not high enough to cause significant interdiffusion between titanium and substrate materials. However, it is high enough to draw carbon atoms, which are abundant (0.85 wt.%) in the steel matrix and very mobile at this temperature, to diffuse through this interface. Evidently, the highly active titanium atoms in the interlayer act as "sinks" for carbon atoms. The benefits of this carbon diffusion, through the substrate–titanium interlayer, on the adhesive strength presumably are the increase of interface contact and stronger interface bonding by closing the vacancy gap at this interface. Interestingly, this knowledge, that uses the diffusion of interstitial elements (*e.g.* carbon, oxygen) of substrate materials to react with active metals (*e.g.* titanium, niobium, tantalum, tungsten) at the interface region for promoting good adhesion, has been widely applied in the field of making reliable metal–ceramic joints for many years [16, 17].

According to the adhesion results, high substrate temperature seems to have the greatest effect in promoting the adhesive strength and several factors may be responsible for it. Figure 7 illustrates that the gap does not exist in this sample. It is reasonable to postulate that, at high substrate temperatures, the surface diffusivities of atoms and the mobility of arriving adatoms are high enough to achieve a thermodynamically stable interface. Moreover, high population of small TiC grains near the substrate–interlayer interface region also indicates that high nucleation density at this interface may be another factor enhancing the adhesive strength of this interface. However, the interlayer–TiN interface of this sample becomes irregular in shape which is generally believed to make interface crack propagation more difficult [11].

## 5. Conclusions

(1) The strength of scratch adhesion: ($T_{sub} = 500$ °C) $\geqslant$ ($T_{sub} = 100$ °C + $T_{ann} = 500$ °C) $\geqslant$ ($T_{sub} = 300$ °C) > ($T_{sub} = 100$ °C) > ($T_{sub} = 500$ °C, without a titanium interlayer).

(2) A disastrous internal stress problem was experienced for the TiN coating deposited at 500 °C without a titanium interlayer, whereas, the internal stresses were not a problem for the samples deposited with a titanium interlayer.

(3) The size of TiN columns of both samples deposited at 500 °C (with and without a titanium interlayer) was approximately the same, which indicated that the existence of the titanium interlayer did not increase the nucleation density of TiN.

(4) A gap was discovered at the substrate–titanium interface of the as-deposited sample ($T_{sub} = 100$ °C). It caused a poor interface contact and weak chemical bonding at this interface, and consequently weakened its adhesive strength.

(5) High substrate temperature (300 and 500 °C) or post-annealing treatment can eliminate or narrow the gap, and can transform the titanium interlayer to TiC interlayer via the diffusion of carbon atoms through the substrate–titanium interface. Consequently, the adhesion is promoted by better chemical bonding and contact at this interface.

## Acknowledgment

The work was supported by the Tribology Program, Energy Conversion and Utilization Technologies Division, U.S. Department of Energy under Contract W-31-109-ENG-38.

## References


1 D. S. Rickerby and P. J. Burnett, *Thin Solid Films, 157* (1988) 195.
2 J.-E. Sundgren, *Thin Solid Films, 128* (1985) 21.
3 J.-E. Sundgren, A. Rockett and J. E. Green, *J. Vac. Sci. Technol., A4* (1986) 2770.
4 D. M. Mattox, in R. F. Bunshah (ed.), *Deposition Technologies for Films and Coatings*, Noyes Publications, Park Ridge, NJ, 1982, p. 63.
5 R. L. Hatschek, *Am. Mach., 127* (1983) 129.
6 S. Vuorinen, E. Niemi and A. S. Korhonen, *J. Vac. Sci. Technol., A3* (1985) 2445.
7 J. A. Thornton, in R. F. Bunshah (ed.), *Deposition Technologies for Films and Coatings*, Noyes Publications, Park Ridge, NJ, 1982, p. 231.
8 R. F. Bunshah, in R. F. Bunshah (ed.), *Deposition Technologies for Films and Coatings*, Noyes Publications, Park Ridge, NJ, 1982, p. 139.
9 J. W. Edington, *Interpretation of Transmission Electron Micrographs*, MacMillan, London, 1975, p. 65.
10 A. Erdemir and C. C. Cheng, Paper presented at the *International Conference on Metallurgical Coatings*, San Diego, CA, 17–21 April 1989.
11 D. M. Mattox, in K. L. Mittal (ed.), *Adhesion Measurements of Thin Films, Thick Films and Bulk Coatings, ASTM STP 640,* ASTM, PA, 1978, p. 54.
12 P. A. Steinmann, Y. Tardy and H. E. Hintermann, *Thin Solid Films, 154* (1987) 333.
13 M. K. Hibbs, B. O. Johansson, J.-E. Sundgren and U. Helmerson, *Thin Solid Films, 122* (1984) 115.
14 J. E. E. Baglin, *Mater. Sci. Eng., B1* (1988) 1.
15 D. S. Rickerby, G. Eckold, K. T. Scott and I. M. Buckley-Golder, *Thin Solid Films, 154* (1987) 125.
16 C. I. Helgesson, *Ceramic to Metal Bonding*, Boston Technical Publishers, Cambridge, MA, 1968.
17 J. T. Klomp, in J. A. Pask and A. G. Evans (ed.), *Ceramic Microstructures '86,* Plenum, NY, 1988, p. 307.

*Surface and Coatings Technology*, *39/40* (1989) 377 - 385 

# REFLECTION–DIFFRACTION X-RAY DIFFRACTION DEPTH PROFILING OF REACTIVELY SPUTTERED HIGHLY ORIENTED TiN ON MgO

C. S. HELRICH, R. C. BUSCHERT and A. J. KROPF

*Turner Laboratory, Goshen College, Goshen, IN 46526 (U.S.A.)*

C. ERNBERGER and T. SMITH

*CTS Corporation, 905 West Boulevard North, Elkhart, IN 46514 (U.S.A.)*

(Received March 16, 1989)

## Summary

Graze angle reflection–diffraction is a new X-ray diffraction (XRD) technique with the surface sensitivity required for XRD depth profiling. We have constructed a graze angle diffractometer allowing control of graze angle to within $\pm 0.012$ mrad ( $\pm 0.0006$ °). This device has been used to study a very highly oriented TiN film reactively sputtered onto (100) MgO. The sample was of interest because it was deposited under conditions giving low resistivity *and* low stress (around 1.4 GPa). The diffractometer successfully obtained diffraction data from the surface passive layer on TiN. The passive layer has been identified as an oxynitride. In addition, we have identified two types of crystallites in the film: epitaxial grains under tensile stress and a certain volume fraction of rotated and highly compressively stressed grains.

## 1. Introduction

X-ray diffraction (XRD) is a powerful technique at the disposal of the materials scientist. As long as the material is not amorphous, information can be obtained regarding grain size, phase distribution, preferred orientation, texture and, from lattice parameter variation, strains from which micro- and macro-stresses may be deduced. Thin films, especially highly textured ones, present problems for the usual Bragg–Brentano diffractometer geometry. A recent paper by Flynn and Waychunas describes some of these problems and a diffractometer they have built to improve the situation [1].

It is well known from destructive techniques such as transmission electron microscopy (TEM), fracture sections, Auger and X-ray photoelectron spectroscopy (XPS) depth profiling etc., that grain size, phase distribution, stress and texture in physical vapor deposition (PVD) films vary from the interface to the film surface. In addition to 'bulk' analysis of thin films, it

0257-8972/89/$3.50 

would be of great benefit to be able to depth profile thin films by XRD. Graze angle diffraction with graze angles in the low milliradian range is capable of the surface sensitivity required for the study of passive films that dictate the behavior of friction, wear, contact resistance, wettability, and adhesion in thin film PVD applications ranging from machining to microelectronics. In this paper we report on a preliminary investigation of utilizing the reflected diffraction wave recently discussed by Afanes'ev [2] and Golovin *et al.* [3] and first reported by Marra *et al.* [4] for X-ray depth profiling. The reflection–diffraction technique can yield information from the top few tens of angstroms of the film surface to the substrate interface. Diffracting planes normal to the film plane may be studied to gain a direct measure of the inplane compressive stress as a function of depth without relying on uncertain poisson ratios in highly textured films. In this paper we report on a diffractometer we have constructed to permit the study of the reflected diffracted wave obtained in graze angle diffraction. We report preliminary results on a highly textured 3000 Å TiN film reactively sputtered onto (100) single crystal MgO. MgO was chosen as a substrate because of the similarity of its crystal properties to those of TiN, *i.e.* lattice parameters of 0.4240 nm (TiN) and 0.4213 nm (MgO), thermal expansion coefficients of $9.5 \times 10^{-6}\ K^{-1}$ (TiN) and $14 \times 10^{-6}\ K^{-1}$ (MgO) and identical sodium chloride crystal structures. The sample was of interest because of its very low bulk resistivity (22.17 $\mu\Omega$ cm) compared to the lowest bulk resistivity on record for TiN (18.5 $\mu\Omega$ cm) for single crystal TiN deposited at high temperature on MgO by Johansson *et al.* [5]. Using similar deposition conditions for the MgO sample studied in the present work, we were able to obtain bulk resistivities as low as 21.17 $\mu\Omega$ cm on common microscope slides. Stress levels (as determined by curvature measurements on titanium and 430 stainless coupons) of less than 1.5 GPa have also been obtained. Low stress, low resistivity films have heretofore been difficult or impossible to obtain [6 - 8]. Control of stress and resistivity is necessary if PVD coatings are to be used to their full advantage. We attribute the low resistivity and stress results to our unique magnetron design which allows high power at low voltage along with a large flux of low energy ion bombardment. This paper will concentrate on the reflection–diffraction technique. Results indicate that reflection–diffraction depth profiling is possible, but improvements in signal-to-noise via high flux X-ray sources and more sensitive detectors are required before the technique can be practical.

## 2. Experimental procedure

The XRD geometry is shown schematically in Fig. 1. The diffractometer was built on a standard GEXRD base, and the X-ray tube and horizontal location of the detector correspond to those of the GEXRD. Rotation and axial positioning of the sample are accomplished by a standard goniometer and micrometer. This device provided variation of the graze angle to within

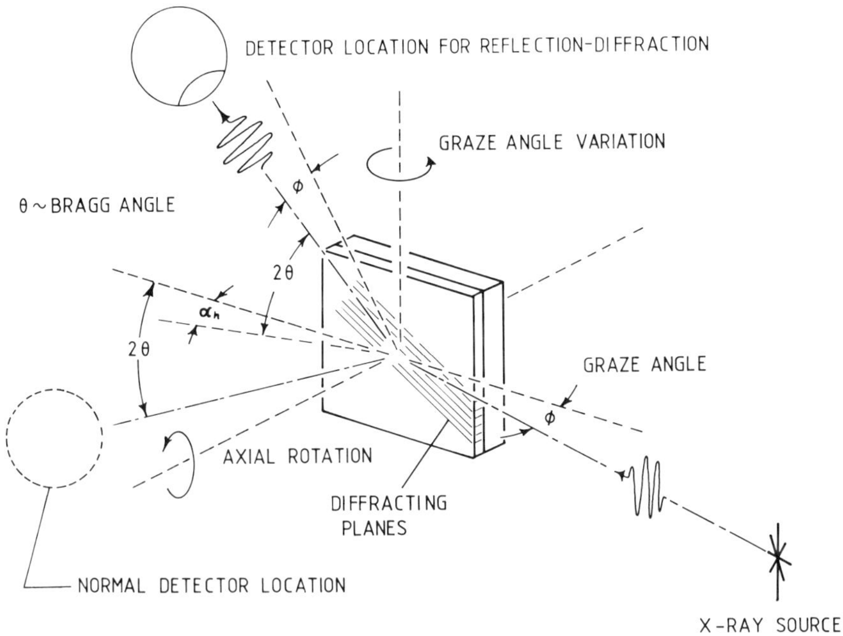


Fig. 1. Schemetic drawing of thin film diffractometer (TFD). Detector location for standard Bragg–Brentano mode is indicated by dashed lines. Vertical detector location is that for reflection–diffraction. Sample may be rotated about and translated along the horizontal axis. Sample rotation about the vertical axis fixes graze angle Φ.

$\pm 0.012$ mrad ( $\pm 0.0006°$) and detector rotation in the vertical as well as the horizontal planes. The beam collimator is of special design providing a constant beam centerline and beam width adjustment to within $\pm 3\ \mu$m at the sample location. Location of the detector in both horizontal and vertical planes is accomplished by stepper motors. The graze angle is also stepper motor controlled.

Figure 2 shows a detail of the surface of the film and the incident X-ray beam. The beam graze angle is Φ. The diffracting planes are shown by lines in the film surface and in the cross-section of the film. These are perpendicular to the film surface. In addition to the incident beam, there are four additional beams: (a) the reflected beam at angle Φ, (b) the penetrating beam, which is an extension of the incident beam, (c) the internal diffracted beam and (d) the reflected–diffracted beam at angle Φ. These are indicated in Fig. 2 by the corresponding lower case letters. Beams (b) and (c) cannot be detected, since they penetrate the film. Beam (a) is of no interest. Beam (d) is the beam which has undergone diffraction with the planes being studied and leaves the surface at an angle which, for exact Bragg diffraction, is equal to the graze angle [2]. By varying the graze angle Φ the penetration depth of the beam into the film can be varied. In this way the structure of the film as well as that of the interface and the substrate can be studied selectively. In the

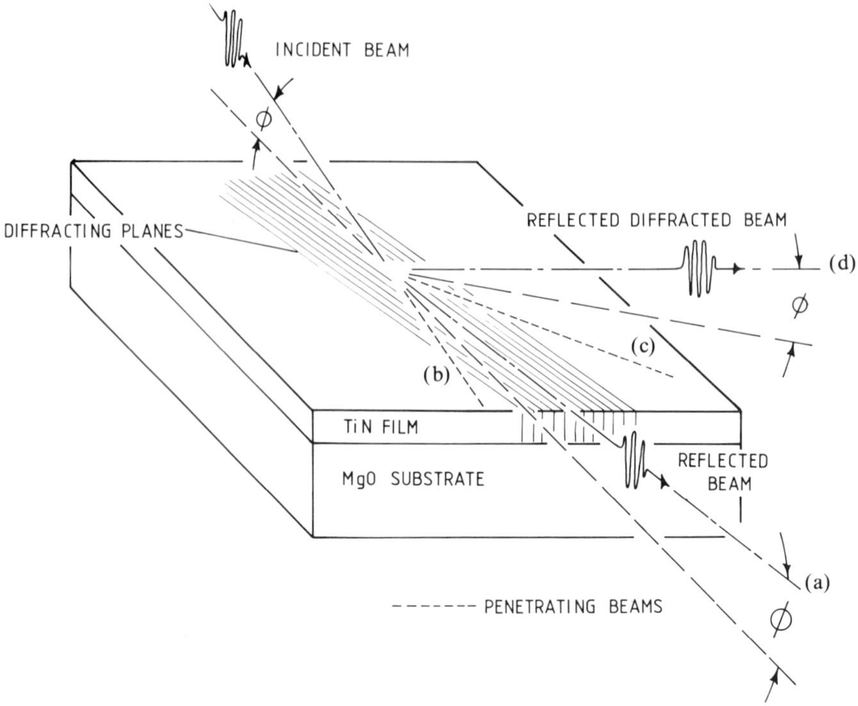


Fig. 2. Schemetic drawing of the beams present in graze angle diffraction. These are described in the text.

studies reported here all data except those resulting from diffraction from the edge of the sample are results of measurements involving this reflected diffracted beam (d). Data were gathered digitally using an IBM PC-based system. All software was written in-house. Iron radiation was used in all measurements reported here.

The films discussed in this paper were deposited by reactive direct current sputtering in a system having a base pressure of less than $5 \times 10^{-9}$ Torr. Partial pressure control during deposition was achieved by a closed loop partial pressure control system described by Sproul and Tomashek [9]. A high purity titanium target was installed in a unique magnetron of CTS Corporation design that allows high rate deposition at pressures in the low $10^{-4}$ Torr range. Target voltage is low (less than 300 V) and the magnetron allows high flux (greater than 1 mA cm$^{-2}$) of low energy (less than 25 V) ion bombardment during deposition. The substrate was radiantly heated to approximately 250 °C and r.f. etched for 4 min at a forward power of 250 W and a pressure of 5 m Torr in ultrahigh purity argon prior to deposition. The substrate-to-target distance was 10 cm, and the samples were scanned $\pm 3.5$ cm either side of the target centerline during deposition. Radiofrequency biasing to the desired direct current sample voltage was utilized.

Film thickness was determined by profilometry. Bulk resistivity measurements were accomplished with a four-point probe having a probe spacing of 0.050 in.

## 3. Results

The Fink index lists the strongest diffraction peak as (002) in both TiN and MgO. These planes were also perpendicular to the surface of the sample studied and were, therefore, ideal for reflection–diffraction studies. These (002) planes could also be used for diffraction from the edge of the sample (*i.e.* $\Phi = 0$) and could be used for alignment of the sample. A typical diffraction pattern obtained with the X-ray beam incident on the edge of the sample is presented in Fig. 3. In this pattern, four peaks, which we shall designate as A, B, C and D are visible. This same cluster of peaks was obtained upon rotating the sample 180 ° about the (100) axis. This diffracted radiation was only visible within a small solid angle centered on $2\theta$ for the (002) planes of MgO and TiN (54.3 - 55.8 ° for Fe $K\alpha$ radiation). By scanning the X-ray beam over the sample edge, it was easy to identify the peak C as that obtaining from the MgO substrate. In what follows we shall show that peaks A and B arise from TiN crystallites and their oxides epitaxial with the MgO substrate. Peak D, we conclude, arises from the (002) planes of a set of TiN crystallites distinct in their D spacing from those producing peak A.

To determine that the peaks A, B, and D arise from crystallites located

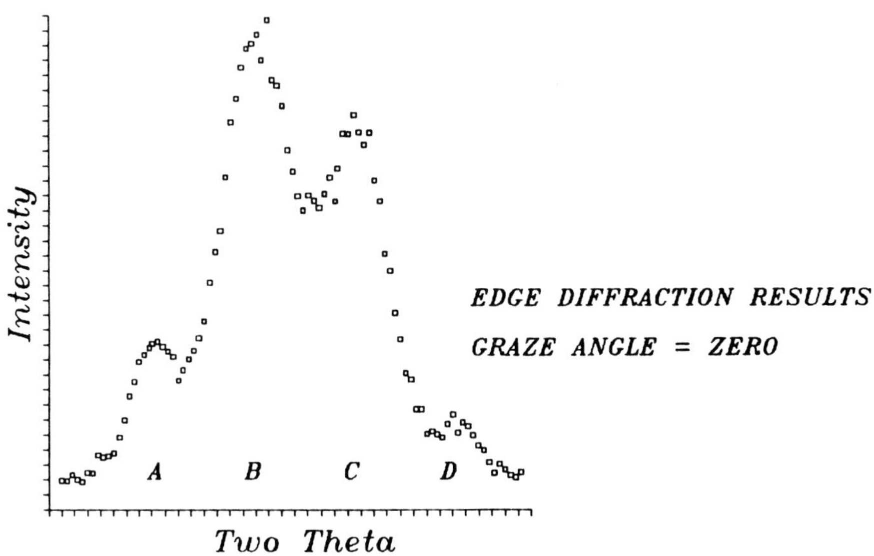


Fig. 3. Bragg scan with beam incident on sample edge ($\Phi = \theta$). Sample rotation such that all peaks are visible in single scan, *i.e.* the Bragg condition is met exactly for none of the peaks. The true Bragg angles are A = 54.01, B = 54.47, C = 54.81, D = 55.21.

in the TiN film or in the vicinity of the film–substrate interface, two separate measurements were performed. In the first, the sample was extracted from the beam in steps of 3 $\mu$m using the axis micrometer and the relative intensities of the peaks studied. It was observed that peaks A and B persisted with a constant ratio of intensities while peak C decreased and was finally lost when the beam intercepted the film and almost no substrate. In the second measurement, the beam centerline was placed within 40 $\mu$m of the film surface and the beam width was varied from 300 to 125 $\mu$m with the sample oriented so that peaks C and D were simultaneously clearly visible. Here the beam intercepted the film and a decreasing amount of the substrate. As the beam width decreased the peak C decreased in intensity while D remained relatively constant.

To determine more precisely the location and identity of crystallites producing the peaks A, B, and D, and to demonstrate depth profiling capability, measurements of the reflected diffracted beam were undertaken for values of the graze angle between 4.07 and 48.31 mrad. This encompassed the region between the surface of the film and a depth of 3000 Å into the film. Estimates of the depth of beam penetration are based on measurements of the specularly reflected beam (beam (a) in Fig. 2) and geometrical optics. Figure 4 presents the results of such measurements. At a graze angle of greater than 4 mrad the

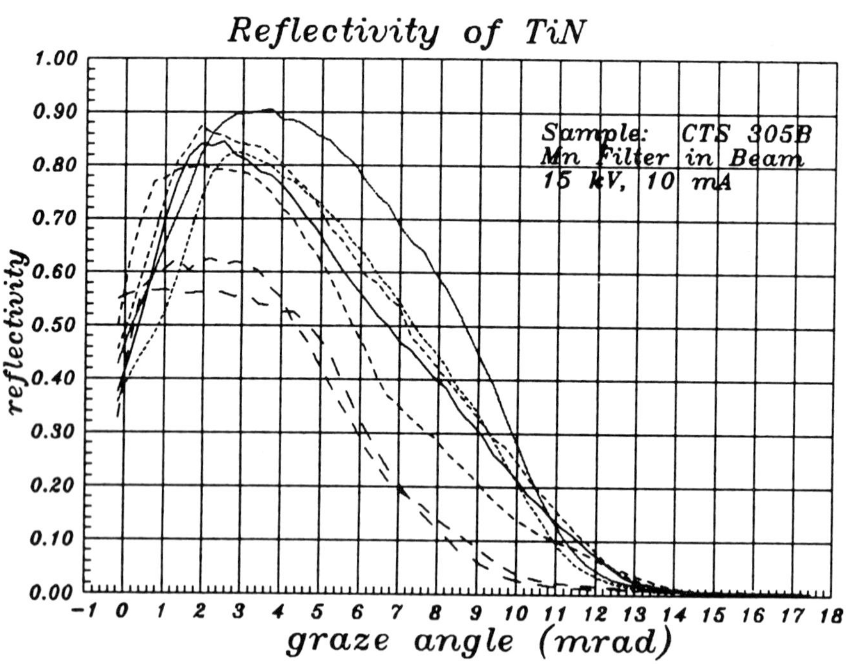


Fig. 4. Total reflection curves for TiN. These are measurements of the (normalized) total intensity of the specularly reflected beam (beam (a) in Fig. 2) as a function of graze angle $\Phi$ for various collimator settings. To be noted is that the penetration of the X-rays into the sample begins at approximately 4 mrad.

beam begins to penetrate the film. The value of 4 mrad was used to determine the index of refraction which, together with tabulated data for the absorption coefficient, formed the basis for the optical calculation. The depth of penetration was defined as the point at which the beam intensity had attenuated to $e^{-1}$ of its original value.

The results of these reflection–diffraction measurements are presented in Fig. 5. Each plot is a $2\theta$ Bragg scan. The graze angle and depth of penetration are presented beside each Bragg scan. Points to be noted in these Bragg scans are: (i) only peak B is present in scans for which the penetration is very shallow and only the surface planes contribute to the diffraction, (ii) as penetration increases and diffraction is increasingly dominated from planes in the central portion of the film, peaks A and D, whose presence is lacking in scans at the surface, increase in intensity and (iii) peak C, which has already been identified as resulting from diffraction from the MgO substrate, is first visible at a depth of penetration of 3000 Å.

Observation (iii) corroborates the profilometer film thickness measurements of 3000 Å. Observation (i) indicates that only the crystallites responsible for peak B are present on the surface. These crystallites we believe to be a titanium oxynitride having the same NaCl structure but with oxygen replacing some of the nitrogen in TiN. To what depth these crystallites are

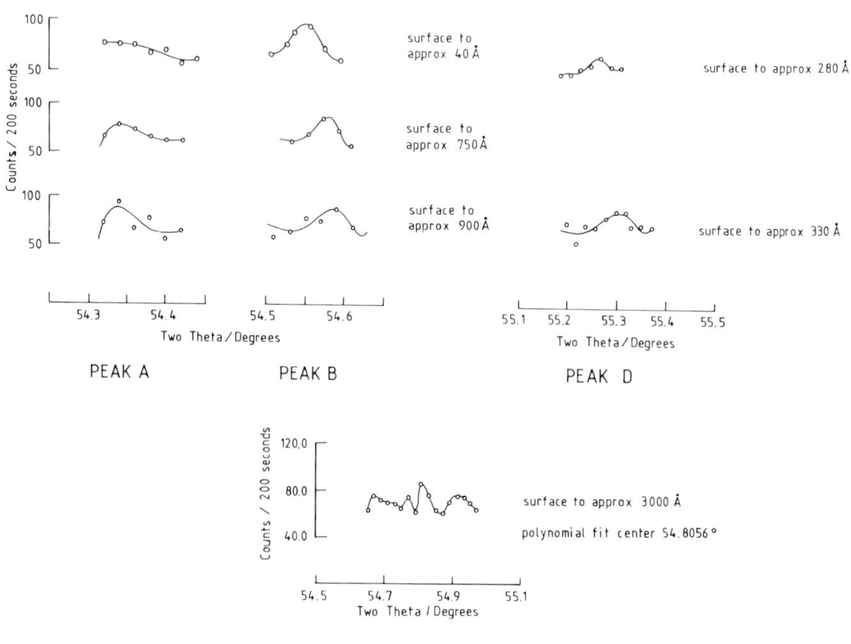


Fig. 5. Reflection–diffraction Bragg scans for the peaks A, B, C and D. The profiles for A, B, and D are least square polynomial fits and that for C is a cubic spline in which all points are included. In each case the sample was first rotated to satisfy very nearly the Bragg condition for the peak of interest.

present cannot be deduced from the present data, since diffraction from a surface layer of any thickness would appear at each graze angle. The remaining peaks A and D are the result from planes in the bulk of the film, which we assert must then arise from TiN crystallites. A thorough survey of planes in the TiN lattice shows that (002) reflections are the only possible candidates for the measured Bragg angle. Neither diffraction from $K\beta$ nor $K\alpha_1$–$K\alpha_2$ splitting accounts for the observed differences in Bragg angle for peaks A and D ($2\theta = 54.01$ for A and 55.21 for D). Because the lattice parameter for peak A (4.27 Å) is greater than that for MgO and because peak A is more intense than peak D, we tentatively assign peak A to (002) planes epitaxial (zero axial rotation) with the substrate. Beyond the contention that peak D arises from TiN because of its absence from the surface, an exact identity of the responsible crystallites is not possible. We return to this point in Section 4.

## 4. Discussion

Graze angle reflection–diffraction studies of a 3000 Å TiN film reactively sputtered onto the (100) face of single crystal MgO have been made. Depth profiling has been accomplished by varying the graze angle.

At the surface we detect the presence of a TiN phase epitaxial with peak A but having distinctly lower d spacings. This implies that a titanium oxynitride to TiN gradient is not present. Rather there is a distinct change from oxynitride crystallites to nitride. To what depth these oxynitride crystallites are present cannot be determined from the experiments reported here. Even a thin layer will provide a diffraction peak at all graze angles. However, one of us (CNE) has previously characterized the surface passive layer as an oxynitride as opposed to an oxide based on XPS binding energy assessments [10, 11]. The present results lend further support to the oxynitride model.

The origin of the enigmatic peak D remains speculatory. The absence of this peak from the surface diffraction data indicates that TiN crystallites in the bulk of the film are its source. However, this identification makes the considerable difference in lattice parameter from that of the A crystallites (4.18 Å for D and 4.27 Å for A) difficult to understand. The accuracy with which we can measure deviations in the axial rotation angle does not permit a definitive statement regarding orientation of these D crystallites. Presumably because of the small volume fraction of these crystallites, it has been very difficult to perform reflection diffraction measurements of them and, therefore, a definitive statement regarding their exact location is accordingly difficult.

The authors believe that clear difference in $d$-spacings must result from one of two reasons: composition (broadly defined as impurity, stoichiometry, and distribution of point defects) modulation and/or rotation of otherwise compositionally homogeneous crystallites, causing lattice strains. Unfortunately, the present measurements alone cannot distinguish small rotations and, of course, cannot speak to the composition modulation issue. The origin of peak D remains enigmatic.

In summary, reflection–diffraction XRD has been shown to have the surface sensitivity required for the study of the top few surface layers comprising passive films. Using the technique on highly oriented TiN on MgO, we have determined the passive layer to be an oxynitride as opposed to an oxide. In addition, an unexpected result was obtained in that two sets of distinctly different grains were observed in the bulk of the film.

To date both theoretical and experimental efforts involving the reflection–diffraction technique have concentrated on studies of single crystals. To our knowledge the present work is the first attempt to use this technique to study oriented polycrystalline films.

## References


1 P. A. Flinn and G. A. Waychunas, *J. Vac. Sci. Technol., B6 (6)* (1988) 1749.

2 A. M. Afanas'ev and M. K. Melkonyan, *Acta Crystallogr., Sect. A, 39* (1983) 207 - 210.

3 A. L. Golovin, R. M. Imamov and S. A. Stepanov, *Acta Crystallogr., Sect. A, 40* (1984) 225 - 228.

4 W. C. Marra, P. Eisenberger, A. Y. Cho, *J. Appl. Phy., 50* (1979) 6927.

5 B. O. Johansson, J. E. Sundgren, J. E. Greene, A. Rockett and S. A. Barnett, *J. Vac. Sci. Technol., A3 (2)* (1985) 303.

6 D. M. Mattox, R. E. Cuthrell, C. R. Peeples, P.L. Dreike, *Surf. Coat. Technol., 36* (1988) 117.

7 H. P. Kattelus, J. L. Tandon, C. Sala and M. A.Nicolet, *J. Vac. Sci. Technol., A4 (4)* (1986) 1850.

8 C. N. Ernsberger, A. J. Perry, L. P. Lehman, A. E. Miller, A. R. Pelton and B. W. Dabrowski, *Surf. Coat. Technol., 36* (1988) 605.

9 W. D. Sproul and J. A. Tomashek, *U.S. patent 4,428,811* (1984).

10 C. Ernsberger, J. Nickerson, T. Smith, A. E. Miller and D. Banks, *J. Vac. Sci. Technol., A4 (6)* (1986) 2784.

11 C. Ernsberger, J. Nickerson, A. E. Miller and J. Moulder, *J. Vac. Sci. Technol., A3 (6)* (1985) 2415.

# THE RESIDUAL STRESS IN TiN FILMS DEPOSITED ONTO CEMENTED CARBIDE BY HIGH-RATE REACTIVE SPUTTERING

A. J. PERRY and M. JAGNER

*GTE Valenite Corporation, 1711 Thunderbird, Troy, MI 48084 (U.S.A.)*

W. D. SPROUL and P. J. RUDNIK

*Basic Industry Research Laboratory, Northwestern University, 1801 Maple Avenue, Evanston, IL 60201 (U.S.A.)*

(Received March 8, 1989)

## Summary

The lattice parameters, lattice strain distribution and residual stress on different planes in TiN films deposited onto cemented carbide by reactive sputtering are studied by X-ray diffraction methods as a function of sputter target input power at a constant bias voltage of $-100$ V. The residual stress shows remarkably large differences between the values measured on different planes in a given film. The form of the various $\sin^2\psi$ *vs.* lattice parameter plots (sspp) also shows differences and can, for example, be linear for the (422) peak but shows $\psi$ splitting for the (220) peak. These results together with differences in the behavior of the lattice parameters and diffraction peak broadening allow two distinct types of behaviors to be isolated. They are not understood but are conjectured as being related to plastic flow, lattice defects and possibly even ultramicrocracking on specific lattice planes.

## 1. Introduction

Residual stress measurements are now carried out routinely on thin films, of for example TiN and ZrN, using the X-ray diffraction (XRD) $\sin^2\psi$ method [1 - 8]. The plots of lattice parameter *vs.* $\sin^2\psi$ (sspp) are generally determined for a high $2\theta$ diffraction peak such as the (422) and are frequently linear so that a residual stress value can be derived.

As usual, the XRD diagram itself allows the lattice parameters to be derived for the individual planes and be plotted out as a function of the Nelson–Riley or the $3\Gamma$ crystallographic parameter. The widths of the diffraction peaks can be used to determine a value of the strain distribution from a Williamson–Hall plot if the lattice is not strained in a uniform way [6, 7].

Some studies have indicated that deviations from linear Nelson–Riley plots can occur, specifically that the {111} family of planes in TiN can have lattice parameters greater than the average in some samples [6 - 9] or less

0257-8972/89/$3.50 

than average in others [10] and in Ti(CN) also [11]. In addition the {200} family has been observed with less than average lattice parameters in films of both TiN and ZrN [10, 12]. Further, the sspp measured on different planes [12] of a ZrN film (only under a differential thermal contraction stress) are not self consistent in that $\psi$ splitting is found on both (400) and (222) planes but not on high index planes which have linear sspp.

The present work reports the first part of a study of films made by high-rate reactive planar magnetron sputtering. Specifically, the input power to the magnetron is taken as the variable, which affects the deposition rate, with the bias voltage on the substrate maintained constant.

## 2. Experimental procedure

The present films had a thickness of about 5 $\mu$m and were made following standard methods as discussed elsewhere [13]. They were deposited with a magnetron power which was stepped through the range of 2.3 - 10 kW. The substrate bias was held constant at $-100$ V. The texture, average deposition rate and composition as determined by Auger electron spectroscopy methods are given in Fig. 1.

The substrate chosen was a C-3 cemented carbide (composition WC–3% Co–0.3% TaC). The advantage is that the thermal expansion coefficient is

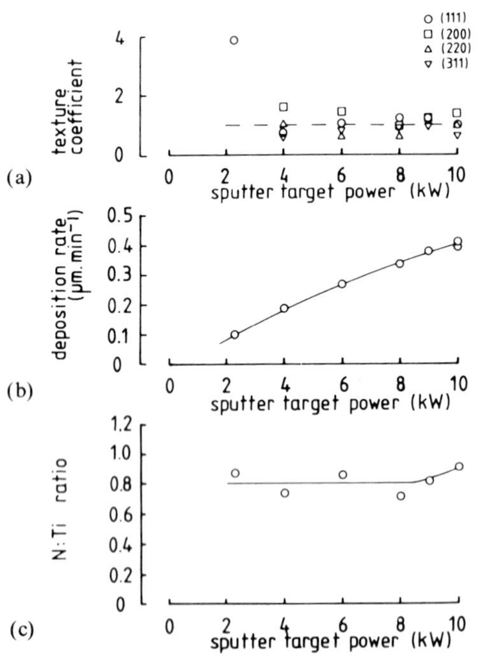


Fig. 1. Texture coefficients, referred to the four peaks indicated (a), average deposition rate (b) and nitrogen to titanium ratio of the present films as a function of the magnetron power input level (c).

about $3 \times 10^{-6}\,°C^{-1}$. This is lower than that of TiN, so that any compressive residual stresses found in the films must be caused by the deposition process. The disadvantage is that diffraction peaks from the WC in the substrate overlap with some of the peaks from the TiN film so that the latter have to be deconvoluted. At higher sputter target powers, the diffraction peaks at high $2\theta$ values have very low intensity and are very broad. The error in the deconvolution is potentially large. In addition it is possible that the peaks are not gaussian, as was found in a previous study [12]. A further consequence of the diffraction peak overlap is that only the (220), (400) and (333)–(511) peaks could be used for residual stress measurements, together with the (422) peak at the lower target power levels where it had not yet been broadened enough to overalp with a WC substrate diffraction peak.

## 3. Results

Typical Williamson–Hall and Nelson–Riley plots are shown in Fig. 2. For convenience in the present discussion, full straight lines have been drawn through all the data points from the individual samples using the method of least squares. This allows the differences to be determined between the measured points and the interpolated values. These differences are

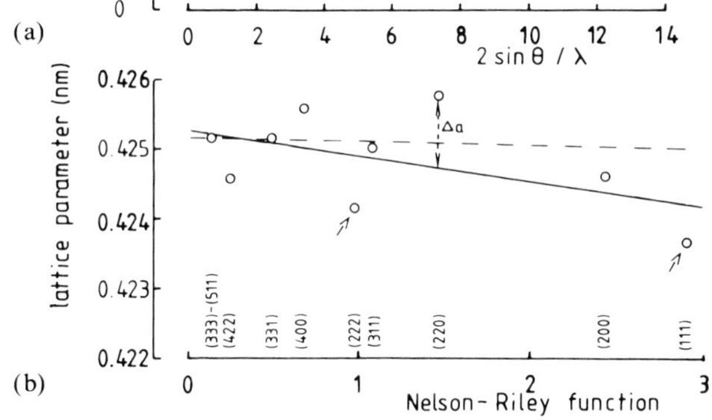


Fig. 2. Williamson–Hall plot showing the difference Δβ between the measured broadening and the interpolated broadening for a typical plane as discussed in the text (a) and Nelson–Riley plot of the lattice parameters showing the difference Δα between the measured and the interpolated values (full curve) for a typical plane as discussed in the text (b). The dashed curve shows the Nelson–Riley extrapolation if the (111) and (222) data (arrowed) are omitted. Both sets of data are taken from the film made with a sputter target power input of 6 kW.

termed $\Delta\beta$ and $\Delta\alpha$ respectively and are indicated in Figs. 2a and 2b respectively for the specific peaks shown. The resulting differences were then considered in terms of behavioral trends for all the peaks studied. It was found that certain peaks showed differences which fell regularly outside the $\pm\sigma$ probability limits for each sample. These were the (111), (220), (222), (400) and (333)–(511) diffraction peaks. The results for these peaks are given in Figs. 3a and 3b respectively together with those from the (311) which is typical of the remaining peaks which showed no such behavioral trends.

The lattice parameter data in Fig. 2b indicate that the values for the (111) and (222) planes (arrowed) are low, as found in other studies [10, 11]. A method of least squares straight line drawn through the remaining data is horizontal (dashed in Fig. 2b) indicating that no stress gradient is present through the film.

The residual stress studies returned sspp which were frequently non-linear. Typical results are shown in Figs. 4a and 4b. As shown, the sspp from the

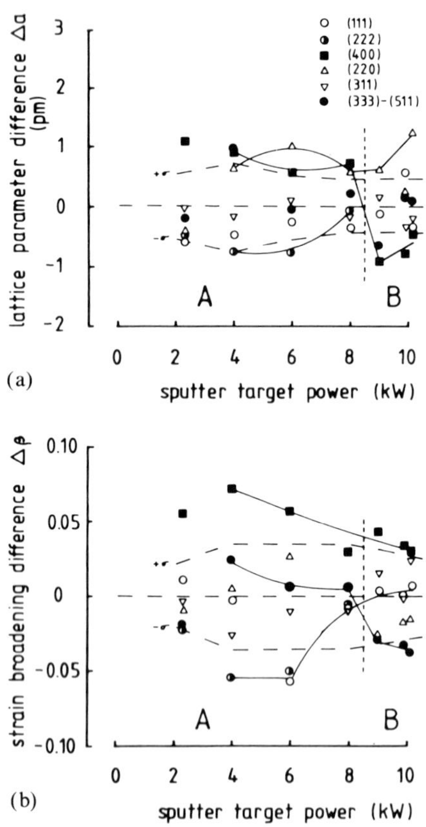


Fig. 3. Lattice parameter differences $\Delta\alpha$ (a) and peak broadening difference $\Delta\beta$ as a function of sputter target power (b). The differences are between the measured values and values interpolated by a method of least squares straight line fit to all the data for a given sample. The corresponding $\pm\sigma$ probabilities are included.

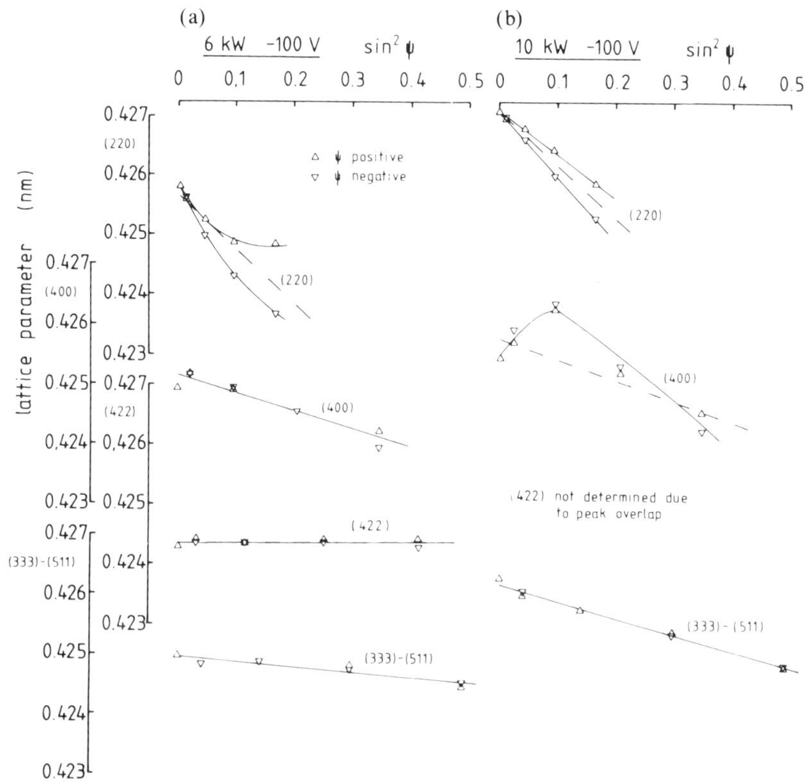


Fig. 4. Typical $\sin^2\psi$ plots (sspp) from two of the present samples indicating the two types of behavior which was found. The (422) diffraction peak could not be used at a sputter target power above 8 kW because peak broadening led to overlap with a WC substrate XRD peak.

(220) peak always showed $\psi$ splitting whereas those from the (422) and (333)–(511) peaks were always linear. The (422) peak could only be studied up to a target power of 8 kW because the peaks in films made at higher powers were so broad that they overlapped a WC substrate diffraction peak as mentioned above. A difference in behavior was found in the (400) peak the sspp of which was linear at lower target powers but markedly curved at higher powers. To allow the data sets to be compared, a straight line was fitted to all sspp data using the method of least squares, as indicated in Fig. 4. A residual stress value and an intercept lattice parameter value was derived in each case, assuming values of 640 GPa, 0.20 and 0.42386 nm for the Young's modulus, Poisson's ratio and the equilibrium lattice parameter respectively. The results are given in Fig. 5 following the convention that data from sspp which are non-linear (and thus not of quantitative significance) are enclosed in parentheses if the sspp is curved and square brackets if it is $\psi$ split.

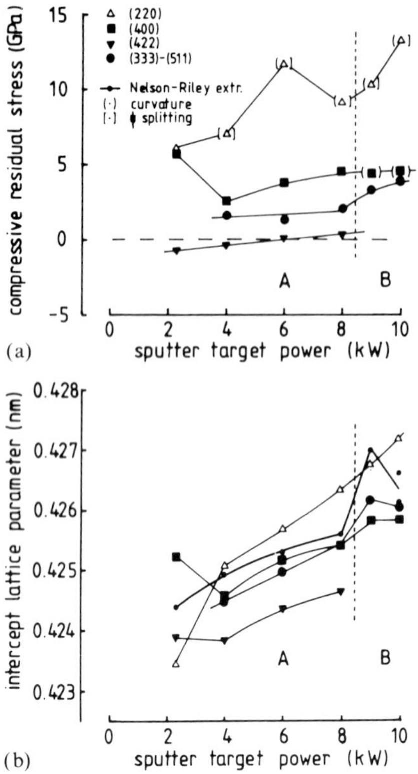


Fig. 5. Residual stress results (a) and intercept lattice parameters as a function of sputter target power for the present sample (b). The residual stress data enclosed in parentheses and square brackets refer to curved and $\psi$ split sspp respectively and are thus not quantitative.

## 4. Discussion

The most remarkable results in the present work are the very marked differences in the level of residual stress measured on the different lattice planes in a given film (even when they have linear sspp) as indicated in Fig. 5a. Indeed, the (422) planes in some samples indicate that a tensile residual stress is operating on them even though the remainder are under high compressive residual stress. Further, some lattice planes, such as the (220), show $\psi$ splitting or curvature of the sspp, while the remainder of the sspp remains linear in any given film. As TiN is nearly elastically isotropic [8, 11], such differences cannot be due to differences in the X-ray elastic constants for the different planes. Stress anisotropy might also be suspected in reactively sputtered films. It has been shown that ZrN films made in the same equipment as the present series and under similar conditions showed no such effect [12]. The differences have to be attributed to the crystallographic condition of the film caused by neutral and ion bombardment during the deposition process. The effects are so large, however, that the following discussion has to be regarded as purely tentative, indeed, speculative, pending more detailed studies which need to be carried out using complementary methods such as transmission electron microscopy, and positron annihilation, etc., to classify the lattice defects which are present.

Study of the data on the lattice parameter and peak broadening differences (Figs. 3a and 3b) and the compressive residual stress (Fig. 5a) together with the behavior of the sspp, allow their separation into two distinct types of behavioral pattern, termed A and B respectively here. These are as follows.

Type A is found at target powers of 8 kW and below and is characterized by lower deposition rates, and linear sspp on all planes, except for the (220) plane which showed both $\psi$ splitting and curvature. The (220) planes also have high lattice parameters but only an average level of strain broadening of the diffraction peaks. In contrast, the (111) and (222) planes show low lattice parameters (*i.e.* negative $\Delta\alpha$ values) and lower than average strain broadening. This behavior is not seen, however, with the (333)–(511) peak, and it is possible that stacking faults [14] on the (111) family of planes are causing this difference in behavior. The lattice parameters, as well as the strain broadening of the (400) planes, are high.

Type B behavior is found at the higher sputter target power of 9 kW and above. In addition to the marked curvature of the sspp of the (400) plane observed in these films, its measured lattice parameters are now very low, *i.e.* $\Delta\alpha$ is negative in contrast to Type A behavior, although the high strain broadening remains. The behavior of the (111) and (333)–(511) planes has also changed (the (222) peak could not be deconvoluted reliably from an overlapping WC substrate peak in these films) because the former has now only average values of both $\Delta\alpha$ and $\Delta\beta$, whereas the latter has a negative $\Delta\beta$, *i.e.* narrow diffraction peaks.

As the power supplied to the sputtering target is raised, the lattice of the TiN film expands and the residual stress level increases. Large differences in the lattice parameters and strain broadening are present in all films, and the (220) planes show a condition of low stress in all samples. At the higher powers, there is a change in the condition of the film which is indicated by the change in the behavior of the (400) planes. There are thus three major effects which need to be understood, namely the origin on the $\psi$ splitting on the (220) planes and the marked curvature of the sspp of the (400) planes found in films made at higher target powers, and finally the large residual stress differences found on the different planes in all films.

Very little transmission electron microscopy work has been reported to date which could allow a more complete interpretation of the present observations. Ernsberger *et al.* [15] have found extremely fine microcracking on the (220) planes in a film made at 5.5 kW and $-100$ V bias, *i.e.* under conditions comparable with the present Type A behavior. Helmersson *et al.* [14] report evidence for slip and stacking faults in TiN deposited at 0.14 $\mu$m min$^{-1}$ with a target power of 1.9 kW onto high speed steel. In later work using the same deposition equipment but operated at 1.8 kW with a deposition rate maintained at 0.14 $\mu$m min$^{-1}$, Petrov *et al.* [16] found a transition in behavior: at a bias of $-80$ V and below, grain boundary porosity was found with defects within the grains. At a bias of $-120$ V and above, no grain boundary porosity was seen but a large increase in defect population was found, so great indeed that the defects were tangled. The differences were attributed [14] to the effect of argon ion bombardment in addition to titanium ions during the deposition process. The study covered a range of substrate temperatures including a value of 300 °C which is comparable with the substrate temperature in the current work. No defects associated with the (400) planes have been reported.

The difference in lattice condition in films of Type A and B also appear to affect their physical properties. It was found in earlier work [17] that the color and reflectance of films corresponding to Type B (*i.e.* made at $-100$ V bias and 10 kW target power) are very prone to changing with time and are more sensitive to heat-treatment than is the case with films made at 8 kW target power, *i.e.* Type A.

In view of the behavior of the (220) planes, it is clear that the condition of the residual stress is far from simple biaxial compressive in the plane of the film. As the films are all nearly randomly oriented, the differences in residual stress should not be due to a preferential orientation of planes relative to the plane of the film. It is more plausible that the differences in the measured residual stress values on the different lattice planes reflect their reaction to an anisotropic residual stress field and is a condition resulting from some form of deformation, *e.g.* on the (220) planes.

It is possible that any slip leading to stress relief on the (111) family of planes has led to the observed low $\Delta\alpha$ and $\Delta\beta$ values. The presence of stacking faults would cause the various $\{111\}$ XRD peaks to behave differently, *i.e.* the (111) peak as opposed to the (333)–(511) peak.

In view of the work reported by Ernsberger *et al.* [15], it is tempting to attribute the $\psi$ splitting on the (220) planes to plastic flow and ultramicrocracking which would cause marked peak broadening as the cracking (and concomitant stress relief) would occur only on some of the planes in a crystal leaving a state of high stress on others. This behavior would be in constrast to that of the $\{111\}$ family of planes where plastic deformation and recovery are accompanied by narrowing of the diffraction peaks.

At high sputter target power levels, the major change in behavior is the transition from a linear sspp in the (400) peak to a markedly curved one. This is accompanied by a fall in the measured lattice parameter. It is conceivable that the defect tangles reported by Petrov *et al.* [16] at high bias levels may have their counterpart here and that these, in some way as yet not understood, have an effect on the $\{200\}$ family of planes. It is worth considering that films grown under such conditions might exhibit a stress gradient through the thickness of the film as the grain structure may coarsen with thickness and enhance intergrain stress fields. However, the lack of such curvature in the sspp of the (333)–(511) peak effectively negates such a possibility.

## 5. Conclusion

Films of TiN reactively sputtered onto cemented carbide substrates at high rates with a substrate bias of $-100$ V and a target power in the range of 2.3 - 10 kW, have a high residual stress in the plane of the film, the measured value of which depends very much upon the crystallographic plane chosen for its study. The behavior of the $\sin^2\psi$ plots (sspp) also differs markedly in a given film where the (220) planes always show $\psi$ splitting, even though the remaining planes may show linear sspp but of differing slopes. There is a behavioral change at sputter target powers between 8 and 9 kW below which the (400) planes show linear sspp and above which a markedly curved sspp together with a reduced lattice parameter. These results, as well as the behavior of the lattice parameter and the strain broadening, are conjectured as being associated with plastic flow, lattice defects such as stacking faults and dislocation loops, and possibly even ultramicrocracking. No detailed analysis is possible at the present time.

## References


1 L. Chollet, H. Boving and H. E. Hintermann, *J. Mat. Energy Syst.*, *6* (1985) 293.
2 L. Chollet and A. J. Perry, *Thin Solid Films*, *123* (1985) 223.
3 A. J. Perry and L. Chollet, *J. Vac. Sci. Technol.*, *A4* (1986) 2801.
4 J. Yoshizawa, Z. Kabeya and K. Kamada, *J. Nucl. Mat.*, *122/123* (1984) 1315.
5 H. Suzuki, H. Matsubara, A. Matsuo and K. Shibuki, *J. Jpn. Inst. Metal*, *49* (1986) 773.
6 D. S. Rickerby, *J. Vac. Sci. Technol.*, *A4* (1986) 2809.
7 D. S. Rickerby, B. A. Bellamy and A. M. Jones, *Surf. Eng.*, *3* (1987) 138.
8 T. Hirsch and P. Mayer, *Surf. Coat. Technol.*, *36* (1988) 729.
9 A. J. Perry and J. Schoenes, *Vacuum*, *36* (1986) 149.
10 A. J. Perry, *Thin Solid Films*, *170* (1989) 63.
11 R. Y. Fillit and A. J. Perry, *Surf. Coat. Technol.*, *36* (1988) 647.
12 A. J. Perry and M. Jagner, *Thin Solid Films*, *171* (1989) 197.
13 W. D. Sproul, P. J. Rudnik and M. E. Graham, *Surf. Coat. Technol.*, *39/40* (1989) 355.
14 U. Helmersson, J.-E. Sundgren and J. E. Greene, *J. Vac. Sci. Technol.*, *A4* (1986) 500.
15 C. Ernsberger, A. J. Perry, L. P. Lehman, A. E. Miller, A. R. Pelton and B. W. Dabrowski, *Surf. Coat. Technol.*, *36* (1988) 605.
16 I. Petrov, L. Hultman, U. Helmersson, J.-E. Sundgren and J. E. Greene, *Thin Solid Films*, *169* (1989) 299.
17 A. J. Perry, M. Georgson and W. D. Sproul, *Thin Solid Films*, *157* (1988) 255.

*Surface and Coatings Technology*, *39/40* (1989) 397 - 408 

# HIGH-TEMPERATURE X-RAY DIFFRACTION STUDIES ON PHYSICAL VAPOUR DEPOSITED TiN

D. S. RICKERBY, S. J. BULL, A. M. JONES, F. L. CULLEN and B. A. BELLAMY

*Materials Engineering Centre, Harwell Laboratory, UKAEA, Oxfordshire OX11 0RA (U.K.)*

(Received March 23, 1989)

## Summary

There is an increasing awareness of the importance of the interrelationship between coating microstructure and properties; however, most of the measurements in the literature relate to room temperature properties. High-temperature X-ray diffraction (XRD) studies confirm the importance of microstructure in dictating the properties of physical vapour deposited (PVD) TiN thin films, but in addition illustrate that these changes in coating properties are not usually reversible. Evidence will be presented on the changes in the levels of residual stress and microstrain broadening which arise in PVD TiN films as a function of annealing temperature. These XRD results are discussed in the light of the structural models proposed for PVD TiN thin films.

## 1. Introduction

In the field of surface engineering there is a growing awareness that the interrelationship between the properties of thin films and coating microstructure, through variations in process-type and deposition conditions, is of crucial importance if 'optimum' coatings are to be selected for a whole range of applications which span the microelectronic and engineering industries [1 - 4]. To date, the criteria used for selection, such as hardness, level of coating–substrate adhesion and internal stress, resistivity etc., are based on measurements made at room temperature. However, there is evidence from the published literature which illustrates that many of the properties of physical vapour deposited (PVD) and chemical vapour deposited (CVD) coatings change with operating temperature. For example, in the case of PVD TiN tempering at elevated temperatures (up to 900 °C) results in a reduction in the levels of internal stress and hardness [5 - 7] which are accompanied by changes in the colour of the nitride [8]. Similar reductions in stress and the coefficient of resistivity have also been reported by Ernsberger *et al.* [9] for sputtered TiN films annealed at temperatures as low as 300 °C.

0257-8972/89/$3.50 

Although *in-situ* high-temperature stress measurements have been reported for PVD TiN [9] and CVD WSi and tungsten films [10], the flexure methods used in these investigations only allow the determination of an integrated stress value averaged through the coating. Consequently the changes in coating microstructure which accompany high-temperature annealing could only be determined from post facto analysis. This paper considers TiN coatings deposited by sputter ion plating for a range of substrate bias voltages. By using high-temperature X-ray diffraction (XRD) techniques the changes in coating microstructure which accompany annealing at elevated temperatures are characterized in terms of a variation in peak position and shape of the TiN reflections. The results obtained from these studies are discussed in terms of the structural models which exist for vapour deposited coatings [11 - 13].

## 2. Experimental procedure

### *2.1. Sputter ion plating (SIP)*

The TiN coatings were produced by SIP [14], which is a soft vacuum direct current sputtering process developed at the Harwell Laboratory of the United Kingdom Atomic Energy Authority. Further details of the process can be found in the literature [15]. For this study the TiN coatings were deposited to a thickness of 3 - 5 $\mu$m with a range of substrate bias voltages onto austenitic stainless steel (1.4541 SS) at a constant deposition temperature of 500 °C.

### *2.2. XRD and scanning electron microscopy (SEM)*

The $\sin^2\psi$ and peak broadening measurements were made using a Harwell-designed APEX goniometer which was interfaced with an IBM PC AT control microcomputer using Harwell 6000 series electronics. For high-temperature XRD studies (up to 1000 °C in vacuum or in a controlled gas mixture) a high-pressure/temperature attachment was incorporated onto the APEX goniometer (manufactured by GTP Engineering, Swindon, U.K.) and measurements of peak position and peak shape were carried out using a position sensitive detector (angular range 10° of $2\theta$) using Cu $K\alpha$ radiation. In all cases, high angle diffraction peaks were used to ensure high precision in the X-ray stress measurements and both positive and negative $\psi$ inclinations were used to check for the presence of stress gradients and shear stresses; fuller experimental details regarding the peak fitting programs [16], the computer controlled diffractometer [17] and the X-ray techniques used at Harwell can be found elsewhere [13].

Both fracture cross-sections and the surface topography of as-deposited and annealed coatings were examined by SEM using a Hitachi S520 microscope operating at 20 - 30 keV.

## 3. Lattice distortions in PVD coatings

Before describing the findings of the present study it is important to consider the physical interpretations behind the various lattice distortions which have been identified in PVD deposits, summarized in Fig. 1 [11].

(1) Type I macrostrains can result from two effects: (a) the presence of a stress in the plane of the coating [18 - 20] and (b) the incorporation of elements such as argon and high concentrations of point defects [12] which expand the host lattice (effectively in solid solution). Compressive stresses give rise to an increase in interplanar spacing (unit cell dimensions) of the TiN coating which are relaxed on dissolution of the underlying substrate, indicating that differences in thermal expansion coefficient between coating and substrate are primarily responsible for Type I distortions in PVD coatings [13].

(2) Type II microstrains are a reflection of the high concentrations of point defects, dislocations etc., which exist in vapour deposited coatings since deposition takes place under conditions which are far removed from equilibrium. In addition to strain broadening ($\beta_\varepsilon$) there is a contribution to the Type II microstrains from grain size broadening ($\beta_s$), and these can be separated by the use of Hall plots as described in ref. 21. It should be noted that the maximum tempering temperatures considered in the present study correspond to a homologous temperature ($T/T_m$) of 0.30 and consequently any changes in peak broadening reflect variations in the strain contribution to the Type II distortions; it is generally recognized that $T/T_m$ should exceed about 0.5 in order to bring about changes in grain size on high-temperature annealing [22].

(3) Type III macrostrains are now generally recognized as being the result of structural or yield anisotropy and are characterized by a peak shift accompanied by X-ray line broadening, and these strains vary with deposition rate, coating thickness and deposition conditions (substrate bias) for PVD coatings [11, 13, 18, 23]. Similar strains have been observed in bulk materials which have been subjected to a particular kind of plastic deformation in the

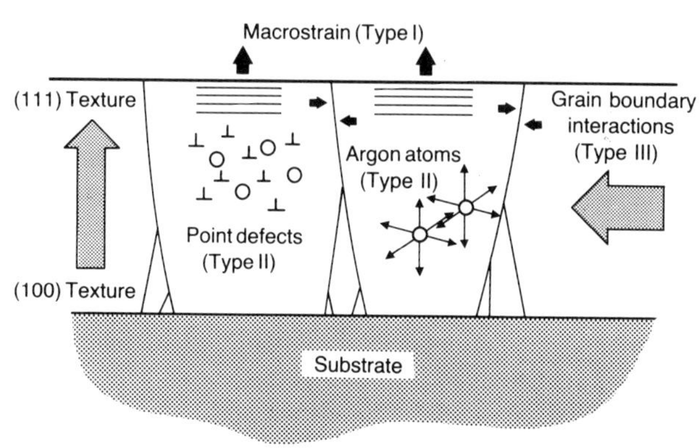


Fig. 1. Schematic diagram of the lattice distortions identified in PVD coatings.

volume averaged by the X-ray beam, and these strains (stresses) have been termed pseudo-macrostrains [24].

The structural model shown schematically in Fig. 1 has been used previously to explain the high lattice parameter of PVD TiN coatings and the variation in unit cell dimensions of the nitride with deposition rate and substrate bias voltage [12]. The purpose of this work is to investigate the general validity of this model with regard to explaining the changes in Types I, II and III lattice distortions which result on high temperature annealing of PVD TiN.

## 4. Results and discussions

### *4.1. Type I macrostrain measurements*

The application of a substrate bias during the deposition of a coating has a profound effect on the growth and resultant microstructure of PVD films [4], and Fig. 2 shows fracture cross-sections through unbiased and biased sputter ion plated TiN films. These two different types of microstructure have been chosen since they represent the extremes of behaviour which might be expected for PVD films. The unbiased coating (Fig. 2a) shows an open columnar structure (a zone-I-type structure [22]) which results in a coating of low hardness and poor wear resistance [25]. Such an open microstructure cannot support any thermal contribution to film stress as these are easily accommodated at the free surfaces which exist perpendicular to the coating–substrate interface, and overall levels of stress are low in unbiased coatings [11]. Consequently, the measured values for the various types of lattice distortion described in Section 3, and shown schematically in Fig. 1, are very low for such open microstructures. In contrast the biased film appears much more dense, the individual columns being less well defined (a

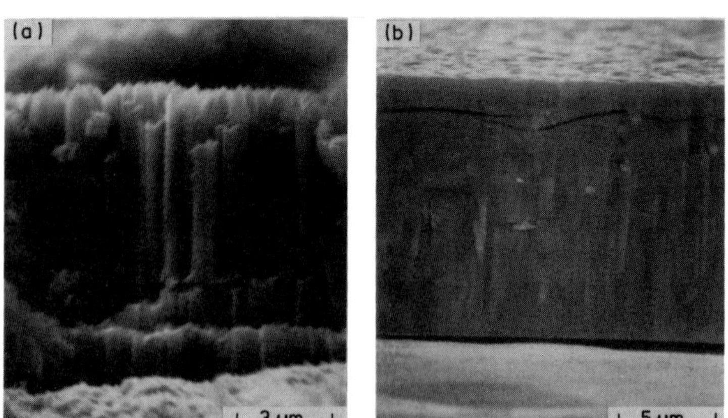


Fig. 2. Secondary electron images of fracture surfaces of TiN coatings deposited onto austenitic stainless steel: (a) no applied substrate bias; (b) biased at $-35$ V.

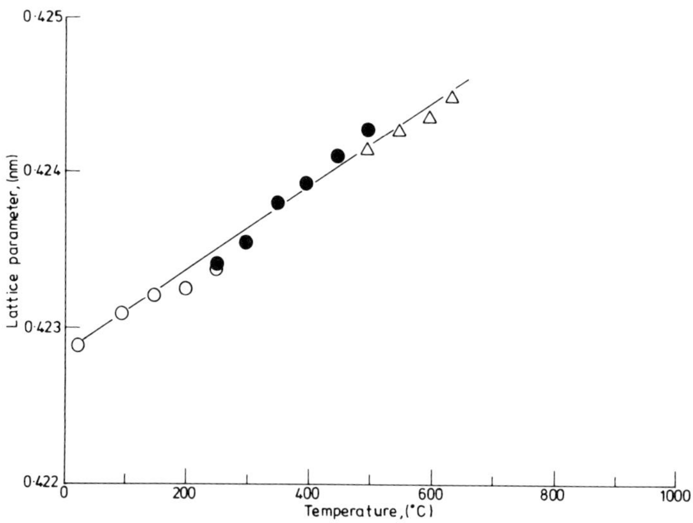


Fig. 3. Lattice parameter *vs.* annealing temperature for a TiN coating deposited at 500 °C onto austenitic stainless steel with no applied substrate bias.

zone-T-type structure [22]), and these possess a high hardness and have excellent wear resistance [25] and also give rise to Types I, II and III lattice distortions.

For unbiased TiN deposited onto austenitic stainless steel, Fig. 3 shows that the lattice parameter increases linearly with temperature, with a coefficient of linear expansion calculated to be $8.5 \times 10^{-6}\ K^{-1}$; a value which is in excellent agreement with that determined by X-ray techniques from sintered compacts ($\alpha = 8.1 \times 10^{-6}\ K^{-1}$ [26]) but slightly lower than that generally given for bulk TiN ($\alpha = 9.4 \times 10^{-6}\ K^{-1}$ [27]). In comparison, Fig. 4 shows that the 'apparent' coefficient of linear expansion for biased TiN varies with temperature, being about $2.5 \times 10^{-6}\ K^{-1}$ in the interval 23 - 500 °C and increasing to $9.5 \times 10^{-6}\ K^{-1}$ between 800 and 900 °C. These changes in lattice parameter for both biased and unbiased TiN are reversible upon cooling back to room temperature. The variation in residual stress with temperature was also monitored for both types of microstructure using the (422) reflection for the biased coating, and due to extreme texture the (333) reflection for unbiased TiN. For the open unbiased microstructure the stress levels were invariant with temperature, always being slightly compressive (less than 200 MPa), whilst those in the fully dense biased coating decreased with increasing temperature to be zero in the interval 800 - 900 °C (see Fig. 5).

### 4.2. *Type II microstrain measurements*

By combining information on the breadth of the X-ray reflection with those of peak shift it is possible to gain a better understanding of the structure of PVD TiN films. Figure 6 shows that there is a dramatic reduction in the breadth of the (422) TiN X-ray reflection with temperature for fully

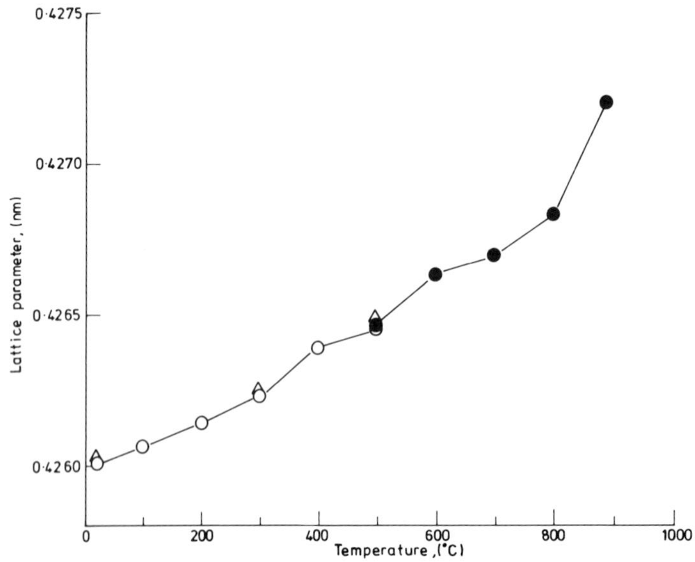


Fig. 4. Lattice parameter *vs.* annealing temperature for a TiN coating deposited onto austenitic stainless steel at a temperature of 500 °C with a substrate bias voltage of −60 V to a thickness of 4 μm. Note that the complete thermal cycle consists of room temperature to 500 °C hold for 15 h (○), 500 °C to 900 °C and back to 500 °C hold for 15 h (●), and finally 500 °C back to room temperature (△).

dense (biased) TiN which is particularly marked above the deposition temperature of 500 °C. On cooling back to room temperature there is a slight increase in microstrain broadening over the value measured at 900 °C but in all cases the levels remain below those observed for as-deposited material. By way of contrast, there was no apparent change in lattice strain with temperature

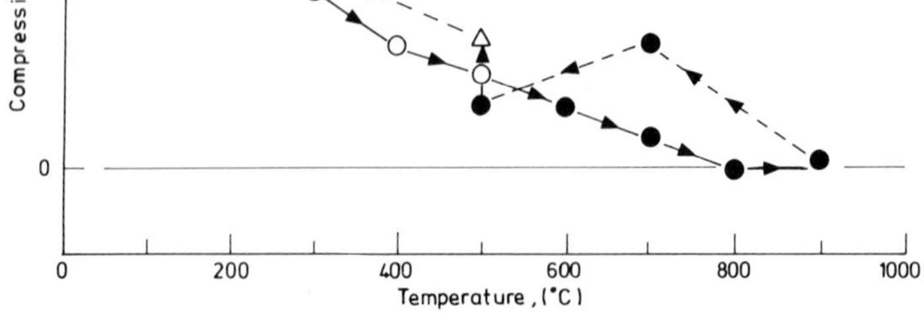


Fig. 5. Variation of compressive residual stress with annealing temperature. Sample as in Fig. 4, arrows indicate direction of temperature changes as detailed previously.

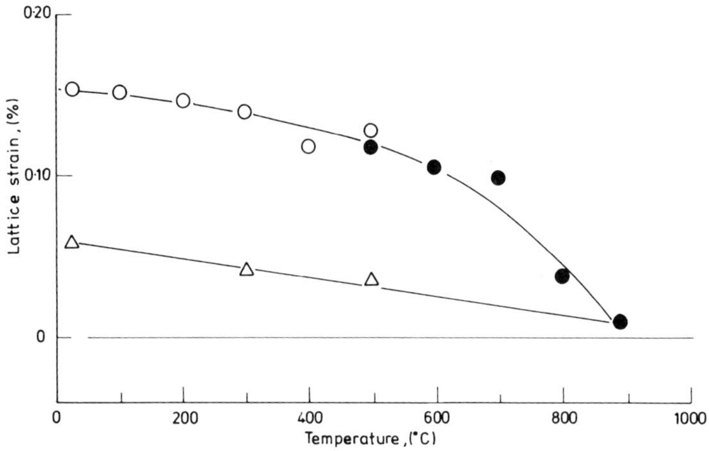


Fig. 6. Variation in lattice strain for the (422) peak of TiN as a function of annealing temperature; symbols as in Fig. 4.

for unbiased TiN, which also showed a much lower level of microstrain broadening than those observed in Fig. 6 for fully dense coatings, confirming our previous observations [11]. By using the method of Hall [21] the microstrain data in Fig. 6 can be separated into two parts, one due to grain size ($\beta_s$) and the other to strain ($\beta_\varepsilon$). Table 1 shows in the case of biased TiN that the changes in lattice strain in Fig. 6 result from variation in $\beta_\varepsilon$, and are not the result of grain coarsening brought about by annealing above the deposition temperature, although Wendler [28] has reported such effects for metastable $Ti_2N$–TiN films tempered at 900 °C and above. Figure 7 contrasts the microstructure of biased TiN before (Fig. 7a) and after (Fig. 7b) the thermal cycle illustrated in Fig. 6, and what is immediately apparent is the opening up of the columnar microstructure after heating. According to the structure

TABLE 1

Contributions to Type II lattice distortions as a function of annealing temperature for PVD TiN films deposited onto austenitic stainless steel with a substrate bias of −60 V (see also Fig. 6)

| *Annealing temperature* | *Stress* (MPa) | *Strain broadening* ($\beta_\varepsilon$) (%) | *Size broadening* ($\beta_s$) (nm) |
|---|---|---|---|
| Room temperature (as-deposited) | −6400 | 0.15 | 80 ± 20 |
| 500 °C (corresponding to the deposition temperature) | −2200 | 0.12 | 62 ± 20 |
| 890 °C | −130 | 0.01 | 45 ± 18 |
| Room temperature after a complete thermal cycle up to 890 °C | −6800 | 0.06 | 56 ± 20 |

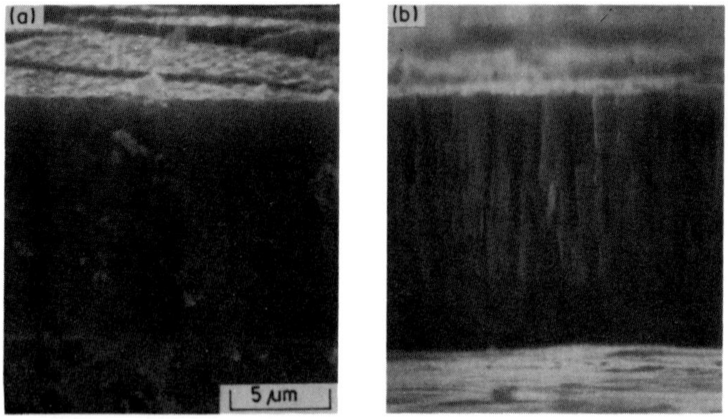


Fig. 7. Secondary electron images of fracture surfaces of TiN coatings deposited onto austenitic stainless steel with a substrate bias of −60 V. (a) As-deposited microstructure; (b) after a thermal cycle indicated by Fig. 4.

model depicted in Fig. 1 such grain boundary relaxations would account for the variation in lattice strain shown in Fig. 6 as the biased coating tends towards the more open columnar structure of an unbiased deposit.

### *4.3. Type III macrostrain measurements*

One feature which is commonly reported for PVD TiN films is that the unit cell dimensions of the nitride determined for each *(hkl)* reflection can vary, and that these deviations are a sensitive function of internal stress and coating thickness [13], deposition rate [23], argon incorporation [29] and nitrogen occupying tetrahedral sites in the TiN lattice [30]. These pseudo-macrostrains are completely relaxed on dissolution of the underlying substrate with an accompanying decrease in X-ray line broadening [11], and for ion plated TiN coatings after tempering for 1 h at 900 °C [12]. Rickerby *et al.* [13] reviewed the possible contributions to these Type III lattice distortions and conclude that the interactions between the columnar units which comprise the coating may give rise to yield anisotropy which can cause peak broadening accompanied by a peak shift. Figure 8 illustrates the type of behaviour often shown by PVD films where, in this particular example, the (111) family of planes are displaced from a line drawn through the remaining experimental points. After the thermal cycle used in this study, room temperature up to 900 °C and back to room temperature in intervals of 100 °C, these pseudo-macrostresses are completely relaxed and the lattice spacing of the (111) planes move towards its "equilibrium" value. Like the changes reported here for lattice parameter and X-ray line broadening these movements in the (111) planes are gradual at first but become more evident above 500 °C, the deposition temperature for these films (see Fig. 9). The reduction in lattice spacing of the (111) planes for temperatures above 500 °C contrasts strongly with the expansion observed for the Type I macrostrain distortion shown in Fig. 4, and this fact is attributed to the relaxation of the grain boundary

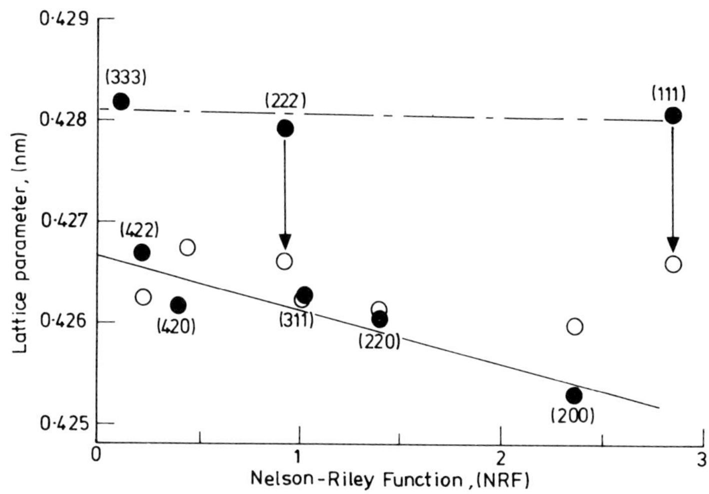


Fig. 8. Lattice parameter ($a_{hkl}$) against the Nelson–Riley Function for TiN coatings deposited onto austenitic stainless steel with a substrate bias of $-90$ V to a thickness of approximately 2.5 $\mu$m. ●, as-deposited coating; ○, as-deposited coating after a thermal cycle indicated by Fig. 4. Note the relaxation of the Type III pseudo-macrostrains on annealing to 900 °C.

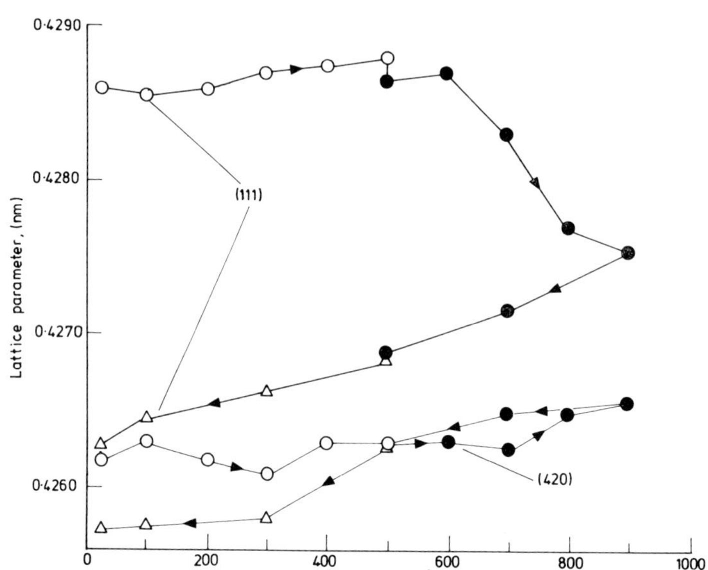


Fig. 9. Variation of lattice parameter with annealing temperature for the (111) and (420) family of planes of TiN. Note the large hysteresis in the behaviour of the (111) planes compared to that of the (420) planes.

interactions, in agreement with the microstructural changes which accompany thermal cycling as shown in Fig. 7.

## 5. Discussion

The microstructure of PVD films, and hence properties such as hardness, effective Young's modulus and level of internal stress [4], varies with annealing (service) temperature and these microstructural changes are not reversible for dense TiN coatings deposited with a substrate bias. Probably the most important parameter is the packing density of the individual columnar units which comprise the coating, and this is a sensitive function of temperature, substrate bias and coating-substrate combination [4]. For the case of TiN deposited onto stainless steel (*i.e.* $\alpha_f < \alpha_s$), on cooling from deposition temperature thermal stress acts to bring the columnar units closer together and a compressive stress is set up (see Fig. 10). This may then act to densify the coating further, resulting in a very dense microstructure such as that shown in Fig. 2b, and giving rise to the various types of lattice distortion shown schematically in Fig. 1. However, at low bias voltages the intrinsic or growth stress component is reduced, and by virtue of the more open microstructure at deposition temperature (reduction in packing density) substantial thermal mismatch stresses cannot be generated in such microstructures, and no Type I distortions are observed on annealing at elevated temperatures. In such cases, since the coating is effectively decoupled from the underlying substrate, the coefficient of thermal expansion of the nitride is close to that of bulk TiN as each column can act independently of its neighbours.

For biased coatings which show a high packing density, and again for the situation $\alpha_f < \alpha_s$, above the deposition temperature the thermal stress is tensile in nature. This can cancel the compressive growth and thermal stresses which are set up in the coating at room temperature, resulting in a low level of residual stress at high temperature as the individual columns are

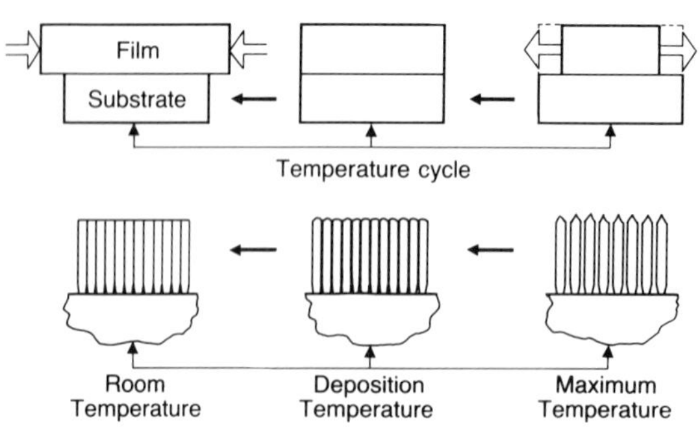


Fig. 10. The effect of thermal mismatch stresses on the microstructure of PVD coatings for the situation $\alpha_f < \alpha_s$.

pulled apart effectively decreasing their packing density. Clearly at low annealing temperatures the coating cannot expand independently of the substrate, a fact which is reflected in an 'apparent' coefficient of linear expansion of $2.5 \times 10^{-6}\ K^{-1}$. With increasing temperature, the coating becomes effectively decoupled from the substrate due to a reduction in compressive residual stress and column packing density, and consequently the expansivity of the nitride increases to that expected for unconstrained bulk TiN ($\alpha = 9.4 \times 10^{-6}\ K^{-1}$). On cooling back to room temperature the thermal stresses will act to increase the packing density by closing the intercolumnar regions shown in Fig. 7b and the final stress levels and coefficients of linear expansion measured after thermal cyling reflect the degree by which these microstructural changes are reversible. Support for these microstructural changes on thermal cycling comes from the hardness hysteresis observed by Knight [31] in PVD coatings after thermal cycling up to 800 °C. For the films studied here there is a significant reduction in coating hardness after thermal cycling due to a reduction in packing density of the columnar units, and in their ability to transmit stresses to neighbouring regions.

## 6. Conclusions

High-temperature XRD can give useful information regarding the microstructural changes which accompany the tempering of TiN films. For the case of $\alpha_f < \alpha_s$, above the coating deposition temperature the substrate causes tensile stresses to be superimposed onto the compressive growth stresses which are present in PVD coatings and these act to reduce the levels of macrostrain and microstrain broadening in TiN films. These changes in X-ray lattice distortions (classified as Types I, II and III) can be correlated with changes in the microstructure of the nitride coatings as well as to variations in the coating's physical properties such as hardness as reported by Quinto *et al.* [7], and are explained by a simple structural model developed for PVD coatings.


## Acknowledgment

This work forms part of the Underlying Programme of the United Kingdom Atomic Energy Authority. Copyright of this paper belongs to the United Kingdom Atomic Energy Authority.


## References


1 J.-E. Sundgren, B.-O. Johansson and S.-E. Karlsson, *Thin Solid Films, 105* (1983) 353.
2 J.-E. Sundgren and H. T. G. Hentzell, *J. Vac. Sci. Technol., A4* (1986) 2259.
3 D. S. Williams, F. A. Baiocchi, R. C. Bearsto, J. M. Brown, R. V. Knoell and S. P. Murarka, *J. Vac. Sci. Technol., B5* (1987) 1723.
4 D. S. Rickerby and P. J. Burnett, *Thin Solid Films, 157* (1988) 195.

5 D. T. Quinto, G. J. Wolfe and P. C. Jindal, *Thin Solid Films, 153* (1987) 19.
6 A. J. Perry and L. Chollet, *Surf. Coat. Technol., 34* (1988) 123.
7 D. T. Quinto, A. T. Santhanam and P. C. Jindal, *Mat. Sci. Eng., A105/106* (1988) 443.
8 A. J. Perry, *Thin Solid Films, 146* (1987) 165.
9 C. Ernsberger, A. J. Perry, L. P. Lehman, A. E. Miller, A. R. Pelton and B. W. Dabrowski, *Surf. Coat. Technol., 36* (1988) 605.
10 Y. Shioya, K. Ikegami, M. Maeda and K. Yanagida, *J. Appl. Phys., 61* (1987) 561.
11 D. S. Rickerby, *J. Vac. Sci. Technol., A4* (1986) 2309.
12 D. S. Rickerby, A. M. Jones and A. J. Perry, *Surf. Coat. Technol., 36* (1988) 631.
13 D. S. Rickerby, A. M. Jones and B. A. Bellamy, *Surf. Coat. Technol., 37* (1989) 111.
14 J. P. Coad and R. A. Dugdale, *Proc. Int. Conf. on Ion Plating and Allied Techniques*, CEP Consultants, Edinburgh, 1979, p. 186.
15 D. S. Rickerby and R. B. Newbery, *Vacuum, 38* (1988) 161.
16 A. M. Jones, *AERE Report R12275* (available from HMSO, London) 1986.
17 F. L. Cullen, *AERE Report M3668*, 1988.
18 L. Chollet and A. J. Perry, *Thin Solid Films, 123* (1985) 223.
19 L. Chollet, H. Boving and H. E. Hintermann, *J. Mat. Energy Syst., 6* (1985) 293.
20 A. J. Perry and L. Chollet, *J. Vac. Sci. Technol., A4* (1986) 2813.
21 G. K. Williamson and W. H. Hall, *Acta. Metall., 1* (1953) 22.
22 J. A. Thornton, *J. Vac. Sci. Technol., A4* (1986) 3059.
23 A. J. Perry and J. Schoenes, *Vacuum, 36* (1986) 149.
24 B. D. Cullity, *Adv. X-ray Anal., 20* (1976) 259.
25 S. J. Bull, D. S. Rickerby, T. Robertson and A. Hendry, *Surf. Coat. Technol., 36* (1988) 743.
26 MCIC Report 1/2, 5.3.4-6 (1979).
27 H. Holleck, *J. Vac. Sci. Technol., A4* (1986) 2661.
28 B. Wendler, *Thin Solid Films, 141* (1986) 223.
29 J. P. Noel, D. C. Houghton, G. Este, F. R. Shepherd and H. Platner, *J. Vac. Sci. Technol., A2* (1984) 284.
30 J.-E. Sundgren, B.-O. Johansson and S.-E. Karlsson, *Thin Solid Films, 80* (1981) 77.
31 J. C. Knight, in K. C. Ludema (ed.), *Wear of Materials*, ASME, 1989, p. 575.

*Surface and Coatings Technology, 39/40* (1989) 409 - 419 

# THREE-DIMENSIONAL GRAIN SIZE CONTROL FOR SIMULTANEOUS IMPROVEMENT OF HARDNESS AND IMPACT ADHESION IN (Ti,Al)N FILMS ON HIGH SPEED STEEL

KYU TAEK HYUM, HUNG HOON JOO and BONG HEE HAHN

*Department of Metallurgical Engineering, Seoul National University, 151-742 Seoul (Korea)*

(Received March 8, 1989)


## Summary

(Ti,Al)N films deposited on HSS by biased activated reactive evaporation showed two competitive mechanical properties, hardness and impact indentation. Effective grain size was determined by X-ray diffraction line broadening. At constant bias of $-1$ kV, increasing nitrogen pressure up to 4 mTorr increased grain size and adhesion, but gave a lower Knoop hardness of KHN 1000. By Rockwell C impact indentation analysis, the failure mode of the film and substrate was used to determine the adhesion. In the range of the $(Ti_2Al)N$ and $(Ti_{0.5}Al_{0.5})N$ dual-phase film structure, a decrease in nitrogen pressure decreased the grain size of the film, which showed a high Knoop hardness of KHN 2500 but poor impact adhesion, being incapable of plastic deformation of the substrate. On the basis of the above results, we designed the deposition process to control the grain size along the thickness direction to obtain both good adhesion at the interface of film and HSS substrate and high Knoop hardness at the rest of the film.


## 1. Introduction

Titanium based hard coatings have been investigated by many groups for more than 10 years [1 - 3]. The principal efforts have been aimed at increasing the hardness through the use of the solid solution hardening mechanism.

These efforts were based on the premise that tool life extension is mainly affected by mechanical hardness. Wear mode can be classified into three areas: abrasive wear, delamination wear and chemical wear.

TiN has a very low friction coefficient and low stability under a severely corrosive environment. At the leading edge of a cutting tool, high enough temperatures for diffusion of the cutting material into the protective coating are observed which decreases tool life. This is not closely related to the mechanical stability of hard coating. Aluminum can form a stable and dense oxide when heated during the cutting operation and reduces interdiffusion.

0257-8972/89/$3.50 

Its atomic size ($r = 1.43$ Å) can fit into the TiN crystal structure at substitutional sites or interstitial sites. For this reason, among various competitive alloying elements, aluminum has been studied as a second element in the TiN system [4]. The (Ti,Al)N system has four crystal structures. Face-centered cubic (f.c.c.) $(Ti_{0.5}Al_{0.5})N$, hexagonal close-packed (h.c.p.) $Ti_{1.5}AlN$, h.c.p. $Ti_2AlN$ and h.c.p. $Ti_3AlN$ ($CuTiO_3$ type) [5]. $Ti_{0.5}Al_{0.5}N$ is a substitutional solid solution of aluminum with titanium sites and the others are stoichiometric compounds [6]. Many research groups have deposited this system with dual target magnetron sputtering, arc evaporation. In this study, biased activated reactive evaporation was used for deposition, and the nitrogen gas pressure and substrate temperature were adopted as the main experimental variables.

The effects of the two process variables were investigated for tool wear on coated high speed steel substrates.

## 2. Experimental procedure

A schematic diagram of our plasma deposition system is shown in Fig. 1. A 4 kW electron beam gun and thermal evaporator were used for titanium

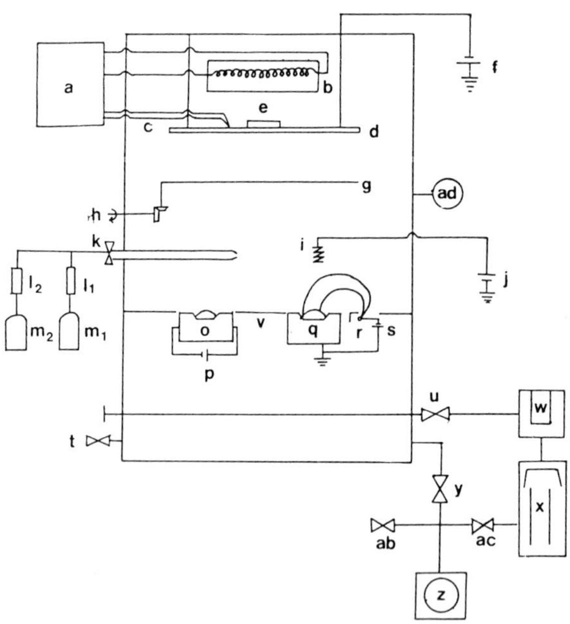


Fig. 1. Schematic diagram of the experimental apparatus. (a) substrate heating controller, (b) hallogen lamp, (c) thermocouple, (d) substrate holder, (e) substrate, (f) regulated DC generator, (g) shutter, (h) rotary feedthrough, (i) ARE probe, (j) regulated DC generator, (k) needle valve, ($l_1$) rotameter, ($l_2$) rotameter, ($m_1$) high purity nitrogen, ($m_2$) high purity argon, (o) tungsten boat, (p) high current generator, (q) water-cooled copper crucible, (r) electron beam filament, (s) high voltage generator, (t) leak valve, (u) main valve, (v) pressure barrier, (w) liquid nitrogen trap, (x) diffusion pump, (y) roughing valve, (z) rotary pump, (ab) foreline leak valve, (ac) foreline valve, (ad) shultz gauge.

and aluminum evaporation. The system was evacuated under $2 \times 10^{-5}$ Torr by a conventional mechanical pump system. On the substrate, a negative bias up to $-2$ kV direct current can be supplied through a high-resistance load resistor and the substrate current was monitored during deposition. The auxiliary anode was 1 cm above the electron-beam heated molten pool and biased at $+100$ V with a 1.5 A discharge current. A high speed steel plate with dimensions 20 mm × 20 mm × 1.2 mm was used as the substrate. After mechanical polishing, TCE and alcohol supersonic detergent cleaning the substrate was handled and finally heated up to 50 °C to release absorbed detergent liquid [7]. After full evacuation down to $2 \times 10^{-5}$ Torr, substrates were argon-sputter cleaned at 40 mTorr with a $-1.4$ kV substrate bias for 10 min. Electron beam current and acceleration voltage were 130 mA and $-4$ kV. At the time of deposition, a pressure baffle retains high vacuum state around the electron beam gun. After vacuum cooling the substrate to room temperature, X-ray diffraction was used for phase identification, grain size measurement and lattice parameter determination. Scanning electron microscope (SEM) micrography and Knoop hardness measurement under 15 gf load were carried out. For the adhesion test, Rockwell C-type impact indentation with 150 kgf was made on film-substrate composite [8].

## 3. Results and discussion

### *3.1. Phase identification and preferred orientation*

The nitrogen pressure was varied over the range $3 \times 10^{-4}$ to $4 \times 10^{-3}$ Torr and the substrate temperature was 250 - 500 °C. From X-ray diffraction, a $Ti_2AlN$ single phase was observed under $4 \times 10^{-4}$ Torr nitrogen pressure. At nitrogen pressures above $4 \times 10^{-4}$ Torr, $Ti_2AlN$ and $(Ti_{0.5}Al_{0.5})N$ coexist.

Figures 2 and 3 are X-ray diffraction results for $3 \times 10^{-4}$ Torr and $1 \times 10^{-3}$ Torr. There are two competitive preferred orientations in $Ti_{0.5}Al_{0.5}N$, (200) and (111) textures. In the case of TiN, the (200) texture developed as incident ion flux into substrate increased. $Ti_{0.5}Al_{0.5}N$ is the alloy phase formed when the nitrogen pressure is over $4 \times 10^{-4}$ Torr, so aluminum is substituted into the normal titanium lattice site. In Fig. 4 the preferred orientation seems to change from (111) to (200) as the nitrogen pressure increases. This can be explained by the model of Lee [9], in which crystal surfaces with higher ledge density (200) grow faster over a certain critical incoming ion flux *i.e.* nitrogen pressure in this experiment.

### *3.2. Effective crystallite size by X-ray diffraction*

Effective crystallite size was measured by the integral breadth method, removing instrumental line broadening by use of a standard quartz sample. All specimens had an average grain size over a few hundred ångstroms (Figs. 5 and 6).

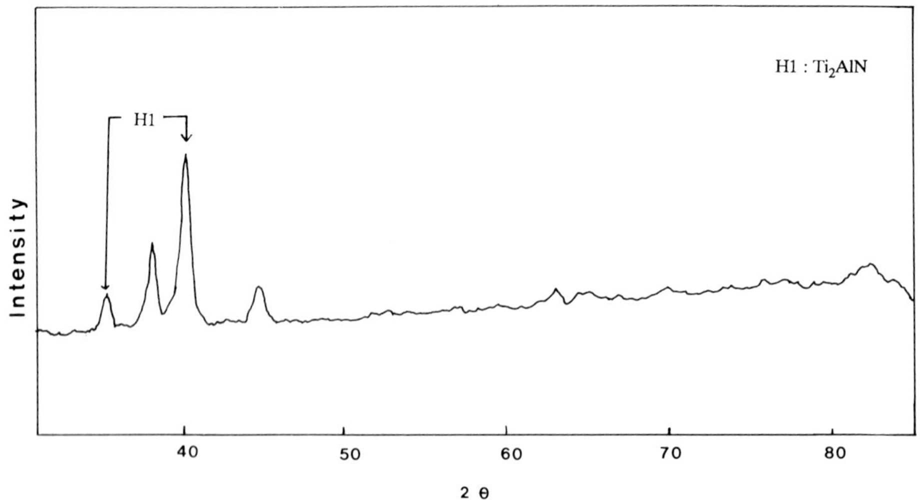


Fig. 2. X-ray diffraction of (Ti,Al)N deposits. $T_s = 500$ °C, $P(N_2) = 3 \times 10^{-4}$ Torr.

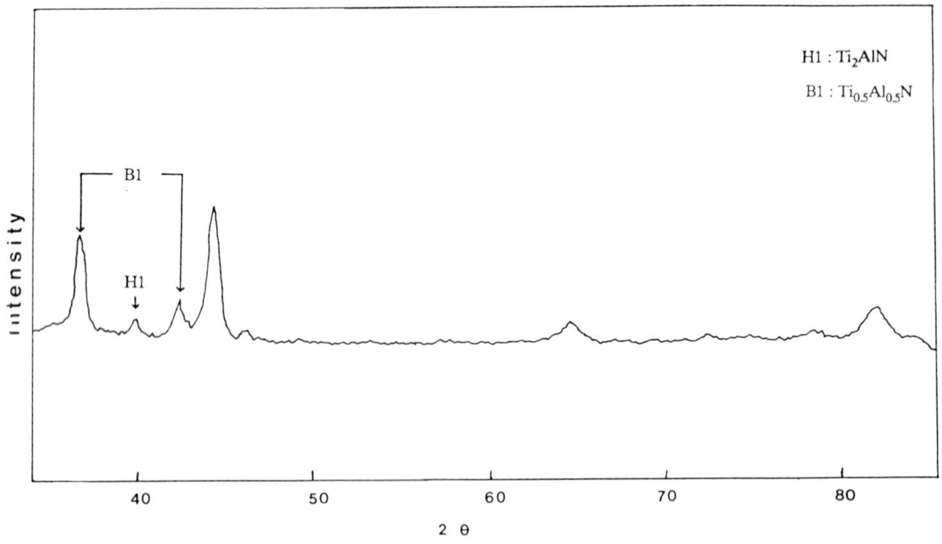


Fig. 3. X-ray diffraction of (Ti,Al)N deposits. $T_s = 300$ °C, $P(N_2) = 1 \times 10^{-3}$ Torr. Fixed variable: $T_s = 400$ °C.

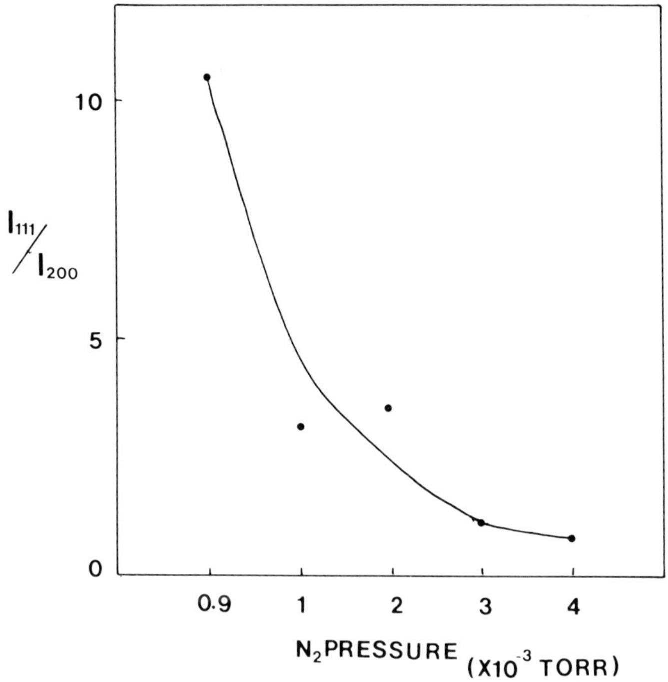


Fig. 4. Variation of integrated intensity ratio of (111) to (200) with nitrogen pressure.

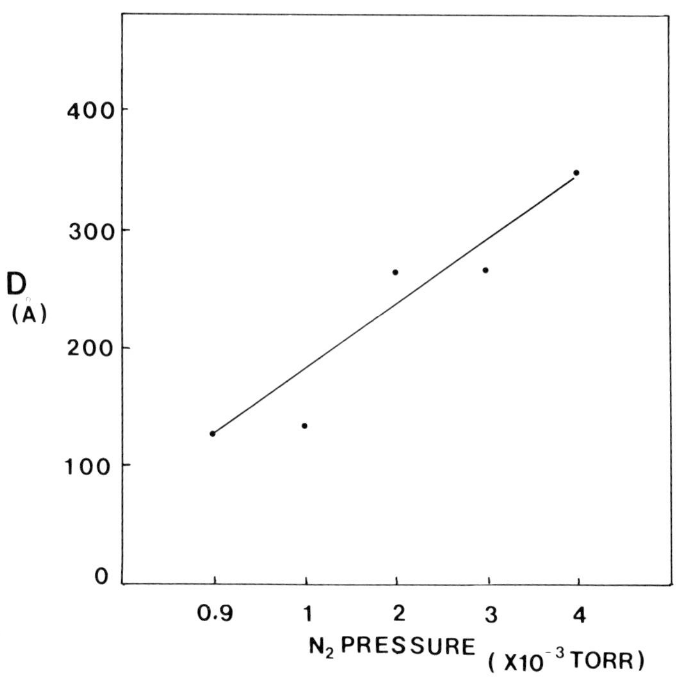


Fig. 5. Variation of effective crystallite size of deposits with nitrogen pressure. Fixed variable: $T_s = 400$ °C.

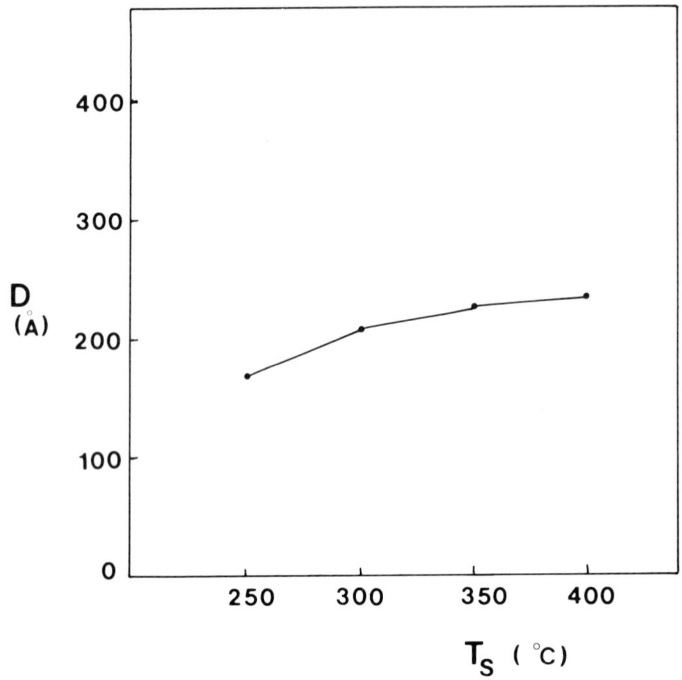


Fig. 6. Variation of effective crystallite size of deposits with substrate temperature (ARE).

In this experiment, the temperature of the substrate was held under 400 °C. This is about $0.25T_m$ of (Ti,Al)N. A higher nucleation site density than that of the unbiased sample was obtained by incoming energetic ions. Our specimens have a thickness of 3 - 4 μm.

### *3.3. Microstructure and microhardness*

Figure 7 shows the surface of (Ti,Al)N deposits by SEM analysis. As nitrogen pressure decreases, the surface gets denser. The observed dome size is not the real crystallite grain size but an X-ray broadened one.

The upper limit in this method is a few tenths of a micrometer. The microhardness variation in (Ti,Al)N deposits is shown in Fig. 8. Under $4 \times 10^{-4}$ Torr nitrogen pressure, the film is single phase $Ti_2AlN$ and the hardness value increases as nitrogen pressure increases. Over this boundary pressure of $4 \times 10^{-4}$ Torr, microhardness decreases with increasing nitrogen pressure. In this region, a two-phase mixture of $(Ti_{0.5}Al_{0.5})N$ and $(Ti_2Al)N$ was observed. However, the microhardness of $(Ti_2Al)N$ is reported to increase as nitrogen pressure increases. The $(Ti_{0.5}Al_{0.5})N$ phase increased its content, so the microhardness decreased.

Concerning the effect of substrate temperature, Fig. 9 shows microhardness variation by temperature increase. Above 350 °C, microhardness started to decrease due to the larger grain size induced by thermal diffusion. This phenomena can be explained as follows. (Ti,Al)N system has its microstructure

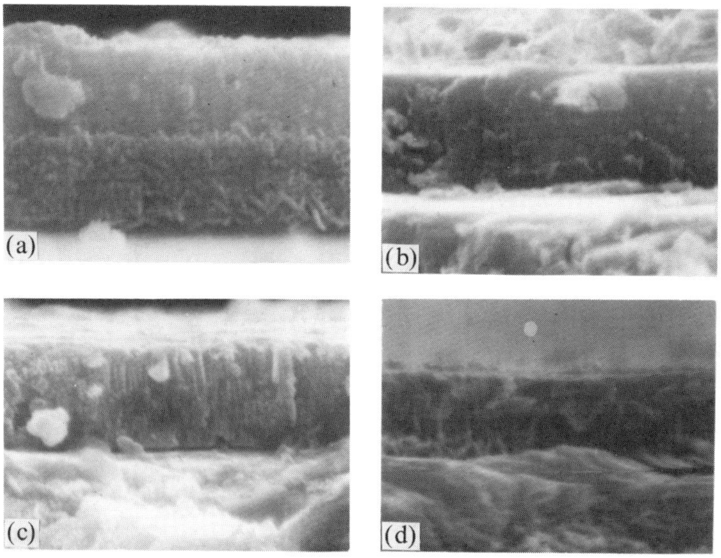


Fig. 7. Scanning electron micrographs of the fracture cross section of (Ti,Al)N deposits ($\times 15\,000$). a, $P(N_2) = 1 \times 10^{-3}$ Torr; b, $P(N_2) = 2 \times 10^{-3}$ Torr; c, $P(N_2) = 3 \times 10^{-3}$ Torr; d, $P(N_2) = 4 \times 10^{-3}$ Torr.

evolution boundary at 0.25 $T_m$ (around 350 - 400 °C) and under 350 °C, ion bombardment during deposition increased nucleation site density hence gave a smaller grain size. Over this temperature, substrate bias effects appeared to enhance surface diffusion, so that a larger grain size and lower hardness would be obtained if a Hall–Petch type deformation behavior is supposed.

### *3.4. Adhesion*

There are many ways of testing adhesion between film and substrate. A good testing method should imitate the real adhesion failure environment. In this experiment, we used an impact indentation test using a Rockwell C-type indent with a 150 kgf load. Films of good adhesion should show no flakes or cracks around the indentation. On the basis of this interpretation, Fig. 10 shows typical impact indentation failure modes. Failure modes for various nitrogen pressure are shown in Fig. 11. As pressure increased, adhesion was better, although the microhardness decreased, this is due to the increase of the softer phase ($Ti_{0.5}Al_{0.5}N$) and the larger grain size. Therefore, we can have an idea of how to control nitrogen pressure from an initially high to a low final pressure so that there is a decreasing grain size distribution with thickness. Figure 12 shows impact indentation failure of film deposited at 300 °C with varying nitrogen pressure from $4 \times 10^{-3}$ to $9 \times 10^{-4}$ Torr. It gave a hardness of 2300 KHN with good adhesion.

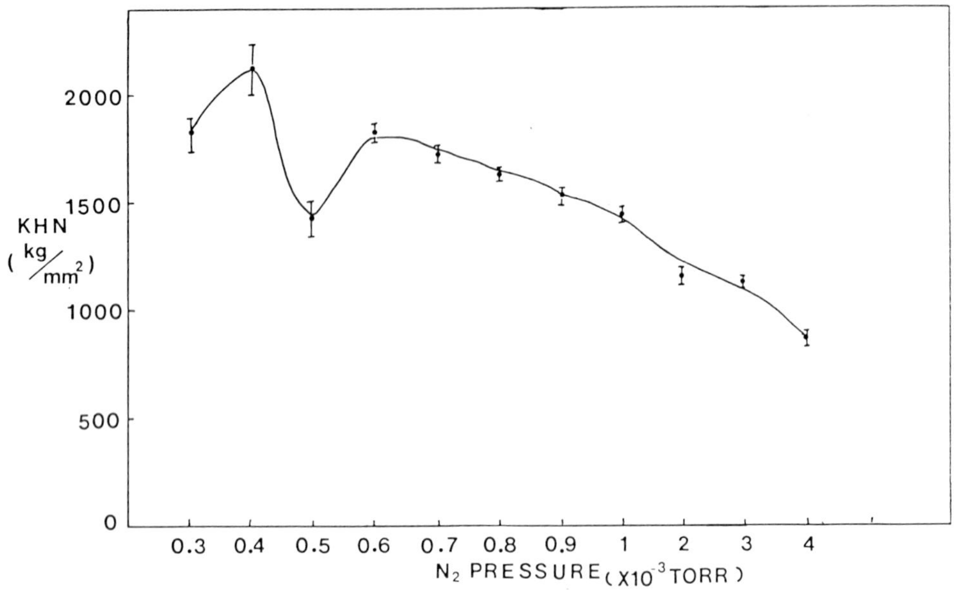


Fig. 8. Variation of microhardness with nitrogen pressure. Fixed variable: $T_s = 300$ °C.

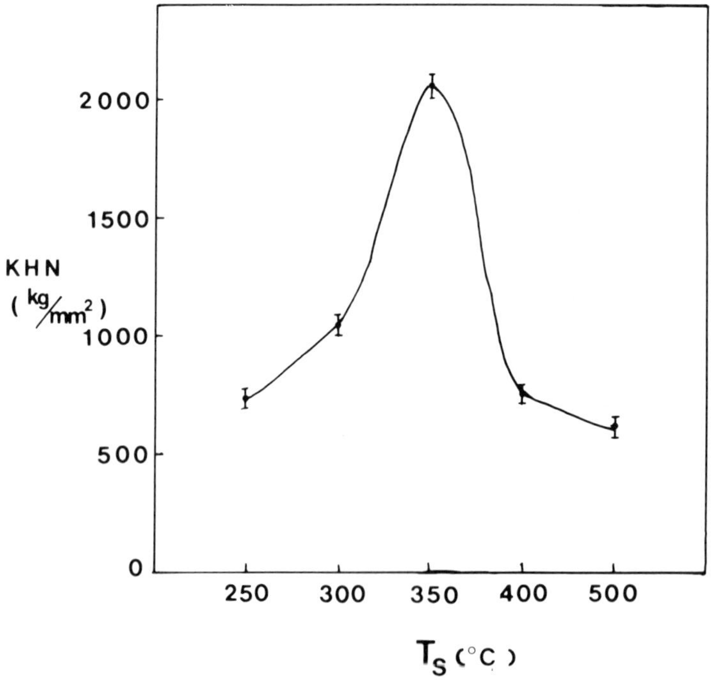


Fig. 9. Variation of microhardness with substrate temperature. Fixed variable: $P(N_2) = 3.0 \times 10^{-3}$ Torr.

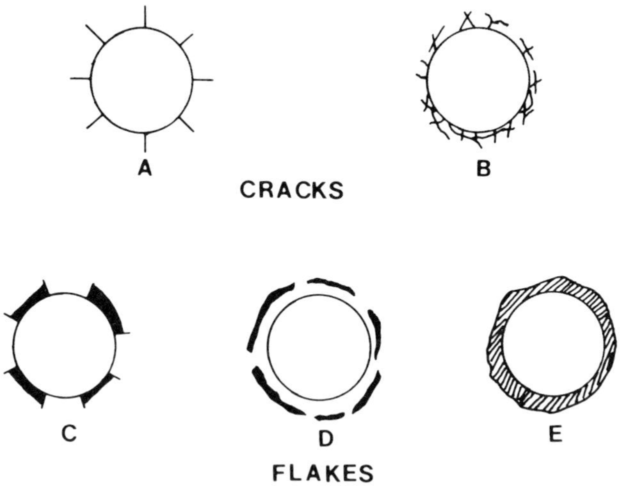


Fig. 10. Expected failure modes around Rockwell-C indentation.

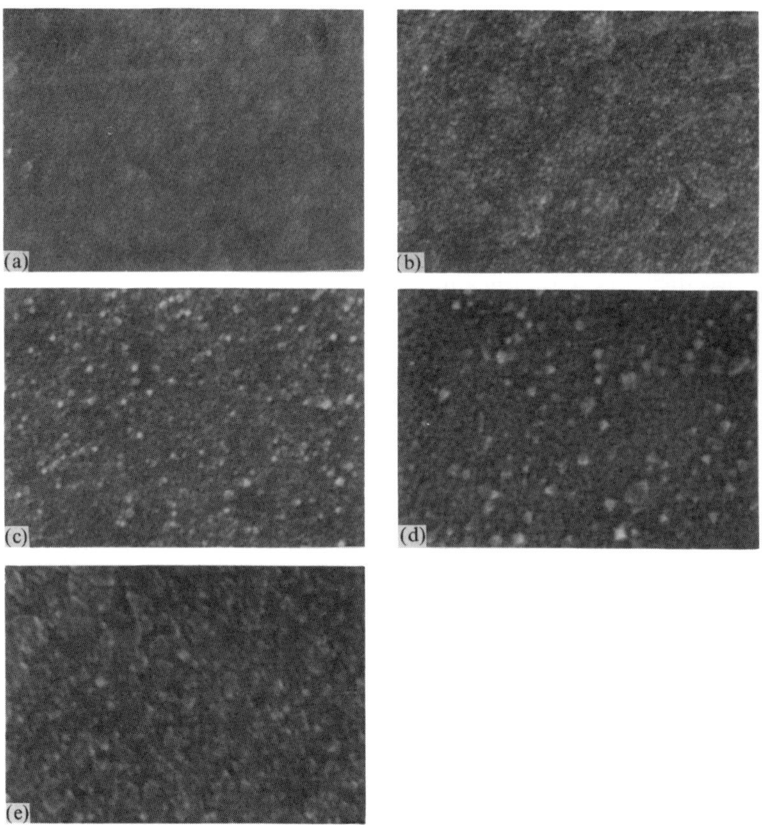


Fig. 11. Scanning electron micrographs of the surface of (Ti,Al)N deposits ( × 15 000). Fixed variable: $T_s = 400\,^\circ C$. a, $P(N_2) = 0.9 \times 10^{-3}$ Torr; b, $P(N_2) = 1 \times 10^{-3}$ Torr; c, $P(N_2) = 2 \times 10^{-3}$ Torr; d, $P(N_2) = 3 \times 10^{-3}$ Torr; e, $P(N_2) = 4 \times 10^{-3}$ Torr.

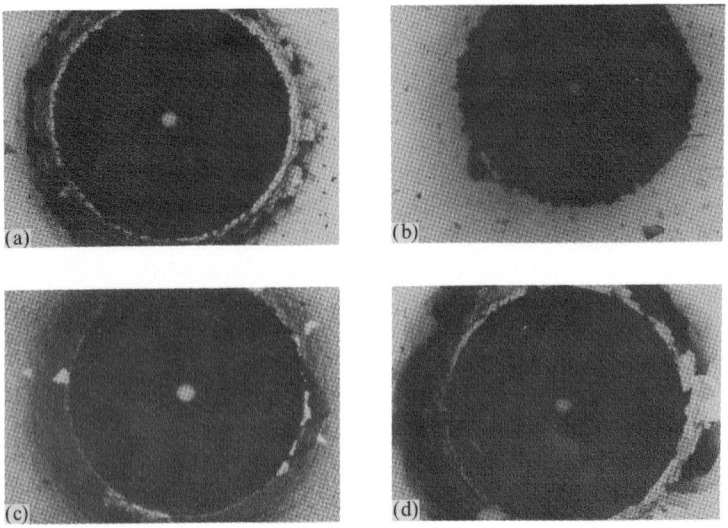


Fig. 12. Optical micrographs showing cracks and flakes around indentation ($\times 100$). Fixed variable: $T_s = 300$ °C. a, $P(N_2) = 9 \times 10^{-4}$ Torr; b, $P(N_2) = 1 \times 10^{-3}$ Torr; c, $P(N_2) = 2 \times 10^{-3}$ Torr; d, $P(N_2) = 3 \times 10^{-3}$ Torr.

## 4. Conclusions

(1) For substrate temperatures of 250 - 500 °C, $Ti_2AlN$ single-phase deposits for nitrogen pressure lower than $4 \times 10^{-4}$ Torr and ($Ti_{0.5}Al_{0.5}N$), $Ti_2AlN$ dual-phase deposits were obtained over $4 \times 10^{-4}$ Torr.

(2) For fixed substrate temperatures of 250 - 500 °C, the $Ti_{0.5}Al_{0.5}N$ (200) preferred orientation developed over the $Ti_{0.5}Al_{0.5}N$ (111) preferred orientation as nitrogen pressure increased.

(3) X-ray determined effective crystallite size increased with increasing nitrogen pressure and substrate temperature.

(4) Better adhesion but lower hardness was obtained at higher nitrogen pressures.

(5) To obtain both good adhesion and high hardness, decreasing nitrogen pressure from an initial value of $3 \times 10^{-3}$ Torr to final value of $1 \times 10^{-3}$ Torr gave good results.


## Acknowledgments

We have been aided in this work by helpful discussions with many individuals; we record our thanks to them all, especially coworkers within the X-ray laboratory and Materials Testing Center at Seoul National University. We also acknowledge Won Kyu Park at Gold Star Research Center and Dr. Jung Hwan Jun at RIST.

## References


1 R. F. Bunshah, *Deposition Technologies for Films and Coating*, Noyes Publication, p. 11.
2 H. A. Jehn, S. Hofmann and W. D. Münz, *Thin Solid Films, 153* (1987) 45 - 53.
3 H. Holleck, *J. Vac. Sci. Technol., 4(6)* (1986) 2661.
4 J. C. Schuster, *J. Solid State Chem., 53* (1984) 260.
5 W. D. Münz, *J. Vac. Sci. Technol., 4(6)* (1986) 2717.
6 J. E. Sundgreen and H. T. G. Hentzell, *J. Vac. Sci. Technol., A4* (1986) 2259.
7 R. Buhl, H. K. Pulker and E. Moll, *Thin Solid Films, 80* (1981) 265.
8 O. Knotek, W. D. Münz and T. Leyendecker, *J. Vac. Sci. Technol., A5* (1987) 2173.
9 D. N. Lee, *J. Mat. Sci.*, in press.

*Surface and Coatings Technology*, *39/40* (1989) 421 - 434


# INFLUENCE OF CRYSTALLOGRAPHIC ORIENTATION, RESIDUAL STRAINS, CRYSTALLITE SIZE AND MICROHARDNESS ON EROSION IN ZrN COATING

J. A. SUE and H. H. TROUE

*Union Carbide Corporation, 1500 Polco Street, Indianapolis, IN 46224 (U.S.A.)*

(Received March 9, 1989)

## Summary

ZrN coatings with a preferred orientation in (111) and (200) are produced with the physical vapor deposition (PVD) arc evaporation process. The microhardness of the coating is found to depend strongly on structural parameters such as crystallographic orientation, microstrain, macrostrain and crystallite size. The erosion rate of the coating to 50 $\mu$m alumina particles at 20° and 90° impact is presented as a function of these structural parameters and microhardness. Scanning electron microscopy examinations reveal that the eroded surface morphology varied significantly with its microhardness. Erosion mechanisms of the coating in response to the oblique and normal impacts are characterized.

## 1. Introduction

During the last few years, TiN has widely gained acceptance in industry for erosion protection as well as for cutting- and forming-tool applications. Material researchers have continued to explore better materials for more demanding operations. ZrN is a hard, refractory compound with good chemical stability and high melting temperature (2980 °C). The compound also exhibits high hardness up to 1000 °C [1]. Combined, these properties have made ZrN a promising protective coating for anti-wear, -corrosion and -erosion applications.

Physical vapor deposition (PVD) processes, including reactive sputtering [2, 3], ion plating [4] and cathodic arc evaporation [5 - 6] have successfully produced ZrN coatings. To date, coating properties evaluated have been limited to reflectivity [8, 9], resistivity [2], hardness [3, 7, 8, 10, 11], Young's modulus [12], thermal stability [13] and bond strength [3, 7]. No systematic evaluation relating properties to the structure parameters has been reported. Furthermore, with regard to practical applications, ZrN coated tools have shown significant performance advantages over TiN coated tools in a drilling test [8]; however, a conflicting result has also been reported in cutting tests

0257-8972/89/$3.50 

[14, 15]. The inconsistency in cutting performance, however, is likely related to differences in the basic properties of the coating and its deposition process. Clearly, a determination of the structure–properties relationship with wear and erosion behavior of ZrN is needed.

In the present study, microhardness of PVD arc evaporation ZrN coatings is evaluated as a function of the crystallite size, lattice strains and crystallographic orientation. Correlation of the erosion rate with material and structural parameters is described.

## 2. Experimental procedure

### *2.1. Coating deposition*

ZrN coatings were deposited on various substrates, such as AISI 304, AISI 410 and 17-4 PH stainless steel, Ti-6Al-4V and Inconel 718, using the Union Carbide PVD arc evaporation system. The principles of the deposition apparatus and process have been described previously [5, 16 - 19].

The substrate (1.5 mm × 25 mm × 50 mm) to be coated was mechanically polished to a surface finish of 0.04 - 0.1 $\mu$m Ra and then was cleaned ultrasonically in a bath of methanol. The zirconium target (purity, 95.5%), 64 mm in diameter, was mounted on the chamber wall facing the substrates. The target-to-substrate distance was between 250 and 400 mm. Before deposition, the chamber was evacuated to a pressure below $7 \times 10^{-4}$ Pa and then backfilled with argon (purity, 99.99%) to 0.7 Pa. The substrate was sputtered to remove surface contaminants using a negative bias of 1 kV direct current. Subsequently, the coating deposition was carried out in an Ar–$N_2$ gas mixture of 1.3 - 4.8 Pa. The purity of $N_2$ was 99.998%. During the deposition, the pressure was measured with a capacitance manometer and was kept constant. Substrates were biased from $-50$ to $-250$ V. The deposition temperature, with an accuracy within $\pm 20$ °C, was measured using an optical pyrometer. Typically, the deposition rate was from 3 to 7.2 $\mu$m h$^{-1}$ and the coating thickness varied between 20 and 30 $\mu$m.

### *2.2. Microstructure examination and microhardness measurement*

The cross-sectional microstructure, surface morphology and erosion scars of the coatings were examined by optical and scanning electron microscope (SEM) analysis. Samples for microstructure and microhardness evaluation were metallographically polished through 120, 240, 320 and 600 SiC grit papers, followed with 1 $\mu$m diamond paste. The hardness was measured on the coating surface using a Vickers hardness tester with a load of 0.1 kg for a loading time of 15 s. At least 15 measurements were taken for each sample. The diagonal length of indentation was measured on optical micrographs at 1000 times magnification and then converted to the Vickers hardness number.

### 2.3. X-ray diffraction (XRD) analysis

XRD techniques using Cu $K\alpha$ radiation were employed to determine phases, preferred orientation, macrostrain, microstrain and crystallite size of the coating. The preferred orientation of the coating either in $\langle 111 \rangle$ or $\langle 200 \rangle$ was obtained and was ranked quantitatively with the intensity ratio of (111) to (200) diffraction, $I(111)/I(200)$. For a ZrN powder, the magnitude of $I(111)/I(200)$ is equal to unity. Therefore, the coating exhibiting a $\langle 111 \rangle$ preferred orientation has $I(111)/I(200)$ greater than unity. Conversely, the coating exhibiting a $I(111)/I(200)$ ratio less than unity has a $\langle 200 \rangle$ preferred orientation.

The macrostrain of the coating is a measure of the change of lattice spacings from the strain-free spacing via a shift of diffraction lines. The macrostrain of the coating in $[hkl]$ is given by $\varepsilon_{\perp}(hkl) = (d_{hkl} - d_0)/d_0$, where $d_{hkl}$ is the $(hkl)$ lattice spacing and $d_0$ is the strain-free $(hkl)$ lattice spacings. The $d_0$ was obtained from a powder coating sample which was mechanically removed from the substrate.

The integral breadth crystallite size and microstrain were determined by Lambda Research Incorporated (Cincinnati, OH). The (111) and (222) diffraction peak profiles were obtained on a horizontal Bragg–Brentano focusing diffractometer using graphite-monochromated Cu $K\alpha$ radiation. The line broadening was assumed to be solely attributed to reduced crystallite size and microstrain in the lattice and the imperfection of the instrument. Annealed ZrN powder with a grain size greater than several microns was used to determine the instrumental broadening.

The measured peak profiles were first corrected for polarization and geometric factors. The Rachinger method [20] was used to separate the $K\alpha$ doublet. The $K\alpha_1$ peak profiles were corrected for instrumental broadening by the Stokes method [21]. The two orders of the diffraction peak profile were analyzed using a program adopted from the DIFFAN program by Wagner [22]. The data was processed using the Enzo–Parrish [23] background estimation method and a seven point moving cubic smoothing of the Fourier coefficients. The Fourier series used to approximate the peak profiles was truncated at the first occurrence of a negative real transformation coefficient.

### 2.4. Erosion test

The erosion resistance of ZrN coatings was determined by impacting with angular $Al_2O_3$ particles at 20° and 90° impact angles. The test apparatus was set up according to recommended guidelines of the American Society of Testing Materials [24]. Compressed air with a pressure of 276 kPa was used to deliver alumina particles (nominal particle size of 50 $\mu$m) through an alumina nozzle (diameter 5 mm). The nominal particle velocity was 60 m s$^{-1}$ and the nozzle-to-specimen distance was maintained at 100 mm. In each test, 0.8 kg and 0.2 kg of alumina particles was used for 20° and 90° erosion, respectively. The erosion rate was measured in terms of the volume loss of the coating per unit gram of the erodent used in the test. The accuracy of the erosion rate measurement was estimated to be within $\pm 10\%$.

## 3. Results and discussion

### 3.1. Microstructure and crystallographic orientation

XRD analysis revealed a B-1 NaCl structure for ZrN coating with a preferred orientation of the (111) or the (200) plane parallel to the substrate surface. ZrN (111) and ZrN (200) represent the coatings with a preferred orientation in ⟨111⟩ and ⟨200⟩ respectively. Both coatings had a light yellow gold color with slightly more yellow color observed in ZrN (111). The difference is probably associated with crystallographic orientation and surface topography of the coating.

The crystallographic orientation of ZrN was dependent on bias voltage and chamber pressure. Johansen *et al.* [7] observed the formation of ZrN (200) at 3.3 Pa chamber pressure and −10 V bias voltage in an arc evaporation process. Jehn *et al.* [11] reported that the reactively sputtered ZrN (200) formed at 0.1 Pa and less than 350 °C. In the present work, at 2.4 Pa ZrN (200) was obtained at a negative bias voltage less than 100 V. At voltage greater than this value ZrN (111) was observed. Consequently, ZrN (200) was produced with relatively low deposition temperatures, less than 400 °C.

Scanning electron microscope (SEM) micrographs show the surface morphology of ZrN (111) and ZrN (200) coatings which were deposited at 560 °C and 390 °C respectively (Figs. 1a and 1b). Also shown are their XRD spectra. The surface structure of both coatings consisted of many dimples (depressions), which were formed along the substrate surface contours and defects. The surface structure of these two coatings was different. The difference was in the dimple size. The dimples of ZrN (111) were relatively larger and well diffused compared with that of ZrN (200), due to higher fluxes of positive ion bombardment during the ZrN (111) coating growth.

The as-polished cross-sectional microstructure of both ZrN (111) and ZrN (200) were featureless. Little or no porosity or inclusions were observed. SEM micrographs of the etched microstructure are shown in Fig. 2. Both coatings exhibited a columnar structure in which the columns grew nearly perpendicular to the substrate surface. The obvious difference between ZrN (111) and ZrN (200) was in the column size. Within columns, there existed fine crystallites with a dimension of approximately 26 nm and 15 nm for ZrN (111) and ZrN (200) respectively (Figs. 2a and 2b). Typically, the crystallite and column size of ZrN (111) was larger than that of ZrN (200). This is in good agreement with its deposition temperatures. The smaller crystallites and columns were formed at lower deposition temperatures.

### 3.2. Crystallographic orientation and microhardness

Figure 3 shows that the microhardness of ZrN coatings is a function of crystallographic orientation. The higher value of $I(111)/I(200)$ indicates the higher degree of ⟨111⟩ preferred orientation. Also shown in Fig. 3 is the microhardness variation of TiN with $I(111)/I(200)$. ZrN coatings had much higher microhardness than the bulk material (bulk hardness value 1500 kgf mm$^{-2}$). The maximum microhardness occurred in the coating with a

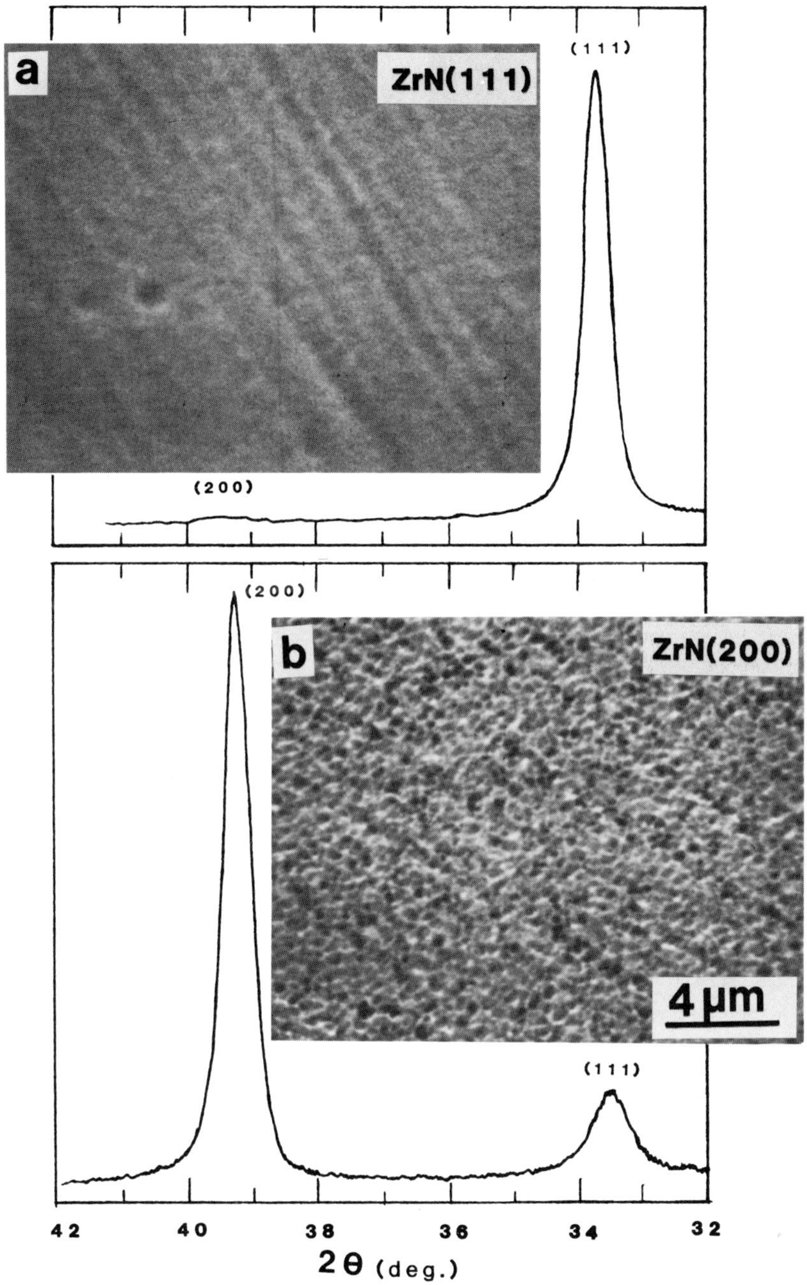


Fig. 1. SEM micrographs of surface morphology and XRD spectrum of typical (a) ZrN (111) and (b) ZrN (200) coatings (magnification marker applies to both micrographs).

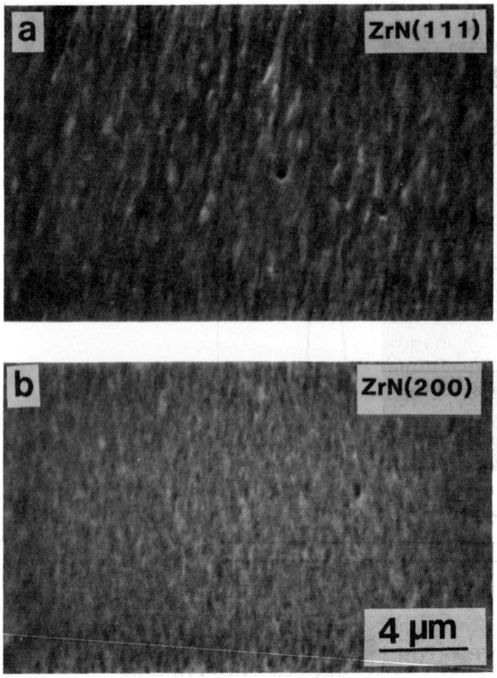


Fig. 2. SEM electron micrographs of etched cross-sectional surfaces of (a) ZrN (111) and (b) ZrN (200) showing columnar microstructure, etched with Kroll's reagent for 60 s (magnification marker applies to both micrographs).

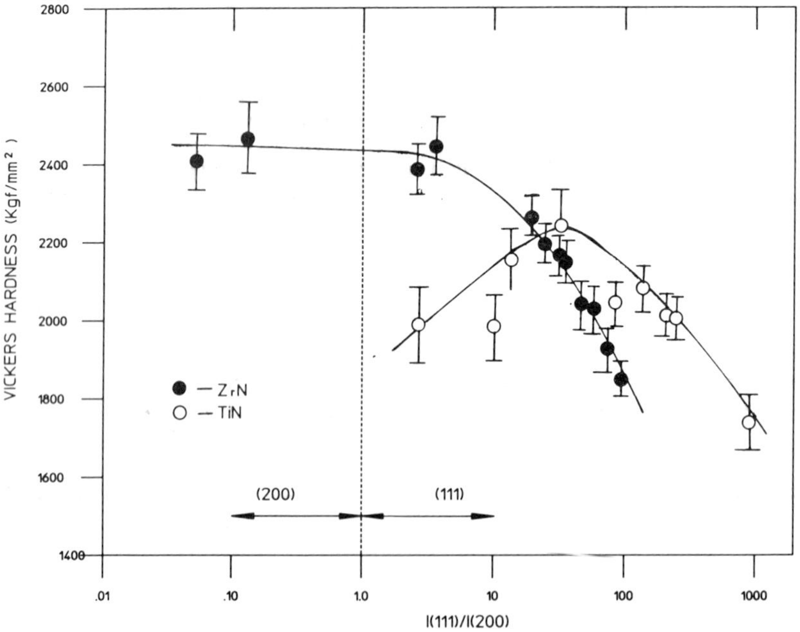


Fig. 3. Microhardness of ZrN and TiN coatings is a function of crystallographic orientation.

⟨200⟩ preferred orientation or with relatively low degree of ⟨111⟩ preferred orientation. This was due primarily to finer crystallites and higher macrostrain as discussed in the following sections. Over a range of $I(111)/I(200)$ from 10 to 100, microhardness of ZrN decreased with increasing $I(111)/I(200)$. Similar behavior occurred in TiN with the exception of $I(111)/I(200) < 80$, in which porosity and inclusions have a profound influence on coating hardness [25]. As mentioned earlier, pores and inclusions were negligible in ZrN. This may account for the fact that the effect of ⟨111⟩ preferred orientation on microhardness is pronounced, even at relatively low $I(111)/I(200)$ values.

### *3.3. Macrostrain and microhardness*

Residual macrostrains of the coatings were measured from the change of lattice spacings which was shown by a shift of diffraction lines. The measured macrostrains $\varepsilon_\perp(hkl)$ which are normal to the coating surface, are attributed to the intrinsic stress and thermal stress acting in the coating plane (parallel to the coating–substrate interface). The former results from the growth processes, depending primarily on deposition parameters; the latter arising from thermal mismatch between the coating and the substrate relates closely to deposition temperature and substrate materials. Marked effects of these stresses on measured macrostrains similar to that shown in TiN previously [26] were observed in ZrN. However, stress relaxation occurred in ZrN when deposited on relatively ductile substrates at more than 600 °C.

With the assumption of biaxial stresses acting in the coating plane, the residual macrostrain within the coating plane $\varepsilon_\parallel$ is proportional to $\varepsilon_\perp$ via Poisson's ratio effect, $\varepsilon_\parallel = -(1-\nu)\varepsilon_\perp/2\nu$, where $\nu$ is Poisson's ratio of the coating. Therefore, the residual macrostress in the coating plane can be easily calculated from either $\varepsilon_\perp$ or $\varepsilon_\parallel$. However, without knowing the specific values of Young's modulus and Poisson's ratio, the discussion is mainly confined to $\varepsilon_\perp$ in this work. For an expansion of lattice spacings in $[hkl]$, $\varepsilon_\perp(hkl)$ is positive and the residual stress in $(hkl)$ is compressive, and vice versa.

Highly anisotropic lattice strains were found in both ZrN (111) and ZrN (200) coatings (Table 1). The maximum macrostrain occurred in [111], and the minimum was in [200], despite the preferred orientation of the coating. Table 1 also shows that the higher the order of ⟨*hkl*⟩, the higher was the magnitude of macrostrain, *e.g.* $\varepsilon_\perp(222) > \varepsilon_\perp(111)$. A plot of microhardness *vs.* macrostrain

TABLE 1

Variations of macrostrain with crystallographic orientation in ZrN (200) and ZrN (111) coatings

| *Coating* | $I(111)/I(200)$ | $\varepsilon_\perp(hkl) = (d_{hkl} - d_0)/d_0$ (%) | | | | | |
|---|---|---|---|---|---|---|---|
| | | (111) | (200) | (311) | (222) | (400) | (331) |
| A | 0.06 | 0.97 | 0.22 | 0.63 | — | 0.30 | 0.70 |
| B | 0.14 | 0.94 | 0.20 | 0.82 | — | 0.32 | 0.81 |
| C | 27 | 0.54 | 0.10 | 0.27 | 0.63 | — | — |
| D | 75 | 0.48 | 0.22 | 0.41 | 0.57 | — | — |

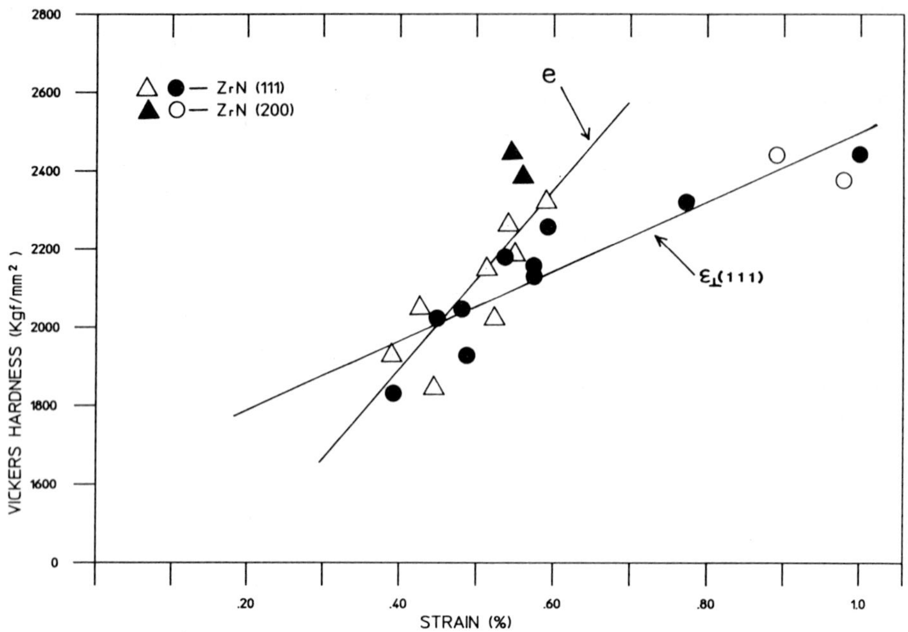


Fig. 4. Microhardness of ZrN coatings as a function of macrostrain $\varepsilon_{\perp}(111)$ and microstrain $e$.

$\varepsilon_{\perp}(111)$ showed a straight line with a correlation coefficient $r = 0.94$ (Fig. 4). A similar relationship would be expected between microhardness and residual compressive stress of the coating.

*3.4. Crystallite size, microstrain and microhardness*

The crystallite size and microstrain of ZrN coatings were determined using the integral breadth analysis based on the line broadening of the (111) and (222) diffractions. The crystallite size represents the dimension between the coherently diffracting domains which are generally smaller than the grain size. It has been reported [10, 27] that the PVD hard coatings exhibit a relatively high concentration of dislocations. The microstrain is attributed primarily to such imperfections. The magnitude of microstrain $e$ is indicative of the width of the microstrain distribution.

The microhardness of the coating was found to increase with increasing the peak width at half maximum (PWHM) of the (111) diffraction line. The combined effects of crystallite size and microstrain on the microhardness of the coatings were clearly evident, but no apparent relationship between crystallite size and microstrain was observed. Figure 4 also shows that microhardness increased with increasing microstrain. This behavior has been observed in steels [28] as well as nitrided steels [29]. Although there is a considerable scatter in the data in the linear plot of microhardness *vs.* microstrain ($r = 0.72$), it was certain that microstrain had greater influence on microhardness than macrostrain $\varepsilon_{\perp}(111)$.

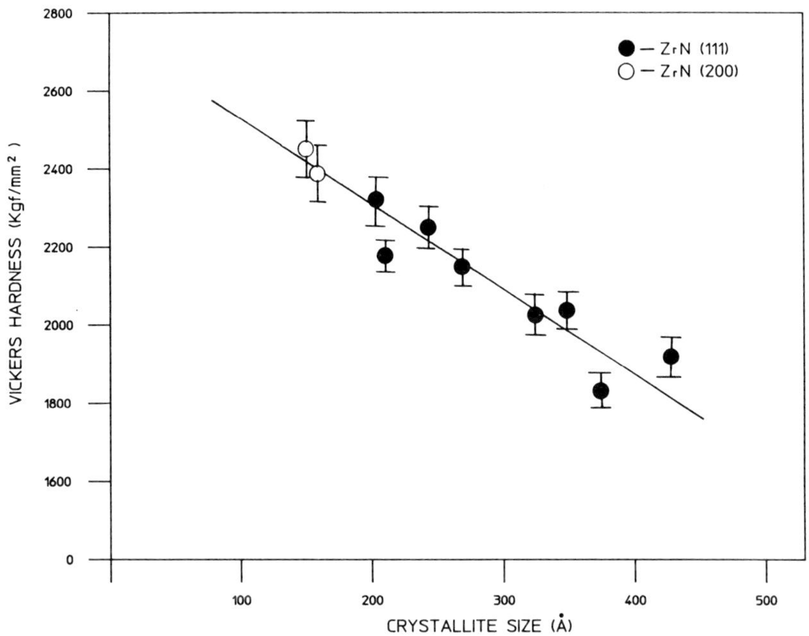


Fig. 5. A linear relationship of microhardness with crystallite size for ZrN coating.

An excellent fit to a straight line ($r = 0.95$) between hardness and crystallite size is shown in Fig. 5. The microhardness of the coating increased with decreasing crystallite size, from 1850 kgf mm$^{-2}$ at 42.5 nm to 2400 kgf mm$^{-2}$ at 14.5 nm. The relatively higher hardness in ZrN (200), compared with that of ZrN (111), was attributed to the finer crystallites. Since the amount of crystallite boundaries is inversely proportional to crystallite size, the hindrance to dislocation motions by crystallite boundaries may account for the increase in hardness of the coating having small crystallites.

*3.5. Erosion*

The erosion behavior of materials is very complex and depends on the properties of the target material as well as the impacting particles. In addition, the shape, velocity and impact angle of the impacting particles also play an important role. Hardness is one of the most important material properties related to wear and erosion behavior. As mentioned above, the microhardness of the ZrN coating depended strongly on crystallographic orientation, microstrain, macrostrain $\varepsilon_{\perp}(111)$ and crystallite size. The effects of these structural parameters on erosion are interrelated. It is not easy to distinguish between them.

A slight decrease in erosion rate with increasing $I(111)/I(200)$ value was noticed when the crystallite size and $\varepsilon_{\perp}(111)$ were kept constant. A linear relationship between erosion rate and microstrain with $r = -0.77$ was observed at 20° impact, but poor correlation ($r = -0.35$) was evident at 90° impact. A plot of the erosion rate against $\varepsilon_{\perp}(111)$ yielded no meaningful correlation. However, if crystallite size was considered, it was apparent that the erosion rate decreased linearly with increasing $\varepsilon_{\perp}(111)$ within the scatter of the data, for crystallite sizes greater than approximately 30 nm. The reduction in erosion rate was probably due to higher hardness as a result of higher $\varepsilon_{\perp}(111)$. As crystallite size decreased to less than approximately 20 nm, the erosion rate was nearly independent of $\varepsilon_{\perp}(111)$. This behavior was noticed at both 20° and 90° impacts. One point to be noted is that there was a transition of erosion behavior, similar to that reported in TiN [26] from a uniform erosion to the lateral cracking mechanism at $\varepsilon_{\perp}(111) > 1.6\%$.

Figure 6 shows that the erosion rate increased markedly with increasing crystallite size with $r = 0.82$ and $r = 0.78$ at 20° and 90° impacts respectively. Generally, ZrN (200) had better erosion performance than ZrN (111). Since the hardness of the coating depended strongly on the crystallite size (Fig. 5), the erosion rate decreased linearly with increasing hardness as

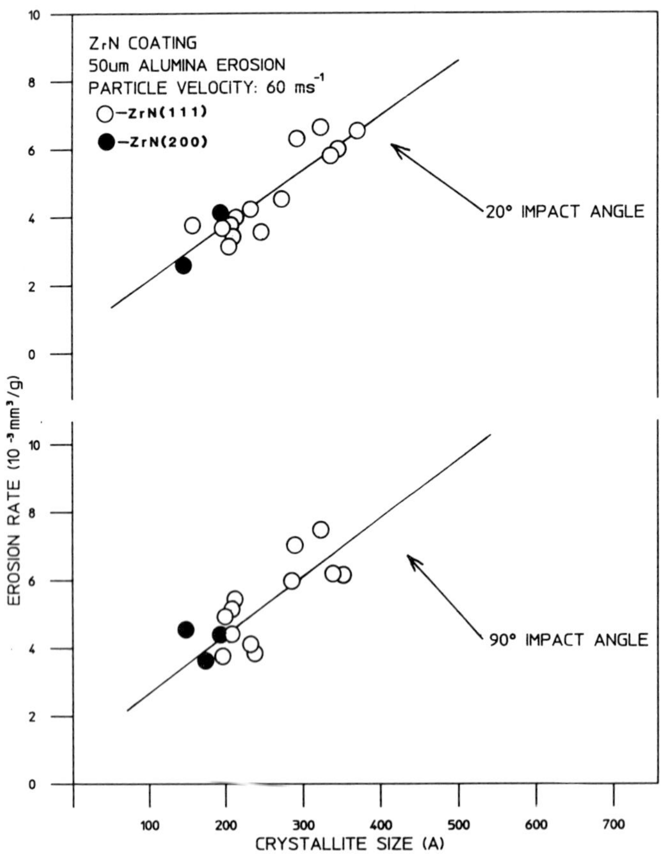


Fig. 6. Erosion rate *vs.* crystallite size plot for ZrN coatings at 20° and 90° impact angle.

expected (Fig. 7). The linear correlation coefficient was −0.77 and −0.82 for 20° and 90° impacts respectively.

SEM examinations revealed the eroded surfaces of typical ZrN (111) and ZrN (200) coatings at 20° and 90° impacts (Fig. 8). The crystallite sizes of ZrN (111) and ZrN (200) in Fig. 8 were 37 nm and 16 nm; macrostrain $\varepsilon_{\perp}(111)$ was 0.36% and 0.83%; and microhardness was 1840 kgf mm$^{-2}$ and 2390 kgf mm$^{-2}$ respectively; the microstrain of both coatings was approximately 0.46%. The hardness of alumina erodent was approximately 2000 kgf mm$^{-2}$. At 20° impact, a gouging mode of plastic deformation prevailed in the eroded surfaces of both coatings. A substantial amount of plastically deformed material appeared as a rim ahead of and around the scars. At 90° impact, eroded surfaces of both coatings consisted of chisel-like crushing scars, chipping and, to some degree, plastic deformation. In ZrN (111), fragments of impacting alumina particles less than 10 μm were observed embedded in the eroded surface, and highly plastically deformed material appeared around the embedded particles. Some intercolumnar fracture was observed in the chisel-like

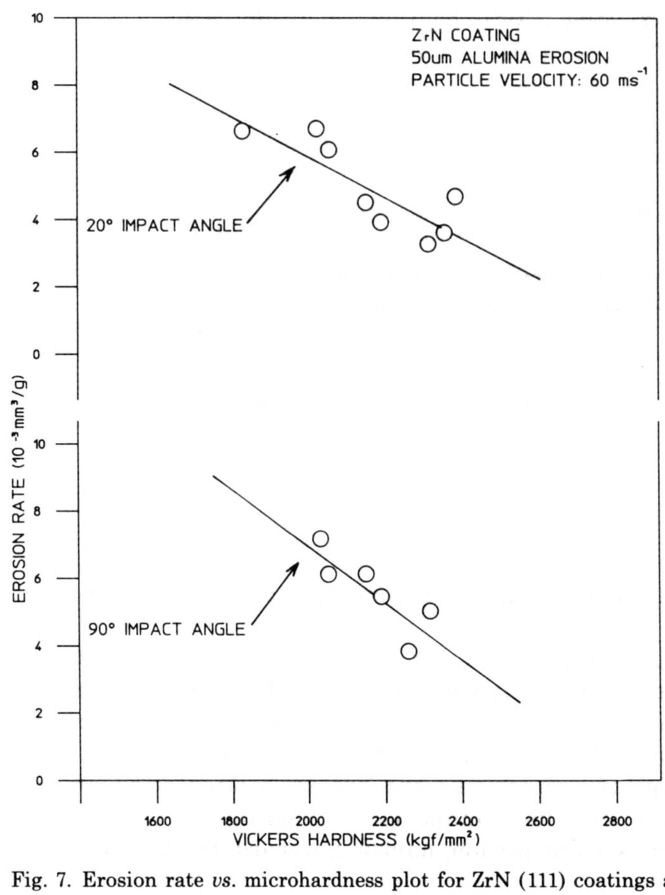


Fig. 7. Erosion rate *vs.* microhardness plot for ZrN (111) coatings at 20° and 90° impact angle.

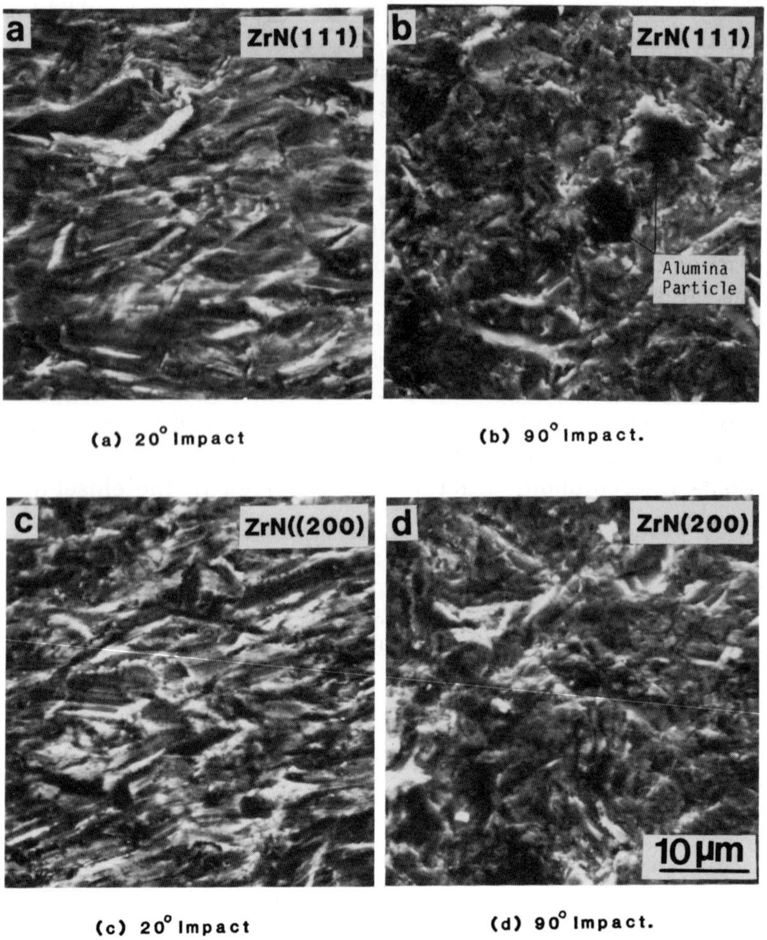


Fig. 8. SEM micrographs of eroded surfaces of ZrN (111) and ZrN (200) coatings at (a) and (c) 20° impact angle and (b) and (d) 90° impact angle (magnification marker applies to all micrographs).

scars of ZrN (200), in contrast to the faceted surface in ZrN (111). Overall, brittle fracture was prevalent on the eroded surfaces at 90° impact in both coatings.

Comparison of eroded surfaces of ZrN (111) and ZrN (200) at 20° and 90° impacts (Figs. 8a–d) shows that the grainy nature of eroded surfaces was more distinct in ZrN (200). The size of eroded scars at both impact angles reflected, to some extent, the significance of hardness effects. Overall, the effects of crystallite size, strains and preferred orientation on erosion are exhibited through their influence on hardness. The erosion processes in ZrN coatings were predominantly governed by hardness and crystallite size.

## 4. Conclusions

Both ZrN (111) and ZrN (200) coatings, which are deposited with the PVD arc evaporation process, consist of columnar structure with crystallite size from 14 to 42 nm. The crystallite size depends essentially on deposition temperatures. The microhardness of the coating depends on crystallographic orientation, microstrain, macrostrain $\varepsilon_{\perp}(111)$ and crystallite size. The microhardness is found to decrease with increasing $I(111)/I(200)$ value when it is greater than 10. The microstrain and macrostrain $\varepsilon_{\perp}(111)$ have an influence on the value of hardness, which increases with increasing microstrain and macrostrain $\varepsilon_{\perp}(111)$. The crystallite size has a predominant effect on the value of microhardness, which is inversely proportional to crystallite size. Typically, the value of the microhardness of ZrN (200) is higher than that of ZrN (111). This is attributed primarily to the finer crystallites in ZrN (200).

The erosion behavior of the coating was evaluated with 50 $\mu$m alumina particles and 60 m $s^{-1}$ particle velocity at 20° and 90° impacts. Generally, ZrN (200) has significantly better erosion performance than ZrN (111). The erosion process is predominantly governed by crystallite size and microhardness at both impact angles. The degree of (111) preferred orientation, microstrain and macrostrain $\varepsilon_{\perp}(111)$ have little influence on erosion rate when compared with the crystallite size. Effects of these structural parameters on erosion are exhibited through their influence on microhardness. SEM examinations of eroded surfaces revealed that brittle fracture and gouging with substantial plastic deformation are the primary mechanisms for coating erosion at 20° and 90° impacts.

## Acknowledgments

The authors are grateful to Mr. R. D. Johnson for preparation of the coatings, Mr. R. E. Newman for performing XRD measurements and Mrs. N. T. Pham for carrying out the alumina erosion test. We also thank Dr. R. C. Tucker Jr., for helpful discussions and Mr. P. S. Prevey III of Lambda Research Incorporated for doing the integral breadth of line broadening analysis.

## References

1 W. D. Sproul, *J. Vac. Sci. Technol., A4(6)* (1986) 2874.
2 R. G. Duckworth and I. H. Wilson, *Thin Solid Films, 63* (1979) 289.
3 W. D. Sproul, *Thin Solid Films, 107* (1983) 141.
4 K. Salmenoja and A. S. Korhonen, *Vacuum, 36* (1986) 33.
5 J. A. Sue and H. H. Troue, *U.S. patent application 905,510* (1986).
6 J. A. Sue and H. H. Troue, *U.S. patent application 911,510* (1986).
7 O. A. Johansen, J. H. Dontje and R. D. Zenner, *Thin Solid Films, 153* (1987) 75.
8 P. C. Johnson and H. Randhawa, *Surf. Coat. Technol., 33* (1987) 53.

9 J. C. Francois, G. Chassaing, P. Gravier, R. Pierrisnard and A. M. Bonnot, *Thin Solid Films, 127* (1985) 205.
10 D. T. Quinto, G. J. Wolfe and P. C. Jindal, *Thin Solid Films, 153* (1987) 19.
11 H. A. Jehn, J. H. Kim and S. Hofmann, *Thin Solid Films, 36* (1988) 715.
12 E. Torok, A. J. Perry, L. Chollet and W. D. Sproul, *Thin Solid Films, 153* (1987) 37.
13 M. Ostling, S. Nygren, C. S. Peterson, H. Norstrom, R. Buchta, H.-O. Blom and S. Berg, *Thin Solid Films, 145* (1986) 81.
14 J. M. Molarius, A. S. Korhonen, E. Harju and R. Lappalainen, *Surf. Coat. Technol., 33* (1987) 117.
15 W. D. Sproul, *Surf. Coat. Technol., 33* (1987) 133.
16 A. A. Snapper, *U.S. patent 3,625,848* (1971).
17 L. P. Sablev, N. P. Atamansky, V. N. Gorbunov, J. I. Dolotov, V. N. Lutseenko, V. M. Lunev and V. V. Usov, *U.S. patent 3,793,179* (1974).
18 L. P. Sablev, J. I. Dolotov, L. I. Getman, V. N. Gorbunov, E. G. Goldiner, K. T. Kirshfeld and V. V. Usov, *U.S. patent 3,783,231* (1974).
19 A. A. Snapper, *U.S. patent 3,836,451* (1974).
20 W. A. Rachinger, *J. Sci. Instrum., 25* (1948) 254.
21 A. R. Stokes, *Proc. Phys. Soc., B61* (1948) 382.
22 C. N. J. Wagner, *Technical Report No. 15, NONR Contract 609,* (1966) 43.
23 S. Enzo and W. Parrish, *Adv. X-ray Anal., 27* (1983) 37.
24 *ASTM G 76-83.*
25 J. A. Sue and H. H. Troue, *Surf. Coat. Technol., 33* (1987) 169.
26 J. A. Sue and H. H. Troue, *Surf. Coat. Technol., 36* (1988) 695.
27 I. Penttinen, J. M. Molarius, A. S. Korhonen and R. Lappalainen, *J. Vac. Sci. Technol., A6(3)* (1988) 2158.
28 R. E. Marburger and D. P. Koistinen, *Trans. Am. Soc. Met., 53* (1961) 743.
29 R. Delhez, Th. H. de Keijser and E. J. Mittemeijer, *Surf. Eng. 3(4)* (1987) 331.

*Surface and Coatings Technology*, *39/40* (1989) 435 - 444



# CrN AND (Ti,Al)N COATINGS DEPOSITED BY THE STEERED ARC AND RANDOM ARC TECHNIQUES

E. ERTÜRK, H.-J. HEUVEL and H.-G. DEDERICHS

*Interatom GmbH, Friedrich-Ebert-Strasse, 5060 Bergisch Gladbach 1 (F.R.G.)*

(Received March 19, 1989)

## Summary

This paper introduces CrN and (Ti,Al)N coatings deposited using the random arc and steered arc techniques.

When compared with TiN, CrN coatings are characterized by the fact that they can be deposited as thick, very dense coatings of up to 60 $\mu$m. They exhibit high hardness values of up to 2900 HK 0.05, very good toughness, relatively smooth coatings and, even in the range of high coating thickness, lower friction coefficients. At high temperatures they show improved oxidation behaviour in the 400 - 750 °C range, better corrosion resistance in the salt spray test and higher deposition rates. These coatings promise successful applications, in particular in the abrasive wear sector.

Like many new coatings, the (Ti,Al)N coatings can be produced using the arc technique at high deposition rates. A loss of aluminium takes place during coating phase. Maximum hardness values of up to 3400 HK 0.05 are attained for coatings with 12 at.% Al. With increasing aluminium content there is an increase in roughness, which is lower on the whole in steered arc coatings.

For uninterrupted cutting (turning), thick (Ti,Al)N random arc coatings on hard metal are even superior to CVD multilayer coatings. In the case of interrupted cutting (milling) with HSS tools, the (Ti,Al)N coatings, in particular the steered arc coatings, are superior to TiN.

## 1. Introduction

In the course of the 1980s, ion-plated TiN coatings have proved that they offer good protection against wear, especially for tools. Today these coatings are successfully used all over the world. However, application cases exist in which TiN coatings are less promising. For instance, these include severe abrasive loads such as those frequently found in the paper, textile and plastic industry. Under these conditions, great importance is not only attached to the hardness and adhesion of the coating but also to its thickness as a reservoir against abrasive loading. However, TiN coatings

0257-8972/89 Elsevier Sequoia/Printed in The Netherlands

can seldom exceed 10 $\mu$m because of the stresses in the coating. Furthermore, problems are often encountered with the corrosion protection of thin TiN coatings.

CrN, a hard material which can be easily deposited, promises better results in this case. Up to now it has been reactively deposited using magnetron sputtering [1 - 3] from chromium cathodes and using low-voltage arc evaporators [4] from a crucible. These papers consider CrN coatings with higher oxidation resistance than TiN and with hardnesses of between 1500 and 2500 HV at higher deposition rates.

In the past years (Ti,Al)N has become popular as a hard coating for tools [5 - 9]. Service life advantages over TiN-coated tools under certain machining conditions, in particular at higher cutting speeds, are discussed.

These two promising new hard materials were deposited using the arc technique.

## 2. Experimental procedures

The arc coatings investigated were produced using Interatom PVD 20 arc equipment with one steered arc cathode (diameter 125 mm) and one random arc cathode (diameter 63 mm).

Prior to coating, the samples were first subjected to vapour degreasing with supporting ultrasonic systems in trichlorotrifluoroethane. They were then ultrasonically cleaned in two aqueous alkaline baths and rinsed with water. This was followed by three-stage spot-free water displacement (Freon TDA 35 X, Freon TF, Freon TMS).

The following investigations for characterization were performed on the coated sample.

The coating thickness was determined with a Sloan Dektak 3030 surface profiler at the step from the uncoated to the coated surface and by means of the X-ray fluorescence method.

Surface roughness was established using a brush analyser. The average surface roughness $R_z$ was determined according to DIN 4768. The surface quality and morphology of the coatings were analysed using a scanning electron microscope. The CSEM Revetest automatic scratch tester was used to determine the critical load of the coatings.

Standard scratch conditions were: diamond tip radius $R = 0.2$ mm, scratching speed $dx/dt = 10$ mm min$^{-1}$, loading rate $dL/dt = 100$ N min$^{-1}$, sensitivity 1.2 scale units.

The Knoop microhardness measurement was used on base materials and coated material because it has advantages over the Vickers microhardness measurement in the case of thin hard coatings.

The friction coefficient was determined with a CSEM Tribometer in the ball–disc arrangement at a normal force of 4.9 N and a grinding speed of 0.1 m s$^{-1}$ over a measured section of 500 mm.

Corrosion resistance of TiN- and CrN-coated samples were examined in spray tests with sodium chloride according to DIN 50021-SS.

The constitution of coatings and cathode materials was determined by quantitative electron beam microanalysis (ESMA) [10].

The oxidation behaviour of TiN- and CrN-coated samples in air was determined by differential thermal analysis (DTA) in the temperature range between 480 and 750 °C.

## 3. Results

### *3.1. CrN*

The deposition rate of the grey-metallic CrN coatings is higher than for TiN coatings for both the random arc and the steered arc technique. As already shown for TiN the deposition rate is higher for the random arc system at the same arc current. However, by increasing the arc current the same deposition rates are attained for both techniques (Table 1).

A special feature of the CrN arc coatings is that coating thicknesses which are one order of magnitude higher can be attained with relatively few problems up to a value of 60 μm. Despite the higher coating thickness, adhesion of the coatings is adequate for technical applications (Table 2).

The dependence of the surface roughness on the thickness is apparent from Fig. 1.

CrN coatings are still relatively smooth despite the very high thicknesses. The structure of a CrN coating is illustrated in Fig. 2. The 30 μm thick coating is exceptionally dense and exhibits an almost amorphous structure without a privileged direction. Magnifications show a particularly fine-grained structure. Individual small droplets are also present in the hard material structure.

Another important technical feature of hard coatings is their toughness. According to Palmquist, a Vickers diamond indentor was used to examine the toughness of CrN coatings.

TABLE 1

Deposition rate and roughness of random and steered arc CrN coatings

| | | *Random arc technique* | *Steered arc technique* |
|---|---|---|---|
| Deposition rate | ($\mu m\ h^{-1}$) | 12 | 12 |
| Surface roughness | $R_a$ (μm) | 0.02 | 0.02 |
| before coating | $R_z$ (μm) | 0.11 | 0.12 |
| Surface roughness | $R_a$ (μm) | 0.05 | 0.04 |
| after coating | $R_z$ (μm) | 2.05 | 1.62 |

Substrate material, HSS Mat.-No. 1.3207; coating thickness, 5 μm; arc current: random arc 60 A, steered arc 120 A.

TABLE 2

Comparison of random arc TiN and CrN coatings

| | | *TiN* (4 μm) | *CrN* | |
|---|---|---|---|---|
| | | | (4 μm) | (20 μm) |
| Hardness (on cemented carbide) (HK 0.05) | | 3300 | 2600 | 2900 |
| Critical load (N) | | 75 ± 3 | 50 ± 5 | 50 ± 2 |
| Friction coefficient | At room temperature against 100 Cr6 | 0.65 | 0.57 | 0.57 |
| | At 600 °C against HSS | 0.47 | 0.39 | 0.47 |
| Wear of ball (100 Cr6) in ball and disc test ($10^{-14}$ m$^3$ m$^{-1}$ N$^{-1}$) | | 3.6 | 2.0 | 3.5 |

Figure 3 compares the coating area around the indentation. The TiN coating on hard metal is relatively brittle and chips widely under this load while no chipping is exhibited by the CrN coatings, not even by the much thicker one. Only fine cracks proceeding from the corners of the indentation can be detected. The fact that CrN coatings show greater toughness than TiN coatings points to a more favourable state of stresses, which allows thicker

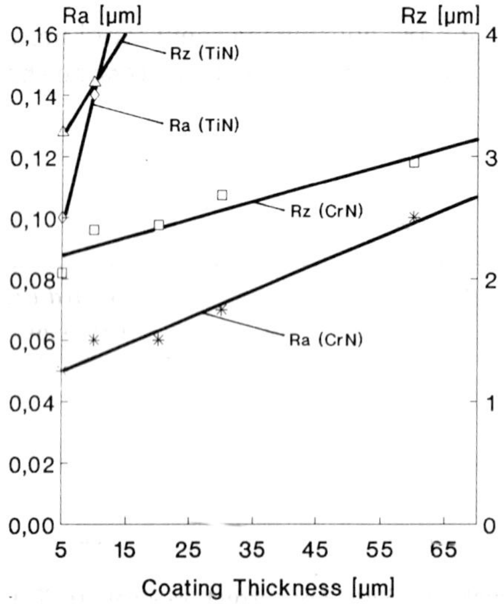


Fig. 1. Influence of coating thickness on surface roughness (random arc technique).

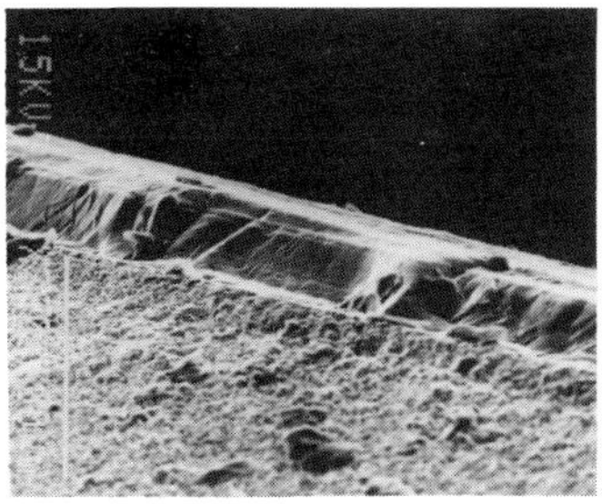

300 x

Fig. 2. Structure of a 30 $\mu$m thick CrN random arc coating on cemented carbide (SEM) (bias voltage −100 V, magnification 150 ×).

TiN, 4 µm    CrN, 4 µm    CrN, 30 µm

Fig. 3. Influence of coating material and coating thickness on toughness of hard coatings (Vickers diamond indentation with 50 kgf, substrate material cemented carbide, bias voltage −100 V, magnification 120 ×).

Fig. 4. Toughness of a 60 $\mu$m thick CrN coating on HSS substrate (Vickers diamond indentation with 50 kgf, bias voltage −100 V, magnification 120 ×).

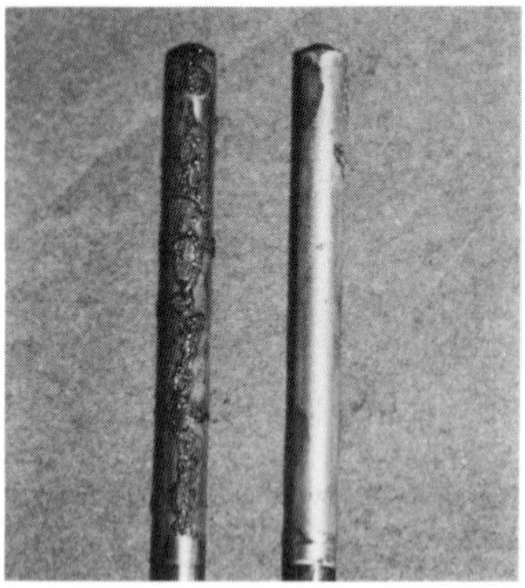


Fig. 5. Corrosive attack in the salt spray test DIN 50021 (a, TiN; b, CrN).

CrN coatings to be deposited with relatively few problems and at higher deposition rates than TiN coatings.

Even a 60 μm thick coating on HSS substrate is quite ductile under load (Fig. 4). These CrN arc coatings, which are unusually thick for ion-plated hard coatings, are capable of absorbing high loads and can be used in technical applications.

In conjunction with the possible high coating thicknesses, CrN coatings improve the corrosion resistance of the coated substrates. Figure 5 shows the different corrosive attack on two coated specimens after 8 days in a salt spray test according to DIN 50021. When compared to the thin TiN coating, the attack on the specimen with a 20 μm thick CrN coating is much slighter. As

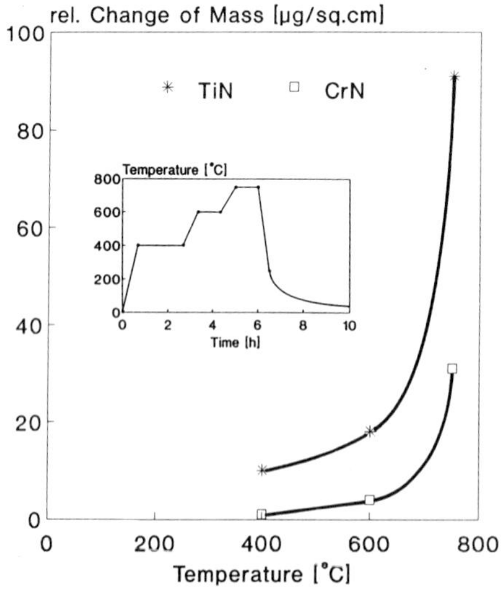


Fig. 6. Oxidation behaviour of TiN and CrN random arc coatings in air (substrate material, stainless steel; coating thickness, 5 μm).

apparent from already corroded points, absolute corrosion protection cannot be achieved with thick CrN either. However, in many application cases, the corrosion resistance which is attained may be adequate.

Finally the higher resistance of the CrN coatings to oxidation is technically interesting. Figure 6 shows the increase in mass of TiN and CrN coatings at temperatures of between 400 and 750 °C.

### *3.2. (Ti, Al)N*

In the case of (Ti, Al)N coatings, the properties are very dependent on the aluminium content in the coatings. Table 3 shows the average roughnesses as a function of the aluminium content of (Ti, Al)N coatings of the same thickness manufactured using the random arc and steered arc techniques. An increase in aluminium content causes deterioration of the surface quality of arc coatings. This is a result of the elevated droplet discharge of arc coatings in the case of metals having lower melting points (*e.g.* aluminium). The steered arc coatings are significantly smoother.

When compared with TiN, the deposition rates increase with increasing aluminium content in the case of (Ti, Al)N. The steered arc coatings are inferior to random arc coatings at the same arc current. However, it is important to remember that, in the steered arc technique, the larger cathodes lead to more uniform distribution of the coating thickness over a much larger area [11] (Table 4).

With reference to the arc technique, the (Ti, Al)N coatings exhibit a lower aluminium content than in the cathode. The higher degree of ionization of titanium (50 - 90% as compared with aluminium at 50 - 60%) and the preferred resputtering of aluminium are mechanisms which lead to an increasing aluminium loss or titanium enrichment in the coating with increasing bias voltage, (Table 5). The random and steered arc techniques do not differ significantly with regard to the aluminium loss.

The structure of the (Ti, Al)N coatings is similar for both arc variants. Densification of the structure takes place with increasing bias voltage. As shown in Fig. 7, compact dense coatings are typical.

TABLE 3

Influence of aluminium content on surface roughness $R_z$ of (Ti, Al)N coatings

| *Aluminium content in the coating* (at.%) | *Roughness $R_z$* (μm) | |
|---|---|---|
| | *Random arc* | *Steered arc* |
| 0 | 3.2 | 2.4 |
| 12 | 4.3 | 3.8 |
| 34 | 4.9 | — |

(Coating thickness, 4 - 4.5 μm; bias voltage, −100 V.)

TABLE 4

Deposition rates with one cathode

| | Deposition rate ($\mu$m h$^{-1}$) | |
|---|---|---|
| | *Random arc* | *Steered arc* |
| TiN | 13 | 5 |
| $Ti_{0.88}Al_{0.12}N$ | 40 | 13 |

(Cathode–substrate distance, 150 mm; $P_{n2} = 5 \times 10^{-3}$ mbar; evaporator current, 80 A)

TABLE 5

Loss of aluminium during coating phase

| | | | | |
|---|---|---|---|---|
| Aluminium content in cathode (at.%) | 25 | 35 | 50 | 65 |
| Aluminium content in coating (at.%) (random arc) | 12 | 26 | 34 | 45 |

Bias voltage, −100 V.

TABLE 6

Critical load of TiN and (Ti, Al)N coatings

| | | | | | | |
|---|---|---|---|---|---|---|
| Aluminium content in cathode (at.%) | 0 | 12 | 16 | 34 | 35 | 45 |
| Critical load (N) random arc | 82 | 54 | 29 | 26 | — | 21 |
| Critical load (N) steered arc | 72 | 69 | — | — | 38 | — |

(Substrate material, HSS Mat.-No. 1.3207; coating thickness, 4.5 - 5.5 $\mu$m.)

In all (Ti, Al)N coatings, X-ray investigations can only determine a single phase, which can be identified as a (Ti, Al)N mixed phase. The preferred orientation is in the 111 plane. A reduction in size of the cubic cell of the (Ti, Al)N mixed phase is observed with increasing aluminium content in the coatings [12].

The hardness values for the (Ti, Al)N coatings are higher than those for TiN. Knoop hardness values between 3100 and 3400 HK 0.05 were measured for $Ti_{0.88}A_{10.12}N$ coatings for both arc variants. The higher values were generally determined for steered arc coatings.

Ball and disc friction tests with a ball of the material 100 Cr6 and coated HSS discs provided a friction coefficient of 0.68 for $Ti_{0.88}A_{10.12}N$ random arc

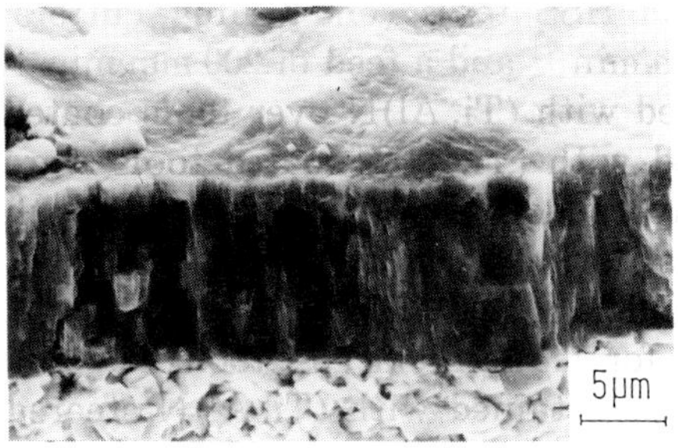


Fig. 7. Fracture surface of a (Ti, Al)N random arc coating (SEM).

coatings while a value of 0.65 was measured for the comparable TiN random arc coating.

It is difficult to determine the critical load with increasing aluminium content due to the increasing roughness of the coating surface. Table 6 must therefore be interpreted with care.

Finally, (Ti, Al)N coatings must prove themselves in the machining test. Figure 8 illustrates the results of a turning test with hard metal indexable inserts. The arc (Ti, Al)N coating with a thickness of 12 $\mu$m was produced in a random arc facility of type PVD 20 at the Technical University of Aachen using a cathode composition of Ti–Al of 75–25 at.%. The comparison of the lifetimes to just before the cutting break shows that the (Ti, Al)N coating is superior to a modern CVD multilayer coating (coating thickness, 13 $\mu$m). An increase in the aluminium content (*e.g.* based on cathode compositions of 50:50) leads to a reduction of the life.

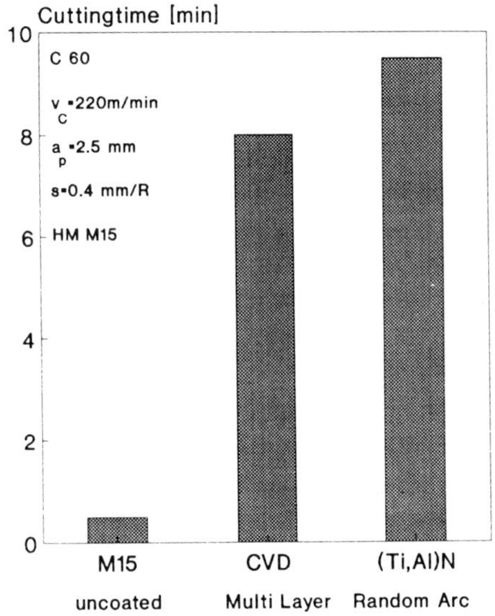


Fig. 8. Tool life of uncoated and coated indexable inserts in turning tests [12].

Similarly, milling of 42 CrMo4 with HSS tools (end-milling cutters, diameter 10 mm) at a cutting speed of 50 m $min^{-1}$ and a feed of 200 mm $min^{-1}$ shows the clear superiority of tools coated with (Ti, Al)N over tools coated with TiN. The milling distances attained with (Ti, Al)N-coated tools were approximately twice as long. The wide divergence of machining results with (Ti, Al)N-coated tools obtained in the past for magnetron sputtering cannot be observed at present in the experimental field for (Ti, Al)N arc-coated tools. The divergences are in the same range as for TiN. The largest tool life gains (which are considerable in some cases) are attained with (Ti, Al)N steered arc-coated tools. On completion of the tests in the experimental field the results will be published in statistically validated form.

## References

1 O. Knotek *et al.*, *High Temp.–High Press.*, *18* (1986) 435 - 442.
2 A. Aubert *et al.*, *Thin Solid Films*, *126* (1985) 61 - 67.
3 R. Gillet, A. Aubert and A. Gaucher, *Met. Technol.*, *10* (1983) 115 - 118.
4 G. Cholvy, J. L. Derep and M. Gantois, *Thin Solid Films*, *126* (1985) 51 - 60.
5 O. Knotek, M. Böhmer and T. Leyendecker, *J. Vac. Sci. Technol. A4* (1986) 2695 - 2700.
6 H. A. Jehn *et al.*, *J. Vac. Sci. Technol.*, *A4* (1986) 2701 - 2705.
7 O. Knotek, W. D. Münz and T. Leyendecker, *Leybold-Heraeus Publication 11-S08.2,* October 1986.
8 G. Hakason, J.-E. Sundgreen *et al.*, *Thin Solid Films*, *153* (1987) 55 - 65.
9 H. Freller and H. Hässler, *Thin Solid Films*, *53* (1987) 67 - 74.
10 O. Knotek, W.-G. Burchard, P. Karduk and T. Leyendecker, *Beitrag elektronenmikroskop. Direktabh. Oberfl. 19* (1986) 211.
11 E. Ertürk, H.-J. Heuvel, and H.-G. Dederichs, *Surf. Coat Technol.*, *39/40* (1989) 455.
12 O. Knotek, M. Atzor, F. Jungblut and H. J. Prengel, *Surf. Coat. Technol.*, *39/40* (1989) 445.

*Surface and Coatings Technology, 39/40* (1989) 445 - 453 

# WEAR RESISTANCE OF ARC-EVAPORATED AND MAGNETRON-SPUTTERED COATINGS ON CEMENTED CARBIDES

O. KNOTEK, M. ATZOR, F. JUNGBLUT and H.-G. PRENGEL

*Institut für Werkstoffkunde B, RWTH Aachen (F.R.G.)*

(Received March 9, 1989)

## Summary

The refinement of the physical vapour deposition (PVD) process has created the necessary conditions for controlled production of thin film systems with defined properties. Complex film systems such as (Ti,Al)N and (Ti,Al,V)N have particularly high potential as wear resistant functional coatings for highly-stressed tools. These films are deposited using the two highly-sophisticated PVD processes of ion plating with a magnetron sputtering source and ion plating with an ARC source. Methods of attaining defined film properties by altering process parameters are indicated.

Production of the films is directly linked with analysis or the development of suitable analytic techniques. Film properties such as thickness, microhardness and adhesion are optimized for maximum wear resistance. X-ray microstructure analyses are used to determine the phases and their crystalline structures and orientations. Finally, scanning electron microscope images document the structure and formation of the films.

The wear resistance of (Ti,Al)N- and (Ti,Al,V)N-coated indexable tips is demonstrated by means of machining tests under high stresses. The wear mechanisms are analysed and implications for further optimization of the films discussed.

## 1. Introduction

Physical vapour deposition (PVD) coating technology is gaining increasing importance in the production of hard wear-resistant coatings on carbide tools, owing to its process advantages, *e.g.* flexibility in terms of coating materials and deposition at temperatures which are low by comparison with those in chemical vapour deposition (CVD) coating [1, 2]. Industrial realization of PVD coating for carbide tools is the aim [3].

In the effort to meet the constantly increasing performance specifications for coated tools, optimization of coating processes is paralleled by the development of complex hard material coatings with improved properties

0257-8972/89/$3.50 

[4, 5]. Multicomponent TiN-based hard material compounds can be produced by substituting other metals (*e.g.* aluminium, vanadium) for the titanium atoms [6, 7]. The pre-conditions for the mixed phases of hard materials according to the Hume–Rothery rule are that the atomic radii are comparable and the lattice structures involved in the formation of the compounds are isotypic [8].

In the present study, complex hard material coatings in the Ti–Al–N and Ti–Al–V–N systems were deposited on carbide indexable tips, using the PVD processes of ion plating with a magnetron sputtering source (MSIP) and ion plating with an arc vaporization source (ARC). Metallographic, X-ray and electron microscope analyses of the coatings were supplemented by machining tests (turning tests) to determine the wear resistance of the (Ti,Al)N and (Ti,Al,V)N coated carbide indexable tips.

## 2. Experimental procedure

Complex (Ti,Al)N and (Ti,Al,V)N hard material coatings were deposited on SNGN 120408 sintered carbides of the ISO M15 group (82 wt.% WC, 11 wt.% (Ti,Ta,Nb)C, 7 wt.% Co).

These multicomponent hard coatings were formed using a Leybold–Heraeus Z400 coating plant (MSIP process) and an Interatom PVD 20 coating plant (ARC process), with nitrogen as a reactive atmosphere. A stable, economically-viable coating process (coating rate 6 - 10 $\mu$m) was achieved in both cases. In the MSIP process, a stable plasma was invariably triggered when a voltage was applied to the target in an argon atmosphere. The ARC process is characterized by a high degree of ionization of the vaporizing metal (approximately 80 - 90%), so that no extraneous gas (*e.g.* argon) is required [2]. A stable arc at the target is achieved by a single triggering operation.

In both these PVD processes, the composition of the compounds can be altered simply by changing the composition of the target and by modifying the reactive-atmosphere partial pressure and substrate bias parameters.

The Ti–Al–V concentration of the targets employed (pyrometallurgical, powder metallurgical and mechanical production) is indicated in Fig. 1. In the tests, the ternary and quaternary hard material films deposited reactively from these targets are sub-divided into ARC(Ti,Al,V)N, ARC(Ti,$Al_x$)N, aluminium-rich MSIP(Ti,Al,V)N and vanadium-rich MSIP(Ti,Al,V)N.

Initial basic results are produced by metallographic analysis of the coating thickness, hardness and adhesion. X-ray microstructural analysis is then used to determine crystalline structure and orientations and to identify the phases.

Scanning electron microscope (SEM) analysis is employed to document the structure and formation of the coatings. Apart from energy dispersive X-ray (EDX) analysis, quantitative electron beam microanalysis (ESMA) [9] yields data on coating composition.

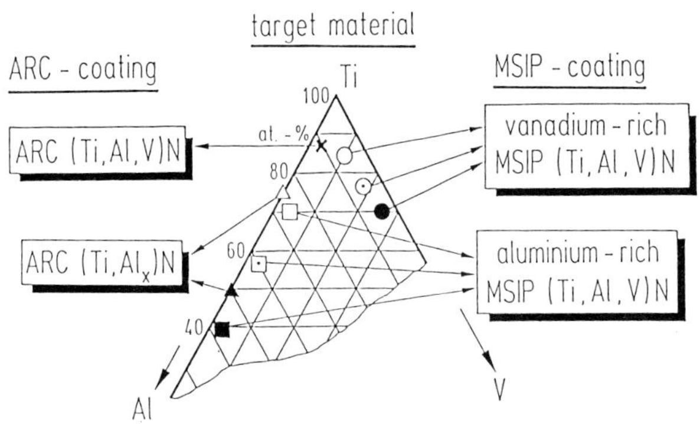


Fig. 1. Target compositions for the production of complex (Ti,Al)N and (Ti,Al,V)N hard material coatings.

The wear resistance of MSIP(Ti,Al,V)N- ARC(Ti,Al,V)N- and ARC (Ti,Al$_x$)N-coated carbide indexable tips is determined in machining tests under high stresses.

## 3. Results

### *3.1. MSIP(Ti,Al,V)N-coated sintered carbide*

Aluminium-rich MSIP(Ti,Al,V)N coatings exhibit the typical violet colouring of (Ti,Al)N, whereas vanadium-rich MSIP(Ti,Al,V)N coatings have a yellow–gold colour resembling that of TiN. Coating adhesion analysis using the scratch test indicated critical loads in excess of 70 N for all MSIP(Ti,Al,V)N coatings. There were no noticeable differences in coating adhesion for different coating compositions in the ultra-hard range of interest for wear resistance.

Peak hardnesses of 3000 HV 0.05 are measured. If aluminium-rich MSIP(Ti,Al,V)N coatings are deposited sub-stoichiometrically, there is a distinct drop in hardness to 1500 HV 0.05 for these aluminium-rich films. Vanadium-rich MSIP(Ti,Al,V)N coatings achieve their greatest hardness (3000 HV 0.05) at 4 at.% V in the coating. Particularly noticeable here are the flat curves with no pronounced hardness maximum, indicating non-critical stoichiometric behaviour.

X-ray structural analyses of the complex MSIP(Ti,Al,V)N coatings reveal the face-centered cubic (f.c.c.) lattice structure of TiN. Influenced by the alloying elements aluminium and vanadium, the X-ray spectrum is displaced towards the higher angles, greatly reducing the size of the MSIP(Ti,Al,V)N cell as compared to the TiN cell. The deviation of the TiN cell lattice parameter is plotted against aluminium and vanadium content in Fig. 2. The contraction of the cell (lattice parameter difference $\Delta\bar{a}$) is limited to a maximum alloying-elements/titanium ratio of 1.1 in the case of aluminium-rich MSIP(Ti,Al,V)N. The lattice parameter difference is then approximately 0.008 nm. If this saturation limit of the cubic cell is exceeded by

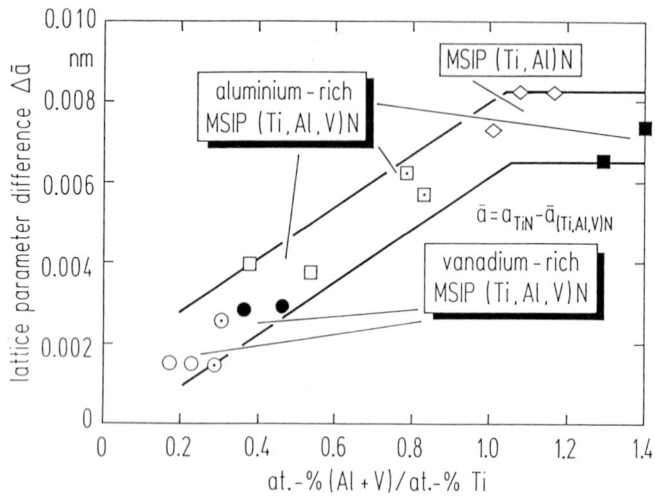


Fig. 2. Lattice parameter difference of complex MSIP(Ti,Al,V)N and MSIP(Ti,Al)N coatings as a function of the (Al + V)/Ti ratio.

further increasing aluminium content, no further reduction of the lattice parameter is observable. Under X-ray analysis, no second phase apart from the cubic (Ti,Al,V)N mixed phase is demonstrable at these high aluminium contents of 30 at.%.

The wear resistance of MSIP(Ti,Al,V)N-coated carbide indexable tips is investigated in machining tests (Fig. 3). Aluminium-rich MSIP(Ti,Al,V)N coatings display good wear behaviour at low aluminium contents. Excessive aluminium content (greater than 30 at.%) leads to partially lower crater wear resistance despite the good results obtained in metallographic analyses. This result may be ascribed to possible reaching or crossing of the solubility limit for aluminium and vanadium in the TiN lattice. This confirms the metastable character of aluminium-rich MSIP(Ti,Al,V)N coatings already observed in vacuum annealing tests at roughly 1000 °C.

The performances of vanadium-rich MSIP(Ti,Al,V)N-coated carbide indexable tips are clearly documented in the machining test. Crater wear is less than on the CVD TiC–TiN reference specimen. MSIP(Ti,Al,V)N with a high vanadium content exhibits particularly good thermal stability and diffusion resistance under the high temperature loads of the machining process, with correspondingly little crater wear, but an arbitrary increase in vanadium content of MSIP(Ti,Al,V)N coatings is limited by the associated increase in flank wear.

### 3.2. *ARC(Ti,Al,V)N- and ARC(Ti,$Al_x$)N-coated sintered carbide*

No problems are presented by ARC reactive processing of the complex hard material coatings of the Ti–Al–V–N and Ti–Al–N systems under investigation. ESMA analysis indicated that stoichiometric hard material

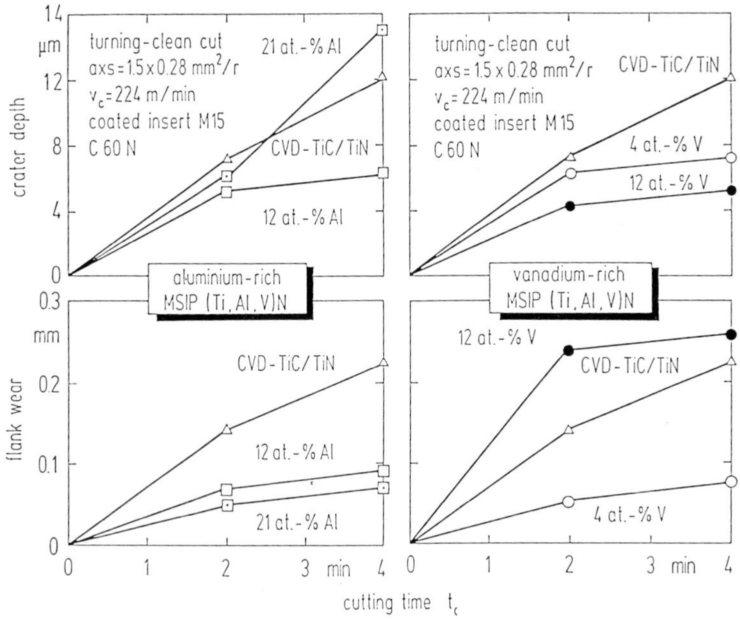


Fig. 3. Crater and flank wear on MSIP(Ti,Al,V)N-coated carbide indexable tips.

compounds are deposited across a wide range of reactive gas volume settings, virtually irrespective of $N_2$ partial pressure.

ARC(Ti,Al,V)N and ARC(Ti,$Al_x$)N coatings have good adhesive properties on carbide substrates. Critical loads of more than 75 N were measured in the scratch test. The ARC coatings in the test were produced to a coating thickness of 6 - 14 μm. A hardness increase for ARC(Ti,Al,V)N and ARC(Ti,$Al_x$)N coatings is observable in microhardness measurements when the bias voltage is raised (Fig. 4).

Above a bias voltage of 200 V, ARC(Ti,Al,V)N coatings achieve coating hardnesses of 3000 HV 0.05. ARC(Ti,$Al_x$)N coatings show comparatively lower hardnesses. A decline in microhardness is observable when the aluminium content in the TiAl target is increased from 25 to 50 at.%. Whereas ARC(Ti,$Al_x$)N coatings produced with a $Ti_{0.75}Al_{0.25}$ target attain a hardness of 2700 HV 0.05 at high bias voltages, microhardness values ranging from 1800 to 2200 HV 0.05 are measured for similar coatings deposited with a $Ti_{0.5}Al_{0.5}$ target.

X-ray analyses confirm the cubic lattice structure of the TiN cell, with little change in the size of the cell and the (111) plane as the preferred growth plane, for both ARC(Ti,Al,V)N and ARC(Ti,$Al_x$)N coatings. Figure 5 shows the lattice parameters as a function of the bias voltage. The lattice parameters of the ARC(Ti,Al,V)N mixed phase correspond to that of TiN in the ASTM register, while the lattice parameter of the ARC(Ti,$Al_x$)N mixed phase tends towards lower values, especially when coatings are deposited at low bias voltages. This reduced cubic cell in ARC(Ti,$Al_x$)N coatings produced

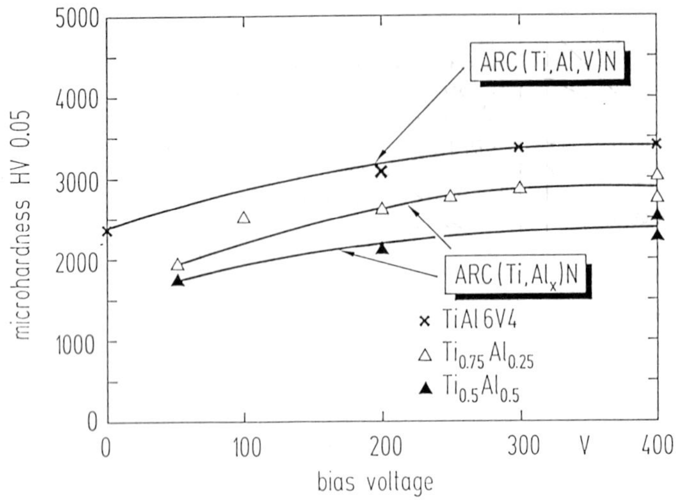


Fig. 4. Microhardness of ARC(Ti,Al,V)N and ARC(Ti,$Al_x$)N coatings as a function of bias voltage.

with a $Ti_{0.5}Al_{0.5}$ target corresponds to the lattice parameter size of MSIP(Ti,Al)N and aluminium rich MSIP(Ti,Al,V)N coatings (*cf.* Fig. 2).

The increase in the lattice parameter of the (Ti,$Al_x$)N mixed phase when the bias voltage is raised is based on a reduction in the aluminium content of the coating (Fig. 6). Since titanium has a higher ionization level (roughly 80%) than aluminium (roughly 50%), more titanium ions condense on the substrate at higher bias voltages, altering the Ti/Al ratio in

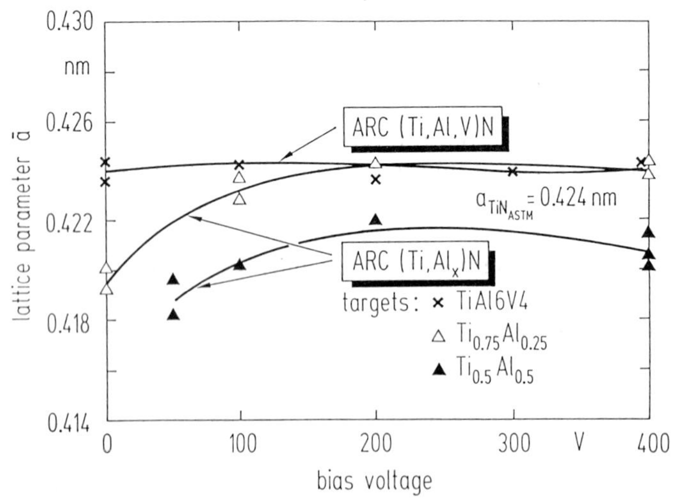


Fig. 5. Lattice parameters of ARC(Ti,Al,V)N and ARC(Ti,$Al_x$)N coatings as a function of bias voltage.

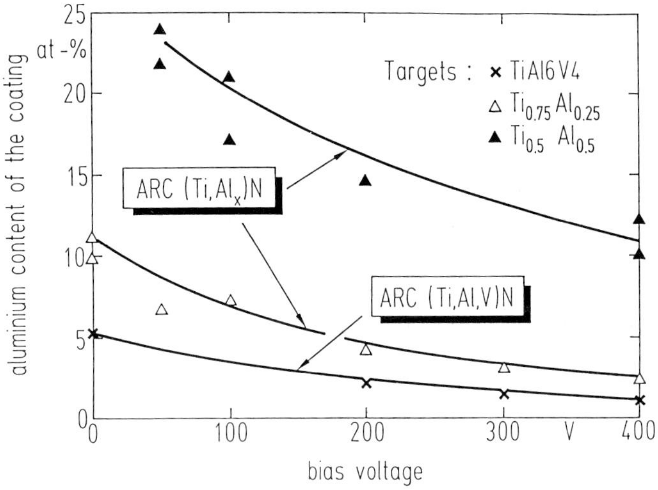


Fig. 6. Aluminium content of ARC(Ti,Al,V)N and ARC(Ti,$Al_x$)N coatings as a function of bias voltage.

uninfluenced stoichiometric deposition. ESMA analyses document this stoichiometric metal–metalloid composition and simultaneously confirm EDX studies used to confirm the titanium, vanadium and aluminium concentrations in the coating. The aluminium concentration in the complex ARC(Ti,$Al_x$)N coatings is particularly easy to modify, and can be adjusted within a concentration range from 22 to 8 at.% in the coating, depending on the bias voltage, if a $Ti_{0.5}Al_{0.5}$ target is employed.

Where a bias voltage is applied, ARC(Ti,Al,V)N and ARC(Ti,$Al_x$)N coatings grow in a dense, compact structure, as indicated by the SEM images in Fig. 7.

The performance and wear resistance of ARC(Ti,Al,V)N- and ARC(Ti,$Al_x$)N-coated carbide indexable tips were examined in a tool life

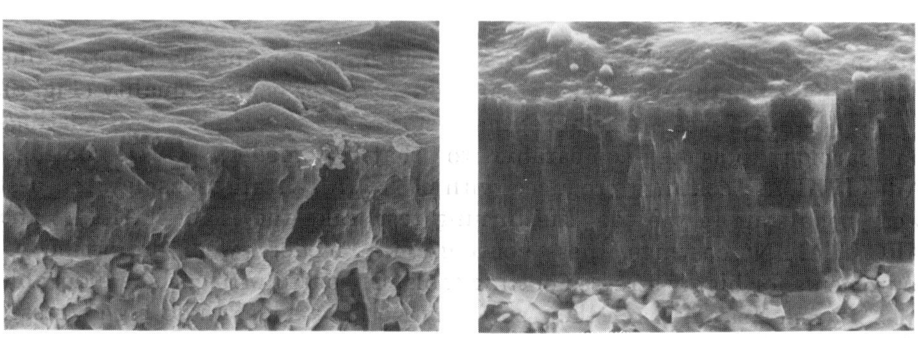

ARC(Ti,Al,V)N ARC(Ti,Alx)N 5μm

Fig. 7. Rupture structure of ARC(Ti,Al,V)N and ARC(Ti,$Al_x$)N coatings on carbide.

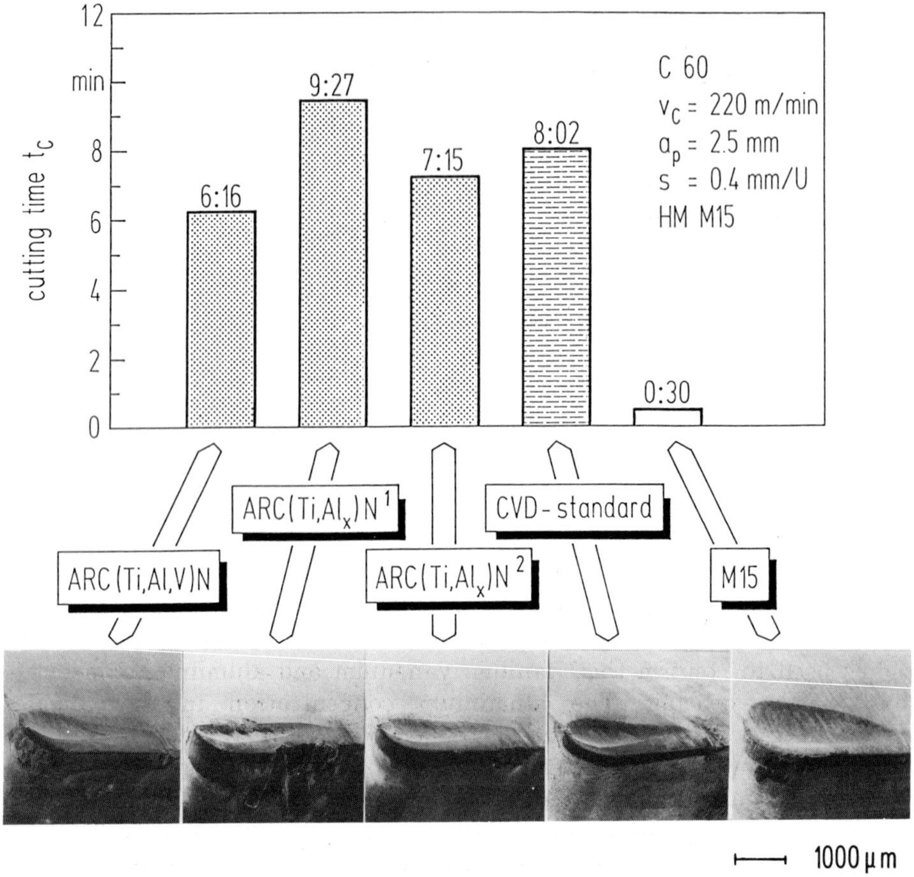


Fig. 8. Edge-holding of ARC(Ti,Al,V)N- and ARC(Ti,$Al_x$)N-coated carbide indexable tips (coating thickness: 10 - 12 µm). ARC(Ti,$Al_x$)$N^1$ with $Ti_{0.75}Al_{0.25}$, ARC(Ti,$Al_x$)$N^2$ with $Ti_{0.5}Al_{0.5}$.

turning test. The tool life of the coated tool up to a point shortly preceding cutting edge fracture was selected as a criterion for coating quality (Fig. 8).

In terms of wear resistance, the performance of complex ARC coatings on carbide tools is comparable to that of modern CVD coatings. ARC(Ti,$Al_x$)$N^1$ coatings produced with a $Ti_{0.75}Al_{0.25}$ target display the best wear behaviour in the tool life turning test. Like the ARC(Ti,Al,V)N and ARC(Ti,$Al_x$)$N^2$ coatings, these coatings meet modern wear protection requirements for highly-stressed cutting tools. Uncoated M15 carbide indexable tips, by comparison, reach their performance limit under these high stresses after a machining time of as little as 30 min.

The wear images of the tool cutting edge exhibit pronounced crater wear in addition to wear on the flank (VB = 0.2 - 0.3 mm). Maximum crater depths of 60 - 80 $\mu$m were measured.

## 4. Conclusions

Complex hard material coatings of the Ti–Al–N and Ti–Al–V–N systems possess considerable potential as wear-resistant functional coatings on highly-stressed carbide tools. These multicomponent hard material coatings are deposited on carbide indexable tips using the two modern PVC processes of ion plating with a magnetron sputtering source (MSIP) and with an arc source (ARC).

The influence of aluminium and vanadium on the properties of MSIP coatings was examined. Machining tests (turning tests) with MSIP(Ti,Al,V)N coated carbide indexable tips confirm the differing influences of vanadium and aluminium. It was demonstrated that a higher vanadium content clearly promotes stability and diffusion resistance, increasing the wear resistance of the face. Aluminium-rich MSIP(Ti,Al,V)N coatings possess high abrasion resistance, as documented by the good flank wear results.

The development of complex coating systems was confirmed by the performance and wear resistance of ARC(Ti,Al,V)N and ARC(Ti,Al)N coatings on carbide tools. Coating properties can be modified according to requirements via the bias voltage applied to the substrate. The edge-holding of ARC(Ti,Al,V)N- and ARC(Ti,Al)N-coated indexable tips is comparable to that of modern CVD coated indexable tips, as demonstrated under high stresses in the tool life turning test.

## References

1 O. Knotek, W. D. Münz and T. Leyendecker, *J. Vac. Sci. Technol.*, *A5* (1987) 2173–2179.
2 E. Ertürk and H.-J. Heuvel, *Thin Solid Films*, *153* (1987) 135 - 147.
3 P. Rödhammer, K. Kailer and A. Mairhofer, *SURTEC Berlin*, 1987.
4 H. Holleck, *Z. Werkstofftech.*, *17* (1986) 334 - 341.
5 U. König, *Surf. Coat. Technol.*, *33* (1987) 91 - 103.
6 O. Knotek, T. Leyendecker and F. Jungblut, *Thin Solid Films*, *153* (1987) 83 - 90.
7 H. Freller and H. Haessler, *Thin Solid Films*, *153* (1987) 67 - 74.
8 W. Hume-Rothery and G. V. Rayner, *The Structure of Metals and Alloys*, Institute of Metals, London, 1969.
9 O. Knotek, W.-G. Burchard, P. Karduck and T. Leyendecker, *Beitr. electronenmikroskop. Direktabb. Oberfl.*, *19* (1986) 211.

*Surface and Coatings Technology, 39/40* (1989) 455 - 464 

# COMPARISON OF THE STEERED ARC AND RANDOM ARC TECHNIQUES

E. ERTÜRK, H.-J. HEUVEL and H.-G. DEDERICHS

*Interatom GmbH, Freidrich-Ebert-Strasse, D-5060 Bergisch Gladbach 1 (F.R.G.)*

(Received March 19, 1989)

## Summary

A comparison of arc coatings produced by the steered arc and random arc techniques is given.

By application of a movable permanent magnet, the arc spot can be steered and forced to follow a predetermined path. This results in smaller craters and a clear reduction of molten cathode areas. Consequently, the steered arc technique can prevent larger droplets, such as those typical of the random arc technique. With regard to coating properties such as hardness, critical load and structure, no essential differences are indicated for both techniques. Both arc variants are suitable for successful application in wear protection. The selection of the appropriate system depends on the technical requirements.

Finally, the new steered arc technique gives the advantage of using segmented cathodes. Multicomponent coatings of various composition from the same target and multilayers, which can both be produced by the steered arc technique with segmented cathodes, signify an enormous industrial potential.

## 1. Introduction

The arc technique has become more and more important in recent years and is now used around the world. Hard coatings with particularly good adhesion and high density are deposited on many parts at relatively low temperatures.

In particular, the possibility offered by this technique of being able to work with independently functioning cathodes of any shape, size and number, and at any position, has led to both diverse batch system concepts up to 3 m in length and, more recently, to in-line systems exceeding 10 m in length.

The only problem with the arc technique concerns the droplets which are hurled off the arc cathodes as molten particles and integrated into the coatings, adversely affecting the surface roughness of the arc coatings.

0257-8972/89/$3.50 

For this reason, considerable efforts have been made in the last few years to reduce the droplet concentration in the arc coatings and to improve the surface roughness.

The steered arc technique is the most effective way of obtaining arc coatings with fewer droplets.

## 2. Considerations of droplet formation

Several researchers have already considered the formation of droplets [1 - 2]. However, the exact processes leading to droplet formation are still not clear. Possible formation mechanisms are as follows.

(a) The impact of ions on liquid craters expels molten material radially from the edges in all directions. However, only the generation of droplets which are smaller than the observed crater diameters can be explained with this assumption.

(b) However, many droplets are larger than the craters. Consequently, they cannot be traced back to the craters. In this case, it is interesting to note that large liquid areas are present on the cathode surface. As a result of the overlapping of craters they are all the more distinctive the slower the arc spot moves. This means that the velocity of the arc spot has a decisive influence on the formation of widespread molten areas and the ejection of large droplets. The processes in the cathode spot take place so quickly (in the order of 10 ns) that thermal equilibrium (in the order of 100 ns) only adjusts long after the arc has "jumped away". Consequently, larger areas are still melted by thermal conduction after leaving the cathode spot. In the course of this the entire liquid surface starts to move in a shock-wave manner as a result of the abrupt cessation of the ion pressure. This wave movement of the liquid surface can cause droplets to form, the diameter of which far exceeds the average crater size [3]. The following considerations on minimization of droplet ejection are drawn from droplet formation as described here.

(i) The volume of molten material produced during the life of a spot must be kept low, *i.e.* small craters (low arc current) and short spot life which can be caused by arc instabilities (*e.g.* as a result of external movable magnetic fields).

(ii) The arc path should be as elongated as possible to prevent widespread melting of the cathode surface (magnetic guidance). At the same time, efficient cooling of the cathode is essential.

(iii) A high melting point of the cathode material reduces the mean radius of craters and a low thermal conductivity limits the extension of molten areas.

Experimental results confirm that droplet ejection can be reduced by each of the points.

## 3. Modified process

It is known that most of the droplets are deposited on the substrate during the heating phase of the standard process (titanium bombardment in high vacuum). Therefore, an attempt was made to run this critical phase in such a way that the generation of droplets is either prevented by applying a different heating mechanism, such as glow discharge, or that the droplets are reduced in size by adding nitrogen during this phase (the modified process).

Nitrogen in the chamber causes poisoning of the cathodes, *i.e.* the formation of a compound which melts at higher temperatures (here TiN), resulting in smaller craters on the cathode. As shown in Fig. 1, the maximum droplet size clearly shifts towards smaller diameters in this process variant. This causes a clear reduction in the roughness of the arc coatings [4].

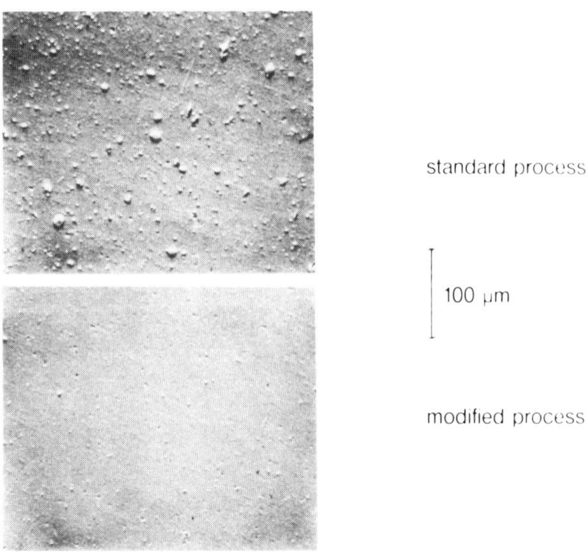


Fig. 1. Surfaces of two different random arc coatings (TiN).

TABLE 1

Surface roughness and critical load of TiN coatings manufactured by the standard or modified process

| | *Uncoated* | *Standard process* | *Modified process* |
|---|---|---|---|
| Surface roughness $R_a$ (µm) | 0.03 | 0.10 | 0.06 |
| Surface roughness $R_z$ (µm) | 0.25 | 2.02 | 1.34 |
| Critical load (N) | — | 53 | 52 |

Substrate material: HSS (1.3207) hardness 61 HRC.
Coating thickness: 5 - 6 µm.
Coating parameter: $P_{N_2} = 5 \times 10^{-3}$ mbar, $V_B = 200$ V, J = 60 A.

However, this type of process guidance also leads to different conditions on the interface of the substrate to be coated and can cause adhesion problems.

Table 1 compares the standard and the modified process. It is evident that the modified process reduces the roughness of the arc coatings by 30 - 40%, in this case without detectable loss of adhesion. The heating period is increased as a result of the decreasing substrate bias current.

## 4. Steered arc technique

The literature shows that magnetic fields can considerably reduce droplet concentrations. In the form realized by Ramalingham [6], a permanent magnet is positioned directly behind the arc cathode so that its magnetic field lines penetrate the arc cathode and exert a Hall force on the arc as current-carrying conductor. This Hall force is also exerted on the ions which form directly in front of the cathode spot (a distance of approximately one mean free path) as a result of the impact of electrons on evaporated cathode material. Some of these ions form the positive cloud in front of the cathode and are attracted by the cathode again. They are therefore overcome by the Hall force and arrive with circular deflection in the vicinity of the arc spot. This new impact point is heated locally. It emits electrons and prepares the location of the new arc spot.

These processes are presented schematically in Fig. 2. Finally, in the case of a round magnet, the arc spot moves along a circular track (Fig. 3)

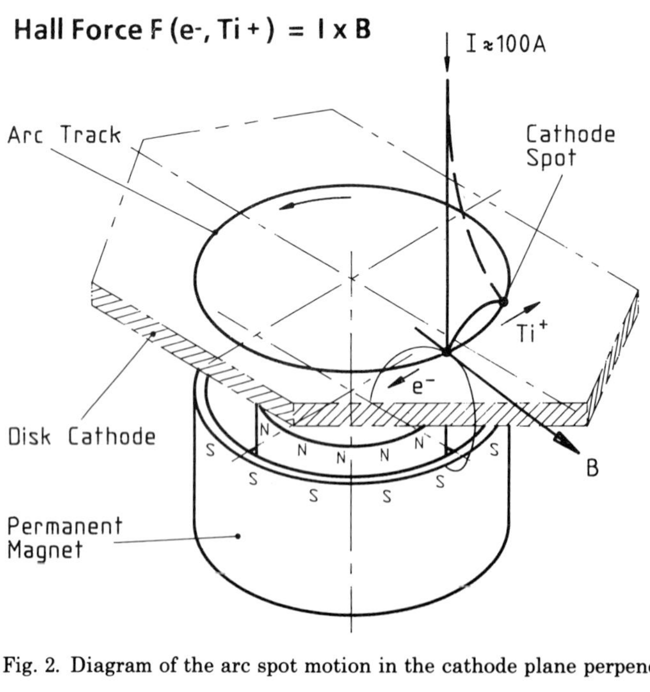


Fig. 2. Diagram of the arc spot motion in the cathode plane perpendicular to the magnetic field.

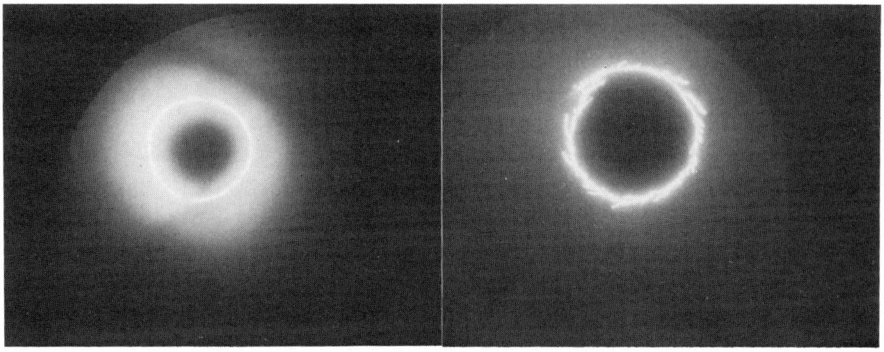

Fig. 3. Examples of steered arc tracks.

which is defined in such a way that the vertical component of the magnetic field is zero here. The cathode thickness, the prevailing external gas pressure [7] and the current strength are all variables which, for specified magnets, influence the control quality and, in the most unfavourable case as shown in Fig. 3 (right), permit too large random movement of the arc transverse to the specified track. Low cathode thickness, low process pressure and low current strength are optimal.

Any arc tracks can be realized by using differently shaped magnets or by superpositioning several magnets (Fig. 4). The background for the selection of a specific track form is the most uniform possible erosion of the cathode surface in conjunction with the mechanical movement of the magnet (or magnets). The life of a spot is also shortened by the continuous movement

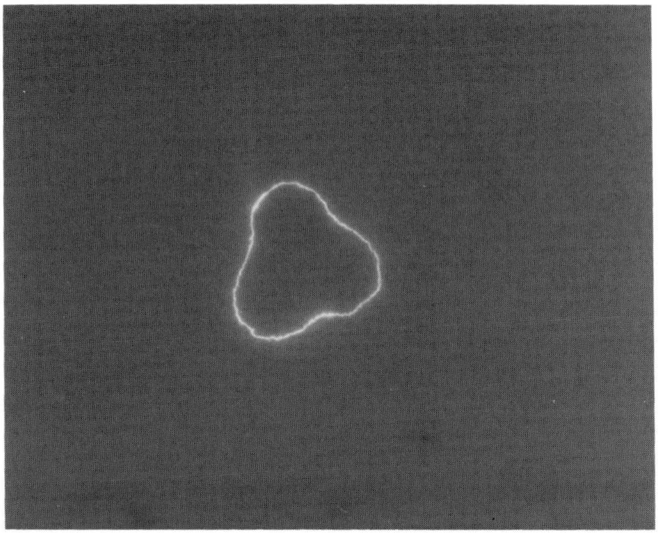

Fig. 4. Example of a steered motion of the arc spot.

of the magnetic field (as discussed in Section 2), *i.e.* smaller craters form and the arc movement is accelerated).

## 5. Comparison of the random and steered arc techniques

In order to see the influence of the moved magnet on droplet formation in the steered arc technique, the magnet was removed and the polished steered arc titanium cathode was operated with a boron nitride edge for arc confinement in the random arc version. As illustrated in Fig. 5 (left), medium-sized craters form here and, in some cases, generate interlinked large molten zones on the cathode. In compliance with the considerations in Section 2, these locations are sources of large droplets which, when integrated into the coatings, have an adverse effect on the roughness.

Very different behaviour is shown by the steered arc cathode operated with mobile permanent magnets in Fig. 5 (right). The average crater size is much smaller. Furthermore, most of the arc craters are directionally positioned one behind the other. Overlapping of these small craters is generally avoided and if it occurs, their small sizes do not lead to large molten areas.

The formation of large droplets can therefore be effectively avoided with the steered arc technique.

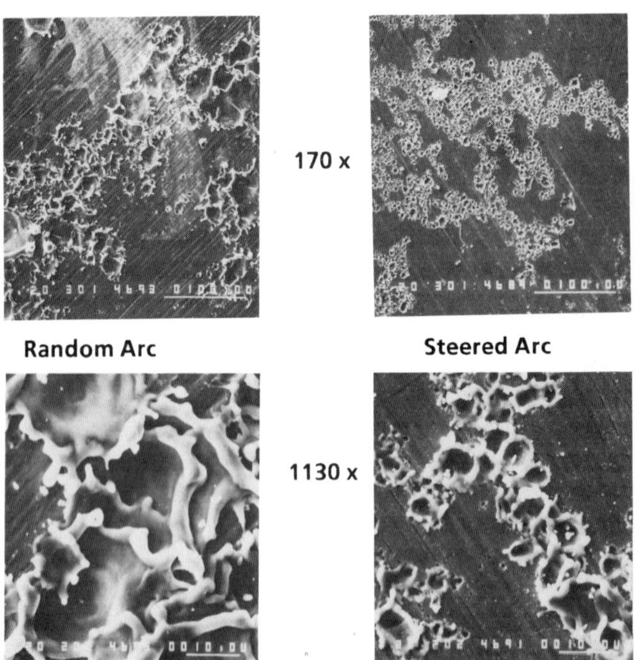


Fig. 5. Erosion craters for steered and random arc (arc current, 120 A; cathode material, titanium).

However, for industrial use the steered arc technique must be compared with the random arc technique which has been used for years now in the variant with stationary magnets for confinement of the arc spot on the cathode surface (in the following denoted by random arc*).

An Interatom PVD 20 system was used for a direct comparison of the properties of the two arc variants. It was fitted with a steered arc evaporator and a random arc evaporator equipped with a stationary magnetic field (Fig. 6).

Electrically connected specimens of the same material, with the same surface treatment and roughness, were subjected to the same surface cleaning. The spaces in front of the two cathodes (diameter of steered arc cathode, 125 mm; diameter of random arc* cathode, 63 mm) were shielded from each other with a sheet of metal to prevent superpositioning of the two vapour clouds. The coating processes were then performed automatically at the same pressure and symmetrical gas distribution, and with the same arc current of 80 A.

First, the erosion rate of the two cathodes was measured as weight loss per time unit and compared to the deposition rate of the TiN process (Table 2). It is evident that the erosion and deposition rates are much higher for random arc* coatings. However, this is combined with the disadvantage of larger droplets.

This is apparent from Fig. 7, in which the roughnesses of the equally thick random* and steered arc coatings are compared. As a result of the lower deposition rate, the coating process had to be prolonged in the case of the steered arc coatings in order to obtain the same coating thickness. This disadvantage of the steered arc coatings can be eliminated by higher currents up to 200 A. However, this adversely affects the surface quality. It is clear that the $R_z$ values of the steered arc coatings are approximately 30% lower.

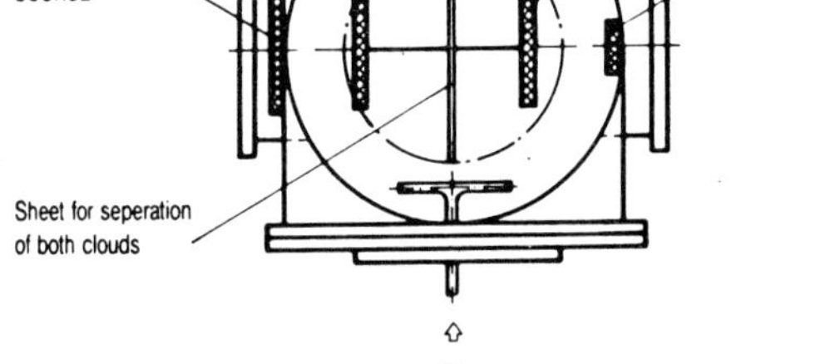


Fig. 6. Configuration of the arc sources.

TABLE 2

Comparison of the erosion and deposition rates

| | *Random arc** | *Steered arc* |
|---|---|---|
| Erosion rate Δg (mg $min^{-1}$) | 230 | 110 |
| Deposition rate s (μm $h^{-1}$) | 13 | 5 |

Cathode material: titanium.
Substrate–cathode distance: 150 mm.

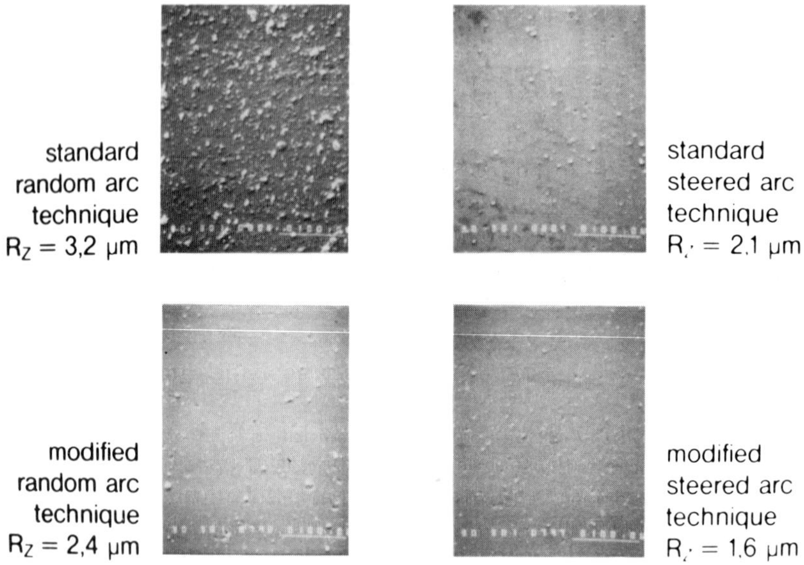


Fig. 7. Surfaces of TiN random arc and steered arc coatings (scanning electron microscope; $U_B = -100$ V; $P_{N_2} = 5 \times 10^{-3}$ mbar; coating thickness, 4.5 μm).

The lowest $R_z$ values are obtained by combining the modified process (as described in Chapter 3) with the steered arc technique. The $R_z$ value of the substrate prior to coating was 1.1 μm on average.

However, the surfaces show that even the smoothest arc coatings are not free of droplets and craters. There is no sign of a completely droplet-free arc coating at present (unless the surfaces to be coated are turned away from the cathodes. This method is sometimes used but is not of great importance either from a technical and/or economic point of view).

Nevertheless it is known that spatter formation also occurs sometimes in competitive processes such as evaporation from a crucible. Therefore, the absolute absence of droplets is not necessary for most wear protection applications. In this case, if special technical requirements exist, it is sufficient to minimize the number and/or size of the droplets by means of suitable actions, such as modified process guidance, other heating techniques and the steered arc technique.

TABLE 3

Critical load $L_c$ (N) of TiN and $Ti_{0.88}Al_{0.12}N$ coatings

| *Coating* | *Random arc** | | *Steered arc* | |
|---|---|---|---|---|
| | *Standard* | *Modified* | *Standard* | *Modified* |
| TiN | $82 \pm 6$ | $78 \pm 8$ | $72 \pm 12$ | $88 \pm 6$ |
| $Ti_{0.88}Al_{0.12}N$ | $54 \pm 4$ | $64 \pm 3$ | $69 \pm 10$ | $85 \pm 6$ |

Substrate material: HSS Mat.-No. 1.3207.
Coating thickness: 4.5 - 5.5 μm.

The hardness values of the TiN coatings produced are in the range of 3000 - 3100 HK 0.05 independent of the arc variant. Apart from the droplets, there is very little difference in the structure of the coatings either. The scanning electron microscope shows the known, particularly dense structure of arc coatings [4, 8 - 9]. Finally, both processes are arc processes, for which the energy and ionization considerations, and the resulting advantages with regard to the formation of hard coatings, apply to an equal extent.

The critical loads of the TiN and $Ti_{0.88}Al_{0.12}N$ coatings generated with the two techniques and process variants are shown in Table 3. The values of (Ti,Al)N, which are lower on the whole, need not necessarily lead to poorer adhesion. The increasing roughness of the arc coatings with increasing aluminium content [9] complicates the interpretation of these results. The higher measured values for the smoother coatings of the modified process, in which better adhesion is not to be expected, strengthen the suspicion that the scratch values are adversely influenced in arc coatings with numerous droplets. On the whole, it can be stated with certainty that TiN and (Ti,Al)N coatings with very good adhesion can be produced with the two arc variants and their process variants.

## 6. Multilayer and multicomponent coatings

The steered arc technique offers the possibility of depositing multicomponent layers with defined composition. For instance, by using segmented cathodes with segments of pure metals and by predetermined variation of location and residence time of the arc spot on the segments, coatings with various compositions are attainable. The deposition of films with controlled composition profiles has been shown by Jewsbury *et al.* [10] for the Ti–Ni system.

By means of the sequence shown in Fig. 8, the steered arc spot can also be guided over a segmented cathode as a very efficient erosion tool in order to obtain multilayer coatings of type $A + A_xB_y + B + \cdots$ and/or to produce graded interfaces.

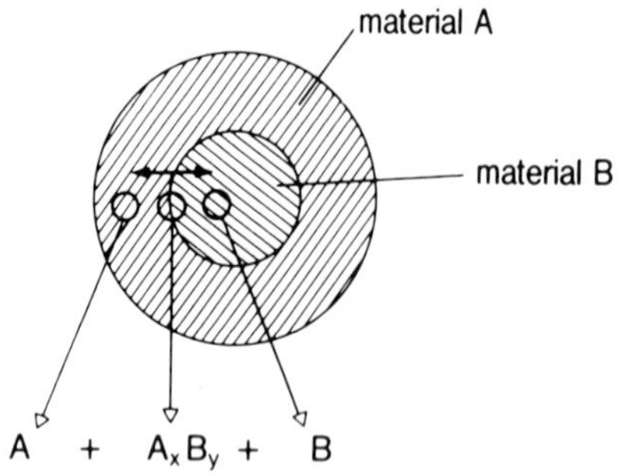


Fig. 8. Possibilities for use of segmented cathodes in the steered arc technique.

## 7. Conclusions

Through application of a movable permanent magnet, the arc spot can be controlled and forced to remain on closed tracks. This results in smaller craters, the directional and in most cases overlapping-free movement of the arc spot and a clear reduction of large molten cathode areas. Consequently, the steered arc technique can prevent larger droplets, such as those which are typical of the random arc technique. Furthermore, it can shift the maximum droplet size distribution towards lower values.

The modified process (input of nitrogen in the phase of heating with ion bombardment) causes a reduction of the roughness both in the random and in the steered arc version.

A comparison of arc coatings produced by both techniques does not indicate any essential differences with regard to coating properties such as hardness, critical load and structure. It can be concluded that both arc variants are suitable for successful application in wear protection. The selection of the appropriate system is conditional on the technical requirements. Finally, the new steered arc technique gives the advantage of using segmented cathodes. Multicomponent coatings of various composition from the same target and multilayers, which can both be produced by the steered arc technique with segmented cathodes, signify an enormous industrial potential.

## References

1 J. E. Daalder, *Physics, 104C* (1981) 91.
2 R. L. Boxmann and S. Goldsmith, *Surf. Coat. Technol., 33* (1987) 153.
3 M. Mack, personal communication, 1988.
4 E. Ertürk, H.-J. Heuvel and H.-G. Dederichs, *1st PSE*, Garmisch Parkenkirchen, 1988.
5 A. M. Dorodnov *et al., J. Appl. Mech. Techn. Phys., 22* (1981) 28.
6 S. Ramalingham *et al., U.S. patent 4,673,477* (1987).
7 B. Y. Molzhes and V. A. Nemchinskil, *Sov. Phys. Techn. Phys., 30* (1985) 136.
8 E. Ertürk and H.-J. Heuvel, *Thin Solid Films, 153* (1987) 135.
9 E. Ertürk, H.-J. Heuvel and H.-G. Dederichs, *Surf. Coat. Technol., 39/40* (1989) 435.
10 P. Jewsbury, S. Ramalingham and R. F. Chang, to be published.

*Surface and Coatings Technology*, *39/40* (1989) 465 - 474 

# THE USE OF PROCESS MODELLING FOR OPTIMUM DESIGN OF REACTIVE SPUTTERING PROCESSES

S. BERG, M. MORADI, C. NENDER and H.-O. BLOM

*Institute of Technology, Uppsala University, Box 534, S-751 21 Uppsala (Sweden)*

(Received March 10, 1989)

## Summary

The reactive sputtering process has been successfully described in a quite simple mathematical model in some recent articles. By the use of such a model it has been possible to predict the partial pressure of the reactive gas, the rate and also the composition of the deposited film. In such a complex process it is not easy to experimentally verify all aspects of the model. Numerous experiments have, however, been carried out which support the validity of the model. In this article we will apply this model to investigate the problem of *target poisoning*. We will calculate the composition of the deposited film, the deposition rate and the poisoning of the target surface. We will mathematically show that by introducing a pressure gradient in the processing chamber it is possible to obtain compound film formation while minimizing the target poisoning effect. This effect has earlier been experimentally observed by several authors. These theoretical findings explain quite satisfactorily the expected improvement in performance of the reactive sputtering of compound films, as far as deposition rate is concerned. However, the most favourable conditions only exist at non-stable processing operating points. To be able to operate the process at optimum conditions a feedback control system for the partial pressure of reactive gas *vs.* gas mass flow has to be incorporated.

## 1. Introduction

The reactive sputtering process is quite complex. It is not always an easy task to optimize processing conditions. In practice this optimization is often done by performing tedious trial and error experiments. By introducing some computer simulations of the processing conditions into the process planning work, the number of unnecessary experiments may be drastically reduced. For this purpose we have recently developed a quite general mathematical model for the reactive sputtering process [1, 2]. The model is based on well known physical parameters like pressure, mass flow, pumping speed, sputtering yield, current density, effective areas, etc. With the help of this

0257-8972/89/$3.50 

model it is possible to theoretically study how the change in a single parameter affects the total processing conditions. This single parameter dependence may be difficult to observe experimentally since most of the parameters involved are, in a complex way, coupled to each other.

The goal in reactive sputtering is normally to try to deposit a compound film at high deposition rate. However, to achieve compound film formation the concentration of the reactive gas in the processing chamber has to reach a certain level. At this concentration level it is unavoidable that a chemical reaction between the elemental target atoms and the reactive gas starts to take place at the target surface. A compound film will then be formed at the surface of the target. The sputtering yield of this compound film is normally much lower than the sputtering yield of the elemental target atoms. Therefore, a drastic drop in the deposition rate is observed at this concentration level of the reactive gas. This points out the conflicting situation. One may not be able to form a compound film at the substrates without also forming the low sputtering yield film on the target surface.

By the use of the reactive sputtering model previously developed it is possible to theoretically simulate the results of different processing conditions. Used in an imaginative way, we believe that such a tool can be very useful as a pre-study before setting up new industrial reactive sputtering fabrication lines. The simulation may give valuable indications about the total performance of the intended process. To our experience it also gives very useful hints about possible feed-back control possibilities. Predicted variations in the involved processing parameters may serve as processing control signals in the experimental set up.

In this article we will focus on the problem of keeping the target surface "non-reacted" while forming a fully reacted film on the substrate surface. By doing so the sputtering rate may be kept at a high level, related to the sputtering yield of the elemental target rather than being related to the low sputtering yield of the formed compound. Indications of suitable processing conditions that satisfy these requirements will be presented.

## 2. Basic reactive sputtering model

A reliable and experimentally verified reactive sputtering model has been presented quite recently [1, 2]. We will refer to this model throughout this article. Figure 1 shows a schematic of the assumed reactive sputtering system. The model operates in the following way.

A mass flow of reactive gas is fed into the system. This input mass flow will be "consumed" by three separate processes.

(1) The fraction $\theta_t$ of the target surface $A_t$ has reacted with the reactive gas and formed a compound material at the surface. Reactive gas will, however, still be gettered at the non-reacted fraction $(1-\theta_t)$ of the target.
(2) In a similar way the fraction $\theta_c$ of the deposited area $A_c$ has also reacted

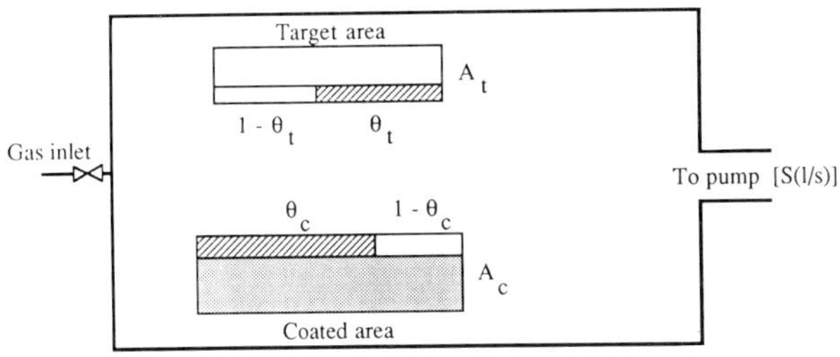


Fig. 1. Schematic of reactive sputtering system. $A_t$ and $A_c$ are the effective areas of the target and coated areas respectively. $\theta_t$ and $\theta_c$ are the surface coverages of the same areas.

with the reactive gas. Also here gettering of reactive gas will take place at the non-reacted fraction $(1-\theta_c)$ of this surface.

(3) The remaining reactive gas will set up a partial pressure $p$ in the vacuum chamber. The system pump (speed $S$) will remove the mass flow $q$ ($q = pS$) of the reactive gas.

In steady state the incoming flow of reactive gas will be the sum of the three "gettering" mechanisms described above. The sputtering takes place by the inert ion currently density $J$ uniformly bombarding the target surface $A_t$. The sputtering rate $R$ from the target is

$$R = J\{S_f\theta_t + S_e(1-\theta_t)\} \tag{1}$$

where $S_f$ is the sputtering yield of the compound film formed on the target and $S_e$ is the sputtering yield of the elemental target material. Normally the condition $S_e > S_f$ holds. Equation (1) illustrates in a simple way that the rate $R$ will drop as $\theta_t$ increases.

$\theta_c$ is a measure of the composition of the reactively sputter deposited film. If $\theta_c$ approaches unity a compound film will be deposited at the area $A_c$. If $\theta_c \ll 1$, elemental unreacted target atoms are deposited.

A theoretical way of searching for a solution to the problem of sputtering a compound film at a high sputtering rate is to find processing conditions that allow $\theta_t$ to be close to zero (high rate) while at the same time keeping $\theta_c$ close to unity (compound).

## 3. Calculations of $\theta_c$ and $\theta_t$

The theoretically obtained results for a normally operated reactive sputtering process are shown in Fig. 2 (from ref. 2). The solid line represents calculated behaviour. The path B–C in the curve represents non-stable operation conditions. Experimental observations normally follow the dotted lines. This demonstrates the well known hysteresis curve. We have concluded [1] that the width of the theoretically obtained S-shaped solid curve represents the width of the hysteresis region. Moreover, several authors [3, 4] have experimentally verified the existence of the theoretical S-shaped type curve

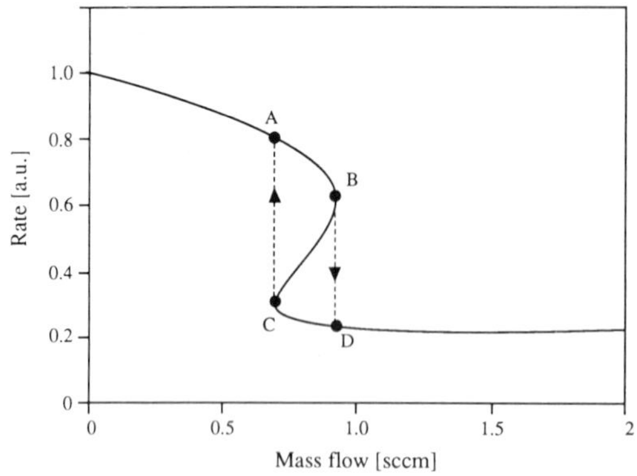


Fig. 2. Calculated sputtering rate *vs.* nitrogen mass flow.

shown in Fig. 2. This verification can only be done if the processing equipment is supplied with a fast feedback control system that enables accurate automatic control of the partial pressure *vs.* mass flow of the reactive gas. With a feedback control system it is possible to experimentally operate the reactive sputtering process also at the non-stable points on the B–C part of the curve shown in Fig. 2. This has been demonstrated to be valid for reactive sputtering in a nitrogen ambient; however, Affinito and Parson [4] reported that their feedback system did not work in an oxygen ambient.

We will now investigate the behaviour of the two important parameters $\theta_c$ and $\theta_t$ corresponding to the processing conditions used in Fig. 2. Results from calculations of $\theta_c$ and $\theta_t$ are shown in Fig. 3. In Fig. 3 we have inserted the processing points A–B–C–D corresponding to the same processing points shown in Fig. 2. It is interesting to notice that $\theta_c$ and $\theta_t$ behave quite differently. There is not a simple relationship between the two parameters. However, we can observe that $\theta_c > \theta_t$ holds for all possible processing points.

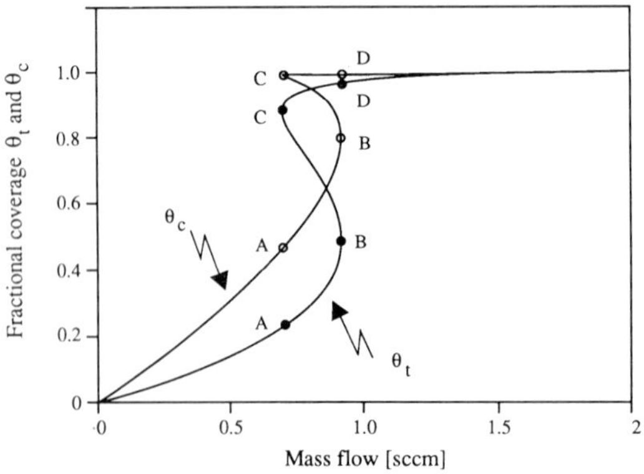


Fig. 3. Calculated fractional coverage $\theta_t$ and $\theta_c$ of the target area and coated area respectively *vs.* mass flow of reactive gas. Corresponding processing points for the two curves are indicated with the same letter.

We can conclude from Fig. 3 that to make $\theta_c$ close to 1.0 (compound deposition) we need *at least* to operate the process somewhere between B and C (corresponding to $0.8 < \theta_c < 1.0$) on the processing curve. However, in this region $\theta_t > 0.5$. This corresponds to 50% compound formation (target poisoning) on the target surface. Therefore we expect a quite low deposition rate, as also can be seen in Fig. 2. The two requirements (high $\theta_c$ and low $\theta_t$) cannot be fulfilled at the same time during the processing conditions related to Figs. 2 and 3.

## 4. Pumping speed effects

The pumping speed $S$ of the system pump has been suggested to have a major influence of the reactive sputtering process. In fact the hysteresis effect may be eliminated if the system pump exceeds a certain pumping speed level [5 - 7]. This may also be predicted by the reactive sputtering model used in our calculations [8]. The effect of changing the pumping speed $S$ on $\theta_t$ is shown in Fig. 4.

The positions of the turning points B (as defined in Fig. 3) exhibit a small shift towards larger values of $q$ for increasing pumping speed. This effect has a minor influence on the processing conditions.

The positions of the turning points C in Fig. 4 also shift towards larger values of $q$. This shift, however, will cause a dramatic impact on the processing conditions. At a certain critical pumping speed (around 400 - 500 l s$^{-1}$), the S-shape of the curve are eliminated. This means that the hysteresis effect will disappear at this pumping speed level.

The $\theta_c$ curves exhibit a similar dependence on the pumping speed as the $\theta_t$ curves.

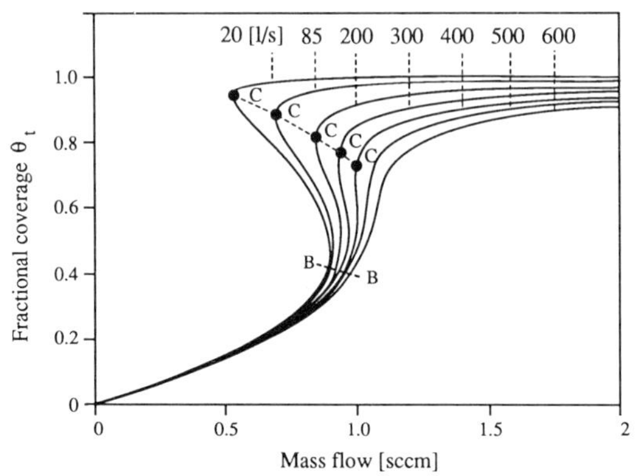


Fig. 4. Calculated fractional coverage of target area *vs.* mass flow of reactive gas for different pumping speeds.

## 5. Eliminating the hysteresis effect

Operating the reactive sputtering process under conditions that do not create the hysteresis effect, as described above, may be advantageous due to the easy control of film composition. We will investigate if it is also favourable from the point of obtaining high sputtering rate. Figure 5 shows the results from calculation of the rate *vs.* mass flow for the same conditions as shown in Fig. 2 but increasing the pumping speed from 85 to 600 l $s^{-1}$. This increase in pumping speed will completely eliminate the hysteresis effect.

Figure 6 shows the calculated values for $\theta_c$ and $\theta_t$ corresponding to Fig. 5. The values of $\theta_t$ closely follow the values of $\theta_c$. From this result we conclude that the target poisoning ($\theta_t$) will not be eliminated or reduced in the hysteresis-free operation. This mode of operation therefore does not offer any advantages concerning the possibility of obtaining high rate deposition and at the same time obtaining compound film formation.

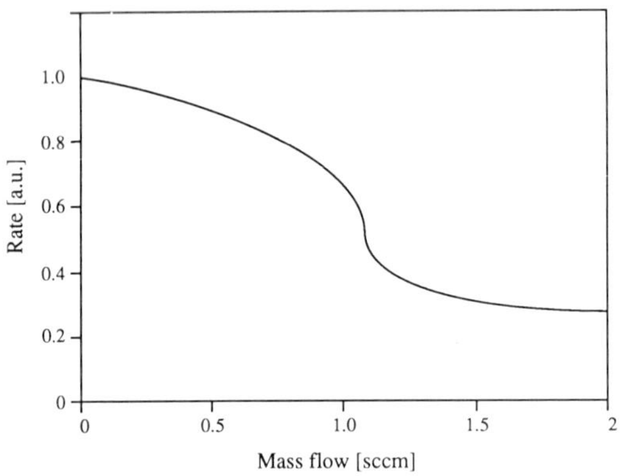


Fig. 5. Calculated sputtering rate *vs.* mass flow of reactive gas for a pumping speed of 600 l $s^{-1}$.

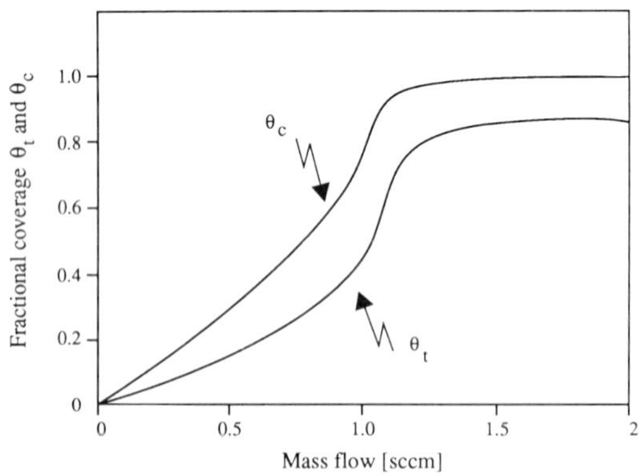


Fig. 6. Calculated fractional coverage of target and coated areas *vs.* mass flow of reactive gas for the same pumping speed as in Fig. 5.

## 6. Introducing a pressure gradient

One obvious way of decreasing the poisoning of the target is to introduce the reactive gas very close to the substrate. In this way a different partial pressure of the reactive gas could be created at these two positions. This technique has been used experimentally by several authors [9, 10]. However, no results have been published from an experimental set up that combines the pressure gradient and a feedback control system that enables operation also on any of the non-stable processing points on the B–C region on the solid line in Fig. 2.

We will investigate the operating points B shown in Fig. 3 for both curves $\theta_c$ and $\theta_t$. At the conditions shown in Fig. 2, $\theta_c = 0.8$ and $\theta_t = 0.45$ at the processing points B. The purpose of introducing the pressure gradient is to increase the value of $\theta_c(\mathrm{B})$ and at the same time decrease the value of $\theta_t(\mathrm{B})$. To simulate the condition of preferentially introducing the reactive gas to the substrate we will introduce a different partial pressure of this gas at this surface. The partial pressure at the substrate surface $A_c$ will be set to $p_s$

$$p_s = kp \tag{2}$$

where $k > 1$ for a higher pressure at the substrate and $k < 1$ for a higher pressure at the target.

The target and the entrance to the pump will remain at the partial pressure $p$. Figure 7 shows results from calculating $\theta_c$ and $\theta_t$ for different values of the pressure gradient coefficient $k$. Apart from the pressure gradient the same conditions are valid as in Fig. 3.

To visualize the favourable effect of the pressure gradient all points B are joined together for both the $\theta_c$ and $\theta_t$ curves. It can be observed that as $k$ increases the values of $\theta_c$ at the points B on the $\theta_c$ curves increase, while the values of $\theta_t$ at the points B on the $\theta_t$ curves decrease. These theoretically calculated results predict that we will approach ideal reactive sputtering

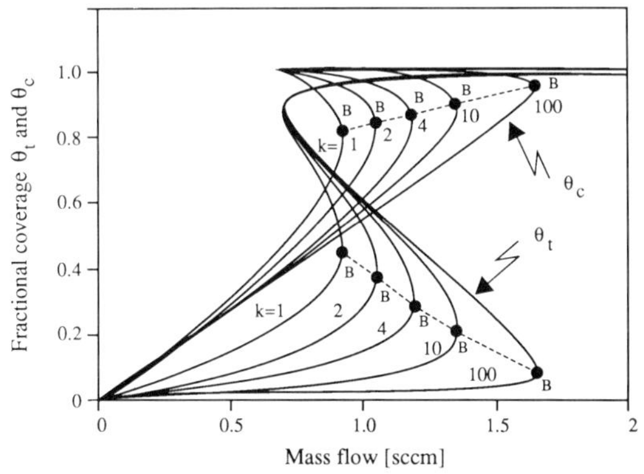


Fig. 7. Calculated fractional coverage of target and coated areas *vs.* mass flow of reactive gas for different pressure gradient coefficients $k$.

conditions for high values of $k$. Thus by introducing the gradient coefficient $k$ into the reactive sputtering model it is in fact possible to predict earlier experimental findings [9, 10].

However, it must be pointed out that theoretically this favourable condition only exists at the turning points B on the $\theta_c$ and $\theta_t$ curves. As was explained in Fig. 2 all reactive sputtering systems not equipped with a feedback control system will avalanche from the non-stable point B to the stable point D. If this happens for high values of $k$, both $\theta_c$ and $\theta_t$ will approach unity and no positive effect of the pressure gradient will be expected. However, in a real reactive sputtering system several second order effects not included in the model may contribute to the overall result. In practice it may therefore be worthwhile to investigate operation points close to B in the search for favourable deposition conditions. The theoretical predictions shall only be taken as "promising" starting points in the search for optimum conditions.

## 7. Experiments

Reactive sputtering experiments were performed in a diode r.f. sputtering equipment. Titanium was sputtered in a mixture of argon and nitrogen. The details of the experimental set-up have been described elsewhere [11].

In order to obtain a larger partial pressure of the reactive gas at the substrate than at the target we introduced the reactive gas through a pipe in the form of a ring close to the substrate. Small holes were drilled at the lower side of the ring. In this way the gas was directed down towards the substrate table. This separate gas inlet system was very simple. We could not measure the pressure at the substrate table position. However, it seems reasonable to assume that we will obtain some pressure gradient with this experimental set-up.

To further study the influence of the pressure gradient we also moved the ring close to the target. In this position we will obtain a higher partial pressure of the reactive gas at the target than at the substrate ($k < 1$).

Measurements of the sputtering rate *vs.* the mass flow of the reactive gas were recorded. A relative value of the sputtering rate was obtained by measuring the optical emission signal from atomic titanium at 4113 Å [11]. Results from the two extreme experiments (ring at target and ring at substrate) are shown in Fig. 8. As a comparison the theoretically calculated curves for $k = 2$ (ring at substrate) and $k = 0.5$ (ring at target) are shown in Fig. 9.

The pressure conditions in the actual reactive sputtering chamber are far more complicated than the conditions assumed in the simple theory. Despite this fact it is interesting to notice that there is a surprisingly good similarity between the set of curves shown in Figs. 8 and 9. Calculations using the theoretical reactive sputtering model predict (Fig. 9) that the hysteresis will disappear if the target is deliberately poisoned. This is also verified by the

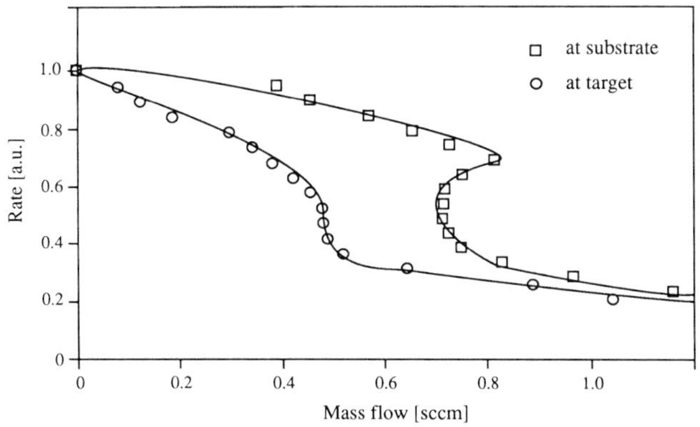


Fig. 8. Experimentally observed relative rate *vs.* mass flow of reactive gas for two positions of the gas inlet.

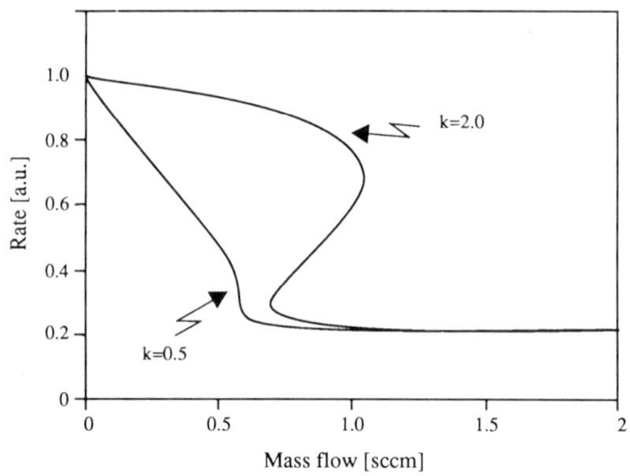


Fig. 9. Calculated sputtering rate *vs.* mass flow of reactive gas for $k = 2$ representing an increased partial pressure at the coated area, and $k = 0.5$ representing a lower nitrogen partial pressure at the coated area.

experimental observations (Fig. 8). This result gives further support to the validity of this reactive sputtering model.

As indicated by Figs. 8 and 9 we did not succeed in creating a large pressure gradient in our reactive sputtering system. We therefore are not in the position to verify our theoretical calculations for very high values of the pressure gradient coefficient ($k > 2$). However, we believe that our reactive sputtering model has proved to satisfactorily predict that it is possible to perform compound film deposition at a high sputtering rate if a large pressure gradient can be established between the target and the substrate surface. The major physical mechanisms involved in the reactive sputtering process are thus quite satisfactorily described by the model.

## 8. Conclusions

The performance of the reactive sputtering process has been theoretically analyzed by the use of a mathematical model. It was found that the elimination of the hysteresis effect did not offer any advantages as far as the deposition rate is concerned. A large hysteresis effect may favour the conditions required for high rate sputtering of compound films. In particular, the creation of a pressure gradient in the processing chamber establishes very favourable processing conditions. However, these conditions theoretically only exist at non-stable processing operating points. To be able to operate the process at the optimum conditions a feedback control system for the partial pressure of reactive gas *vs.* gas mass flow thus has to be incorporated. By the use of this system it should be possible to perform reactive sputtering of compound films at a rate limited by the high sputtering yield of the target material rather than being limited by the low sputtering yield of the compound material.

These general remarks are of course only valid in systems similar to the theoretical system described in the text. For example, in magnetron sputtering systems with non-uniform sputter erosion rates from the target, modifications of the basic simple assumptions in the model have to be done. However, we believe that this can be done quite simply. We have demonstrated that the basic model could very easily be modified by introducing the pressure gradient coefficient $k$ into the basic equations. The theoretical results from this simple modification closely resemble experimentally observed results.

We therefore very strongly believe that our reactive sputtering model will give realistic indications of reactive sputtering performance for a large variety of different operating conditions. The key factor is only to introduce relevant modifications to the basic equations so that the basic simple mode more closely resembles the actual processing conditions.

## References

1 S. Berg, T. Larsson, C. Nender and H.-O. Blom, *J. Appl. Phys., 63(3)* (1988) 887.
2 S. Berg, H.-O. Blom, T. Larsson and C. Nender, *J. Vac. Sci. Technol., A5(2)* (1987) 202.
3 A. F. Hmiel, *J. Vac. Sci. Technol., A3(3)* (1985) 592.
4 J. Affinito and R. R. Parsons, *J. Vac. Sci. Technol., A2(3)* (1984) 1275.
5 S. Kadlec, J. Musil and H. Vyskocil, *J. Phys. D, 19* (1986) L187.
6 T. Serikawa and A. Okamoto, *Thin Solid Films, 101* (1983) 1.
7 M. E. Sjöstrand and A. G. Thelin, *Proc. 8th Int. Conf. on Vacuum Metallurgy*, 1985.
8 T. Larsson, H.-O. Blom, C. Nender and S. Berg, *J. Vac. Sci. Technol., A6(3)* (1988) 1832.
9 G. Este and W. D. Westwood, *J. Vac. Sci. Technol., A2(3)* (1984) 1238.
10 F. Jones, *J. Vac. Sci. Technol., A6(6)* (1988) 3088.
11 S. Berg, T. Larsson and H.-O. Blom, *J. Vac. Sci. Technol., A3* (1986) 594.

*Surface and Coatings Technology*, *39/40* (1989) 475 - 486


# ACTIVATED REACTIVE ION PLATING (ARIP)

E. MOLL, R. BUHL, H. K. PULKER and E. BERGMANN
*Balzers AG, FL-9496 Balzers (Liechtenstein)*
(Received March 9, 1989)

## Summary

Activated reactive ion plating (ARIP) processes are characterized by the following features: (1) the substrate is subjected to a flux of ions before and during film formation, (2) a reactive gas is incorporated into the growing film, and (3) the evaporating material is activated because it is either the cathode or the anode of an arc discharge such as in thermionic arc evaporators, hot hollow cathode discharge evaporators or cathode spot arc evaporators. ARIP processes are used to produce well adhering, dense nitride, carbide and oxide films at relatively low substrate temperatures. They are in commercial use to produce films for both wear protection and optical applications.

## 1. The basis of activated reactive ion plating (ARIP) processes

### *1.1. Ion plating*

The term "ion plating" was introduced by Mattox [1] for processes which include ion bombardment of the substrate and of the growing film. Historically, Berghaus [2] was probably the first to propose the exposure of a negatively biased substrate to a plasma in order to get "perfect structure and adhering strength".

A large number of process variants including triode and tetrode configurations have been developed. Not only evaporation sources are used for ion plating. Bias sputtering and some types of plasma chemical vapor deposition (CVD) also come under the above mentioned definition. An interesting review was published by Matthews [3]. In the following we will use the term ion plating according to Mattox's definition in which ion plating is exclusively related to effects produced on the substrate, the interface and the growing film. In fact, many effects of ion bombardment can also be understood without taking the method used for ion production into consideration.

In sputter etching — the substrate preparation step preceding deposition — ion bombardment removes material from the substrate surface. However, it does not create a clean surface because the coating material previously deposited on the substrate holders is also sputtered off and

0257-8972/89/$3.50 

redeposited on the new batch of substrates. Etching plus redeposition produces a very thin interface layer made of a mixture of substrate and film material which improves the adhesion of the film.

Ion bombardment during deposition causes cascades of atomic collisions in the growing film. The recoiled or displaced atoms cause a kind of continuous atomic mixing and also enhance surface migration. This results in the filling up of voids and in smoothed or graded grain boundaries. The well-known structural zone diagram proposed by Thornton [4] could be modified for ion plating by replacing the axis "argon pressure" by "ion bombardment energy per molecule" (Fig. 1). Another result of ion bombardment is a high concentration of point defects. These are frozen in the structure under growth conditions that are comparable to rapid quenching. High compressive intrinsic stress is the direct consequence of these defects.

In wear protection applications density, high adhesive strength and compressive stress are features underlying the huge success of these thin brittle films [5]. Adhesive strength is needed to compensate the buckling forces caused by the high compressive intrinsic growth stress, and this stress is needed as a mechanical prestressing which is relaxed by the mechanical load placed on the films in actual use. The consequence of this prestressing is the fact that as long as the deformation of the substrate remains in the elastic range such loads do not damage the film.

A thermal compressive stress is added when the thermal expansion coefficient of the film is less than that of the substrate and when the film was deposited at a temperature higher than or equal to the maximum expected operating temperature. This thermal stress also relaxes in use when the

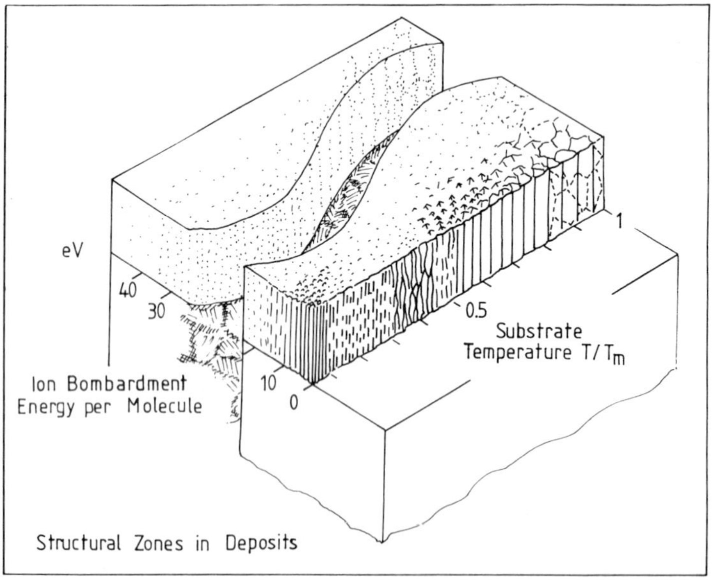


Fig. 1. Structural zones in deposits.

operating temperature is higher than room temperature, which is normally the case.

Optical films produced in processes without ion bombardment show properties associated with low mobility of the condensing atoms and molecules due to their low thermal energies of between 0.1 and 0.2 eV. As no current can flow in or on an insulating substrate, ion bombardment of such substrates or films requires that exactly the same number of positive and negative charges impinge onto every point of the surface in order to achieve charge balance.

Plasma assisted ion plating makes use of the negative self-biasing which not only occurs on the surface of thin insulating substrates placed on an r.f. electrode. In d.c. plasmas, alone the big difference between the mean velocities of the electrons and the ions in the plasma causes a self-bias potential of more than 10 V with respect to the plasma. This bias accelerates positive ions towards the substrate surface [6]. Their energies are usually not sufficient to cause sputtering, but they are nevertheless at least 50 times higher than the thermal energies of the vapor atoms and molecules and higher than crystalline binding energies.

### *1.2. Reactive evaporation*

Dissociation is a problem with compound films. Even in evaporation, which is the gentlest physical vapor deposition (PVD) process, chemical compounds are dissociated to a certain extent. Due to their low sticking coefficient, gaseous components can be pumped off, resulting in substoichiometric composition of the deposited film.

In reactive evaporation the gaseous components are continuously replaced. Because of the high reactivity of oxygen, only oxide films have been successfully produced industrially by reactive evaporation [7]. Oxides or sub-oxides are used as evaporation materials and the oxygen partial pressure in the chamber is stabilized at about $10^{-4}$ mbar using a controlled gas inlet valve. The actual oxidation takes place to a great extent on the substrate surface. According to Ritter [8], the chemisorption rate of oxygen is the critical factor for the completion of the reaction.

Nevertheless, optical coatings produced by reactive evaporation alone are often still slightly substoichiometric and thus slightly absorbing. These films have rough surfaces and a columnar or spongy microstructure with large void volume and great internal surface area. As a consequence of the low density, the refractive indices of these films are considerably lower than the values for the bulk oxides. These films absorb water vapor and other gases from the atmosphere which change their refractive index and other physical properties. Their adhesion to the substrate is poor and their abrasion resistance and hardness are low.

It is possible to improve most of these properties by heating the substrates to about 300 °C before reactive evaporation. Substrate heating has, in fact, become a standard procedure in this process. It has, however, the undesirable "side-effects" of producing coarser film microstructures and higher surface roughness.

### *1.3. Activated reactive evaporation*

The term "activated reactive evaporation" (ARE) was created by Bunshah in 1972 [9]. Here, "activation" is a synonym for the increase of chemical reactivity through ionization, dissociation and excitation in a gas discharge. An activated and at the same time reactive deposition method was described as early as 1932 by Berghaus [10]. Because even traces of non-oxidized metal atoms or metal suboxides in the film cause absorption [11], activation of oxygen [12] is an important improvement in the production of films for highly sophisticated optical components such as laser mirrors with low absorption losses and high damage thresholds.

In these early gas activation methods, the oxygen ions were generated in glow discharge ion sources, *e.g.* a cold hollow cathode inside a quartz tube [13]. Today we know that reactivity can be improved to a much higher extent by a high density of metal ions. This can be achieved using the evaporation source as the anode or cathode of a discharge. In this way, not only oxides but also nitrides, carbides and borides can be synthesized thus opening the door for the production of hard materials. In the following text we will concentrate our attention on processes in which the evaporation source acts as the anode or cathode of a concentrated discharge, *i.e.* an arc discharge. A famous example is the ARE process proposed by Bunshah [9, 14]. In ARE processes dissociation is so predominant and activation so powerful that it makes little difference whether the compound or the pure metal is used as the evaporation material.

### *1.4. ARIP*

Experience has shown that bombarding the growing film with predominantly film forming species of atoms has many adantages over bombardment with inert gas species. Therefore the combination of ion plating with those ARE processes in which the coating material is activated using anodic or cathodic evaporation sources results in a new type of process yielding superior results. We call it "activated reactive ion plating" (ARIP). ARIP includes not only the biased activated reactive evaporation (BARE) process [15], which was used as early as 1975 at Sumitomo Electric [16], but also the three most important industrial PVD processes for the deposition of wear resistant coatings. They are all based on arc discharges, *i.e.* gas discharges in which most of the electrons are generated by electron emission from a hot cathode spot. After ignition, this hot spot may be maintained by ion bombardment heating (self-sustaining arc) or it may be initiated, maintained and localized by an independent energy source such as a heated filament (thermionic arc). The following three arc vapor sources with anodic or cathodic evaporation material were especially developed for the combination of high vapor activation and high vapor ion bombardment of the substrates:

(1) Thermionic arc → anodic source
(2) Hot hollow cathode discharge → anodic source
(3) Cathode spot arc → cathodic source

## 2. ARIP processes for wear resistant coatings

CVD has been the traditional method for the deposition of thin, wear resistant films. It is limited to cemented carbide tools, as only these substrates can withstand the high CVD process temperatures around 1000 °C. In the late seventies, ARIP processes became an alternative industrial technology. As these are PVD methods, no thermal equilibrium has to be taken into consideration. Thus, wear resistant coatings can now also be deposited on steel at temperatures below the annealing temperature, *i.e.* without soft annealing the substrate. High speed steel (HSS) is usually coated at about 500 °C. Recently TiN coatings can be ion plated at substrate temperatures as low as 250 °C with great success [17].

TiN is by far the most popular wear resistant coating. One of the reasons is the yellow color which facilitates controls of the finished film and its wear performance. However, this advantage was not our predominant argument when we started to develop TiN coatings for HSS tools. The more important reasons were the previous success of TiN coatings deposited by CVD on cemented carbide tools and, with respect to the steel substrate, the approximate correspondence of the Young's moduli and the thermal expansion coefficients. It does not seem to be easy to find a general purpose tool coating better than TiN.

Ten years ago, we assumed that PVD would be the coating technology for steel, whereas CVD would continue to be an adequate coating technology for cemented carbide tools. However, today there is a strong tendency to also use ARIP for hard metal substrates because it does not round the cutting edge of the substrate and the substrate is not embrittled as it is in CVD. Therefore, the danger of breaking the edges of the tool during interrupted cutting is minimized. Another advantage is that steel tools with hard metal tips can be coated without destroying the brazed joint.

New coatings will be developed for special tools or for tools to cut special materials. Balzers recently introduced the TiCN coating BALINIT B to the market. It combines the higher hardness (compared with TiN) of TiC with the lower friction coefficient (compared with TiC) of TiN. The results of this combination are outstanding when this new coating is used for cutting abrasive materials and aluminum alloys with a high percentage of silicon. The color of TiCN is noticeably darker than that of steel. Other possible coatings for the cutting of abrasive materials or for very high cutting speeds could be made of ternary or quaternary materials such as TiAlN or TiAlVN [18, 19].

### *2.1. Thermionic arc ion plating*

Thermionic arc ion plating was invented at Balzers in 1977 and, right from the beginning, mainly applied for the production of titanium nitride films on HSS substrates [20, 21]. All important details are shown in Fig. 2. The cathode of a non-self-sustaining arc discharge is a resistance heated filament situated in a separate chamber under a relatively high noble gas

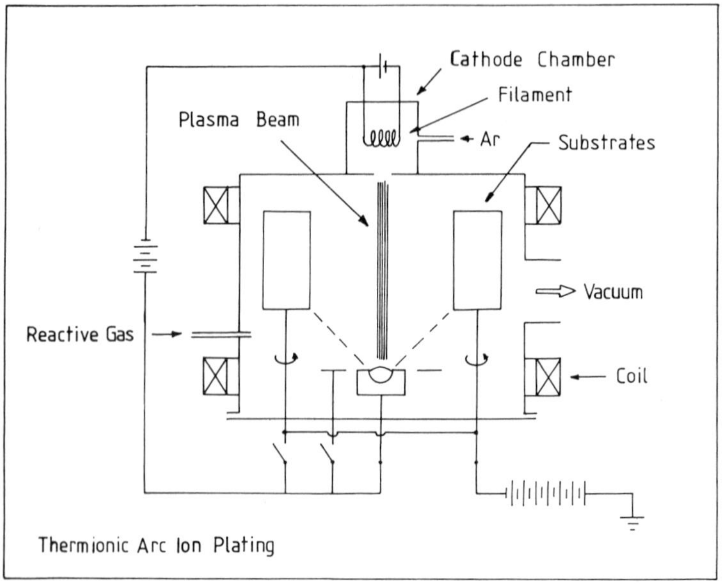


Fig. 2. Thermionic arc ion plating.

pressure. The arc is transferred into the evaporation chamber to different anodes, depending on the process step. A small aperture mounted between the cathode chamber and the evaporation chamber confines the arc plasma.

Heating is the first operation carried out inside the high vacuum coating system. The thermionic arc that will later be used for etching and evaporation also serves as the heater. This is achieved by making the substate holders and thus the substrates themselves the anode of the arc discharge. Sufficient uniformity can be achieved by magnetic dispersion [22].

Etching is performed using the thermionic arc in a triode arrangement. The substrates are immersed in the dense plasma of an arc discharge between the cathode and a metal electrode inside the evaporation chamber (which acts as the anode in the etching step) and are bombarded by argon ions drawn from this plasma. The great advantage of this method is that the plasma boundaries perfectly follow the contours of geometrically complex substrates because, according to the Child–Langmuir law, the width of the dark space is less than 1 mm when the plasma density is high and bias voltage low. In addition, the low noble gas pressure in this method ensures that the mean free paths are much longer than the structural details of the substrates, which allows excellent penetration. Control is easy and there are only minor problems with mixed loads. The etching voltage can be adjusted independently to regulate the etching intensity. No high voltage power supply is necessary.

After etching, the crucible is made the anode of the arc discharge. A magnetic field guides the arc plasma in a straight line to the anodic crucible. When the power density is high enough, certain materials such as titanium will be evaporated and then partially ionized. A negative d.c. bias is applied to the substrates to achieve the ion bombardment just described for etching with all its advantages. The substrate current is approximately proportional to the deposition rate because of the high proportion of vapor ions. At high evaporation rates, the arc discharge takes place in the metal vapor instead of in the noble gas. This results in a concentration of the plasma beam, *i.e.* the formation of an anodic spot.

An even more flexible system is obtained by combining an electron beam source with a thermionic arc source (Fig. 3) [23]. This can be done by insulating the crucible of the electron beam gun and then using it as the anode of the arc discharge. Such systems are used in box coaters for depositing TiN coatings on parts with delicate relief structures or for the deposition of those compounds which cannot be evaporated by thermionic arc evaporation alone.

In another commercial system (Balzers BAP 800) this dual beam technique can produce extremely dense oxide coatings at low substrate temperatures for optical applications as will be shown in Section 3 of this paper.

*2.2. Hollow cathode discharge (HCD) ion plating*

In hot HCD evaporation, as well as in the thermionic arc process, an arc discharge is anchored anodically on the crucible. In contrast to the thermionic arc, the HCD arc is a self-sustaining discharge, because the hollow cathode is heated (and sputtered!) by ion impact after an r.f. start up. The HCD evaporator for ion plating was invented by Morley in 1968 [24]. The use of an HCD device in a reactive process (chromium carbides and nitrides)

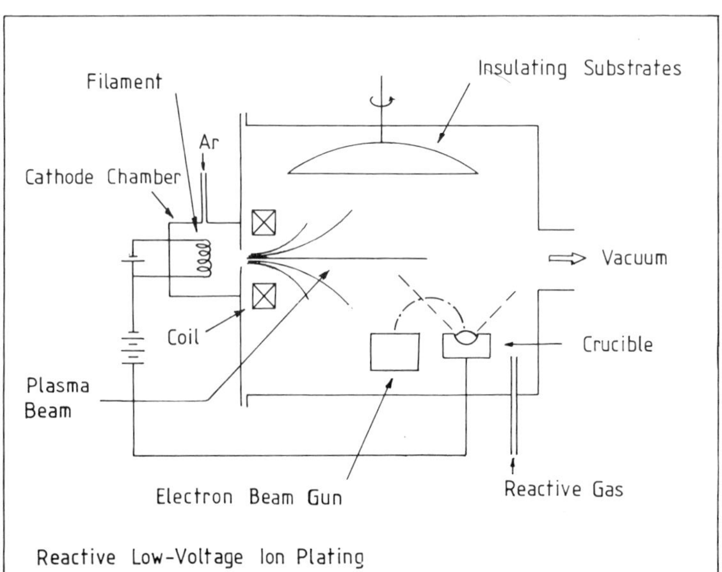


Fig. 3. RLVIP.

was first reported by Komiya and Tsuruoka [25]. A principle diagram of the commercial Ulvac systems is shown in Fig. 4. A small replaceable hollow cathode is situated in front of the anodic crucible so that it shadows the substrates above as little as possible. Before deposition the substrates are prepared by radiant heating and glow discharge etching.

### *2.3. Cathodic arc plasma deposition*

As early as 1962, the cathode spot arc was used to produce thin films from a variety of metals [26]. The high vapor ionization rate, the occurrence of multiply charged ions and the comparatively high energy of the emitted ions were reported by several authors, among them Plyutto *et al.* in 1965 [27]. Industrial use of cathode spot arc evaporation was started in the USSR where it is called Pusk [28, 29].

A detailed review on cathodic arc plasma deposition was recently published by Randhawa [30]. Here we will only mention some points of interest for comparison with other ARIP processes: the cathode spot is a very dense plasma cloud hanging above an extremely small pool of molten metal which emits very large quantities of vapor, most of which is ionized during its passage through the plasma cloud. Currents up to 100 A can be drawn through a single spot. When 1 coulomb is passed through the arc, about $10^{-7}$ kg of cathode material is evaporated no matter what metal it is made of. This shows the particular suitability of the cathode spot arc for the evaporation of refractive metals.

Figure 5 shows a cathode spot arc set-up. Unlike the processes with anodic crucibles, arc cathodes can be oriented in any direction. They can also

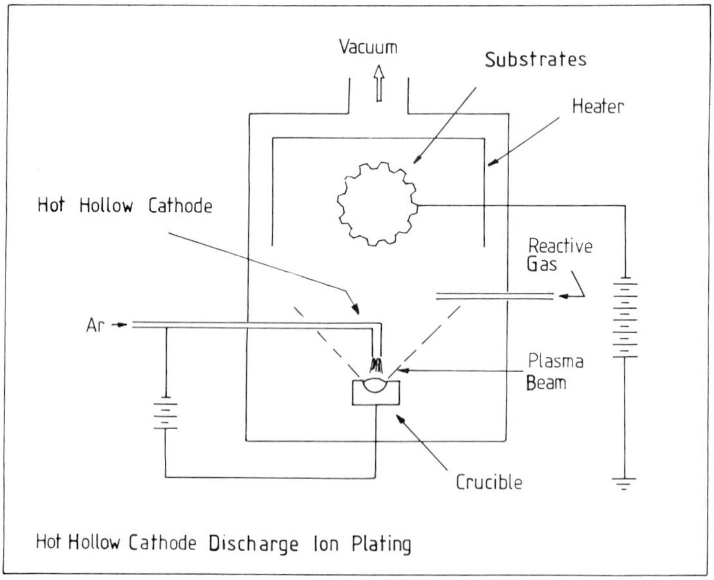


Fig. 4. Hot HCD ion plating.

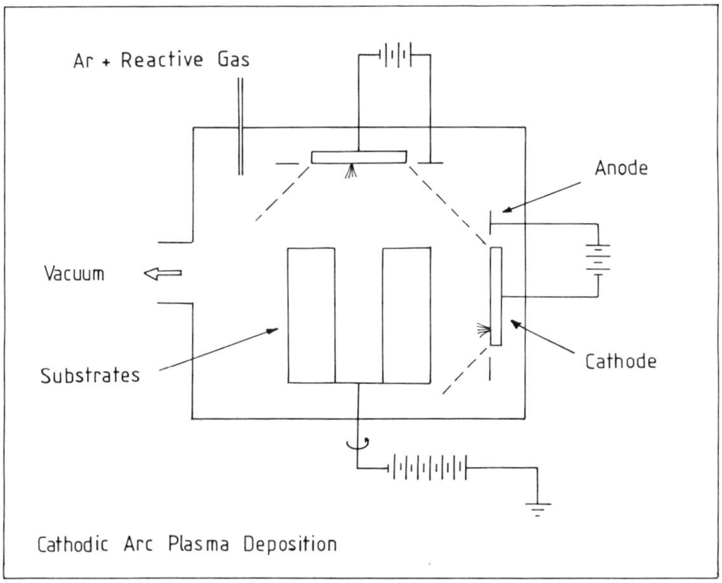


Fig. 5. Cathodic arc plasma deposition.

be used to deposit alloys or hot-pressed mixtures of metal powders from a single source. The composition of the coatings sometimes differs from that of the cathode, *i.e.* the coating material. Corrections can be achieved by changing the bias voltage applied to the substrate [31].

Bombardment with cathode material ions is the method most frequently used to heat and etch the substrates. These must be biased at a sufficiently high level to avoid cathode material deposition. Sometimes glow discharge and/or radiant heating boost or replace the metal ion bombardment.

The vapor pressure in the plasma cloud can cause splashing of the molten metal. The amount of droplets emitted has been shown to be higher when the melting point and/or the vapor pressure at the melting point is lower. During reactive deposition of nitride films, it was observed that increasing the nitrogen partial pressure reduces the amount of droplets emitted [32]. Therefore, if ion bombardment heating and etching is used, most of these microdroplets are already deposited on the substrates during the heating and etching phase at low pressure.

The positive effect of nitrogen is probably due to the clearly recognizable increase in speed of the randomly moving spot arc. In fact the spot should be kept moving at high speeds. A magnetic field can be used to speed up the motion and prevent the spots from moving randomly around the surface of the cathode. Further increased performance was achieved by steering the arc exactly on a closed path [33].

Another promising possibility appears to be a plasma-optical separator that guides electrons as well as positive ions on a buckled trajectory through which droplets cannot pass. Thus only ions reach the substrate while droplets splash onto the chamber walls [34].

## 3. An ARIP process for optical coatings

Reactive low-voltage ion plating (RLVIP) [6, 35] is a relatively new ARIP process for the deposition of oxide and nitride coatings onto unheated insulating substrates such as glass. It uses the self-biasing of insulating substrates mentioned in Section 1.1 as well as the dual beam technique with electron beam evaporator and thermionic arc mentioned in Section 2.1.

The BAP 800 automatic box-type ion-plating system was especially designed for RLVIP, but it can also be used for conventional reactive evaporation, allowing alternating processes without requiring any modification of the plant. It is equipped with two special 270°-type electron beam evaporators. The substrate holder is electrically insulated. An oscillating quartz crystal monitor is used to control the deposition rate and film thickness [36]. An optical thickness control also is available.

When used for RLVIP, the evaporation materials (metals or suboxides) form electrically conducting melts. Neither substrate heating nor any other special pretreatment is needed. A typical gas composition is $5 \times 10^{-4}$ mbar of argon plus $1 \times 10^{-3}$ mbar of oxygen. Quarterwave films in the visible are deposited within two to three minutes. A film thickness uniformity of $\pm 1\%$ across the 800 mm diameter substrate holder can be achieved when a static distribution shield is used.

Electron microscope investigations of oxide films ion plated onto unheated substrates showed amorphous and fine grained polycrystalline structures. All films produced by RLVIP are very dense and often without columnar microstructures. The surfaces were found to be very smooth compared with films deposited by conventional evaporation [37].

Infrared, ultraviolet and Rutherford backscattering measurements of films deposited by ion plating showed a stoichiometric composition and the absence of incorporated water. The high density hindered water absorption from humid atmospheres. The films are therefore especially suited as protective final layers [38 - 41]. With the exception of $TiO_2$, all coatings tested were found to have a higher Knoop microhardness than glass. Excellent adherence of single-layer films and film systems on glass substrates was found in scratch tests. Abrasion resistance values (eraser test) and environmental stability tests according to US MIL-C-675 B indicate very high film quality [42].

Ion plated oxide films are fully oxidized. No absorption can be measured with simple photometric intensity methods in the high transmittance range. Photometric and ellipsometric measurements of the refractive indices of the films gave values close to those of the bulk materials. In all cases, the refractive indices were much higher than those of evaporated films. Various values are listed in Table 1 for comparison. The values of single-layer films were reproduced with an accuracy of better than 1%. Homogeneity of the

TABLE 1

Optical properties of ion-plated films [42]

| *Film material* | *Evapor. material* | *Coating time (s)* | *Refractive index (550 nm)* | *Extinction coefficient* |
|---|---|---|---|---|
| $Nb_2O_5$ | $Nb_2O_5$ | 120 | 2.40 ± 1% | $<5 \times 10^{-4}$ (450 nm) |
| $Ta_2O_5$ | $Ta_2O_5$ | 250 | 2.23 ± 1% | $<10^{-3}$ (400 nm)<br>$k = 4 \times 10^{-6}$ [43] |
| $ZrO_2$ | ZrO | 120 | 2.20 ± 1% | $<10^{-3}$ (380 nm)<br>$k = 2.3 \times 10^{-5}$ [43] |
| $HfO_2$ | Hf | | 2.17 ± 1% | $<10^{-3}$ |
| $Si_3N_4$ | Si | 180 | 2.06 ± 1% | $<10^{-3}$ (400 nm) |
| $Y_2O_3$ | Y | | 1.95 ± 1% | $<10^{-3}$ |
| $Al_2O_3$ | Al | 180 | 1.66 ± 1% | $<5 \times 10^{-4}$ (250 nm) |
| $SiO_2$ | Si | | 1.485 | $<10^{-3}$ |
| $SiO_xN_y$ | Si | | 1.5 - 2.0 | $<10^{-3}$ |

refractive index was good for films up to thicknesses of 7/4 (equal to 550 nm). The values for thicker films were slightly higher. As a consequence of the intrinsic compressive stress, the films showed slight birefringence. The lack of water vapor sorption and desorption phenomena caused by the dense microstructure resulted in very stable refractive indices. During controlled heating of various multilayer interference filters in a photometer to at least 130 °C, no measurable shifts in the position of the edge or of the transmittance band occurred. The optical characteristics remained constant even during repeated heat treatment cycles. Heating $ZrO_2-SiO_2$ and $Ta_2O_5-SiO_2$ multilayers on glass for several hours at 400 °C and subsequent submersion in water for 3 days produced no change in the optical and mechanical properties.

Comparing RLVIP with ion-assisted evaporation and ion beam sputtering, we found that the RLVIP film properties are comparable, but have more in common with ion beam sputtered films than with films obtained by ion assisted deposition. Nevertheless, we do not expect the applications for RLVIP and the two ion beam processes to overlap. Whereas ion beam processes are powerful tools for basic parameter investigations, optimization studies or for the deposition of low-loss oxide films on a limited number of small to medium size substrates [44], the new LVRIP process is fast and optimized for high volume production. We expect that RLVIP will become a large scale production technique for high quality optical coatings.

## 4. Conclusion

ARIP processes have become the predominant industrial method for depositing nitrides, carbides and oxides on three-dimensional substrates such as tools, machine parts, surgical prostheses or as decorative coatings. ARIP processes are becoming more and more important for optical applications.

## References


1 D. M. Mattox, *Electrochem. Technol.*, *2* (1964) 295.
2 B. Berghaus, *British Patent 510,933* (priority 1937).
3 A. Matthews, *J. Vac. Sci. Technol.*, *A3(6)* (1985) 2354.
4 J. A. Thornton, *J. Vac. Sci. Technol.*, *11* (1974) 666.
5 E. Moll, to be published in *Proc. IPAT-89*, Geneva, CEP Consultants Ltd., Edinburgh, 1989.
6 E. Moll, H. K. Pulker and W. Haag, *U.S. Patent 4,619,748* (priority 1985).
7 M. Auwärter, *U.S. Patent 2,920,002* (priority 1952).
8 E. Ritter, *J. Vac. Sci. Technol.*, *3* (1966) 225.
9 R. F. Bunshah, *U.S. Patent 3,791,852* (priority 1972).
10 B. Berghaus, *German Patent 683 414* (priority 1932).
11 H. K. Pulker, *Coatings on Glass*, Elsevier, Amsterdam, 1987, p. 262.
12 W. Heitmann, *Appl. Optics*, *10* (1971) 2414.
13 J. Ebert, *Proc. SPIE*, *325* (1982) 29.
14 R. F. Bunshah and A. C. Raghuram, *J. Vac. Sci. Technol.*, *9(6)* (1972) 1385.
15 R. F. Bunshah, *Thin Solid Films*, *107* (1983) 21.
16 M. Kobayashi and Y. Doi, *Thin Solid Films*, *54* (1978) 67.
17 H. Rudigier, E. Bergmann and J. Vogel, *Surf. Coat. Technol.*, *36* (1988) 675.
18 O. Knotek, M. Atzor and H.-G. Preugel, *Surf. Coat. Technol.*, *36* (1988) 265.
19 H. Holleck, *Surf. Coat. Technol.*, *36* (1988) 151.
20 E. Moll, H. Daxinger, *U.S. Patent 4,197,175* (priority 1977).
21 R. Buhl, H. K. Pulker and E. Moll, *Thin Solid Films*, *80* (1981) 265.
22 E. Moll, *U.S. Patent 4,555,611* (priority 1982).
23 R. Buhl, E. Moll and H. Daxinger, *U.S. Patent 4,448,802* (priority 1981).
24 J. R. Morley, *U.S. Patent 3,562,141* (priority 1968).
25 S. Komiya and K. Tsuruoka, *J. Vac. Sci. Technol.*, *13(1)* (1976) 520.
26 M. S. P. Lucas, C. R. Vail, W. C. Stewart and H. A. Owen, *Am. Vac. Soc. Trans.*, *2* (1962) 988.
27 A. A. Plyutto, V. N. Ryzhkov and A. T. Kapin, *Sov. Phy. JETP*, *20* (1965) 328.
28 L. P. Sablev *et al.*, *U.S. Patent 3,793,179* (priority 1971).
29 A. I. Grigorov, A. M. Dorodnov and M. D. Kiselev, *Tekhnol. Automobilestr.*, *12* (1978) 10.
30 H. Randhawa, *Thin Solid Films*, *167* (1988) 175.
31 H. Freller and H. Haessler, *Surf. Coat. Technol.* *36* (1988) 219.
32 I. I. Aksenov, I. E. Konovalov, E. E. Kudryavtseva, V. V. Kunchenko, V. G. Padalka and V. M. Khoroshikh, *Sov. Phys. Tech. Phys.*, *29* (1984) 893.
33 S. Ramalingam, C. B. Qi and K. Kim, *U.S. Patent 4,473,477.*
34 I. I. Aksenov, V. A. Belous, V. G. Padalka and V. M. Khoroshikh, *Sov. J. Plasma Phys.*, *4* (1978) 425.
35 H. K. Pulker, W. Haag, M. Buehler and E. Moll, *Proc. IPAT-85*, Munich, CEP Consultants Ltd., Edinburgh, 1985, p. 299.
36 J. Edlinger, M. Buehler and H. K. Pulker, *Proc. Int. Symp. on Trends and New Applications in Thin Films*, Strasbourg, 1987, Vol. II, p. 735, supplém. à la revue "le vide, les couches minces", 1987, 235.
37 H. K. Pulker, W. Haag, M. Buehler and E. Moll, *J. Vac. Sci. Technol.*, *A3(6)* (1985) 2700.
38 H. K. Pulker, M. Buehler and R. Hora, *Proc. SPIE*, *678* (1986) 110.
39 H. K. Pulker, J. Edlinger and M. Buehler, *Proc. Int. Symp. on Trends and New Applications in Thin Films*, Strasbourg, 1987, Vol. II, p. 741, supplém. à la revue "le vide, les couches minces", 1987, 235.
40 H. K. Pulker, J. Edlinger and M. Buehler, *Proc. IPAT 87*, Brighton UK, CEP Consultants Ltd., Edinburgh, 1987, p. 371.
41 M. Buehler, J. Edlinger, H. K. Pulker, M. Reinhold, B. Bendow and O. El Bayoumi, *Proc. Halide Glass Symp.*, Hyatt Regency, Monterey, CA, 1987.
42 R. Hora, *Balzers Application Report BB 800 039.*
43 A. Saxer, *ENEA/Casaccia*, Roma, to be published.
44 H. A. Macleod, *Proc. Int. Symp. on Trends and New Applications in Thin Films*, Strasbourg, 1987, Vol. I, p. 43, supplém. à la revue "le vide, les couches minces", 1987, 235.

*Surface and Coatings Technology*, *39/40* (1989) 487 - 497



# REACTIVE DEPOSITION OF TiN FILMS USING AN UNBALANCED MAGNETRON

S. KADLEC and J. MUSIL

*Institute of Physics, Czechoslovak Academy of Sciences, Na Slovance 2, 180 40 Prague 8 (Czechoslovakia)*

W.-D. MÜNZ

*Leybold AG, Wilhelm-Rohn-Strasse 25, Postfach 1555, D-6450 Hanau 1 (F.R.G.)*

G. HÅKANSON and J.-E. SUNDGREN

*Thin Film Group, Department of Physics, Linköping University, S-58183 Linköping (Sweden)*

(Received March 29, 1989)

## Summary

The paper presents discharge characteristics for a sputtering device operated as a conventional magnetron (CM) or as an unbalanced magnetron (UM). In the UM, with a strong magnetic field of several millitesla reaching the substrates, a double-site-sustained discharge (DSSD) was observed. The UM with the DSSD operation transports a current $I_s$ to substrates which is more than 50% of the magnetron current for a wide range of substrate–target distances $d_{S-T}$ up to 150 mm at a pressure $p_T = 5$ Pa and low substrate bias $U_s = -100$ V.

Microstructure, composition, crystallographic structure, mechanical properties and colour are reported for TiN films reactively sputtered by the UM at $d_{S-T} = 200$ mm, $p_T = 5$ Pa, for different $U_s$ values between $-5$ and $-100$ V and substrate current densities up to 6 mA cm$^{-2}$. Dense stoichiometric TiN films with a microhardness of 2270 kg mm$^{-2}$ and critical load $L_c = 64$ N were obtained at low bias $U_s = -40$ V.

## 1. Introduction

It is well known that ion bombardment of growing films can strongly influence their microstructure and, consequently, their physical properties. Magnetron sputtering is now a widely used deposition technique. Its essential disadvantage is a strong decrease of the substrate ion current $I_s$ with increasing distance of substrates from the magnetron target $d_{S-T}$.

In principle there are two possible ways to increase the ion current density $i_s$ on substrates in magnetron sputtering, *i.e.* by (1) additional gas ionization, for instance by a hot cathode electron beam [1] or a hollow

0257-8972/89/$3.50 

cathode arc electron source [2] or (2) a magnetic confinement of the plasma, for instance by an unbalanced magnetron (UM) [3 - 8].

It is well known [4] that in sputtering systems equipped with a UM high ion currents can flow to substrates. This paper presents results of a systematic study of discharge characteristics for a UM equipped with two electromagnets. The device was also used to demonstrate the possibility of reactive sputtering of TiN films at large distance $d_{S\text{-}T}$ and high pressures $p_T = 5$ Pa.

## 2. Experimental procedure

The experimental apparatus used for investigation of magnetron discharge characteristics is shown schematically in Fig. 1. The vacuum chamber, a non-magnetic stainless steel cylinder with inner diameter 255 mm and length 450 mm, was evacuated by an oil diffusion pump rated at 2000 l s$^{-1}$. The circular planar magnetron with a titanium target (diameter 124 mm × 9 mm) made of titanium (purity, 99.5%) was mounted in one side of the vacuum chamber. A heatable substrate holder with diameter 130 mm, placed against the magnetron, could be negatively d.c. biased up to $U_s = -600$ V. The substrate-target distance $d_{S\text{-}T}$ could be changed from 10 to 300 mm.

The UM consists of a conventional magnetron (CM) and an external magnetic coil $C_2$ (400 turns; see Fig. 1). The magnetic field of the CM was produced by the internal coil $C_1$ (1200 turns).

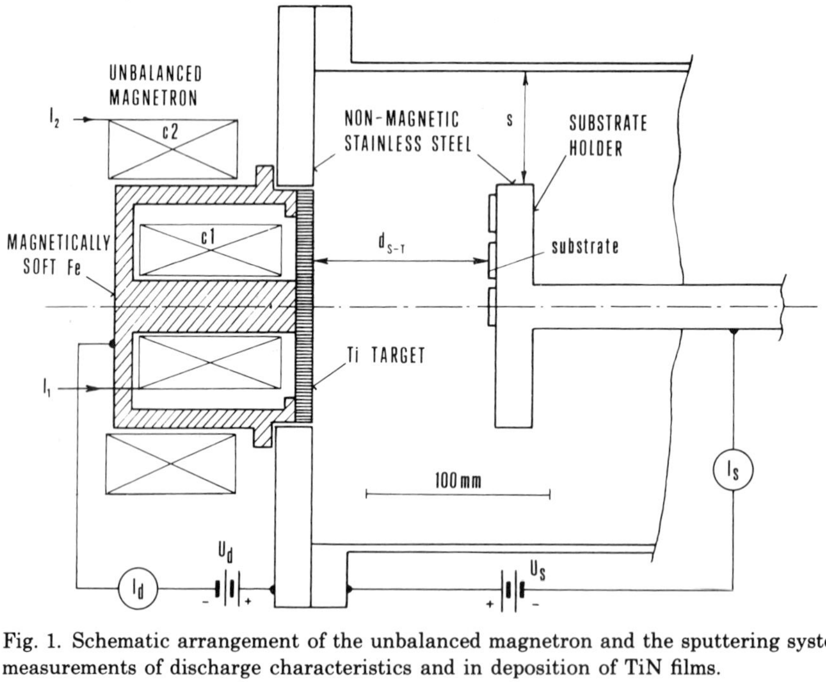


Fig. 1. Schematic arrangement of the unbalanced magnetron and the sputtering system used in measurements of discharge characteristics and in deposition of TiN films.

## 3. Discharge characteristics of magnetron sputtering systems

In a deposition device equipped with a UM, the discharge between the magnetron cathode and the anode can be influenced not only by the gas pressure p, the discharge power $P_d = U_d I_d$ and the magnetic field configuration, controlled, for example, by the currents $I_1$ and $I_2$, but also by the presence of the substrate holder biased to the potential $U_s$ respective to the anode and the distance $d_{S-T}$. Therefore, interrelations between all the above mentioned parameters and, especially, their influence on the substrate current $I_s$ have been studied [7]. The main features of the discharge characteristics are displayed in Figs. 2 - 4.

Figure 2 shows $I_s$ as a function of the current $I_2$, *i.e.* the external magnetic field strength. Here, the current $I_1$ necessary to maintain constant discharge current $I_d = 1$ A and the discharge voltage $U_d = 500$ V is also

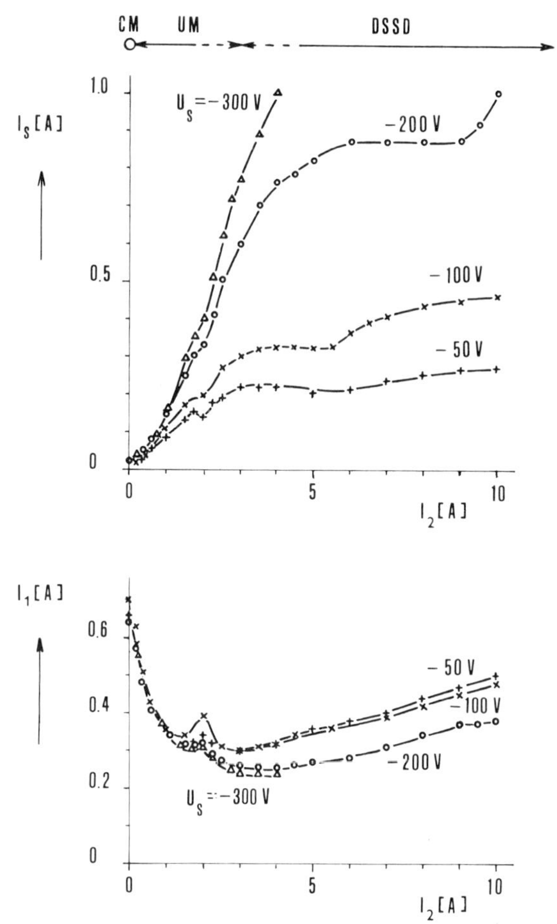


Fig. 2. The substrate ion current $I_s$ and the current $I_1$ as functions of the current $I_2$ for different values of substrate bias $U_s$. Constant parameters: $U_d = 500$ V, $I_d = 1$ A, $d_{S-T} = 70$ mm, $p_{Ar} = 5$ Pa.

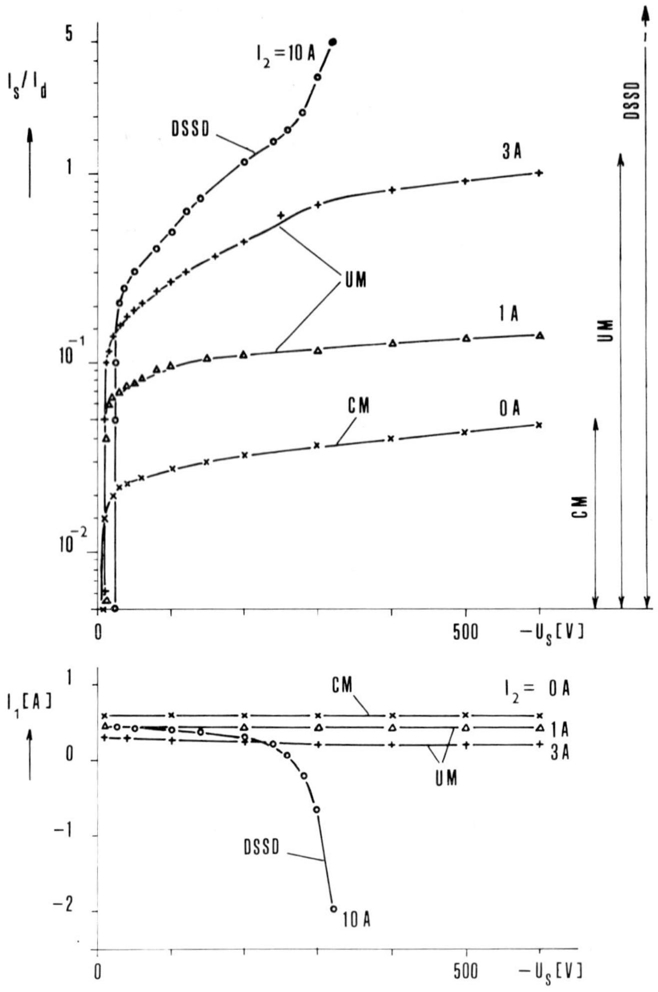


Fig. 3. The ratio $I_s/I_d$ and the current $I_1$ as functions of the substrate bias $U_s$ for different values of $I_2$. Constant parameters: $U_d = 500$ V, $I_d = 0.2$ A, $d_{S\text{–}T} = 70$ mm, $p_{Ar} = 5$ Pa.

shown. First, the current $I_1$ decreases, but starting when $I_2$ is around 3 A it increases with increasing of $I_2$. According to the magnitude of the external magnetic field ($I_2$), we can distinguish three types of magnetron discharges:

(1) Conventional magnetron (CM); this corresponds to the point $I_2 = 0$ and is characterised by relatively low values of $I_s$.
(2) Unbalanced magnetron (UM); this exists in the range $0 < I_2 \lessapprox 3$ A, where $I_1$ decreases and $I_s$ increases with increasing $I_2$.
(3) Double-site-sustained discharge (DSSD); this exists for $I_2 \gtrapprox 3$ A, where $I_1$ increases with increasing $I_2$ and $I_s$ is either almost independent on $I_2$ (for low substrate bias $-U_s \lessapprox 200$ V) or increases with $I_2$ and achieves high values of $I_s \approx I_d$ (for higher bias $-U_s > 200$ V).

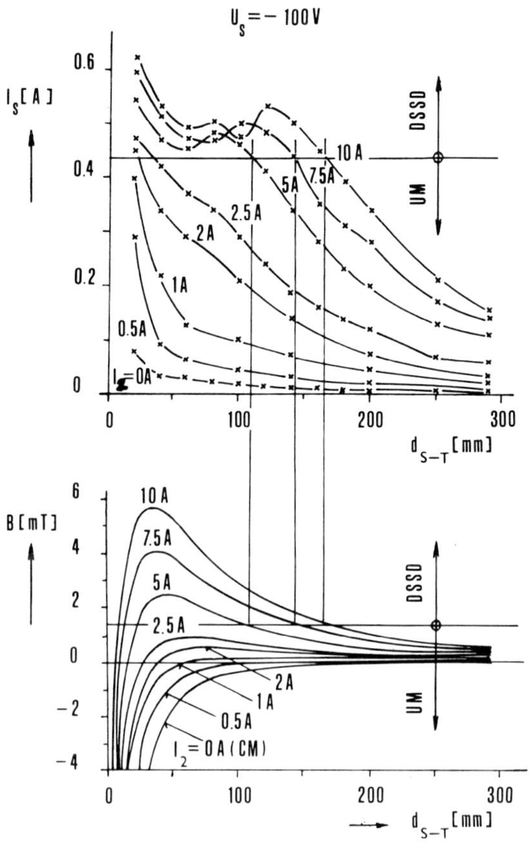


Fig. 4. The substrate ion current $I_s$ and the magnetic induction $B$ on the magnetron axis as functions of the distance $d_{S-T}$ for different values of $I_2$. Constant parameters: $U_d = 500$ V, $I_d = 1$ A, $p_{Ar} = 5$ Pa, $U_s = -100$ V.

The magnetron operating in a strong external magnetic field ($I_2 \gtrsim 3$ A) can completely confine the plasma between the magnetron target and the substrates. In this sputtering system the discharge can be sustained on both the magnetron target and the substrate sites.

The difference in discharges of CM, UM and DSSD is also clearly seen in Fig. 3. Here the ratio $I_s/I_d$ is displayed as a function of the negative substrate bias $U_s$. This function is rather flat for the CM and the UM operated at low $I_2 < 3$ A, *i.e.* in relatively low external magnetic fields. A steep increase of the ratio $I_s/I_d$ is, however, observed for the UM operated in the transition to the DSSD ($I_2 \approx 3$ A) and especially for DSSD operated in strong magnetic fields ($I_2 > 3$ A).

High values of $I_s \approx 5I_d$ were achieved in the DSSD at $U_s = -320$ V (see Fig. 3). When the negative bias $U_s$ was increased and the discharge became more substrate-site-sustained, the current $I_1$ in the inner coil of magnetron had to be decreased to keep $U_d$ and $I_d$ constant, and $I_1$ even reached negative values ($I_2 = 10$ A, $-U_s > 265$ V). This was evidence that the DSSD can be operated even without the CM magnetic field ($I_1 = 0$). Conversely, CM and UM exhibit almost no or only weak dependence of $I_1$ on $U_s$ and always have to be operated with some component of the CM magnetic field.

Also very interesting is the dependence of $I_s$ on $d_{S\text{–}T}$ for CM, UM and DSSD (see Fig. 4). In sputtering systems equipped with CM and UM the current $I_s$ decreases with increasing $d_{S-T}$. A high and almost constant ion current $I_s$ is transported over large distances $d_{S\text{–}T}$ for DSSD.

For comparison, the axial component of the magnetic induction $B$ on the magnetron axis is also shown in Fig. 4. It can be seen, that $I_s$ begins to fall when the induction $B$ weakens to about 1.4 mT.

In our experimental arrangement the separation distance $s$ between the anode (chamber cylinder) and the outer field lines connecting the cathode with the edge of the substrate holder is about 50 mm. The Larmor radius of electrons $r_L$ [9] is $r_L = s = 50$ mm for $B = 1.4$ mT and $E = 500$ eV. This is really the highest energy which electrons can acquire in our plasma, because the target potential $U_d = 500$ V with respect to the anode. In a sputtering system with DSSD the plasma electrons move along the magnetic field lines between the cathode and the substrates on helical paths like in the Penning discharge.

The sputtering system with DSSD can be operated under different conditions. In the case when the substrate bias is low ($-U_s \lesssim 100$ V) the discharge is magnetron-site-sustained, for higher biases ($100 \text{ V} \lesssim -U_s \lesssim 600$ V) it is double-site-sustained and for high substrate biases ($-U_s \gtrsim 600$ V) it can be substrate-site-sustained. The last case was verified in a special experiment [7] carried out at $I_s = 0.2$ A, $d_{S\text{–}T} = 70$ mm, $p_{Ar} = 3$ Pa, $I_1 \doteq -2$ A, $I_2 = 10$ A. It was found that $I_d$ is nearly constant over a broad range of $U_d$ between 100 and 1000 V, and considerably lower than $I_s$, *i.e.* $I_d = 0.04 < I_s = 0.2$ A. At the same time the ion current $I_s = 0.2$ A was extracted to substrates even at $I_d = 0$ and $-U_s < 1000$ V. This demonstrated that the discharge was substrate-site-sustained and the magnetron cathode served as an ion collector only.

The greatest advantage of UMs, especially when operated in the DSSD mode, over CM sputtering lies in the fact that they provide much higher ratios $\nu_i/\nu_m$ of the flux of bombarding ions $\nu_i$ to the flux of arriving metal atoms $\nu_m$ at substrates. A comparison of ratio $\nu_i/\nu_m$ for CM, UM and DSSD sputtering of titanium is given in ref. 10. The ratio $\nu_i/\nu_{Ti}$ increases with increasing magnetic field from about 0.5 in CM to about 2 in UM and to about 10 in DSSD, for $U_s = -100$ V. The ratio $\nu_i/\nu_m$ shows a weak dependence on $d_{S\text{–}T}$ for CM and UM but considerably increases with increasing $d_{S\text{–}T}$ for DSSD. Such high ratios of $\nu_i/\nu_m$ strongly influence the microstructure and the properties of thin films. Typical values of $\nu_i/\nu_m$ in commercial ion plating systems range between 5 and 10 [11].

## 4. Properties of TiN films sputtered at large distances and high pressures

The unbalanced magnetron described above was used in the high substrate ion current density mode to study properties of TiN films deposited at a high total pressure $p_T = 5$ Pa and a large distance $d_{S-T} = 200$ mm. These conditions are rather unfavourable for CM sputtering. TiN films sputtered by CM at $p_T = 5$ Pa, $d_{S-T} = 150$ mm and $U_s = -100$ V are black, porous and soft [12].

TiN films were deposited at different values of $U_s$ between $-5$ and $-100$ V on ground HSS and electropolished SS under the following conditions: $T_s = 500$ °C, $I_d = 3$ A, $U_d = 600$ V, $\phi_{N_2} = 8.4$ s.c.c.m., $\phi_{Ar} = 40$ s.c.c.m., $p_T = 5$ Pa, $d_{S-T} = 200$ mm, $I_2 = 10$ A, $I_1 \doteq 0.5$ A. Table 1 summarizes the deposition parameters, argon content, X-ray data, mechanical properties and colour of deposited TiN films. It can be seen that the DSSD provides high $i_s$ up to about 6 mA cm$^{-2}$, high $\nu_i/\nu_{Ti}$ up to about 8 for $U_s = -100$ V and the conditions chosen. These values correspond to an average energy up to 0.8 keV per arriving titanium atom. Figure 5 shows scanning electron microscope (SEM) micrographs of fracture cross sections and the surface morphology of selected films. The SEM micrographs were produced with a Cambridge S360 instrument.

The film properties systematically varied with increasing substrate bias $-U_s$. The main features of this variation can be summarized as follows:

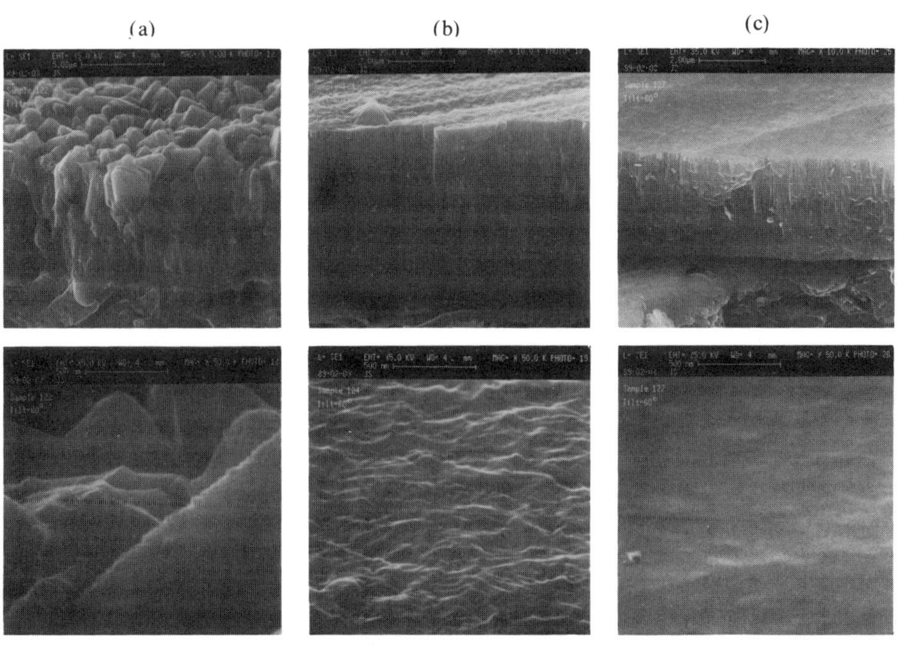


Fig. 5. High resolution SEM micrographs of fracture cross-sections (a, ×5000; b and c, ×10000) and top views of the film surfaces (×50000) of selected TiN samples deposited by DSSD at (a) $U_s = -30$ V, (b) $U_s = -40$ V and (c) $U_s = -100$ V.

TABLE 1

Dependence of deposition conditions, X-ray data, argon content, mechanical properties and colour on bias voltage $U_s$ for TiN films prepared in a DSSD sputtering system

| *Deposition parameters* | | | | | *X-ray results* | | | | | | *Properties* | | | | | |
|---|---|---|---|---|---|---|---|---|---|---|---|---|---|---|---|---|
| $-U_s$ (V) | $i_s$ (mA cm$^{-2}$) | $h$ (μm) | $a_D$ (nm s$^{-1}$) | $\frac{v_i}{v_{Ti}}$ | $I_{hkl}$ | | | $a_0$ | | $\beta$(111) (deg in 2Θ) | $Ar$ (at.%) | $HV_{10}$ (kg mm$^{-2}$) | $L_c$ (N) | *Colour* | | |
| | | | | | (111) | (200) | (220) | (111) (nm) | (200) (nm) | | | | | $L^*$ | $a^*$ | $b^*$ |
| 5 | 1.8 | 11.5 | 1.60 | 2.2 | 0.52 | 1.71 | 0.77 | 0.4235 | 0.4235 | 0.21 | 0 | — | — | 41.4 | 4.5 | 24.3 |
| 30 | 3.0 | 11.2 | 1.56 | 3.7 | 1.34 | 0.91 | 0.75 | 0.4243 | 0.4241 | 0.26 | 0.06 | 522 | — | 45.9 | 2.1 | 16.3 |
| 40 | 3.5 | 7.3 | 1.01 | 4.3 | 2.96 | 0.03 | 0.01 | 0.4284 | 0.4258 | 0.33 | 0 | 2270 | 64 | 70.2 | 2.6 | 44.2 |
| 50 | 4.0 | 2.7 | 0.87 | 4.9 | 2.32 | 0.68 | 0 | 0.4283 | 0.4268 | 0.36 | 0.19 | — | 39 | 74.8 | 2.6 | 45.4 |
| 60 | 4.4 | 6.1 | 0.85 | 5.4 | 2.97 | 0.02 | 0.01 | 0.4277 | 0.4251 | 0.39 | 0.10 | 2370 | 36 | 77.4 | 0.3 | 43.9 |
| 70 | 5.1 | 5.6 | 0.78 | 6.2 | 2.96 | 0.03 | 0 | 0.4277 | 0.4258 | 0.40 | 0.22 | 2420 | 30 | 77.7 | 0.4 | 43.1 |
| 83 | 6.0 | 5.2 | 0.79 | 7.3 | 2.95 | 0.05 | 0 | 0.4277 | 0.4263 | 0.41 | 0.14 | 2470 | 25 | 78.9 | 0.8 | 43.0 |
| 100 | 6.3 | 4.9 | 0.68 | 7.7 | 2.96 | 0.03 | 0.01 | 0.4285 | 0.4270 | 0.51 | 0.84 | 2470 | 30 | 78.7 | 0.3 | 41.5 |

Independent deposition parameters: $p_T = 5$ Pa, $d_{S\text{-}T} = 200$ mm, $T_s = 500$ °C, $U_d = 600$ V, $I_d = 3$ A, $\phi_{Ar} = 40$ s.c.c.m., $\phi_{N_2} = 8.4$ s.c.c.m., $I_2 = 10$ A, $I_1 \doteq 0.5$ A.

(1) The films deposited at $-U_s \leqslant 30$ V are relatively thick and seem to be porous, which also indicates low density. The films consist of sharp-edged, rather large grains separated by voids (Fig. 5(a)). The X-ray analysis shows mixed (200) + (111) + (220) textures, lattice parameters determined from all reflections close to the bulk value 0.424 nm for stoichiometric TiN and narrow diffraction peaks, indicating a low density of defects, low intrinsic stresses and larger grain sizes in films. Large grain sizes up to 100 nm evaluated by transmission electron microscope (TEM) analysis have also been previously found for $Ti_{0.5}Al_{0.5}N$ films which were deposited under substrate biases below $-30$ V. The study showed also intergranular voids [13].

The microhardness of the porous TiN films was very low or almost not measureable. The same is true for the critical load $L_c$ evaluated by scratch testing. Due to the rough surface morphology the films appeared dull grey in colour. This results also in low brilliance values $L^*$ (Table 1). Energy dispersive X-ray (EDX) analysis was carried out using the Magic V program [14]. Low-biased films exhibited quite a high amount of gaseous impurities, up to 15 at.% O and 5 at.% C. Secondary neutral mass spectroscopy (SNMS) depth profile analysis using the Leybold INA 3 system showed that the oxygen and carbon concentration decayed within approximately 0.2 $\mu$m of the outer surface region. At 0.6 $\mu$m below the film surface the impurity level of the porous films was still higher by a factor of three (4 at.% for each concentration) as compared to the compact films. For all films, either porous or dense, the N/Ti ratio measured by EDX or SNMS was unity within the accuracy of the measurements, if the measurements were carried out far enough below the surface level. For the measurements of the N/Ti ratios both the EDX and the SNMS analyser were calibrated by a stoichiometric thin film single crystal TiN sample.

(2) All films deposited at $U_s$ between $-40$ and $-100$ V are dense, compact and consist of fine fibrous grains (Fig. 5(b) and (c)). It is surprising that a negative bias voltage increase from 30 to 40 V is sufficient to densify the microstructure as demonstrated in Fig. 5(b) in comparison with Fig. 5(a). With increasing bias voltage the deposition rate decreases (see Table 1). This is mainly due to densification, especially in comparison with the low biased deposits, but probably also due to resputtering in the case of higher bias values. Although the argon pressure during deposition was comparatively high, argon incorporation of approximately 0.84 at.% was only observed for a bias value of $-100$ V.

The texture of the films is also clearly influenced by the increase in bias voltage. An increase of (111) orientation was observed, whereas the intensity of the (200) lines was decreasing and the (220) orientation was only detected at the lowest bias values (Table 1). Similar tendencies for the (111) and (200) reflections have been reported [15] for the ion plating process with evaporation sources.

These effects are accompanied by an enlargement of the lattice parameter $a_0$ as determined from all the diffraction peaks. However, the denser films exhibit significantly higher $a_0$ values, as determined from the (111) reflections, than those determined from the (200) reflections. Also observed is an increase of the full width $\beta$ at half maximum intensity, obtained after subtracting instrumental broadening effects. The results for the (111) signals are shown in Table 1. Both the increase of $a_0$ and $\beta$ with bias indicate mainly a higher density of defects, macrostress and microstrain in the films [13, 16 - 19].

It is interesting to see also that the microhardness values, which increase with $-U_s$, have rather moderate values, namely 2270 - 2470 kg mm$^{-2}$ as compared with values up to 3000 kg mm$^{-2}$ previously reported [13, 18 - 20].

The critical load $L_c$ of dense TiN films is decreased if bias voltage increases. This can be explained by the increase of the formation of stress. Similar results have been reported, *e.g.* in ref. 19.

Finally, the influence of the bias voltage on the surface morphology is discussed. As $-U_s$ increases the surface becomes smoother (Fig. 5(b) and (c)) and shallow dimples, typical of ion plated TiN films rather than for those sputtered by CM [16], are observed (Fig. 5(c)). This may be caused by a decrease in grain size as observed for $Ti_{0.5}Al_{0.5}N$ films [13], or by some kind of surface smoothing effects due to the intensive ion bombardment [21] at $-U_s > 50$ V and $v_i/v_m$ values of 5 or less. Colour changes show analogue trends, *i.e.* the increase of brilliance $L^*$, and the decrease of redness $a^*$ and yellowness $b^*$ with increasing bias $-U_s$. This can be related not only to the increase of surface smoothness, but also to the expansion of the lattice parameters [22].

## 5. Conclusion

In sputtering systems equipped with UMs, high ion currents $I_s$ can be transported to substrates, which are even greater than the magnetron current $I_d$. If the magnetic field of UM reaching substrates is sufficiently strong (several mT), the discharge strongly differs from that in a weak magnetic field. Such a discharge can be sustained at both the magnetron and substrate sites, and so it is called the double-site-sustained discharge (DSSD). UM, working in a regime of DSSD, can also be operated with $I_1 = 0$ (the internal electromagnet ($C_1$) is switched off) or even with $I_d = 0$. Moreover, in the sputtering system with DSSD $I_s$ virtually does not decrease with $d_{S\text{–}T}$ up to 200 mm.

It was found that the sputtering system with DSSD allows the production of dense, compact, bright golden TiN films with the microhardness greater than 2200 kg mm$^{-2}$ and good adhesion (a high critical load up to 64 N) even at pressures as high as 5 Pa and $U_s \approx -40$ V and in the range of target to substrates distanced $d_{S\text{–}T} = 200$ mm. The latter compares with distances typical for ion plating systems with a low energy electron beam or arc evaporation sources. Nevertheless, the pressure and bias ranges used for sputtering in this study are not common for electron beam and arc techniques. A comparison study on the properties of sputtered and ion plated TiN films is given in ref. 11.

## References

1 R. Adachi and K. Takeshita, *J. Vac. Sci. Technol.*, *20* (1982) 98.
2 J. J. Cuomo and S. M. Rossnagel, *J. Vac. Sci. Technol.*, *A4* (1986) 393.
3 T. Hata, E. Noda, O. Marimoto and T. Hada, *Appl. Phys. Lett.*, *37* (1980) 633.
4 B. Window and N. Savvides, *J. Vac. Sci. Technol.*, *A4* (1986) 196.
5 A. G. Spencer, K. Oka, R. W. Lewin and R. P. Howson, *Vacuum*, *38* (1988) 857.
6 T. Fukami, F. Shintani and M. Naoe, *Thin Solid Films*, *149* (1987) 373.
7 S. Kadlec and J. Musil, Research Report, *Institute of Physics CSAS*, *no. 055/88*, November 1988.
8 J. Vyskočil, J. Musil, S. Kadlec and W.-D. Münz, in E. Brozeit *et al.* (eds.), *Plasma Surface Engineering*, DGM Informations Gesellschaft, Oberasel, 1989, Vol. 1, pp. 661 - 668.
9 F. F. Chen, *Introduction to Plasma Physics*, Plenum, New York, 1974.
10 J. Musil, S. Kadlec, J. Vyskočil and V. Poulek, *Surf. Coat. Technol.*, *39/40* (1989) 301.
11 W.-D. Münz, J. Bartella, G. Håkansson, J.-E. Sundgren, D. McIntyre and J. E. Greene, *presented at 16th ICMC*, San Diego, CA, *April 17 - 21, 1989.*
12 S. Kadlec, J. Musil, W.-D. Münz and V. Valvoda, *Proc. 7th Int. Conf. on Ion and Plasma Assisted Technologies*, Geneva, Switzerland, CEP Consultants, Edinburgh, 1989, p. 100.
13 G. Håkansson, J.-E. Sundgren, D. McIntyre, J. E. Greene and W.-D. Münz, *Thin Solid Films*, *153* (1987) 55.
14 J. W. Colby, *Quantex Instruction Manual*, Kevex, Foster City, CA, 1980.
15 H. Rudigier, E. Bergmann and J. Vogel, *Surf. Coat. Technol.*, *36* (1988) 678.
16 D. T. Quinto, G. J. Wolfe and P. C. Jindal, *Thin Solid Films*, *153* (1987) 19.
17 I. Petrov, L. Hultman, U. Helmersson, J.-E. Sundgren and J. E. Greene, *Thin Solid Films*, *169* (1988) 299.
18 D. S. Rickerby and P. J. Burnett, *Thin Solid Films*, *157* (1988) 195.
19 P. J. Burnett and D. S. Rickerby, *Thin Solid Films*, *154* (1987) 403.
20 W. D. Sproul, P. J. Rudnik and C. A. Gogol, *Thin Solid Films*, *171* (1989) 171.
21 D. M. Mattox and G. J. Kominiak, *J. Vac. Sci. Technol.*, *9* (1972) 528.
22 A. J. Perry, *Thin Solid Films*, *146* (1987) 165.

*Surface and Coatings Technology, 39/40* (1989) 499 - 506 

# ADVANCES IN PARTIAL-PRESSURE CONTROL APPLIED TO REACTIVE SPUTTERING

W. D. SPROUL and P. J. RUDNIK

*Basic Industry Research Laboratory, Northwestern University, 1801 Maple Avenue, Evanston, IL 60201-3135 (U.S.A.)*

C. A. GOGOL and R. A. MUELLER

*Leybold Inficon, Inc., 6500 Fly Road, East Syracuse, NY 13057 (U.S.A.)*

(Received March 8, 1989)

## Summary

High-rate reactive sputtering (HRRS) requires rapid and careful control of the reactive gas partial pressure to achieve high deposition rates and to maintain stoichiometry in the coating. HRRS of TiN was first achieved with feedback control of the $N_2$ utilizing a differentially pumped mass spectrometer system to sense the $N_2$ partial pressure. Recently, a new instrument called the optical gas controller (OGC) that operates at sputtering pressures became available for sensing the partial pressure of the gases in the sputtering atmosphere. The OGC was used during the reactive sputtering of $TiO_x$ with partial-pressure control, and the $TiO_x$ hysteresis loop exhibited a wide negative slope region which is much larger than the one found during the reactive sputtering of $TiN_x$. Partial-pressure control, compared to mass flow control during oxide reactive sputtering, leads to better control of the process and to a 3.5 times higher deposition rate for highly resistive oxide coatings. With partial-pressure control, the anatase form of $TiO_2$ has been produced with lattice parameters of $a_0 = 3.780$ Å and $c_0 = 9.610$ Å.

---

## 1. Introduction

Reactive sputtering is a popular means for producing coatings of compounds on an industrial scale. It is a process that is conceptually simple. Metal is sputtered from a pure target, and sufficient reactive gas is added to the process chamber to form the desired compound at the substrate. Any facility with sputtering equipment can easily convert its equipment to make reactive films.

The transformation of reactive sputtering into an economically attractive manufacturing process is, however, a balancing act of atomic proportions. To form the stoichiometric compound, the arrival rate of metal atoms must be matched by an appropriate arrival rate of the desired reactive gas

0257-8972/89/$3.50 

atoms at the substrate. If these arrival rates are not balanced, the resulting film will not be of the desired composition. By controlling the partial pressure of the reactive gas in the region of the substrate, it is possible to maintain the required arrival rate of gas atoms so that, when they are combined with the arriving metal atoms, the proper material phase is produced [1].

The partial pressure of the reactive gas should be controlled closely so that wide variations in pressure are not experienced. Even short periods of low pressure result in gas deficits that will produce several monolayers of metal rich composition, while a short term overabundance will lead to lower deposition rates and possibly the formation of an undesired phase. These regions of non-stoichiometric composition will reduce the performance of the coating and generally are the cause of early coating and possibly part failure.

Maintaining a constant metal arrival rate is not always easy. A well-conditioned metal target in a pure argon atmosphere will have a constant metal removal rate if the power to a magnetron sputtering target is constant. This constant metal removal rate at the cathode translates into a constant transport rate to the substrate and a constant flux of available metal.

When the reactive gas is added to the process chamber during sputtering, it reacts with the substrates, chamber walls and target [2]. When there is sufficient reactive gas to form the stoichiometric compound at the substrate, there usually is sufficient gas to form a compound on the surface of the target. The sputtering rate from the compound on the target is less than that from the pure metal surface. When this transition occurs, the overall sputtering rate drops rapidly. If the reactive gas is supplied at a constant rate, the partial pressure of the reactive gas rises rapidly as the rate drops because less gas is being consumed in the process. If the flow of reactive gas is reduced after the partial pressure has risen to a high level, the partial pressure will not follow the same trajectory on the way down as it did on the way up, and the so-called reactive sputtering hysteresis loop is formed. This hysteresis effect makes it impossible to produce certain compositions in the coating if the mass flow control of the reactive gas is used [2, 3].

The severity of the hysteresis effect can be reduced by increasing the pumping speed of the system such that the quantity of gas removed by the pumps is much greater than that consumed chemically [4, 5]. In this way, the destabilizing pressure swings at the target are greatly reduced when the target changes from the metallic to the compound mode. Increasing the pumping speed can eliminate hysteresis, but it is costly in terms of the expense of added pump capacity and increased gas consumption. The added gas flow does have the advantage of reducing contamination from vessel outgassing and leaks through dilution.

Through partial-pressure control of the reactive gas it is possible to produce all material compositions in spite of the hysteresis effect. If the reactive gas partial pressure is held constant at the same time that the power to the sputtering target is held constant, a balance between consumption and availability of the reactive gas is maintained [1]. The partial pressure sets the flux of gas atoms at every surface. If the partial pressure is controlled, the availability of that gas will be controlled. If there is a process disturbance, such as an arc on the target, a partial-pressure controller will momentarily reduce flow to maintain constant partial pressure. Once the plasma is re-established (after the arc is quenched) and the metal is being sputtered at the full rate, the flow will once again be increased to maintain the desired partial pressure. In partial-pressure control, there is inherent stability. The removal of material from the target is nearly constant except for perturbations such as arcs, and at the substrate, the metal and gas atoms arrive in the proper ratio to produce the stoichiometric compound.

Partial-pressure control requires a species selective means of monitoring the gases in the process chamber in real time. The most frequently used piece of equipment is the quadrupole mass analyzer (QMA), which has the ability to separate gases by their mass ratios and which generally provides a unique signal for each gas present. Good partial-pressure control is achieved when an adequate signal-to-noise ratio is obtained by the analyzer in a time frame that is short enough that the flow can be adjusted before the inherent process instabilities take the partial pressure too far from equilibrium.

The location of the partial-pressure transducer is important for minimizing the response time of the system sufficiently to achieve good control of the partial pressure. Ideally, the transducer should be placed as close to the reactive gas manifold as possible, but, due to the differences in operating pressure for sputtering and the QMA (typically $8 \times 10^{-3}$ Torr for sputtering and $1 \times 10^{-5}$ Torr or less for the QMA), it is necessary to differentially pump the QMA system through an orifice. The size of the needed pumping system often prohibits the placement of the QMA right next to the sputtering plasma, but fortunately transport of partial pressure information to other regions of the chamber takes place very rapidly at sputtering pressures. At a sputtering pressure of $10 \times 10^{-3}$ Torr, it is estimated that it will take a gas atom approximately 0.09 s to traverse a 50 cm distance. Delays in signal acquisition time can be introduced at the pressure reduction orifice if there is insufficient pumping speed between the orifice and the QMA, and a poorly designed orifice can change the apparent composition of the sputtering atmosphere. Care must be taken to avoid these problems.

Recently, a new instrument for partial-pressure control, called the optical gas controller (OGC) [6, 7], has been introduced which operates at pressures up to $100 \times 10^{-3}$ Torr without having to be differentially pumped. Because pressure reduction is not necessary, the distortions that can be introduced by orifices and capillaries are not present. The OGC sensor is placed directly in the sputtering chamber and is exposed to the sputtering atmosphere. This favorable location is extremely important for measuring condensable gases that may not be detected due to adsorption on the extra surfaces of a QMA pressure reduction apparatus. This adsorption–release process can distort readings for several hours until equilibrium is reached. In addition, the elimination of a differential pumping system means that the OGC system is not subject to periodic pump failures that may occur.

The operation of the OGC is based on electron impact emission spectroscopy (EIES) [8]. An electron gun provides 500 $\mu$A of primary electrons at 100 V that are used to excite the gas molecules. The subsequent de-excitation of the gas molecules is responsible for the release of a photon. These photons are typically in the visible and ultraviolet region of the electromagnetic spectrum and have precise wavelengths that are characteristic of the species. The photons relating to a specific species are selected by thin film optical filters and then converted into an electrical signal proportional to the density of the gas by a photomultiplier.

In a recent study, both an OGC and a QMA system were used during the high-rate reactive sputtering of TiN [9]. The QMA was located at one end of the sputtering chamber, whereas the OGC sensor was placed on top of the sputtering target and communicated with the sputtering plasma through a small tube located along the long axis of a rectangular magnetron sputtering target. Both the QMA feedback control system and the OGC were able to control the high-rate reactive sputtering process for TiN, but some interesting effects were observed with the OGC. When power was applied to the sputtering target, the partial pressure of the argon and nitrogen, as measured by the OGC, appeared to drop, whereas they remained constant at the QMA. As power is applied to the sputtering target, there are collisions between the sputtered atoms, energetic ions and neutrals, and the gas atoms. These collisions lead to gas heating and rarefaction. The number density of gas atoms in front of the target is reduced [10, 11], but the pressure is not reduced when the metal atoms in the transport are considered. Since the number density of the gas is reduced in front of the target and since the temperature of the gas is cooled between the plasma and the OGC sensor, there is an 'apparent' pressure drop measured by the OGC or any other 'density sensing' measurement scheme.

This apparent pressure drop has serious consequences for reactive sputtering. As the target power is increased, the density of gas atoms in front of the target is decreased whereas the number of metal atoms is increased [9]. In order to maintain the proper arrival ratio of reactive gas atoms to metal atoms to achieve the stoichiometric composition, it is necessary to increase the reactive gas partial pressure for each target power increase. Thus, each power level will have its own unique reactive gas partial pressure to maintain the correct coating composition. The OGC, by virtue of not requiring pressure reduction, is able to detect this condition better.

Since the OGC and the QMA are so effective in controlling the reactive sputtering of nitrides, it was decided to determine how well these instruments would work for the reactive sputtering of titanium oxide. The remainder of this paper describes the results from that work.

## 2. Experimental procedures

All reactive sputtering work was carried out in an MRC in-line sputtering system, the operation of which has been described previously [9]. Only direct current power was used, both for the target and for the bias. Power to the target was set at 8 kW, and the bias on the substrate was $-100$ V. Total system pressure was $8 \times 10^{-3}$ Torr. In this work, partial-pressure control of the oxygen reactive gas (purity, 99.999%) was by the OGC, but it is also possible to control the oxygen with the QMA feedback control system.

The thickness of steps in the coating on the glass slide substrates was measured with a Dektak surface profilometer. The bulk resistance of the films was measured with Fluke 73 digital multimeter, and this resistance was converted to a resistivity once the coating thickness was known. X-ray diffraction analysis of the coatings was done in a Scintag XDS 2000 PAD V diffractometer with Cu $K\alpha$ radiation.

## 3. Results and discussion

Most reported reactive sputtering work has been accomplished with mass flow control of the reactive gas (see ref. 12, for example). For the Ti–O system, mass flow control produces a typical hysteresis curve, as is shown in Fig. 1 for the reactive sputtering of $TiO_x$. There is little change in the partial pressure of the reactive gas until the flow reaches approximately 40 sccm, but

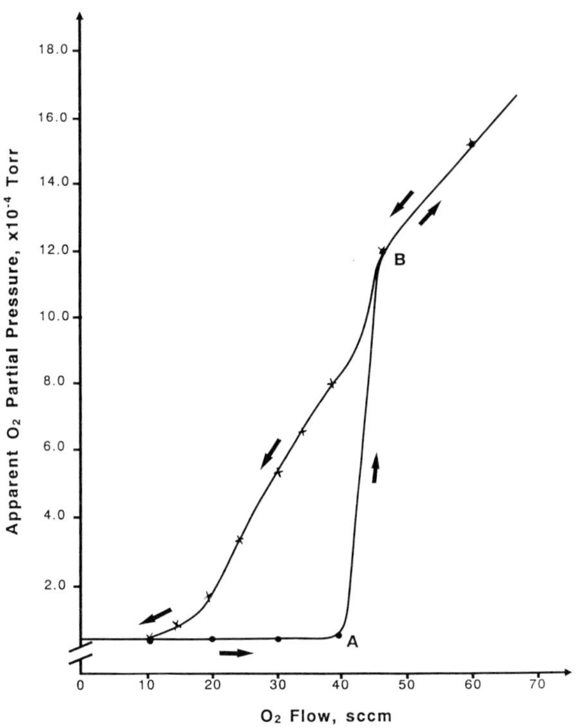


Fig. 1. The hysteresis curve for the reactive sputtering of $TiO_x$ at a target power of 8.0 kW and mass flow control of the $O_2$ reactive gas. The partial pressure of the $O_2$ was measured with the OGC.

at that flow the partial pressure starts to change rapidly. It jumps from a partial pressure of $5 \times 10^{-5}$ Torr at point A to $1.2 \times 10^{-3}$ Torr at point B, and the deposition rate drops from 0.33 $\mu$m min$^{-1}$ at point A to 0.017 $\mu$m min$^{-1}$ at point B. With mass flow control, it is not possible to operate between points A and B. This is the so-called forbidden zone of reactive sputtering.

With partial-pressure feedback control of the reactive gas during the reactive sputtering of $TiO_x$, the shape of the hysteresis curve, as shown in Fig. 2, is quite different from the one produced with mass flow control (see Fig. 1). As the partial pressure is increased beyond the nose of the hysteresis curve at point A in Fig. 2, the flow of the reactive gas actually decreases. This phenomenon has been observed during the reactive sputtering of the nitrides [3, 9, 13], but even at high nitride deposition rates, the drop in the flow of the nitrogen is only 1 - 2 sccm. For reactive sputtering of $TiO_x$, the reduction in the flow of oxygen is very large. It drops from a peak flow of 48 to 28 sccm before the flow begins to increase again at point B in Fig. 2.

There are several reasons for this large drop in reactive gas flow, and they all relate to the reduction in sputtering rate of the metal from the magnetron sputtering target. When the reactive gas is initially introduced into the sputtering chamber, all of the reactive gas is either consumed in the

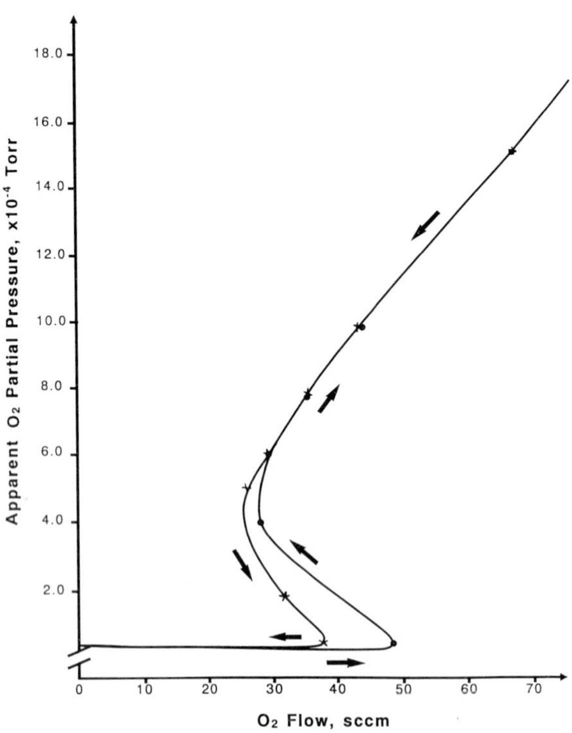


Fig. 2. The hysteresis curve for the reactive sputtering of $TiO_x$ at a target power of 8.0 kW with OGC partial-pressure control of the $O_2$ reactive gas.

reaction with the sputtered metal or it is pumped away. As the partial pressure increases, an oxide starts to form on the outer edges of the magnetron racetrack region where the metal erosion rate is not as great as it is in the center of the racetrack. When this oxide forms, the deposition rate starts to slow down, and less oxygen flow is needed to maintain the partial pressure. At the same time as more oxygen enters the sputtering plasma, the number of oxygen ions increases, and the sputtering rates drop even more because oxygen ions are not as efficient at sputtering as the argon ions are. Finally, as the oxide builds up on the target, the number of electrons given off by the oxided surface increases substantially because the secondary electron emission coefficient is greater for the oxide than for the metal. This increase in electrons leaving the cathode cannot be separated electrically from argon ions arriving at the cathode. This change dramatically reduces the sputtering efficiency because both the ion current and plasma impedance are reduced.

The resistivity of the oxide films produced at different points along the partial-pressure controlled hysteresis curve changes by many orders of magnitude. At partial pressures up to point A in Fig. 2, the material is still conductive. A film produced at point A at a partial pressure of $5 \times 10^{-5}$ Torr was reddish brown in color and had a resistivity of 3700 $\mu\Omega$ cm whereas one made at a partial pressure of $8 \times 10^{-4}$ Torr with a deposition rate of 0.06 $\mu$m $\min^{-1}$ was clear in color and had a resistivity in excess of $10^6$ $\Omega$ cm, which is the limit of the instrument used to measure these films.

The resistivity of the coatings produced by mass flow control also changed greatly when the flow was increased. As the flow of oxygen was increased up to point A in Fig. 1, the coatings remained conductive, and the color of the coatings varied from metallic to brown. A coating sputtered at a flow of 35 sccm had a resistivity of 7800 $\mu\Omega$ cm and a brown color, and the deposition rate at this point was 0.33 $\mu$m $\min^{-1}$. Once the oxygen flow reached 40 sccm, the partial pressure jumped up to point B, and the deposition rate dropped to 0.017 $\mu$m $\min^{-1}$. The resistivity of the coating made at this point was again greater than $10^6$ $\Omega$ cm, and the coating, although clear in color, was deposited at less than one third the rate of partial-pressure control.

Even though the resistivity of the coatings made by the two different control methods can be quite large, the deposition rate for two techniques for high resistivity materials is different. In a production environment, the deposition rate is an important factor. As long as two techniques can produce the same coating with the same properties, the one with the higher deposition rate will be used in order to save money.

The clear color of the highly resistive coatings made in this study should be contrasted with what Scherer and Wirz [12] reported. They showed that their $TiO_x$ coatings had a transmission of only 30% in the visible spectrum. This result would indicate that the coatings were not clear and that they had a distinct tint to them. It is estimated that the highly resistive $TiO_x$ coatings made in this study have a transmission in the visible spectrum of at least 80%.

X-ray diffraction of the coating, made by partial-pressure control at an oxygen partial pressure of $8 \times 10^{-4}$ Torr, produced a pattern with many sharp peaks. The coating is crystalline, and comparison of the peaks in this pattern with the Powder Diffraction File for the various $TiO_x$ materials shows that the coating is the anatase form [14] of $TiO_2$. This result compares favorably with what Schiller *et al.* [15] reported for their reactively sputtered $TiO_x$ films. The lattice parameters for this $TiO_2$ coating are $a_0 = 3.780$ Å and $c_0 = 9.610$ Å.

## 4. Conclusions

Partial-pressure control during reactive sputtering is a very effective way to produce compound coatings. Real-time control of the reactive sputtering process has been accomplished with the aid of quadrupole mass spectrometers, but there are drawbacks with this instrument. During reactive sputtering, the mass spectrometer must be differentially pumped through an orifice which can lead to time delays and signal distortions. Recently, a new instrument called the optical gas controller (OGC), which does not have to be differentially pumped, has been introduced for measuring the partial pressure of gases, and the small size of the sensor allows it to be placed very close to the sputtering plasma. The OGC was used during the reactive sputtering of $TiO_x$ for the partial-pressure control of the oxygen. The partial pressure–flow hysteresis loop made under partial-pressure control exhibited a very large negative slope region. $TiO_2$ coatings were produced at an oxygen partial pressure of $8 \times 10^{-4}$ Torr, and this material was the tetragonal anatase form of $TiO_2$ with lattice parameters of $a_0 = 3.780$ Å and $c_0 = 9.610$ Å. Partial-pressure control leads to 3.5 times higher deposition rates for highly resistive materials than does mass flow control.

## References


1 W. D. Sproul, *Surf. Coat. Technol., 33* (1987) 73.
2 S. Berg, H.-O. Blom, T. Larsson and C. Nender, *J. Vac. Sci. Technol. A, 5* (1987) 202.
3 A. F. Hmiel, *J. Vac. Sci. Technol. A, 3* (1985) 592.
4 T. Serikawa and A. Okamoto, *Thin Solid Films, 101* (1983) 1.
5 T. Larsson, H.-O. Blom, C. Nender and S. Berg, *J. Vac. Sci. Technol. A, 6* (1988) 1832.
6 Manufactured by Leybold Inficon, Inc., East Syracuse, New York, USA.
7 C. A. Gogol and R. Mueller, *Surf. Coat. Technol., 36* (1988) 939.
8 C. A. Gogol, *U.S. Patent 4,692,630,* September 8, 1987.
9 W. D. Sproul, P. J. Rudnik, and C. A. Gogol, *Thin Solid Films, 171* (1989) 171.
10 D. W. Hoffman, *J. Vac. Sci. Technol. A, 3* (1985) 561.
11 S. M. Rossnagel, *J. Vac. Sci. Technol. A, 6* (1988) 19.
12 M. Scherer and P. Wirz, *Thin Solid Films, 119* (1984) 203.
13 R. McMahon, J. Affinito and R. R. Parsons, *J. Vac. Sci. Technol. A, 2* (1984) 1275.
14 JCPDS Power Diffraction Card File Number 21-1272.
15 S. Schiller, G. Beister, W. Sieber, G. Schirmer and E. Hacker, *Thin Solid Films, 83* (1981) 239.

*Surface and Coatings Technology*, *39/40* (1989) 507 - 520 

# HARD DECORATIVE COATINGS BY REACTIVE PHYSICAL VAPOR DEPOSITION: 12 YEARS OF DEVELOPMENT

B. ZEGA*

*Battelle-Europe, Geneva Research Centres, 7, Route de Drize, Carouge, Geneva (Switzerland)*

(Received March 8, 1989)

## Summary

Hard nitrides of group IVb metals can display not only gold–yellow, but also other fashionable colors as bronze, brown, violet, dark-blue and almost black by small variations of their stoichiometrical composition. The interest in this type of coatings for millinery goods such as watch cases, bracelets, lighters, jewels etc., is growing. The exact reproduction of tints remains delicate because of the extreme sensitivity of the human eye for small color differences. This fact limits the number of deposition methods available for decorative purposes to those where precise control of deposition parameters is possible. After a short review of the historical development of these coatings, their optical properties and other quality requirements, factors influencing color and various possibilities of process monitoring in reactive plasma-assisted physical deposition methods are discussed.

## 1. Introduction

The origin of efforts for improving the aesthetic appearance as well as the resistance to deterioration of various metallic objects is lost in the mists of time. The optimal combination of all desired properties is generally difficult to obtain, *e.g.* gold coatings are unrivalled for their beautiful tint and chemical inalterability, but have a poor resistance to wear. By alloying gold with one or more other metals, its hardness can be increased, but not sufficiently for resisting wear as a consequence of friction by tissues, cleaning products etc.

At the beginning of this century, chemists and physicists became intrigued by the peculiar properties of carbides and nitrides of transition metals. Simple curiosity at the beginning, they have soon been recognized as extremely interesting new materials. Apart from the present boom in their use as hard, wear resistant coatings on tools and machine parts, their

*Present address: 67, Rue de Lyon, 1203 Genève, Switzerland.

0257-8972/89/$3.50 

decorative aspect, combined with durability, has opened a new market for millinery goods, coated by this low cost, abrasion resistant "gold".

We know that optical properties of gold cannot be imitated; no other material has its high reflectivity in the long-wave part of the spectrum. In consequence, the initial efforts for obtaining high quality yellow nitrides for decorative purposes have been much less successful than in other fields of application. More recently, other colors than gold–yellow, such as brown, violet, blue–black or even completely black, obtained by slight modifications of the composition, have come into fashion, and the interest for this type of coatings is rapidly growing.

The number of contributions to the knowledge of synthesis and properties of pure nitrides of titanium and its homologues, accumulated during the last two decades, is impressive. Much less has been published on their optical properties, and almost nothing on other than yellow colored compositions. The intention of this paper is, therefore, to devote some more space to hard coatings displaying other colors than gold–yellow. Many presently known technological methods can be used for the deposition of such coatings. The most important PVD methods will be reviewed, in particular from the point of view of control of the deposition parameter, which is essential for the exact reproducibility of hues, where the eye is an extremely severe judge.

Although decorative hard coatings represent, by their economical importance, a relatively small part of the total activity in the field, the investments related to this scope are tentatively estimated to amount to 30 - 50 million dollars, growing at a rate of more than 15% per year.

## 2. Short historical review

The oldest reference mentioning titanium nitride in powder form dates from 1912 [1]. During the following two decades, various nitrides were synthesized from the gas phase [2, 3]. During the period from 1948 to 1953, the interest for these compounds suddenly increased, and more systematic research was done on chemical vapor deposition (CVD) and on the powder synthesis [4 - 7]. Kieffer's team in Vienna has the merit of having systematically explored the ternary systems of isomorphous phases TiC–TiN–TiO, ZrO–ZrN–ZrC and HfO–HfN–HfC [8 - 11], being also the first to observe various colors appearing in the case of the titanium system.

The relatively high temperatures needed to obtain coatings by the CVD method limits their application to a relatively restricted number of substrate materials. The method has nevertheless found wide application in the coating of cemented carbide tools in the 1960s and 1970s [12].

Looking for lower temperature processes, the physical vapor deposition (PVD) methods have been examined since the beginning of the 1970s. Apparently, TiN and TiC coatings were first obtained by triode sputtering from sintered targets of the same composition [13]. Due to the low deposition rate and expensive ceramic targets, it became obvious that "reactive" methods, starting from metallic titanium, are much more interesting. In this way, reactive ion-plating was used [14, 15], as well as activated reactive deposition [16], activated reactive ion-plating [17], hollow cathode ion-plating [18], triode reactive sputtering [19], sputter-ion plating [20], cathodic arc deposition [21, 22] and reactive magnetron sputtering [23].

Some of these methods have been developed up to industrial scale. Three of them have been particularly successful in coating cutting tools, *i.e.* biased activated reactive evaporation, cathodic arc deposition and reactive magnetron sputtering.

Watchmakers were the first group interested in the decorative aspect of hard colored nitrides. The first patent in this field seems to be from Suwa Seikosha in 1973 [24], but it appears that the technique for carrying out such coatings was not yet mastered at that time. The same is probably true for the next patent (Koehli and Glavan from March 1974 [25]). A real avalanche of patents came from Japan in 1976–78 (Suwa Seikosha, Nishida, Citizen Watch [26 - 28, 30]). This period coincides with development work done at Battelle-Geneva [15], Balzers [29] and Leybold Heraeus [31], and can be considered as the real beginning of the activity in the field. More recently, it was followed, between many others, by Vac-Tech, using cathodic arc deposition.

## 3. Optical properties of group IVb metal nitrides

The oldest known reference giving reflectance spectra of Ti–TiN–TiC and Zr–ZrN–ZrC systems is from Knosp and Goretzki [32]. In ref. 15, spectra of ion-plated TiN coatings have been compared with these data, as well as with pure gold and some standard alloys. Their coordinates in the standard tristimulus system have also been calculated. This work has been continued with more refined techniques by Perry, in particular on the less known HfN [33]. The influence of deposition parameters such as nitrogen partial pressure, deposition rate, substrate temperature and possible substrate bias, as well as aging and tempering effects, have been demonstrated by a number of authors [34, 35].

In all these studies, the attention was focused on the gold–yellow hues. In the ref. 8, significant deviations of color from gold–yellow to brown, violet, reddish and grey were described in the TiC–TiN–TiO ternary system. This was observed on bulk samples obtained by high-temperature sintering, and consequently very close to the 50:50 composition. According to this data, all compositions within the binary system TiN–TiO are yellow; by admixing some TiC, the color goes through brown and blue toward metallic grey. Somewhat different results are obtained by PVD, because of the low substrate temperature, non-equilibrium compositions with large deviations from stoichiometry are readily obtained. Compositions where the sum of nitrogen and oxygen is less than 50% are light yellow or even metallic white, while a small excess of these two components gives dark tones with increasing optical absorption.

X-ray diffraction spectra indicate that no real solid solubility takes place at low-temperature syntheses. Diagrams show the degradation of the initially well crystallized films by the widening of lines as a consequence of over-stoichiometric composition by nitrogen or oxygen addition. Precise determination of crystalline parameters or the detection of new phases is difficult in this case. It is supposed that a large excess of nitrogen or oxygen was not incorporated into the basic cubic structure, but deposited on defect sites of a highly perturbed lattice and in grain boundaries, thus increasing the optical absorption.

Figure 1 represents a series of reflectance spectra of ZrN- and TiN-based coatings, obtained by reactive magnetron sputtering (spectra taken with a

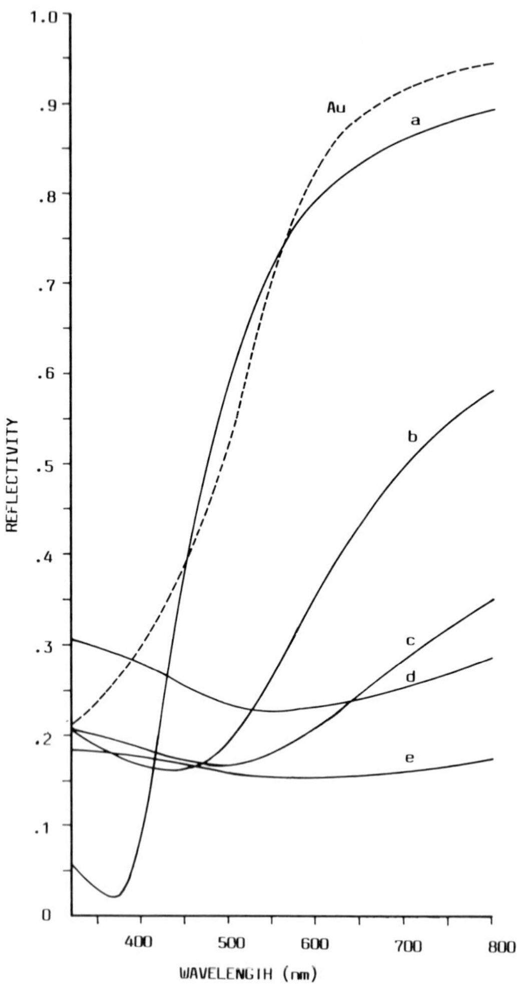


Fig. 1. Reflectivity spectra of some decorative hard coatings: (a) gold–yellow ZrN; (b) bronze-colored $TiN_{1+x}$; (c) dark-brown Ti(N,O); (d) and (e) bluish-black Ti(N,O) and black Ti(N,O,C). Dotted line: 24-carat gold.

Perkin–Elmer Lambda 9 spectrophotometer). Sample (a) is a closely stoichiometric, gold–yellow ZrN; increasing amounts of nitrogen and oxygen have been introduced in the sputtering atmosphere in the case of next, TiN-based samples (b, c and d). Two effects are observed: the surplus of nitrogen reduces the reflectivity on the long-wavelength side; the coatings become darker yellow (old gold), then bronze colored. The reflectance minimum, situated in the near UV for yellow samples, is shifted to longer wavelength, as has already been observed in ref. 36. Oxygen addition combined with excess nitrogen acts in the same direction, reducing the red-side reflectivity to very low values. Due to the displacement of the minimum, the blue-side reflectance increases, neutralizing the red and yellow component and giving blackish, or even bluish, tints. Carbon additions act essentially by further lowering the total reflectance, giving still darker black (sample e).

More recently, some other binary and ternary systems have been explored, where titanium is substituted by aluminum and vanadium [37, 38]. Blue and violet colors have been alluded to, but no detailed information about potential decorative applications is available.

As already noticed by Perry, the exact identification of the color of highly reflective, mirror polished samples is difficult. The reproducibility of spectroscopical measurements depends on illumination and viewing conditions. To stress the point, the human eye, especially that of an expert in decorative applications, is extremely sensitive to small variations in hue, in such a way that even the most refined measurements may not be able to define a given color with a sufficient precision.

## 4. Mechanical and other quality needs

While color is an essential characteristic of any decorative coating, hardness is, in our case, the second one. Unfortunately, a compromise has often to be accepted when matching both aspects. This is particularly true when we move away from the gold–yellow tints toward dark coatings, obtained owing to the over-stoichiometric composition. From very high values (over 2500 VH) in case of yellow coatings, the hardness can be as low as 1000 VH for very dark ones. This is still acceptable as protection from wear by friction of cloths or by conventional cleaning products.

Other criteria appear to be more important for acceptance as an efficient coating: the first is obviously adhesion, which has to be "ideal", *i.e.* no peeling should appear, even on deep scratching. Enough has been said about means for improving adhesion in PVD methods, particularly in plasma-assisted ones, that we shall not come back to this point. The second and third criteria do not depend on the coating alone, but also on the substrate. The hardest coating will behave poorly when deposited on a soft metallic substrate, because of the very limited resistance to shock and scratching. Simultaneously, problems with corrosion proofing may arise; coatings may not be continuous and pin-hole free as deposited, or become discontinuous after being exposed to mechanical damage. Acceptance criteria vary between countries and companies. They are very severe in Switzerland, where more or less standardized tests are practised.

An ideal case is represented by the combination of a cemented carbide substrate (watch case) with a hardness of 1500 VH or more, but of a dull grey color, coated by a gold–yellow TiN or ZrN, even harder than this (2500 - 3000 VH). This solution is unfortunately very expensive and not very often applied.

Another approach was experimented in 1978 at Battelle-Geneva. Alloyed steel substrates were first plasma nitrided, to obtain a hardness gradient within the first 100 $\mu$m beneath the surface, and then the TiN or ZrN coating deposited, increasing further the hardness and the corrosion resistance. Difficulties were encountered at that time with damage of the initially polished surface during the nitriding cycle. The work of the Helsinki team [39] has demonstrated the advantages of low pressure plasma nitriding in an intensified glow discharge, but no further data are available concerning the potential application of this technique to decorative hard coatings.

Another case concerns brass and similar soft alloys: direct deposition of hard coatings on these types of substrates results in dull coatings and in poor protection from scratching and shock. A hard chromium undercoat with a thickness of many microns, deposited by well known electrochemical methods, gives an acceptable solution.

In order to make economy of real gold platings on medium quality goods, an undercoat of yellow nitride is first deposited and the initial thin PVD gold layer is later reinforced electrochemically. Adhesion and corrosion problems are often observed in this combination, and manufacturers use their own tricks to avoid them.

## 5. Some technological aspects of PVD processing

### *5.1. General*

The most convenient deposition methods for group IVb nitrides are reactive, for the obvious reason of the refractory nature of these compounds. Base metals are relatively inexpensive and can be easily delivered in the appropriate forms (crucible charges, rods, planar targets etc.). Their vapor pressure is also relatively high, as well as the chemical reactivity.

As mentioned in Section 1, several PVD methods have been successfully applied to the synthesis of hard nitrides in form of coatings, but not all perform in the same manner, due to the mechanism governing any reactive deposition process.

It is generally accepted that the major part of the chemical reaction between the metal and the reactive gas happens on the substrate surface. The composition of the coating will then result from the impingement ratio between the gas molecules and the metal atoms. The effective impingement rate includes an activation factor, specific for each type of molecule and varying considerably from one (activated) deposition method to another. In addition, some contribution is due to reactions in the homogeneous (vapor) phase, where reactive collisions are frequent in the activated and partially ionized atmosphere. The contribution of these reactions on the source itself can be neglected in some cases (as in electron beam evaporation and arc erosion), but not in others (sputtering).

From these considerations we can deduce that many parameters will influence the composition of the deposit, such as the reactive gas partial pressure, the degree of its activation, the evaporation rate of the base metal, the degree of activation of the metallic vapor, the source-to-substrate distance and the substrate temperature.

In their paper, published in 1984, Bunshah and Deshpandey [40] extensively reviewed phenomena appearing in so-called "plasma activated PVD" methods. The comparison between two extensively used methods, ARE (activated reactive evaporation)and RS (reactive sputtering), presented in their paper gives a pronounced advantage to the ARE process. In the present paper, some complementary arguments are brought to these conclusions, according to the present state of knowledge.

In the following Sections, three of the most important methods used for the deposition of hard coatings are analysed from the point of view of the ease of parameter control, essential for the reproducibility of colors in the case of decorative applications.

### *5.2. Plasma-assisted methods, based on thermal evaporation*

In this Section we shall consider all methods using thermal evaporation by electron beam in the widest sense of the word: high energy (10 - 20 keV) and low energy beam (few 100 eV or less), in combination with additional activation systems. These can be (i) a positively biased grid, as in ARE, (ii) an additional electron emitter (triode ion-plating) or (iii) the r.f. excitation of plasma etc.

In all cases, the substrate can be negatively biased or not, but this is not essential from the point of view of source control.

One characteristic of electron beam heated sources is the difficulty in monitoring the evaporation rate by simply modulating the total input power. The crucible charge, beam focusing etc., also strongly influence the evaporation rate. In a reactive gas atmosphere, an additional factor is the presence of a nitride or oxide skin floating on the liquid metal and limiting the evaporating surface. The deposition process must in consequence be monitored; this can be done in two ways, *i.e.* (i) by maintaining the reactive gas flow constant and controlling the metal evaporation rate with the heating power or (ii) by measuring the effective metal evaporation rate and adapting the gas flow accordingly.

For the process control, three essentially different principles can be used: the first one is based on the deposition rate control, using, for example, continuous thickness measurement devices (such as quartz crystal balances) and differentiating the obtained signal variation in time; the second method directly measures the evaporation rate or, more exactly, the metal vapor density over the source (by collecting ions created by the primary electron beam in the beam–vapor collision zone or even by additional ionization of the vapor by a special electron beam). A third method can be taken into consideration, which seems to be particularly well suited to the control of phenomena appearing in plasma activated reactive processes, using optical emission and/or absorption spectroscopy.

It is well known from current practice that serious reliability problems exist when some of these methods are used in the presence of intensely ionized atmospheres, either because of excessive heating of the quartz crystal or because of parasitic signals from plasma. The spectroscopic method is supposed to give the correct information needed for monitoring the reactive deposition process by indicating the concentration ratios between reacted and unreacted species in the chamber atmosphere. In the simplest case, *i.e.* the concentration of metallic vapor over the source, the evaporation rate can be monitored in this way. It has been demonstrated by many authors (*e.g.* Schiller *et al.* [37]; Enjouji *et al.* [38], Poitevin *et al.* [39] and Ricard *et al.* [40]) in various cases of reactive PVD that a (almost) linear relationship exists between the incident power (or deposition rate) and the intensity of emission lines from the metallic vapor. Complications originating from the influence of the total gas pressure and from interference between the numerous emission lines present in gaseous discharge spectra somewhat limit the usefulness of this method for practical applications.

### *5.3. Deposition from cathodic arc sources*

By its basic mechanism, a cathodic arc source can be compared to thermal evaporation; its behavior has many similarities with "flash" evaporation and also with the low energy electron beam evaporation. Very restricted areas of the target are instantaneously heated and the material is transformed from its solid state to vapor at a temperature level where the high vapor pressure makes it literally explode. As this flash of vapor is produced by a low energy, high intensity electron beam, concentrated in the cathodic spot, a high degree of ionization and excitation is present.

From the mechanism governing the cathodic arc in vacuum we know that cathodic spots multiply proportionally to the total current. The average evaporation rate is also proportional to the total current. Owing to the "flash" type of evaporation from the cold target, there is practically no opportunity for compound formation on its surface. As a consequence, monitoring of the reactive deposition process is possible by simply controlling the total power input, as claimed by Randhawa *et al.* [22], provided the influence of the magnetic field present is also considered, as has been demonstrated by Aksenov *et al.* [45].

Optical emission spectroscopy has also been applied to cathodic arc deposition methods, with the restrictions previously mentioned.

*5.4. Deposition by high-rate (magnetron) sputtering*

From the process control point of view, sputtering combines both advantages mentioned in the previous Section, concerning cathodic arc deposition, *i.e.* the sputtering rate is, in principle, almost linearly proportional to the discharge current (or better: to the total power dissipated on the cathode) and the target remains (relatively) cold during the deposition.

The essential difference lies in the considerably slower local erosion of the surface of the target than in the case of the vacuum arc method; consequently, reactive gas ions bombarding the surface are able to form a compound layer, having very different, much lower sputtering rate. As in high-rate (magnetron) sputtering the erosion rate is also inhomogeneous due to the graded distribution of the ion current, a non-linear behavior appears: whether the power (or the reactive gas pressure) has been increased or decreased during the deposition cycle, the target's characteristics will stabilize either on the ascending or on the descending branch of the hysteresis loop, circumscribing a zone of instability.

In order to control the composition of the coating, we have to be able to stabilize the working conditions of the cathode or to monitor the real metal erosion rate. This problem has been approached in different ways. McMahon [45] finds that the voltage controlled mode is better suited to reactive magnetron sputtering, in his case of AlN, while Schiller and coworkers [37] recommend the plasma emission spectroscopy previously discussed for monitoring the metal sputtering rate. Hmiel [46] controls the reactive gas pressure by a mass spectrometer, while Sproul [47] introduces the reactive gas in short pulses, circumscribing in this way the hysteresis loop to a very narrow surface around the desired working point.

Various more or less usable geometrical solutions have also been proposed, such as the introduction of the reactive gas near the substrate, and the protection of the cathode area by screens and by the argon flow etc. These methods are either too complicated to handle under production conditions or not adaptable to large coating systems. An important contribution to the understanding of instability in reactive sputtering, is that of the Uppsala team (Berg and coworkers [49 - 51]). Their conclusions are similar to the empirical solution of Danroc *et al.* [52], who recommend the use of high capacity pumps and/or having the reactive gas inlet near the pumping orifice. In this manner, the hysteresis disappears at the expense of higher reactive gas consumption, which is obviously acceptable. The same authors also refer to the influence of the degree of target erosion. A new planar target behaves differently from a deeply eroded one, and the latter is easier to stabilize than the former. This is essentially due to the modified magnetic field intensity and geometry near the target surface, and consequently to a different plasma confinement. This also coincides with observations on focusing effects of deeply eroded targets and, to some extent, with the Rossnagel's study [53] on the gas density effects in magnetrons.

As all these phenomena can influence the final composition of reactively deposited coatings, we conclude that sputtering parameters cannot be fixed once for ever, but have to follow the evolution of the target during its life. The introduction of microprocessor-controlled deposition systems makes this possible.

Apart from the history of the target, the environment influences the composition to some extent, revealing itself in the shade of the resulting color. The large surface of the chamber walls is one such factor, remembering one or more previous runs by its more or less pronounced capacity of gettering reactive gases.

In conclusion, considered from the standpoint of reproducibility of composition and, therefore, of color, each of the three most important groups of deposition methods possesses specific advantages and drawbacks. Those related to thermal evaporation allow high evaporation rates, limited essentially by the efficiency of activation, but also suffer from evaporation rate monitoring problems, when high precision is needed.

Vacuum arc deposition is attractive from many standpoints. The progress obtained recently on this method makes it very promising for decorative applications, in particular because of the easy control of parameters.

However, on the subject of reactive magnetron sputtering, considerable knowledge has been accumulated during the last decade by a large number of researchers. This gives a somehow wrong impression of much wider complexity.

As a criterion for selection in industrial scale applications, the geometry of the deposition chamber can be the deciding factor. Coaters based on thermal evaporation are essentially batch systems. Good utilization of the space surrounding the source is not easily compatible with the "in-line" concept. The vapor beam, though perceptibly scattered in the plasma, nevertheless possesses a certain degree of directionality. Rotating substrate holders become unavoidable.

In the case of the cathodic arc method, the source already has a larger surface, and this is especially true for sputtering, where target dimensions, as well as their geometry, can be adapted to particular tasks. Simultaneous deposition from two sides is easily feasible and the integration of the deposition chamber into "in-line" coaters is possible.

### *5.5. Substrate – plasma interaction effects*

In Sections 5.2. and 5.3, phenomena governing the efficiency of the metal vapor sources have been discussed. As mentioned, the most important part of the reaction between the deposited metal and the reactive gas occurs on the substrate surface. In plasma-activated or plasma-assisted deposition methods, some interaction between energetic particles from the plasma and the arriving metallic atoms is unavoidable, and it can be further induced, becoming one of the principal factors determining the final properties of the coating.

Some of these aspects have been studied since the early applications of the "bias sputtering", particularly by Mattox, who introduced the notion of "ion plating". All these interactions cannot be modulated as independent variables, since they depend on the source working conditions (X-ray, UV and visible radiation from the heated source, neutral thermalized or reflected energetic particles). However, the main influence comes from charged particles, whose flux toward the substrate can be controlled by an appropriate bias.

The first consequence of the presence of plasma between the source and the substrate is the appearance of a negative space charge near the substrate due to the electron mobility being higher than that of ions. An electrically floating metallic substrate will acquire a negative bias of a few tens of volts (autobias). This bias is not sufficient for extracting many ions from the positive space charge region. Apart from this situation, two possibilities exist: the substrate can be positively or negatively biased. In the first case, an electronic current is collected, whose effect is limited essentially to the substrate temperature increase. No other favorable influence on the deposit has been observed. Collected current is particularly important in magnetron sputtering, where it can reach almost 100% of the target current by simply grounding the substrate. The negative bias, however, provokes substrate bombardment by positive ions from the plasma region, and this has much more important effects.

As is generally recognized [54], many ion-bombardment effects can be observed:

(1) desorption and sputtering from the substrate surface,
(2) penetration and entrapment of gas atoms in the coating,
(3) (re)sputtering from the substrate and consequent reduction of the effective deposition rate,
(4) generation of defects in the growing film and as a result modification of its texture.

While the first point plays an important role at the beginning of the deposition, improving the adhesion, the last point is undoubtedly the most important from the standpoint of the final appearance of the coating (color, brilliance). Working conditions (pressure, temperature) when depositing on various metallic objects are often such that a typical columnar structure appears, particularly by sputtering, as stated by Thornton [55]. One of the consequences of this type of texture is the microscopical roughness, which is approximately proportional to the coating thickness. In the case of decorative coatings, which are generally thinner than those deposited, for example, on cutting tools, this phenomenon is not very pronounced, but appears when the fresh surface is touched with fingers. Fingerprints cannot be wiped away by simple rubbing, because of traces of greasy deposits in the microscopical depressions between single columnar crystals. This roughness, representing in the worst case up to 10% of the total coating thickness, can be drastically reduced under conditions where equiaxial texture is favored. It is well known that intense low energy ion bombardment, as in ion plating, creates favorable conditions for such a texture and thus more brilliant surfaces.

### 5.6. *"Doping" of nitride coatings*

In order to move away from the gold-colored stoichiometric nitrides, additional quantities of nitrogen, oxygen or even carbon have to be incorporated into the coating.

When nitrides only are deposited, the final nitrogen content is roughly proportional to the partial pressure of the gas present. To obtain darker colored oxynitrides, the simplest way is to start with a yellow nitride and to add some oxygen. The affinity of both titanium and zirconium for this gas is markedly higher than for nitrogen, the first gas is more quantitatively gettered by the metal vapor than the second. One result is that relatively small quantities of oxygen are needed.

Carbon can obviously not be introduced in elementary form. One possibility is using gaseous hydrocarbons, but this is not very recommendable. Partial splitting of hydrogen from these molecules leaves free radicals behind. The probability that such radicals react between themselves, creating new organic molecules, is higher than the formation of metal carbides. A series of organic syntheses takes place in the plasma, seriously contaminating the chamber and condemning the next experiment to fail in the reproducation of color. A partial solution to this problem lies in the use of carbon oxides, unavoidably giving oxycarbonitrides, or even of carbon–halogen compounds, provided the pumping unit is compatible with these gases.

Precise quantitative prescriptions for obtaining various colors are evidently impossible. Parameters adapted to every coating system have to be found out experimentally by trial and error.

Doping by partially substituting the target metal by other elements, giving rise to numerous new possibilities, lies beyond the scope of this paper.

## 6. Conclusions

Over more than a decade of development in the field of hard coatings, obtained by PVD, an impressive amount of experimental as well as fundamental work has been accomplished, essentially owing to their impact on the metal machine-tool market. The parallel application of colorful transition metal nitrides as decorative coatings has induced sensitive measurements of their optical properties and the necessity to introduce more refined process control methods in order to warrant the exact reproduction of colors. This last requirement limits the number of presently known deposition methods to two or three, and in spite of the clear preference given to (activated and biased) ion plating for cutting tool coating, magnetron sputtering still remains the most widespread method for decorative applications, but is entering into competition with the cathodic arc method.

Further development trends point toward new binary or ternary compositions, giving new colors and potentially even better mechanical and chemical properties.

## References


1 O. Ruff and O. Treidel, *Berichte*, *45* (1912) 1364.
2 A. E. van Arkel and J. H. de Boer, *Z. Anorg. Chem.*, *148* (1925) 345.
3 K. Moers, *Z. Anorg. Chem.*, *198* (1931) 243.
4 F. M. Pollard and P. Woodward, *J. Chem. Soc.*, *2* (1948) 1709.
5 J. E. Campbell, C. F. Powell, D. H. Nowicki and B. W. Gonser, *J. Electrochem. Soc.*, *96* (1949) 318.
6 A. Münster and K. Sagel, *Z. Electrochem.*, *57* (1953) 564.
7 P. Schwarzkopf and R. Kieffer, *Refractory Hard Metals*, MacMillan, 1953.
8 G. Neumann, R. Kieffer and P. Ettmayer, *Monatsh. Chem.*, *103* (1972) 1130.
9 K. Constant, R. Kieffer and P. Ettmayer, *Monatsh. Chem.*, *106* (1975) 823.
10 K. Constant, R. Kieffer and P. Ettmayer, *Monatsh. Chem.*, *106* (1975) 973.
11 R. Kieffer, W. Wruss, K. Constant and H. Habermann, *Monatsh. Chem.*, *106* (1975) 1349.
12 W. Ruppert, in J. M. Blocher and J. C. Withers (eds.), *Chemical Vapour Deposition*, The Electrochemical Society, New York, 1970, p. 443.
13 G. Mah, C. W. Nordin and J. F. Fuller, *J. Vac. Sci. Technol.*, *11* (1974) 371.
14 D. Carmichael and D. Chambers, *U.S. Patent 3,756,193* (1973).
15 B. Zega, M. Kornmann and J. Amiguet, *Thin Solid Films*, *9* (1977) 577.
16 R. F. Bunshah and A. C. Raghuram, *J. Vac. Sci. Technol.*, *9* (1972) 1385.
17 E. Moll and H. Daxinger, *Patent DE 28 23 876* (1977).
18 J. R. Morley and H. R. Smith, *J. Vac. Sci. Technol.*, *9* (1972) 1377.
19 D. G. Teer and B. Kirkham, *Proc. Conf. on Ion Plating and Allied Techniques*, CEP Consultants Ltd., Edinburgh, 1979, p. 35.
20 R. A. Dugdale, *Proc. Conf. on Ion Plating and Allied Techniques*, CEP Consultants Ltd., Edinburgh, 1977, 177.
21 A. I. Grigorov *et al.*, *Patent DE 29 02 142* (1978).
22 H. Randhawa, P. C. Johnson and R. Cunningham, *J. Vac. Sci. Technol.*, *A6* (1988) 2136.
23 P. J. Clarke, *J. Vac. Sci. Technol.*, *14* (1977) 141.
24 Suwa Seikosha KK, *U.K. Patent 1427-674* (March 1973).
25 W. Koehli and B. Glavan, *Swiss Patent Application 2962/74* (March 1974).
26 Nishida N., *14 Patent Applications JA 066320 - 066336* (June 1977).
27 Nishida N., *U.K. Patent 2000-812* (June 1977).
28 Suwa Seikosha KK, *Patent J5 4062-183* (October 1977).
29 H. K. Pulker and H. Daxinger (Balzers), *U.K. Patent 2010-919* (December 1977).
30 Citizen Watch KK, *Swiss Patent 631-040* (May 1977).
31 W. D. Münz and G. Hessberger, *Industrial R&D*, September 1981, 130.
32 K. Knosp and H. Goretzki, *Z. Metallkd.*, *60* (1969) 587.
33 A. J. Perry, *Thin Solid Films*, *157* (1988) 255, mentioning previously published papers of the author since 1983.
34 H. Takei, J. Minami and A. Shimizu, *Met. Finish.*, *4* (1983) 59.
35 K. Y. Ahn, M. Wittmer and C. Y. Ting, *Thin Solid Films*, *107* (1983) 45.
36 E. Vogelzang, J. Sjollema, H. J. Boer and J. Th. M. De Hosson, *J. Appl. Phys.*, *61* (1987) 4606.
37 H. Jehn, S. Hofmann, V.-E. Rückborn and W.-D. Münz, *J. Vac. Sci. Technol.*, *A4* (1986) 2701.
38 O. Knotek, T. Leyendecker and F. Jungblut, *Thin Solid Films*, *153* (1987) 83.
39 A. S. Korhonen, E. H. Sirvio and M. S. Sulonen, *Thin Solid Films*, *107* (1983) 387.
40 R. F. Bunshah and C. V. Deshpandey, *J. Vac. Sci. Technol.*, *A3* (1985) 553.
41 S. Schiller, U. Heisig, K. Steinfelder and J. Schtrümpfel, *Thin Solid Films*, *63* (1979) 369.
42 K. Enjouji, K. Murata and S. Nishikawa, *Thin Solid Films*, *108* (1983) 1.
43 J. M. Poitevin and G. Lempérière, *Thin Solid Films*, *120* (1984) 223.
44 A. Riccard, H. Michel, P. Jacquot and M. Gantois, *Thin Solid Films*, *124* (1985) 67.

45 I. I. Aksenov, Ju. P. Antufiev, V. G. Bren, V. G. Padalka, A. I. Popov and B. M. Horoshikh, *J. Techn. Phys.*, *51* (1981) 303.
46 R. McMahon, J. Affinito and R. R. Parsons, *J. Vac. Sci. Technol.*, *20* (1982) 376.
47 A. F. Hmiel, *J. Vac. Sci. Technol.*, *A3* (1985) 592.
48 W. D. Sproul, *Thin Solid Films*, *118* (1984) 279.
49 S. Berg, H.-O. Blom, T. Larsson and C. Nender, *J. Vac. Sci. Technol.*, *A5* (1987) 202.
50 S. Berg, T. Larsson, C. Nender and H.-O. Blom, *J. Appl. Phys.*, *63* (1988) 887.
51 T. Larsson, H.-O. Blom, C. Nender and S. Berg, *J. Vac. Sci. Technol.*, *A6* (1988) 1832.
52 J. Danroc, A. Aubert and R. Gillet, *Surf. Coat. Technol.*, *33* (1987) 83.
53 S. M. Rossnagel and J. J. Cuomo, *Vacuum*, *38* (1988) 73.
54 G. Carter and D. G. Armour, *Thin Solid Films*, *80* (1981) 13.
55 J. A. Thornton, *Ann. Rev. Mater. Sci.*, *7* (1977) 239.

*Surface and Coatings Technology*, *39/40* (1989) 521 - 529 

# THERMAL SURFACE MODIFICATION BY ELECTRON BEAM HIGH-SPEED SCANNING

S. SCHILLER and S. PANZER

*Forschungsinstitut Manfred von Ardenne, Dresden (G.D.R.)*

(Received April 17, 1989)

## Summary

Highly productive techniques of thermal surface modification require the use of areal temperature fields that are restricted to a thin surface layer. They can be generated by means of an electron beam scan with deflection speeds in a range of a few kilometers per second or scan frequencies of about 100 kHz. A raster scan which allows an isothermal energy transfer to moving workpiece surfaces was developed for surface hardening. Typical hardness depths are of the order of 0.5 - 1.2 mm. At a beam power of 10 kW the areal throughout is around 10 $cm^2$ $s^{-1}$. This new technique is used successfully by some enterprises in the G.D.R. Apart from the qualitative advantages, it is the financial advantages which determine its industrial application.

## 1. Introduction

Many development centres all over the world have recently dealt with the application of high-energy beams for thermal surface modification. In most of the cases laser beams have been used for the energy transfer. However, in the world-wide euphoria over lasers, the outstanding properties of electron beams for these tasks were often neglected or utilized only to a very low extent [1]. In the following contribution we therefore present our results about thermal surface modification with electron beams. It was the basic objective of this work to develop a highly efficient energy transfer technique and to create the instrumental preconditions needed for such a task. As to the aspects of process engineering, it was the implementation of the methods compiled in Fig. 1 that stands in the limelight; giving top priority to the solid-phase process of martensitic hardening [2].

## 2. Energy transfer in the high-speed scan technique

Highly productive techniques in thermal surface improvement call for areal temperature fields that must be restricted to the surface layer of the

0257-8972/89/$3.50 

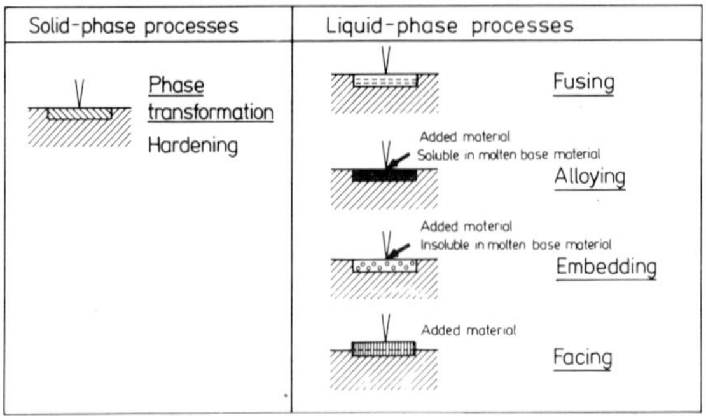


Fig. 1. Possibilities for thermal surface modification of metal parts by electron beams or laser beams.

component to be treated. As indicated in Fig. 2, their generation is, as a rule, possible with the aid of an energy beam of adequate cross-section, or by cyclic areal scanning by means of a focused beam. However, even with electron beams, the energy transfer technique with adequate cross-section can be implemented by sophisticated means only within close limits. The raster scan technique offers the big advantage that the energy transfer and deposition within the scanned surface, as well as the shape and dimensions, solely depend on the beam deflection parameters. At a first glance, it has the drawback that the temperature field produced at the instantaneous point of beam action exhibits a hot spot due to the fact that energy conversion takes place in a small volume element that is fixed by the focal spot size and the depth of electron penetration. This results in a timed scan-bound temperature oscillation. With a finite beam power the said oscillation vanishes only if the beam scan rate goes to infinity. In practice, however, it is fully adequate if the temperature oscillation is kept low compared with the process temperature required. Figure 3 shows the estimated temperature jump $\Delta T$ as a function of the scan rate $v_s$ for various ratios of the beam power $P$ to the focal spot extension $d_{Fx}$ referenced to an electron beam of 60 keV and steel. It should be pointed out here that the focal spot extension $d_{Fy}$ in the direction of the scan line has no bearing on this temperature jump. The typical example of an

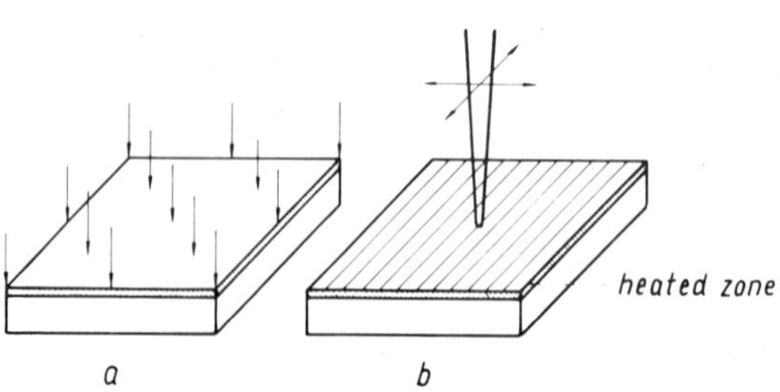


Fig. 2. Heating of workpiece surface by energy beams: (a) beam action with areal cross-section; (b) areal scanning with a focused beam.

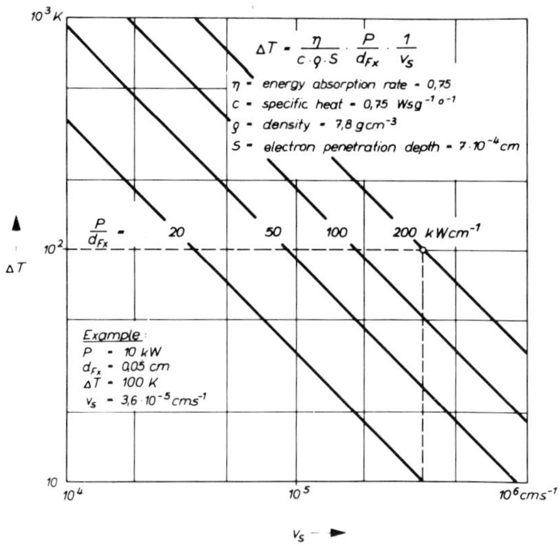


Fig. 3. Temperature jump $\Delta T$ on the point of beam impingement for scanning a steel surface with an electron beam of 60 keV at a rate $v_s$ for various ratios of the beam power $P$ to the focal spot extension $d_{Fx}$ normal to the direction of scan.

electron beam of 10 kW with a spot diameter of 0.5 mm given here reveals that a scan rate of $10^5$ cm s$^{-1}$ (1 km s$^{-1}$) or less is needed to obtain a temperature jump of $\Delta T \leq 100$ °C. With a deflection amplitude of some centimeters such a high-speed procedure corresponds to a scan frequency of about 100 kHz.

From the relation for $\Delta T$ plotted in Fig. 3 it can be seen that, at a given scan rate and beam power, a reduction of the temperature jump can be attained by using a greater electron penetration S and/or a greater focal spot extension $d_{Fx}$ (normal to the direction of the scan line). As the depth of electron penetration rapidly increases with the electron energy, an electron energy of 100 - 150 keV proves to be favourable. Moreover, it is of advantage to use an elliptical focal spot extended normal to the direction of scan. The extension of the elliptical focus should be equal to the spacing of the scan lines that are passed successively in time. At a corresponding elliptic focal spot the greater spot extension $d_{Fx}$ has no adverse consequences with regard to the timed sharpness of the energy transfer field.

With the martensitic hardening of strip-like surface regions as an example, the inherent advantages of an energy transfer via the high-speed scan technique shall be explained in more detail. The surface region of the workpiece to be hardened is passed through the stationary field of energy transfer at a constant feed rate. Figure 4 shows the temperature–time dependence of a surface location obtained when two energy transfer fields of different power density distribution are being passed. The case of constant power density shown in Fig. 4a, for instance, corresponds to laser beam hardening according to the focusing hood method, or to electron beam

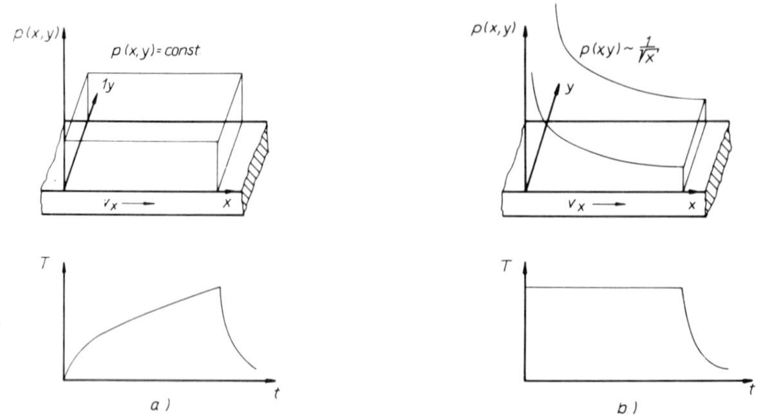


Fig. 4. Power density distribution $p(x, y)$ and associated temperature–time cycle $T(t)$ for a workpiece surface passing the energy transfer range at a speed $v_s$ (schematic). (a) Power density $p(x, y)$ = constant. (b) Surface–isothermal energy transfer $p(x,y) \approx (x)^{-1/2}$.

hardening in the HSS technique performed at a raster of equidistant scan lines. Figure 4(b) shows the ideal case of an isothermal energy transfer as obtained theoretically by contacting a temperature bath. The associated power density distribution shows a typical ski-jump profile. Here it is only the HSS technique which allows us to implement such a case in good approximation. Figure 5 shows the associated raster scan. For minimizing the temperature oscillation, the focal spot extension has to be matched to the variable spacing of the scan lines. The pattern of this raster scan is independent of the required hardening depth, the width of the hardening track and the feed rate of the workpiece.

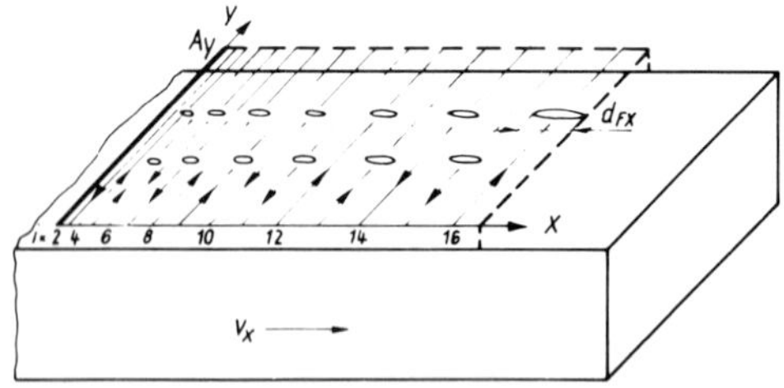


| $i$ | 1 | 2 | 3 | 4 | 5 | 6 | 7 | 8 | 9 | 10 | 11 | 12 | 13 | 14 | 15 | 16 |
|---|---|---|---|---|---|---|---|---|---|---|---|---|---|---|---|---|
| $\frac{x_i}{A_x} \cdot \frac{1}{1024}$ | 2 | 10 | 26 | 50 | 82 | 122 | 170 | 226 | 290 | 362 | 442 | 530 | 626 | 730 | 842 | 962 |
| $\frac{d_{FXi}}{A_x} \cdot \frac{1}{100}$ | 0,4 | 1,2 | 2,0 | 2,7 | 3,5 | 4,3 | 5,1 | 5,9 | 6,6 | 7,4 | 8,2 | 9,0 | 9,8 | 10,6 | 11,3 | 12,1 |

Fig. 5. Location of $y$-deflection tracks in the $x$–$y$ scan field of the beam allocated to the workpiece during isothermal heat transfer as well as relative numerical values of $x_i/A_x$ and $d_{Fxi}$ for $N = 16$.

## 3. Electron guns

Thermal surface modification makes similar demands on the electron guns as electron beam welding. This particularly holds for the electron energy and the beam power. In the implementation of the HSS technique, however, difficulties with these guns did occur because of the limited frequency of the beam deflection — being too low by one order of magnitude — and the missing capability for focal spot shaping. Based on experience gained in the field of electron beam micromachining, a new type of electron gun was developed. In addition to thermal surface modification and electron beam welding, these guns may further be used for electron beam engraving and similar tasks. This gun is shown schematically in Fig. 6. A new feature compared with the welding gun is an electron–optical system arranged within reach of the focusing lens. It consists of a high-speed deflection array and, for focal spot shaping, a quadrupole and a dynamic lens. Figure 7 shows the actual design of these guns. With an acceleration voltage of 60 kV, they are rated for a maximum beam power of 5, 10 and 20 kW. An electron gun of 150 kV was developed for a maximum beam power of 50 kW.

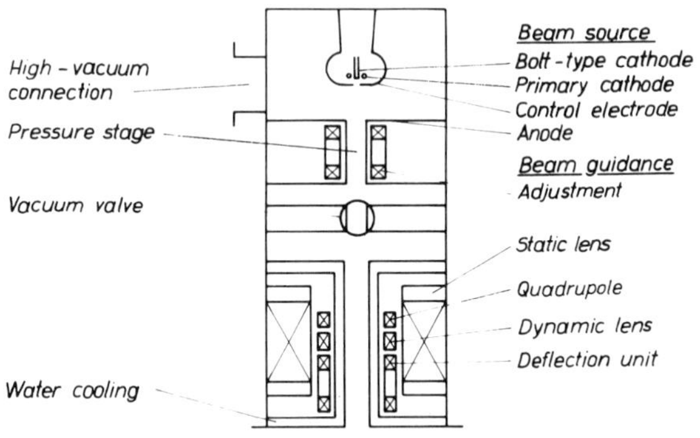


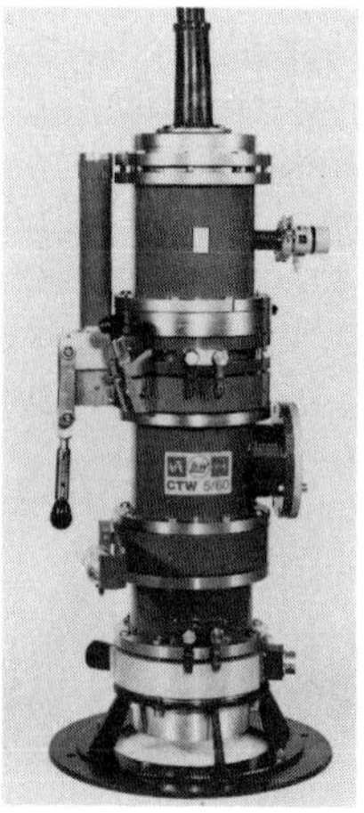

Fig. 6. Schematic diagram of an electron beam gun for thermal surface modification and welding.

Fig. 7. Actual design of a Model CTW 5/60 electron beam gun for thermal surface modification and welding.

## 4. Results and applications

A highly efficient method for surface hardening was developed that is based on an areal isothermal energy transfer with high-speed scan technique. Since 1985 this method has increasingly been used in some G.D.R. factories on an industrial scale. Being suited for hardening depths to 1.5 mm, it has

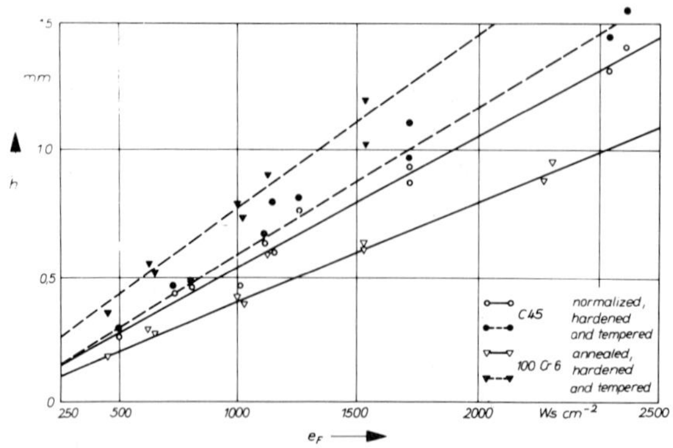


Fig. 8. Hardness depth $h$ as a function of the surface energy $e_F$ transmitted through isothermal energy transfer at about the melting temperature for two different materials and various material states.

found its main applications where a hardening depth of 0.5 - 1.2 mm must be obtained. In practice, all transformation hardenable steel grades and pearlitic cast iron can be hardened in this way. Energy transfer occurs at surface temperatures that are just below the melting temperature of the materials concerned, *i.e.* without incipient fusing of the surface. The duration of the energy transfer and the power density required to attain a particular hardening depth depend on the material composition and on the thermal pretreatment employed. Figure 8 shows the relation between the hardening depth and the energy transfer $e_F$ per unit areas as required for two materials and different thermal pretreatment states. By way of these rather extreme examples this illustration clearly reveals the weak and the strong dependence of the hardening depth $h$ on the thermal pretreatment states concerned. For all materials investigated there exists a linear relationship between hardening depth and transfered surface energy. Depending on the material in question a surface energy between 1400 and 2400 W cm$^{-2}$ is needed to obtain a hardening depth of 1 mm.

Referenced to an available beam power of 20 kW, and a hardened layer of 1 mm, the typical energy expenditure of 2000 W cm$^{-2}$ results in an areal throughput of 10 cm$^2$ s$^{-1}$. The costs, fixed chiefly by plant amortization, then lie within a range of 0.5 cents cm$^{-2}$ hardened surface.

Figure 9 shows the cross-sectional profile of a hardening track. Typical of this process is the constant hardening depth over the track width. The ascent of the hardened interfacial layer towards the surface is limited to a narrow marginal zone that approximately corresponds with the hardening depth. Up to now, hardening tracks up to 70 mm in width have been used. The rather broad tracks and the constant hardening depth are a boon compared with spherical hardening profiles attained by means of other energy transfer techniques.

As indicated in Fig. 10, different temperature cycles are passed at various distances from the surface. According to the materials used, different transfor-

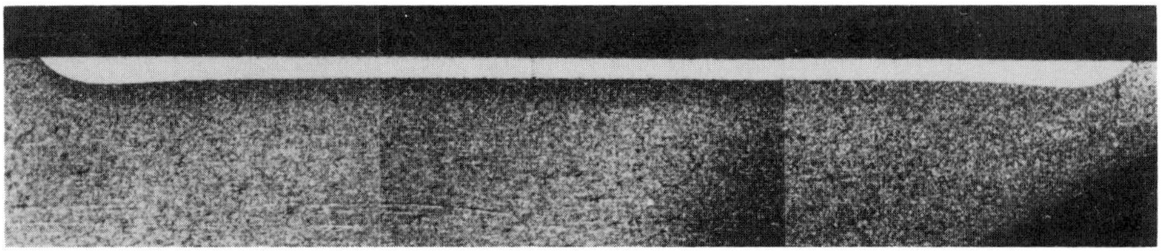

Fig. 9. Cross-sectional profile of a hardening track. Material: Ck 67, refined; track width 20 mm, hardening depth 0.4 mm, transferred energy density $e_F = 600\ \mathrm{W\,s\,cm^{-2}}$, areal throughput $F = 5\ \mathrm{cm^2\,s^{-1}}$, isothermal energy transfer.

mation products may therefore occur at the various distances from the surface. In particular, at a hardening depth greater than 1 mm it is very likely that coarse-grain growth occurs near the surface of hypoeutectic materials. High-alloyed hypereutectic materials, however, tend towards an excessively high residual austenite content if the hardening depth is greater than 1 mm.

For a great many load conditions of the surface it proves to be of advantage that substantial compression stresses usually occur in the hardening zones; typically on the order of 300 - 600 Pa.

Contrary to the customary opinion that beam hardening techniques are suitable for finish hardening we have found surface deformations of about 1% of the hardening depth (Fig. 11). However, these deformations are not caused by the specific energy transfer as such but have to be attributed to the different specific volumina of the initial and final products in the hardened layer. Whenever high demands have to be made on the surface quality, grinding is therefore required after hardening in each case.

The most important benefits of electron beam hardening compared with conventional hardening techniques are (i) the low energy input into the component and therefore a rather low hardening distortion, (ii) the low energy expenditure and (iii) the high process confidence. With these characteristics the HSS technique also allows us to treat parts hitherto inaccessible to hardening for reasons of warping, *i.e.* cutting tools and similar

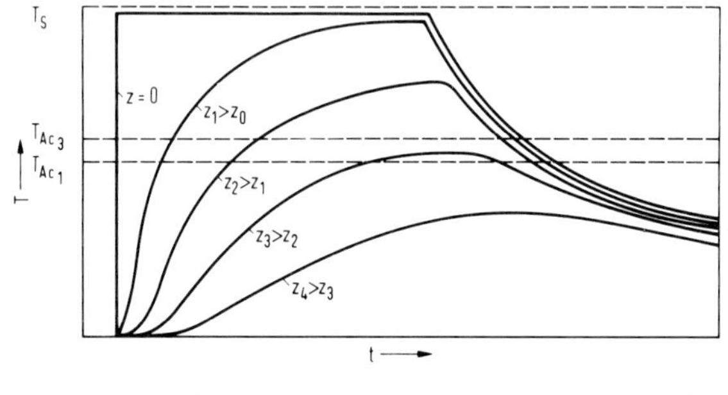


Fig. 10. Passed temperature cycles $T(t)$ for various distances $z$ to the surface for isothermal energy transfer at about the melting temperature $T_s$ (schematic).

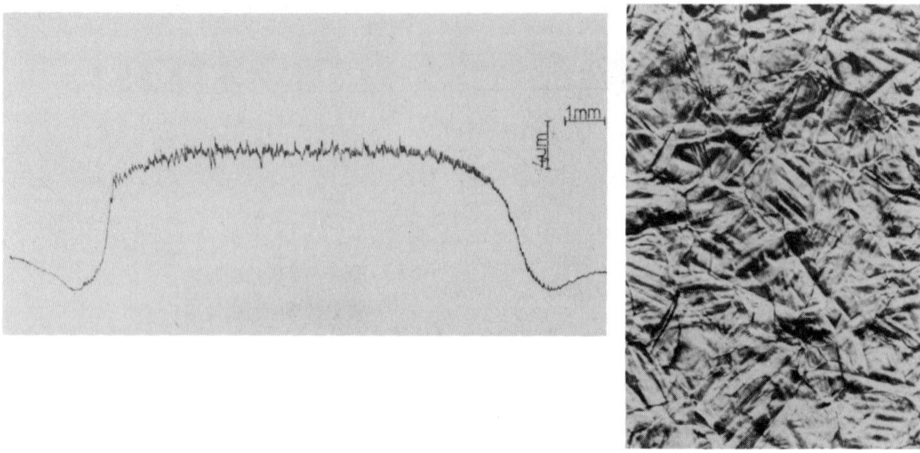


Fig. 11. Surface deformation in the region of the hardness track.

objects. However, most of its applications accrue from the profitable plant economy. For many machine elements the expenses to be spent in the implementation of electron beam hardening are often lower than those of conventional hardening techniques. Figure 12 shows the development with regard of the assortment of electron beam hardened parts in a machine-tool factory experienced since this technique was introduced. By utilizing the capabilities of electron beam hardening in full, the assortment of parts processed in the said factory will embrace about 110 different items. The "historical" first example, the quill of a machine tool, is shown in Fig. 13. Apart from a simplified production flow, savings in material costs were obtained by using a less expensive heat-treatable steel grade in place of the hitherto employed carburizing steel. This is accompanied by a cut in grinding allowance to about one tenth of the former value, and a reduction in energy expenditure from 130 to 19 kW h per component.

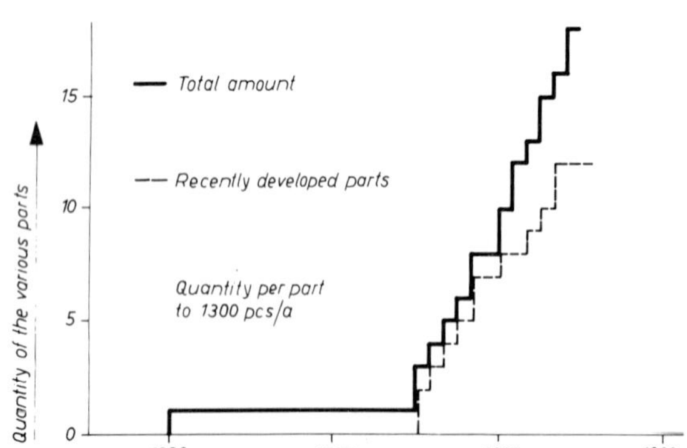


Fig. 12. Industrial application of electron beam hardening in the parent firm of Werkzeugmaschinenkombinat "Fritz Heckert", Karl-Marx-Stadt.

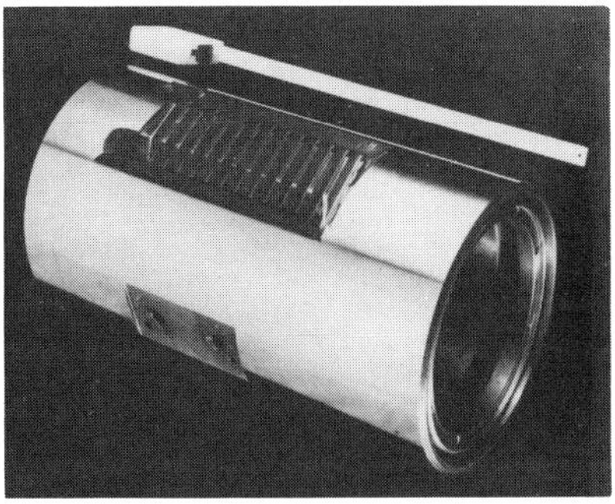

Fig. 13. Quill of a milling machine surface hardened by electron beams. Hardness depth: 0.7 mm, toothed region and plane surfaces not hardened.

Fig. 14. Electron beam plant for welding and hardening of parts in machine tool building.

Very good preconditions for the introduction of this technique are obtained in cases where electron beam welding has been used already. Figure 14 shows a production plant in which welding and hardening are performed over equal periods of time. Although initially intended only for electron beam welding, this plant was also converted to accomplish electron beam hardening in the HSS technique; an endeavour that resulted in extra costs of about 5% referenced to the total investment costs. Meanwhile, plants are under development that are intended for electron beam hardening exclusively.

## 5. Conclusions

With the advent of the isothermal energy transfer in high-speed scan technique, surface hardening with electron beams has been introduced successfully on an industrial scale. The capability to attain energy transfer fields up to about 100 mm × 100 mm, and the use of high beam powers up to 50 kW give the preconditions for high plant productivity and economical benefits compared with competing hardening techniques. The software-controlled energy transfer confers a high flexibility to this technique and results in a plant configuration similar to that of a machine tool. The incorporation of electron beam processing equipment into flexible machining centers is now a task to be dealt with in the immediate future.

## References

1 M. von Ardenne *et al.*, *Effekte der Physik und ihre Anwendungen*, VEB Deutscher Verlag der Wissenschaften, Berlin, 1988, pp. 55 - 63.
2 S. Schiller and S. Panzer, *HTM*, *42* (1987) 293 - 300.

*Surface and Coatings Technology, 39/40* (1989) 531 - 538 

# TECHNICAL NOTE: DEVELOPMENT OF CORROSION-RESISTANT DECORATIVE COLORED STAINLESS STEEL COILS BY AN IN-LINE DRY-COATING PROCESS

I. ITO, T. KOMORI, Y. OIKAWA and M. ONOYAMA

*Hikari Works, Technical Department, Nippon Steel Corporation, 3434 Shimata, Hikari 743 (Japan)*

M. HASHIMOTO and S. ITO

*R&D Laboratories-1, Nippon Steel Corporation, 1618 Ida, Nakahara-ku, Kawasaki 211 (Japan)*

(Received March 22, 1989)

## Summary

Dry-coating technologies have been widely used in the field of semiconductors and electronics. However, the technologies are rarely applied to coat or treat steel surfaces. Basic types of research have been conducted, particularly on the coating materials and process for several years. Based on the results, a new coil-coating pilot facility using three coating processes (ion plating, sputtering, plasma chemical vapor deposition (CVD)) in tandem for the first time was developed and installed at the Hikari Works of Nippon Steel Corporation.

In particular, technologies to connect the three different processes for continuous operation were developed, and the optimum coating treatment was established for corrosion resistance and decorative color. The development of the differential vacuum chamber enabled the connection of the three coating processes with different vacuum pressures on a single continuous line. This in-line dry-coating process can be utilized to apply not only single-layer films (TiN, $SiO_x$ etc.) but also multiple-layer films. Thus, it meets a great variety of market needs.

In this paper, the deposition of TiN and TiC films by ion plating, $Al_2O_3$ and chromium films by sputtering and $SiO_x$ films by plasma CVD on stainless steel sheet in coil form will be introduced, together with the results of thickness uniformity, adhesivity, corrosion resistance and other properties.

Decorative TiN coatings on bright annealed surfaces of type 304 stainless steel by ion plating should be uniform in thickness with a gold color, the tone of which can be easily tuned by varying the ratio of the titanium nitrogen composition. It also gives excellent abrasion resistance.

$SiO_x$ coatings on bright annealed surfaces of type 430 stainless steel by plasma CVD have excellent corrosion resistance against $Cl^-$ attack, to the degree of uncoated type 304 BA. They also have outstanding electrical resistance for use as an insulating film.

0257-8972/89 Elsevier Sequoia/Printed in The Netherlands

$Al_2O_3$ coatings on BA surfaces of type 304 stainless steel formed by sputtering give rise to an interference color, so that any desired color from blue to pink can be obtained by changing the thickness of the coating. $Al_2O_3$ films possess not only high electrical resistance but also high radiant capability for far-IR radiation.

Thus, a new in-line dry-coating process is capable of coating a variety of materials on stainless steel sheet in coil form, and also producing highly functional steel surfaces.

---

## 1. Introduction

Various types of surface treatment techniques have been utilized for stainless steels, *e.g.* the chemical conversion coating oxide (INCO) method, electroplating, organic coating, thermal spraying and cladding. However, highly flexible coating techniques are needed to meet the great variety of surface needs, *i.e.* not only corrosion resistance but also decorative color, electrical resistance, etc. Thus, a new type of in-line coating technique for stainless steel sheet in coil form was developed. Utilizing ion plating, sputtering and plasma CVD, this process can coat both metals and ceramics on steel surfaces. This paper deals with the details of the new process and introduces the result for coated materials in terms of the various properties, thickness uniformity and workability, etc.

## 2. Experimental procedure

The substrate materials used in this study were commercially available stainless steels type 304 and 430 and Nippon Steel Corporation grade YUS-180 (19Cr–0.3Ni–0.4Cu–0.4Nb–0.01 C,N) stainless steel. The surface of the materials were mainly of bright annealed sheet of thickness 0.5 mm. Specimens were pre-treated before deposition by (1) ion bombardment and (2) electrolysis in nitric acid [1]. Ion bombardment primarily eliminates surface contaminations, resulting in the high adhesivity of deposited layers. The nitric acid treatment chemically stabilizes the pre-existing passive film on stainless steels.

The developed in-line dry-coating process consists of three different coating systems, *i.e.* ion plating, sputtering and plasma CVD, together with an ion bombardment system.

The difference in vacuum conditions in each processing chamber requires that a differential vacuum pressure chamber be installed between the sputtering and plasma CVD chambers.

A coil pay-off chamber and a coil take-up chamber are located at both ends of the pressure chamber (Fig. 1).

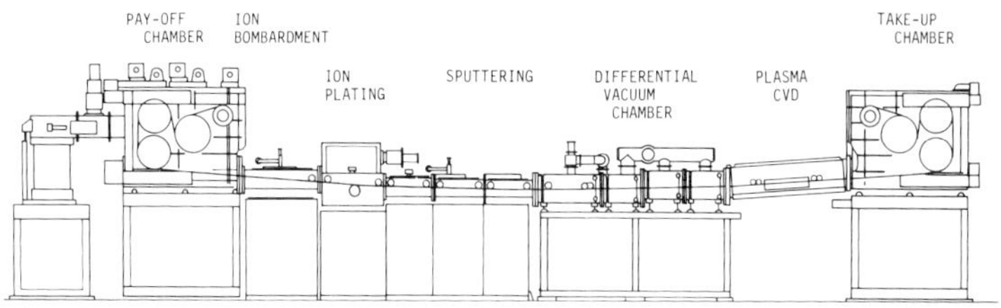


Fig. 1. The in-line dry-coating process.

TABLE 1

Deposition conditions

| *Coating species* | *Process* | *Conditions* |
|---|---|---|
| TiN | Ion plating | HCD current, 280 A × 2; substrate temperature, 260 °C; operated pressure, $1 \times 10^{-4}$ Torr; web speed, 0.4 m min$^{-1}$ |
| TiC | Ion plating | HCD current, 250 A × 2; substrate temperature, 260 °C; operated pressure, $1 \times 10^{-4}$ Torr; web speed, 0.4 m min$^{-1}$ |
| $Al_2O_3$ | Sputtering | r.f. power, 5 kW; substrate temperature, 200 °C; operated pressure, $2 \times 10^{-4}$ Torr; web speed, 0.1 m min$^{-1}$ |
| Chromium | Sputtering | r.f. power, 5 kW; substrate temperature, 200 °C; operated pressure, $2 \times 10^{-4}$ Torr; web speed, 0.1 m min$^{-1}$ |
| $SiO_x$ | Plasma CVD | r.f. power, 35 W; substrate temperature, 250 °C; operated pressure, $2 \times 10^{-1}$ Torr; web speed, 0.08 m min$^{-1}$ |

TABLE 2

Methods for determining coating properties

| | |
|---|---|
| Thickness uniformity | Coating thickness (Dektak II-A, Ellipsometry) |
| Structure and chemical composition | X-ray analysis<br>XPS |
| Workability | Tensile test<br>Bending test |
| Wear resistance | Wear resistance tester |
| Corrosion resistance | Atmospheric corrosion test at sea shore (9 weeks)<br>Actual exposure test inside of indoor swimming pool (9 weeks)<br>Modified salt spray test (0.5% NaCl + 0.2% $H_2O_2$, 24 h) |
| Electrical property | Insulation resistance using contact method |

The ion-plating process utilizes a hollow cathode discharge gun and crucibles for vaporizing and ionizing the metals.

The sputtering process in this study utilizes r.f. magnetron excitation between the substrate and the target (457 mm × 325 mm in size) located 60 mm apart.

The plasma CVD process adopts capacitively coupled r.f. plasma which can be applied to a width of approximately 370 mm with a maximum power of 300 W.

Details of the deposition conditions of each coating species are shown in Table 1. After coating, surface properties of each coating film were evaluated in terms of coating thickness uniformity, chemical composition, wear resistance, insulation resistance and corrosion resistance (see Table 2).

## 3. Results and discussions

### *3.1. In-line dry-coating process*

Ion plating, sputtering and plasma CVD can be applied successively in this continuous line process.

The general specifications for the stainless steel sheet in coil form are as follows: thickness 0.1 - 0.5 mm, width 370 mm, length 300 m (maximum), coil weight 250 kg (maximum). The web is allowed to run at the speed of 0.01 - 0.4 m $min^{-1}$, facing downwards to the deposition sources. Supporting rolls are installed on the substrate web from the back to protect it from damaging the active surface. Slits and differential pressure chambers between processing chambers give independent pressure control and perfect protection from cross-contamination of reactive gases. This technology has made it possible to produce an excellent functional surface finish on stainless steel sheet in coil form without defects.

### *3.2. Thickness uniformity of coated stainless steel coils*

The coating thickness of TiC, TiN and chromium was measured with the precision stylus method (Dektak II-A), whereas that of $Al_2O_3$ and $SiO_x$ was measured by the ellipsometry method.

Typical results for thickness uniformity in the direction of the web width are shown in Fig. 2. Figure 2 shows the highly uniform surface thickness of these five coatings, especially $SiO_x$-coated stainless steel sheet in coil form. The best results indicated that scatter was less than 2.5%.

### *3.3. The structure and chemical composition of thin films*

The structure and chemical composition of films were measured by X-ray analysis and X-ray photoelectron spectroscopy (XPS) respectively. X-ray analysis shows that chromium, TiC and TiN are all crystalline, whereas $Al_2O_3$ and $SiO_x$ show very diffuse diffraction peaks indicating a non-crystalline or amorphous structure. $Al_2O_3$ is essentially stoichiometric and the O/Al ratio is between 1.5 and 1.6. $SiO_x$ shows a slight oxygen deficiency and the O/Si ratio changes from 1.3 to 1.7 depending on the details of the coating conditions. The surface of the metallic chromium contains large amounts of oxygen presumably due to the air-formed oxide film.

| Species \ Location | ① | ② | ③ | ④ | ⑤ |
|---|---|---|---|---|---|
| $Al_2O_3$ (SP) | 0.19 | 0.26 | 0.21 | 0.21 | 0.18 |
| Cr (SP) | 0.25 | 0.22 | 0.21 | 0.21 | 0.18 |
| SiOx(P-CVD) | 0.21 | 0.21 | 0.21 | 0.21 | 0.23 |
| TiC (IP) | 0.28 | 0.26 | 0.30 | 0.34 | 0.29 |
| TiN (IP) | 0.27 | 0.24 | 0.18 | 0.22 | 0.20 |

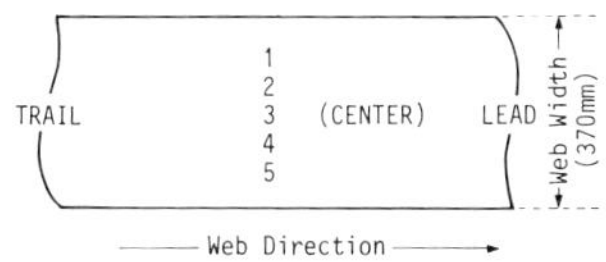


Fig. 2. Thickness uniformity of coated stainless steels.

### 3.4. *Workability*

When the ceramic-coated stainless steels undergo severe deformation, the formation of microcracks is inevitable since ceramics are generally brittle. However, the results of tensile tests of TiN-coated type 304 stainless steel show that the films are essentially unchanged below 5% elongation. Above this elongation, however, scanning electron microscope (SEM) analysis shows the formation of microcracks, which cause a slight change in color.

### 3.5. *Wear resistance*

The wear resistance of stainless steels coated with ceramics and metal was assessed by scrubbing the surface with felt. In this experiment, a solution

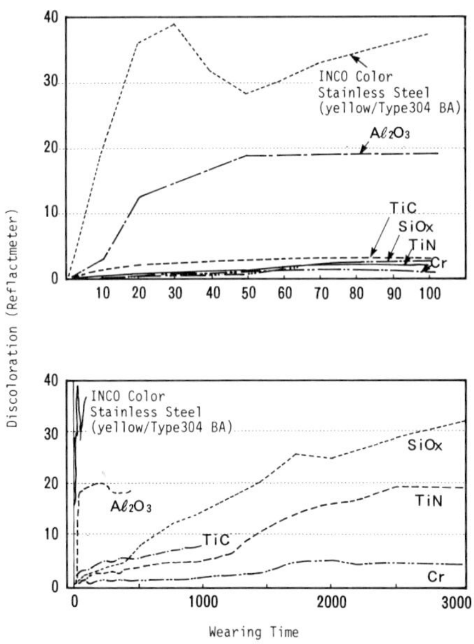


Fig. 3. Wear resistance of coated stainless steels.

containing alumina was continuously supplied to the surface of the specimens. Figure 3 summarizes the results and clearly shows the superiority of these ceramics or metal coated stainless steels compared to the INCO color stainless steel. In particular, Cr and TiN are superior to the others.

### *3.6. Corrosion resistance*

In this study, the corrosion behavior of stainless steels coated with ceramics and metals was assessed by the following corrosion tests: (1) an atmospheric corrosion test at the sea shore, (2) actual exposure test inside of an indoor swimming pool and (3) a modified salt spray test (0.5% NaCl plus 0.2% $H_2O_2$) [2].

Figure 4 shows the results of the atmospheric exposure test at the sea shore. The corrosion resistance of stainless steels was much improved in all cases by coating with these ceramics or metals. In particular the $SiO_x$-coated type 430 BA coated by the Plasma CVD process shows excellent corrosion resistance.

It is interesting to note that the degree of corrosion degradation depends on the location of the bend in the specimen. Region 4 shows the most severe corrosion damage. One of the reasons is that this region stays wet for a long period of time due to its geometrical positioning limitations compared with the other locations. This can be seen from the fact that a few stains of 5 - 10 mm diameter were observed in these specimens. These spots correspond to the location of dried droplets. This result is also consistent with our previous study [3], where the chemistry of a thin water film on metal surfaces was shown to be the important factor in these atmospheric corrosion tests.

$SiO_x$ and chromium coatings are also beneficial for improving corrosion resistance, especially under oxidizing conditions such as in the modified salt spray tests using $H_2O_2$ and a swimming pool containing hydrogen gas as sterilizer. However, TiN-coated stainless steels suffer extensive corrosion

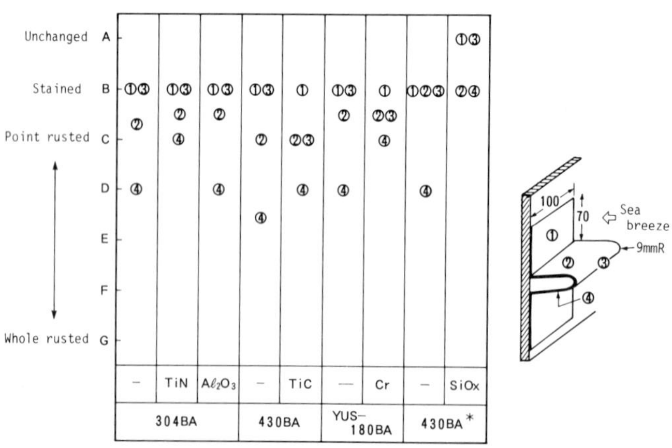


Fig. 4. Result of atmospheric exposure test at sea shore.

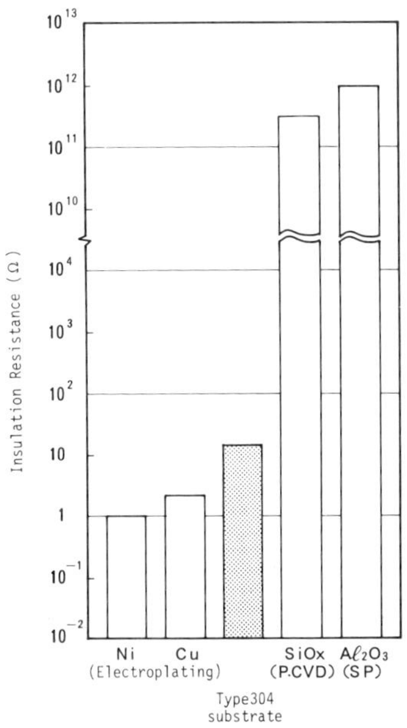


Fig. 5. Insulation resistance of coated stainless steels.

damage under these oxidizing conditions due to the lack of chemical stability of these nitrides against oxidation. In this case, the corrosion product was mainly composed of $TiO_2$ and was detected at the tarnished spots on the surface after modified spray testing. However, coating with TiN makes type 304 stainless steels as resistant as super stainless steels such as 25Cr–4Mo–4Ni ferritic steels in the case of pitting corrosion. Therefore, it is important to make the correct material selection, especially in severe environments.

### *3.7. Electrical property*

Electrically insulated stainless steels can also be produced by coating with dielectric materials such as $SiO_x$ and $Al_2O_3$. Figure 5 shows that these coatings provide excellent electrical insulation when measured by the contact method.

## 4. Conclusions

A new in-line dry-coating process for ceramic and metal coatings utilizing ion plating, sputtering and plasma CVD has been developed and applied to stainless steel coils for additional surface functions such as excellent corrosion and wear resistance as well as electrical insulation.

The following is a summary of these results.

(1) In-line dry-coating process. Technologies that connect processes for continuous operation were developed, and optimum conditions for carrying

out coating treatment were established. Incorporating the three coating processes into a single continuous line made it possible to perform coating treatments that apply single-layer films and as well as multiple-layer films.

(2) Characteristics of the coated stainless steel coils

| | |
|---|---|
| $Al_2O_3$/type 304 BA by sputtering | Interference decorative color, insulation resistance. |
| Chromium/YUS 180 BA by sputtering | Pure metallic color, wear resistance, corrosion resistance. |
| $SiO_x$/type 430 BA by plasma CVD | Corrosion resistance, insulation resistance. |
| TiN/type 304 BA by ion plating | Material decorative color, wear resistance. |
| TiC/type 304 BA by ion plating | Material decorative color. |

Great interest has been expressed in the field of thin film coatings and surface modification of metallic materials, due to its potential application in broad engineering fields. In order to meet these requirements it will be necessary to develop a more large-scale and efficient continuous-coating system in the very near future.

## References

1 H. Omata, S. Ito and T. Murata, *U.S. Patent 4,612,095.*

2 M. Nakata, I. Ito, M. Onoyama, H. Inagaki, *Proc. Joint Symp. on Corrosion Science and Technology, JSCE, 1981*, Vol. 28, p. 18.

3 M. Hashimoto, S. Miyajima, W. Ito, S. Ito, T. Murata, T. Komori, I. Ito and M. Onoyama, *Program of 9th Int. Conf. on Vacuum Metallurgy, 1988,* Vol. 8, p. 139.

*Surface and Coatings Technology, 39/40* (1989) 539 - 548



# TECHNICAL NOTE: AGGLUTINATION PROCESS FOR COATING CERAMIC PARTICLES

S. DALLAIRE and B. CHAMPAGNE

*Industrial Materials Research Institute, National Research Council Canada, 75 De Mortagne Blvd., Boucherville, Quebec J4B 6Y4 (Canada)*

J. HOJO

*Department of Applied Chemistry, Faculty of Engineering, Kyushu University, 6-10-11 Hakozaki, Higashi-ku, Fukuoks-shi 812 (Japan)*

(Received March 8, 1989)

## Summary

In the fabrication of high performance metal–ceramic composites and plasma-sprayed coatings containing dispersed ceramic particles, the use of metal-coated ceramic powders can be expected to ensure better bonding between constituents and provide a more uniform microstructure than the mixing of ceramic and metal powders could do. Various coating methods have been used to coat ceramic powders. However, these methods are limited in terms of coating materials and thicknesses. A new inexpensive method based on an agglutination process is proposed to coat different kinds of ceramic particles (40 - 100 $\mu$m) with various metallic elements. The process variables are discussed with examples concerning $Al_2O_3$ and SiC. It was found that there is no limitation in terms of ceramic cores and metal coatings and that it is possible to produce thick coatings depending upon the nature of metal powders and ceramic particles.

## 1. Introduction

Composite powder particles comprising of a central ceramic core coated with a metal layer are well known, and in commercial use in the plasma and flame spraying fields as well as in the composite fabrication field. A suitable metal coating prevents ceramic powders that do not melt, such as BN or SiC, from decomposition when heated in a flame [1]. It also protects carbide powders from oxidation and decomposition when they are heated to melting point [1]. The spraying of metal-coated ceramic powders has given coatings with higher hardness, better bond strength and higher wear resistance than achieved by the conventional blend of metal and ceramic powders [2, 3]. The use of metal-coated ceramic particles should provide high performance composites with the uniform microstructure they require [4, 5].

0257-8972/89/$3.50 

So far, the metal coating of ceramic particles has been mainly done by the following techniques: (1) chemical plating from the aqueous solutions of metal salts [3, 5 - 8], (2) physical vapour deposition [9] and (3) chemical vapour deposition [10]. The first technique is only possible with metals that can be reduced such as nickel, cobalt and molybdenum, whereas the two others need volatile and reactive metal sources. Moreover, all these techniques are limited to thin coatings.

An agglutination technique [11] has been developed to circumvent the severe drawbacks of the above-mentioned techniques. It consists of agglutinating fine metal powder on ceramic cores covered with binder to produce composite particles with a thick metal coating. This process is described here with examples concerning the coating of $Al_2O_3$ and SiC with various kinds of fine metal powders. The effects of some process variables such as the size of the ceramic particles, the shape of the metal powders, the amount of binder and the addition of a surfactant are discussed.

## 2. Description of the process

The agglutination coating technique is partly derived from the tumbling agglomeration of powders [12]. Binder-coated ceramic particles and fine metal powders are loaded into a cylinder tilted at an angle to the horizontal axis of rotation (Fig. 1). The tumbling ensures good mixing and a uniform

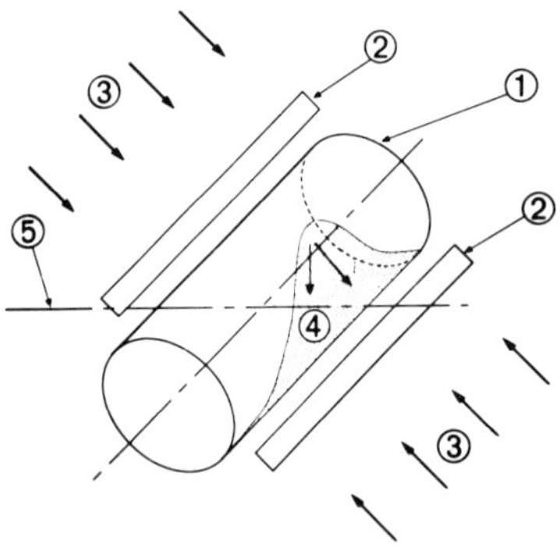


Fig. 1. Schematic representation of the agglutination apparatus: 1, tilted cylinder; 2, heating elements; 3, cooling medium; 4, load; 5, rotation axis.

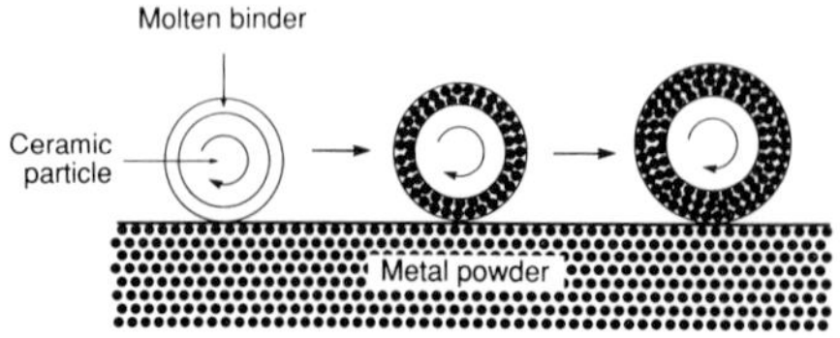


Fig. 2. Mechanism of agglutination coating.

movement of ceramic particles and metal powder. Upon heating the cylinder, the binder reaches a viscous state which allows the formation of a fine powder coating on the ceramic cores (Fig. 2). In a subsequent step, heat treatment can be carried out to remove the binder and to sinter the powder layer which is then coherently bonded to the ceramic core.

## 3. Experimental procedure

The agglutination was initially carried out in a cylinder made from acrylic glass heated and cooled down with an air blower in order to observe the tumbling of powders. A cylinder with a higher thermal conductivity heated by electric bands was thereafter preferred. A temperature controller regulates the power delivered to the bands to avoid binder overheating.

Angular particles of $Al_2O_3$ with a mean size of 260 $\mu$m and SiC with different mean sizes of 40, 90 and 770 $\mu$m were used as cores. Aluminum, chromium, cobalt, molybdenum and tungsten powders were used as coating powders. These metal particles are in the size range of 5 - 10 $\mu$m as measured with an HIAC particle size analyser PA-720 (Pacific Instruments). Polyethylene glycol (PEG) (molecular weight 3350) with a melting range between 54 and 80 °C was used as binder.

The general procedure for covering cores with the binder consists of loading the cylindrical container with the binder and cores. The container is rotated and heated until the binder becomes viscous. Then the mixture is cooled and the agglomerated particles are broken up by adding balls to the container. An alternative method for coating cores with the binder consists of dissolving the binder in a solvent (acetone), mixing the solution with the cores, evaporating the solvent and then crushing the resulting agglomerates. This procedure is longer, more complex and more hazardous than the former procedure.

Binder-coated ceramic cores and fine metal powder with or without addition of a surfactant are charged into the cylinder. The cylinder is then simultaneously heated and rotated leading to the agglutination process. Typical process parameters are listed in Table 1.

After cooling, metal-coated ceramic cores were separated from the free metal powder by sieving. The proportion of the coating was determined by

TABLE 1

Process parameters used for coating ceramic particles

| | |
|---|---|
| Content of binder | 4 wt.% (based on ceramic cores) |
| Load of metal powder | 20 - 60 vol.% (Me/(Me + ceramic)) |
| Filling of cylinder | 5 - 10% |
| Rotation speed | 40 - 80 rev min$^{-1}$ |
| Tilt angle | 30° or 45° |
| Temperature | 50 - 85 °C |
| Heating time | 300 - 600 s |

weight measurements. Particle shape and appearance of coatings were observed by scanning electron microscope (SEM) analysis.

## 4. Results and discussion

### *4.1. Agglutination coating of $Al_2O_3$ with various metal powders*

$Al_2O_3$ cores (mean size of 260 μm) were coated with aluminum, chromium, nickel, cobalt, molybdenum and tungsten powders. The metal powder load was kept constant on a volume basis in order to make comparisons on the efficiency of the coating process. Table 2 summarizes the type of metal powders used and the results obtained. Both flaky and spherical aluminum powders covered the whole surface of $Al_2O_3$ particles. The coating volume obtained from flaky powders was smaller than that obtained from spherical ones. Cobalt, chromium and molybdenum powders yielded a fairly large volume of metal coating. Since chromium powder contained many large particles, the surface of the coating was rough. The volume of the nickel coating was small and the coating was partial whereas the volume of the tungsten coating was large and the coating was nearly perfect.

Figure 3 shows the surface of aluminum-coated $Al_2O_3$ particles with flaky and spherical aluminum powders. The flaky Al powder covers the surface of $Al_2O_3$ core well in spite of the small coating volume. It seems that the lamella form of the powder prevents the binder from agglutinating more powder. Up to 63 vol.% of the spherical aluminum powders can be agglutinated on to $Al_2O_3$ cores.

The following coating efficiencies of metal powders in decreasing order was obtained: Al(sphere) > Cr = W > Co = Mo > Al(flake) > Ni. The wettability of metal powders by the viscous binder is the main factor explaining the difference in efficiencies. The efficiency not only depends upon the shape and the size of coating powders but also upon the size of ceramic cores.

TABLE 2

Coating of $Al_2O_3$ with various metal powders

| *Metal* | *Powder shape* | *Weight mean diameter* (μm) | *Metal coating* $Me/(Me + Al_2O_3)$ (vol.%) | *Coating appearance* |
|---|---|---|---|---|
| Al | flaky | 10 | 14 | perfect |
| Al | spherical | 8 | 31 | perfect |
| Ni | spherical | 9 | 5 | partial |
| Cr | irregular | 6 | 25 | perfect |
| Co | agglomerate | 7 | 16 | perfect |
| Mo | irregular | 7 | 18 | perfect |
| W | agglomerate | 5 | 25 | perfect |

Load: $Me/(Me + Al_2O_3) = 32$ vol.%; PEG content, 4 wt.%.

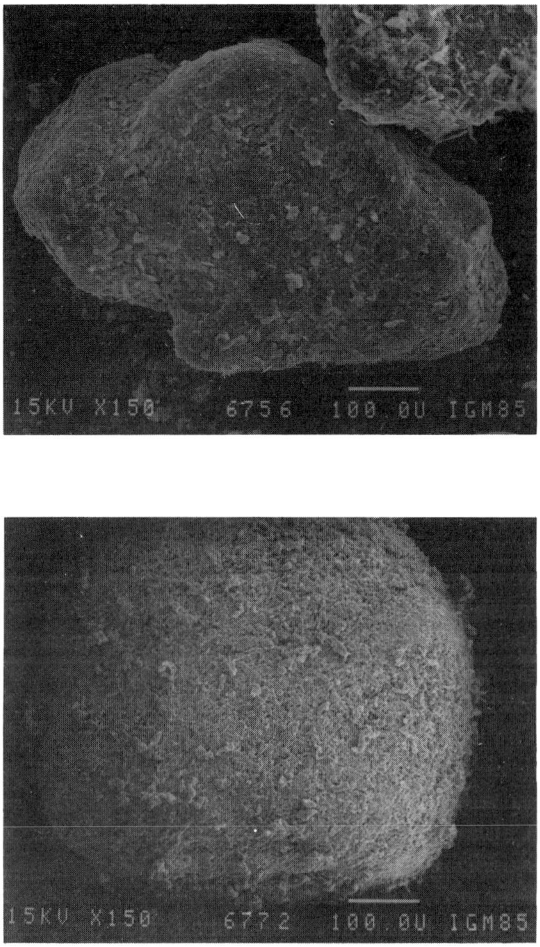


Fig. 3. Scanning electron micrograph of aluminum-coated $Al_2O_3$ particles: (a) with flaky aluminum powder (11 vol.%); (b) with spherical aluminum powder (63 vol.%).

*4.2. Effect of the size of ceramic cores*

Results on the nickel and aluminum powder coating of $Al_2O_3$ and SiC particles are reported in Table 3. Because SiC particles have a larger size than $Al_2O_3$ particles, they agglutinated more metal powder and the resulting coating was nearly perfect. A decrease of aluminum powder coating on SiC particles with a decrease of SiC particle size is also observed. Therefore, large particles are easier to coat than small ones. This is likely due to the higher kinetic energy of coarse particles during their tumble which produces higher compressive and collision forces that improve the powder coating.

*4.3. Effect of additives to metal powders*

As mentioned previously, the nickel coating was the most difficult to perform. The effect of additives (PEG and aluminum stearate) was thus

TABLE 3

Coating of metal powder on particles of various size

| *Ceramic particle* | *Weight mean diameter* ($\mu$m) | *Metal* | *Load*[a] (vol.%) | *Coating*[a] (vol.%) | *Coating appearance* |
|---|---|---|---|---|---|
| $Al_2O_3$ | 260 | Ni | 32 | 5 | partial |
| SiC | 770 | Ni | 31 | 19 | nearly perfect |
| SiC | 770 | Al | 55 | 55 | perfect |
| SiC | 90 | Al | 55 | 49 | perfect |
| SiC | 40 | Al | 55 | 39 | perfect |

[a]Me/(Me + ceramic); PEG content = 4 wt.%.

investigated to enhance the growth of the coating on $Al_2O_3$. A PEG addition to the metal powder did not result in an improvement: the nickel coating remained imperfect and, moreover, large nickel agglomerates were formed. However, the addition of aluminum stearate increased the thickness of the coating as shown in Fig. 4. The addition of aluminum stearate to the nickel powder is effective only up to 0.2 wt.% and no further increase of the nickel coating was observed above this value.

With no addition of aluminum stearate, many coating defects were observed on the edges of irregular shape cores (Fig. 5). The edges are indeed prone to collisions and friction. On the contrary, the aluminum stearate-doped nickel powder perfectly covered irregular $Al_2O_3$ cores, even on their edges as shown in Fig. 5.

The effect of the addition of aluminum stearate to the metal powder is noteworthy. Aluminum stearate is a surfactant and, when the molecules are adsorbed on the metal powder, the long carbon chains inhibit the agglomeration of the metal powder and help to spread the core surface with metal powders.

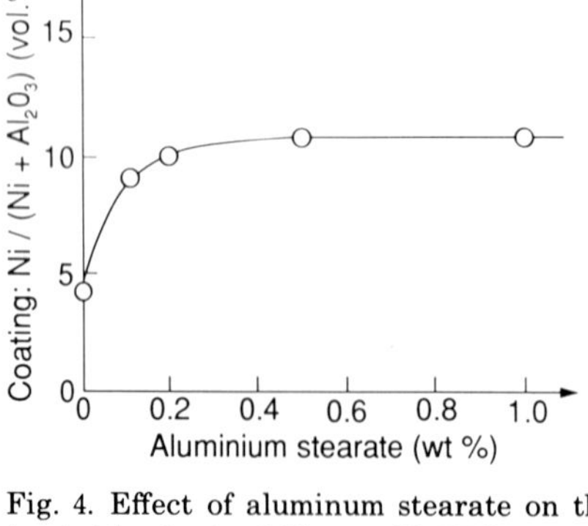


Fig. 4. Effect of aluminum stearate on the agglutination of nickel powder on $Al_2O_3$ particles (nickel load: 16 vol.%, 4 wt.% PEG).

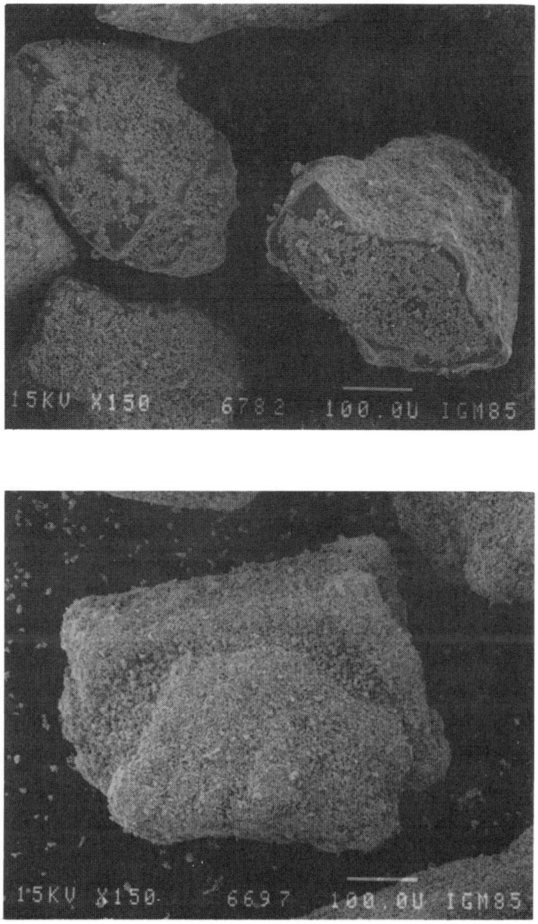


Fig. 5. Typical appearance of nickel-coated $Al_2O_3$ particles: (a) undoped nickel powder; (b) aluminum stearate-doped nickel powder.

### 4.4. *Effect of PEG content on the coating volume*

The influence of PEG content on the coating volume was investigated with aluminum and nickel and $Al_2O_3$ cores. The variation of aluminum coating volume with PEG content is shown in Fig. 6. The volume of aluminum coating remarkably increases up to about 1 wt.% PEG, and at this point most of the aluminum particles were consumed to form the coating. The variation of nickel coating volume with PEG content is somewhat different, as shown in Fig. 7. The volume increases almost linearly with the PEG content to reach 13 vol.%. A large proportion of coating powder (up to 5 vol.%) was not agglutinated on $Al_2O_3$ cores even with a PEG content as high as 7 wt.%.

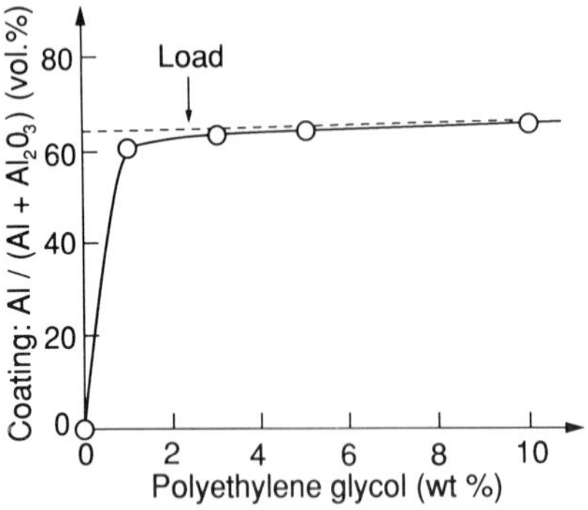


Fig. 6. Effect of PEG content on the agglutination of aluminum powder on $Al_2O_3$ particles (aluminum load: 64 vol.%, spherical aluminum).

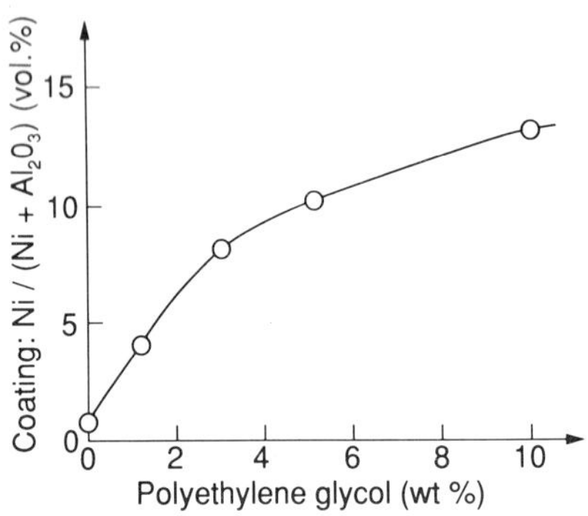


Fig. 7. Effect of PEG content on the agglutination of nickel powder on $Al_2O_3$ particles (nickel load: 18 vol.%, 1 wt.% aluminum stearate-doped nickel powder).

### *4.5. Effect of the load in the rotating cylinder*

The effect of the load on the coating volume was investigated to determine to what extent a coating could be built up around ceramic cores. The influence of the load of aluminum powder on the coating volume formed around $Al_2O_3$ particles is shown in Fig. 8. The coating volume increases almost proportionally with the aluminum load up to 60 vol.%. Aluminum loads below 20 vol.% result in the agglomeration of $Al_2O_3$ particles, thus indicating that the PEG content (4 wt.%) is excessive in this case.

The influence of the charge of nickel powders on the coating volume formed around $Al_2O_3$ cores is shown in Fig. 9. An increase of the nickel load leads to a greater coating thickness for aluminum stearate-doped nickel powder whereas a decrease in thickness is observed for undoped powder. The addition of a surfactant has also a benefic effect for the nickel powder since no nickel agglomerates are formed for a load above 5 vol.%, and no tendency for $Al_2O_3$ particles to stick together is observed for a load below 5 vol.%.

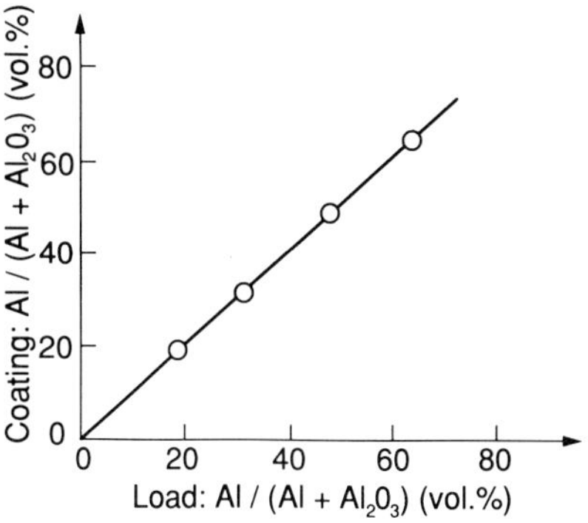


Fig. 8. Effect of the aluminum powder load on the agglutination of aluminum powder on $Al_2O_3$ particles (containing 4 wt.% PEG).

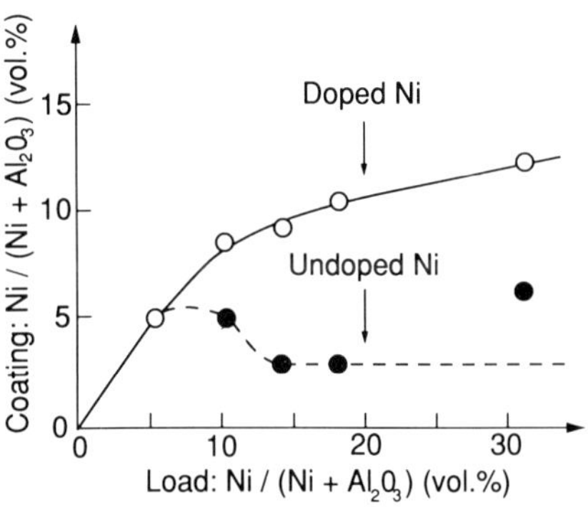


Fig. 9. Effect of the nickel powder load on the agglutination of nickel powder on $Al_2O_3$ particles (containing 4 wt.% PEG): ○, 1 wt.% aluminum stearate-doped nickel powder; ●, undoped nickel powder.

### *4.6. Effect of other process variables*

The influence of other process variables, *i.e.* tilt angle of the cylinder, rotation speed and kind of adhesive binder, were also investigated. The effect of these variables was not significant in the present study. However, it is worth mentioning that the temperature must be controlled and kept as close as possible to the melting point of the binder in order to avoid the release of binder from cores.

## 5. Conclusion

The agglutination coating method has been found to be very effective for coating ceramic particles with various kinds of metal powders. This was exemplified with the coating of $Al_2O_3$ and SiC. It was found that the thickness of the metal coating depends on the nature and shape of the metal powder and that the most important variable is the proportion of binder. It was also pointed out that a surfactant can play a fundamental role in preventing the self-agglomeration of fine metal powders and in helping to spread them on the ceramic particles. This clearly indicates that the mechanisms responsible for the agglutination also depend upon the surface chemistry properties of the materials involved.

The agglutination coating method is simple and can be used to coat particles of a few tens of micrometers suitable for thermal spraying or for fabricating metal-bonded ceramics. It is, in principle, applicable to other kinds of materials.

## Acknowledgments

One of the authors (J. Hojo) would like to thank the Natural Science and Engineering Research Council of Canada for financial assistance. The authors also wish to thank H. Levert for his technical help and S. F. Turcotte for reviewing the manuscript.

## References

1 F. N. Longo and M. S. Patel, *U.S. Patent 3,655,425*, 1972.
2 J. D. Reardon, R. Mignoma and F. N. Longo, *Airline Plating and Metal Finishing Forum*, Society of Automotive Engineers Inc., Warrendale, PA, 1982, p. 19.
3 W. Kunda, *7th Plansee Seminar Preprints, Reutte, Austria*, 1971.
4 H. W. Lavendel, *Bull. Am. Ceram. Soc.*, *53* (1974) 797.
5 A. C. D. Chaklader and K. R. Linger, *Composites*, *7* (1976) 239.
6 B. Meddings, W. Kunda and V. N. Mackiw, *Proc. Int. Conf. on Powder Metallurgy*, Interscience Publishers, New York, 1961, p. 775.
7 Y. Horie, T. Kanbe, Y. Kumagaya and S. Isobe, *Bosei Kanri*, *31* (1987) 162.
8 C. P. Vijayan, J. Hojo, J. J. Heckler, B. Champagne and S. Dallaire, *Electroless Nickel Conf. IV*, Gardner Publications Inc., Cincinnati, OH, 1985, p. 17.
9 Yu. A. Pavolv, I. V. Blinkov, K. A. Myagkov and A. V. Manukhin, *Poroshk, Metall.*, *8* (1984) 1.
10 C. M. Pleass and D. G. Schimmel, *Proc. Int. Conf. Chemical Vapor Deposition*, American Nuclear Society, Hinsdale, 1972, p. 513.
11 J. Hojo, S. Dallaire and B. Champagne, *U.S. Patent 4,788,080*, 1988.
12 N. Claussen and G. Petzow, *Proc. 3rd European Powder Metallurgy Conf. Suppl.*, 1971, p. 225.

*Surface and Coatings Technology, 39/40* (1989) 549 - 564



# THE EFFECT OF TARGET–SUBSTRATE COUPLING ON REACTIVE DIRECT CURRENT MAGNETRON SPUTTERING

S. SCHILLER, U. HEISIG, CHR. KORNDÖRFER, J. STRÜMPFEL and P. FRACH

*Forschungsinstitut Manfred von Ardenne, Zeppelinstrasse 7, 8051 Dresden (G.D.R.)*

(Received April 17, 1989)

## Summary

From a consideration of the reactive gas flow rates and of the reactive gas partial pressure a criterion for process stability in reactive direct current magnetron sputtering is derived. By a discussion of this stability criterion the methods of self-stabilization and stabilization through high pump speed are deduced. If the gas discharge is considered as a getter pump with a reactive gas pressure dependent pumping rate $S_{Dis}(p_R)$, the stability criterion includes a term $-dS_{Dis}/dp_R$ which describes the degree of target-to-substrate coupling (TSC). From this the method of stabilization through reduction of TSC, *i.e.* through decoupling, is deduced. The relationship of process stability, TSC and film properties is discussed and exemplified by the system $CrSi–O_2$. Coating of large areas and plasma-assisted deposition call for strong TSC, which may be contradictory to the stability criterion. An active closed-loop control circuit, *e.g.* using the plasma emission intensity signal (PEM control), has to be employed for process stabilization.

## 1. Introduction

The fundamentals of reactive direct current magnetron sputtering, especially its regime and stabilization, have been described in many papers and summaries [1 - 3]. Different methods for process stabilization have been worked out empirically, *i.e.* stabilization through (i) installation of high-capacity pumps [4], (ii) decoupling of target and substrate processes ("gas separation") [5, 6], (iii) self-stabilization [7] and (iv) active closed-loop control via a plasma emission monitor (PEM) [8].

Recent papers have dealt with theoretical investigations into processes which cause instabilities [9] and with the specification of an empirical stability criterion.

It is the objective of this contribution to derive the said stability criterion in detail, and to link it with the degree of target-to-substrate coupling (TSC). The effect of TSC on the film properties shall be exemplified by the system $CrSi–O_2$. With regard to ion-assisted reactive coating at a

0257-8972/89/$3.50

strong TSC, reasons are given why the active PEM control will be the method of choice wherever the said stability criterion cannot be satisfied.

## 2. Stability and target–substrate coupling

### *2.1. Stability criterion*

In an arrangement for reactive direct current magnetron sputtering a reactive gas flow $f_{\mathrm{R,\,tot}}$ is admitted, for example, via dosage valves or mass flow controllers in addition to the inert gas used. In the process chamber, the reactive gas flow $f_{\mathrm{R,\,tot}}$ is split up into a portion $f_{\mathrm{R,\,P}}$ to the pumps having a pump speed $S_{\mathrm{P}}$, and a portion $f_{\mathrm{R,\,Dis}}$ that is consumed in the discharge. The portion $f_{\mathrm{R,\,Dis}}$ comprises (i) the share of the reactive gas converted into the chemical compound on the substrate, (ii) the share of the reactive gas converted on the target and (iii) the share of the reactive gas gettered on other surfaces (*e.g.* those of the plasma shield that confine the discharge). Hence

$$f_{\mathrm{R,\,tot}} = f_{\mathrm{R,\,Dis}} + f_{\mathrm{R,\,P}} \tag{1}$$

With $f_{\mathrm{R,\,P}} = p_{\mathrm{R}} S_{\mathrm{P}}$ this gives

$$f_{\mathrm{R,\,tot}} = f_{\mathrm{R,\,Dis}} + p_{\mathrm{R}} S_{\mathrm{P}} \tag{2}$$

where $p_{\mathrm{R}}$ is the partial pressure of the reactive gas in the process chamber. In industrial plants this approximately corresponds to the partial pressure of the reactive gas that acts on the substrate.

The reactive gas flow consumed in the discharge depends on both the partial pressure of the reactive gas and the sputter or condensation rate $a_{\mathrm{C}}$

$$f_{\mathrm{R,\,Dis}} = f_{\mathrm{R}}(p_{\mathrm{R}}, a_{\mathrm{C}})$$

From eqns. (1) and (2) the reactive gas pressure then becomes

$$p_{\mathrm{R}} = \frac{f_{\mathrm{R,\,tot}} - f_{\mathrm{R,\,Dis}}(p_{\mathrm{R}})}{S_{\mathrm{P}}} \tag{3}$$

Figure 1 shows the behaviour of $p_{\mathrm{R}}$ when the reactive gas flow $f_{\mathrm{R,\,tot}}$ is increased. At a critical flow rate $f_{\mathrm{R,\,tot}} = f_{\mathrm{M\text{-}R}}$ one observes a sudden increase in $p_{\mathrm{R}}$. Phenomenologically this behaviour can be described by the transition from the metallic mode to the reactive mode. The instable region is termed the transition mode [3]. This response is caused by the drop in sputter rate owing to the growing coverage of the target surface with reaction products (target poisoning) when the reactive gas flow increases [1, 10]. In consequence the reactive gas consumption in the discharge is further reduced and thus the partial pressure of the reactive gas is increased. However, this rise in $p_{\mathrm{R}}$ further enhances the target coverage; a positive feedback effect that is responsible for the sudden increase in $p_{\mathrm{R}}$. From the differentiation of eqn. (2) it follows that

$$\frac{\mathrm{d}p_{\mathrm{R}}}{\mathrm{d}f_{\mathrm{R,\,tot}}} = \left(\frac{\mathrm{d}f_{\mathrm{R,\,tot}}}{\mathrm{d}p_{\mathrm{R}}}\right)^{-1} = \left(S_{\mathrm{P}} + \frac{\mathrm{d}f_{\mathrm{R,\,Dis}}}{\mathrm{d}p_{\mathrm{R}}}\right)^{-1} \tag{4}$$

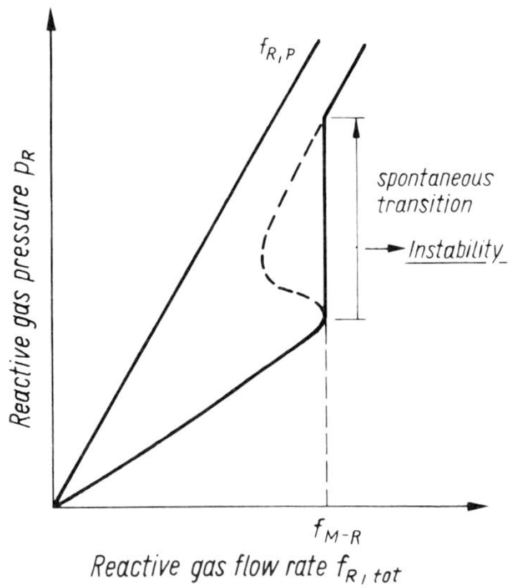


Fig. 1. Schematic behaviour of reactive gas pressure $p_R$ with increasing reactive gas flow rate $f_{R,\,tot}$ and flow rate to pump $f_{R,\,P}$.

The discharge will be unstable if the expression in brackets is equal to or less than zero because, without additional measures, the section of the curve with infinite or negative slope (dotted line in Fig. 1) cannot be passed. From this results the stability criterion

$$S_P > -\frac{df_{R,\,Dis}}{dp_R} \tag{5}$$

This agrees with the relationship given by Spencer *et al.* [4].

In the following sections the various methods used for stabilization of the reactive magnetron discharge are derived from the stability criterion discussed (eqn. (5)).

### 2.2. Self-stabilization

Independently of the pumping capacity installed (*i.e.* even at $S_P = 0!$), relationship (5) is always satisfied if the reactive discharge can be operated in such a manner that

$$\frac{df_{R,\,Dis}}{dp_R} > 0$$

is obtained throughout the working range, *i.e.* if an increase in the partial pressure of the reactive gas results in an increase in the reactive gas flow consumed by the discharge. With all reactive systems this holds in a small section of this characteristic (for small values of $f_{R,\,tot}$). For some particular systems such as Si–O or Al–N this behaviour can be extended over the total working range: a mode termed self-stabilization. For instance, it can be

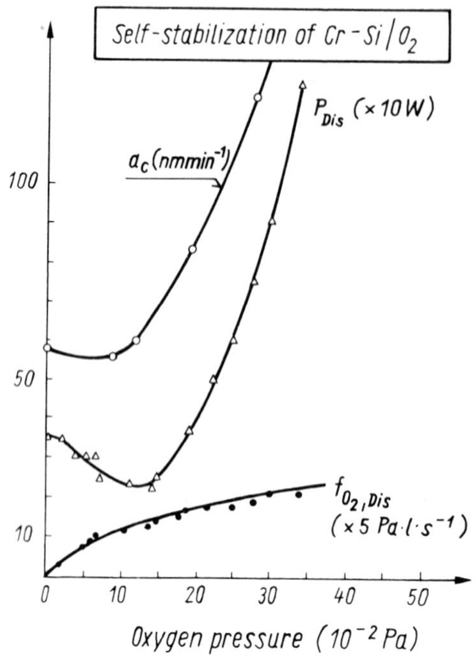


Fig. 2. Discharge power $P_{Dis}$, condensation rate $a_C$ and oxygen flow rate consumed by discharge $f_{O_2, Dis}$ *vs.* oxygen partial pressure $p_{O_2}$ with self-stabilized sputtering of Cr–Si–$O_2$ (discharge voltage 600 V, target diameter 156 mm, target–substrate distance 50 mm).

implemented if reactive sputtering of CrSi occurs in an Ar–$O_2$ mixture. Self-stabilization occurs if CrSi is sputtered within a practically closed reaction chamber, and if the source operates in the constant-voltage mode [7]. Here, the drop in sputter rate caused by an increase in the partial pressure of the reactive gas is compensated by an increasing discharge power; a procedure in which the reduction of the discharge impedance at higher values of $p_R$ is utilized in a dedicated manner.

Figure 2 shows (i) the reactive gas flow $f_{R, Dis}$ consumed in the discharge, (ii) the discharge power $P_{Dis}$ and (iii) the condensation rate $a_C$ as a function of the partial pressure $p_R$ of the reactive gas. The share $f_{R, Dis}$ of the $O_2$ flow spent in the discharge increases monotonously with the rise in oxygen partial pressure.

### *2.3. Stabilization through high pump speed*

In ref. 4 the conclusion has been drawn from eqn. (5) that, for a given sputtering assembly, the installation of a vacuum pump with a higher pump speed for the reactive gas improves the stability of the discharge. It is further shown by the example of the InSn–$O_2$ system that doubling the installed pump speed allows us to satisfy criterion (5) throughout the working range.

It has already been pointed out in ref. 4 that, for strongly reactive systems such as Ti–$O_2$, stabilization through the installation of an adequately high pump speed is practically not possible. The stability range can only be shifted towards higher values of $f_{R,\,tot}$ and $p_R$. For reasons of economy, stabilization through an installed high pump speed gives no practical solution for industrial plants in which high sputter rates have to be matched on extended target–substrate surfaces.

### 2.4. *Stabilization through decoupling*

For a given pump speed $S_P$ the right-hand term of eqn. (5) allows us to derive the requirements to be satisfied by a sputtering arrangement so as to meet the stability criterion. With discharge burning the sputtering arrangement can be taken as a getter pump for the reactive gas with the pumping rate

$$S_{Dis}(p_R) = \frac{f_{R,\,Dis}}{p_R} \tag{6}$$

With eqn. (6) the stability criterion (5) then becomes

$$S_P > -S_{Dis} - \frac{dS_{Dis}}{dp_R}\, p_R = S_C(p_R) \tag{7}$$

where $S_C(p_R)$ is a function which characterizes the stability performance of the system.

In industrial plants the pumping rate of the discharge $S_{Dis}$ obtained at a reactive gas flow $f_{R,\,tot} < f_{M-R}$ is large compared with the installed pump speed (*i.e.* $S_P \ll S_{Dis}$). As a consequence the following approximation holds for the stability criterion

$$\frac{dS_{Dis}}{dp_R} > -\frac{S_{Dis}}{p_R} \tag{8}$$

Here $dS_{Dis}/dp_R$ describes the change in pumping rate of the sputtering arrangement caused by an alteration in the partial pressure of the reactive gas near the substrate. With reactive sputtering systems the pumping rate of the discharge reduces with increasing reactive gas partial pressure

$$\frac{dS_{Dis}}{dp_R} < 0$$

The stability criterion (8) therefore becomes

$$\left|\frac{dS_{Dis}}{dp_R}\right| < \frac{S_{Dis}}{p_R} \qquad \text{for } \frac{dS_{Dis}}{dp_R} < 0 \tag{9}$$

Criterion (8) calls for a sufficiently high pumping rate $S_{Dis}$ or for a small decrease in the pumping rate of the discharge as the partial pressure of the residual gas increases. By causality, however, the pumping rate of the arrangement depends on the metal vapour stream, *i.e.* on the sputter rate at the target. This is why the term $dS_{Dis}/dp_R$ is also a measure of the coupling of the processes which take place on the target and on the substrate. As a consequence, eqn. (8) becomes a criterion with respect to the degree of coupling needed for stabilization. Thus eqn. (9) indicates that stabilization can be attained by a reduced coupling between target and substrate or, in other words, by decoupling. In this respect empirical measures have been introduced into the design of reactive sputtering arrangements.

(i) Reduced ratio of the effective target surface to the condensation surface.

(ii) Enhanced target-to-substrate distance (TSD).

(iii) Arrangement of additional getter surfaces between target and substrate, *e.g.* plasma shields.

(iv) Partial separation of substrate space and target space by means of a shutter with a slit aperture which reduces the metal vapour stream to the substrate, and admission of the reactive gas near the substrate ("gas separation") [3, 5].

The fundamental action of these measures is expressed by the stability criterion of eqn. (9). These measures for decoupling are used for stabilization in most of the industrially employed reactive sputter techniques, either individually or in combination with a high pump speed.

### *2.5. Stability, degree of coupling and film properties obtained*

The original objective in reactive coating process engineering is to deposit films with dedicated properties efficiently and at a high rate, rather than the creation of a stable discharge over a wide working range. The methods for process stabilization derived from the stability criterion make demands on the regime which are sometimes contradictory to this basic objective. This shall be critically discussed in more detail.

(i) Stabilization through installation of high-capacity pumps results in excessively high costs with respect to the vacuum system to be used in arrangements with large target and substrate surfaces. For stable sputtering of titanium in an Ar–$O_2$ mixture by means of a plasmatron of length about 2.2 m, for example, the active pump speed must be of the order of 4000 l $s^{-1}$.

(ii) Self-stabilization is restricted to some combinations of target material and reactive gas only, *e.g.* CrSi–$O_2$ or Al–$N_2$.

(iii) All measures introduced for stabilization through decoupling of target and substrate processes are accompanied by a reduced condensation rate on the substrate. For certain applications, this allows us to exert a positive influence on the film structure, *e.g.* on resistive films. In most cases, however, such a reduced rate is a serious economical drawback. This refers particularly to large-area coating.

(iv) In addition, decoupling results in a reduced plasma density near the substrate which, in turn, considerably reduces the capabilities for plasma-enhanced deposition.

A great many applications of reactive direct current magnetron sputtering, especially for large-area coating and plasma-enhanced deposition, call for process engineering facilities that go far beyond the framework of practicable stabilizing methods based on the said stability criterion.

In Section 3, an example shall be given to demonstrate that pronounced differences in the degree of coupling may result in films which exhibit distinctly varying features.

## 3. Properties of Cr–Si–O films obtained at different degrees of target–substrate coupling

Reactive direct current magnetron sputtering has been used to deposit Cr–Si–O layers with different degrees of oxidation at comparable stationary rates. Two coating arrangements were employed which differ with respect to the degree of TSC [7]. The main features of both arrangements are compiled in Table 1. In Form A the coupling is so poor that a stable reactive discharge can be obtained through decoupling even at a low pumping capacity of $S_P < 60\ l\ s^{-1}$. In Form B, self-stabilization is employed to obtain a stable discharge.

TABLE 1

Fundamental features of sputter arrangements used for CrSi–$O_2$ deposition with different degrees of TSC

| | *Form A* *set-up with poor coupling* | *Form B* *set-up with strong coupling* |
|---|---|---|
| Target diameter (mm) | 160 | 160 |
| Target-substrate distance (mm) | 110 | 55 |
| Rate reduction by slit aperture (%) | 80 | — |
| Effective anode | annular around the target | annular outside the vapor stream close to the substrate |
| Total pressure (Pa) | 0.4 | 0.4 |
| Discharge power (W) | 800 | 600 - 1000 |
| Pump speed $S_P$ ($l\ s^{-1}$) | 55 | 55 |
| Mean steady-state rate (nm $min^{-1}$) | 350 | 320 - 360 |
| Carrier speed (rev $min^{-1}$) | 6 | 0.6 |
| Sublayer thickness per revolution (nm) | $\approx 1$ | $\approx 100$ |
| Ratio of coating time to exposure time in the residual gas | 1:75 | 1:7.5 |
| Stabilizing method | decoupling | self-stabilization |
| Floating substrate potential (V) | −3 | −30 to −220 |

A substantial influence of the various TSCs on reactive film deposition is exerted by the floating potential which, in Form B, is greater by more than one order of magnitude. This can be attributed to the very high plasma density close to the substrate obtained at a very pronounced TSC.

From the characteristic of the sheet resistivity, plotted as a function of the oxygen concentration (measured by Rutherford backscattering), it can be seen that, at the same oxygen content, samples produced with a strong coupling exhibit a higher sheet resistance compared with those obtained in the decoupled state (Fig. 3).

X-ray photoelectron spectroscopy was used to investigate the silicon 2p and chromium 2p peaks in order to obtain information about the bonding states. This shall be demonstrated by the less complicated silicon spectra. Figure 4 shows a superposition of the silicon peaks of three sample layers with different oxygen concentrations. Here, the peak assigned to elemental silicon at 99 eV shows the highest share of elemental silicon to be in sample A3 (produced in the decoupled state), although its oxygen content of 46.8

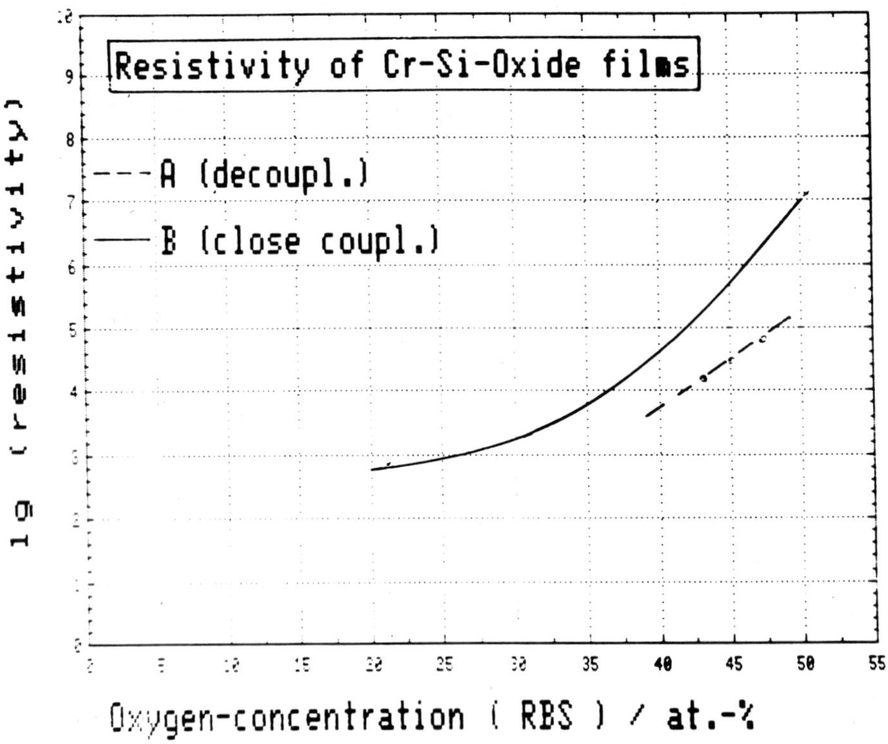


Fig. 3. Resistivity *vs.* oxygen concentration of CrSi–$O_2$ films with (A) decoupled and (B) closely coupled deposition.

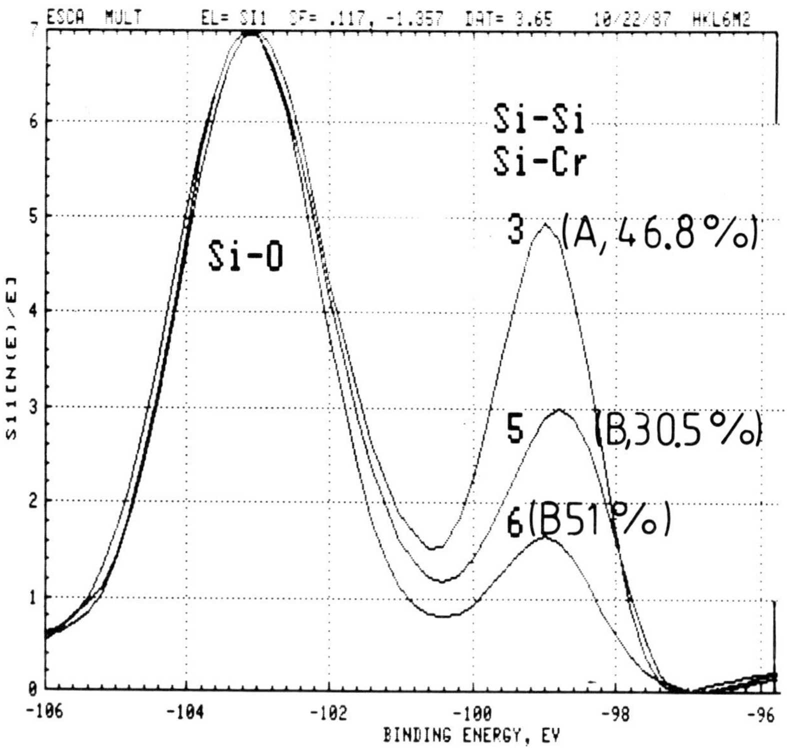


Fig. 4. Peak height of 99 eV silicon element peak with normalized $SiO_2$ peak *vs.* oxygen concentration of CrSi–$O_2$ films with (A) decoupled and (B) closely coupled deposition.

at.% is much higher than in the obviously more heavily oxidized sample B5 (produced with a strong coupling) with an oxygen content of 30.5 at.%. Above all, the more pronounced oxidation of the silicon obtained films produced at a strong coupling becomes manifest in the higher intensity ratio of the $SiO_2$ peak to the elemental silicon peak (Fig. 5). The chromium spectra exhibit a similar behaviour.

The results obtained can be summarized as follows. Compared with the films deposited in the Form A arrangement in a decoupled mode, and referred to the same oxygen content, the CrSi oxide films produced in the Form B arrangement at a close coupling between target and substrate reactions are characterized by (i) a higher resistivity, (ii) a higher share of chromium oxides of a high oxidation number ($Cr^{3+}$, $Cr^{6+}$) and (iii) a higher share in oxidized silicon ($SiO_2$).

All in all it can be concluded that Form B, with its strong TSC, gives a higher reactivity. Therefore, it will be the system of choice for reactive high-rate sputtering, *i.e.* for an efficient chemical conversion of the highest possible amount of target material in the compound layer aimed at.

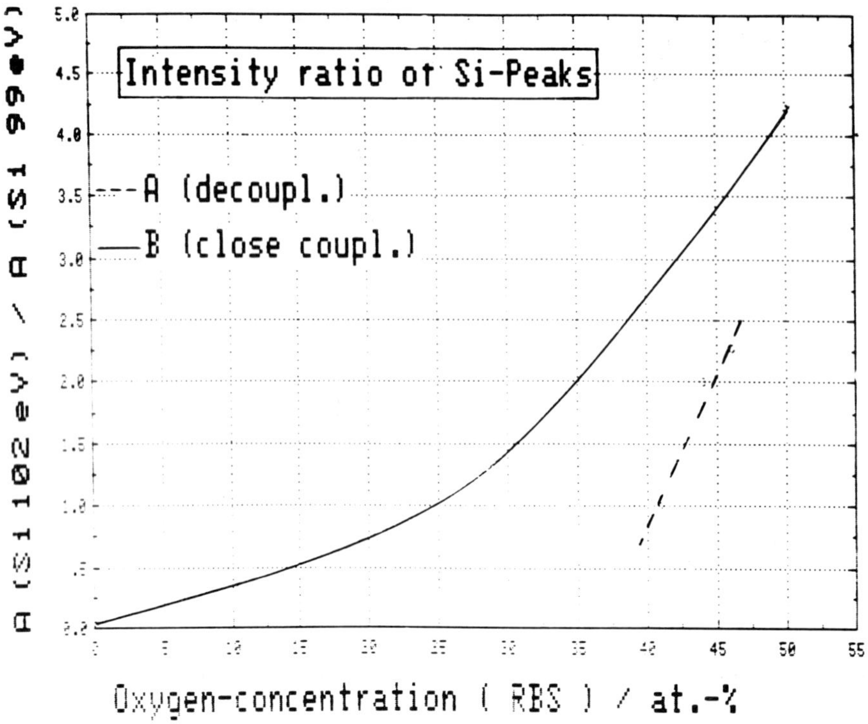


Fig. 5. Area ratio of the 102 eV $SiO_2$ peak to the 99 eV silicon element peak *vs.* oxygen concentration of CrSi–$O_2$ films with (A) decoupled and (B) closely coupled deposition.

## 4. Stabilization through active control with non-fulfilled stability criterion

The accomplishment of the basic objective to carry out reactive coating with maximum throughput and a minimum of installed pumping speed as well as with dedicated plasma action on the film growth is, as a rule, tied in with a non-observance of the stability criterion (5) in a part of the reactive working range. To obtain the process stability required it is therefore necessary to use an active control loop which operates as follows. A suitable measurand is used to measure the discharge state and to compare it with a given setpoint. With the aid of a control element (*e.g.* a piezo valve) the reactive gas intake $f_{R,\,tot}$ is adjusted so that the actual value is equal to the setpoint of the measurand. The emission intensity of a characteristic spectral line of the target material is employed as a suitable measurand for the discharge state. With industrial plants this line intensity $\varphi$ is a unique function of the metal sputter rate [8].

On the target, the reactive gas causes a drop in the sputter rate as mentioned already in Section 2.1. The partial pressure $p_R$ of the reactive gas (measured close to the substrate surface) determines the chemical conversion of the condensing metal vapour stream and affects the metal sputter rate on the target to an extent that is the greater, the closer the coupling becomes between the target and substrate processes. Because of this TSC the measured line intensity $\varphi$ does not only cover the metal vapour stream, but indirectly also covers the partial pressure of the reactive gas. Thus a close coupling is a substantial prerequisite to stabilize the discharge via PEM control within the critical region in which the stability criterion is not satisfied, and to control the chemical conversion in the deposit. On the one hand, close coupling is the reason for the non-fulfilment of the stability criterion and, on the other hand, it is a precondition for the applicability of active PEM stabilization.

The schematic layout of a stabilizing arrangement with PEM control is given elsewhere [3, 8]. The behaviour of a PEM stabilized reactive direct current discharge for sputtering titanium in an Ar–$O_2$ mixture is illustrated in Fig. 6. A system with a 610 mm × 160 mm titanium target, a TSD of 50 mm and a plasma shield with an aperture of the target size was used. The reactive gas flows $f_{R,\,tot}$, $f_{R,\,Dis}$ and $f_{R,\,P}$ *vs.* the partial pressure of the reactive gas are plotted in Fig. 6 in order to check the stability in the stoichiometric working point. The arrow indicates the working point for the stoichiometric deposition of $TiO_2$ films. The pumping speed of the vacuum pump installed amounts to $S_P = 100\ l\ s^{-1}$. At the working point the pumping rate of the discharge is $S_{Dis} = 550\ l\ s^{-1}$ and the gradient of the reactive gas flow consumed by

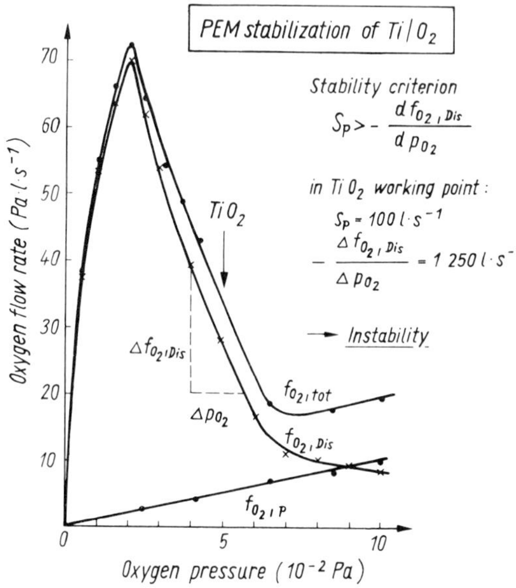


Fig. 6. Total oxygen flow rate $f_{O_2,\,tot}$, oxygen flow to discharge $f_{O_2,\,Dis}$ and to pump $f_{O_2,\,P}$ *vs.* oxygen pressure in PEM stabilized sputtering titanium (target 610 mm × 160 mm, discharge power 5 kW, argon pressure 0.4 Pa, target–substrate distance 50 mm, pump speed $S_P = 100\ l\ s^{-1}$).

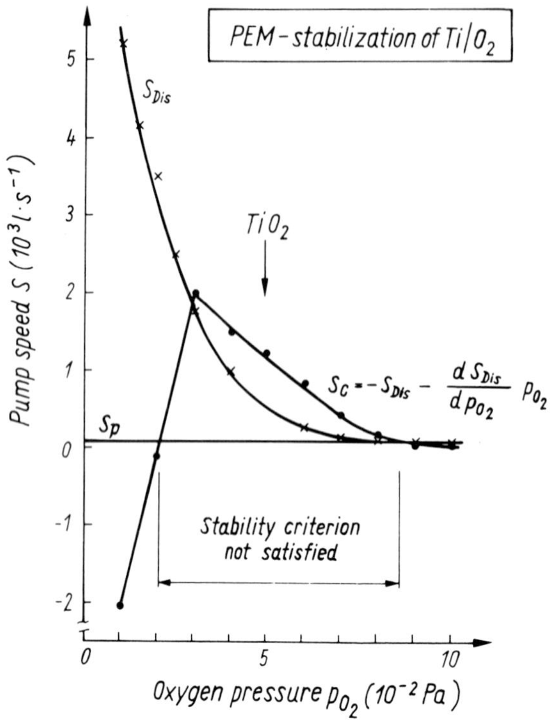


Fig. 7. Pump speed of vacuum pump $S_P$, of the discharge $S_{Dis}$ and characteristic function $S_C$ *vs.* oxygen partial pressure in PEM stabilized sputtering titanium (target 610 mm × 160 mm, discharge power 5 kW, argon pressure 0.4 Pa, target–substrate distance 50 mm).

the discharge is $df_{R,\,Dis}/dp_R = -1250\ l\ s^{-1}$ so that the stability criterion $S_P > -df_{R,\,Dis}/dp_R$ is not satisfied. In other words, the pump speed installed ($100\ l\ s^{-1}$) is too low for stabilization. To attain a stable discharge at this very point by means of the pumps installed, their capacity should be greater than $1250\ l\ s^{-1}$.

If, however, the stability criterion has to be satisfied at a given pumping speed $S_P$, the degree of coupling of the system must be reduced in accordance with eqn. (7). To this end, the pumping capacity $S_{Dis}$ has been plotted in Fig. 7 *vs.* the partial pressure $p_R$ of the reactive gas. In the stoichiometric working point the degree of coupling, $-dS_{Dis}/dp_R$, amounts to $35\,000\ l\ s^{-1}$ Pa. However, it follows from eqn. (7) that coupling has to be reduced to $-dS_{Dis}/dp_R < 12\,000\ l\ s^{-1}$ Pa for obtaining stability by matching the arrangement.

The active PEM control, however, allows all working points, inclusive of the point for stoichiometric $TiO_2$, to be stabilized in a given arrangement at a high degree of coupling.

## 5. Application of active PEM stabilization to the reactive deposition of TiN films

The production of TiN hard layers meeting the demands for wear-resistive tool coatings requires the use of reactive plasma-assisted deposition. On the one hand, the reactive discharge has to be stabilized to attain a defined film stoichiometry and, on the other hand, the film growth must be influenced to obtain a fine crystalline structure by means of a dedicated plasma action. This plasma action is absolutely necessary and determines the selection of adequate process stabilizing methods. A high plasma density at the substrate calls for a high degree of TSC. Therefore we consider the active PEM control to be the method of choice.

The behaviour of the TiN discharge was investigated in an arrangement for tool coating with the aid of a large-area sputtering source and a rotary cage (target 1200 mm × 160 mm, rotary cage diameter 700 mm, TSD = 40 mm). The process chamber was evacuated by mechanical pumps. The reactive gas was distributed by a manifold close to the sputter source. The titanium line intensity (501 nm) was measured in the immediate vicinity of the target. A piezo valve controlled the reactive gas flow $f_{N_2,tot}$ by means of the PEM controller.

To investigate the behaviour of the Ti–N system the set-point of the line intensity was continuously reduced from $\varphi = 100$ in pure argon to $\varphi = 20$ (poisoned target). With this procedure it is no problem to also stabilize those working points in the region where the stability criterion has not been satisfied. The characteristics $f_{N_2,tot}(p_{N_2})$ and $f_{N_2,P}(p_{N_2})$ were measured and the curves $f_{N_2,Dis}(p_N)$ were derived. This was performed for different values of both the discharge power $P_{Dis}$ and the TSC.

For an arrangement with strong TSC it is evident from Fig. 8 that, as expected, the critical reactive gas flow $f_{M\text{-}R}$ increases with the power, *i.e.* $f_{M\text{-}R}(28\text{ kW}) = 1.9 f_{M\text{-}R}(14\text{ kW})$. Moreover, it shifts towards higher values of the nitrogen pressure. In accordance with the enhanced sputtering rate of metals, the stoichiometric working point concurrently shifts towards higher values of $f_{N_2,tot}$ and $p_{N_2,Dis}$. An increase in discharge power simultaneously means an increase in target power density. In this way, the partial pressure of the reactive gas has a lower influence on the sputter rate, which becomes manifest in a reduced coupling parameter $-dS_{Dis}/dp_R$ (from 250 000 l s$^{-1}$ Pa for 14 kW to 110 000 l s$^{-1}$ Pa at 28 kW in the working point for TiN). The relative position of the stoichiometric working points on the curve $f_{N_2,tot}(p_{N_2})$, however, is maintained in the region being unstable in compliance with the criterion (5).

To reduce the degree of coupling by a simple alteration of the arrangement, a diaphragm with a slit aperture of 50 mm was arranged between the target and substrate in the vicinity of the latter. For a sustained discharge with and without the diaphragm the function $S_C(p_{N_2}) = -S_{Dis} - (dS_{Dis}/dp_{N_2})\, p_{N_2}$ has been plotted over $p_{N_2}$ in Fig. 9. From the stability criterion (7) it follows that the discharge is stable in the pressure ranges where the pump

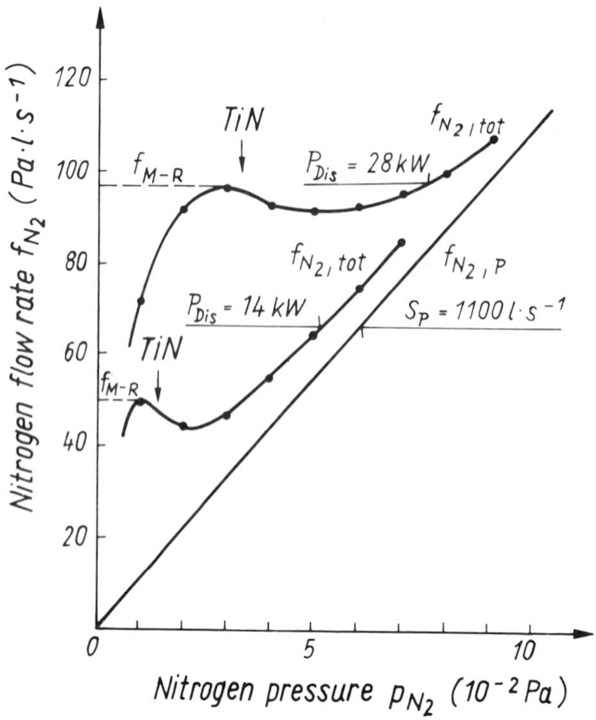


Fig. 8. Nitrogen flow rate $f_{N_2}$ *vs.* nitrogen partial pressure $p_{N_2}$ with variation of discharge power $P_{Dis}$.

speed of the vacuum pump is greater than $S_C(p_{N_2})$. The characteristic function $S_C(p_{N_2})$ contains the coupling parameter $-dS_{Dis}/dp_{N_2}$ which determines the stability in the range of progressive target poisoning. The stabilizing effect of a reduced TSC is also evident from the reactive gas flow rates plotted as a function of the $N_2$ partial pressure (Fig. 8). Here the critical flow $f_{M-R}$ is shifted towards higher values of $f_{N_2,\,tot}$ and $p_{N_2}$.

With strong TSC, the stoichiometric working point lies in the unstable region, and with poor coupling in the stable working range. For the coating system considered here, a mere reduction in the TSC therefore suffices to adjust a stable working range in which TiN can be deposited in a stoichiometric composition. Films produced at a poor TSC are suitable for use as barrier layers in electronics or as decorative layers. However, such a poor TSC does not allow for the plasma action required to obtain the structure needed for hard layers. Determining the pumping speed needed for stabilizing the discharge we obtained for the strong coupled arrangement $S_P > 1900\ l\ s^{-1}$ and for the reduced coupling a value of $S_P > 1600\ l\ s^{-1}$.

Using active PEM control at a high degree of TSC a stable discharge will be obtained independent of the installed pumping speed down to a value

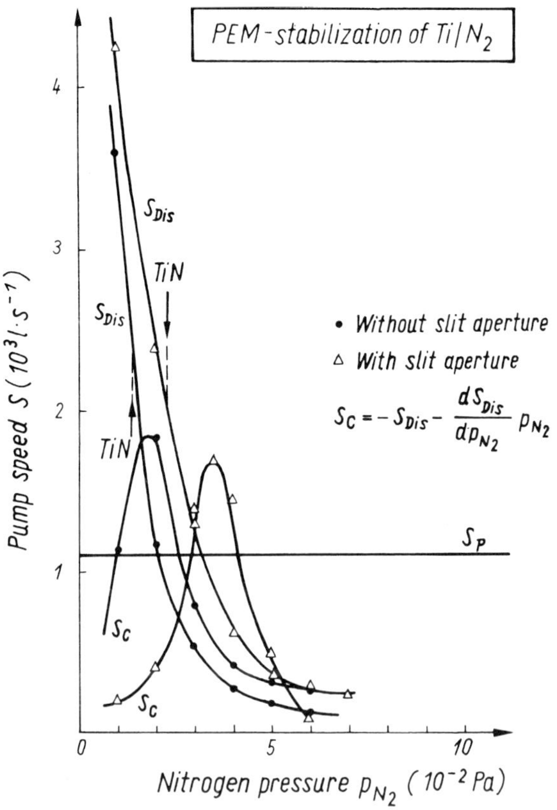


Fig. 9. Pump speed of vacuum pump $S_P$, of the discharge $S_{Dis}$ and characteristic function $S_C$ with and without slit aperture *vs.* nitrogen partial pressure in PEM-stabilized sputtering of Ti–$N_2$ (target 1200 mm × 160 mm, discharge power 14 kW, argon pressure 1.6 Pa, target–substrate distance 40 mm).

of $S_P = 300\ l\ s^{-1}$. At the same time this creates the prerequisites for ion-assisted coating. Referred to the length of the erosion track a power density of 177 W cm$^{-1}$ and a substrate bias of −200 V give an ion current density of 9 mA cm$^{-2}$ on the substrate. This high value is a measure of the high plasma density on the latter. Layers deposited through reactive high-rate sputtering under the conditions described meet all applicative demands on wear-resistive tool coating.


## Acknowledgments

The authors are gratefully indebted to Dr. R. Grötzschel for his assistance in RBS measurements and to Dr. K. Haupold for his XPS investigations carried out with special care and outstanding knowledge.

## References


1 S. Schiller, U. Heisig, G. Beister, K. Steinfelder, J. Strümpfel, Chr. Korndörfer and W. Sieber, *Thin Solid Films, 118* (1984) 255.
2 J. Affinito and R. R. Parsons, *J. Vac. Sci. Technol. A, 2* (1984) 1275.
3 S. Schiller, U. Heisig, Chr. Korndörfer, G. Beister, J. Reschke, K. Steinfelder and J. Strümpfel, *Surf. Coat. Technol., 33* (1987) 405.
4 A. G. Spencer, R. P. Howson and R. W. Lewin, *Proc. 6th Intern. Conf. IPAT*, Brighton, 1987, p. 1.
5 H. Frey and G. Kienel, *Dünnschichttechnologie*, VDI Verlag, Düsseldorf, 1987, p. 128.
6 S. Maniv, C. J. Miner and W. D. Westwood, *J. Vac. Sci. Technol. A, 1* (1983) p. 1370.
7 S. Schiller, U. Heisig, Chr. Korndörfer and K. Haupold, *Proc. 7th Intern. Conf. IPAT*, Geneva, 1989.
8 S. Schiller, U. Heisig, K. Steinfelder, J. Strümpfel, A. Friedrich and R. Fricke, *Proc. 6th Intern. Conf. IPAT*, Brighton, 1987, p. 23.
9 T. Larsson, H.-O. Blom, C. Nender and S. Berg, *J. Vac. Sci. Technol. A, 6* (1988) 1832.
10 J. A. Thornton, *Thin Solid Films, 80* (1981) 1.

*Surface and Coatings Technology*, *39/40* (1989) 565 - 572 

# TECHNICAL NOTE: A MAGNETRON SPUTTER ION-PLATING SYSTEM

D. G. TEER

*D. G. Teer Coating Services Limited, 285 Hartlebury Trading Estate, Hartlebury, Worcestershire DY10 4JB (U.K.)*

(Received March 12, 1989)

## Summary

A new type of magnetron sputter ion-plating system has now been fully developed. High rate deposition at low substrate temperatures is now possible. High ion bombardment of the samples during deposition is achieved at low bias voltages giving coatings with good adhesion and structure. Independent tests on coated drills demonstrate the effectiveness of the coatings in reducing wear under accelerated drilling conditions.

## 1. Introduction

A new type of system for the deposition of hard coatings on cutting tools has previously been reported [1]. At that time the development of the system was incomplete and a number of projected modifications and improvements were given. Those modifications and others have now been incorporated, and are described below, together with the general characteristics of the system, and typical deposition parameters. The results of various tests on the coatings are also included.

## 2. Summary of the equipment as described previously [1]

The chamber is a relatively long, narrow cylinder. The dimensions are 36 in × 12 in diameter. Samples are mounted from an insulated, rotating holder running along the axis of the chamber. Three magnetron sources of target size 30 in × 4 in are mounted symmetrically in the chamber walls. The pumping system is a $700\,\mathrm{l\,s^{-1}}$ oil diffusion pump backed by an $18\,\mathrm{m^3\,h^{-1}}$ rotary pump. The small, almost totally usable chamber volume, fast pumping and large source area result in fast cycle times. The cylindrical symmetry results in near uniform deposition rates on all samples and all sample

0257-8972/89/$3.50 

surfaces, and also in uniform plasma conditions. This means that all samples are in the same deposition and plasma environment resulting in coatings of uniform thickness, stoichiometry and quality on all samples and sample surfaces.

As expected, excellent coatings were obtained with the equipment as originally constructed. However, the deposition rate was limited by the power that could be applied to the magnetrons. The magnetrons were indirectly cooled and if more than 2.5 kW was applied to each magnetron, then the sample temperature could exceed 500 °C with the risk of softening of high speed steels. This was clearly unacceptable and the magnetron electrode design was changed to incorporate direct cooling of the target. This modification allowed much higher powers to be applied to the magnetrons without the risk of softening the samples. The changes in the sources and the consequences of these changes are now described in more detail.

## 3. Modified sources

The sources fitted originally were indirectly cooled magnetrons, in which the target was clamped against the copper front plate of the magnetron electrode which was in direct contact with the water cooling system. This arrangement initially allowed a maximum power to the electrode of about 2.5 kW. However, during operation the target could distort slightly and this reduced the efficiency of the thermal contact and therefore the maximum allowable power was also reduced. Although the use of soft backing inserts of copper and graphite gave some improvement it was decided to modify the electrode so that the targets were in direct contact with the cooling water. This arrangement necessitates breaking a water seal when changing targets and is less convenient than with the indirectly cooled system. The use of spare electrodes already fitted with new targets circumvents this inconvenience.

The directly cooled electrodes have been used at 10 kW power and this is limited only by the availability of more powerful supplies. Tests on shorter electrodes of similar design indicate that 15 kW operation should be possible.

The higher power to the magnetrons results in higher deposition rates, and rates of more than 1 $\mu$m in 10 min are routinely achieved under production conditions. This means that total cycle times of about 1 h are possible. Under these conditions 0.25 in drills attain a temperature of about 450 °C. The main source of heating in the original system was radiant heat from the targets. This has been considerably reduced and now other forms of heating such as ion bombardment come into consideration.

It has been found that good TiN coatings can be deposited at temperatures under 250 °C simply by reducing the power to the magnetrons. In this type of system the ion current density to the samples depends on several factors but the most important factor is the power applied to the magnetrons. This is now discussed in the next section.

## 4. Ion current density to the samples

In ion-plating systems the ion bombardment of the samples during deposition is all important in obtaining desirable coating structures. In the original ion-plating systems [2, 3] the source of the ions was the abnormal glow discharge created by applying a high negative voltage to the samples in an atmosphere of argon gas. Ionization efficiency is very low [4] in this type of system and the main source of ions in all successful modern systems used for hard coatings is no longer a simple abnormal glow discharge. Additional positive electrodes may be used to attract electrons either from an electron gun source used for evaporation [5] or from a hot filament [6]. In many successful systems the main source of ions is the vapour source itself, a low voltage electron beam gum [7, 8], an arc [9] or the overlapping plasmas from two facing magnetrons [10].

In the present system it is easy to demonstrate that the level of ion current density to the samples is strongly influenced by the power applied to the magnetron electrodes. Thus, in order to reduce sample heating during deposition it is necessary only to reduce the power to the magnetrons, which simultaneously reduces the ion bombardment and any radiant heating. Good TiN coatings have been deposited at temperatures below 250 °C at reduced magnetron powers.

Freller and Lorenz [11] have shown that it is relatively easy to produce dense structures on sample surfaces where the coating flux arrives at normal incidence, while high ion current densities are required to prevent open, columnar structures on oblique surfaces. In order to increase the ion current density in the present system the magnetic fields of the magnetron were modified to produce unbalanced magnetrons [12]. As shown in Fig. 1, the unbalanced magnetrons gave significantly higher ion currents than the

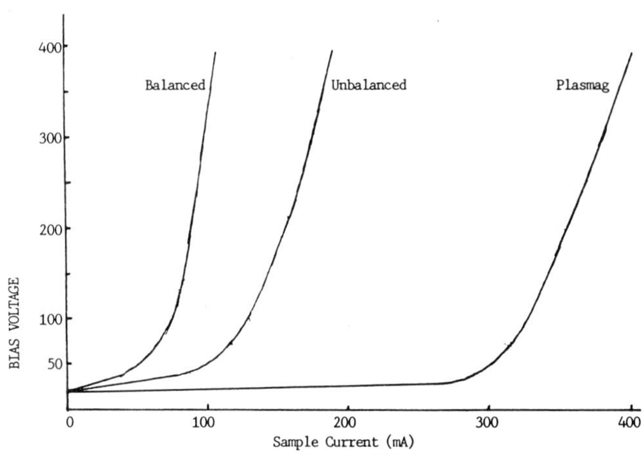


Fig. 1. Bias voltage–current relationship for balanced magnetrons, unbalanced magnetrons and PLASMAGS all operating at 4 A.

conventional magnetrons, and further modifications have been carried out to increase the ion current even more. The ion current–bias voltage relationships are shown for conventional magnetrons, unbalanced magnetrons and the modified unbalanced magnetron system that we have called the "PLASMAG" system. It can be seen that high ion currents to the samples are produced even at bias voltages as low as 50 V. This low bias voltage helps in obtaining good coatings at low temperatures. The intense plasma and high ion current to the samples also makes the control of the process, especially the nitrogen gas flow rate, much less critical. However, the introduction of the directly cooled source did change the characteristics of the process, and this is now discussed.

## 5. Control of coating stoichiometry

Magnetron sputtering has many advantages in the deposition of wear resistant coatings such as TiN by reactive processes, but it has one major disadvantage. This is the well known target "poisoning" effect, which necessitates a precise balance between metal sputtering rate and reactive gas flow in order to maintain high deposition rates and stoichiometric coatings.

When using the indirectly cooled targets it was found that a very simple control mechanism could be employed. When the target starts to poison, TiN is formed on the surface. This has a higher electron emission coefficient than titanium causing the effective impedance of the discharge to be reduced. If the magnetron power is applied as constant current, then the power is reduced due to the poisoning, the sputtering rate is also reduced and the poisoning is increased. However, if constant voltage power supplies are used, the power is increased due to poisoning, and the metal sputtering rate is increased, thus reducing the poisoning. It was found that this effect could be employed to control stoichiometry of the coating when using the indirectly cooled targets. Once the correct voltage was set, the nitrogen flow rate could be varied and the current to the magnetrons would change to compensate and to maintain high rate metallic deposition and a stoichiometric coating. A similar system has been described for metal oxide coatings [13].

When directly cooled magnetrons are used it is found that the use of constant voltage power supplies are not sufficient to provide control of target poisoning and coating stoichiometry. This effect has not been studied in detail but it is thought that the lower temperatures of the target surface mean that the reaction is slower, and that the nitrogen pressure causing an increase in discharge impedance has a greater effect than the target poisoning.

A feed-back system, in which the intensity of the characteristic titanium line in the discharge is maintained at a constant level by control of the nitrogen flow, is now used and has been found to be fully reliable.

## 6. Deposition sequence and parameters

The equipment is now as described previously [1] but with the new directly cooled "PLASMAG" sources producing high deposition rates due to higher applied power and giving high ion currents to samples biased at relatively low voltages. A spectrometer based feed back system is used to control the nitrogen flow rate and maintain stoichiometric coatings. A typical deposition sequence used to coat 0.25 in drills is as follows

| | |
|---|---|
| Pump down to $5 \times 10^{-5}$ Torr. | 10 min |
| Ion clean and heat at $3 \times 10^{-3}$ Torr argon pressure and a bias voltage of 1000 V. | 15 min |
| Coat at a total gas pressure of $3 \times 10^{-3}$ Torr. Each of the three magnetrons set to 15 A. Sample bias voltage of 50 V and ion current density of about 2.0 mA cm$^{-2}$. Nitrogen flow rate is automatically set to maintain stoichiometry. | 30 min (for a coating thickness of 3 $\mu$m) |
| Cool down | 10 min |

## 7. Results

Figure 2 shows a typical structure for a TiN coating on a 0.25 in twist drill obtained under the above conditions. The temperature of the drill was below 400 °C during coating. The structure can be seen to be fine columnar. Scratch adhesion tests on cleaned coatings give $L_c$ values in excess of 50 N. The coating has a strong (111) preferred orientation. Figure 3 shows a titanium TiN nitride coating deposited at half the magnetron power and at a

Fig. 2. Fracture cross-section of TiN coating, deposited on a 0.25 in drill under standard conditions, *i.e.* 50 V bias and 15 A applied to magnetrons.

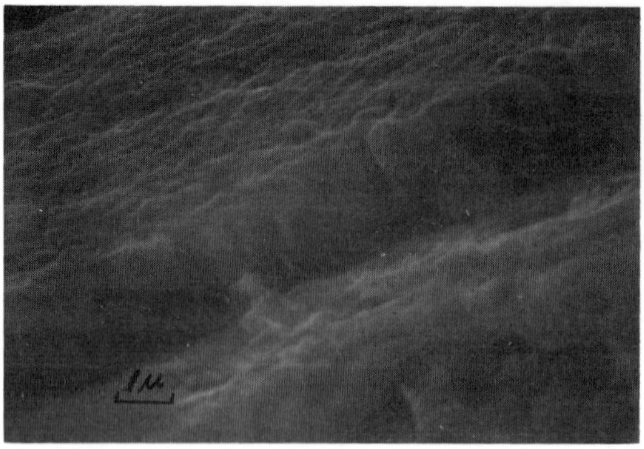


Fig. 3. Fracture cross-section of TiN coating deposited on a 0.25 in drill, with 50 V bias and 7.5 A applied to magnetrons.

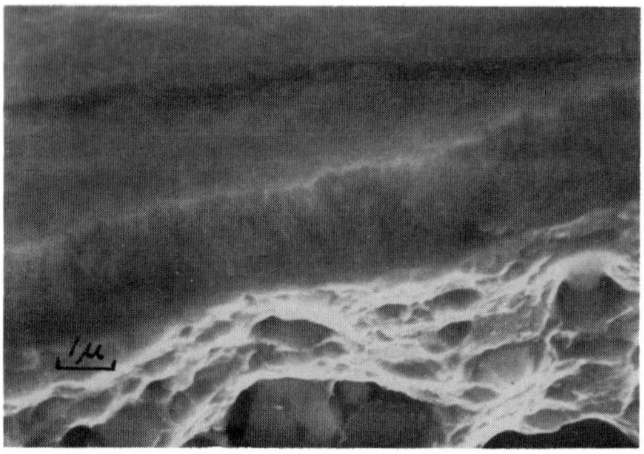


Fig. 4. Fracture cross-section of TiN titanium nitride coating deposited on a 0.25 in drill, with 100 V bias and 15 A applied to magnetrons.

temperature of 250 °C. Figure 4 shows the effect of increasing the bias voltage to 100 V. The change in structure is apparent.

*7.1. Drilling tests*

A batch of 10 drills were tested in the S.K.F. Dormer standard drill test. The drills were coated under the typical conditions given above but with a shorter coating time to give an average thickness of 2.2 μm. The test conditions were

| | |
|---|---|
| Speed | 2250 rev $min^{-1}$ |
| Penetration rate | 657 mm $min^{-1}$ |
| Depth of hole | 3 × drill diameter, breaking through |
| Test material | Tyre steel |

Soluble coolant applied in a flood.
One drill broke under test. The other nine each produced more than 100 holes.

### 7.2. *Alloy coatings*

Samples were rotated round the axis of the cylindrical chamber during coating and each sample had exactly the same geometrical relationship to each of the three magnetrons, therefore if all magnetrons were operating with the same target materials at the same power, then each sample received an equivalent coating flux from each of the magnetrons.

By varying the power to the different magnetrons and by using different target materials alloy coatings can be deposited. The alloy composition is controlled by applying the appropriate power to the magnetrons. Thus alloy nitride and carbide coatings can be deposited using pure metal targets, rather than alloy or "inserted" targets [14]. This allows a range of alloy compositions to be deposited rapidly and conveniently in order to study the optimum alloy composition for a particular application. Also, it is easy to produce graded coatings by varying the magnetron powers during the coating. It may be desirable to start with a pure TiN coating for optimum adhesion but to increase the aluminium content in the outer layers for improved wear resistance at high temperature. The alloy coatings produced in this equipment are described in more detail elsewhere [15]. It is probable that the facility to produce graded coatings will find application in other fields, *e.g.* protective coatings on turbine blades.

### 7.3. *Chamber size*

The success of this method depends on using a relatively small diameter chamber. However, the length can be varied while still maintaining the same deposition and plasma conditions. Versions 300 and 450 mm long have been built and make excellent units for the development of new coatings and coating methods. They have the advantage of simulating exactly the conditions in the full size production unit and can be used to coat practical samples. For instance, a gear cutting hob of diameter 8 in can be coated in the 300 mm unit. The basic design is ideally suited to coating long cylindrical objects such as broaches and a stretched version is now planned.

## 8. Conclusion

The objective of constructing a relatively small inexpensive coater suitable for routine deposition of TiN on small tools has been achieved. High rate deposition at low temperatures is now possible. The geometrical arrangement gives several advantages, including uniformity of coating thickness, composition and quality over all samples and all sample surfaces.

Alloy coatings are easily achieved as are graded coatings. This design concept should find many applications including the coating of long broaches with TiN and coating of turbine blades with MCrAlY.

## Acknowledgments

The author is indebted to Mr. P. D. Boeck and Mr. P. Logan for their help in the construction, testing and development of the equipment and the investigations of coating parameters. He is grateful to Dr. R. Goodwin of S.K.F. Dormer for his cooperation and for supplying the test results on coated drills, and to the U.K. Department of Trade & Industry for a Support for Innovation Grant.

## References

1 D. G. Teer, *Surf. Coat. Technol., 36* (1988) 901.
2 D. M. Mattox, *J. Appl. Phys., 34* (1963) 2493.
3 T. Spalvins, J. S. Przybyszewski and D. H. Buckley, *NASA TND - 3707* (1966).
4 D. G. Teer, *J. Adhesion, 8* (1977) 289.
5 R. Bunshah and A. C. Raghuram, *J. Vac. Sci. Technol., 9* (1972) 1385.
6 A. Matthews and D. G. Teer, *Thin Solid Films, 80* (1981) 41.
7 J. R. Morley and H. R. Smith, *J. Vac. Sci. Technol., 9* (1972) 1377.
8 E. Moll and H. Daxinger, *U.S. Patent 4,197,175,* (1980).
9 P. Hatto and D. G. Teer, *Vacuum, 36* (1986) 67.
10 W. D. Münz, D. Hoffmann and K. Hartig, *Thin Solid Films, 96* (1982) 79.
11 H. Freller and P. Lorenz, *J. Vac. Sci. Technol., 14* (1986) 2691.
12 B. Window and N. Savvides, *J. Vac. Sci. Technol., 14* (1986) 196.
13 S. Schiller, V. Heisig, G. Beister, K. Steinfelder, J. Strumpfel, Chr. Korndorfer and W. Sieber, *Thin Solid Films, 118* (1984) 255.
14 O. Knotek, T. Leyendecker and F. Jungblut, *IPAT Proc.* (1987) 190.
15 D. G. Teer, *Presented at 16th ICMC, San Diego, CA, April 17–21, 1989.*

*Surface and Coatings Technology, 39/40* (1989) 573 - 585 

# ULTRAHIGH CURRENT DENSITY NITROGEN IMPLANTATION OF 440C STEEL FOR IMPROVED ROLLING CONTACT FATIGUE LIFETIMES

F. M. KUSTAS and M. S. MISRA

*Martin Marietta Astronautics Group, Denver, CO 80201 (U.S.A.)*

P. J. WILBUR

*Colorado State University, Ft. Collins, CO 80523 (U.S.A.)*

(Received March 8, 1989)

## Summary

Very high current density $N_2^+$ implantation has been used to improve the rolling contact fatigue (RCF) resistance of 440C steel. Equivalent tempering relations were used to predict allowable thermal exposures that 440C steel could withstand during implantation without sustaining bulk hardness loss which would reduce fatigue properties. A heat transfer model was formulated to calculate the surface temperature attained as a function of implantation parameters. Analytical results indicated that a dual end grip configuration was needed to process RCF rods to prevent excessive thermal exposures. Good correlations were observed between predicted and measured temperatures obtained during implantation processing. Improvements in B10 RCF life (at 5.42 GPa (786 ksi) Hertz stress) by a factor of about 6 were observed for $N_2^+$ implantation of $10^{17}$ ions cm$^{-2}$ at 40 keV using a current density of 1300 $\mu$A cm$^{-2}$. Current densities of this magnitude result in significantly reduced processing times compared to conventional implantation processing (*i.e.* 10 $\mu$A cm$^{-2}$).

## 1. Introduction

Surface modification by directed energy beams (*i.e.* ion implantation) has been successfully used to provide wear and corrosion resistant surfaces. Conventional implantation (*i.e.* using ion beam current densities of about 10 $\mu$A cm$^{-2}$) is, however, costly due to the long beam exposure times required to obtain a large dose. In addition, unless effective cooling techniques are used, tempering of thermally sensitive substrates can occur for these long exposures, resulting in degradation of properties.

A relatively new implantation technique, developed from ion thruster technology [1], is being developed to provide very high current densities (up

0257-8972/89/$3.50 

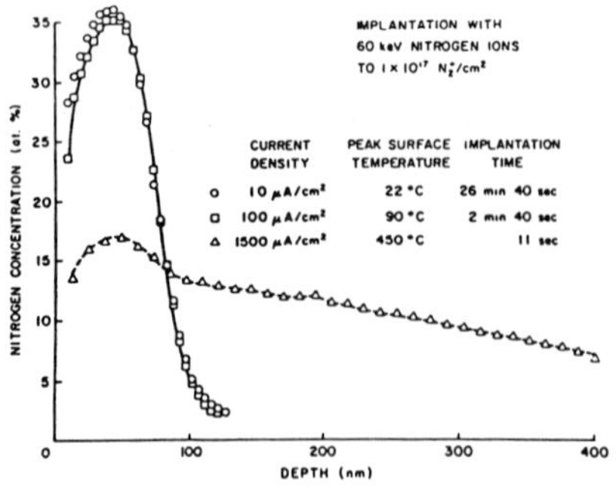


a)

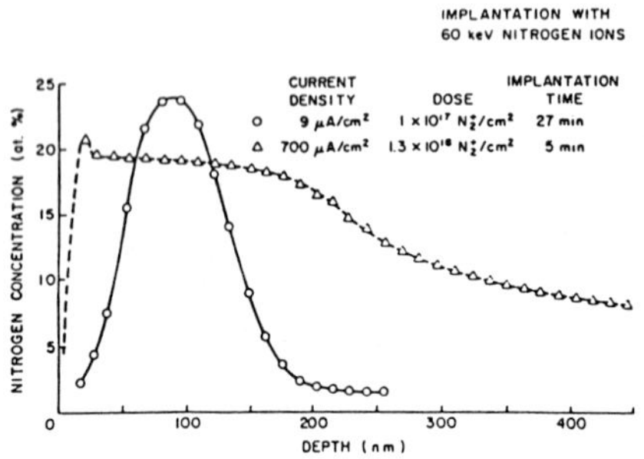


b)

Fig. 1. Enhanced penetration depths from high current density nitrogen implantation processing of (a) 304 SS and (b) 01 tool steel (from ref. 1).

to 1500 $\mu$A cm$^{-2}$). Potential benefits of this technique include: (1) significantly reduced processing time, which is directly proportional to ion beam current density, (2) deeper penetration depths (Fig. 1), speculated to result from radiation-enhanced and thermal diffusion, (3) formation of unique microstructures by controlling penetration depth (by current density), retained concentration (by integrated dose) and induced temperature (by heat sink fixturing), and (4) minimal effect on substrate properties due to short time exposures at relatively high temperatures [2].

Other investigators have observed diffusion effects from implantation of substrates held at high temperatures [3] and from post-implantation anneals [4], although the magnitudes of the profile modifications have been minor compared to those observed from high current density implantations, such as shown in Fig. 1.

In a previous investigation [5], systematic studies were performed to select implantation process parameters for 440C steel which would prevent heating into the tempered martensite embrittlement (TME) range and also prevent a hardness loss of 2 Rockwell C points which both reduce fatigue properties. Nitrogen implantation using these moderate parameters

(107 $\mu$A cm$^{-2}$ ion beam current density) resulted in reproducible improvements in rolling contact fatigue (RCF) lifetimes for 440C steel. In the present study, equivalent tempering relations were used to predict higher allowable thermal exposures (from a given current density) to prevent a hardness loss of 2 Rockwell C points. The combination of high temperatures and high ion beam current densities is speculated to account for enhanced diffusion rates [6]. Subsequent implantations were performed using ultrahigh (up to 1300 $\mu$A cm$^{-2}$) current densities (which did not excessively temper the substrate) to evaluate the effects of this technique on RCF performance.

## 2. Experimental procedure

### *2.1. Test specimens*

An annealed bar of diameter 6.35 cm (2.5 in) of air-melted 440C steel (nominal composition in weight percent: chromium 17.0; carbon 1.08; silicon, manganese 1.0 max; molybdenum 1.0; iron, balance) was rough machined into oversized solid cylindrical rods and heat treated according to the following NASA specification [7]: (1) austenitize, 1054 °C for 1 h; (2) oil quench (50 °C); (3) temper, 163 °C for 1 h; (4) air cool to ambient temperature; (5) liquid nitrogen soak for 0.5 h; (6) temper, 163 °C for 1 h. Specimen surfaces were ground to final dimensions (diameter 0.953 cm, length 8.26 cm) and then buffed to produce a surface finish of 100 nm (4 $\mu$in) root mean square (rms).

### *2.2. Ion implantation*

Implantation processing was performed in the Mechanical Engineering Department of Colorado State University, Ft. Collins, CO, using a broad beam system developed from ion thruster technology [1]. A gaseous nitrogen source emitting primarily molecular $N_2^+$ ions was accurately controlled to maintain uniform dose exposures to within $\pm 10\%$ of the desired levels. High current densities of up to 1500 $\mu$A cm$^{-2}$ can be achieved with this broad beam system, although the highest current density utilized in this study was 1300 $\mu$A cm$^{-2}$. Implantation beam time to process an RCF rod at 1300 $\mu$A cm$^{-2}$ to a total dose of $10^{17}$ $N_2^+$ cm$^{-2}$ was about 1.25 min compared with about 2.67 h for conventional implantation using a current density of 10 $\mu$A cm$^{-2}$. RCF specimens were mounted in a single water cooled heat sink fixture (Fig. 2a) for early tests, but the apparatus was later modified to a dual grip configuration (Fig. 2b) to provide more efficient heat removal. A graphite mask, which was used to provide near-normal beam incidence, provided an included angle of surface exposure of about 60 ° (Fig. 2). RCF specimens were rotated in 5 ° increments instantaneously, with intervals between rotation being 1/72 of the total time duration.

Table 1 shows a test matrix of implantation process parameters which was selected to evaluate several things: (1) the effect of increasing dose at constant current density, (2) the effect of increasing current density at constant dose and (3) investigation of an apparent critical current density for radiation enhanced diffusion effects. All implantations were performed using

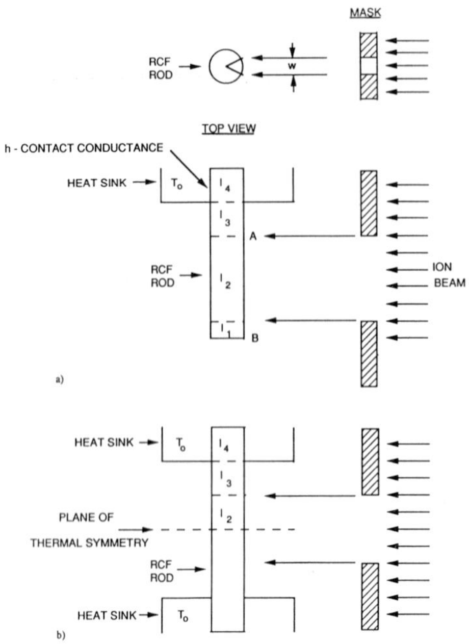


Fig. 2. Heat sink configuration used for implantations of 440C steel RCF rods: (a) single end-grip configuration, (b) dual end-grip configuration.

TABLE 1

Test matrix of implantation process parameters

| *Dose* (ions $cm^{-2}$) | *Current density* ($\mu A\ cm^{-2}$) | | | |
|---|---|---|---|---|
| | 200 | 655 | 1000 | 1300 |
| $5 \times 10^{16}$ | | $\times$ [b] | | $\times$ [b] |
| $10^{17}$ | $\times$ [c] | $\times$ [a, b] | $\times$ [b] | $\times$ [b] |

[a]Single end grip for specimen, $N_2^+$ accelerating energy 40 keV.
[b]Dual end grips for specimen, $N_2^+$ accelerating energy 40 keV.
[c]Dual end grips for specimen, $N_2^+$ accelerating energy 20 keV.

an acceleration energy of 40 keV for $N_2^+$ ions except for a single implantation at 20 keV using a current density of 200 $\mu A\ cm^{-2}$ which was performed to produce a high near-surface nitrogen concentration.

### *2.3. Rolling contact fatigue tests*

RCF tests were conducted using a standard ball–rod type rolling contact fatigue test machine which is described in detail in ref. 8. The 440C steel rod was loaded between three 1.27 cm (0.5 in) diameter hardened 52100 steel balls which had a roughened surface finish of 0.089 $\mu$m (3.5 $\mu$in) roughness average

(RA). Roughened balls were used to obtain a statistically significant number of failures within a suspension time limit of 200 h.

Parameters used in the RCF tests included: (1) 5.42 GPa (786 ksi) calculated maximum Hertz stress, 10 tests per stress level; (2) rotational velocity of 3600 rpm and (3) turbine engine oil lubrication (MIL-L-7808J) at 6 - 8 drops $min^{-1}$. Tests were conducted over the central 5 cm (2 in) zone of the specimen rod to prevent any end effects.

Termination of a test occurred when vibration, sensed by a close-proximity accelerometer, exceeded a pre-set level. The signal from the accelerometer triggered an automatic shut off device. Excessive vibration was produced by the emergence of a fatigue spall on the specimen surface.

### *2.4. Statistical analyses*

Statistical analyses were performed to evaluate the effects of nitrogen implantation on the RCF lifetime of 440C steel. Included was determination of B10 (10% of specimens predicted to fail) and B50 or mean (50% of specimens predicted to fail) Weibull fatigue lifetimes, Weibull slope and confidence levels. The latter parameter is useful in determining the significance of the difference between the nitrogen implanted specimens' and unimplanted specimens' fatigue lifetimes [9]. This confidence number is calculated taking into account three things: the slope of the Weibull line, the sample size and the B10 or mean life of the tested populations.

### *2.5. Surface analysis*

Auger electron spectroscopy (AES) was performed to determine the atomic concentration–depth profiles for selected specimens. Data were taken during argon ion sputter depth profiling and the following sensitivity factors were used to quantify the peak height data: nitrogen (379 eV), 0.23; carbon (272 eV), 0.28 (carbide); oxygen (503 eV), 0.40 [10, 11]. Sputter rates were determined from previous profilometry measurements of crater depths on polished 440C steel specimens [12]. It was found that actual sputter rates on 440C steel were 10% lower than calculated for the $Ta_2O_5$ standard material.

## 3. Results and discussion

### *3.1. Selection of implantation process parameters*

Martensitic steels, such as 440C, are susceptible to mechanical property degradation from excessive thermal exposures. In a previous study [5], it was established that a thermal exposure of 300 °C for 30 min, applied to normal double tempered 440C steel, resulted in a hardness loss of about 2 Rockwell C points, which has been found to be large enough to cause degradation in RCF lifetimes [13, 14].

Using this thermal exposure as a reference baseline, an equivalent tempering relation, such as shown below [15],

$$\frac{\tau_1}{\tau_2} = \exp - \frac{Q}{R}\left(\frac{1}{T_2} - \frac{1}{T_1}\right) \tag{1}$$

where $Q$ is the activation energy for carbon diffusion, $R$ is the universal gas constant, $T_1$ and $T_2$ are the temperatures, and $\tau_1$ and $\tau_2$ are the exposure times at temperatures $T_1$ and $T_2$ respectively, was used to predict other temperature–time combinations that would result in equivalent carbon diffusion distances (using $Q$ for carbon into $\alpha$-Fe [15]) for a treatment of 300 °C for 30 min. These calculated values, shown in Table 2, illustrate the significant effect of temperature on reducing the allowable exposure times. For example, increasing the temperature to 500 °C reduced the allowable exposure time to only 19 s. This set of temperature–time combinations therefore represents an upper limit for allowable thermal exposures to prevent hardness loss in excess of 2 Rockwell C points.

In order to predict the temperature–time history experienced by the RCF rod during implantation processing, a thermal model was developed to calculate the maximum surface temperature at the end of the RCF rod for a single end-grip configuration (Fig. 2a) or at the center of the rod for the dual-end grip configuration (Fig. 2b) [16]. The thermal model used is a simple one-dimensional, unsteady conduction model that follows the development of the numerical models cited in ref. 17. A one-dimensional model, *i.e.* one that considers only temperature variations along the rod axis, was selected as adequate in this case because preliminary experiments suggested that the radial and azimuthal temperature variations experienced in the rod during a typical implantation treatment were small. It was also assumed that the rod temperatures never reached values where radiation would be significant, and because the processing was accomplished in a vacuum, the effects of convective heat losses from the rod could also be neglected. The fact that these heat losses from the rod surface may be considered insignificant makes it permissible to use the one-dimensional model and to prescribe a zero heat flux boundary condition at the bottom of the rod. In order to carry out the

TABLE 2

Calculated thermal exposures equivalent to 300 °C for 30 min exposure (based on equivalent tempering relation)

| *Temperature* (°C) | *Time* (min) |
|---|---|
| 300 | 30 |
| 350 | 7.23 |
| 400 | 2.16 |
| 450 | 0.76 |
| 500 | 0.31 |
| 550 | 0.14 |
| 600 | 0.07 |

numerical analysis, the rod was divided into four thermally coupled zones (labelled $l_1$ through $l_4$ in Fig. 2). In two of these ($l_1$ and $l_3$) transient conduction with no internal generation occurred, while in one ($l_2$) internal heat generation induced by ion beam impingement was added. These three zones were coupled with a final zone of heat conduction through a contact conductance ($h$) to the heat sink at an ambient temperature ($T_0 = 20$ °C). The internal generation power per unit rod length was computed as the product of the ion beam current density ($j$) ion energy ($E$) and mask width ($w$).

Because the plane of thermal symmetry in Fig. 2b represents a zero heat flux boundary, just as the lower boundary at the bottom of the rod in Fig. 2a does, the same model used for the single heat sink configuration can also be used to describe temperature histories for the double heat sink configuration. This is done by requiring $l_1 = 0$ and redefining the other lengths in the manner suggested in Fig. 2b. In this case, solutions to the one-dimensional thermal model above and below the plane of symmetry are identical so it is necessary to determine thermal histories in only half of the rod.

The thermal model was verified experimentally using the test configuration of Fig. 2a and implanting an instrumented (with thermocouples) RCF rod at various beam current densities to a total dose of $10^{17}$ $N_2^+$ cm$^{-2}$ at 40 keV. As shown in Fig. 3, the temperatures measured at the points labeled A and B in Fig. 2a are seen to agree quite well with the computed values when the published thermal properties of 440C steel [17] and a contact conductance $h$ of 250 W m$^{-2}$ are used in the model. This contact conductance cannot be measured directly, but it is considered reasonable [17], and it yields results

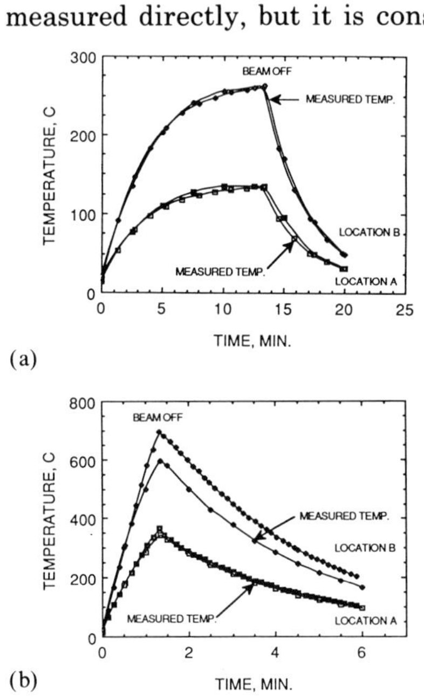


Fig. 3. Computed and measured temperatures for nitrogen implanted RCF rods showing good correlations for (a) 100 μA cm$^{-2}$ current density implantation and (b) 1500 μA cm$^{-2}$ current density implantation.

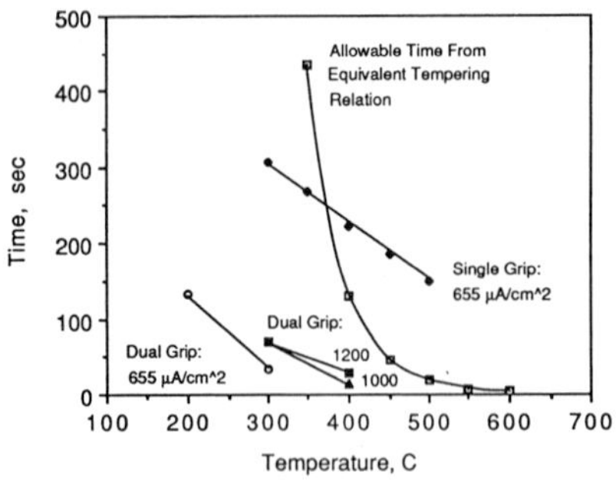


Fig. 4. Comparison of thermal exposures as a function of heat sink fixture for implantation of 440C steel RCF rods showing excessive exposures for single-grip configuration.

that agree with experimental ones over a range of implantation conditions.

Figure 4 shows a graphical representation of the calculated data from Table 2 along with analytically derived (from the heat transfer model) thermal exposures for different current densities and grip configurations. Any time–temperature combinations to the left of the allowable curve does not result in hardness loss of more than 2 Rockwell C points. As observed in Fig. 4, the type of heat sink fixturing is crucial in determining the thermal history experienced by the specimens. For example, early implantations (655 $\mu$A cm$^{-2}$) using the single grip configuration produced exposure times that were above the calculated allowable values for temperatures greater than about 375 °C. Modifying the heat sink configuration reduced the exposure times at elevated temperatures to within the allowable range, as shown for implantations using current densities of 655, 1000 and 1200 $\mu$A cm$^{-2}$. It is also apparent that higher temperatures are attained for higher current density implantations which was expected from the increased applied power. These results are shown for constant implantation conditions of a total dose of $10^{17}$ $N_2^+$ cm$^{-2}$ and accelerating energy of 40 keV. Thus it is apparent that specimen fixturing is very important, using this high current density implantation technique, to prevent excessive substrate tempering and reduction in near surface properties.

### *3.2. Surface analyses*

Large area (matrix and carbides) AES analyses were performed on selected RCF rods to evaluate the effects of ultra high current density on the resultant nitrogen concentration–depth profiles. As shown in Fig. 5, slightly reduced retained concentrations were observed as the current density was increased from 107 (from ref. 5) to 1300 $\mu$A cm$^{-2}$. In a preceding study [5], near-surface AES analyses showed a high near-surface nitrogen concentration, which was independent of dose, and suggested the attainment of sputter limited conditions. A similar effect may have occurred in this study of specimens implanted using ultrahigh current densities. These results were somewhat unexpected, since it has been shown that high (up to 1500 $\mu$A cm$^{-2}$) current density implantations of other steels result in

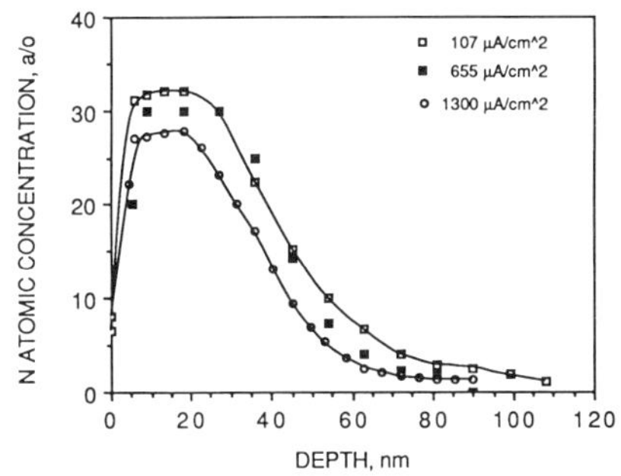


Fig. 5. Large area AES analyses of nitrogen implanted 440C steel RCF rods showing slightly reduced retained nitrogen concentrations for specimens implanted with ultrahigh current densities.

enhanced penetration depths, as shown in Fig. 1 [1]. In these cases, $N_2^+$ implantation of an austenitic matrix, *i.e.* 304 SS at 1500 $\mu$A cm$^{-2}$, resulted in enhanced penetration whereas lower current density (*i.e.* 10, 100 and 300 $\mu$A cm$^{-2}$) implantation profiles were similar. For 01 tool steel, a hard martensitic matrix, a broadened profile was observed for implantation at 700 $\mu$A cm$^{-2}$, although the order of magnitude increase in dose (from $10^{17}$ to $10^{18}$ $N_2^+$ cm$^{-2}$) may have also influenced the profile, but probably not to the extent observed in Fig. 1. From these results, it appears there may be a threshold current density and thus temperature above which diffusion coefficients are increased, which suggests we may have not reached these values for the 1300 $\mu$A cm$^{-2}$ $N_2^+$ implantation of 440C steel. The combination of radiation effects from atomic displacements and induced temperatures is speculated to account for the high diffusion coefficients, as calculated for irradiated Ni in the intermediate temperature region of 200 - 700 °C [6]. For the 440C steel RCF specimens processed with high current densities in this study, the following thermal histories were calculated from the thermal model (using the dual heat sink configuration): 655 $\mu$A cm$^{-2}$, 300 °C for 33 s; 1000 $\mu$A cm$^{-2}$, 400 °C for 15 s; 1300 $\mu$A cm$^{-2}$, 400 °C for 30 s. While these temperatures are within the above range for enhanced diffusion, as calculated for nickel [6], the very short durations may have restricted mass transport since diffusion distance is traditionally proportional to $(Dt)^{0.5}$ [18]. Additional work is underway to establish the singular and synergistic effects of temperature and irradiation on the mobility of nitrogen atoms.

### *3.3. Rolling contact fatigue test results*

Tables 3 and 4 present the B10, B50 RCF lifetimes, lifetime ratios ($N_2^+$ implanted life/unimplanted life), confidence levels and specimen processing details for the $N_2^+$ implanted 440C steel rods. For the B10 RCF lifetimes (Table 3), an increase in lifetimes (and thus life ratios) occurred as the current density was increased from 107 $\mu$A cm$^{-2}$ [5] to 655 $\mu$A cm$^{-2}$ to 1300 $\mu$A cm$^{-2}$ at a constant dose of $10^{17}$ $N_2^+$ cm$^{-1}$. For the lower dose of $5 \times 10^{16}$ $N_2^+$ cm$^{-2}$, an increase in life ratio was also observed as the current

TABLE 3

Summary of B10 RCF results for 440C steel showing improved fatigue lifetimes for specimens nitrogen implanted with ultrahigh current densities

| $N_2^+$ *dose* (ions cm$^{-2}$) | *Accelerating energy* (keV) | *Ion current density* (μA cm$^{-2}$) | *B10 life*[a] (×10$^6$ cycles) | *Life ratio*[b] | *Confidence level* (%) | *Processing details* |
|---|---|---|---|---|---|---|
| Untreated | — | — | 0.38 | 1.0 | — | — |
| $10^{17}$ | 40 | 107 | 1.9 | 1.27 | 88 | single end grip (ref. 5) |
| $10^{17}$ | 20 | 200 | 1.03 | 2.71 | 76 | dual end grip |
| $10^{17}$ | 40 | 655 | 0.13 | 0.34 | — | single end grip |
| $5 \times 10^{16}$ | 40 | 655 | 1.27 | 3.3 | 85 | dual end grip |
| $10^{17}$ | 40 | 655 | 1.51 | 4.0 | 86 | dual end grip |
| $10^{17}$ | 40 | 1000 | 0.26 | 0.68 | — | dual end grip[c] |
| $5 \times 10^{16}$ | 40 | 1300 | 1.75 | 4.61 | 91 | dual end grip |
| $10^{17}$ | 40 | 1300 | 2.20 | 5.8 | 89 | dual end grip |

[a]785 ksi maximum Hertz stress. [b]$N_2^+$ implanted B10 life/unimplanted B10 life. [c]Poor grip contact.

TABLE 4

Summary of B50 RCF results for 440C steel showing reduced magnitude of improvements compared to B10 RCF results (Table 3)

| $N_2^+$ *dose* (ions cm$^{-2}$) | *Accelerating energy* (keV) | *Ion current density* (μA cm$^{-2}$) | *B50 life*[a] (×10$^6$ cycles) | *Life ratio*[b] | *Confidence level* (%) | *Processing details* |
|---|---|---|---|---|---|---|
| Untreated | — | — | 2.0 | 1.0 | — | — |
| $10^{17}$ | 40 | 107 | 5.0 | 1.67 | 97 | single end grip (ref. 5) |
| $10^{17}$ | 20 | 200 | 3.79 | 1.9 | 95 | dual end grip |
| $10^{17}$ | 40 | 655 | 0.77 | 0.39 | — | single end grip |
| $5 \times 10^{16}$ | 40 | 655 | 2.96 | 1.48 | 86 | dual end grip |
| $10^{17}$ | 40 | 655 | 3.04 | 1.52 | 90 | dual end grip |
| $10^{17}$ | 40 | 1000 | 1.26 | 0.63 | — | dual end grip[c] |
| $5 \times 10^{16}$ | 40 | 1300 | 3.94 | 1.97 | 97 | dual end grip |
| $10^{17}$ | 40 | 1300 | 10.99 | 5.50 | 99 | dual end grip |

[a]786 ksi maximum Hertz stress. [b]$N_2^+$ implanted B50 life/unimplanted B50 life. [c]Poor grip contact.

density was increased from 655 to 1300 $\mu$A cm$^{-2}$. These specimens all experienced good thermal contact with both heat sink fixtures except for one instance ($10^{17}$ ions cm$^{-2}$ at 1000 $\mu$A cm$^{-2}$ specimen) where poor thermal contact apparently resulted in excessive tempering of the 440C steel substrate which resulted in reduced RCF lifetimes (*i.e.* B10 life ratio of 0.68). A graphic illustration of excessive tempering was also shown for a rod implanted at 655 $\mu$A cm$^{-2}$ using only a single rod grip fixture where the RCF lifetime was reduced (B10 life ratio of 0.34) compared to untreated 440C steel. This result confirms the predictions of the thermal exposure study, which showed excessive exposures to temperatures above 400 °C are detrimental (Fig. 4). Increasing the $N_2^+$ dose from $5 \times 10^{16}$ to $10^{17}$ ions cm$^{-2}$ at a constant current density, also enhanced the B10 RCF life for current densities of 655 and 1300 $\mu$A cm$^{-2}$ as shown in Table 3.

For the B50 lifetimes (Table 4), smaller improvements in the life ratio were observed compared to the B10 values, except for the highest current density specimen (*i.e.* 1300 $\mu$A cm$^{-2}$). Similar reductions in the B50 lifetimes, as for the B10 lifetimes, were observed for the improperly processed specimens.

The confidence levels shown in Tables 3 and 4 provide an indication of the degree of certainty that the nitrogen implanted specimens' RCF life is actually greater than the untreated baseline material. It is generally regarded that confidence levels greater than 80% indicate the tested population has B10 or B50 lifetimes that are statistically significant. For the B10 lifetimes (Table 3) all of the specimens, except for the lower current density (200 $\mu$A cm$^{-2}$) and lower energy (20 keV) processed specimen, exhibited statistically significant results. Some of these calculated values are even conservative, since several of the test specimens ($5 \times 10^{16}$ $N_2^+$ cm$^{-2}$, 655 $\mu$A cm$^{-2}$; $10^{17}$ $N_2^+$ cm$^{-2}$, 655 $\mu$A cm$^{-2}$, $5 \times 10^{16}$ $N_2^+$ cm$^{-2}$, 1300 $\mu$A cm$^{-2}$) had Weibull slopes greater than 2.0, which is the upper limit for interpolation of the confidence level. For the B50 lifetimes, all of the reported values for the nitrogen implanted specimens are considered statistically significant, since confidence levels greater than 80% were calculated for all of the specimens.

## 4. Conclusions

Equivalent tempering relations were used to predict allowable thermal exposures applied to 440C steel that would not degrade substrate material properties. It was shown that increasing temperature significantly reduced allowable exposure time. For example, an exposure of 19 s at 500 °C produces equivalent carbon diffusion distances to a treatment of 300 °C for 30 min, which reduces material hardness by 2 Rockwell C points. Reductions in hardness of this magnitude reduce RCF performance. A heat transfer model was developed to enable correlation of the thermal history experienced during implantation processing with implantation parameters (i.e. type of heat sink configuration, ion beam current density and accelerating energy).

The use of a dual specimen end grip configuration significantly reduced the time at temperature, resulting in allowable thermal exposures for 440C steel.

Subsequent $N_2^+$ implantation processing at ultrahigh (up to 1300 $\mu$A $cm^{-2}$) current densities, for very short times, produced enhanced (up to 6 times) B10 RCF lifetimes. Increasing the current density, at a constant dose, increased B10 RCF lifetimes and a slight dose dependence was also observed, with higher lifetimes shown as the dose was doubled from $5 \times 10^{16}$ to $10^{17}$ $N_2^+$ $cm^{-2}$. With the exception of one specimen, all of the B10 and B50 lifetimes for the nitrogen implanted specimens are considered statistically significant.

Surface analyses by AES showed slightly lower retained nitrogen concentrations as the ion beam current density was increased. This indicates that considerable reductions in processing time (*i.e.* by up to 150 times based on beam exposure time) can be realized while still maintaining a sufficient nitrogen retention and thus beneficial effects. The reason for the apparent depth independence of current density is unknown at this time, although the very short processing times at elevated temperature may restrict the penetration depths.

Failure examinations of the RCF rods to establish the reasons for lifetime improvement, as a result of implantation processing, and microstructural characterization (by Mössbauer spectroscopy) to establish microstructure–tribological property correlations is planned.

## Acknowledgments

Surface analyses were performed by Dr. S. R. Smith of the Martin Marietta Astronautics Group Failure Analysis Laboratory. This work was supported by a Martin Marietta IR&D project with partial support also provided by National Science Foundation Grants INT 8501157 and MSM 8617811.

## References

1 P. J. Wilbur and L. O. Daniels, The development and application of an ion implanter based on ion thruster technology, *Vacuum, 136 (1–3)* (1986) 5.
2 P. J. Wilbur and W. S. Sampath, The effects of ultrahigh current density ion implantation of bulk microstructure and bulk hardness, in *Proc. of Industry–University Advanced Materials Conf.*, The Metallurgical Society, 1987, p. 47.
3 P. D. Goode and I. J. R. Baumval, The influence of ion implantation parameters on the surface modification of steels, *Nucl. Instrum. Methods, 189* (1981) 161.
4 Th. Barnavon, H. Jaffrezic, G. Marets, N. Monocoffie, J. Tousset and S. Fayeulle, Influence of temperature on nitrogen-implanted steels and iron, *Mater. Sci. Eng., 69* (1985) 531.
5 F. M. Kustas, M. S. Misra, S. R. Smith, P. J. Wilbur and R. L. Thom, High current density ion implantation processing of 440C steel. Part 1 — implantation parameters and rolling contact fatigue improvement, *Surf. Coat. Technol.*
6 H. Wiedersich, Kinetic processes during ion bombardment, *Nucl. Instrum. Methods Phys. Res., B7/8* (1985) 1.

7 NASA Marshall Space Flight Center (MSFC), *Heat Treatment Specification DCN 28433.*
8 D. Glover, in T. C. Hoo (ed.), *Rolling Contact Fatigue Testing of Bearing Steels*, 1982, p. 107.
9 R. B. Abernethy, J. E. Breneman, C. H. Medlin and G. L. Reinman, *Weibull Analysis Handbook*, Government Report AFWAL-TR-83-2079, 1093.
10 L. E. Davis (ed.), *Handbook of Auger Electron Spectroscopy*, Physical Electronics Division, Perkin–Elmer, Eden Prairie, MN, 2nd edn. 1978.
11 J. F. Moulder, Physical Electronics Division, Perkin–Elmer, personal communication.
12 F. M. Kustas and M. S. Misra, *Progress Report R85-48681-002*, Martin Marietta Denver Aerospace, 1985.
13 R. L. Thom, NASA Marshall Space Flight Center, Huntsville, AL, personal communication.
14 F. M. Kustas and M. S. Misra, Analysis of rolling contact fatigue failures of ion implanted 440C steel, *Martin Marietta IR&D Progress Report R85-48681-005*, 1986, p. 16.
15 G. Krauss (ed.), *Principles of Heat Treatment of Steel*, American Society for Metals, Metals Park, OH, 1980.
16 P. J. Wilbur, Colorado State University, Ft. Collins, CO, private communications and unpublished work.
17 J. T. Holman, *Heat Transfer*, McGraw-Hill, New York, 1981.
18 P. G. Shewmon, in *Diffusion in Solids*, McGraw-Hill, New York, 1963, p. 9.

# SURFACE MODIFICATION OF MONEL K-500 BY THE PLASMA SOURCE ION IMPLANTATION PROCESS

M. MADAPURA, J. R. CONRAD, F. J. WORZALA and R. A. DODD

*University of Wisconsin, Madison, WI 53706 (U.S.A.)*

F. C. PRENGER and J. A. BARCLAY

*Astronautics Corporation of America, Madison, WI (U.S.A.)*

(Received March 22, 1989)

## Summary

The monel series of nickel base alloys (Ni–Cu type) are attractive for components like bearings and shafts operating in high magnetizing environments because of their non-magnetic characteristic coupled with superior corrosion and oxidation resistance at extreme service temperatures. Unfortunately, the current bearing technology is based on the use of two classes of materials, 5000 series alloy steels and 400 series stainless steels, both of which are magnetic. Thus, the successful operation of a magnetic device requires a new class of bearing materials. Some of the non-ferrous materials such as monels have not demonstrated superior wear characteristics as bearing materials.

Plasma source ion implantation (PSII), a non-line-of-sight technology for implanting preferably gaseous species, is a new and innovative approach to surface modification of materials. The principal objective of the present investigation was to apply PSII technology to improve the wear characteristics of monels for developing a new class of bearing materials operating in highly magnetic environments.

Monel K-500 alloy samples were nitrogen implanted by the PSII process and the modified layers were characterized by profilometry and optical metallography, low load microhardness testing, scanning Auger microscopy (SAM), pin-on-disk wear testing and scanning electron microscopy (SEM). The results of nitrogen implantation show a significant improvement in wear resistance of this alloy without any perceptible increase in surface hardness.

## 1. Introduction

Surface modification of materials and components through ion beam ion implantation is being used selectively to tailor the surface properties for improved wear, corrosion and oxidation resistance in a variety of non-semiconductor applications. With respect to nickel base alloys, there are a

0257-8972/89 Elsevier Sequoia/Printed in The Netherlands

number of studies [1] dealing with enhancing the corrosion and oxidation resistance, specially in Ni–Cr-type alloys by implanting with yttrium, calcium, zirconium, aluminum, cerium and platinum species.

Plasma source ion implantation (PSII) [2], the newest member of the surface modification processes, is a non-line-of-sight process and is being used for developing modified layers in a variety of materials such as titanium alloys, low-alloy steels, stainless steels, niobium base alloys and high nickel containing steels. PSII has demonstrated (1) efficient implantation of ions to the concentration and depth required for surface modification, (2) dramatic improvement in the life of manufacturing tools in actual industrial application and (3) acceptable dose uniformity on non-planar targets without target manipulation [2 - 4]. Also, the economics of surface modification by the PSII process relative to the conventional ion implantation process indicate a substantial reduction in operating costs by virtue of the enhanced throughput.

The monel series (Ni–Cu type) of nickel base alloys are used for a variety of high temperature and cryogenic applications, and for moving components operating in magnetic environments because of their non-magnetic characteristics. Monel K-500 is an age hardenable Ni–Cr alloy containing aluminum and titanium, is very attractive for applications that require the corrosion resistance of monel 400 but greater strength [5, 6]. This alloy can be used for bearings that operate at extreme service conditions, *e.g.* magnetic refrigeration, provided its wear resistance is significantly improved. As this alloy has aluminum and titanium as strengthening elements, it appears to be a good candidate material for nitrogen implantation by the PSII process. There is hardly any information in the published literature on the nitrogen implantation characteristics of nickel base alloys in general. With these objectives in view, a detailed study on the nitrogen implantation of monel K-500 by the PSII process was initiated to characterize the adhesive–sliding wear resistance, load bearing capacity and corrosion resistance of the modified layers.

In this paper, we will be presenting the recent results of characterization of the modified layers by profilometry, optical metallography, low load Knoop microhardness testing, scanning Auger microscopy (SAM), scanning electron microscopy (SEM) and pin-on-disk wear testing.

## 2. Experimental procedure

Disks of thickness 6 mm and diameter 28 mm of monel K-500 in the age hardened condition were used in this study. The disks were ground and polished to a final finish of approximately 0.05 $\mu$m. The samples were degreased with a detergent and washed with running water, then washed in acetone and dried. The samples were nitrogen implanted by the PSII process. In a typical PSII cycle of implantation, the sample disk was placed on an aluminum pedestal and the surface to be implanted was sputter cleaned with argon. Nitrogen implantation from the plasma was performed at an energy of 50 keV to a dosage of $3 \times 10^{17}$ atoms $cm^{-2}$. During implantation, the process parameters were so chosen to maintain the temperature of the target disk at less than 100 °C.

Optical metallography and profilometry were used to characterize the surface roughness–finish of the implanted surfaces. SAM was used to determine the depth profile of nitrogen after implantation. Low load Knoop microhardness measurements were performed to discern the change in hardness relative to the unimplanted condition.

The wear tests were performed on a laboratory model pin-on-disk wear tester using a ruby ball with a diameter of 3 mm. During wear testing, the sliding speed was maintained constant at $0.02 \, m \, s^{-1}$. The tests were conducted at various loads in the range of 2 - 10 N, to a sliding distance of 23.8 m. The surfaces were lubricated by a thin layer of WD40 oil throughout the test.

To compare the wear rate of untreated and nitrogen-implanted disks, the maximum wear depth (MWD) which is defined as the distance from the original surface to the deepest part of the profile was measured from profilometric traces. An average of four traces were reported. In addition, the cross-sectional area of the wear tracks were measured and the volume loss of material was computed and expressed as volume loss per unit length. SEM was used to study the features of the wear tracks.

## 3. Results

Figure 1 shows the profilometric traces of a sample of monel K-500 before and after implantation. After implantation to a dose of $3 \times 10^{17}$ atoms $cm^{-2}$ at 50 keV energy, the overall surface finish does not change from its pre-implantation value of about 0.05 $\mu$m. However, the surface loses its smooth appearance due to a certain amount of preferential sputtering,

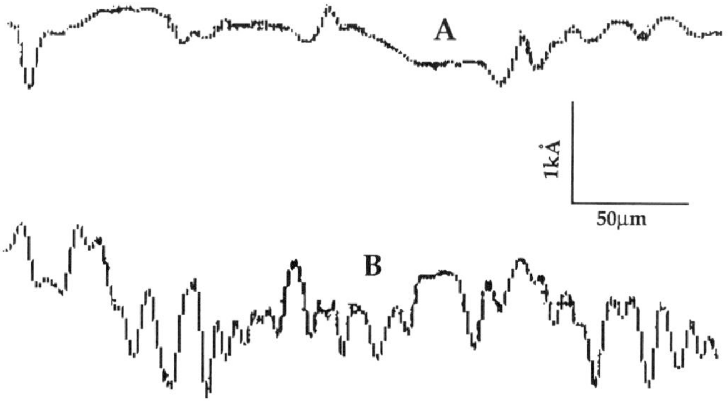


Fig. 1. Profilometer traces of unimplanted (A) and nitrogen implanted (B) surfaces of monel K-500 alloy.

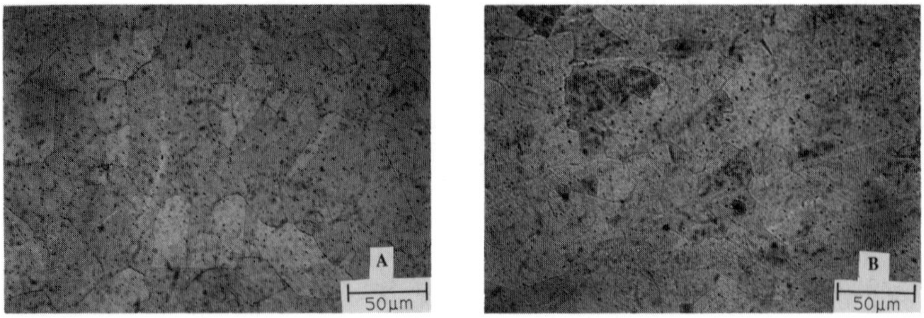


Fig. 2. Optical micrographs of unimplanted (A) and nitrogen implanted (B) surfaces of monel K-500 alloy.

particularly at the grain boundaries. The etching resulting from the implantation process is shown in Fig. 2 and is compared with a chemically etched surface of an unimplanted sample.

Figure 3 is a plot of low load Knoop microhardness *vs.* load for implanted and unimplanted samples of monel K-500 alloy. The average hardness of the implanted sample shows a slight increase relative to the unimplanted sample at all the measured loads. However, the increase does not appear to be significant as most of the error bars of the implanted sample include the unimplanted sample hardness readings.

Figure 4 shows the nitrogen profile measured by SAM. The nitrogen concentration shows a peak value of about 9 at.% which appears to be low for a dosage of $3 \times 10^{17}$ atoms $cm^{-2}$. The analysis of nickel and copper by AES in the present system is complicated by the fact that the ${}^{Ni}I_{848}$ peak overlaps with the ${}^{Cu}I_{849}$ peak. To separate the intensities of nickel and copper from ${}^{Ni+Cu}I_{848}$, the following relationship has been used, $I_{Ni} = I_{Ni+Cu} - 0.302 I_{Cu}$. This assumes that the ratio ${}^{Cu}I_{848}/{}^{Cu}I_{920}$ is constant. Thus, the atomic concentrations obtained by this technique may not be quantitatively accurate and are useful for a relative comparison.

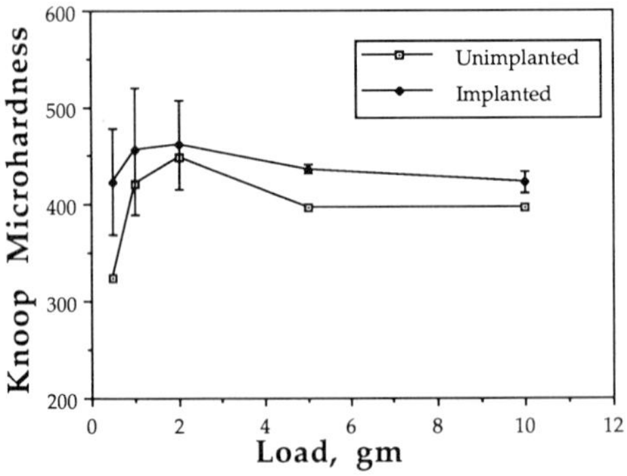


Fig. 3. Variation of Knoop microhardness with load of nitrogen implanted and unimplanted surfaces.

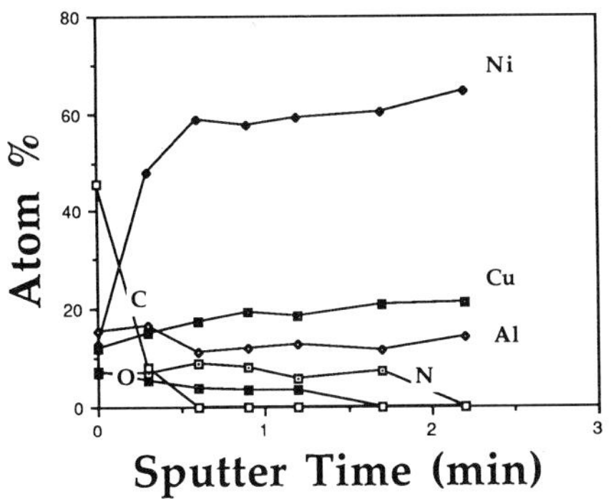


Fig. 4. Auger concentration profile of implanted monel K-500 alloy sample.

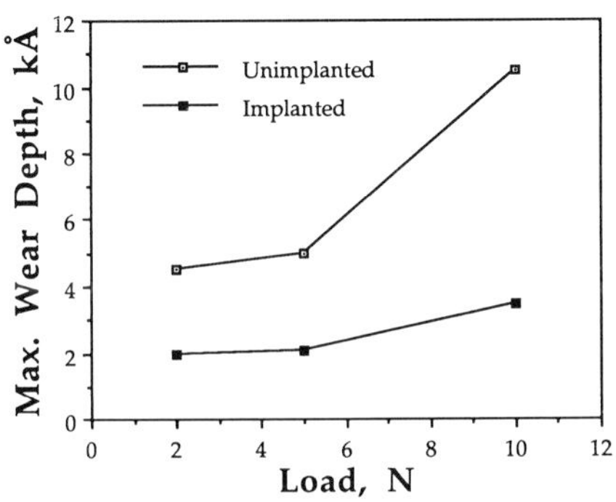


Fig. 5. Plot of wear rate characterized by MWD *vs.* load.

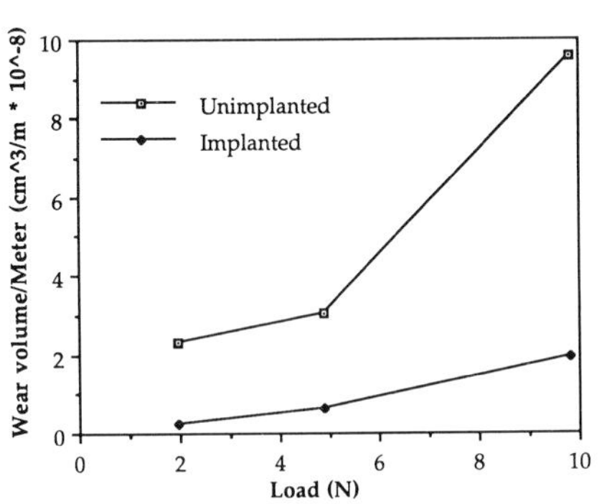


Fig. 6. Plot of wear rate characterized by wear volume per unit length *vs.* load.

The trend of wear rate expressed by the maximum depth of wear track *vs.* load is shown in Fig. 5 for both implanted and unimplanted samples. Figure 6 represents the wear rate in terms of volume of the tracks *vs.* load. These two plots essentially reveal a significant improvement in wear resistance of monel K-500 due to implantation. In the case of the unimplanted

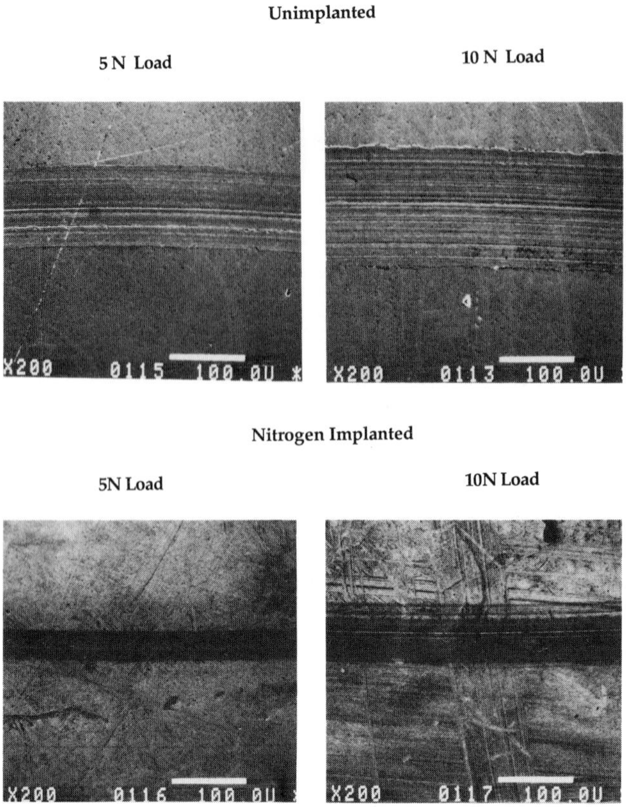


Fig. 7. SEM micrographs of wear tracks of unimplanted and nitrogen implanted samples.

sample, the wear rate increases gradually up to a load of 5 N and increases very steeply beyond. This trend is significantly moderated by nitrogen implantation.

Figure 7 shows the SEM micrographs of wear tracks at loads of 5 and 10 N. These micrographs reveal a significant reduction in wear track width and changes in surface topography of the tracks due to implantation.

## 4. Discussion

The results of nitrogen implantation in monel K-500 alloy show a significant increase in the wear resistance in the load range tested (2 - 10 N) but do not reveal any perceptible increase in the surface Knoop microhardness. A careful study of the wear tracks by SEM and optical metallography indicate severe ploughing and deformation of the material at the tracks in the case of unimplanted material. However, the wear tracks in the case of implanted material do not show any ploughing but show a relatively smooth

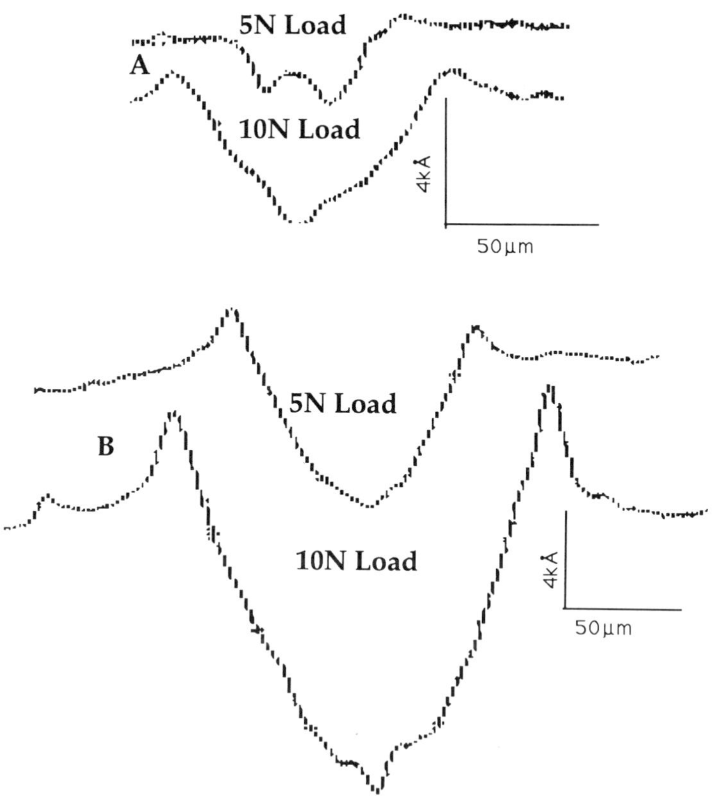


Fig. 8. Profilometer traces of wear tracks showing deformed metal in the case of unimplanted sample (A) and relatively negligible flow of metal near the tracks in nitrogen-implanted sample (B).

appearance. The lack of any pushing or deformation of the material from the tracks is evident from the profilometric traces shown in Fig. 8.

From the studies conducted by a number of researchers [7 - 9] on the wear behavior of nitrogen implanted metals, only a loose correlation between reduced wear and increased surface hardness has been found. There is agreement about the contributing effects of increased surface hardness. It has been experimentally verified by the aforementioned researchers that the best improvements in wear performance occur only when the implantation produces a change in the dominant mode of water.

At present, work is underway to identify the probable mechanisms for the observed wear characteristics in the present system. Experiments have been designed to characterize the implanted surface for amorphous behavior by grazing incidence X-ray diffraction, to identify the stability of the gamma prime precipitate and preferential segregation of major elements in the implanted layer by transmission electron microscopy and measurement of coefficient of friction during wear testing.

## 5. Conclusions

Nitrogen implantation of monel K-500 by the PSII process produces a modified layer with a significant increase in wear resistance compared to the unimplanted alloy.

Low load Knoop microhardness measurements of the implanted layers do not show any perceptible difference compared to the unimplanted surfaces.

The implanted layer shows a peak nitrogen concentration of about 9 at.% which seems to be very low in comparison with the calculated value.

Even though the implanted surfaces show a certain degree of sputter etching, the overall surface finish remains at its pre-implantation value of 0.05 $\mu$m which is highly desirable for precision bearings.

## Acknowledgments

The authors acknowledge the technical assistance of P. Fetherston, J. Firmess, L. Xie and N. Tran in several phases of this investigation. This work was supported by NSF grants DMC-8712461 and CDR-8721545, and by a number of industrial grants.

## References

1 K. Khobaibe and R. C. Krutenat, *Proc. Symp. High Temperature Coatings*, TMS Meeting, Orlando, Florida, 1986, p. 153.
2 J. R. Conrad, J. L. Radke, R. A. Dodd and F. J. Worzala, *J. Appl. Phys.*, *62* (1987) 4591.
3 J. R. Conrad, R. A. Dodd, F. J. Worzala and X. Qiu, *Surf. Coat. Technol.*, *36* (1988) 927.
4 J. R. Conrad, R. A. Dodd, F. J. Worzala, X. Qiu and R. S. Post, *ASM Conf. on Ion Implantation and Plasma-assisted Processes for Industrial Applications*. Atlanta, Georgia, May 1988.
5 *Metals Handbook*, Vol. 3, ASM, 9th edn., 1985, p. 133.
6 *Metals Handbook*, Vol. 9, ASM, 9th edn., 1985, p. 435.
7 W. C. Oliver, R. Hutchings and J. B. Pethica, *Metal. Trans.*, *15A*, (1984) 2241.
8 C. J. Mchargue, *Int. Metals Rev.*, *31* (1986) 49.
9 K. Z. Gahr, *Tribology Series 10, Microstructure and Wear of Materials*, Elsevier, Amsterdam, 1987, p. 416.

*Surface and Coatings Technology*, *39/40* (1989) 595 - 598 

# MECHANISMS OF ATOMIC OXYGEN INDUCED MATERIALS DEGRADATION

H. G. PIPPIN

*Boeing Aerospace, P.O. Box 3999, Seattle, WA 98124–2499 (U.S.A.)*

(Received March 9, 1989)

## Summary

This paper includes discussions of mechanisms by which atomic oxygen may attack materials being flown in low Earth orbit (LEO), suggestions for development of materials intrinsically resistant to the harsh LEO environment and results of an Arrhenius fit to mass loss of Kapton under exposure to thermal atomic oxygen in a laboratory environment. Oxygen atoms are powerful oxidizing agents which can rapidly degrade many types of polymeric materials. The process details depend upon the specific material, its detailed bonding, the thermodynamic stability of potential products, temperature and particle dynamics. With regard to spacecraft, many effects associated with atomic oxygen may be explained simply by ordinary thermal oxidation processes and by the fact that the oxygen has a preferred direction of impingment.

## 1. Introduction

Spacecraft operating in low Earth orbit (LEO) are exposed to an extremely harsh environment. One component of this environment is atomic oxygen, which is produced by dissociation of molecular oxygen by ultraviolet radiation from the sun. Owing to the high velocity required to maintain the spacecraft in orbit, the collision energy for impacts between atomic oxygen and the ram facing (direction of motion) surfaces averages about 5 eV. The flux of oxygen atoms to the surface is a function of orbital altitude, orientation of the surface relative to the flux and phase of the solar cycle.

Examination of external surface materials returned from early shuttle flights showed surface recession and mass loss of organic based materials. Surface analysis by scanning electron microscope revealed a roughened texture on the surface.

The recognition of atomic oxygen as a major contributor [1] to the observed materials degradation has led to large-scale efforts to identify materials which are resistant to oxidation and still have the other properties required for use at LEO.

0257-8972/89 Elsevier Sequoia/Printed in The Netherlands

## 2. Results

Boeing Aerospace has designed and built a space environment screening chamber for testing materials resistance to atomic oxygen [2]. Under an advanced development contract from NASA-MSFC, "Space Station Protective Coating Development", Boeing has measured the mass loss of selected materials at well defined temperatures as a function of oxygen fluence [3].

Data obtained for Kapton recession rates as a function of temperature were fitted to an Arrhenius type equation. The values for the pre-exponential factor and the activation energy are $2400 \times 10^{-24}\,cm^3\,atom^{-1}$ and 0.31 eV respectively. Recession rates are determined by measuring the mass loss of a known geometric surface area, dividing this value by the density of the material and then dividing the result by the fluence of oxygen atoms to the surface. This puts the removal rate of different materials by oxygen atoms on a comparable basis. Fits to measurements of Kapton recession rates at different oxygen kinetic energies, using modified Arrhenius functions, also predict activation energies of about 0.3 eV [4].

## 3. Discussion

The values cited above are "global" activation energies, not for a specific chemical step, and their meaning is subject to interpretation. In the laboratory tests whose results are presented here, the materials were exposed to a flux of thermal oxygen atoms and molecules. The mass loss is due to thermal oxidation and abstraction processes. Observed mass loss of organic based materials on orbit can also be explained by thermal oxidation. Using the value of $3.0 \times 10^{-24}\,cm^3\,atom^{-1}$ for the on-orbit recession rate and the Arrhenius fit from the laboratory data, a thermal oxidation process would require a temperature of about 500 K. This is a reasonable value. The ambient temperature at altitudes of interest ranges between about 900 and 1200 K. The spacecraft surface will range between about 200 and 400 K. The actual surface temperature of the spacecraft, where the reaction is occurring, will be somewhat in-between these values. Any proposed mechanism is still highly speculative. For typical on-orbit conditions, the average lifetime of an atom on a spacecraft surface, regardless of what may happen to that atom, is short in comparison to the average time between arrival of atoms at the same location. Under such conditions of effective oxygen starvation, where the arrival of oxygen at the surface is the rate limiting step, the observed rates of oxidation would not show the expected change as a function of temperature [5]. Data for materials held at different temperatures during a shuttle flight [6] showed no difference within the precision of the measurement.

For materials flown on-orbit, the surfaces exposed directly to atomic oxygen show a roughened texture. This was originally thought to be due to the 5 eV impact energy. However, this textured result has been reproduced in laboratory beam systems of lower energy [7]. This indicates that the texturing is more likely to be a geometric effect.

Many details of the degradation of materials on orbit remain unknown. A comprehensive solution will have to describe the distribution of all energy sources impinging on the surface. Collisions are not particularly efficient at transferring energy and enhancing chemical reaction rates. The 5 eV must be absorbed in the bonds of the polymer and dissipated ultimately as heat. This source of energy may indirectly increase the reaction rate by creating excited states within the polymer but this has not been studied. A further difficulty in determining the mechanism of degradation is that due to the extreme vacuum on orbit, any intermediate state species which are volatile will leave. This means that reaction pathways which have lower activation energies will dominate the process. An oxygen atom which arrives at the spacecraft surface will (1) scatter inelastically and cause no degradation, and (2) adsorb on the surface and remain for a time. An oxygen attached to the surface may migrate to a reactive site and attack the polymer, migrate to another oxygen atom and recombine to form molecular oxygen, or desorb after some period of time on the surface without having undergone a chemical reaction. The relative probabilities of surface migration and desorption are a function of how tightly an atom or molecule is bound to the surface and the temperature of the surface. A model for surface lifetimes and surface migration rates of atomic and molecular species is presented in ref. 5. If the atom desorbs little or no damage will have occurred. The impact energy of the atoms is not great enough to cause sputtering and the energy will be dissipated throughout the many bonds of the polymer and eventually just add to the heat load of the surface (this will be an extremely small contribution). If the atoms recombine the surface material will also have to dissipate the up to 5 eV of energy released by this process.

Given sufficient time any system will approach equilibrium with its environment. Since there is a steady flux of oxygen atoms to the spacecraft surfaces the need is for materials thermodynamically more stable than their potential oxidation products. For organic polymers this is virtually not possible. For inorganic polymers this is a possibility for some fluorinated species. For metal oxide ceramics this condition is met by default. Unfortunately, for particular applications where optical properties, low mass and flexibility are essential requirements, ceramics are not suitable.

The solutions to the materials degradation problems vary with material type. One possibility is to use materials which have bonds stronger than about 5 eV so that recombination of oxygen atoms is favored over degradation pathways. Siloxane to silica conversion is relatively slow and the Si—O bonds formed are about 4.7 eV. The concern with siloxane materials is contamination of nearby surfaces and flexibility. The diatomic species Si—O may also be volatile in the vacuum of space. Materials which have very few thermodynamically stable (relative to the original material) products are also of interest. For the conditions at low earth orbit altitudes such materials are hard to identify.

Materials which have structural advantages so that oxygen atoms cannot approach and attack relatively weak bonds provide the basis for a "kinetic" solution. Energy barriers are established and the decay rate is slowed. A clear example of this is the difference between ordinary hydrocarbons and fluorocarbons ("teflons"). Oxygen atoms may extract hydrogen from a hydrocarbon polymer by direct free radical abstraction and approach the carbon–carbon bonds rather easily. The oxidation products formed, carbon monoxide and/or carbon dioxide, are quite stable and this process is favorable. Fluorocarbons must degrade by different mechanisms because fluorine atoms will not be abstracted by oxygen atoms. The O—F bond strength is only about 2.2 eV whereas the C—F bond strengths are around 3.7 eV and greater. The fluorine atoms are much larger than hydrogen atoms thus oxygen atoms have greater difficulty in getting around the fluorine to attack the polymer backbone. Fluorocarbons are susceptible to UV degradation which will disrupt the polymer structure and may provide additional pathways for oxygen attack. This issue has not been thoroughly investigated.

Limitations exist to the so-called "strong bond materials" solution because oxides of so many species form very strong bonds and the oxidizing agent is in continual supply.

## Acknowledgments

The author would like to thank NASA-MSFC, Dr. Ann Whitaker and Mr. Roger Linton in particular; and The Boeing Company for supporting the work which formed the basis of this paper.

## References

1 L. J. Leger, Oxygen atom reaction with shuttle materials at orbital altitudes, *NASA TM-58246.*

2 H. G. Pippin and R. Carruth, Materials screening chamber for testing materials resistance to atomic oxygen, *Proc 34th Int. SAMPE Symp.* Reno, NV, May 8 - 11, 1989, accepted for publication.

3 R. Linton, A. Whitaker, L. P. Torre and H. G. Pippin, Materials resistance to low Earth orbit environmental effects. *Proc 34th Int. SAMPE Symp.* Reno, NV, May 8 - 11, 1989, accepted for publication.

4 L. J. Leger, S. L. Koontz, J. T. Visentine and J. B. Cross, Laboratory investigation involving high-velocity oxygen atoms, *4th Int. Symp. Spacecraft Materials in Space Environment,* Toulouse, France, September 6 - 9, 1988.

5 W. D. Watson and E. E. Salpeter, Molecule formation on Interstellar Grains, *Astrophys J., 174* (1972) 321 - 340.

6 L. J. Leger, J. T. Visentine, and J. F. Kuminecz, Low Earth orbit atomic oxygen effects on surfaces, *AIAA-84-0548, AIAA 22nd Aerospace Sciences Meeting*, Reno, NV, January 1984.

7 R. C. Tennyson, J. B. French and T. Braithwaite, Atomic oxygen studies on polymers, *IES Symposium*, Williamsburg, VA, November 1988.

*Surface and Coatings Technology, 39/40* (1989) 599 - 605 

# ATOMIC OXYGEN RESISTANT COATINGS ON KAPTON

I. RITCHIE

*ANDUS Corporation, 21019 Osborne St, Canoga Park, CA 91304 (U.S.A.)*

H. B. GJERDE

*Lockheed Missiles and Space Company, 1111 Lockheed Way, Sunnyvale, CA 94089 (U.S.A.)*

(Received March 8, 1989)

## Summary

Coatings on Kapton film have been developed to provide resistance to atomic oxygen attack in a low Earth orbit environment. This material is to be used as the flexible substrate of the solar array panels on the Space Station Platform; and it must be both atomic oxygen resistant and sufficiently flexible to withstand solar array manufacture and deployment.

Several candidate coatings and deposition processes were investigated, including sputtered $SiO_2$, $SiO_2$–fluoropolymer mixtures and indium–tin oxide. Results are presented to show the different failure modes of these coatings. $SiO_2$ coatings were chosen for scaleup to manufacture, and initial results of this phase are presented.

## 1. Introduction

Polymer films have been used extensively on the outer surfaces of space vehicles, mainly as thermal control surfaces. The Space Station Freedom plans to use flexible solar photovoltaic arrays which consist of solar cells bonded to flexible Kapton® printed circuits. Early Space Shuttle flights have however shown that Kapton and other polymers are reactive with the low Earth orbit (LEO) environment of both the Shuttle and the Space Station. [1, 2]. Several experiments were conducted on Shuttle flights and their results attribute the degradation of the materials to attack by atomic oxygen impinging on surfaces at high velocities. Loss rates of several microns of Kapton per week were observed.

The solar photovoltaic array is constructed of a flexible Kapton printed circuit, to which are welded a large number of small rigid solar cells. The array panels are hinged together so that they may be folded for storage, and a lightweight mast is provided to unfold and tension the panels [3]. This design allows significant amounts of Kapton to be exposed to the LEO environment. It is therefore necessary to provide a protective coating on the

0257-8972/89/$3.50 

TABLE 1

Representative atomic oxygen erosion yields are given for several materials

| *Material* | *Erosion yield* ($10^{-24}$ cm$^3$ atom$^{-1}$) | *Reference* |
|---|---|---|
| Al–Kapton | 0.1 | 4 |
| (700Å) $Al_2O_3$–Kapton H | < 0.02 | 5 |
| (125Å) Cr–Kapton H | 0.006 | 6 |
| (1000Å) Cu–sapphire | 0.007 | 6 |
| FEP Teflon | 0.1 | 6 |
| Teflon | 0.5 | 6 |
| ITO–Kapton–Al | 0.01 | 4 |
| PET polyester (Mylar) | 1.5 - 3.9 | 6 |
| Kapton H | 1.5 - 2.8 | 6 |
| (650Å) $SiO_2$–Kapton H | 0.001 | 5 |
| $SiO_x$–Kapton–Al | 0.01 | 4 |

Kapton, so that a reasonable array lifetime can be expected. The mass loss rate which can be tolerated is less than $5 \times 10^{-9}$ g cm$^{-2}$ s$^{-1}$, as measured in a plasma reactor. This mass loss rate is approximately 50 times less than that of uncoated Kapton. The coating should also be sufficiently flexible that it withstands normal handling during array fabrication and deployment.

The erosion yields of many materials which were exposed to atomic oxygen in LEO have been studied and compiled by various authors. Representative values are given in Table 1 for several materials. These results show that polymer films are readily attacked by atomic oxygen, whereas metals are not, and metal oxides have intermediate resistance. Metal coatings clearly cannot be used in this application, and consequently the effort focussed upon metal oxide coatings.

## 2. Coating materials and processes

Initial work in this area was performed at NASA Lewis Research Center, concentrating on ion-beam sputtered $SiO_2$, $Al_2O_3$ and $SiO_2$–PTFE mixtures [7, 8]. This last coating was developed in an attempt to produce a coating which has superior flexibility to a pure metal oxide coating, and yet retains the good resistance to atomic oxygen attack of the metal oxides. Note that the results given in Table 1 show that the fluorinated polymer films are more resistant to atomic oxygen than non-fluorinated polymers.

The work described here concerns three coatings: $SiO_2$, $SiO_2$–fluoropolymer mixtures and non-conductive indium–tin oxide (ITO) coatings. The ITO coating was chosen since it is a common material used on the outer surfaces of satellite thermal control blankets [9] and initial data, presented in Table 1, showed good atomic oxygen resistance.

Reactive magnetron sputtering was chosen as the deposition process, rather than ion-beam sputtering, as it is a well proven industrial deposition

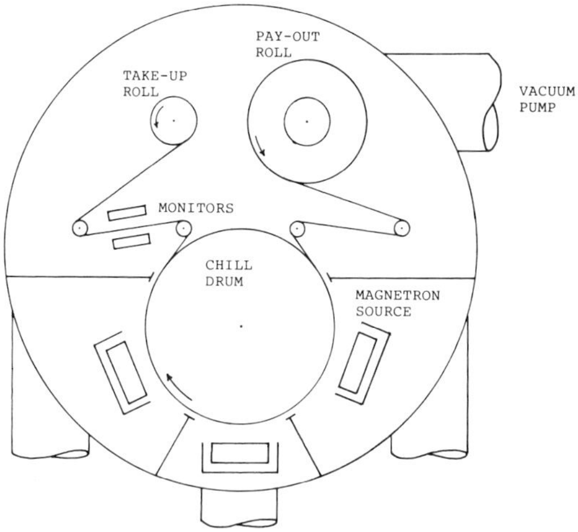


Fig. 1. A cross-section of a sputter roll coater is shown.

process used to coat webs of film such as Kapton [10]. Both d.c. and h.f. sputtering were investigated. Reactive sputtering was chosen to give maximum control over the coating process and to achieve commercially viable deposition rates, and variations on this process which include plasma deposition (introduction of additional silicon into the discharge as silane, to increase the deposition rate) and plasma polymerization were also used.

Figure 1 shows a cross-section of a sputter roll coater. A roll of film is unwound in vacuum, passing over a chill drum which stabilizes the film as it is coated by one or several magnetron sputtering sources, and the coating properties are monitored in-line before the film is rewound. This configuration allows for continuous monitoring and feedback control of the coating process and properties. Various optical and electrical properties can be monitored, depending upon the coating being made.

## 3. Results

### *3.1. Initial experiments*

Initial experiments were done using d.c. reactive sputtering. Samples were made of $SiO_2$ and mixed $SiO_2$–fluoropolymer coatings. The fluoropolymer was deposited by introducing hexafluoropropylene (HFP) into the sputter discharge; effectively a plasma polymerization process, which occurs simultaneously with the reactive deposition. Samples approximately 1000 Å thick on Kapton were tested for atomic oxygen resistance in a commercial plasma asher, which simulates atomic oxygen exposure in a reproducible, though not necessarily accurate, manner. Samples were desiccated for 24 h prior to exposure. Results are shown in Table 2, and compared with a sample made at NASA Lewis Research Center by ion-beam sputtering from a mixed $SiO_2$–

TABLE 2

Ashing rate of initial samples, tested in a commerical plasma asher at NASA Lewis Research Center

| *Material* | *Ashing rate (g cm$^{-2}$ s)* | |
|---|---|---|
| | *Initial* | *After peel test* |
| Kapton | $1.1 \times 10^{-8}$ | $1.3 \times 10^{-9}$ |
| $SiO_2$–Kapton | $7.1 \times 10^{-9}$ | $9.8 \times 10^{-9}$ |
| $SiO_2$(16%FP[a])–Kapton | $5.4 \times 10^{-9}$ | $9.7 \times 10^{-9}$ |
| $SiO_2$(8%FP[a])–Kapton | $4.8 \times 10^{-9}$ | $9.1 \times 10^{-9}$ |
| $SiO_2$(8%PTFE)–Kapton — NASA | $4.2 \times 10^{-10}$ | $5.4 \times 10^{-10}$ |

[a]Fluoropolymer (see text).

PTFE target. These samples were tested at NASA Lewis Research Center. The d.c. sputtering process was very noisy, with frequent arcs, and it produced coatings which were macroscopically grainy with small particles embedded in them, resulting from the discharge arcs. Sputtering conditions could not be established which gave a quiet process and a clear coating. These small particles did not adhere well to the Kapton substrate, which is the reason for the poor, but similar post-peel test results for the d.c. sputtered coatings; the Kapton substrate is exposed, and ashes rapidly.

### *3.2. Process development*

In the next stage of the program coatings were made using h.f. sputtering at 100 kHz in an attempt to eliminate the process arcs and particulate contamination of the coating. Some arcs occurred with the h.f. process, but it was considerably more stable than the d.c process. Addition of the HFP also improved process stability. Coatings of pure $SiO_2$ and $SiO_2$ with a fluoropolymer addition were made, and compared with a dielectric ITO coating made by the standard ANDUS d.c. sputtering process. All coatings were approximately 1000 Å thick, and were made in a machine which coated a 12 in wide web of Kapton substrate.

The coatings were tested for the following properties: adhesion, flexibility, handling characteristics and atomic oxygen mass loss rate. Adhesion was tested by a standard tape test. Flexibility was tested by judgement of the coating's ability to withstand bending under compression around a Teflon coated mandril. Handling characteristics were measured by the ability of the coated film to be fabricated into complete solar array panels; atomic oxygen mass loss was measured after fabrication. Atomic oxygen resistance was tested in the Lockheed Palo Alto Research Laboratory atomic oxygen plasma chamber, which has an atomic oxygen flux of $2.48 \times 10^{16}$ atoms cm$^{-2}$ s$^{-1}$. Samples were desiccated for 24 h before exposure.

The results of these tests are shown in Table 3.

TABLE 3

Ashing rate of samples made with improved processes, tested at Lockheed Palo Alto Research Laboratory

| *Material* | *Adhesion* | *Flexibility* | *Handling* | *Initial mass loss* ($g\ cm^{-2}\ s^{-1}$) |
|---|---|---|---|---|
| Kapton | — | — | — | $2 \times 10^{-7}$ |
| $SiO_2$–Kapton | No removal | Parallel cracks Passes 0.10 in mandril test | Good | $2.1 \times 10^{-9}$ |
| $SiO_2$–FP–Kapton | No removal | Good Passes 0.04 in mandril test | Poor | $6.3 \times 10^{-9}$ |
| ITO–Kapton | No removal | Poor | — | $2.9 \times 10^{-9}$[a] |

[a]Coating turned dark after testing; see text.

The silicon dioxide coating showed excellent atomic oxygen resistance, good handling characteristics and surprisingly good flexibility. It could be flexed around a 0.10 in radius mandril.

The $SiO_2$ fluoropolymer coating showed good atomic oxygen resistance, and it was more flexible than the pure $SiO_2$ coating; it passed the 0.04 in radius mandril test. However, the coating was very soft and complete array panels could not be successfully fabricated.

The ITO coating did not pass the atomic oxygen exposure test. Although the mass loss rate was low, the coating turned dark after exposure and became conductive; indicating that it was reduced to metal by the atomic oxygen beam. This is rather surprising as it was initially a fully oxidized coating, and ITO coatings have been shown to withstand the LEO environment. The coating was not very flexible and was not processed into an array panel.

As a result of these tests $SiO_2$ coatings were chosen to be scaled up for full-sized solar array panel experiments.

### *3.3. Scaleup*

Full-sized solar array panels require wider material than could be produced in the machine used to deposit the coatings described above, so that the process was scaled to a machine which coats a 48 in wide web of Kapton film. The process was eventually successfully transferred to this larger machine, although several difficulties were encountered in the scaleup process.

Firstly, arcing had been present in all smaller scale processes but it had not presented serious problems. However, at the higher powers required to coat the 48 in wide film at a reasonable deposition rate, the process arcs could be quite destructive, both to process control equipment and to the coating. An arc causes a section of dark, reduced silicon oxide coatings to be produced. These sections of unusable coating cause loss of yield during subsequent array manufacturing processes. The arcs had two principal causes: debris from the coating chamber falling upon the sputtering cathode, and the loss of an electrical ground for the sputtering glow discharge due to the deposition and buildup of insulating $SiO_2$ on all surfaces exposed to the sputtering flux. Both of these causes become more severe with time.

Secondly, the $SiO_2$ coatings were deposited on all chamber surfaces exposed to the sputtering flux, and depending upon the local conditions, they were deposited with different degrees of adhesion and under diffeent amounts of stress. Again, as the length of the coating time increased, this problem became more severe; and material delaminated from the chamber walls, causing process arcs and subsequent coating imperfections. Upon chamber venting, large amounts of debris fell from all areas of the chamber exposed to coating flux.

Nevertheless, the process was sufficiently well developed so that coatings several hundred feet long could be successfully made at reasonable deposition rates and with similar coating properties to those of the coatings produced on a smaller scale, with mass loss rates of less than $5 \times 10^{-9}$ g cm$^2$ s, as measured in the Lockheed plasma chamber.

## 4. Conclusion

A deposition process has been developed and successfully scaled up to coat full-size solar photovoltaic array panel substrate material with $SiO_2$ to be used as an atomic oxygen resistant coating. This coating meets all design criteria.

However, process scaleup has identified several problems inherent in the deposition process, most notably undesirable buildup of coating material on the interior of the process chamber, which should be minimized before full scale manufacture commences.

## Acknowledgments

The authors wish to acknowledge valuable discussions with, and experimental assistance of, W. Kittler and P. Maschwitz.

## References

1 L. J. Leger, J. Visentine and B. Santos-Mason, *SAMPE Quarterly*, 18 (1987) 48.
2 J. Dauphin, in J. De Bossu, G. Briens and P. Lissac (eds.), *Looking Ahead for Materials and Processes*, Elsevier, Amsterdam, 1987, p 345.

3 Lockheed Missiles and Space Company, *Product Brochure — SA/10-85.*
4 K. A. Smith, Evaluation of oxygen interaction with materials (EOIM)-STS-8 atomic oxygen effects, *AIAA-85-7021*, 1985.
5 B. A. Banks, M. J. Mirtich, S. K. Rutledge and H. K. Nahra, *Proc. 18th IEEE Photovoltaic Specialists Conf.*, 1985.
6 L. J. Leger, I. K. Spiker, J. F. Kuminecz, T. J. Ballentine and J. T. Visentine, STS Flight 5 LEO effects experiment — Background description and thin film results, *AIAA-83-2631-CP*, 1983.
7 B. A. Banks, M. J. Mirtich, S. K. Rutledge and D. M. Swec, *Thin Solid Films, 127* (1985) 107.
8 J. S. Sovey and B. A. Banks, *U.S. Patent 4,664,980* (1987).
9 R. D. Goldstein, E. M. Brown and L. C. Muldoon, *IEEE Trans. Nucl. Sci., NS-29* (1982) 1621.
10 W. C. Kittler and I. T. Ritchie, *Proc. SPIE, 325* (1982) 61.

# UNDERCUTTING OF DEFECTS IN THIN FILM PROTECTIVE COATINGS ON POLYMER SURFACES EXPOSED TO ATOMIC OXYGEN

SHARON K. RUTLEDGE

*NASA Lewis Research Center, Cleveland, OH (U.S.A.)*

JUDITH A. MIHELCIC

*Cleveland State University, Cleveland, OH (U.S.A.)*

(Received March 21, 1989)

## Summary

Protection for polymeric surfaces is needed to make them durable in the low Earth orbital environment, where oxidation by atomic oxygen is the predominant failure mechanism. Thin film coatings of oxides such as $SiO_2$ are viable candidates to provide this protection, but concern has been voiced over the ability of these coatings to protect when defects are present in the coating due to surface anomalies occurring during the deposition process, handling or micrometeoroid and debris bombardment in low Earth orbit. When a defected coating protecting a polymer substrate is exposed to atomic oxygen, the defect provides a pathway to the underlying polymer allowing oxidation and subsequent undercutting to occur. Defect undercutting was studied for sputter deposited coatings of $SiO_2$ on polyimide Kapton®. Preliminary results indicate that undercutting may be limited as long as the coating remains intact with the substrate. Therefore, coatings may not need to be defect free to give protection to the underlying surface.

## 1. Introduction

In low Earth orbit (LEO) between altitudes of 180 and 650 km, single, neutral oxygen atoms in the ground state are the most predominant species [1]. These atoms impact spacecraft surfaces with energies between 4.2 and 4.6 eV due to the movement of the spacecraft through the environment [2]. Many materials including polyimide Kapton® are oxidized in LEO by the reactive atomic oxygen [2]. The flexible solar array blanket for Space Station Freedom which acts as a support for the solar cells and cell interconnects is composed of this polyimide [3]. Based on Kapton® erosion yields on shuttle flight experiment STS-8, it is estimated that the blanket would undergo structural failure due to material loss in approximately 0.5 years if the blanket were not protected [4], while the desired life for the Space Station

0257-8972/89/$3.50 

Freedom array is 15 years. The most promising protection technique to date is the application of a thin film coating (approximately 500 - 2000 Å) of a metal oxide such as $SiO_2$ to the surface which acts as a barrier to the atomic oxygen, remains somewhat flexible and transparent, and allows the bulk properties of the underlying polyimide to be maintained [2]. One of the major concerns about using a protective coating is the possibility of coating failure due to defects generated in the coating during the coating deposition or array manufacturing process or during space flight by micrometeoroid and debris impact [5]. These defects, usually in the form of pinholes or scratches, are sites where no coating exists to protect the underlying surface. Upon exposure to the LEO environment, the underlying material would be oxidized by atomic oxygen thus creating a void underneath the coating. Of concern is the degree to which the oxidation can spread transversely, undercutting beyond the initial site of the defect. If undercutting continues through atomic oxygen collisional and reactive processes or by growth of the size of the initial defect, then the removal of Kapton® may result in structural failure of the array blanket. Also of concern is whether undercutting observed in ground based atomic oxygen simulation systems would be similar to that observed in space. This paper addresses these concerns by determining the extent undercutting occurs at defect sites due to oxidation by atomic oxygen, and whether the directionality of the atomic oxygen entering the defect plays a role in the extent of undercutting.

## 2. Experimental procedure

### *2.1. Protective coatings and substrates*

All of the materials evaluated in this paper consisted of $SiO_2$ ion beam sputter deposited onto ion beam sputter cleaned Kapton® type HN with a 1000 eV argon ion beam in an Ion Tech dual ion beam system. Pressure during deposition was approximately $10^{-4}$ Torr. Film thickness was approximately 600 Å as measured by a stylus surface profiler (Sloan Dektak IIA). Although much of the data for Kapton® obtained in space was for type H, type H and HN of the same thickness have very similar erosion yields based on observed mass loss rates in plasma ashers ($1.26 \times 10^{-8} \pm 6 \times 10^{-10}$ and $1.37 \times 10^{-8} \pm 6 \times 10^{-10}$ g $(cm^2\, s)^{-1}$ respectively).

### *2.2. Scanning electron microscopy sample preparation*

The $SiO_2$-coated Kapton® samples were mounted to quartz optical flats with double stick Kapton® tape, then a thin film coating of approximately 200 - 300 Å of gold or platinum was applied over the coated surface so that the surface would be conductive enough to be viewed by scanning electron microscopy (Cambridge Stereoscan 200). The conductive coating was applied prior to atomic oxygen exposure.

*2.3. Atomic oxygen simulation*

Two different facilities were used to expose samples to atomic oxygen for testing. A plasma asher (SPI Plasma Prep II), which generated a 13.56 MHz r.f. discharge of air in a partial vacuum (19 - 100 mTorr), was used to produce a plasma containing oxygen and nitrogen ions and atoms in various states. The nitrogen species in the plasma appear to have a negligible effect on the mass loss rate of materials exposed in the plasma as determined in previous experiments [4]. The asher is extremely useful in providing qualitative indication of material survivability. Materials which survive in the asher have been found to survive in LEO. Plasma ashers allow accelerated testing to be performed because they present a much more severe environment due largely to a higher oxygen atom arrival rate. The arrival of the oxygen species to the surface of a material in the asher is from random directions while in space, a directed sweeping arrival would occur due to the rotation of the solar array to track the sun.

A directed beam of oxygen ions was produced by means of an end Hall ion source (Commonwealth Scientific Inc.). Oxygen ions at an energy of approximately 57 eV at a flux of approximately $8.5 \times 10^{16}$ atoms $(cm^2\,s)^{-1}$ were used to expose the samples in a vacuum chamber whose pressure was $10^{-4}$ Torr during operation with a base pressure of $10^{-6}$ Torr. The beam was a mixture of $O^+$ and $O_2^+$. The samples were located approximately 25 cm downstream of the ion source on a water cooled sample holder. A water cooled shield between the cathode filament and the samples was used to reduce the heat arriving at the samples from the filament. This source was used in conjunction with the asher to determine the role of atomic oxygen arrival directionality and energy on undercutting.

## 3. Results and discussion

*3.1. Determining the extent of undercutting*

The amount of undercutting, at defect sites formed during the deposition process, as a function of time was studied by exposing a sample of $SiO_2$-coated Kapton® in a plasma asher and removing it at selected time intervals for observation in a scanning electron microscope. Figure 1 shows a series of photomicrographs of a selected area on the coated surface as a function of the effective fluence (based on Kapton® mass loss area$^{-1}$ and an erosion yield of $3 \times 10^{-24}$ $cm^3$ atom$^{-1}$) incident upon the sample surface in the plasma asher. The light or dark ring surrounding each defect indicates the extent of coating undercutting. This is visible due to charging of the coating surface in the electron beam at points where the coating is no longer in contact with the underlying substrate when the electron beam energy is high enough. By measuring the distance between the center of the defect and the average edge of the undercutting ring, a plot of the undercutting extent *vs.* atomic oxygen fluence could be made (Fig. 2). The plots indicate a slow initial growth in the undercutting radius which passes through a maximum before

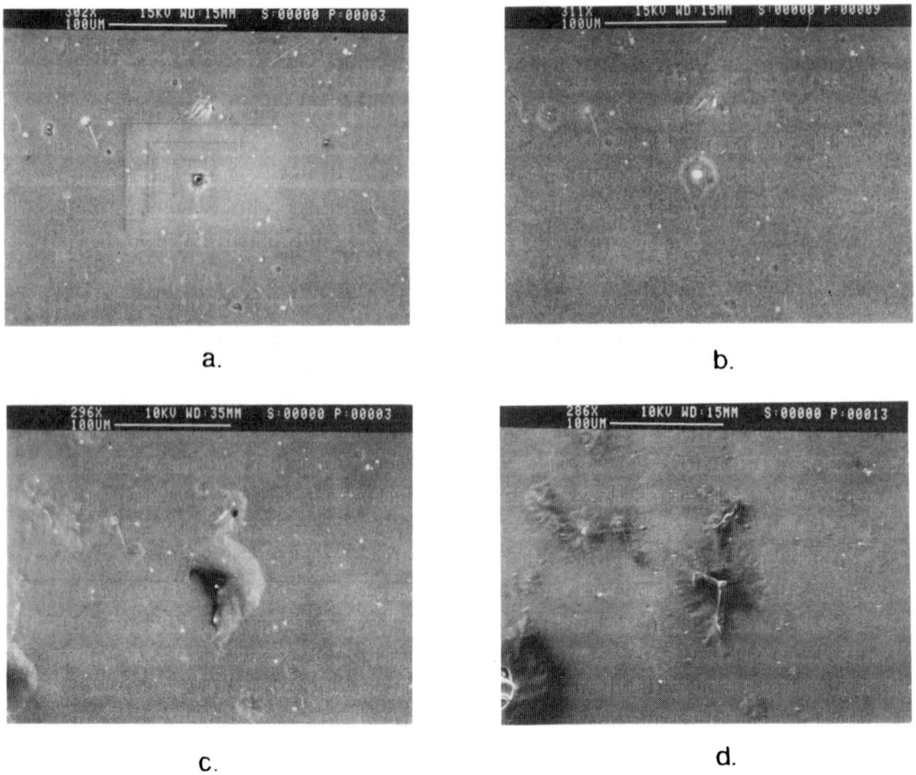


Fig. 1. Scanning electron photomicrographs of $SiO_2$-coated Kapton® exposed in an atomic oxygen plasma asher at various fluence levels: a, $2.95 \times 10^{20}$ atoms $cm^{-2}$; b, $4.6 \times 10^{20}$ atoms $cm^{-2}$; c, $1.0 \times 10^{21}$ atoms $cm^{-2}$; d, $1.57 \times 10^{21}$ atoms $cm^{-2}$.

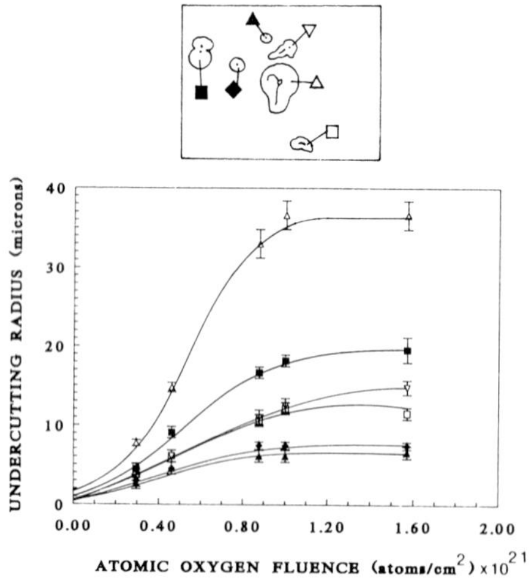


Fig. 2. Undercutting extent of defects on $SiO_2$-coated Kapton® as a function of atomic oxygen fluence during exposure in a plasma environment.

slowing to near zero growth after an effective atomic oxygen fluence of approximately $1 \times 10^{21}$ atoms $cm^{-2}$ for this sample. The photomicrographs appear to show the coating collapsing into the undercut area which could account for the reduction in the growth of the undercutting radius. These types of growth data, however, have been observed for defects which remain open through the duration of the atomic oxygen exposure.

Curve shape may be driven by several factors. Only about 11% of the atomic oxygen which lands on a carbon surface in LEO undergoes a chemical reaction with the surface, the remainder is scattered [6]. Scattering is believed to occur in almost a cosine distribution from the surface [7]. The reaction probability is believed to be energy dependent, with more energetic atoms having a higher reaction probability than those at lower energy [8]. Also since these measurements are made for a two-dimensional surface, the extent of the depth of oxidation into the substrate is not observed, but could account for the initial increasing slope of the undercutting radius curves. A combination of these factors could provide a possible explanation for the growth and cessation of undercutting observed for these samples (Fig. 3).

Initially, the opening in the coating would allow the polymer exposed by the defect to be oxidized but little undercutting would be observed because the material near the surface must be removed to a level where scattering of atomic oxygen under the coating could take place. Once this occurs, the rate

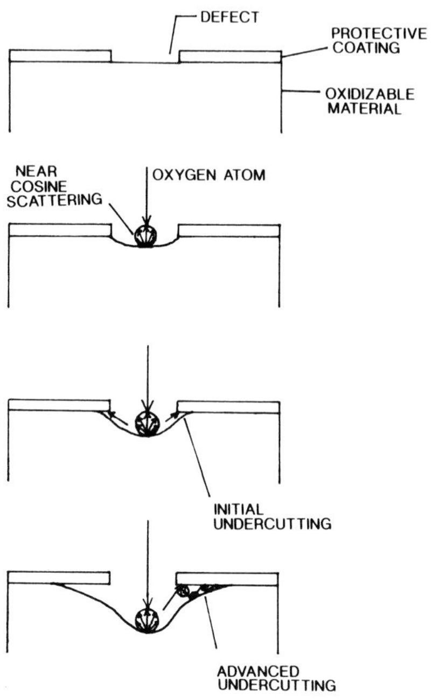


Fig. 3. A possible scenario for undercutting of a defected coating in an atomic oxygen environment.

of undercutting would increase. Some of the atomic oxygen will be scattered back out of the opening, and the probability of this would increase with the size of the coating defect, so for large defects it would be expected that smaller undercutting radii with respect to the size of the defect would be seen with exposure time. As time progresses, the atomic oxygen would have to undergo multiple scattering off of the polymer surface and the underside of the coating in order to allow the undercut region to grow outward. The farther it extends, the less likely it will be for the atomic oxygen to arrive at that point, without having undergone reaction, reassociation or loss out of the defect. More testing is needed at higher atomic oxygen fluence levels to determine the extent of oxidation depth and whether undercutting does terminate as initial testing indicates. Modeling is needed in order to determine which factors are important.

For this sample of coated Kapton®, out of approximately 113 800 defects $cm^{-2}$ visible on the surface, about 21 800 actually showed signs of undercutting. This means that a visual observation of a coated surface is not enough to determine if the coating has atomic oxygen pathways to the Kapton® surface; it must be subjected to an atomic oxygen environment. In low Earth orbit, the micrometeoroid environment has been predicted by the 1987 micrometeoroid model to provide approximately 1 hit $cm^{-2}$ over 15 years for $10^{-12}$ g or larger particles [5]. Debris is predicted to provide less than 1000 hits $cm^{-2}$ over 15 years [5]. The two combined represent approximately 20 times fewer oxidizable defects than are produced by coating and handling. Defects that can occur on the ground seem to be a greater problem, more worthy of addressing.

If intrinsic stresses in the protective coating cause the coating to tear due to lack of support in the area of the undercutting, then one would expect to see a continued growth in the undercut radius [9]. Figure 4 shows an example of a reactively magnetron sputter deposited coating of silicon dioxide on Kapton® that does not continue to protect because the coating has torn as the undercutting progressed allowing more of the underlying

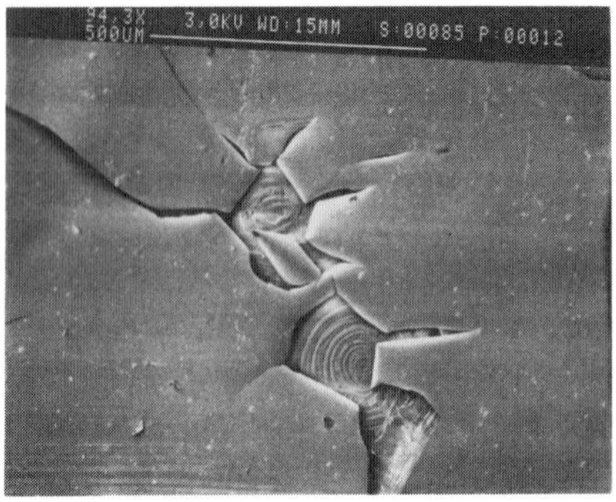


Fig. 4. Reactive magnetron sputter deposited $SiO_2$ on Kapton® after exposure to an equivalent atomic oxygen fluence of approximately $10^{22}$ atoms $cm^{-2}$.

surface to be exposed. This can lead to total failure of the protective coating. Deposition of a low stress or adherent coating is needed to prevent this.

*4.2. Energetic directed vs. a thermal random oriented atomic oxygen arrival*

In the plasma of the asher, the atomic oxygen arrives with low energy (0.1 eV) from many different angles while in space the arrival is closer to a rotated, directed beam at higher energy (4.2 - 4.6 eV). To bound the degree of undercutting one might expect in LEO, data from the plasma asher was compared to that from a 57 eV directed oxygen ion beam exposure. The energetic ions would be expected to have a higher first impact reaction probability than in the plasma asher or LEO. Thus one would expect less undercutting, but deeper erosion of the Kapton® at the site of the defect. Samples of the $SiO_2$-coated Kapton® were exposed to atomic oxygen in the plasma asher and the directed beam for an equivalent fluence of $1.5 \times 10^{21}$ atoms $cm^{-2}$ and $1.2 \times 10^{21}$ atoms $cm^{-2}$ respectively, which corresponds to almost 1 year in the Space Station Freedom constant density orbit [10]. Figure 5 shows the photomicrographs of these samples after exposure.

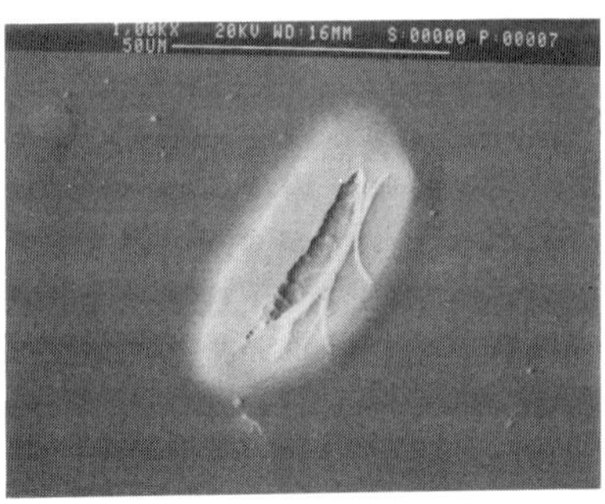

a.

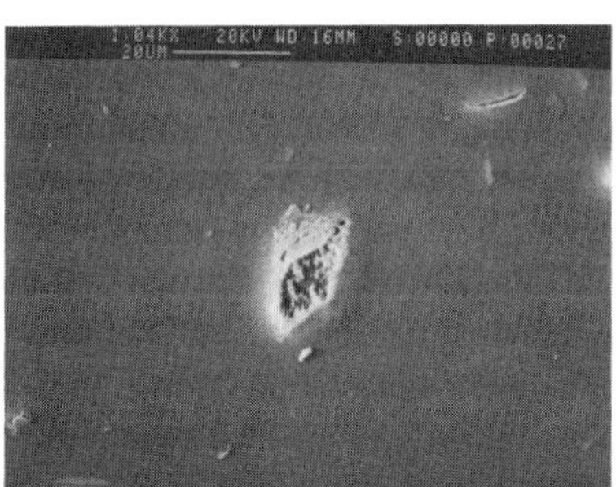

b.

Fig. 5. Photomicrographs of $SiO_2$-coated Kapton® after exposure to atomic oxygen. a, Plasma asher effective fluence of $1.5 \times 10^{21}$ atoms $cm^{-2}$; b, directed beam effective fluence of $1.2 \times 10^{21}$ atoms $cm^{-2}$.

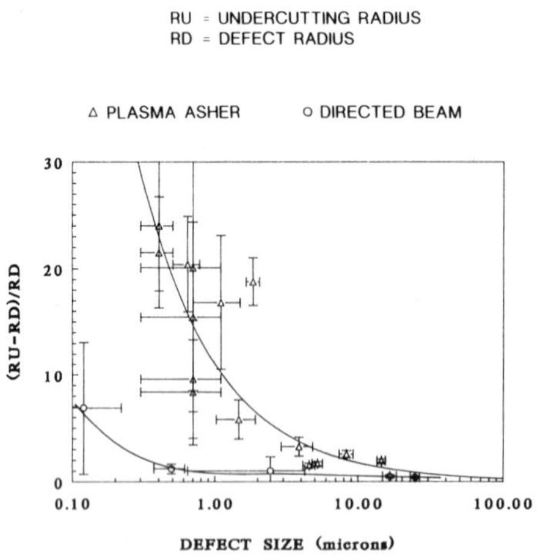


Fig. 6. Extent of undercutting in comparison to the original defect size for plasma asher exposed and directed beam exposed $SiO_2$-coated Kapton®.

Although undercutting is present for the sample exposed in the energetic directed beam (Fig. 5b), the extent of undercutting in comparison to the size of the defect is smaller than for the plasma asher exposed defects. This is illustrated in Fig. 6. In general, the radius of undercutting compared to the defect size is smaller for larger defects in both cases as would be expected. A directed atomic oxygen bean rotating around a surface, which is closer to the arrival in LEO, would tend to produce more undercutting than a fixed directed beam.

## 5. Conclusions

Tests to date have shown that the growth of undercutting appears to be limited with atomic oxygen exposure on coated polyimide surfaces in which the protective coating stays in contact with the substrate. Therefore, the production of defect free coatings, although desirable, is not as important as the ability of the coating to remain intact as undercutting occurs. Thus deposition of low intrinsic stress, tear resistant films may be critical to long term durability in LEO. Large defects do not produce proportionately large areas of undercutting, therefore size should not cause great concern provided the defects are not long and interconnecting which could provide a structurally weak point. Methods of ground evaluation for defects should involve the testing of coupons of the coated material with ground based atomic oxygen plasma devices in order to accurately determine the number of defects produced. The number of defects on the surface produced in the coating process and the mechanical integrity of the coating, are quality control issues extremely pertinent to the survival of polymers to be used on Space Station Freedom and for other LEO space systems if long term durability against attack by atomic oxygen is to be achieved.

## References


1 NOAA, ANSA and USAF, *U.S. Standard Atmosphere*, 1976.

2 B. A. Banks, M. J. Mirtich, S. K. Rutledge and H. K. Nahra, Protection of solar array blankets from attack by low earth orbital atomic oxygen, *Proc. IEEE Photovoltaic Specialists Conf.*, Las Vegas, Nevada, October 21 - 25, 1985.

3 C. R. Baraona, The Space Station Power System, NASA TM-88847, *Fifth Conf. on Photovoltaic Generators in Space*, European Space Agency, Noordwijk, Netherlands, September 30 - October 4, 1986.

4 S. K. Rutledge, B. A. Banks, F. DiFilippo, J. Brady, T. Dever and D. Hotes, An evaluation of candidate oxidation resistant materials for space applications in LEO, NASA TM-100122, *Proc. NASA Workshop on Atomic Oxygen Effects*, Sponsored by NASA JPL, Pasadena, CA, November 10 - 11, 1986.

5 M. J. Mirtich, H. Mark and W. R. Kerslake, The effect of the near Earth micrometeoroid environment on a highly reflective mirror surface, NASA TM-101307, *Proc. the 26th Aerospace Sciences Meeting*, Sponsored by the AIAA, Reno, Nevada, January 11 - 14, 1988.

6 C. K. Purvis, D. C. Ferguson, D. B. Snyder, N. T. Grier, J. V. Staskus and J. C. Roche, *Environmental interactions considerations for space station and solar array design, Preliminary Report*, July, 1985.

7 J. C. Gregory, Interaction of hyperthermal atoms on surfaces in orbit: The University of Alabama Experiment, *Proc. NASA Workshop on Atomic Oxygen Effects*, Sponsored by NASA JPL, Pasadena, CA, November 10 - 11, 1986.

8 D. C. Ferguson, The energy dependence and surface morphology of Kapton degradation under atomic oxygen bombardment, *Proc. of the 13th Space Simulation Conf*, Orlando, Florida, October 8 - 11, 1984.

9 B. Banks and S. Rutledge, Low Earth orbital atomic oxygen simulation for materials durability evaluation, *Proc. 4th Int. Symp. on Spacecraft in the Space Environment*, Toulouse, France, September 6 - 9, 1988.

10 L. Leger, J. Visentine, and B. Santos-Mason, Selected materials issues associated with space station, *Proc. 18th Int. SAMPE Technical Conf.*, October 7 - 9, 1986.

*Surface and Coatings Technology, 39/40* (1989) 617 - 626 

# EVALUATION OF PLASMA-DEPOSITED PROTECTIVE COATINGS FOR MULTIPURPOSE SPACE APPLICATIONS

D. G. ZIMCIK

*Canadian Space Agency, Ottawa, Ontario (Canada)*

M. R. WERTHEIMER

*Ecole Polytechnique, Montreal, Quebec (Canada)*

K. G. BALMAIN and R. C. TENNYSON

*University of Toronto, Toronto, Ontario (Canada)*

(Received March 24, 1989)

## Summary

This paper reports evaluations of protective thin films applied to important polymers used on spacecraft, Kapton® polyimide and graphite–epoxy, that are attacked by atomic oxygen. These films, derived from volatile compounds through microwave glow discharge deposition, include amorphous silicon (a-Si:H) and inorganic silicon compounds (P-SiN and P-$SiO_2$). These materials are shown to provide excellent protection against atomic oxygen attack, as evidenced by the reduction in mass loss compared to unprotected specimens of the same materials after exposure to an atomic oxygen beam source. Scanning electron microscope photomicrographs are presented to confirm the lack of apparent attack of the surface. Results from high-energy (20 keV) electron beam bombardment show that these coatings reduce charge build-up which might otherwise result in electrostatic discharge phenomena. These materials provide promising candidates for coatings for exterior surfaces of spacecraft to protect the underlying materials from the deleterious effects of the space environment.

## 1. Introduction

Space systems operating in low Earth orbit (LEO) environment for long periods of time, such as the planned International Space Station, require materials which are stable in that environment. It is known from Shuttle flight experience that certain organic (*e.g.* polymeric) and metallic materials, used in structural and thermal protection applications, are attacked and eroded by atomic oxygen. Experience has shown that even a short-term exposure to this environment can have harmful effects on spacecraft surfaces, particularly if these are composed of or covered with organic materials [1 - 4].

® indicates a registered trademark of E. I. DuPont de Nemours & Co.

0257-8972/89/$3.50 

Organic polymers are important to the design of present and future generations of spacecraft, as they are used in advanced composites for structural members, and for thermal protection in the form of films and coatings. When organic materials are exposed to the highly oxidizing oxygen atoms, especially at the elevated velocities of spacecraft, the observed result is rapid erosion (loss of mass) and surface roughening. This, in turn, leads to irreversible degradation of the physical characteristics (optical, electrical and mechanical) for which the surface or structural members were designed.

At higher altitudes, charge build-up due to impinging electrons and ions in the Van Allen Belts and solar wind (*e.g.* in geostationary orbit) has other deleterious effects on insulating spacecraft materials such as organic polymers. When charge build-up on an exterior surface reaches a certain critical level, the result is surface discharge arcs or flashover, which can permanently damage the polymer structure and integrity [5]. Furthermore, electrical transients associated with these discharge events can perturb or destroy electronic circuitry aboard the space vehicle. These events have been shown to cause loss or near loss of spacecraft [6].

The present paper reports results of evaluations of thin protective layers [7] applied to important polymers used on spacecraft: Kapton® polyimide and graphite–expoxy, which have been shown to be attacked by atomic oxygen. These protective coatings, derived from volatile compounds through microwave glow discharge, include amorphous hydrogenated silicon (a-Si:H), and inorganic silicon compounds (silicon nitride, P-SiN; oxide, P-$SiO_2$; and oxynitride, P-SiON). These microwave plasma-deposited coating materials evaluated in this study provide an unique opportunity to tailor material properties through the process parameters in the plasma reactor to meet mission specific needs. These materials provide promising candidates for coatings for exterior surfaces of spacecraft to protect the underlying materials from the deleterious effects of the space environment.

## 2. Protective coatings for spacecraft materials

It is possible to apply very thin coatings onto the exposed surfaces of organic spacecraft materials in order to protect these against the harmful effects described above. Experiments with thin layers of polytetrafluoroethylene (PTFE, produced by plasma spraying) [2]; sputtered or vapour deposited indium–tin oxide (ITO) [2]; and a $SiO_2$-polytetrafluoroethylene (PTFE) mixture prepared by ion beam sputtering [8], have been reported by other workers. It was found, in both ground-based laboratory and in-space tests, that these thin protective coatings can act as barriers against oxidation erosion of the underlying polymer.

The object of the present study was to evaluate the performance of thin barrier coatings applied to spacecraft materials by low pressure plasma (glow discharge) deposition (patents pending). Performance characteristics of various coating materials were measured in a simulated space environment to evaluate the resistance to atomic oxygen attack and electrostatic charging. Other performance characteristics including thermal radiative properties and flexibility are reported in refs. 7, 11 and 12.

The use of glow discharges for preparing thin solid films is well known [9 - 13]. A glow discharge is established in a reagent gas flow, which results in a plasma-chemical reaction leading to the desired thin film deposit. Various types of thin film coatings can be prepared, namely

(a) plasma polymers,
(b) inorganic insulating films,
(c) semiconducting and conducting materials.

In category (a), films may be derived from hydrocarbon, fluorocarbon, or organosilicon monomer compounds. Typical category (b) materials are silicon compounds such as plasma silicon dioxide (P-$SiO_2$), silicon nitride (P-SiN) or silicon oxynitride (P-SiON). These may be prepared by reacting silane ($SiH_4$) with an appropriate second gas (or mixture) such as $N_2O$, $NH_3$, $N_2$, etc. Films in category (c) include amorphous, hydrogenated silicon (a-Si:H) or germanium, prepared from $SiH_4$ and $GeH_4$ gas respectively, and metal films prepared from volatile organometallic compounds. Among the above-mentioned materials many can, in principle, provide suitable barriers except those prepared from hydrocarbon reagents which are not stable against oxygen atom attack. One of the most interesting of these materials is amorphous silicon. It has been shown that changes in the processing conditions, for example the addition of gaseous dopants like phosphine or diborane, reagent flow rate, substrate temperature, power density in the plasma and reactor pressure can very significantly alter the electrical conductivity of this material [9, 14].

A series of specimens from categories (a) to (c) was fabricated in this study. The materials, shown in Table 1, included silicon compounds, P-SiN, P-$SiO_2$, P-SiON and a-Si:H; and organosilicon (hexamethyldisiloxane, PP-HMDSO). As shown in ref. 10, all of these films can be prepared using the microwave plasma reactor as very thin, strongly adhering and pinhole-free

TABLE 1
Coatings on Kapton® polymide substrate

| *Specimen* | *Coating* | *Coating conditions*[a] | | | |
|---|---|---|---|---|---|
| | | $T_s$ (°C) | $SiH_4$ (sccm) | *Gas/Flow* (sccm) | *Thickness* (μm) |
| 1 | $SiO_2$ | 250 | 8 | $N_2O$/35 | 0.7 |
| 2 | SiN | 250 | 10 | $NH_3$/35 | 0.5 |
| 3 | a-Si:H | 300 | 10 | — | 0.5 |
| 4 | a-Si:H | 300 | 8 | He/35 | 0.5 |
| 5 | a-Si:H | 300 | 10 | He/35 | 0.7 |
| 6 | HMDSO | 250 | — | HMDSO/10 | 0.7 |

[a]Pressure, 100 m Torr; microwave power; 100 W.

over-coats on various substrate materials including the polymers used in the present study, as well as other polymers such as Mylar®.

A microwave plasma reactor was used to deposit the thin film materials onto Kapton® polyimide and graphite–epoxy composite specimens. Microwave power was coupled into the plasma chamber from a slow-wave applicator (periodic travelling wave structure) and the substrate holder was rotated to minimize any non-uniformities of thickness over the specimen. More details of the apparatus and the plasma coating procedures may be found in refs. 10 - 13.

## 3. Mass loss due to atomic oxygen exposure

Coated specimens of 13 $\mu$m aluminized Kapton® film were exposed in an atomic oxygen beam simulation facility described in detail in ref. 15. Briefly, this facility is comprised of three major components: a microwave induced plasma torch, a sampler-skimmer interface and a vacuum chamber with the associated support electronics. The plasma torch is used to generate a stream of essentially neutral oxygen atoms seeded in a helium gas carrier. The sampler–skimmer system strips off a portion of the lighter gas carrier from the plasma and produces a diverging atomic oxygen beam which is then directed into the connecting vacuum chamber. The vacuum chamber maintains candidate specimens at about $10^{-5}$ Torr to approximate LEO conditions. The input power to the plasma determines the energy level of the incoming atoms, which was held constant for these tests at 2.2 eV. Flux at the specimen location is determined by both input flow rate and specimen position. The flux used in these tests was approximately $8 \times 10^{16}$ atoms (cm$^2$ s)$^{-1}$ which provides a slight acceleration factor on test time over that experienced in a nominal Shuttle orbit.

Each specimen was subjected to a fluence of approximately $3 \times 10^{20}$ atoms cm$^{-2}$, although a second specimen coated with $SiO_2$ was also subjected

TABLE 2
Mass loss measurements

| *Specimen* | *Coating* | *Coating thickness* ($\mu$m) | *Fluence* (atoms cm$^{-2}$) | *Mass loss* ($\mu$g) |
|---|---|---|---|---|
| 1 | $SiO_2$ | 0.7 | $3 \times 10^{21}$ | $0.00 \pm 0.02$ |
| 1a | $SiO_2$ | 0.7 | $3 \times 10^{20}$ | $0.01 \pm 0.02$ |
| 2 | SiN | 0.5 | $3 \times 10^{20}$ | $0.02 \pm 0.02$ |
| 3 | a-Si:H | 0.5 | $3 \times 10^{20}$ | $0.00 \pm 0.02$ |
| 4 | a-Si:H | 0.5 | $3 \times 10^{20}$ | $0.04 \pm 0.02$ |
| 5 | a–Si:H | 0.7 | $3 \times 10^{20}$ | $0.04 \pm 0.02$ |
| 6 | HMDSO | 0.7 | $3 \times 10^{20}$ | $0.00 \pm 0.02$ |
| Kapton® (ref) | — | — | $3 \times 10^{20}$ | $0.43 \pm 0.02$ |

Flux = $8 \times 10^{16}$ atoms (cm$^2$ s)$^{-1}$.

to a higher fluence of $3 \times 10^{21}$ atoms cm$^{-2}$. Duplicate specimens were located inside the vacuum chamber but protected from the oxygen beam to be used as reference for mass loss measurements. Also, specimens were kept in a desiccator for 24 h prior to initial measurements. As expected, the unprotected Kapton® used as a comparative standard to flight experiment data, showed significant mass loss and visible surface change. In fact, for the 13 $\mu$m thick material used, the material over a 1 cm diameter circle in the centre of the beam was completely removed leaving only the aluminum coating on the back. However, the materials covered with all of the candidate protective coatings showed no visible signs of attack by the atomic oxygen beam. Mass loss measurements made using an electronic balance are shown in Table 2. No meaningful loss of mass was measured for any of the coating materials after exposure.

## 4. Surface analysis

Detailed examinations were made of the coated surfaces both before and after oxygen beam exposure using a scanning electron microscope (SEM). Photomicrographs of the surface before and after oxygen beam exposure are shown in Figs. 1, and 2 for coatings of two material types investigated.

(a) (b) (c)

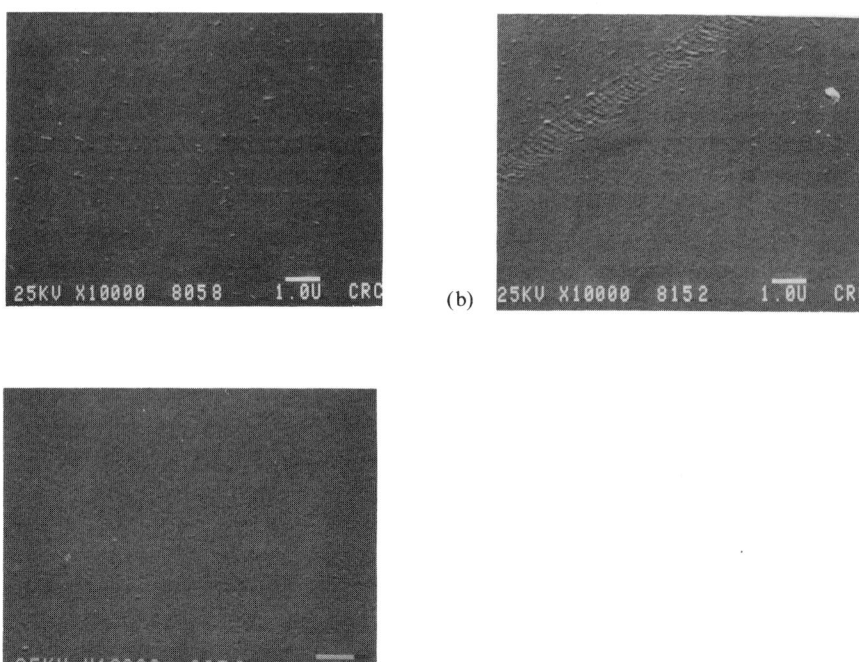


Fig. 1. SEM photomicrographs of P-$SiO_2$ coating on Kapton® before and after oxygen exposure. (a) Before oxygen beam exposure; (b) after oxygen beam exposure ($3 \times 10^{20}$ atoms cm$^{-2}$); (c) after oxygen beam exposure ($3 \times 10^{21}$ atoms cm$^{-2}$).

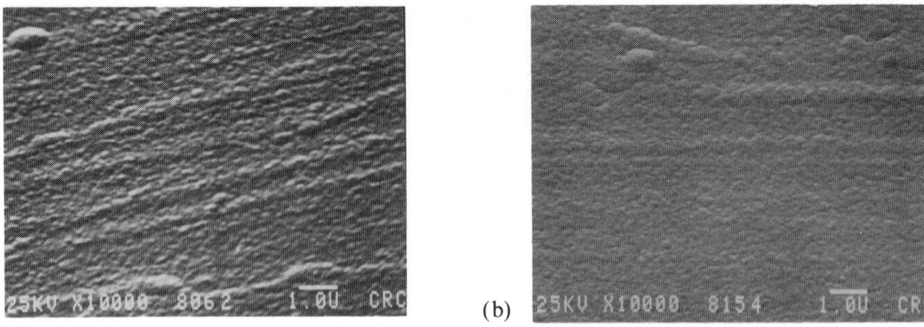


(a) (b)

Fig. 2. SEM photomicrographs of PP-HMDSO coating on Kapton® before and after oxygen exposure. (a) Before oxygen beam exposure; (b) after oxygen beam exposure ($3 \times 10^{20}$ atoms $cm^{-2}$).

Photomicrographs of other materials evaluated were similar. For comparative purposes, a photomicrograph of bare Kapton® exposed in the same facility is shown in Fig. 3. As would be expected from the results of mass loss measurements, no distinguishing features were apparent that would indicate significant erosion of the surface of any of the coating materials. Rather, the surfaces before and after exposure were very similar in appearance indicating little effect due to the oxygen atoms. This is also true of the specimens coated with $SiO_2$ which were exposed to a total fluence of an order of magnitude greater than most past Shuttle flight test data that have been shown to severely erode this material [2, 3]. This confirms significantly improved results over other coating techniques [16] that have reported good results for lower levels of fluence but observed signs of erosion at higher fluence levels due to the presence of pin-holes in the coating. It should be noted that the thickness of the coating was approximately 0.5 - 1 $\mu$m.

The removal of organic materials by oxygen plasma is among the earliest applications of plasma chemistry. This process has also been shown [17] to be an effective screening process for atomic oxygen resistant materials. Results from these works have shown that materials that are attacked by

Fig. 3. SEM photomicrograph of uncoated Kapton® after oxygen exposure.

(a) (b) (c)

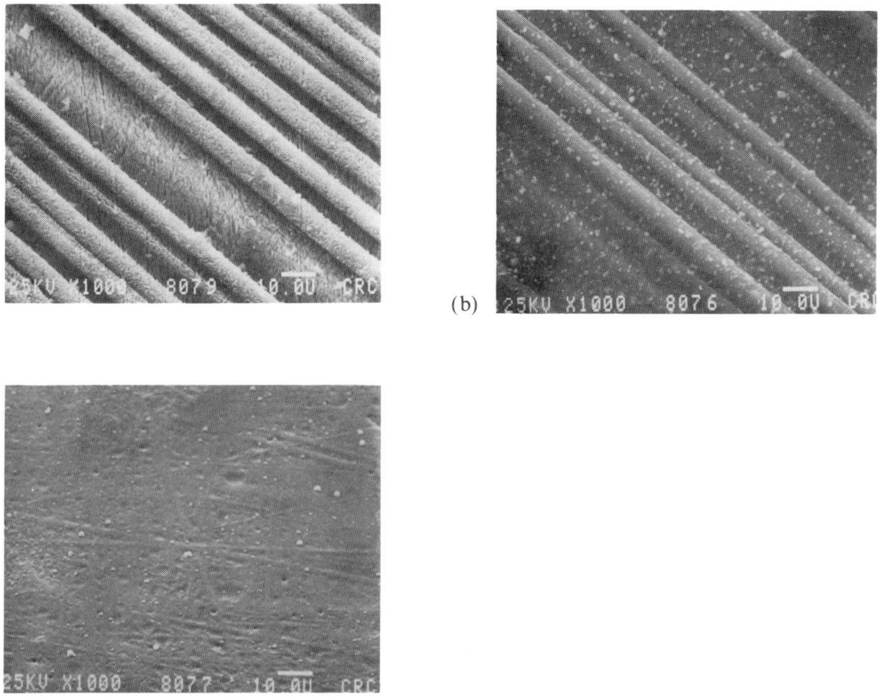


Fig. 4. SEM photomicrographs of graphite–epoxy with and without a-Si:H coating. (a) True space exposure from STS 41-G ($3 \times 10^{20}$ atoms $cm^{-2}$) (without protective coating); (b) after oxygen plasma exposure (without protective coating); (c) after oxygen plasma exposure (with protective coating).

atomic oxygen in Shuttle flight experience, are also attacked by an oxygen plasma. Such an approach was used to evaluate a-Si:H coating for graphite–epoxy. Examinations were made of the coated surfaces both before and after oxygen plasma exposure. The latter treatment occurred in the reactor of ref. 13 using the methodology described in the same reference. SEM photomicrographs taken after oxygen plasma exposure of the surface of graphite–epoxy with and without protective coating are shown in Fig. 4 and compared to bare graphite–epoxy material exposed in space during Shuttle mission STS 41-G as part of the Advanced Composite Materials Exposure to Space Experiment (ACOMEX) [2]. The bare material from both the true space exposure (ACOMEX) and the oxygen plasma erosion are remarkably similar in appearance. The initially "smooth" epoxy surface has been eroded exposing the fibers in a "corduroy" fashion although the apparent attack on the fibers by the oxygen plasma is not as marked. Nevertheless the images confirm the use of an oxygen plasma as a screening method. In contrast, the specimen coated with a-Si:H shows no attack of the original surface of the epoxy verifying the effectiveness of the protective coating.

## 5. Electrostatic discharge testing

To evaluate the tendency to arc-discharge in an energetic electron environment, two coating materials, a-Si:H (two specimens) and P-SiN, were applied to 51 $\mu$m thick Kapton® and tested in an electron-beam apparatus employing 20 keV electrons [18]. Each specimen was mounted behind a 15.9 $cm^2$ aperture in a metal mask and over a metal substrate. The experimental sequence involved a 20 min exposure at each of three current densities 25, 50 and 100 nA $cm^{-2}$. This exposure sequence was terminated early if sufficient data were acquired (four or more discharges). The exposure sequence was performed first on the coated side of the material, and then on the uncoated (reverse) side to provide a comparison or "control" case. When a surface arc discharge occurred, both mask and substrate currents were measured, with the mask current magnitudes typically being 50–80% of the substrate current. In two cases the coatings eliminated arc discharges completely, while in the third case, the coating reduced the arc discharge strength by 50%. Also, the coated specimens exhibited reduced long-term charge retention. The results are summarized in Table 3 and further details are described below.

Two specimens of a-Si:H coating which were slightly different in composition (due to changes in processing parameters) and thickness were evaluated. On the coated side of the first specimen with a coating 1.0 $\mu$m thick, there were no discharges while on the uncoated, reverse side there were three discharges averaging 175 A peak substrate current. For the second specimen with an a-Si:H coating 1.5 $\mu$m thick, there were five discharges averaging 85 A when exposed on the coated side, while on the uncoated, reverse side there were six larger discharges averaging 183 A. Furthermore, the arc patterns on the uncoated side were observed to cover the entire dielectric surface exposed to the electron beam, producing the very strong arcs measured. The coated specimens tended to flatten out under the electron beam due to electrostatic attraction suggesting that some charging was occurring. However, they did not stick to the substrate upon removal from the vacuum, whereas all uncoated specimens stuck to the substrate, indicating long-term charge retention.

TABLE 3
Electrostatic discharge results

| *Specimen*[a] | *Coating* | *Coating thickness* ($\mu$m) | *Discharges* | *Substrate current* (A) |
|---|---|---|---|---|
| 1 | a-Si:H | 1.0 | 0 | — |
| | reverse side | — | 3 | 175 |
| 2 | a-Si:H | 1.5 | 5 | 85 |
| | reverse side | — | 6 | 183 |
| 3 | P-SiN | 0.5 | 0 | — |
| | reverse side | — | 4 | 53 |

[a]Kapton® substrate.

One specimen with a P-SiN coating 0.5 $\mu$m thick was tested. There were no discharges on the coated side, while on the uncoated, reverse side there were four discharges measured. Of these, peak currents over 53 A were recorded for two discharges. The coated side of the specimen exhibited a faint luminescence and tended to flatten against the substrate under the electron beam, indicating slight charging. After removal from the vacuum, the coated material showed no remaining electrostatic attraction and did not stick to the metal substrate.

## 6. Conclusions

Evaluation of thin barrier coatings applied to typical spacecraft polymeric materials using a microwave plasma reactor have been conducted. The coating materials have been shown to be resistant to attack by atomic oxygen while also acting to prevent charge build-up that might lead to harmful electrostatic discharge. SEM photomicrographs of the coated surfaces before and after exposure to an atomic oxygen beam source confirm that the surfaces were not attacked by the beam. In conclusion, these materials provide promising candidates for protective coatings for exterior surfaces of spacecraft to protect the underlying materials from the deleterious effects of the space environment.

## Acknowledgments

The authors gratefully acknowledge the contributions of J. E. Klemberg-Sapieha, J. Noad, Dr. W. D. Morrison and G. Dubois who assisted in the preparation of the coatings and subsequent investigations. This work has been supported in part by research grants from the Natural Science and Engineering Research Council (NSERC), Institute for Space and Terrestrial Science (ISTS) and Fonds Formation de Chercheurs et Aide à la Recherche de Québec (FCAR).

## References

1 D. G. Zimcik, R. C. Tennyson, L. J. Kok and C. R. Maag, *Proc. Third European Symp. on Spacecraft Materials in Space Environment, ESA SP-232,* 1985.
2 D. G. Zimcik, and C. R. Maag, *AIAA J. Spacecraft Rockets, 25* (1988) 162.
3 J. T. Visentine, L. J. Leger, J. F. Kuminecz and I. K. Spiker, *AIAA 23rd Aerospace Sciences Meeting, 85-0415,* 1985.
4 D. G. Zimcik, *CASI J., 33* (1987) 4.
5 K. G. Balmain and W. Hirt, *IEEE Trans, EI-18* (1983) 498.
6 E. R. Schmerling (ed.), *Proc. Conf. on The Aerospace Environment at High Altitude and Its Implications for Spacecraft Charging and Communications, AGARD-CP*-406, 1986.
7 J. E. Klemberg-Sapieha, M. R. Wertheimer and D. G. Zimcik, *Proc. Fourth ESA Int. Symp. on Spacecraft Materials in Space Environment,* 1988.

8 B. A. Banks, M. J. Mirtich, S. K. Rutledge, D. M. Swec and H. K. Nahra, *AIAA 23rd Aerospace Science Meeting*, 1985.

9 J. Mort and F. Jansen, *Plasma Deposited Thin Films*, CRC Press, Boca Raton, FL, 1986.

10 M. R. Wertheimer, J. E. Klemberg-Sapieha and H. P. Schreiber, *Thin Solid Films, 115* (1984) 109.

11 Y. Tessier, J. E. Klemberg-Sapieha, S. Poulin-Dandurand, M. R. Wertheimer and S. C. Gujrathi, *Can. J. Phys., 65* (1987) 859.

12 S. Blain, J. E. Klemberg-Sapieha, M. R. Wertheimer and S. C. Gujrathi, *Can. J. Phys., 67* (1989) 190.

13 B. Lamontangne, A. M. Wrobel, G. Jalbert and M. R. Wertheimer, *J. Phys. D., Appl. Phys., 19* (1987) 844.

14 J. L. Brebner *et al., Can. J. Phys., 63* (1985) 786 - 797.

15 D. Morrison, R. C. Tennyson and J. B. French, *Proc. ESA Fourth Int. Symp. on Spacecraft Materials in Space Environment*, 1988.

16 J. B. Cross, E. H. Lan and C. A. Smith, *Proc. ESA Fourth Int. Symp. on Spacecraft Materials in Space Environment*, 1988.

17 M. McCargo, R. A. Dammann, T. Cummings and C. Carpenter, *Proc. Third European Symp. on Spacecraft Materials in Space Environment, ESA SP-232.* 1985.

18 K. G. Balmain, *Proc. ESA Symp. on Spacecraft Materials in Space Environment, ESA SP-145,* 1979.

*Surface and Coatings Technology, 39/40* (1989) 627 - 636 

# ATOMIC OXYGEN RESISTANT PROTECTIVE COATINGS FOR THE HUBBLE SPACE TELESCOPE SOLAR ARRAY IN LOW EARTH ORBIT

H. D. BURNS, A. F. WHITAKER and R. C. LINTON

*Materials and Processes Laboratory, NASA/Marshall Space Flight Center, Marshall Space Flight Center, AL 35812 (U.S.A.)*

(Received March 9, 1989)

## Summary

The Hubble Space Telescope back-up solar array (BSFR-ATOX) was designed with atomic oxygen resistant materials to replace the original high efficiency cell (HEC) array. Pending qualification of the BSFR-ATOX array, protective coatings for application to the existing HEC array were evaluated. Protective coatings would extend the material life, thereby extending the solar array useful life. Three polysiloxane-type coating materials were evaluated, SWS V-10, DC-100 and CV-1144. Coatings were evaluated with respect to atomic oxygen resistance, outgassing, contamination potential, stability under ultraviolet irradiation, electron–proton irradiation, thermal cycling and long-term storage after application. Atomic oxygen testing included characterization of the atomic oxygen plasma using comparison of Kapton reaction to existing flight data, residual gas analysis and temperature measurements. Results from these tests show DC-100 as a greater contamination concern if used in the vicinity of sensitive optical surfaces, *i.e.* solar cell cover glasses. The SWS V-10 and CV-1144 coatings met all acceptance criteria.

## 1. Introduction

The Hubble Space Telescope (HST) is powered by a high power-to-weight ratio solar array (STSA). This solar array consists principally of silicon solar cells mounted on a flexible polyimide based structural substrate. The European Space Agency (ESA) in cooperation with NASA designed and built the first STSA known as the “high efficiency cell” (HEC) solar array. A 3 year minimum life requirement on this array which resulted from launch timeline considerations led to reassessment of the array materials. The projected launch date for the delayed mission would result in greater orbital degradation due to increased solar activity, *i.e.* increased atomic oxygen (AO) flux which would lead to increased materials erosion.

0257-8972/89/$3.50 

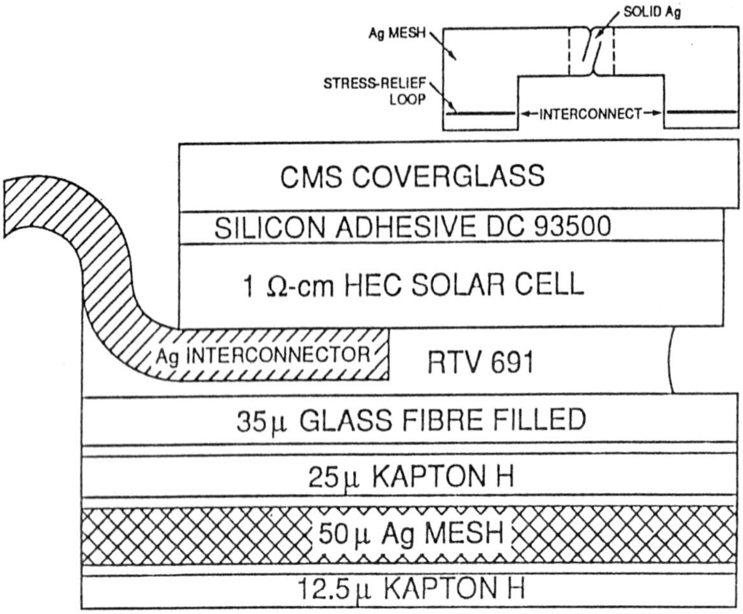


Fig. 1. HST HEC solar array cross-section.

Highly reactive AO in low Earth orbit (LEO) would limit the useful life of the HEC solar array due mainly to its effect on solar cell silver interconnects, the polyimide Kapton blanket substrate on the array backside (see Fig. 1) and composite hinge material.

In early 1987, a coordinated effort was begun to determine the feasibility of applying protective coatings to the existing HEC array pending qualification of the second generation "back surface field reflecting-atomic oxygen resistant" (BSFR-ATOX) array. These protective coatings were to be applied to AO susceptible materials extending the life of these materials, and therefore the life of the existing HEC solar array.

Candidate protective coatings were tested for their space environmental durability and compatibility with the existing HEC solar array. The material acceptance criteria, test methods and test results utilized in evaluating these materials are described.

## 2. Material selection criteria

Candidate coating material selection was primarily based on resistance to the combination of space environmental conditions. Space environmental compatibility requirements included contamination (outgassing), atomic oxygen resistance, ultraviolet (UV) irradiation stability, electron–proton irradiation stability and thermal stability. Coating application processes and coating effects on array deployment were also major considerations.

Outgassing tests were conducted per Marshall Space Flight Center (MSFC) Specification MSFC-SPEC-1443 [1], "Outgassing Test for Non-Metallic Materials Associated with Sensitive Optical Surfaces in a Space Environment," which is a modification of JSC-SP-R-0022A [2].

Candidate coatings must provide atomic oxygen protection for a minimum of 3 years. The coatings were selected to prevent erosion (weight loss) of Kapton and oxidation of the silver interconnects, thereby insuring structural stability and electrical continuity of the solar array.

Additional requirements included coating stability after exposure to HST mission levels of UV and electron–proton irradiation. Thermophysical properties, solar absorptivity ($\alpha_s$) and infrared thermal emissivity ($\epsilon_{IR}$), were measured to ensure minimum change over the array lifetime, since increases in the average array temperature would result in solar array power loss.

Coating stability to thermal cycling over the solar array operating range of $-100$ °C to $+100$°C was also a major consideration. The intrinsic deflection of silver interconnect stress relief loops due to thermal cycling required coating flexibility (elasticity). In addition to space environmental requirements, coating application to the existing array was complicated by the relative inaccessibility of the array's 99 520 silver interconnects requiring protection.

Finally, the coating applied to the array could not affect the performance of the deployment mechanism of the solar array. This requirement considered sticking (blocking) of array panels after long term storage.

## 3. Candidate protective coatings

Three candidate coatings received qualification testing. Other coatings were included in various tests where time and facilities permitted. Materials which received substantial testing included a modified non-curing Dow Chemical dimethyl polysiloxane, DC-100, which will be referred to as G-1B in the test. The second coating was SWS V-10, a curable dimethly polysiloxane. McGhan Nusil's CV-1144-0, a diphenyl dimethyl polysiloxane, was the third candidate.

## 4. Candidate coatings tests

### *4.1. Outgassing–contamination tests*

Candidate coating materials were tested in bulk form initially, then in thin film form (as-used condition) if they failed the previous test. These materials were required to pass a 24 h outgassing test at a maximum of $10^{-6}$ Torr and temperature equal to or above their use temperature. The material was required to have a total mass loss (TML) of 1.0% or less and a collectable volatile condensable material (VCM) content of 0.1% or less VCM was determined using collector plates held at or below the temperature of sensitive surfaces in the vicinity of the material being proposed. Optical witness samples (OWSs) included in the test were magnesium fluoride coated aluminum second surface mirrors or quartz transmission samples held at the same temperature as collector plates to determine effects of outgassing on sensitive optical surfaces. Reflectance or transmission sample respectively, over the appropriate wavelength ranges. Reflectance measurements were made over the vacuum UV (120 - 200 nm) wavelength range utilizing a Seya–Namioka vacuum UV spectrometer with a focal length of 0.5 m. An acceptance level of 3.0% or less change in reflectance at the 121.6 nm wavelength has been used for HST materials used in the vicinity of sensitive optical components.

Spectral transmittance measurements were made using a Beckman DK 2 spectrometer with a Gier–Dunkle integrating sphere over a wavelength range of 250 - 2500 nm so as to overlap the HEC solar cell spectral response.

The possibility of contamination of the solar cell cover glass during protective coating application to the silver interconnect's stress relief loop was a concern. The intercell gap (0.9 mm) was further reduced by overlap from cover glasses. Therefore the effective intercell gap was 0.5 mm.

This consideration led to tests with candidate coatings deliberately applied to solar cell cover glasses then exposed to various combinations of space environmental conditions. Optical transmission properties were measured before and after exposure to determine the contamination potential of the coating materials.

### *4.2. AO exposure testing*

A Tegal Corporation "Plasmod" asher was used for AO exposure of candidate coatings. The Plasmod, which utilized a radio frequency r.f. generated (50 - 100 W) plasma, was run with a Fomblin oil prepped roughing pump. Operating pressure of the exposure chamber was approximately 800 m Torr vacuum while flowing technical grade oxygen with an inlet pressure of 2 - 5 lbf in$^{-2}$.

AO fluence, predicted for the HEC array with adjustments for altitude, orbit, and orientation was distributed to 37.5% of the RAM (velocity vector) fluence for the Kapton backside and 20.2% of RAM for the cell side. The predicted AO fluence, based on a December 1988 launch date, is shown in Table 1.

TABLE 1

Predicted HST mission atomic oxygen fluence (assuming December 1988 launch)

| | *Year 1* | *Year 2* | *Year 3* | *Year 4* | *Year 5* |
|---|---|---|---|---|---|
| RAM fluence ($\times 10^{21}$ cm$^{-2}$) | 0.153 | 0.596 | 1.197 | 2.412 | 3.57 |
| Array sun-side fluence ($\times 10^{21}$ cm$^{-2}$) | 0.031 | 0.120 | 0.242 | 0.487 | 0.721 |
| Array backside fluence ($\times 10^{21}$ cm$^{-2}$) | 0.057 | 0.224 | 0.449 | 0.904 | 1.34 |

Kapton mass loss from Plasmod exposure was compared to calculated mass loss based on published reaction efficiency values [3] as a function of HST orbital fluence to determine equivalent orbital exposure. For example, a 1 h Plasmod exposure proved to be the equivalent of 1 year orbital array backside exposure.

### *4.3. Characterization of oxygen plasma*

The Plasmod was operated at standardized test conditions for characterization testing. A UTI quadrapole residual gas analyzer (RGA) was utilized for chemical characterization of the oxygen plasma. The RGA scan was performed on a gas sample differentially pumped from the Plasmod chamber.

Temperatures of exposed samples in the oxygen plasma were determined using Wahl "Temp-plates," thermocouples and a Hughes Thermal Video System with Probeye infrared viewer.

Kapton and silver interconnect samples were exposed to the plasma intermittently, *i.e.* removed for weighing, requiring compensation for moisture absorption. Cyclic operation reduced heating effects on samples. All coated Kapton samples were run with a Kapton control (uncoated) for normalizing exposure environment variation. Uncoated silver interconnects were exposed with coated interconnects for comparison of coating protection efficiency.

Coated Kapton samples were exposed to AO after other space environmental exposure tests (UV and electron–proton irradiation) to evaluate combined effects on coating protection capability.

### *4.4. Ultraviolet and electron–proton irradiation tests*

Coated Kapton samples were used to determine effects of UV and electron/proton irradiation on candidate coatings. A Varian Cary 2300 spectrophotometer was used for measuring solar absorptivity ($\alpha_s$). Infrared thermal emissivity ($\epsilon_{IR}$) measurements were performed with a portable emissometer. Measurements of $\epsilon_{IR}$ were made over the 250 - 2500 nm wavelength range to overlap the HEC solar cell spectral response range (0.3 - 1.1 $\mu$m).

UV irradiation testing consisted of exposing water cooled samples through a quartz window in $10^{-7}$ Torr vacuum. A 10 kW mercury–xenon lamp was used in an accelerated test (2 solar constant) for 2000 equivalent sun hours (ESH) for G-1B and SWS V-10. The McGhan Nusil CV-1144-0 coating was exposed for 500 ESH.

Electron–proton irradiation was conducted in a diffusion pumped vacuum chamber at $10^{-7}$ Torr with a 1 MeV Van de Graaff ion accelerator. These accelerated exposures generated a maximum integrated particle flux (fluence) of $10^{15}$ particles $cm^{-2}$ in approximately 5.5 min.

### *4.5. Thermal stability tests*

Thermal stability tests included verification that protective coatings applied to solar array materials would not significantly degrade the thermophysical properties ($\alpha_s$, $\epsilon_{IR}$) of the HEC array. A significant change in $\alpha_s$ and $\epsilon_{IR}$ would lead to increased orbital temperatures thereby decreasing solar array power output.

Thermal cycling of the coated solar array interconnects was also a concern. Simulation of the silver interconnect stress relief loop deflection due to thermal cycling (35 $\mu$m) in orbit was accomplished with a device designed for *in-situ* use in the Plasmod. The "Wiggler" as it was called was a ceramic device mounted on a flange which was sealed against the Plasmod internal chamber. Interconnects were mechanically deflected during AO plasma exposure. Mechanical design limited the minimum deflection possible to 70 $\mu$m. Coated and uncoated interconnects were flexed and exposed simultaneously with periodic interruptions for microphotography of the interconnects to document coating protection efficiency and stress relief loop strand breakage due to cycling.

Composite HEC solar array coupons were utilized for thermal cycling under vacuum. The test coupons were cut from an actual HEC solar array blanket. Candidate coatings were applied to the interconnects and the backside Kapton then the coupons were exposed ($-100$ °C to $+100$ °C) for 30 000 cycles (flight equivalent of 5 years) over a 14 month time period with periodic inspection and continuous electrical continuity monitoring.

### *4.6. Coating viscosity – ease of application testing*

Candidate coatings were graded for low viscosity for ease of application. Protective coatings had to be brushed or sprayed on and have the ability to "wet" the surface. Wettability was a consideration since the interconnect's stress relief loop underside was inaccessible. Atomic oxygen could reach the loop underside by reflecting off the blanket surface (RTV 691) underneath [4].

### *4.7. Blocking – storage verification test*

Blocking samples (10 cm × 10 cm) were coated and placed, in pairs with coated surfaces in contact, under a dead load of 75 N to simulate compressive loads of 7500 N m$^{-2}$ on the stowed solar array. The calculated peel force acceptance limit was 1 N. This force was the maximum allowed as the samples were pulled apart at 90° after initial separation at the leading edge.

## 5. Results

### *5.1. Outgassing – contamination test results*

Outgassing tests showed both SWS V-10 and CV-1144-0 met the TML and VCM requirements at 125 °C. G-1B had slightly high values for TML and VCM at 125 °C but passed both criteria when tested at 90 °C. The 90 °C test was adequate since the predicted array maximum temperature was reassessed and determined to be approximately 85 °C.

Outgassing of the G-1B coating left a visible contaminant film on the OWS at both temperatures. Reflectance measurements were not made due to possible contamination of the reflectometer.

Post-test reflectance measurements on the OWS run with SWS V-10 showed a maximum change in reflectance ($\Delta R/R_0$) of $-5.0\%$ at 200 nm and $-3.2\%$ at 140 nm. Measurements at other wavelengths met the $-3.0\%$ acceptance limit. Outgassing of CV-1144-0 caused a $-4.2\%$ $\Delta R/R_0$ at 121.6 nm and greater degradation from 170 to 200 nm.

Results from these initial tests led to direct coating application to solar cell cover glasses or quartz transmission samples for contamination evaluation. No significant changes in transmission were seen after application of the candidate coatings to cover glasses. Coated cover glasses were then subjected to UV irradiation (400 ESH) and measured again. No transmission loss due to UV irradiation was noted for any of the coated cover glasses.

The coated cover glasses were then exposed to AO plasma. Cover glasses with all three coatings had decreases in transmission after AO exposure due to coating drying or "glazing". G-1B showed greater loss of transmission than the other two candidates.

### *5.2. AO exposure results*

RGA scans of the empty Plasmod chamber with oxygen plasma on showed atomic mass units (a.m.u.) interpreted as $O_2$ (a.m.u. 32), (a.m.u. 16), $N_2$ (a.m.u. 28) and $H_2O$ (a.m.u. 18) listed in order of decreasing partial pressures. The a.m.u. 16 peak was not attributed to atomic oxygen alone due to its high reactivity. Fragmentation patterns for molecular oxygen could partially account for this a.m.u. peak. No a.m.u. values greater than 44 were seen. The absence of higher a.m.u. values indicates the exposure chamber was free of contamination.

Temperatures in the oxygen plasma were measured using three methods. A Wahl "Temp-plate" placed under exposed Kapton samples indicated sample temperatures up to 88 °C. Thermocouples placed under exposed Kapton (coated and uncoated) and coated silver interconnects showed temperatures from 72 to 82 °C. Bare, uncoated, silver reached a maximum temperature of 174 °C.

Testing of uncoated silver interconnects showed the characteristic black oxide layer forming almost immediately upon exposure to AO. Considerable oxide buildup was noted after exposure for 1 h. Interconnect mesh holes grew noticeably smaller (oxide growth filled in) after exposure for 3 h. Interconnects properly coated with all three candidates were not visibly oxidized after 3 year equivalent flight exposure (3 h) (see Fig. 2). Some oxidation was noted on interconnects which were not fully coated. Protected interconnects showed a slight weight loss attributed to the coating drying while unprotected interconnects showed weight gain due to oxide growth.

Scanning electron microscope (SEM) inspection of exposed, unprotected interconnects confirmed oxidation of the stress relief loop underside.

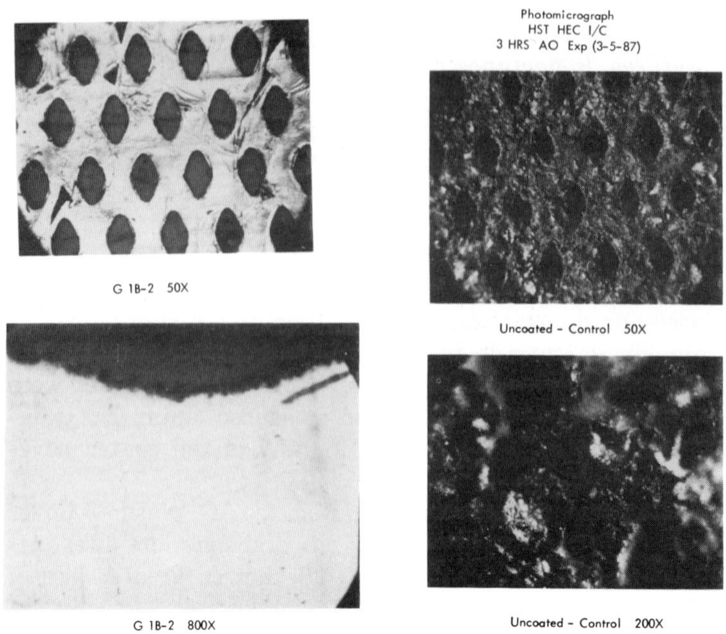


Fig. 2. AO atomic oxygen exposure of coated *vs.* uncoated silver interconnects.

Protective coatings applied to the stress relief loop top appeared to provide protection for the loop underside, *i.e.* very little oxidation was noticed.

Kapton samples of thickness $2 \times 10^{-3}$ in exposed for 3 h (flight equivalent of 3 years) showed an average weight loss of 20 - 25%. SWS V-10 and G-1B coated (4 - 8 $\mu$m) samples had either negligible or no more than 5% weight loss for a similar exposure of 3 h. CV-1144-0 ($0.5 \times 10^{-3}$ in thick) coated specimens provided equivalent protection (maximum loss of 8%). Thinner coatings of all three candidates generally provided less protection.

### 5.3. *Ultraviolet and electron–proton irradiation test results*

Kapton samples coated with SWS V-10 and G-1B showed no significant change in thermophysical properties after 2000 ESH UV exposure. CV-1144-0 samples showed no property changes after a shorter exposure of 500 ESH.

Subsequent exposure of SWS V-10 and CV-1144-0 samples to AO plasma (flight equivalent of 3 years) resulted in less than 1% mass loss.

Electron–proton irradiation tests of the coated Kapton specimen provided no significant evidence of coating degradation for up to 5 year mission flux levels. No measurable changes in the thermophysical properties were detected.

### 5.4. *Thermal stability test results*

Kapton samples showed no significant change in thermophysical properties ($\alpha_s$ and $\epsilon_{IR}$) after application of any of the three candidate coatings.

Further tests were conducted to determine if changes would occur after space environmental exposure of the coatings.

Thermal cycling simulation in conjunction with AO exposure with the Wiggler device induced no silver strand breakage on protected interconnects after 30 000 cycles. Unprotected interconnects suffered substantial strand breakage and embrittlement as well as flaking off of the silver oxide layer.

The strand breakage noted after 30 000 cycles of laboratory bench operation was less than the 50% found after thermal vacuum cycling qualification testing of the original HEC array. This was attributed to the deflection not including the twisting moments which were experienced in thermal cycling.

The limited number of available HEC array blanket specimens restricted full qualification testing to SWS V-10 and G-1B. Acceptance criteria were met for coated coupons after thermal cycling exposure.

### *5.5. Coating viscosity results*

The G-1B coating had a fairly low viscosity and wet the interconnect surface easily. SWS V-10 and CV-1144-0 had moderate wetting capability. These coatings could be diluted, therefore improving wettability. Dilution could result in requirement of multiple coating layers.

### *5.6. Blocking–storage verification results*

Kapton samples coated with SWS V-10 and CV-1144-0 were placed under load after air curing for 3 days. Samples coated with the non-curing G-1B were easily peeled due to the fluidity of the coating. None of the three candidate coatings exceeded the peel test limit of 1 N. Maximum peel forces during the 6 month test were 0.74 N for SWS V-10 (12 weeks), 0.14 N for CV-1144-0 (18 weeks) and 0.02 N for G-1B (14 weeks).

## 6. Conclusions

The lack of time limited the number of candidate coatings substantially tested to three. Since protection from AO was the major concern some coatings were merely screened for their AO resistance and eliminated due to low potential for protection. Other coatings equally resistant to AO exist, but, due to schedule concerns, only three coatings were selected for thorough characterization.

The candidate coatings (SWS V-10 and CV-1144-0) met outgassing requirements for flight. G-1B, which was acceptable at lower temperatures, was less desirable due to greater potential contamination threat due to observed material creepage, particularly above ambient temperature. SWS V-10 and CV-1144-0 were considered acceptable since borderline optical acceptance levels were obtained at elevated test temperatures (125 °C) compared to the HEC array operating temperature (85 °C), and the relative remoteness of HST UV optics.

All three candidate coatings provided adequate AO protection for laboratory exposure equivalent to 3 years of flight exposure at predicted HST conditions.

Thermophysical properties ($\alpha_s$ and $\epsilon_{IR}$) of Kapton were not affected by application of a coating or subsequent exposure to atomic oxygen, UV irradiation, electron–proton irradiation and thermal cycling. Thermal modeling calculations predicted a maximum decrease in power output of the HEC solar array of 3% after 5 years of AO exposure.

Acceptable application techniques were demonstrated with the three coatings. Storage verification or blocking tests revealed no detrimental effects on array deployment after long term storage (6 months).

Successful qualification of the backup BSFR solar array eventually negated the need for the HEC solar array blankets, but CV-1144-0 has been used on selected surfaces of the BSFR array, *e.g.* cushion roller.

## Acknowledgments

The following are acknowledged for their considerable contributions: Dr. Raymond Gause, John Reynolds, Bob DeHaye, Bruce Glick, Joey Norwood, Bobby Cothren, Ed White, Roger Harwell, Ted Edge, Jim Burka, Doug Alexander, Jewell Cargile and Barbara Simms of MSFC. The cooperation of the ESA/ESTEC is appreciated.

## References

1 *MSFC-SPEC-1443*, Outgassing test for non-metallic materials associated with sensitive optical surfaces in a space environment, George C. Marshall Space Flight Center, 1987.

2 *JSC-SP-R-0022A*, General specification, vacuum stability requirements of polymeric materials for spacecraft application, Johnson Space Center, 1974.

3 L. Leger *et al.*, Selected materials issues associated with the space station, *SAMPE Quarterly, 18 (2)* (1987) 48 - 54.

4 A. F. Whitaker, LEO atomic oxygen effects on spacecraft materials — STS-5 results, *NASA TM-86463*, George C. Marshall Space Flight Center, 1984, p. 3.

# EFFECT OF REACTOR LOADING ON ATOMIC OXYGEN CONCENTRATION AS MEASURED BY NO CHEMILUMINESCENCE

N. R. LERNER

*Ames Research Center, NASA, Moffett Field, CA 94035 (U.S.A.)*

(Received March 8, 1989)

## Summary

It has previously been observed that the etch rate of polyethylene samples in the afterglow of an r.f. discharge in oxygen increases with reactor loading. This enhancement of the etch rate is attributed to reactive gas phase products of the polymer etching. In the present work, emission spectroscopy is employed to examine the species present in the gas phase during etching of polyethylene. In particular, the concentration of atomic oxygen downstream from the polyethylene samples is studied as a function of the reactor loading. It is found that the concentration of atomic oxygen increases as the reactor loading is increased. Since atomic oxygen is known to etch hydrocarbons, the increase of etch rate with increased reactor loading is attributed to the increase of atomic oxygen concentration in the vicinity of the sample.

## 1. Introduction

Polymeric materials subjected to the low Earth orbit (LEO) environment have been observed to undergo extensive weight loss and surface recession [1 - 6]. This weight loss is thought to be due to the reaction of atomic oxygen with the polymer surface, rather than simple ablation by the high speed oxygen atoms. This conclusion suggested exposure of polymers to the afterglow of an r.f. discharge in oxygen, where the dominant reactive species present is thought to be $O(^3P)$, might be a suitable method of screening polymers for service in the LEO environment. However, it has been found that the erosion rates of polymers in afterglow reactors show a greater variation with polymer composition than is observed for the same polymers in LEO [1 - 5].

Lerner and Wydeven [7, 8] recently studied the etch rate of hydrocarbon polymers in the afterglow of an r.f. discharge in oxygen. The etch rate was studied as a function of reactor loading, flow rate and distance downstream from the r.f. discharge. The etch rate of polyethylene and polyvinylfluoride was found to increase with increased reactor loading and decrease with

0257-8972/89/$3.50 

increasing gas flow rate. In addition, Lerner and Wydeven found that the etch rate of polypyromellitimide downstream from an r.f. discharge in oxygen, which is normally quite low [7, 9, 10], could be increased by well over an order of magnitude by adding polyethylene to the reactor. These observations suggest that the etching products influence the erosion rate of polymers in afterglow reactors; this effect might be the cause of the large variation in erosion rate of polymers observed in such reactors. The pressure in LEO is $10^{-7}$ to $10^{-8}$ Torr [4, 6]. At such low pressures the etching products arising from surface recession would not be in contact with the surface for a sufficient amount of time to contribute to the rate of surface recession.

Lerner and Wydeven [7, 8] speculated that a species more reactive than $O(^3P)$, such as OH [11], is formed by the interaction of $O(^3P)$ with the products of polymer etching. Emission from OH has been observed during the etching of organic materials [12 - 14]. Therefore, we undertook a spectroscopic study to determine the effect of reactor loading on the species present during the etching of polyethylene downstream from an r.f. discharge in oxygen. Spectroscopic measurements made 9 cm downstream from the r.f. discharge detected emission from excited CO, OH, hydrogen and oxygen, and 29 cm downstream from the discharge (5 cm downstream from the polymer sample) detected only emission from excited OH. Measurements made at the former location indicate that polymer degradation products diffuse toward the discharge, where they interact with the plasma forming excited CO, OH, hydrogen and oxygen. The intensity of atomic oxygen spectral lines increased with an increase in the sample etch rate (reactor loading). This suggested that the total concentration of oxygen atoms might also be increased by increased reactor loading. In order to verify this, the chemiluminescence from the reaction

$$NO + O \rightarrow NO_2 + h\nu \quad (1)$$

was studied downstream from the sample during $NO_2$ titrations of the atomic oxygen concentrations. These measurements indicate that increased loading of the reactor with polyethylene samples increases the concentration of ground state atomic oxygen. It is well known that addition of hydrogen-containing compounds to an r.f. discharge in oxygen causes a significant increase in the concentration of atomic oxygen [15 - 19]. The results presented here are consistent with a mechanism in which the products of degradation of polyethylene diffuse upstream to the discharge, where they cause an increase in the concentration of atomic oxygen.

## 2. Experimental procedure

Figure 1 is a schematic diagram of the reactor used in this study. The reactor was made of quartz. During etching, black cloth was draped over the reactor thus shielding the sample from the ultraviolet radiation produced in the discharge. The reactor was fitted with UV grade fused silica windows

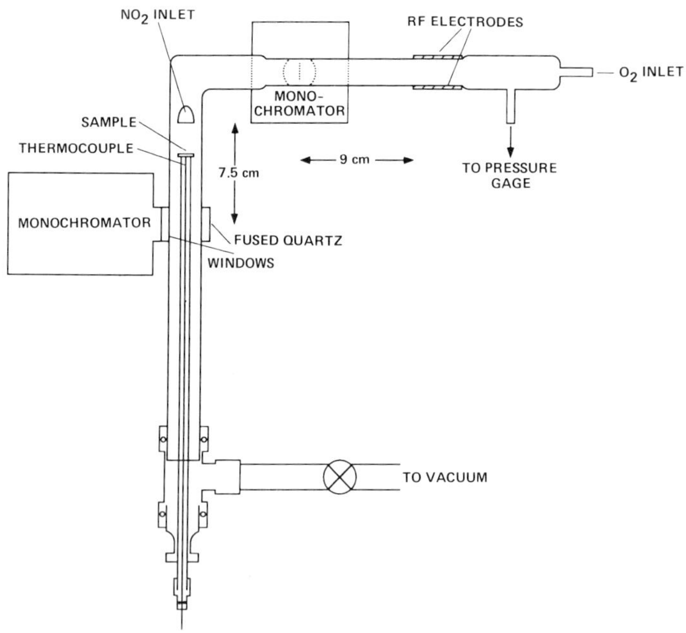


Fig. 1. Experimental arrangement. The two monochromator positions are indicated on the diagram; only one monochromator was used in this work.

which were centered 29 cm from the electrodes (7.5 cm from the $NO_2$ inlet). The net r.f. power into the reactor was 15 W at 13.56 MHz. The reactor pressure was maintained at 0.55 Torr. Samples, and weight loss, gas flow rate and pressure measurement are as described in ref. 7.

The monochromator used in these experiments was an Oriel Model 77257 Grating Monochromator with a photomultiplier detector incorporating a Keithly 416 picoammeter. The sensitivity of the monochromator–photomultiplier tube combination was checked from time to time with a calibration lamp. The emission from the reactor was measured at the two positions indicated in Fig. 1. The entrance slit of the monochromator was positioned 5 mm from the square cross-section tubing at the 9 cm position, and 5 mm from the fused silica windows at the 29 cm position. The emission at the 9 cm location was sufficiently intense that a monochromator slit width of 100 $\mu$m could be used, giving a bandpass of 1 nm. Due to the lower intensity of the emitted radiation at the 29 cm position a monochromator slit width of 400 $\mu$m was used, resulting in a band pass of 4 nm. The slit height was 12 mm.

The concentration of atomic oxygen was measured by $NO_2$ titration. $NO_2$ titration curves were constructed as follows. The spectrometer was set so that it passed radiation at 530 nm. $NO_2$ was added to the reactor at a fixed flow rate, and the intensity of the radiation at 530 nm was recorded. The flow rate of the $NO_2$ was then increased and again the intensity of the radiation was recorded. This procedure was repeated many times until a complete titration curve was recorded. Titration curves were always recorded 29 cm from the electrodes and 7.5 cm from the titration point.

## 3. Results

### *3.1. Emission spectra 9 cm from the plasma*

The emission spectra obtained at the observation point 9 cm from the discharge are displayed in Fig. 2. The main features of the emission spectrum obtained with no sample in the reactor are the emission lines from atomic oxygen and the bands of the second negative ($A^2\Pi_u - X^2\Pi_g$) and Hopfield ($c^4\Sigma_u^- - b^4\Sigma_g^-$) systems of $O_2^+$. Previous workers [13, 14, 16] studying emission spectra during plasma etching of photoresists have not observed the spectrum of $O_2^+$. However, Pearse and Gaydon [20] report the emission bands of this species as a prominent feature of discharges in oxygen. Atomic oxygen is clearly not the only active species in an r.f. discharge in oxygen. The two weak peaks at 337 nm and 316 nm can be attributed to the second positive system of $N_2$, which is present as an impurity. In addition to the above lines and bands, there is a very intense emission of unknown origin at 209 nm. The intensity of this emission was invariant under all reactor conditions. We attribute it to fluorescence from the quartz.

With the addition of a sample to the reactor, emission bands arising from CO and OH are observed together with the $H_\beta$ line. The intensity of the atomic oxygen emission also increases. Clearly, the products of polymer etching are diffusing upstream from the sample to the discharge and affecting the distribution of species produced in the discharge. At constant flow rate, the intensity of the emission arising from CO, OH, hydrogen and oxygen depended linearly upon the rate of mass loss by the samples. The mass loss rate was varied by varying the size of the sample being etched. The slope $S_\lambda$ of the plot of intensity *vs.* mass loss decreases with increasing gas flow rate. Table 1 lists the values of $S_\lambda$ obtained from plots of $I_\lambda$ for emissions from these species *vs.* mass loss rate. Also identified in Table 1 are the wavelengths of the emissions. For all the species considered, the magnitude of $S_\lambda$ decreases with increasing gas flow rate. This decrease can be understood on the following basis: the emission $I_\lambda$ at wavelength $\lambda$ arising from the excited species $X$ is directly proportional to its concentration $[X]$ in the plasma. $[X]$ is proportional to the concentration of its precursor $[R]$ in the plasma. Except at the surface of the sample, $[R]$ is determined by the mass transport equation

$$\frac{d[R]}{dt} = \frac{Dd^2[R]}{dz^2} - \frac{v d[R]}{dz} \tag{2}$$

where $D$ is the diffusion coefficient of $R$, $v$ the gas velocity and $z$ the distance from the sample, with positive $z$ taken in the direction of the gas flow.

The solution to this equation in the steady state, $d[R]/dt = 0$, is

$$[R] = c_1 + c_2 e^{\frac{v}{D}z} \tag{3}$$

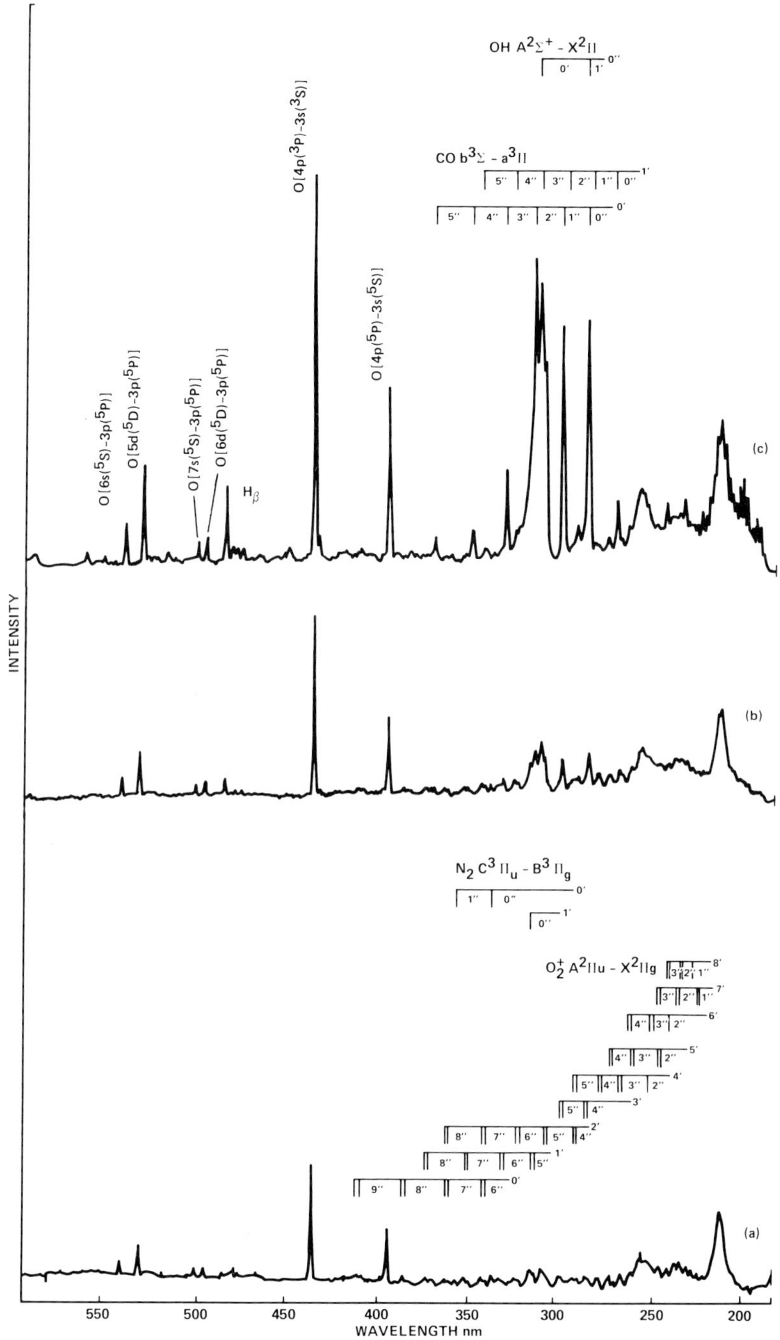


Fig. 2. Emission spectra observed 9 cm from the plasma: (a) spectrum with no sample in the reactor; (b) spectrum with a 3.0 mg sample in the reactor (etch rate, 0.21 mg $h^{-1}$); (c) spectrum with a 10.0 mg sample in the reactor (etch rate, 1.71 mg $h^{-1}$).

TABLE 1

Influence of flow rate and etch rate on emission spectra

| $\lambda$(nm) (species) | $S_\lambda$ | *Flow, F* (s.c.c.m.) | $FS_\lambda$ | $\ln(FS_\lambda)$ |
|---|---|---|---|---|
| **486** | **$20 \pm 3$** | **2.0** | **40** | **3.69** |
| ($H_\beta$) | $8.8 \pm 1$ | 3.0 | 26 | 3.27 |
| | $4.4 \pm 0.2$ | 4.0 | 17.6 | 2.87 |
| | $1.9 \pm 0.1$ | 6.0 | 11.4 | 2.43 |
| **436** | **$35 \pm 10$** | **2.0** | **70** | **4.24** |
| (O) | $27 \pm 3$ | 3.0 | 81 | 4.39 |
| | $17 \pm 3$ | 4.0 | 68 | 4.21 |
| | $8.3 \pm 2$ | 6.0 | 49.8 | 3.91 |
| **395** | **$17.6 \pm 6$** | **2.0** | **35** | **3.56** |
| (O) | $11.6 \pm 1$ | 3.0 | 34.8 | 3.55 |
| | $7.6 \pm 1$ | 4.0 | 30.4 | 3.41 |
| | $4.3 \pm 1$ | 6.0 | 25.8 | 3.25 |
| **309** | **$58.3 \pm 10$** | **2.0** | **117** | **4.76** |
| (CO + OH) | $30 \pm 2$ | 3.0 | 90 | 4.50 |
| | $16 \pm 0.5$ | 4.0 | 64 | 4.16 |
| | $6.6 \pm 0.1$ | 6.0 | 39.6 | 3.67 |
| **295** | **$115 \pm 16$** | **2.0** | **230** | **5.43** |
| (CO) | $31 \pm 4$ | 3.0 | 94 | 4.54 |
| | $14 \pm 1$ | 4.0 | 55 | 4.00 |
| | $3.5 \pm 0.1$ | 6.0 | 21 | 3.04 |
| **283** | **$103 \pm 15$** | **2.0** | **206** | **5.33** |
| (CO) | $36 \pm 5$ | 3.0 | 108 | 4.68 |
| | $14 \pm 1$ | 4.0 | 55 | 4.01 |
| | $3.5 \pm 0.3$ | 6.0 | 21 | 3.04 |

The sample position $z = 0$ is a singular point. At this point $R$ enters the gas stream. In the steady state, the flow of $R$ out of the reactor will be equal to its rate of production due to the etching of the polymer, with the result that downstream from the sample ($z > 0$), $[R] \propto E/F$, where $E$ is the etch rate and $F$ is the flow rate of the gas through the reactor. For $z > 0$, $[R]$ is a constant and $c_2 = 0$. For $z < 0$, we have $c_2 \propto E/F$ and $c_1 = 0$. Since $I_\lambda \propto [X]$ this leads to the chain of proportionalities

$$I_\lambda \propto [X] \propto [R] \propto (E/F)\, e^{-\frac{v}{D}d} \tag{4}$$

where $d$ is the distance between the sample and the plasma. Since $S = I_\lambda/E$, this leads to the result

$$S \propto (1/F) e^{-\frac{v}{D}d} \tag{5}$$

or

$$\ln(SF) = -\frac{v}{D} d + \text{constant} \tag{6}$$

But $v = FP/A$, where $P$ is the pressure and $A$ is the cross-sectional area of the reactor. It follows that

$$\ln(SF) = -\frac{FP}{DA} d + \text{constant}. \tag{7}$$

A plot of $\ln(SF)$ *vs.* flow rate should yield a straight line whose slope is proportional to $1/D$. The data do not conform exactly to a straight line, indicating this model is far too simplistic. However, the spread in $\ln(SF)$ for the $H_\beta$ line of hydrogen is roughly 1/2 of that for the 298 nm band of CO indicating a diffusion coefficient for the precursor of the $H_\beta$ emission roughly twice that for the CO emission. The diffusion coefficient for $H_2O$ at room temperature is also roughly twice that of CO (the diffusion coefficient in $O_2$ for standard conditions for $H_2O$ is 0.282 and for CO 0.183, both given in $cm^2\ s^{-1}$). The spread in the quantity $\ln(SF)$ for the atomic oxygen lines is smaller than that for the $H_\beta$ line. This indicates that the precursor for the excited atomic oxygen has a diffusion coefficient larger than that of $H_2O$. The only species expected to have a diffusion coefficient in oxygen significantly larger than $H_2O$ are atomic hydrogen and $H_2$ (the diffusion coefficient in $O_2$ for $H_2$ is $0.679\ cm^2\ s^{-1}$). Previous workers [14 - 19] have found that the concentration of atomic oxygen in discharges in oxygen was increased by the addition of hydrogen or hydrogen containing compounds to the gas.

### *3.2. $NO_2$ titrations*

The $NO_2$ titration technique has been described in considerable detail by Kaufman [15]. In particular, he has shown that $I_g$, the intensity of the glow from (1) is given by

$$I_g \propto p_{NO} p_O \tag{8}$$

where $p_{NO}$ is the partial pressure of NO and $p_O$ is the partial pressure of atomic oxygen. When the flow rate of $NO_2$ into the reactor at the titration point is less than the flow rate of atomic oxygen past the titration point, all of the $NO_2$ is converted to NO by means of the reaction

$$NO_2 + O \rightarrow NO + O_2 \tag{9}$$

Thus $p_{NO}$ and $p_O$ at locations downstream from the titration point are given by

$$p_{NO} = \frac{f_{NO_2}}{F} P \qquad p_O = \frac{f_O}{F} P = \frac{f_O^* - f_{NO_2}}{F} P \tag{10}$$

where $F$ is the total flow rate of gas past the titration point, $f_{NO_2}$ is the flow rate of $NO_2$ into the reactor, $f_O$ is the flow rate of atomic oxygen past the titration point, $f_O^*$ is the flow rate of atomic oxygen with no flow of $NO_2$ into the reactor and $P$ is the total gas pressure.

We constructed titration curves for our reactor from measurements taken 7.5 cm downstream from the titration point and 5 cm downstream from the sample. The sample holder was in position for all titrations. Figure 3

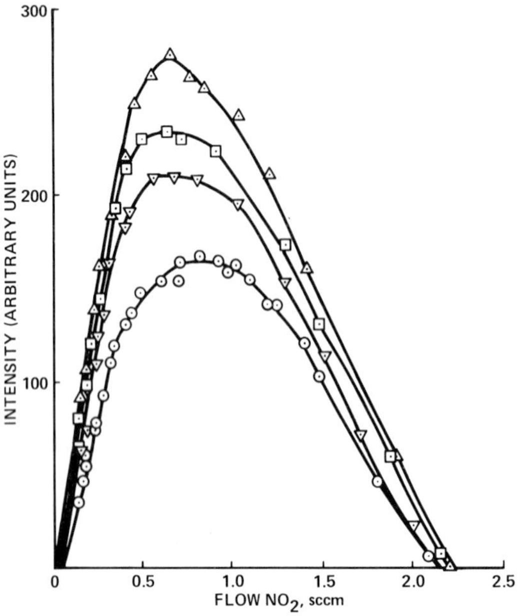


Fig. 3. Plot of the intensity of the chemiluminescent reaction $NO + O \rightarrow NO_2 + h\nu$ *vs.* flow of $NO_2$ into the reactor: ⊙ no sample; ▽ a 5.9 mg polyethylene sample in the reactor; ⊡ a 7.2 mg sample in the reactor; △ a 12.5 mg sample in the reactor.

shows plots of the intensity of the radiation at 530 nm measured with the monochromator centered 7.5 cm downstream from the titration point (5 cm downstream from the sample position) *vs.* flow of $NO_2$ into the monochromator. The most striking feature of these plots is that for a given flow rate of $NO_2$ into the reactor, the intensity of the radiation increases as the size of the sample in the reactor increases, and thus as the etch rate of the sample increases. These results show that the etching products of polyethylene increase the total concentration of atomic oxygen.

## 4. Discussion

Our results clearly indicate that the products of polymer etching are diffusing upstream from the sample to the plasma. In the plasma, they produce excited hydrogen, OH, CO and oxygen. In addition, the concentration of atomic oxygen is increased by the addition of polymer samples to the reactor. Many workers [15 - 19] have observed that the production of oxygen atoms is greatly enhanced when hydrogen or hydrogen-containing compounds are added to an electrical discharge in oxygen. Costa *et al.* [17] estimate that about 300 extra oxygen atoms are produced per hydrogen atom added to the gas. We now show that under our reactor conditions there was a sufficient concentration of hydrogen atoms in the plasma to produce a significant increase in the atomic oxygen concentration. The etch rates we have observed are of the order of 1 mg $h^{-1}$ = 0.016 mg $min^{-1}$ = 0.0012 mmol ($CH_2$ units) $min^{-1}$. If the products are in the form of $H_2O$ and $CO_2$, 0.026 s.c.c.m. of $H_2O$ will be produced. For a flow rate of 6 s.c.c.m. the concentration of $H_2O$ at the sample will be 0.43%. The diffusion coefficient of $H_2O$ is roughly 0.3 $cm^2\ s^{-1}$, at atmospheric pressure and 298 K. Since diffusion coefficients are inversely proportional to pressure, the diffusion coefficient of $H_2O$ will be given by $0.3/P$, with $P$ in atm. The velocity of gas through the reactor is given by $F/AP$, where $A$, the cross-sectional area of the reactor, is about 5 $cm^2$. With a flow rate through the reactor of 6 s.c.c.m. (0.1 s.c.c.s.), we have $v/D = 0.07\ cm^{-1}$. The distance from the sample to the electrodes is about 24 cm. Substituting these values into eqn. (4), we find that a fall-off of 0.19 in the concentration of $H_2O$ should occur between the sample and the plasma, resulting in a concentration of 0.08% for $H_2O$ in the plasma. If each hydrogen atom introduced in the plasma results in 300 oxygen atoms, the concentration of oxygen should be 48%. Actually the catalytic efficiency in producing oxygen atoms of hydrogen containing additives drops off at concentrations much less than 0.08%, so the high value given for the catalytic efficiency is too large. However, this calculation illustrates that backward diffusion to the plasma of the products of polymer etching can result in significant increases in the concentration of atomic oxygen.

## 5. Conclusions

The presence of a polyethylene sample etching downstream from an r.f. discharge in oxygen significantly affects the concentration of atomic oxygen produced in the discharge. In view of earlier research [15 - 19] on the catalytic production of atomic oxygen by hydrogenated additives in discharges in oxygen this is not a surprising result. These results conform to a model in which hydrogen containing products of polymer etching, diffuse upstream to the plasma. This effect may be partly responsible for the strong dependence of surface erosion on polymer structure observed when polymers are etched downstream from an r.f. discharge in oxygen in contrast to what is observed when polymers are exposed to a space environment.

## References

1 A. Garton, P. D. McLean, W. Wiebe and J. Densley, *J. Appl. Polym. Sci., 32* (1986) 3941.
2 A. Garton, P. D. McLean and T. K. Stevenson, *Polym. Preprints, 29* (1988) 104.
3 B. A. Banks, M. J. Mirtich, S. K. Rutledge and D. M. Swec, *Thin Solid Films, 127* (1985) 107.
4 G. S. Arnold and D. R. Peplinski, *AIAA J., 23* (1985) 1621.
5 J. Stecher, *Fifteenth Space Simulation Conf.: Support the Highway to Space Through Testing*, NASA-CP-3015, 1988, p. 51.
6 L. J. Leger and J. T. Visentine, *J. Spacecraft Rockets, 23* (1986) 505.
7 N. R. Lerner and T. Wydeven, *J. Electrochem., Soc. 136* (1989) 1426.

8 N. R. Lerner and T. Wydeven, *J. Appl. Polym. Sci.*, *35* (1988) 1903.
9 S. J. Moss, A. M. Jolly and B. J. Tighe, *Plasma Chem. Plasma Process*, *6* (1986) 401.
10 V. Vukanovic, G. A. Takacs, E. A. Mutuzak, F. D. Egitto, F. Emmi and R. S. Horwath, *J. Vac. Sci. Technol. B*, *6* (1988) 66.
11 J. S. Gaffney and S. Z. Levine, *Int. J. Chem. Kinet.*, *9* (1979) 1197.
12 R. H. Hansen, J. V. Pascale, T. De Benedictis and P. M. Rentzepis, *J. Polym. Sci.*, *A3* (1965) 2205.
13 E. O. Degenkolb, C. J. Mogab, M. R. Goldrick and J. E. Griffiths, *Appl. Spectrosc.*, *30* (1976) 520.
14 E. O. Degenkolb and J. E. Griffiths, *Appl. Spectrosc.*, *31* (1977) 40.
15 F. Kaufman, *Proc. R. Soc. London, Ser. A*, *247* (1958) 123.
16 F. Kaufman and J. R. Kelso, *Proc. 8th Symp. (Int.) on Combustion*, The Combustion Institute, Williams and Wilkins, Baltimore, 1960, p. 230.
17 M. D. Costa, P. Zuliani and J. M. Deckers, *Can. J. Chem.*, *57* (1979) 568.
18 F. Kaufman and J. R. Kelso, *J. Chem. Phys.*, *32* (1960) 301.
19 R. L. Brown, *J. Phys. Chem.*, *71* (1966) 2492.
20 R. W. B. Pearse and A. G. Gaydon, *The Identification of Molecular Spectra, 3rd edn.*, Chapman and Hall, London, 1965, p. 262.
21 W. R. Harshbarger, R. A. Porter, T. A. Miller and P. Norton, *Appl. Spectrosc.*, *31* (1977) 201.

*Surface and Coatings Technology*, *39/40* (1989) 647 - 653


# EFFECTS OF OXYGEN "ASHING" ON INDIUM–TIN OXIDE THIN FILMS

B. N. DE, YONG ZHAO, P. G. SNYDER and J. A. WOOLLAM

*Center for Microelectronic and Optical Materials Research, Department of Electrical Engineering, University of Nebraska, Lincoln, NE 68588-0511 (U.S.A.)*

T. J. COUTTS and X. LI

*Solar Energy Research Institute, 1617 Cole Blvd., Golden, CO 80401 (U.S.A.)*

(Received March 16, 1989)

## Summary

Indium–tin oxide (ITO) films were prepared by d.c. sputtering of ITO in an oxygen atmosphere, and measurements of resistivity, thickness, and optical constants were made. The resistivity was found to increase at first and then decrease substantially with ashing in an oxygen plasma. The basic optical gap is relatively unaffected by ashing, but impurity band absorption increased substantially.

## 1. Introduction

Materials for use in low earth orbit (LEO) in space will be exposed to atomic oxygen, which promotes oxide formation and generally degrades typical material properties [1]. Indium–tin oxide (ITO) is transparent and electrically conducting [2]. It is desirable for use as an antireflecting coating on solar cells and as coatings to reduce spacecraft electrostatic charging [3 - 5]. ITO has a wide optical bandgap (3 eV) and thus is optically transparent over much of the visible (solar) spectrum [2]. It is also potentially useful for coating mirrors used on spacecraft.

Indium and oxygen are not soluble in each other and form only the compound $In_2O_3$. It can be doped with tin, and an oxygen deficiency results in oxygen vacancies and increased electrical conduction by impurity banding effects.

This paper describes the results of a series of experiments in which ITO films were exposed to a series of ashings in a commercial oxygen plasma asher. Films were deposited onto both silicon and quartz substrates for evaluation by variable angle spectroscopic ellipsometry (VASE), X-ray diffraction and electrical resistivity measurements.

0257-8972/89/$3.50 

## 2. Experimental procedure

Ashing was done in a commercial (Plasma Prep II) barrel-type asher manufactured by SPI Corporation. This was capacitively coupled, operating at 13.56 MHz and 100 W at a pressure of 100 $\mu$m of pure oxygen. The chamber was 10 cm diameter and 16 cm long. The presence of atomic oxygen was confirmed by spectroscopic measurement of the plasma discharge emission, in which atomic oxygen "lines" were observed. The oxygen fluence was determined by measuring the weight loss of a thin sheet of Kapton. The loss rate was 0.00014 g $cm^{-2}$ of Kapton per hour of ashing, corresponding to an atomic oxygen fluence of about $4 \times 10^{19}$ $cm^{-2}$ $h^{-1}$. It should be born in mind that the asher only qualitatively simulates the low Earth orbit (LEO) environment. Therefore, measurement of the oxygen fluence in the asher is only an order of magnitude estimate.

Films were nominally 650 Å thick (the antireflecting thickness for visible light for ITO on silicon), prepared by direct current sputtering in an oxygen atmosphere onto unheated substrates. These films were not annealed after deposition. Their microstructure was amorphous. Two ITO samples from two separate runs (samples 240 and 243) were chosen for VASE (using the silicon substrate) and resistivity (using quartz substrates) measurements. Since sample 243 was made at a higher oxygen partial pressure, more oxygen was incorporated inside the film and consequently had a higher sheet resistance and resistivity.

Measurements of electrical resistivity were done by shaping the ITO films into long narrow strips with accurately positioned contacting pads located along the length of the strips. Thus, the resistivity measurements were in the conventional four-probe configuration. Sample shaping was done using steel masks and a miniature sand blaster.

VASE uses a collimated monochromatic beam of linearly polarized incoming light at an angle to the sample normal [6]. The reflected light is generally elliptically polarized, and the state of polarization is determined using a second (rotating) polarizer. The measured quantity is the ratio of the complex Fresnel reflection coefficients. The procedure is to compare the measured data with that expected from an assumed model for the material and its microstructure. The Marquardt algorithm is used in regression analysis in which the parameters of the model are varied until a best fit solution is found.

In the present VASE experiments, the unknown quantities were the film thickness and the real and imaginary parts of the optical constants. Generally, three angles of incidence (60°, 65°, 70°) and a photon energy range of 1.65 - 3.65 eV were used.

## 3. Results

Figure 1 shows the change in resistivity as a function of ashing time, for the previously mentioned samples of ITO on quartz (samples 240 and 243).

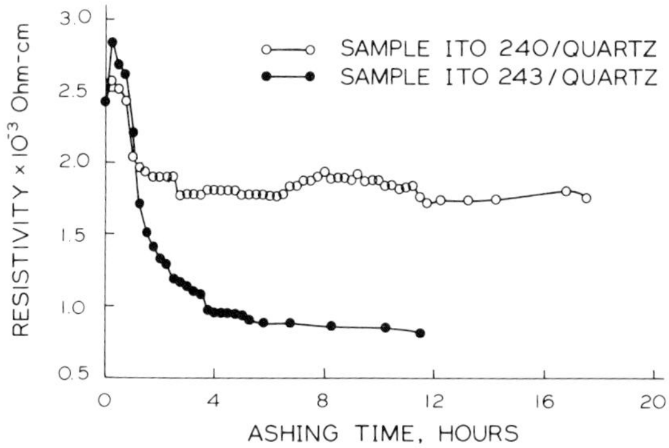


Fig. 1. Electrical resistivity of ITO *vs.* ashing time for two different samples.

The general features of the data to be noted are an initial small increase followed by a substantial decrease. In the case of sample 240, the decrease was to a value of 0.71 times the initial value. For sample 243, the resistivity dropped to a value of 0.34 times its initial value. The final value of resistivity for sample 240 was $1.75 \times 10^{-3}$ ohm cm, and for sample 243, it was $9.17 \times 10^{-4}$ ohm cm.

Figure 2 shows thickness derived from analysis of VASE data. It should be noted that the error bars represent an uncertainty of only one monolayer! Thus, the initial decrease in thickness by 5 Å is possibly real. Notice that the thickness increases monotonically for ashing up to 480 min. In calculating the resistivity in Fig. 1, a constant thickness of 650 Å was assumed. The maximum change in thickness seen in Fig. 2 is only 3%, thus assuming a constant thickness for purposes of resistivity calculation is reasonable.

Figures 3 and 4 show the photon energy dependence of the real (E1) and imaginary (E2) parts of the dielectric function. These data were taken before ashing and after ashing for up to 8 h. Notice the systematic decrease in E1 with ashing, followed by a slight increase between 6 and 8 h. However, in E2,

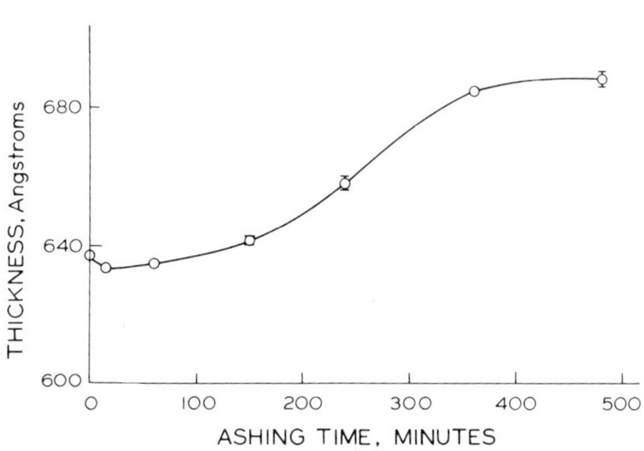


Fig. 2. Ellipsometrically determined thicknesses of ITO *vs.* ashing time.

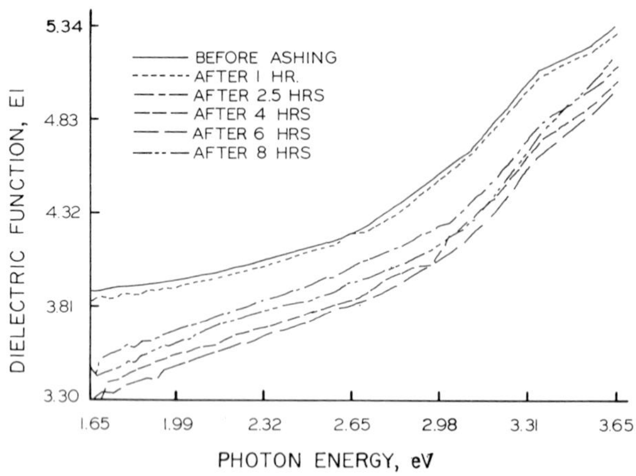


Fig. 3. Real part of the dielectric function *vs.* photon energy.

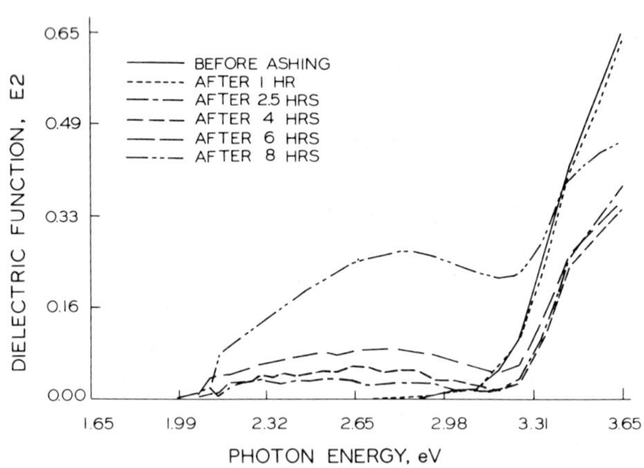


Fig. 4. Imaginary part of the dielectric function *vs.* photon energy.

the trend is monotonic in ashing time over the entire 8 h, with a steady increase in E2 for the spectral region below 3 eV.

The theory of optical effects in materials relates E2 to transitions between energy states, and the associated absorption of photons. The data of Fig. 4 are characterized by two distinct spectral regions; above and below 3 eV. Above 3 eV, the E2 data can be interpreted as being due to absorption by direct transitions from valence to conduction band. The data below 3 eV can be interpreted as being due, at least in part, to transitions from the valence band to an impurity band just below the conduction band, which broadens and densifies with increasing fluence.

Figure 5 shows "Tauc" plots [7, 8] for data on the ashed ITO sample number 205. Tauc plots are used to determine optical gaps in amorphous semiconductors and to investigate the energy dependence of absorption

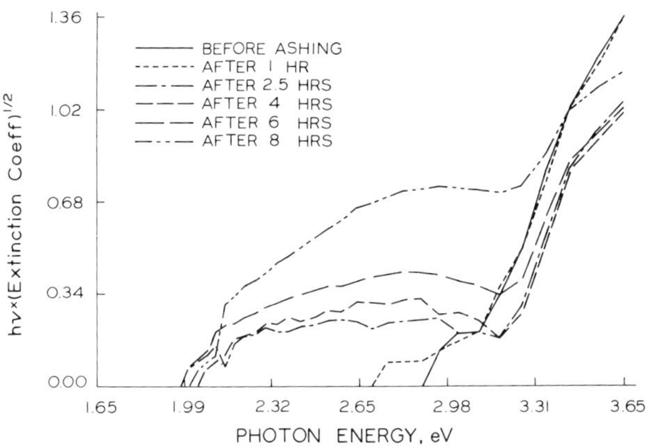


Fig. 5. Tauc plot (see text) for no ashing, and for ashing up to 8 h.

phenomenon. The ordinate is the square root of the photon energy times the absorption coefficient. The linear extrapolation of the high energy data to zero ordinate value gives the optical bandgap. Usually Tauc plots have regions with reasonably straight lines. This is moderately true for energies above 3 eV in Fig. 5. The unashed sample yields a direct gap of 3 eV. For ashing time less than a few hours, the gap increases slightly.

Figure 6 shows the reflectance for ITO on silicon for no ashing, as well as ashing up to 8 h. The general trend is towards a reduced reflectance with ashing. Reflectance depends on both film thickness and film index of refraction. From VASE data, we have shown that both thickness and index are changing with ashing, such that the reflectance decreases.

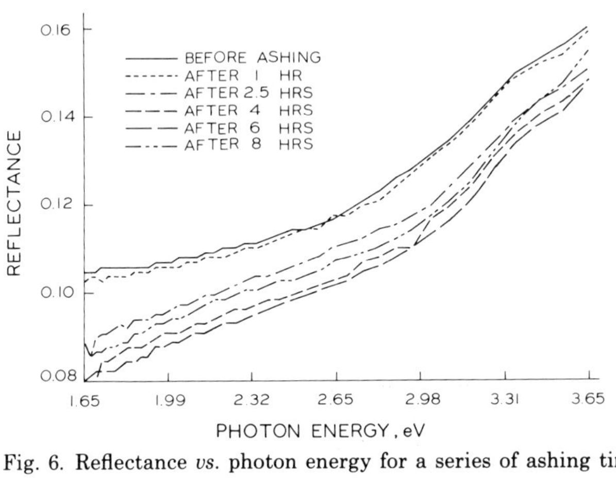


Fig. 6. Reflectance *vs.* photon energy for a series of ashing times.

## 4. Conclusions

Ashing in an oxygen plasma has been found to substantially decrease the electrical resistivity of direct current sputter prepared ITO, but that the effect saturates after approximately 4 h of ashing. The magnitude of the effect is sample dependent, but the functional dependence (a small increase followed by a large decrease, followed by saturation to a constant value) is sample independent. Furthermore, the time constant for saturation appears to be sample independent.

The apparent ITO film thickness increases slightly (of the order of 3%) with ashing. This may be due to the increased volume required as oxygen diffuses into the film, or to the growth of a thin oxide ($SiO_x$) at the interface between the ITO and silicon.

The optical data indicate that at least for short ashing times, the direct optical bandgap is not significantly affected by ashing, but the absorption tail (below 3 eV) increases and broadens with ashing. These results are consistent with the picture of a broad impurity band below the conduction band edge, in which the density of states increases with increasing ashing.

Preliminary investigations of the possible effects of sample heating by the plasma suggest that the films are no hotter than 70 °C, which is not likely to cause annealing [9]. Annealing is known to change the oxygen stoichiometry in $In_2O_3$ and to cause a reduction in electrical resistivity. X-ray scan on an ITO–quartz sample, after 17.5 h of ashing indicated that ITO was still amorphous. This implies that ITO was not annealed.

## Acknowledgments

The research at the University of Nebraska was supported by NASA Lewis Grant NAG3-95. Samples were prepared at SERI, which is supported by the Department of Energy.

## References

1 S. Rutledge, B. Banks, F. DiFilippo, J. Brady, T. Dever and D. Hotes, An evaluation of candidate oxidation resistant materials for space applications in LEO, *NASA Technical Memorandum 100122*, 1986.

2 K. L. Chopra, S. Major and D. K. Pandya, Transparent conductors — a status review, *Thin Solid Films, 102* (1983) 1.

3 T. J. Coutts and S. Naseem, High efficiency indium tin oxide/indium phosphide solar cells, *Appl. Phys. Lett., 46* (1984) 164.

4 T. J. Coutts, X. Wu, T. A. Gessert and X. Li, Direct current magnetron fabrication of indium tin oxide/InP solar cells, *J. Vac. Sci. Technol., A6* (1988) 1722.

5 R. D. Goldstein and L. C. Maldoon, Usage of ITO to prevent spacecraft charging, *IEEE Trans. Nucl. Sci., 29* (1982) 1621.

6 See for example: P. G. Synder, M. C. Rost, G. H. Bu-Abbud, J. A. Woollam and S. A. Alterovitz, Variable angle of incidence spectroscopic ellipsometry: application to GaAs–AlGaAs multiple heterostructures, *J. Appl. Phys.*, *60* (1986) 3293.
7 J. Tauc, R. Grigonovici and A. Vancu, *Phys. Status Solidi*, *15* (1966) 627.
8 T. Datta, J. A. Woollam and W. Notohamiprodjo, *Phys. Rev. B*, *40* 1989 (accepted).
9 A. J. P. Theuwissen and G. J. Declerck, Optical and electrical properties of reactively D.C. magnetron-sputtered $In_2O_3$:Sn films, *Thin Solid Films*, *121* (1984) 109.

*Surface and Coatings Technology, 39/40* (1989) 655 - 665


# AQUEOUS CORROSION RESISTANT PLATINUM AND STAINLESS STEEL COATINGS

R. P. WALTERS

*U.S. Bureau of Mines, Department of the Interior, 1450 Queen Avenue, SW, Albany, OR 97321 (U.S.A.)*

(Received March 9, 1989)

## Summary

Laser consolidation of platinum coatings electrodeposited from molten cyanide electrolytes has been studied. Laser processing was investigated as a means of consolidating platinum coatings on molybdenum substrates. Direct current deposited platinum coatings were coated with platinum black to enhance laser coupling. The platinum coatings were then melted with a Nd–YAG laser. An electrochemical technique was used to evaluate the corrosion resistance of the processed samples in a 1 N $HNO_3$ environment. Results indicated that laser processing could produce a pore-free platinum coating.

Using a particle occlusion technique followed by heat treatment, thick coatings of stainless steel were produced. Chromium particles (averaging 3 $\mu$m in diameter) were suspended in a ferrous–nickelous sulfate electrolyte and occluded in the electrodeposited Fe–Ne matrix. Subsequent heat treatment at 1100 °C produced the ternary alloy. Corrosion rates equivalent to those for bulk stainless steel were measured when the ternary alloy coatings were tested in a boiling nitric acid solution.

## 1. Introduction

Materials-related problems in the mining and metallurgical industries add to production costs and can limit the operational efficiency and safety of many processes. Typically, these problems are related to the performance of the materials in wear and corrosive environments. Often the solution to these problems involves the use of conventional alloys which contain large quantities of strategic metals, such as platinum and chromium, or the use of alloys which do not possess the appropriate physical or chemical resistance. The Bureau of Mines has investigated the possibility of conserving both platinum and chromium by producing corrosion resistant platinum and stainless steel coatings.

0257-8972/89 Elsevier Sequoia/Printed in The Netherlands

Electrodeposited platinum coatings prepared from aqueous baths are limited to internally stressed thin deposits. The aqueous coating process is inefficient, expensive, and highly sensitive to impurities [1, 2]. Bureau research has focused on producing platinum-group metal coatings from molten cyanide electrolytes on refractory metals as well as on more common materials of construction. A summary of the plating methods developed by the Bureau for producing platinum-group metal coatings from molten cyanides appears in a 1984 Bureau of Mines Information Circular [3]. If platinum-coated materials are to be used in real environments, they must be adherent, thick, stress-free and most importantly pore-free.

The Bureau of Mines has investigated the use of pulse plating to produce pore-free, corrosion resistant coatings [4]. Pulse deposition was found to significantly reduce the measured corrosion rates compared to direct current deposited coatings. However, a pore-free coating was not obtained for the pulse parameters tested.

Another method which could eliminate porosity is the melting of the surface with a laser. An examination was made of the effect of laser surface glazing thick platinum coatings with a pulsed Nd–YAG laser. Provided that the coatings are consolidated without melting or otherwise affecting the substrate, laser glazing should produce coatings equivalent to bulk platinum. A rapid electrochemical test was used to evaluate the effectiveness of the laser processing technique. Samples were held at an anodic potential in nitric acid where the substrate would be rapidly dissolved and the platinum coating unaffected. The use and limitations of this anodic polarization method are discussed.

It has been known for many years that insoluble particles suspended in an electrolyte during plating can be occluded in the electrodeposited metal. In recent years, particle-occlusion methods have been used advantageously to produce a variety of electrodeposited composite coatings [5 - 8]. Because chromium bearing alloys electroplated by conventional processes are thin, highly stressed and can only be produced at low efficiencies, particle-occlusion methods have been considered as an alternate method of introducing chromium in an alloy coating. In this procedure, metal powder is occluded in an electrodeposited metal matrix and heat treated to form an alloy. Bazzard and Boden [9, 10] reported on preparing Ni–Cr alloy coatings by occluding chromium powder in a nickel matrix and then heat treating at 800 - 1050 °C. Kilgore [11] and Williams [12] suggested that a particle-occlusion method could be used to produce stainless steel coatings, but their reports did not contain experimental results. Research conducted by the Bureau of Mines [13, 14], however, has provided experimental evidence of the feasibility of this method. The results of experiments to produce corrosion resistant Fe–Ni–Cr ternary alloy coatings using the particle-occlusion, heat-treatment method are discussed in this paper. Stainless steel electroforms and coatings were evaluated for their corrosion resistance in a boiling nitric acid environment.

## 2. Experimental procedure

### 2.1. Electrodeposition

The experimental apparatus for electrodepositing platinum, which includes a helium-filled, controlled environment glovebox and plating bath, have been previously described [15]. Coatings were prepared on TZM (0.41 titanium, 0.10 zirconium, balance molybdenum) substrates, with the sample preparation described in ref. 16. A potentiostat–galvanostat was used to control deposition potential. Platinum coatings used for the laser consolidation study were deposited under potentiostatic control at either one-half or the full diffusion limited current. The experimental apparatus used to deposit the composite coatings has been described in a recent Bureau publication [17]. Coatings were prepared on cold-drawn AISI 1020 carbon steel and on Sn–Bi mandrels. Three types of test specimens were prepared for determining the corrosion resistance of the heat-treated Fe–Ni–Cr alloys. The first type was an electrodeposited composite coating that had been physically removed from the carbon steel substrate prior to heat treatment. In this form, corrosion resistance of the coating served as a standard for evaluating the effectiveness of the various heat-treatment procedures used to homogenize the coatings. This procedure eliminated adverse effects caused by coating–substrate interactions. The second type had a substrate which was comprised of a carbon steel rod mechanically butt joined to a 304 stainless steel rod. The 304 stainless steel end extended out of the electrolyte during electrodeposition. A nickel barrier strike coating and the particle occluded composite coating bridged the surface between the 304 stainless steel rod and the carbon steel rod. Because the carbon steel was effectively encapsulated with corrosion resistant material, it was possible to evaluate the overall protective quality of the heat-treated coatings using this type of sample. The third type of specimen was an electroformed shape from which a tin–bismuth mandrel was removed by melting prior to heat treatment. Corrosion resistance of free standing, thick, and intricately shaped alloys was determined with the electroformed shapes.

### 2.2. Laser processing

Parameters for the pulsed Nd–YAG laser are summarized in Table 1. The laser treatment was performed at a commerical installation by mounting

TABLE 1

Laser processing parameters for surface melting with pulsed Nd–YAG laser

| *Case* | *Power density* ($kW\ cm^{-2}$) | *Pulse duration* (ms) | *Pulse frequency* (cycles $s^{-1}$) | *Incident power* (J) |
|---|---|---|---|---|
| 1 | 12 | 0.8 | 10 | 30.4 |
| 2 | 13 | 0.95 | 8, 9 and 10 | 39.5 |
| 3 | 12.7 | 1.25 | 6 | 50 |

Wavelength, 1.6 μm; beam spot size, 0.2 cm; area of focused beam, 0.0314 $cm^2$; table velocity, 0.6 cm $s^{-1}$; inert cover gas, argon; pressure, ambient.

specimens on a computer-controlled table which allowed horizontal motion in both the $x$ and $y$ directions. Sample surfaces were processed in an argon atmosphere. The gas cover was maintained by using a gas shield box with a fixed opening [18]. Overlaps of 1 mm of the 2 mm beam were used on successive passes of the laser to process the entire surface. Samples were TZM substrates, about 2 cm × 2 cm, coated with 12.5 - 27.5 $\mu$m of platinum. Before laser processing, an additional 3 $\mu$m of platinum black was plated from an aqueous bath to enhance laser coupling.

### 2.3. *Corrosion resistance tests*

The evaluation of laser melted coatings was conducted in 1 N $HNO_3$ at room temperature in an oxygen saturated solution. TZM coated samples were placed in a holder with 0.636 $cm^2$ of exposed surface area. Once in the test solution, the samples were maintained at 0.750 V SCE with a potentiostat. Current readings were taken after a steady state was reached, usually within 5 min.

Corrosion testing of the stainless steel coatings was conducted to ensure that uniform diffusion of chromium had occurred and that a stainless alloy was produced. Test specimens were exposed to boiling 65% $HNO_3$ solution for as many as five 48 h stages, the standard corrosion test for chromium segregation in commercial stainless steels (the Huey test).

### 2.4. *Characterization of coatings*

Scanning electron microscope (SEM) analysis and energy dispersive X-ray analysis (EDX) were used to examine the surfaces of all coatings and the distribution of the chromium powder within the electrodeposited matrix of the composite coatings. Image analysis was performed using a SEM equipped with an automated image analysis system.

Optical micrographs of cross-sections were obtained using standard techniques. Platinum coatings were electroetched in an aqueous solution containing 10 wt.% KCN with an alternating current. The best results were obtained by maintaining 2.5 V for 1.5 - 3 min. The stainless steel coatings were electrolytically etched for 30 s in a 50 g $l^{-1}$ cupric sulfate solution to reveal the grain structure.

## 3. Results and discussion

### 3.1. *Platinum coatings*

Studies were conducted on the use of laser surface melting of platinum coatings to determine the feasibility of using lasers to consolidate porous coatings. The objective was to determine if laser processing could be adjusted to effectively consolidate the coatings without damaging the substrate. A rapid test method was used, based on the potentiostatic selective dissolution of the substrate material, to determine if the platinum coatings were consolidated. Anodic polarization scans of bulk TZM and platinum were conducted

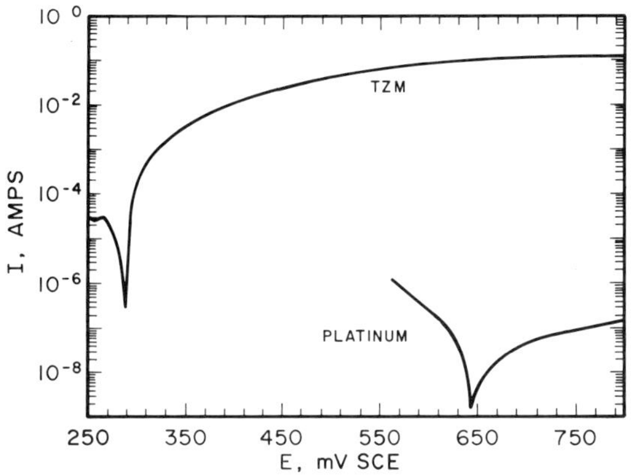


Fig. 1. Anodic polarization of TZM and platinum in oxygen saturated 1 N $HNO_3$ solution. Surface area, 0.636 $cm^2$, scan rate, 1 mV $s^{-1}$.

(Fig. 1) to determine the potential at which TZM would anodically dissolve at high current densities while oxidation on platinum occurred at very low rates. A potential of 0.750 V SCE was chosen since TZM was on a current dissolution plateau and the reaction was diffusion limited. The TZM dissolution current density was six orders of magnitude greater than the platinum current density at this potential.

Severe undercutting occurred for samples which exhibited large pores. This resulted in larger measured currents for samples with very porous coatings. Greater contributions to the measured current also occurred when alloying had taken place in the coating during the laser processing. Both of these additional contributions to the measured current would simply provide additional reason for not considering the laser parameters further.

The current readings obtained on all samples tested are summarized in Table 2. Pore area calculations were made assuming that the sum of the products of the fractional area of each material, platinum or TZM, and its current density equalled the total measured current. It was also assumed that the area of platinum was equal to the total area exposed minus the area of TZM. This resulted in the equation:

$$A_{\text{TZM}} = (I_{\text{Total}}/j_{\text{TZM}} - j_{\text{Pt}}) - (A_{\text{Total}}j_{\text{Pt}}/j_{\text{TZM}} - j_{\text{Pt}}) \qquad (1)$$

where $A$ is the area, $I$ is the current and $j$ is the current density.

Selected samples were examined with automated SEM–image analysis and optical techniques to determine the amount of pore area. Molybdenum elemental discrimination, EDX, was used in conjunction with image analysis in order to count only those areas which had dissolution products present.

A linear fit of the measured current *vs.* the measured pore area provided a check on eqn. (1). The value determined from the fit for the slope was $5.2 \pm 0.5$ as compared with a calculated value of 5.1 and for the intercept was $0.028 \pm 0.029$. The error associated with the determination of the intercept

TABLE 2

Results of potentiostatic polarization of laser processed platinum coating in oxygen saturated 1N $HNO_3$

| *Sample* | *Thickness,* ($\mu$m) | *Laser processing parameters* | *Measured current* (A) | *Calculated pore area* ($cm^2$) | *Measured pore area* ($cm^2$) |
|---|---|---|---|---|---|
| Pt | — | — | $8.0 \times 10^{-8}$ | $2.6 \times 10^{-10}$ | 0 |
| Mo | — | — | $1.2 \times 10^{-1}$ | $6.4 \times 10^{-1}$ | $6.4 \times 10^{-1}$ |
| 1 | 14 | — | $2.8 \times 10^{-2}$ | $1.4 \times 10^{-1}$ | $3.5 \times 10^{-1}$ |
| 2 | 22 | — | $2.8 \times 10^{-4}$ | $1.4 \times 10^{-3}$ | $5.2 \times 10^{-4}$ |
| 3 | 25 | — | $4.5 \times 10^{-3}$ | $2.3 \times 10^{-2}$ | $5.4 \times 10^{-3}$ |
| 4 | 24 | — | $8.0 \times 10^{-6}$ | $4.1 \times 10^{-5}$ | $3.7 \times 10^{-5}$ |
| 5 | 102 | — | $4.0 \times 10^{-6}$ | $2.0 \times 10^{-5}$ | — |
| 6 | 109 | — | $3.7 \times 10^{-6}$ | $1.9 \times 10^{-3}$ | — |
| 7 | 19 | case 1 | $8.0 \times 10^{-4}$ | $4.1 \times 10^{-3}$ | — |
| 8 | 25 | case 1 | $5.0 \times 10^{-8}$ | $-1.5 \times 10^{-7}$[a] | — |
| 9 | 25 | case 1 | $8.5 \times 10^{-7}$ | $3.9 \times 10^{-6}$ | — |
| 10 | 25 | case 1 | $3.0 \times 10^{-8}$ | $-2.6 \times 10^{-7}$[a] | — |
| 11 | 25 | case 1 | $3.4 \times 10^{-7}$ | $1.3 \times 10^{-6}$[a] | — |
| 12 | 17 | case 2 | $2.3 \times 10^{-2}$ | $1.2 \times 10^{-1}$ | — |
| 13 | 19 | case 2 | $8.3 \times 10^{-4}$ | $4.1 \times 10^{-3}$ | — |
| 14 | 21 | case 2 | $8.0 \times 10^{-3}$ | $4.1 \times 10^{-2}$ | — |
| 15 | 23 | case 2 | $2.6 \times 10^{-2}$ | $1.3 \times 10^{-1}$ | — |
| 16 | 25 | case 2 | $1.5 \times 10^{-2}$ | $7.7 \times 10^{-2}$ | — |
| 17 | 31 | case 2 | $1.5 \times 10^{-7}$ | $3.6 \times 10^{-7}$[a] | — |
| 18 | 38 | case 3 | $2.5 \times 10^{-4}$ | $1.3 \times 10^{-3}$ | — |

[a] Calculated areas have little significance since measured currents are approximately pure platinum current.

was due to the lack of data at low values of current and area. The excellent agreement between the fitted data and eqn. (1) justify the original assumptions. Equation (1) was therefore used to calculate the pore areas given in Table 2. There are two major limitations to the use of a technique which measures current assuming that the base of all pores are polarized to the same potential. These limitations must be considered since they prevent an absolute determination of a pore free coating. First, there is a detection limit where the contribution of the TZM substrate to the total current becomes indistinguishable from the current of the platinum surface. This is equal to the value of pore area for platinum from Table 2 and corresponds to $2.56 \times 10^{-10}$ $cm^2$ for the detection limit of exposed substrate. Therefore, if a single circular pore 0.2 $\mu$m in diameter were present it could not be measured by this technique.

Second, and more importantly, the electrolyte resistance prevents the substrate in all of the pores from being polarized to the same potential [19]. If a value of 3.2 $\Omega$ $cm^{-3}$ [20] is used for the resistance of the nitric acid, an average coating thickness of 25 $\mu$m, and a current of one order of magnitude greater than that of platinum, the cross sectional area of a pore can be calculated which will have a significant potential drop due to IR [21]:

$$\Delta\psi = (t/\kappa A)I \tag{2}$$

where $\Delta\psi$ is the difference in potential, $t$ is the thickness of the coating, $\kappa$ is the specific conductivity, $A$ is the area of the pore and $I$ the current. If we use the worst case where all of the current must flow through one pore, then the pore must be larger than 0.4 $\mu$m in diameter in order to polarize the substrate to within 0.06 V of 0.750 V SCE.

These simple, worst case calculations indicate that the minimum diameter that could be detected by this technique was on the order of 0.5 $\mu$m, which was adequate to demonstrate the feasibility of laser processing for the consolidation of platinum coatings.

The results of the current testing indicated that the laser processing parameters which produced the best surface were case 1, Table I. The effectiveness of laser processing appears to be very sensitive to the thickness of the coatings, with as little as 8 $\mu$m difference producing radically different results. Future efforts in this area will require some attention to surface pretreatment in order to provide a uniform cross section to the laser.

The inability of the potentiostatic technique to detect all size pores limits the use of this method and prevents an absolute determination of the porosity. In spite of the limitations of the technique, it is an effective rapid screening test.

### *3.2. Stainless steel coatings*

To produce a practical, corrosion-resistant coating from an Fe–Ni matrix containing chromium powder, the coating must be heat-treated. The heat-treatment steps must be effective in dispersing the chromium throughout the deposit without causing detrimental interaction betweeen the substrate and coating. This interaction may include migration of carbon from the substrate to the coating and subsequent formation of harmful carbides or diffusion of the iron, nickel and chromium constituents into the substrate which could alter the coating composition.

Composite Fe–Ni coatings containing chromium particles were heat treated at 1100 °C in order to produce stainless steel type coatings. After heat treating, the coatings were of a uniform matrix Fe–Ni–Cr ternary alloy and were austenitic. A nickel diffusion barrier was needed to reduce carbon migration from the substrate during the heat treatment. Electron microprobe studies of coating cross-sections showed that although interdiffusion between the major coating elements and substrate occurred, the nickel barrier effectively prevented carbon migration into the coating.

After heat treatment an average grain size of 3 - 5 $\mu$m was observed, representing an ASTM grain size between 13.5 and 14. Figure 2 shows the microstructure of a typical coating after heat treatment at 1100 °C for 8 h. These coatings are considered to be fine grained since typical hot rolled bar or sheet steel have ASTM grain sizes of 6. The fine grain size is believed to

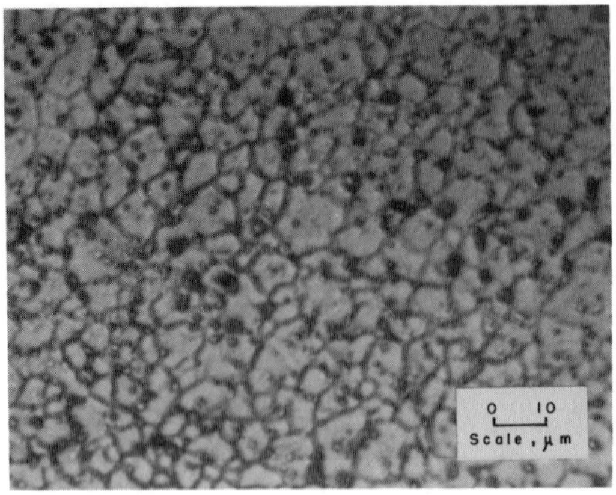


Fig. 2. Cross-sectional microstructure of a heat-treated stainless steel coating. The coating has been etched to reveal the grain boundaries.

be due to the chromium particles acting as nucleation sites during the heat treatment.

As stated previously, the coating exhibits an austenitic, stainless steel-like microstructure after heat treatment. As with conventional austenitic stainless steels, the coating's corrosion resistance is strongly affected by its carbon content. In the specific case of 304 stainless steel, a significant decrease in corrosion resistance to nitric acid is observed when the level of carbon in the metal exceeds 0.2 wt.%. The effectiveness of the heat-treatment procedure was measured by boiling the electrodeposited metal in 65% $HNO_3$ for five 48 h periods. This test (Huey test) was designed to identify chromium segregation in stainless steels. Nitric acid will preferentially dissolve those areas in the electrodeposit that have a lower chromium content. In conventional stainless steels, these areas are created by localized chromium carbide precipitation at the grain boundaries.

*3.2.1. Heat-treatment effects*

Coatings containing nominally 27 wt.% Ni and 17 wt.% Cr were physically removed from the cathode, vacuum-treated at 1100 °C and quenched in water. The coatings were removed from the substrate before heat treating to eliminate the possibility of carbon diffusion from the substrate and to prevent the results of the test from being affected by acid attack of the carbon steel substrate. The heat treated coating was quenched to prevent localized chromium carbide precipitation.

Heat-treatment times of 4 - 16 h at 1100 °C produced low corrosion rates (Table 3). The best corrosion resistance was obtained with a vacuum heat treatment of 8 h at 1100 °C.

Vacuum heat treatment is not required to obtain low corrosion rates. When the coatings were heat treated under a nitrogen atmosphere, there was no significant change in their corrosion rates.

TABLE 3

Effects of heat-treatment time and furnace pressure on coating homogenization

| *Time* (h) | *Pressure* (Torr) | *Corrosion rate* (mm $year^{-1}$) |
|---|---|---|
| 4 | $1 \times 10^{-6}$ | $4.1 \pm 1.3$ |
| 8 | $1 \times 10^{-6}$ | $1.3 \pm 1.0$ |
| 16 | $1 \times 10^{-6}$ | $4.0 \pm 0.7$ |
| 8 | 760[a] | $2.0 \pm 0.5$ |

[a]Nitrogen atmosphere.

### *3.2.2. Protection of carbon steel*

The integrity (protective quality) of the Fe–Ni–Cr alloy coating was demonstrated by exposing several coated specimens, 36 wt.% Ni and 21 wt.% Cr, to boiling 65% $HNO_3$ solution. The electrolyte composition and heat-treatment conditions that had produced a low-porosity, crack-free, homogenous alloy coating [17] were used to prepare the specimens. A 50 $\mu$m thick nickel barrier and a 250 $\mu$m thick coating protected the substrate during a 240 h test in the $HNO_3$ solution. Corrosion rates averaged $0.6 \pm 0.4$ mm $year^{-1}$ with some rates as low as 0.2 mm $year^{-1}$. Figure 3 illustrates the uniformity of corrosion of the coating after exposure to the acid environment.

### *3.3. Electroforms*

The alloy coating method worked well for preparing electroforms because relatively thick (up to 750 $\mu$m) deposits were possible. A variety of small objects including a crucible of 35 ml capacity as well as square and cylindrical shapes were prepared on Sn–Bi alloy mandrels. The electroforms were removed from the mandrels, heat-treated to homogenize the metal and

Fig. 3. Fe–Ni–Cr alloy coating after exposure to $HNO_3$ solution. This 53Fe–26Ni–21Cr alloy coating was boiled in 65% nitric acid for 240 h. The average corrosion rate was 0.6 $mm^{-1}$ $year^{-1}$.

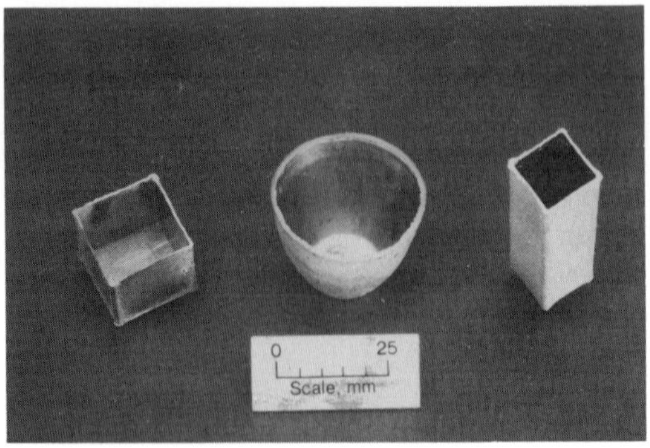


Fig. 4. Heat-treated Fe–Ni–Cr alloy electroforms.

quenched to prevent carbide segregation at the grain boundaries and the formation of eta phase. The electroforms could be polished to produce a smooth surfaced part in its final shape (Fig. 4).

The results of 65% $HNO_3$ corrosion tests conducted for 96 h on several electroforms, 28 wt.% Ni and 19 wt.% Cr, indicated good uniformity of alloy composition throughout the object. After heat treating four of the square- and crucible-shaped objects and cutting each vertically into half-sections, the average corrosion rate for the eight sections was $1.0 \pm 0.6$ mm year$^{-1}$. The corrosion rates of the two half-sections comprising each electroform were within 0.3 mm year$^{-1}$ of each other.

## 4. Conclusions

Platinum coatings electrodeposited from molten cyanide electrolytes were consolidated by laser surface melting. By utilizing a potentiostatic technique, the feasibility of consolidating platinum coatings with a pulsed Nd–YAG laser was demonstrated. Coatings around 25 $\mu$m thick were processed without damage to the substrate material, and anodic polarization currents were similar to pure platinum. The potentiostatic dissolution technique does not provide an absolute measure of porosity; however, the method can be effective for initial screening tests.

Coatings which had corrosion resistance equivalent to stainless steel were produced by using a combined particle occlusion-heat treatment procedure. The technique was applicable to the production of electroforms as well as protective coatings.

## References

1 A. J. Foster, *Electroplat. Met. Finish.*, *28* (1975) 5.
2 D. Schlain, F. X. McCawley and G. R. Smith, *BuMines RI 8249* (1977) 22.
3 R. P. Walters and D. R. Flinn, *BuMines IC 8974* (1984) 17.

4 M. J. Lynch, R. P. Walters, and D. R. Flinn, *BuMines RI 8829* (1983) 18.
5 J. P. Celis and J. R. Roos, *Rev. Coat. Corros.*, *5* (1 - 4) (1982) 1.
6 J. R. Roos and J. P. Celis, Is the Electrolytic Codeposition of Solid Particles a Reliable Coating Technology? Paper 0-1 in *Dispersion Coatings I: Technical Proc. 71st Ann. Tech. Conf. of the American Electroplaters Society, New York, July 16 - 19, 1984*), American Electroplaters Society, Winter Park, FL, 1984, 14 pp.
7 N. Pushpavanam, G. Vardajan, S. Krishnamurthy and A. Shenoi, *Electroplat. Met. Finish.*, *27* (1974) 10.
8 P. K. Sinha, N. Dhamansayan and H. K. Chakrabarti, *Plating*, *60* (1973) 55.
9 R. Bazzard and P. J. Boden, *Trans. Inst. Met. Finish.*, *50* (1972) 63.
10 R. Bazzard and P. J. Boden, *Trans. Inst. Met. Finish.*, *50* (1972) 207.
11 C. R. Kilgore, *Prod. Finish. (Cincinnati)*, *27* (1963) 34.
12 R. V. Williams, *Electroplat. Met. Finish.*, *19* (1966) 92.
13 J. E. Allison, Jr. and G. R. Smith. *Technical Proc. 71st Ann. Tech. Conf. of the American Electroplaters' Society, New York, July 16 - 19, 1984*, American Electroplaters' Society, Winter Park, FL 1984, 14 pp.
14 G. R. Smith and J. E. Allison, Jr., *U.S. Patent 4,601,795*, July 22, 1986.
15 D. R. Flinn and C. L. Manger, *BuMines RI 8656* (1982) 16.
16 R. P. Walters, M. J. Lynch, and D. R. Flinn, *Plating Surf. Finish.*, *7* (1983) 91.
17 J. E. Allison, Jr. and G. R. Smith, *BuMines RI 9213* (1989) 13.
18 C. R. Molock, R. P. Walters, and P. M. Fabis, *J. Electrochem. Soc.*, *134* (1987) 289.
19 M. Clarke, *Trans. Inst. Met. Finish.*, *51* (1973) 150.
20 C. D. Hodgman (ed.), *Handbook of Chemistry and Physics*, Chemical Rubber Co., Cleveland, OH, 1959, p. 2603.
21 J. O'M. Bockris and A. K. N. Reddy, *Modern Electrochemistry*, Vol. 1, Plenum, New York, 1970, p. 355.

*Surface and Coatings Technology*, *39/40* (1989) 667 - 674 

# CHEMICAL STABILITY OF PHOSPHATE CONVERSION COATINGS ON COLD-ROLLED AND ELECTROGALVANIZED STEELS

W. J. VAN OOIJ and A. SABATA

*Department of Chemistry, Colorado School of Mines, Golden, CO 80401 (U.S.A.)*

(Received March 9, 1989)

## Summary

The attack of phosphate conversion coatings on cold-rolled (CRS) and electrogalvanized (EG) automotive steels in high and low pH solutions has been studied. Hopeite (HP) and phosphophyllite (PPH) crystals dissolve completely at pH below 3. HP also dissolves in alkali, whereas PPH is converted to a mixed oxide. The phosphate crystals have a pronounced texture which is influenced by the texture of the substrate. Certain orientations are more stable at high pH than others. There is a strong effect of the temperature on the chemical stability of the phosphate at high pH. The presence of NaCl increases the rate of phosphate attack at high pH. The mechanism of attack at high pH is a solid state reaction forming $Zn(OH)_2$ which is stabilized by iron, manganese and other ions resulting in the formation of complex zinc hydroxy compounds.

## 1. Introduction

The phosphate conversion coating on automotive steels prior to the application of the paint film is assumed to be an important factor in the cosmetic corrosion performance of the total system. The phosphate layers on CRS consist of a mixture of hopeite (HP) and phosphophyllite (PPH) which has been optimized with respect to stability in an alkaline environment, *e.g.* by the addition of small amounts of nickel and manganese. It has been conclusively demonstrated that underfilm corrosion in painted CRS in the presence of an electrolyte is associated with high pH conditions [1]. However, the use of modern precoated steels, which are zinc- or zinc alloy-coated, may modify the mechanism of corrosion propagation under paint films [2]. Although these mechanisms have not yet been studied in great detail, the same phosphating processes are used for precoated steels as for cold-rolled steels (CRS).

The objective of the studies described here is to compare the behavior and properties of phosphate coatings on electrogalvanized steel (EG) with

0257-8972/89/$3.50 

those on CRS. In the first part of this study it was shown that in modern phosphate processes the incorporation of nickel and manganese in the bath results in improved chemical stability of the phosphate crystals on CRS and EG, especially at high pH [3]. It was also observed that the presence of nickel and manganese retards the rehydration of $Zn_3(PO_4)_2{\cdot}2H_2O$ to the originally present hopeite, zinc phosphate tetrahydrate, upon exposure to water. Such a rehydration process is detrimental to paint adhesion. We now report on the effects of temperature, presence of $Na^+$ and $Cl^-$ ions, and texture on the rate of phosphate dissolution in alkali. For the evaluation of the performance of painted automotive steels, cyclic corrosion tests are frequently used which involve a wet cycle at elevated temperature and exposure to $Na^+$ and $Cl^-$ ions. It has further been reported that certain crystallographic orientations of hopeite on EG are more resistant to alkali than others [4].

## 2. Experimental procedure

Phosphated test panels were obtained from Parker + Amchem. The CRS panels were spray phosphated (Granodine 902, coating weight $1.39\ g\ m^{-2}$). The EG panels were spray phosphated (Granodine 902, coating weight $2.22\ g\ m^{-2}$) or dip phosphated (Granodine 958, coating weight $2.8\ g\ m^{-2}$). The spray process was nickel-containing whereas the dip process contained nickel and manganese. The coating on CRS was a 75/25 mixture of HP and PPH, whereas those on EG consisted of hopeite only. The nickel and manganese contents were around 2 wt.% and 4 wt.% respectively, as per EDX. By XRD no crystalline form of these elements could be detected.

The effect of pH on phosphate attack was studied by immersion of $1.5 \times 1.5\ cm^2$ samples in stirred solutions of HCl or NaOH at 25 °C. Unless noted otherwise, the cut edges of the samples were not covered. Following treatment, the samples were rinsed in water, blown dry and analyzed away from the edges to avoid the effects of local corrosion cells at the edges. This was done by SEM–EDX using a JEOL JX 840 instrument with Tracor Northern EDX system TN-5500. XRD was performed on a Rigaku-200 system using Cu $K_\alpha$ radiation.

## 3. Results and discussion

### 3.1. *Effect of temperature*

Figure 1 shows the phosphate dissolution at pH 11 as a function of temperature. At room temperature no dissolution was observed under these conditions [3]. The phosphate on CRS retains its original morphology at all temperatures, despite the complete removal of $PO_4^{3-}$ ions. The reaction product is a complex iron zincate–hydroxide [3]. This reaction is strongly accelerated by the temperature. The two phosphates on EG behave in a

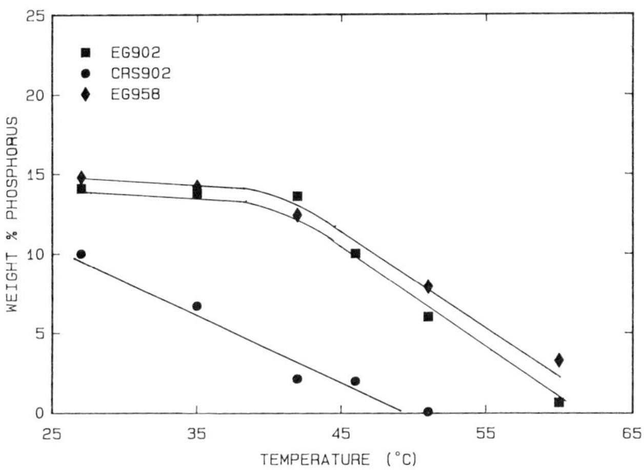


Fig. 1. Effect of temperature on phosphate dissolution in a stirred solution of pH 11. Plotted are the residual amounts of wt.% P after 30 min treatments, as per EDX.

different manner. They seem to be stable up to about 40 °C. Above this temperature they lose their $PO_4^{3-}$ ions at approximately the same rate as CRS. The Ni/Mn containing coating is slightly more resistant than the nickel-containing phosphate.

It has recently been reported that contrary to expectation, the weight loss from phosphated CRS is higher than from EG upon immersion in pH 12 solutions for 30 min [5]. This was explained by the redeposition of ZnO in the pores of the hopeite. In our experiments the residual weight percent phosphorus was determined by EDX, not the total weight loss.

SEM of the completely dissolved EG-902 showed that very little residue was left after treatment at 60 °C. However, the treated EG-958 had retained its original morphology almost completely, as is shown in Fig. 2. By XRD, only very small amounts of ZnO could be detected. It seems likely that this residue after attack is an amorphous form of $Zn(OH)_2$. It is known that $Zn(OH)_2$ decomposes readily to crystalline ZnO, but it is stabilized by small amounts of Mn ions [6]. Such a manganese-stabilized residue was not observed after a 30 min treatment at room temperature and higher pH (12.7) which completely dissolves the coating, except for manganese and nickel which are probably converted to oxides or hydroxides [3].

Although the mechanism by which manganese stabilizes $Zn(OH)_2$ is not understood, it is conceivable that a mixed zincate–hydroxide is formed here, too, as was proposed for PPH-containing phosphates [3]. In these cases, the products formed, iron or manganese zincate–hydroxide, are stable at very high pH, whereas $Zn(OH)_2$ formed from HP in the absence of iron or manganese is not stable above pH 12. It has also been observed in zinc corrosion studies that zinc forms a wide range of complex hydroxy salts which are very insoluble and stable [2].

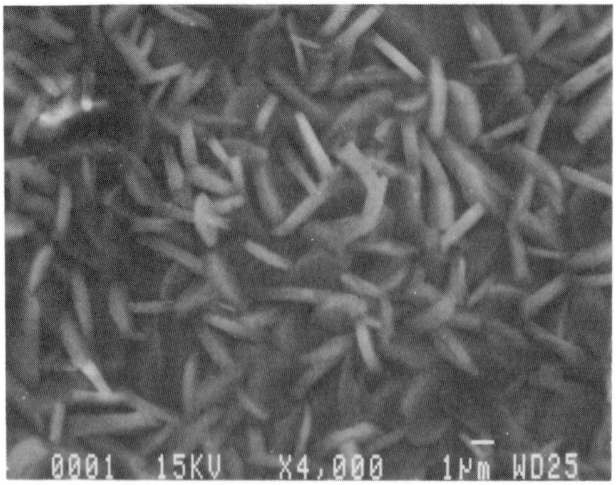


(a)

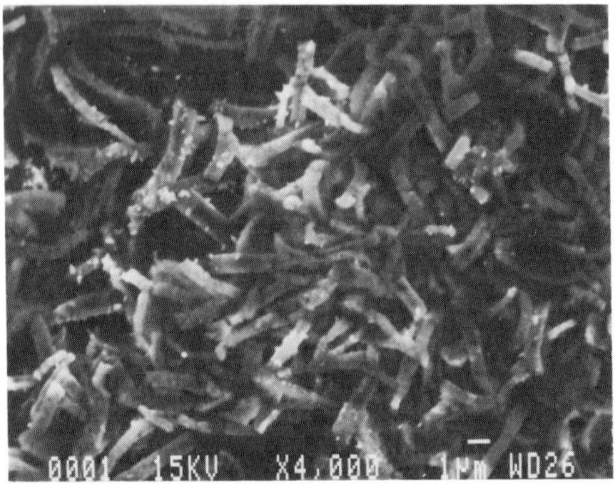


(b)

Fig. 2. Secondary electron image of EG-958 sample before (a) and after (b) treatment as in Fig. 1 at 60 °C.

### *3.2. Effect of sodium chloride*

Figure 3 shows the effect of the presence of NaCl on the phosphate dissolution at pH 12 for 30 min at room temperature. For both EG and CRS, the rate of phosphate leaching is accelerated with increasing concentration of NaCl. For CRS-902 the morphology of the crystals is completely retained as in other high pH treatments of mixtures of HP and PPH. Probably the same iron zincate–hydroxide is formed, since no $Cl^-$ could be detected in the reaction product. For EG-902 the crystal morphology was also retained, as opposed to treatments without the presence of NaCl which would result in the same degree of attack, *e.g.* at the same pH for a longer time, or at higher pH for the same exposure time as in Fig. 3. Although around 30% of the original phosphorus content was still present after the attack in 0.5N NaCl,

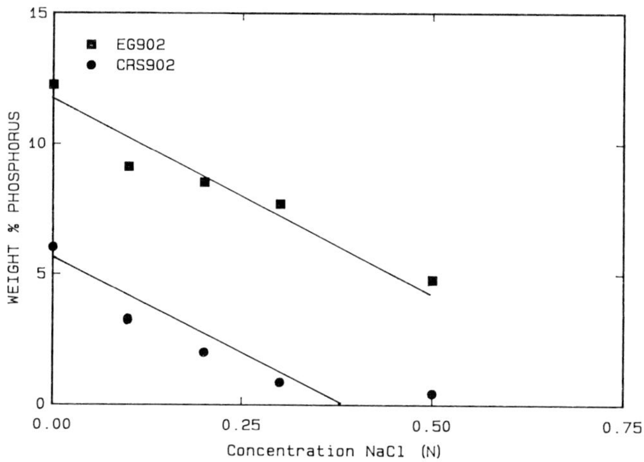


Fig. 3. Effect of NaCl on phosphate dissolution in a stirred solution of pH 12; residual amount of wt.% P (EDX) after 30 min treatments.

XRD could no longer detect any hopeite crystals. Only very small amounts of ZnO were identified. EDX did not detect any chlorine in the reaction product.

In normal corrosive environments, $Na^+$ and $Cl^-$ ions are always present. These results show that they can modify corrosion mechanisms and products. The exact mechanism is not understood, but the retention of the original morphology suggests that here, too, the $PO_4^{3-}$ ions are leached out of the hopeite crystals in a solid state exchange reaction between phosphate and hydroxyl ions. This reaction seems to be accelerated by the presence of NaCl. The result is the formation of $Zn(OH)_2$ which is stabilized here, too, probably by the formation of a mixed amorphous sodium zincate–zinc hydroxide [7].

### *3.3. Effect of phosphate texture*

Figure 4 shows the change in texture with pH for EG-902 treated for 30 min at room temperature. In the pH region 3 - 12.5, phosphate dissolution was only partial and Fig. 4 indicates that certain crystallographic orientations are more stable than others. At high pH, the (311) plane is more stable than the (020) plane, whereas in the low pH range, the latter is more stable; however, the effect here is less pronounced. These results show that $OH^-$ and $H^+$ ions attack certain planes preferentially, based on their atom distribution. This observation has important implications for further improvement of the total system if certain phosphate textures can be promoted by controlling the texture of the substrate. A relationship between texture of the phosphate layer and that of the metal substrate has recently been reported [4].

The same effect shown in Fig. 4 was observed for the EG-958 system.

### *3.4. Effect of corrosion cells on phosphate stability*

In the experiments described so far, phosphate dissolution was a chemical reaction. Analyses were only done in the center of the samples where the

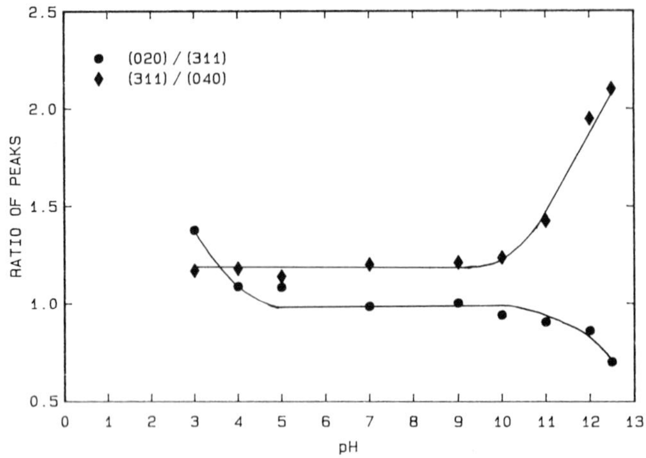


Fig. 4. Change in texture of phosphate on EG-902 with pH. Immersion in stirred solutions at 25 °C for 30 min.

effect of the local corrosion cells at the edges were assumed to be absent. In these cells the steel exposed in the cut edges is the cathode. The zinc layer is the anode which dissolves with propagation from the edges. Upon exposure of EG-902 samples for a longer length of time, *i.e.* 11 days at pH 12, the corrosion at the edges was clearly visible. The center of the samples was not corroded. The phosphate here showed very little attack whereas at the edges it had completely been converted to ZnO which in turn had been converted to zinc hydroxy carbonate, as was shown in XRD.

In similar experiments with CRS-902, *i.e.* exposure to pH 12 for 11 days, many localized corrosion cells were set up all over the sample surface. In the anodic areas the phosphate crystals were attacked and converted to ZnO. In the cathodic high pH areas the phosphate ions were leached out completely. The crystals had been converted to the iron zincate–hydroxide as described in the dissolution experiments and the original phosphate morphology had been retained.

Two experiments were done with EG-902 to investigate how the presence of corrosion cells affect the phosphate dissolution. In Fig. 5 a backscattered electron image shows the distribution of anodes (dark areas) at the phosphate surface of a panel that had been exposed to pH 11 for one day. The cut edges were taped in order to remove the effect of these exposed surfaces. Many circular anodes are formed where the phosphate is not attacked, but covered with ZnO–$Zn(OH)_2$ in pores where the zinc substrate dissolves anodically and then precipitates. The remainder of the panel acts as the cathode where oxygen is reduced, generating hydroxyl ions which increase the pH. The crystals in these areas were attacked and had lost 50% of their original phosphorous content. In a second experiment, an EG-902 sample with taped edges was treated in the same manner but the phosphate crystals were removed by scraping near the edges, exposing a bare zinc surface. The

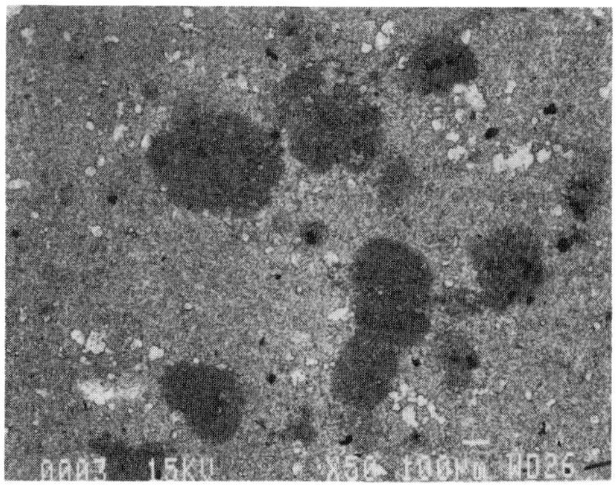


Fig. 5. Backscattered electron image of EG-902 sample with taped edges after exposure to pH 11 solution for one day (25 °C).

anode/cathode ratio thus obtained was 3. In this sample no local anodes, as shown in Fig. 5, were formed and the degree of attack and phosphate loss of the remaining phosphate layer was almost identical to that in the first experiment.

These experiments demonstrate that phosphate crystals on CRS and EG are very sensitive to pH changes occurring in corrosion reactions. Also, the surface of the crystals can sustain a cathodic corrosion reaction. Further, they are porous and zinc can dissolve and precipitate as corrosion products in these pores. The corrosion products formed in these corrosion reactions appear to be the same as those formed in our model dissolution experiments, as described here and elsewhere [3].

## 4. Conclusions

(1) Hopeite crystals are very sensitive to low and high pH. At low pH they dissolve completely, whereas at high pH they are converted to zinc hydroxide. This product is normally unstable. It decomposes to ZnO or it dissolves as zincate ions.

(2) In the presence of certain ions, $Zn(OH)_2$ is stabilized by forming a mixed zincate–hydroxide compound containing these ions. These ions are $Fe^{2+}$, $Mn^{2+}$ and probably $Na^+$. The former two are present in modern phosphating processes.

(3) The stability of hopeite at high or low pH can be improved by promoting certain crystallographic orientations of the crystals. Such textures depend on the texture of the metal substrate.

(4) Anodic and cathodic sites can develop on phosphated surfaces of steel and zinc when immersed in an electrolyte. Phosphate coatings are porous, allowing zinc dissolution from the substrate at local anodes.

## References

1 C. R. Shastry and H. E. Townsend, *Corrosion, 45* (1989) 103.
2 W. J. van Ooij, A. Sabata and G. Ström, *Prog. Org. Coat.*, in press.
3 W. J. van Ooij and A. Sabata, *J. Coat. Technol.*, in press.
4 T. Yoshihara and H. Okita, *Trans. ISIJ, 23* (1983) 984.
5 J. P. Servais, B. Schmitz and V. Leroy, *Mat. Performance, 27* (1988) 56.
6 Y. Miyoshi, J. Oka and S. Maeda, *Trans. ISIJ, 23* (1983) 974.
7 H. Leidheiser, Jr., and A. Sommer, *Semiannual Report*, Center for Surface and Coatings, Lehigh University, 1984.

# HERMETIC COATINGS FOR HEAVY METAL FLUORIDE GLASSES

D. W. REICHER, M. S. MURILLO and J. R. McNEIL

*University of New Mexico, Center for High Technology Materials, Albuquerque, NM 87131 (U.S.A.)*

P. A. SMITH, S. Q. YU and G. J. COLLINS

*Engineering Research Center, Colorado State University, Fort Collins, CO 80523 (U.S.A.)*

I. D. AGGARWAL and B. HARBISSION

*Naval Research Laboratory, Washington, DC 20375 (U.S.A.)*

(Received March 17, 1989)

## Summary

Heavy metal fluoride glasses constitute a group of materials with promising optical properties, but which are extremely fragile environmentally. We have investigated methods of protectively coating optical elements made of a specific type of heavy metal fluoride glass (ZBLAN). Dielectric coatings were reactively sputtered using a direct current magnetron with concurrent ion bombardment (ion-assisted deposition (IAD)) onto planar substrates. These coatings were tested by exposing the coated substrate to deionized water and determining the concentration of fluorine in the water after specified time intervals. Results indicate that reactively sputtered dielectric materials provide substantial protection from aqueous attack and that IAD can improve the performance of some materials. This is shown to be related to the film packing density.

## 1. Introduction

Heavy metal fluoride (HMF) glasses constitute a class of relatively new optical materials that have many desirable properties, including a wide range of transparency and low optical dispersion. The fact that this material is a glass allows it to be molded or easily drawn into fibers. There is particular interest in this material for use in optical fibers in the mid infrared wavelength region [1].

A major drawback of the material is its susceptibility to damage by water. Previous work has been done investigating the reaction of HMF material with water as well as applying hermetic coatings to the material [2, 3]. Another problem is that the material cannot be heated to temperatures

0257-8972/89/$3.50 

in excess of 150 °C without damage from crystallization. This precludes use of normal film deposition techniques which require substrate temperatures of 250 - 300 °C to produce films of good optical and mechanical properties.

In this study three materials, $ZrO_2$, MgO and AlN, were reactively sputtered onto HMF substrates using a direct current magnetron. In addition, ion-assisted deposition (IAD) techniques were applied. The substrate material was ZBLAN (56% $ZrF_4$, 14% $BaF_2$, 6% $LaF_3$, 4% $AlF_3$, 20% NaF). Several test methods were used to evaluate the films. The goal of the effort was to identify materials that could be used as hermetic coatings for HMF optical fibers that might be coated using a cylindrical magnetron.

## 2. Film deposition

The method for producing the films was reactive direct current magnetron sputtering. This method has the advantages of being fairly simple, requiring inexpensive equipment, being easily scalable and easy to automate. Targets for the magnetron were fabricated from aluminum, magnesium and zirconium. The films were deposited in a vacuum system that also contained a Kaufman ion source. The substrate holder was 15 cm from the magnetron and 15.5 cm from the ion source. A quartz crystal monitor was used to monitor the film thickness.

Ion bombardment from the Kaufman source was applied during deposition. Ion assisted deposition has been used to improve optical and mechanical properties of films in applications such as the present one where the substrate cannot be heated [4]. IAD has provided films with higher refractive indices, lower extinction coefficients, and improved adhesion and abrasion resistance. These effects can be attributed to a higher packing density of the film material and improved stoichiometry promoted by IAD. A densified film would be expected to provide improved hermetic protection of the HMF material.

Relevant deposition parameters are shown in Table 1. The films were deposited on HMF substrates of diameter 1.5 in with ion bombardment current densities of 0, 5, 15 and 45 $\mu A\ cm^{-2}$ and energy of 300 eV. The gas

TABLE 1

Deposition conditions

| | *Material* | | |
|---|---|---|---|
| | *AlN* | *MgO* | *$ZrO_2$* |
| Gas mixture (Torr) | $7 \times 10^{-4}$ $N_2$ | $0.5 \times 10^{-4}$ Ar $+6.5 \times 10^{-4}$ $O_2$ | $0.5 \times 10^{-4}$ Ar $+6.5 \times 10^{-4}$ $O_2$ |
| Rate (Å $s^{-1}$) | 0.5 | 0.3 | 0.2 |
| Film thickness (Å) | 2400 | 2400 | 1800 |

pressures were the same for films produced with and without IAD. In the case of AlN, pure nitrogen was used for the ion source discharge; for the $ZrO_2$ and MgO films, equal amounts of argon and oxygen were used in the ion source discharge to minimize erosion of the cathode filament. During deposition the substrates were rotated to provide films of uniform thickness. Because of this, the flux of particles from both the magnetron and the ion source striking the substrate varied by approximately 30% within one rotation. During preparation of the substrate for coating, care was taken to avoid exposure to moisture.

## 3. Test methods

Environmental testing of thin films is typically qualitative and usually, to some extent, subjective. Common tests such as abrasion, humidity and adhesion usually are evaluated on a pass or fail basis, and the criteria for evaluation tend to be loosely defined.

In this investigation more quantitative methods to evaluate the hermetic properties of protective thin films applied to heavy metal fluoride materials were used. One technique involved immersing the substrate in water and measuring the amount of fluorine leached from the substrate. The coated substrates were placed in holders made of nylon which were designed to hold approximately 15 $cm^3$ of deionized water. An O-ring seal confined the area exposed to water to the coated surface of the substrate. At different time intervals the water was removed from the holders and the $F^-$ concentration was measured using a standard fluorine probe. The water was then replaced in the holders. The amount of fluoride compounds leached from the substrate is proportional to the $F^-$ concentration of the water.

Optical absorption at 2.8 $\mu$m due to the OH stretch band was used as a measure of hydration. This technique has been used in studies involving leaching of bare HMF substrates [5]. In our investigation, the usefulness of the spectral data is somewhat diminished due to the inability to separate the hydration of the substrate from the hydration of the film. However, as discussed below, the level of fluorine concentration and the degree of absorption at 2.8 $\mu$m correlated quite well.

Spectral analysis of the films in the visible region was used to determine the index of refraction and extinction coefficient of the films at 560 nm. Using the refractive index, it is possible to determine the packing density of the film. The extinction coefficient is an indicator of the stoichiometry of the film. This data was taken using $SiO_2$ samples coated at the same time as the HMF samples.

## 4. Experimental results

The results of the fluorine concentration test on a bare HMF substrate are shown in Fig. 1. As can be seen, the fluorine concentration in water

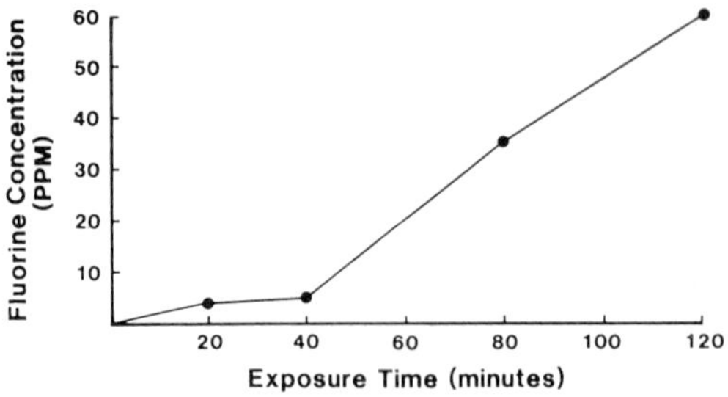


Fig. 1. Fluorine concentration *vs.* time for an uncoated HMF substrate.

increases rapidly. In Figs. 2, 3 and 4 the results of the same test on the coated substrates are shown. It can be seen that the coated substrates are leached at much lower rates. In most cases even after 196 h of direct water exposure, the fluorine concentration is much less than that illustrated in Fig. 1. Of the three materials, the MgO has the lowest average level of fluorine concentration, the AlN concentration is slightly higher, and the $ZrO_2$ is much higher. The results do not show a consistent correlation between ion bombardment flux and hermetic properties for the $ZrO_2$ and MgO films. For the AlN films there may not appear to be a relation between the level of bombardment and

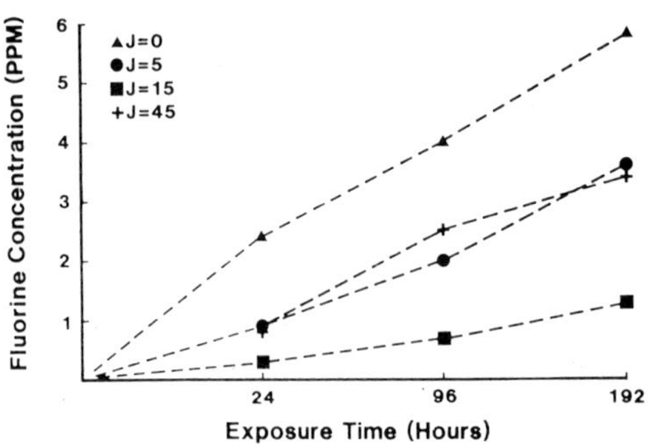


Fig. 2. Fluorine concentration *vs.* time for HMF substrates coated with AlN films.

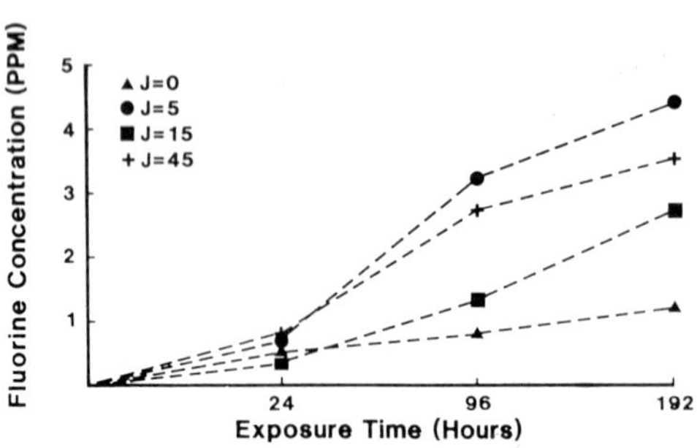


Fig. 3. Fluorine concentration *vs.* time for HMF substrates coated with MgO films.

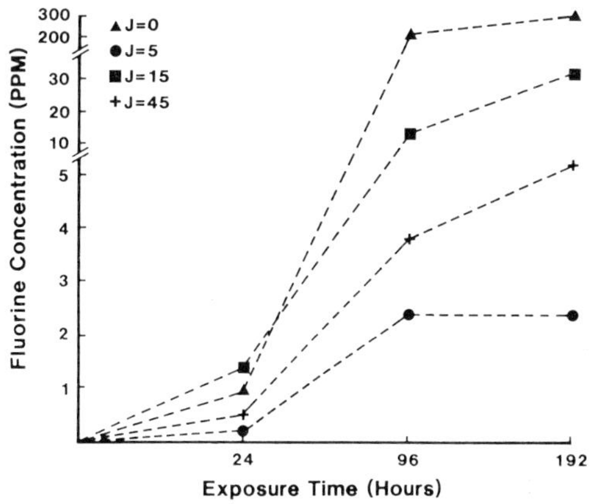


Fig. 4. Fluorine concentration *vs.* time for HMF substrates coated with $ZrO_2$ films.

the fluorine concentration, however the fluorine concentration does decrease for increased flux levels up to 15 $\mu$A cm$^{-2}$ and then increases for the next higher level of ion flux. As will be discussed later, this trend is consistent with previous work. Spectrographic data agree quite well with fluorine concentration, in that samples with higher fluorine concentration have greater optical absorption at 2.8 $\mu$m.

Figures 5, 6 and 7 show the index of refraction and extinction coefficients for the three film materials as a function of bombardment ion

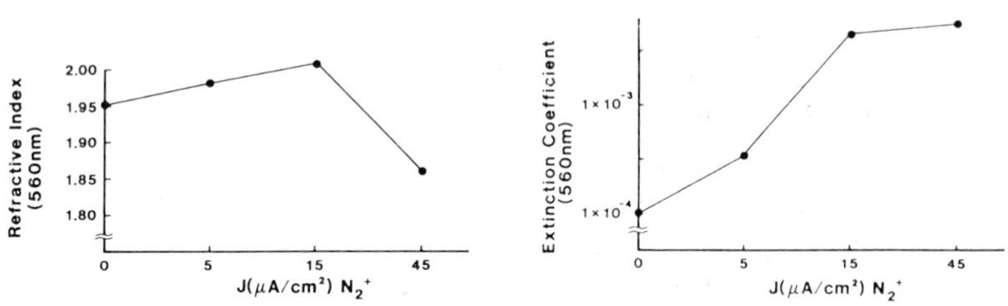


Fig. 5. Refractive index and extinction coefficient of AlN films as a function of $N_2^+$ ion current density.

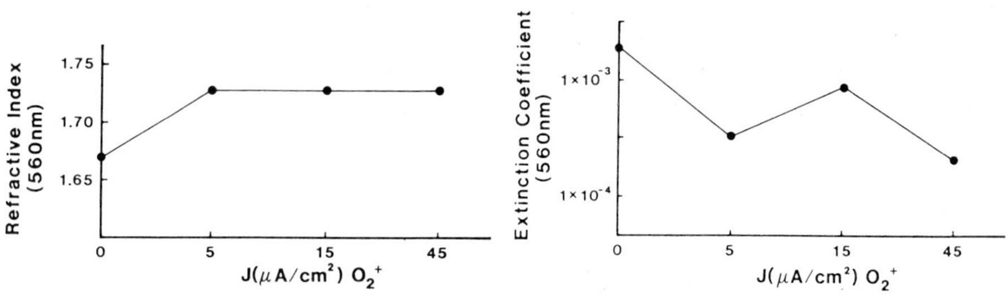


Fig. 6. Refractive index and extinction coefficient of MgO films as a function of $O_2^+$ ion current density.

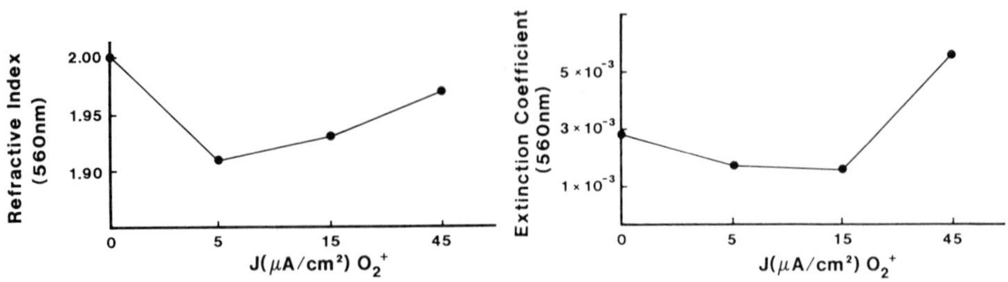


Fig. 7. Refractive index and extinction coefficient of $ZrO_2$ films as a function of $O_2^+$ ion current density.

current density. The bulk indices of the materials at 589 nm are 1.73 for MgO, 2.15 for AlN and 2.15 for $ZrO_2$. The MgO films attained refractive index values of close to bulk value for certain deposition conditions. The AlN films and $ZrO_2$ films did not exhibit bulk index under any of the deposition conditions.

After the termination of water exposure, the substrates coated with AlN and $ZrO_2$ had numerous cracks and pinholes in the films. It can be surmised that these flaws were the major failure mechanism for AlN and $ZrO_2$ films. The MgO films had fewer flaws, but were uniformly thinner over the exposed area. This was probably caused by dissolution of the MgO film material, which is soluble in water. For MgO dissolution is the most likely failure mechanism and this material may not prove useful in applications requiring direct exposure to a humid environment.

As discussed above, there are reasons to believe that IAD should improve the film properties. In prior work using IAD for electron beam evaporated films, it has been observed that films densify until a critical level of ion flux is reached. For flux levels greater than this critical level, the film becomes less dense [6]. We have discussed this behavior in terms of the incorporation probability of the bombarding ions; and how this probability is dependent upon ion species, energy and flux level [7]. The AlN films were observed to follow this trend. Measurements of the refractive index of the films at 560 nm indicate an increase in refractive index with increasing ion flux to levels of 15 $\mu$A cm$^{-2}$. At greater levels of bombardment a decrease in the index of refraction is observed. As can be seen from Fig. 4, the fluorine concentration exhibits a similar trend. The experimental results do not show any consistent correlation between ion bombardment flux provided by the Kaufman source and hermetic or optical properties for the $ZrO_2$ and MgO films.

## 5. Discussion

From the experimental results it is not obvious why certain of the films provide better hermetic barriers than others. It is also not clear why the MgO and $ZrO_2$ films do not respond to IDA as the AlN does.

As mentioned above, we have assumed that a denser film will provide a superior hermetic barrier. The density of the film can be quantified in terms of the packing density $p$, which is the ratio of the solid volume of the film to the total volume of the film. A film with no voids would have a $p$ equal to one. Maxwell–Garnett theory [8] can be used to determine $p$. Using the relation

$$\frac{n_e^2 - n_s^2}{n_e^2 + 2n_s^2} = p_f \frac{n_x^2 - n_s^2}{n_x^2 + 2n_s^2}$$

where $n_e$ is the refractive index of the film, $n_x$ is the refractive index of the voids (here assumed to be 1.16), $n_s$ is the refractive index of the bulk material and $p_f$ is the volume fraction of voids. All of the indices of refraction may be complex. The value of 1.16 for $n_x$ is chosen as an average of the indices of water (1.33) and air, assuming water occupies 50% of the void volume. Then $p$ is given by

$$p = 1 - p_f$$

The results of this calculation using the data in Figs. 5, 6 and 7 are shown in Fig. 8. Since both the real and imaginary parts of the refractive index are known, it is possible to estimate the effects of changes in stoichiometry. The MgO films have the highest packing density, while the AlN and $ZrO_2$ have lower packing densities. For AlN the film with the highest packing density has the lowest fluorine concentration which is consistent with the premise that a higher packing density improves the hermeticity of the film.

The differences observed in applying IAD to magnetron deposition compared with what we have observed using electron beam evaporation may be explained as follows. In previous investigations [9], the effects of energetic neutral atoms emitted by the magnetron have been analyzed. The origin of these energetic particles is believed to result from negative ions generated near the cathode being subsequently neutralized in the discharge. Prior to being neutralized, the ions are accelerated away from the cathode. Time of flight mass spectroscopy has been employed to detect significant numbers of neutral oxygen atoms generated by the magnetron. However, in the case of nitrogen very few neutral atoms were detected. It has been proposed that the large

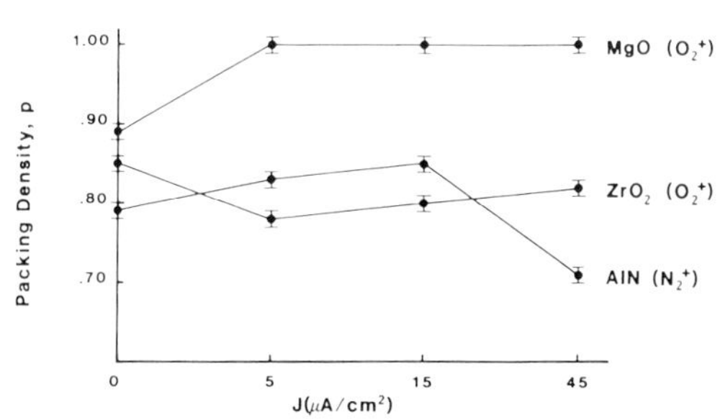


Fig. 8. Packing density of MgO, AlN and $ZrO_2$ films as a function of ion current density.

positive electron affinity of oxygen leads to a large production of $O^-$ ions, but that the negative electron affinity of nitrogen makes $N^-$ difficult to form [10].

In our investigation it is likely that with an oxygen magnetron discharge the large amount of high energy oxygen atoms created would produce bombardment of the substrate at a much higher flux level than that provided by the Kaufman source. However, this is not the case in using nitrogen in the magnetron discharge; there would be very little energetic neutral flux from the magnetron discharge, and the ion flux provided by the Kaufman source would be effective in altering film properties.

## 6. Conclusion

The three film materials in this investigation all provided substantial protection from moisture to the HMF substrate. Reactive direct current magnetron sputtering appears to be an effective technique of film deposition for this purpose. The data indicate that hermeticity is related to packing density. Only in the case of AlN did there appear to be an improvement in film hermeticity promoted by IAD. This may be a result of the energetic particle flux created by the magnetron in an oxygen discharge.

## References

1 D. C. Tran, G. H. Sigel and B. Bendow, *J. Lightwave Technol., LT-2* (1984) 566.
2 Robert H. Doremus *et al.*, *Mater. Sci. Forum, 5* (1985) 291.
3 J. J. McNally, G. A. Al-Jumaily, J. R. McNeil and B. Bendow, *Appl. Opt. 25* (1986) 1973.
4 F. L. Williams, J. J. McNally, G. A. Al-Jumaily and J. R. McNeil, *J. Vac. Sci. Technol., A5* (1987) 2159.
5 D. Tregoat *et al.*, *Mater. Sci. Forum, 5* (1985) 335.
6 F. L. Williams, D. W. Reicher, C. B. Juang and J. R. McNeil, *J. Vac. Sci. Technol., A7* (1989) 2286.
7 F. L. Williams, *et al.*, presented at the *Materials Research Society*, Fall Meeting (1988), Session A1.3.
8 A. Feldman, *Proc. SPIE, 821* (1988) 129.
9 G. Este and W. D. Westwood, *J. Vac. Sci. Technol., A5* (1987) 1892.
10 K. Tominaga, S. Iwamura, Y. Shintani and O. Tada, *Jpn. J. Appl. Phy., 22* (1983) 418.

*Surface and Coatings Technology, 39/40* (1989) 683 - 690 

# EFFECT OF DEPOSITION VARIABLES ON THE PROPERTIES OF MOLYBDENUM SULFIDE FILMS PREPARED BY THE ACTIVATED REACTIVE EVAPORATION TECHNIQUE

H. SHIN, H. J. DOERR, C. V. DESHPANDEY, B. DUNN and R. F. BUNSHAH

*School of Engineering and Applied Science, University of California, Los Angeles, CA 90024-1595 (U.S.A.)*

(Received March 13, 1989)

## Summary

Non-crystalline molybdenum sulfides ($MoS_x$, $2 \leq x \leq 3$) are excellent cathode materials for the secondary lithium batteries. Molybdenum sulfide films were therefore synthesized by the activated reactive evaporation (ARE) technique which involved evaporating molybdenum using an electron beam source in an $H_2S$ plasma. The composition of the films showed wide variation depending on the deposition variables such as deposition rate, anode current, working pressure and gas flow rate. Deposition rate was the most dominant factor controlling the composition of the films where desorption of sulfur from the films was a limiting step. Sulfur adsorption was enhanced due to reaction of condensing molybdenum and sulfur atoms to form molybdenum sulfide. However, after reaching saturation (*i.e.* no more sulfur atoms are available to form $MoS_x$) molybdenum atoms were embedded as excess atoms, leading films to be called sulfur deficient films in a chemical composition sense. The second dominant factor controlling the film composition was ARE anode current. Sulfur content in the film increased at low anode currents, where thermodynamically reactive HS was produced (as opposed to sulfur and hydrogen at high anode current) hence enhancing the formation of molybdenum sulfide.

## 1. Introduction

Ambient temperature secondary lithium batteries are the newest and most convenient electrochemical cells [1, 2]. The cell performance strongly depends on the cathode material, which is required to be an electronic and ionic conductor. Non-crystalline molybdenum sulfides such as a-$MoS_3$ [3, 4], a-$MoS_2$ [4, 5] and a-$Mo_2S_5$ [4] are known to be promising candidates due to their high intercalation ratio of Li/Mo. Synthesis of molybdenum sulfide films was therefore explored by the activated reactive evaporation (ARE) technique, which provides the desired flexibility to obtain films with a wide

0257-8972/89 Elsevier Sequoia/Printed in The Netherlands

compositional range. The sulfur content of the film could be varied by manipulating deposition variables such as deposition rate, anode current, working pressure and gas flow rate. The effect of these variables on the sulfur content of the films was investigated using the electrical resistivity of the film which changes monotonically with sulfur content. The substrate temperature kept constant around 200 °C to obtain a non-crystalline structure.

## 2. Experimental procedures

### 2.1. Synthesis of thin films

The schematic diagram of the deposition system used in this study is shown in Fig. 1. The films were synthesized by evaporating molybdenum (purity, 99.99%) using a rod fed electron beam gun in $H_2S$ plasma. Flow rate of $H_2S$ gas was controlled by an electronic flowmeter (Tylan Model FC-260). The plasma was generated with a filament–anode geometry. The working pressure of the chamber was measured by a Convectron gauge (Granville Phillips Model 275) calibrated for $H_2S$ gas and was adjusted via pressure control plate opening. Films were deposited onto silica glass substrates located 12 in above the evaporation source and beneath the water-cooled copper holder. The temperature of the substrate during the deposition run was monitored using a chromel–alumel thermocouple and was kept around 200 °C (by adjusting the water flow) to obtain non-crystalline films.

### 2.2. Characterization of film properties

Molybdenum sulfide is difficult to analyze quantitatively by energy dispersive X-ray analysis (EDX) due to the peak overlap of S $K$ and Mo $L\alpha$ lines. Electron microprobe analysis (CAMECA-MICROBEAM) clearly separates the Mo $L\alpha$ and S $K$ peaks at $\sin\theta = (61797 \pm 175) \times 10^{-5}$ and

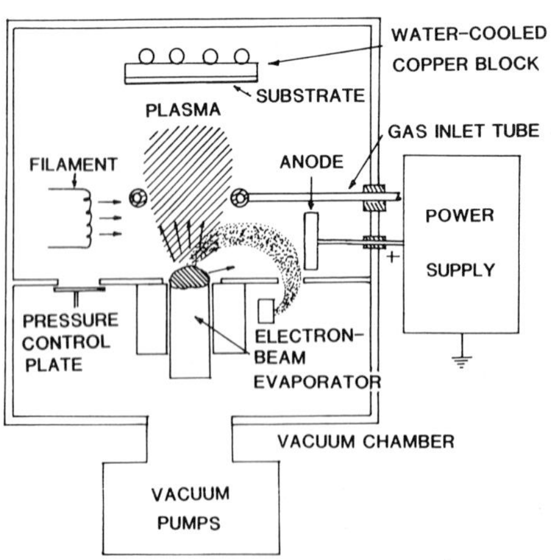


Fig. 1. Schematic diagram of ARE system.

$(61385 \pm 215) \times 10^{-5}$ respectively, after diffraction from PET crystal. Electron microprobe analysis was therefore used to determine the composition of the films. $MoS_2$ powder purchased from ESPI was used as a standard. The relative intensity ratio of peak of film to that of standard was measured and converted to the actual concentration within accuracy of $\pm 2\%$.

Sheet resistance ($R_\square$) of the film was measured using a Magnetron Instrument four-point probe tester (model M-700) and the thickness ($t$) of the film was measured using a surface profilometer (Sloan DEKTAK IIA). Room temperature resistivity was calculated using the equation $\sigma = R_\square t$ and averaged from at least eight measurements.

## 3. Results

The effect of the deposition variables was examined by changing one variable at a time while keeping the other variables fixed within the experimental limits. Figure 2 shows the room temperature resistivity variation against atomic percent of sulfur measured by electron microprobe analysis and illustrates that resistivity increased monotonically with increasing sulfur content. The room temperature resistivity measured by four-point probe tests reflects the sulfur content of the film and was used as a reference value to study the effect of changes in the deposition variable.

Among the deposition variables in the ARE technique, deposition rate was the most dominant variable in the deposition process. The deposition rate was measured *in situ* by a gold-coated quartz crystal using the thickness monitor. Figures 3 and 4 show variation of resistivity with deposition rate at $H_2S$ gas flow rates of 20 and 25 sccm respectively. It is observed that for a given flow, the resistivity increases with deposition rate, passes through a maximum and then decreases with further increase in deposition rate. It is also seen from Figs. 3 and 4 that the trend of resistivity variation with

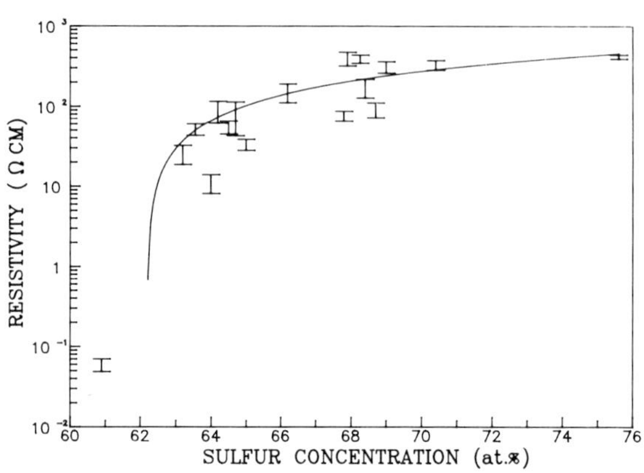


Fig. 2. Room temperature resistivity *vs.* atomic percent of sulfur of the film.

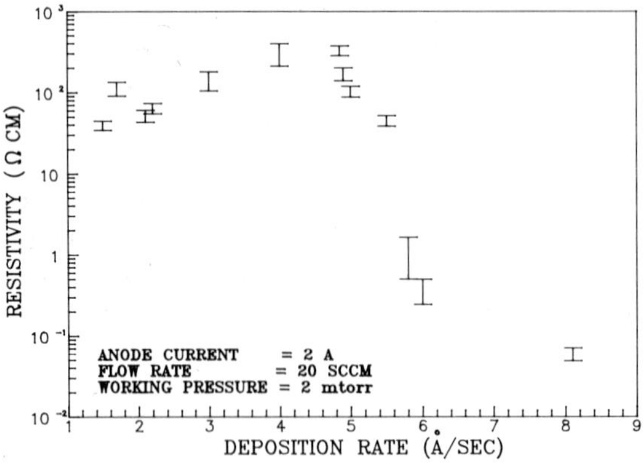


Fig. 3. Room temperature resistivity *vs.* deposition rate at flow rate of 20 sccm.

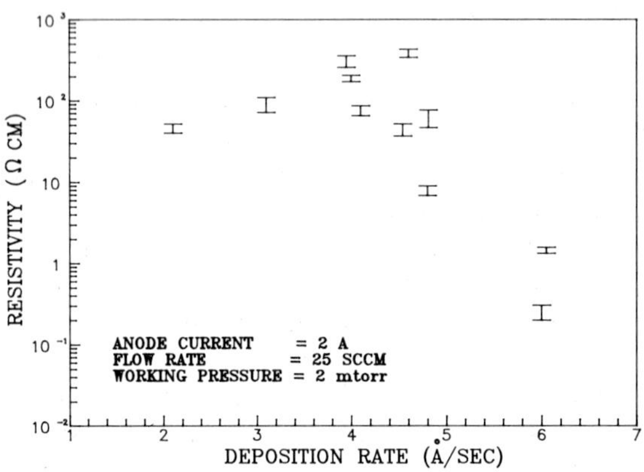


Fig. 4. Room temperature resistivity *vs.* deposition rate at flow rate of 25 sccm.

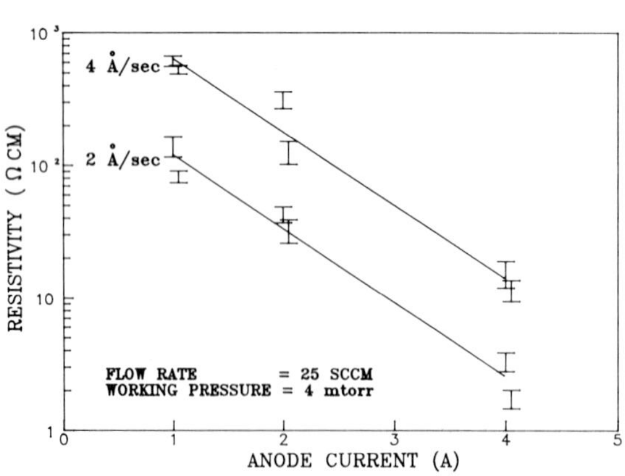


Fig. 5. Room temperature resistivity *vs.* anode current at deposition rate of 2 and 4 Å $s^{-1}$.

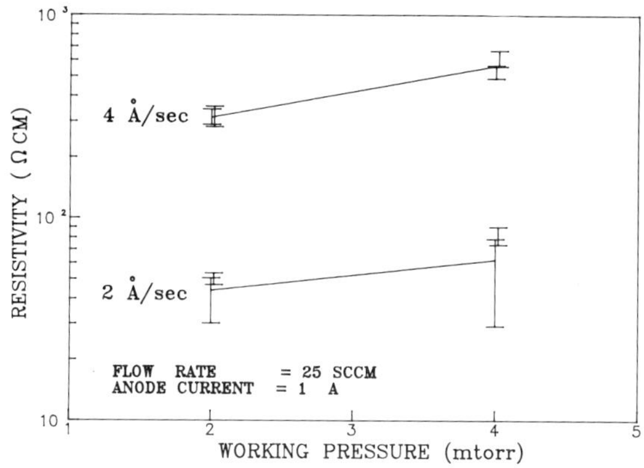


Fig. 6. Effect of working pressure on the room temperature resistivity at $I$ (anode) = 1 A.

deposition rate is similar for different flow rates. However, the resistivity maxima shifts to higher deposition rate with increase of flow rate.

Figure 5 shows the dependence of resistivity on anode current. It is observed that for both of the deposition rates (2 and 4 $Å^{-1}$ s), resistivity (and thus sulfur content) decreases as the anode current increases.

The effect of working pressure during deposition on the resistivity of the films was not as marked. The resistivity increases slightly with increasing pressure as shown in Fig. 6, which shows the variation in resistivity of the films with working pressure for a deposition rate of 2 and 4 Å $s^{-1}$, a flow rate of 25 sccm and anode current of 1 A.

## 4. Discussion

The observed variation in sulfur content and resistivity of the films with the deposition variables discussed above can be quantitatively explained as follows.

### *4.1. Sulfur content and resistivity vs. deposition rate*

The adsorption of sulfur may be a major factor determining sulfur content in the film citing a model proposed by Winters *et al.* [6] for the composition of sputtered multicomponent thin films and successfully applied to the $MoS_2$ compound by Reichelt and Mair [7]. Gas incorporation into a film is due to the sorption of molecules, atoms and ions produced by dissociation of $H_2S$ gas. They assumed that sulfur atoms, except for the energetic atoms which penetrate the lattice, equilibrate thermally with the surface and are absorbed with weak binding energy $E_{ad}$. These atoms are mobile and diffuse on the surface in a two-dimensional Brownian motion. Since adsorption energy may be assumed to be related to the sublimation energy, the desorption of molybdenum is negligible due to its high sublimation energy. Molybdenum atoms either react with sulfur or are embedded as excess atoms into the growing film. If the desorption of sulfur is a limiting step, the atomic ratio $\eta = \mathrm{S/Mo}$ in the films is given by [7]

$$\eta = \frac{\mathrm{S}}{\mathrm{Mo}} = \frac{R_{\mathrm{Sv}} - (N_{\mathrm{s}}/\tau_{\mathrm{s}})}{R_{\mathrm{Mo}}} = \frac{R_{\mathrm{Sv}} - N_{\mathrm{s}}\nu \exp(-E_{\mathrm{ad}}/kT)}{R_{\mathrm{Mo}}} \tag{1}$$

where $R_{\mathrm{Sv}}$ is the impingement rate of sulfur (molecules, atoms and ions) on the surface, $N_{\mathrm{s}}$ is the number of surface site $\mathrm{cm}^{-2}$ occupied by sulfur adatoms, $\tau_{\mathrm{s}}$ is the mean residence time of sulfur, $R_{\mathrm{Mo}}$ is the deposition rate of molybdenum, $\nu$ is the atomic frequency of sulfur and $E_{\mathrm{ad}}$ is adsorption energy of sulfur. Because of the low sticking coefficient of sulfur atoms, the desorption of sulfur is a limiting step in this process. If the sulfur adatoms react with molybdenum adatoms on the film surface during the Brownian motion, the desorption of sulfur is greatly suppressed. Therefore, it is more reasonable that $E_{\mathrm{ad}}$ in eqn. (1) related to $\tau_{\mathrm{s}}$, should include the $\Delta G$ of molybdenum sulfide compound formation. Increasing the deposition rate only increases the molybdenum flux while the total rate of sulfur incorporation is nearly constant since the other deposition variables are fixed. Referring to Figs. 3 and 4, $\eta$ is initially increasing with deposition rate because $\tau_{\mathrm{s}}$ is increasing and $N_{\mathrm{s}}$ is decreasing. However, the slope is not steep because $R_{\mathrm{Mo}}$ is also increasing. After reaching the maximum, the enhancement of sulfur adsorption reaches saturation (*i.e.* no more sulfur atoms are available to form $\mathrm{MoS}_x$) and molybdenum atoms start to be embedded as excess atoms into the growing film. It leads the films which consist of $\mathrm{MoS}_x$ with excess molybdenum atoms to be called sulfur deficient films in a chemical composition sense.

The maximum shift at higher flow rate can be understood from the fact that, for a larger number of sulfur atoms due to higher flow rate, more molybdenum atoms are needed for the saturation of molybdenum sulfide formation. However, the shift is not marked.

### 4.2. Anode current and working pressure

Dissociation rate of a gas reactant in a plasma is given by [8]

$$R_{\mathrm{d}} = n_{\mathrm{e}}(2/m_{\mathrm{e}})^{1/2}[p] \int E^{1/2} f(E)\sigma(E)\,\mathrm{d}E \tag{2}$$

where $R_{\mathrm{d}}$ is the dissociation rate of gas reactant, $n_{\mathrm{e}}$ is electron density, $m_{\mathrm{e}}$ is electron mass, $[p]$ is concentration of reactant, $E$ is electron energy, $f(E)$ is a normalized distribution function and $\sigma$ is the cross-section for the reaction. As anode current was increased by increasing anode voltage, $n_{\mathrm{e}}$ and $E$ in eqn. (2) would also increase with increasing anode current. Therefore, the dissociation rate increases as the anode current increases, which enhances the complete dissociation to atomic species. However, if the anode current is low (corresponding to about 70 V anode voltage), dissociation rate is low and thus molecular species such as HS would be formed [9]. It was expected that

TABLE 1
Gibb's free energy change of the reaction of hydrogen sulfides and molybdenum at 500 K [10]

| *Reactions* | $\Delta G$ (cal mol$^{-1}$) |
|---|---|
| (a) $Mo + 2S \rightarrow MoS_2$ | −74 190 |
| (b) $Mo + 2HS \rightarrow MoS_2 + H_2$ | −111 750 |
| (c) $Mo + 2H_2S \rightarrow MoS_2 + 2H_2$ | −14 270 |

sulfur content would increase if the anode current increased. However, Fig. 5 shows that the resistivity and sulfur content decrease as anode current increases. This discrepancy can be explained in terms of the following two factors: gas decomposition products and thermodynamic data of molybdenum sulfide formation. First, the low anode current product such as HS may have a higher sticking coefficient than sulfur atoms and hence increases $R_{Sv}$ and $\eta$. Secondly, as in the case of the effect of deposition rate, $\Delta G$ of formation of molybdenum sulfide should be considered in sulfur impingement rate ($R_{Sv}$). Table 1 summarizes the Gibb's free energy of the possible reactions at 500 °K (near the substrate temperature) [10]. Table 1 clearly shows that the driving force of the reaction of molybdenum with HS is larger than that with sulfur atom and thus the $R_{Sv}$ greatly increases by the formation of molybdenum sulfide from HS.

As the working pressure increases, the number of sulfur atoms in the chamber increases, which increases the impingement rate ($R_{Sv}$ in eqn. (1)) of sulfur atoms as shown in Fig. 6.

## 5. Conclusions

The ARE process was used to deposit $MoS_x$ films with compositions containing from 61 to 75 at.% s. Deposition rate was the most dominant deposition variable. Electrical resistivity exhibited a maximum at a deposition rate of 4 - 5 Å s$^{-1}$, depending on a flow rate. Desorption of sulfur is a rate limiting step and was suppressed by the molybdenum atoms to form molybdenum sulfide. However, after reaching saturation molybdenum atoms are embedded as excess atoms leading films, which consist of $MoS_x$ with excess molybdenum atoms, to be called sulfur deficient films in a chemical composition sense. Anode current is the second dominant factor which affects the film composition. Resistivity–sulfur content increases as anode current decreases, since the low anode current product, HS, other than atomic sulfur is more reactive with molybdenum to form a compound with. In both cases, $\Delta G$ of formation of molybdenum sulfide contributes significantly.

## Acknowledgment

This work was supported by a grant from Army Research Office (contract DAAG 29-85-K-0170).

## References


1 D. W. Murphy and P. A. Christian, *Science*, *205* (1979) 651.

2 R. Selim and P. Bro, *J. Electrochem. Soc.*, *121* (1974) 1467.

3 A. J. Jacobson, R. R. Chianelli, S. M. Rich and M. S. Whittingham, *Mater. Res. Bull.*, *14* (1979) 1437 - 1448.

4 J. J. Auborn, Y. L. Barbeiro, K. J. Hanson, D. M. Schleich and M. J. Martin, *J. Electrochem. Soc.*, *134* (1987) 580.

5 A. J. Jacobson, R. R. Chianelli and M. S. Whittingham, in *Proc. 3rd Int. Conf. Chem. Uses Molybdenum*, Ann Arbor, Michigan, 1979, pp. 209 - 212.

6 H. F. Winters, D. L. Raimondi and D. E. Horne, *J. Appl. Phys.*, *40* (1969) 2996.

7 K. Reichelt and G. Mair, *J. Appl. Phys.*, *49* (3) (1978) 1245.

8 A. T. Bell, in J. R. Hollahan and A. T. Bell, (eds.), *Techniques and Applications of Plasma Chemistry*, Wiley, New York, 1974, p. 31.

9 A. Cornu and R. Massot, *Compilation of Mass Spectral Data*, Heyden & Son Ltd., 1966, p. 1-B.

10 K. C. Mills, *Thermodynamic Data for Inorganic Sulfides, Selenides and Tellurides*, Butterworth, London, 1974.

*Surface and Coatings Technology, 39/40* (1989) 691 - 700


# CHEMICAL VAPOR DEPOSITION COATINGS IN FLUIDIZED BED REACTORS

A. SANJURJO, M. C. H. McKUBRE and G. D. CRAIG

*SRI International, 333 Ravenswood Avenue, Menlo Park, CA 94025–3493 (U.S.A.)*

(Received March 6, 1989)

## Summary

A coating technique based on chemical vapor deposition (CVD) in a fluidized bed reactor (FBR) is described. To coat an article, we load the coating material as a powder in an FBR and fluidize it using an inert gas. The system is heated to operating temperature, and vapors of HBr are mixed with the fluidizing gas. The sample to be coated is suspended on top of the bed or immersed into the bed. The reaction of the HBr with the powder in the bed produces halide species of the coating material that can deposit a coating on the substrate by a disproportionation or reduction reaction. Using this technique, we deposited titanium, silicon, zinc, TiN and $Si_3N_4$ coatings on substrates such as copper, steel, silica and graphite.

## 1. Introduction

Metallic and ceramic coatings can be applied by a variety of techniques including physical vapor deposition, chemical vapor deposition (CVD), sputtering, dip coating, slurry painting, pack cementation and thermal spraying. Each technique has advantages and disadvantages that dictate the major area of application. CVD is particularly well suited for many applications, including deposition of semiconductors, metallic interconnects, hard coatings and corrosion resistant coatings for aqueous and high temperature environments. Typically, a gas mixture, *e.g.* $TiCl_4$ and $NH_3$, is passed over a substrate heated to high temperature (1000 °C or greater) and a film (in this case, TiN) deposits on the substrate. For a current review of typical systems and applications, see ref. 1.

One factor that may limit the use of CVD is cost, including capital equipment cost and operating cost. Other possible limitations include the need to operate at high temperatures, which may degrade the properties of the substrate, and the difficulties in maintaining a constant temperature on the substrate and a constant composition of the gas phase on all parts of the substrate, especially if the substrate has a convoluted geometry. Methods have been devised to circumvent these limitations. For example, the use of

0257-8972/89/$3.50 

plasma assisted CVD allows deposition of coatings at much lower temperatures than those used with conventional CVD, but this technique requires working at low pressures, which results in higher coating costs. The homogeneity of the coatings can be improved by modifying the fluodynamics of the gas around the piece to be coated and careful design of the heating unit, but this approach may result in expensive, one-of-a-kind systems, and low utilization of the reactants.

Fluidized bed reactors (FBR) have been used to coat particles. For example, particles in a bed can be coated with pyrolytic graphite by using a hydrocarbon gas at temperatures ranging from 1200 to 2000 °C [2]. Kaae [3] codeposited HfC and SiC by injecting simultaneously $HfCl_4$, $CH_3 SiCl_3$ and $H_2$ in an FBR kept at 1300 °C. The same author discusses that FBR could also be used for coating larger objects by immersing them in the bed [3]. A similar approach has been disclosed in U.S. Patent 4,623,400 [4]. A gas mixture containing vapors of the coating precursors (such as $TiCl_4$ and $NH_3$) is injected in an FBR containing inert particles such as $Al_2O_3$ kept at 1000 °C. The piece to be coated is immersed in (and heated by) the bed. The use of an FBR with its fast heat transfer permits heating the piece to be coated at uniform temperature, and the result is a uniform coating. It should be noted that most of the coating material is deposited on the inert particles that form the fluidized bed, and that the deposition temperature is similar to that of conventional atmospheric CVD.

This paper describes a coating process that uses an FBR to take advantage of the high rates of heat and mass transfer, but the particles in the bed are the source (rather than the sink) for the precursor vapors. Powders of metals are used as the fluidized particles and when reacted with the vapors of HBr they generate *in-situ* volatile sometimes unstable precursors that can be used to coat at much lower temperatures than with previous process approaches. These species either by themselves or mixed with externally introduced gases, can react with, or disproportionate on the substrate to form the desired coating. For example, if titanium or a titanium-containing material such as TiN is to be deposited, titanium particles are used. Although the halide species introduced could have been $TiCl_4$, HCl or any halide that can react with the bed to form titanium halide species, we used HBr in this work. When we used titanium powders and HBr, we obtained a titanium coating. When $NH_3$ was also added through the top of the reactor, we obtained a TiN coating on articles located above the bed.

It is the purpose of this paper to describe the basis of the process and to illustrate with a few applications how it can be applied.

## 2. Experimental procedure

The coating system consists of an FBR (Fig. 1), a gas delivery unit, and a mass spectrometer to determine the composition of the gas phase. In Fig. 1, the reactor is made of quartz, although other inert, heat-resistant materials

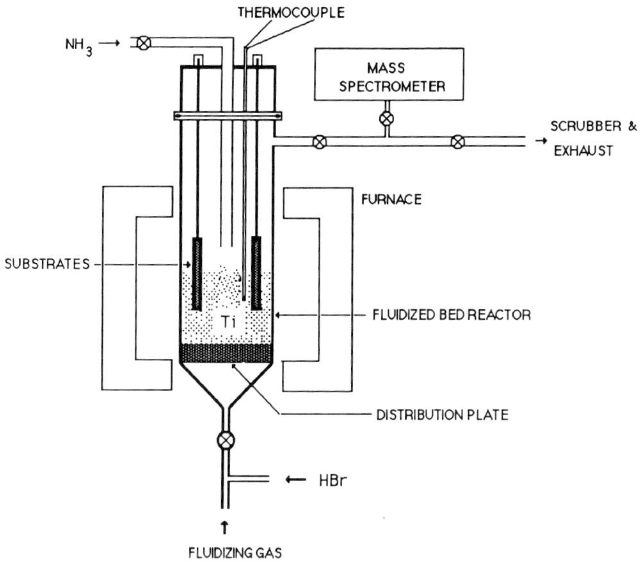


Fig. 1. FBR for coating TiN on steel bars.

TABLE 1

Coating experiments in FBR-CVD

| *Run* | *Substrate* | *Bed* | *Substrate temperature* (°C) | *Time* (min) | *Comments* |
|---|---|---|---|---|---|
| 1 | Cu wire | Si | 750 | 120 | Wire (1 mm OD) in bed totally siliconized. Wire above bed perfectly coated |
| 2 | Cu tube | Si | 600 | 15 | Very thin coating ( < 1 $\mu$m) |
| 3 | Cu tube | Si | 600 | 37 | Porous free, homogeneous, thick coating (2 - 3 $\mu$m) |
| 4 | Cu tube | Ti | 750 | 45 | Porous free homogeneous coating (3 - 5 $\mu$m) |
| 5 | Steel bar | Ti | 750 | 45 | Porous free, with TiC interface (1 $\mu$m) |
| 6 | Silica tube | Ti | 750 | 45 | Mirror coating |
| 7 | Steel bar | Ti/$NH_3$ | 850 | 75 | TiN homogeneous continuous (3 $\mu$m) |
| 8 | Silica tube | Ti/$NH_3$ | 850 | 75 | Gold color TiN, mirror coating (1 $\mu$m) |
| 9 | Steel bar | Zr | 900 | 120 | Homogeneous, continuous coating with ZrC interface ( < 1 $\mu$m) |
| 10 | Graphite fibers | Si | 600 | 150 | Continuous, coating of amorphous $Si_3N_4$ |

could be used. The coating material (or at least a component) was loaded as a powder through a detachable top in the FBR. At the start of a coating run, the sample to be coated was introduced through the top and the system was purged for 5 min with an inert fluidizing gas, typically argon. The system was then heated by an external furnace to operating temperature, and the sample to be coated was suspended on top of the bed or immersed into the bed. Vapors of HBr were mixed with the inert fluidizing gas and reacted with the particles of the bed to produce halide species of the coating material.

The experiments were performed using fluidized beds of silicon, titanium and zirconium powders. Several substrates were used during each coating run. The substrate materials used included copper wires and tubing, steel bars, quartz tubes and graphite fibers. The coating conditions, materials used, and results obtained are summarized in Table 1. A detailed description of each type of coating is given in the next section.

## 3. Results

### *3.1. Silicon on copper articles*

Metallurgical grade silicon powder with particle sizes ranging from 250 to 500 mm was loaded in a 5 cm ID, 40 cm high fluidized bed reactor (Fig. 1) to form a bed 10 cm high (in repose). The bed was fluidized by a 1 atm. argon flow with an approximate linear flow velocity of 10 cm $s^{-1}$ in the bed. Vapors of HBr were mixed with the argon flow to obtain a partial pressure of 7 Torr (kPa). Copper tubes were then immersed in the fluidized bed reactor, which was kept at temperature by external heating. The conditions and results of the coating experiment are listed as runs 1, 2 and 3 in Table 1. At low temperatures (600 °C), a thin silicon layer was formed, as determined by scanning electron microscopy (SEM) and energy dispersive X-ray analysis (EDAX). At high temperatures or long processing times an interface phase of copper silicides is also formed (Fig. 2). The corrosion resistance of the coated samples in aqueous environments with high concentrations of $Cl^-$ ions was studied by impedance techniques [5]. We found that the impedance at the electrolyte coating interface was a factor of 2 greater than that of uncoated samples, indicating a clear increase in corrosion resistance.

### *3.2. Titanium on copper articles*

The experimental setup was similar to the one above, except that titanium particles were used. The coating conditions and results are listed in Table 1. A Ti coating was deposited, as determined by X-ray diffraction analysis, SEM (Fig. 3) and EDAX. The interface impedance between the $Cl^-$ solution and the coated samples was approximately 10 times that of uncoated samples.

### *3.3. Titanium on steel and silica*

As in the previous experiments we loaded the FBR with titanium particles and injected HBr vapors in argon. The substrates were steel rods

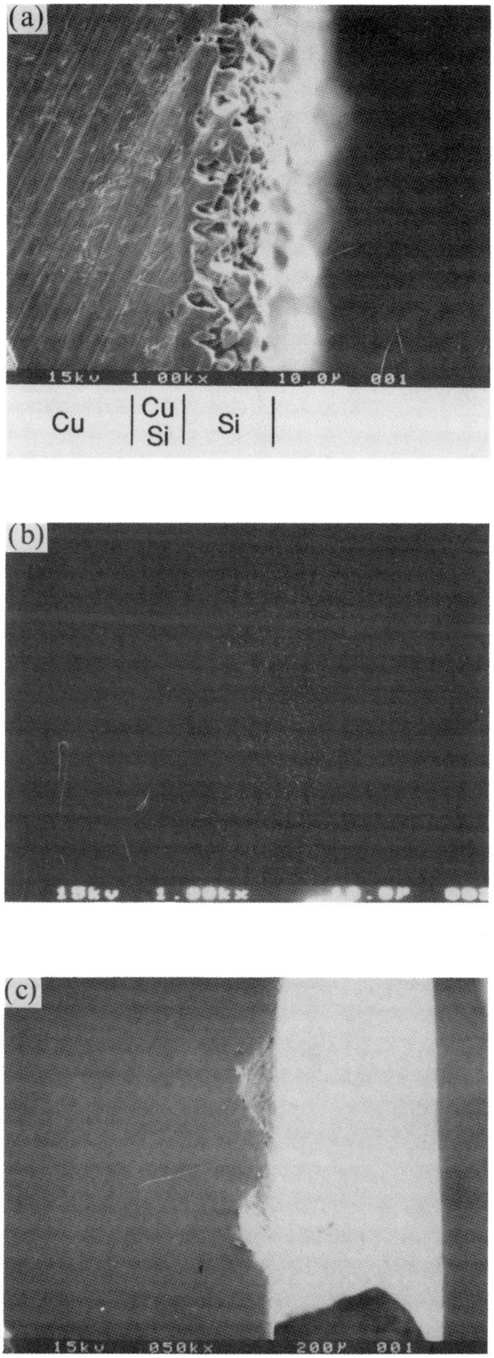


Fig. 2. Silicon coating on copper substrate: a, cross-section showing the coating; b, silicon distribution as determined by X-ray analysis; c, indentations on original copper coated with silicon showing the conformation of the coating to the substrate surface.

External coating layer: Cu 50%, Ti 44%, Si 6%
Intercoating layer: Cu 96.5%, Ti 3.5%

Fig. 3. Titanium on copper substrate.

and silica tubes. The coating conditions and results are listed in Table 1. SEM and X-ray analysis showed that a thin interface layer (about 0.1 $\mu$m) of TiC was formed on steel substrates.

### *3.4. TiN on steel and silica*

The setup described above was used, but $NH_3$ was injected through the top of the reactor and the substrates were located above the bed. The coated samples were gold colored, and the presence of the TiN coating was confirmed by X-ray diffraction analysis, SEM and EDAX. The interface impedance between the coated steel and the aqueous chloride solution was approximately 30 times that of the interface between uncoated controls and the chloride solution.

### *3.5. Zirconium on steel*

The design was the same used for coating with silicon or titanium, but a bed of zirconium particles was used. A zirconium coating with a ZrC interface layer between substrate and coating was formed, as determined by X-ray diffraction analysis and SEM.

### *3.6. Coating of carbon fibers with $Si_3N_4$ films*

Silicon particles (metallurgical grade, 325 mesh) were fluidized in a 5 cm ID, 50 cm long fluidized bed reactor (FBR) by an $N_2$ flow. A felt of graphite fibers was placed immediately above the bed. The height of the fluidized bed was about 10 cm in repose. A 20 $cm^3\ min^{-1}$ flow of HBr gas was mixed with the $N_2$ and co-injected into the bed. The silicon bed was preheated to 600 °C. This temperature was selected to maximize the partial pressure of bromosilane present in the gas phase of the FBR. Ammonia gas at a flow rate

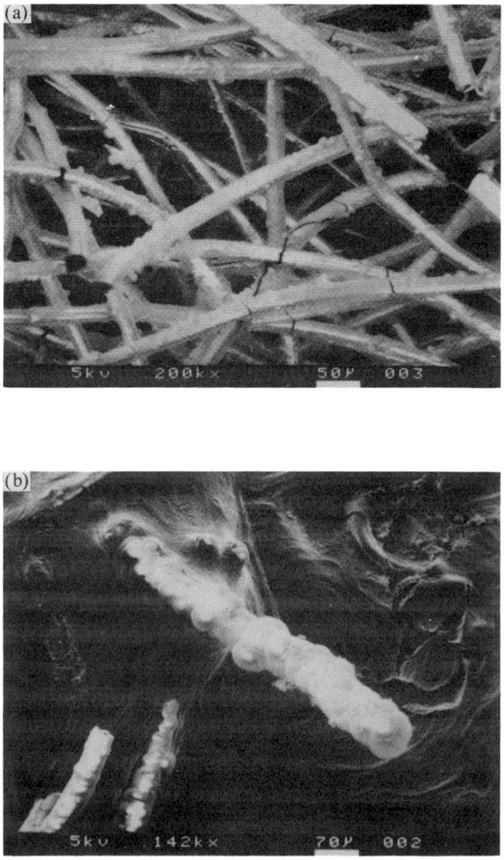


Fig. 4. $Si_3N_4$ coated graphite fibers: a, $Si_3N_4$ coating on graphite fibers; b, detail of the coated fiber.

of flow 30 - 100 $cm^3 min^{-1}$ was fed through the top of the reactor so that the $NH_3$ and bromosilanes mixed in the hot zone above the bed. The idea was to allow the bromosilanes and subbromides ($SiBr_x$ where $x < 4$) to react directly with $NH_3$ before either one could decompose to more stable compounds. The zone above the felt was kept at 1200 °C.

The graphite fibers were coated with a grayish, hard, compact brittle material. SEM and EDAX studies confirmed the presence of the silicon nitride coating (Fig. 4). The cracks and bare fiber area visible in Fig. 4a were generated on purpose by mechanical means to show the thickness of the coating. Figure 4b shows the modular growth observed on some fibers.

## 4. Discussion

We will now discuss in some detail the chemistry involved in producing silicon coatings using the bromosilane chemistry. This example is used to illustrate the basic concept as well as some of the advantages of this method of coating. Coatings or other elements and compounds can, of course, be produced by this technique, and other halosilane chemistries can also be used depending on the needs.

When $H_2$ is used as a carrier gas, the gas is mixed with $SiBr_4$ vapors, and the mixture is passed through a bed of silicon particles, Si–Br–H species are produced. Thermodynamic equilibrium estimates can be used to determine the composition of the gas phase [6] and the relative temperatures of the bed and the substrate needed for deposition to occur. Figure 5 shows the curves obtained experimentally by determining the composition of the gas over the FBR using mass spectrometry. These experimental values and estimated values agree well, which indicates that the equilibrium conditions were reached in the reactor at temperatures ranging from 700 to 800 °C. Other subhalides with short lifetimes may also have been formed, but they did not survive the distance through the sampling port to the mass spectrometer. Nevertheless, they may be important for the deposition process.

When HBr vapor was used instead of $SiBr_4$, similar bromosilane and subbromide species were formed, but at significantly lower temperatures. Furthermore, we observed that coatings could be deposited at lower temperatures. When HBr vapors mixed with the fluidizing gas reach the bed of silicon

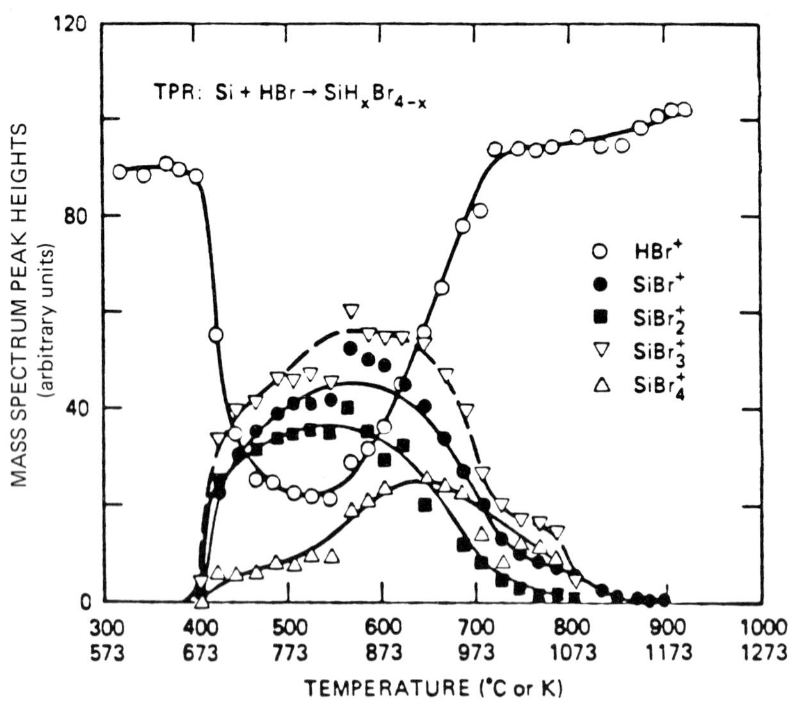


Fig. 5. Effect of temperature on composition of the gas on the top of the bed. Gas feed $H_2$–$SiBr_4$(Br/H = 0.0063).

particles, they react according to the following equations:

$$Si + 4HBr = SiBr_4 + 2H_2 \quad (1)$$

$$Si + 3HBr = SiBr_3H + H_2 \quad (2)$$

$$Si + 2HBr = SiBr_2H_2 \quad (3)$$

Figure 6 shows the experimentally determined pressures of these bromosilanes as they exit the FBR which occur at temperatures between 400 and 700 °C when the total pressure is 1 atm.

We also postulate the formation of subhalides intermediates ($SiBr_x$ where $x < 4$) according to equilibria such as

$$SiBr_4 + Si = 2SiBr_2 \quad (4)$$

In effect, when this equilibrium is established in high vacuum in a Knudsen cell, $SiBr_2$ is present at 600 °C and is the most abundant species at 900 °C [7]. Other similar subhalides, such as $SiF_2$ and $SiCl_2$ have also been detected at high temperatures [8 - 10] but their lifetimes are of the order of milliseconds to seconds at temperatures near 1000 °C [11]. As the temperature decreases, the equilibrium is displaced to the left by disproportionation of the $SiX_2$ (where X is a halide) to deposit solid silicon and regenerate $SiX_4$. The fact that we can coat substrates that are kept at a lower temperature than the bed seem to be consistent with the existence of these subhalide species at atmospheric pressure. From the linear flow velocities in the reactor and the location of the deposits, we estimated that the lifetime of these species (at low partial pressures and 600 °C in 1 atm argon) is less than 100 ms.

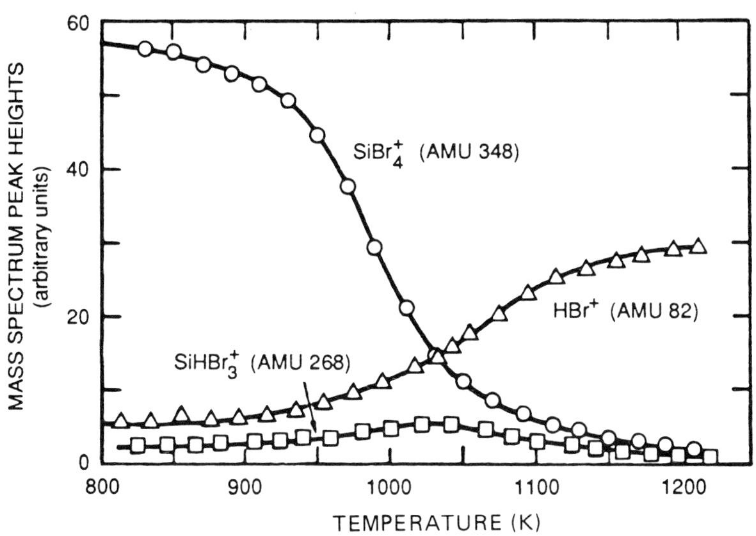


Fig. 6. Effect of temperature on composition of the gas on the top of the bed. Gas feed HBr.

In contrast, coatings deposited on a substrate kept hotter than the bed are attributable to bromosilane species containing H such as $SiHBr_3$. As the temperature increases, the Si/X ratio in the gas phase decreases if the H/X ratio is less than 1, as predicted by thermodynamic estimates [6].

The rates of deposition can be high ($\mu m\ min^{-1}$) or low ($\mu m\ hr^{-1}$) on copper at 750 °C and 600 °C respectively. The rates obtained with the other systems were in some cases lower, most probably because the temperature for formation of the transport species was not optimized as was the case for the silicon species.

## 5. Conclusions

Chemical vapor deposition, fluidized bed technology and subhalide chemistry has been combined to coat metallic and ceramic substrates such as copper, steel, silica, and graphite with metallic and ceramic coatings of silicon, zirconium, TiN and $Si_3N_4$. Subhalide and element–halide–hydrogen species act as transport agents carrying the coating material from the bed to the sample to be coated. Because the piece to be coated is in close contact with the bed, even very unstable, short lived species can contribute significantly to the coating. They provide a low temperature path for the transport of mass. By using reactive vapors such as HBr, these transport species can be produced at much lower temperature that is possible with fully halogenated species such as $TiCl_4$ that are thermodynamically and kinetically much more stable. The overall result is that deposition is achieved at atmospheric pressure and at lower temperatures than with conventional chemical vapor deposition. The technique should be applicable to coating with other materials including carbides, borides and sulfides.

## References

1 D. P. Stinton, T. M. Besmann and R. A. Lowden, *Ceram. Bull.*, *67* (2) (1988) 350–355.
2 J. Gilleray, R. L. R. Lefeure, M. S. T. Price and J. P. Thomas, in J. Blocker, H. Hietermann and L. Hall (eds.), *Proc. 5th Int. Conf. on CVD*, The Electrochem. Society, 1975, p. 727.
3 J. L. Kaae, *Ceram. Eng. Sci.*, *9* (1988) 1159–1168.
4 J. Japka, R. Staffin and S. Bhatia, *U.S. Patent 4,623,400* (November 18, 1986).
5 D. D. Macdonald and M. C. H. McKubre, Corrosion of materials, in J. R. Macdonald (ed.), *Electrochemical Impedance Methods*, Wiley-Interscience, New York, 1987.
6 L. P. Hunt and E. Sirtl, *J. Electrochem. Soc.*, *120* (1972) 806–811.
7 Unpublished results.
8 L. Troost and P. Hantefeuille, *Ann. Chim. Phys.*, *7* (1976) 452.
9 H. Schafer and J. Nickl, *Z. Anorg. Allg. Chem.*, *274* (1953) 250.
10 C. Kempter and C. Alvarez-Tostado, *Z. Anorg. Allg. Chem.*, *290* (1957) 239.
11 C. D. Graham *et al.*, *Final Report ERDA/SE/EC (11–1)-2721/FR/76/1* (1976).

*Surface and Coatings Technology*, *39/40* (1989) 701 - 709 

# ELECTROCHEMICAL CORROSION STUDIES OF ALLOYS PLASMA SPRAYED WITH $Cr_2O_3$

A. A. ASHARY and R. C. TUCKER, Jr.

*Union Carbide Coatings Service Corporation, Indianapolis, IN 46224 (U.S.A.)*

(Received March 16, 1989)

## Summary

Interconnected porosity in a thermal spray coating may act as a crevice which may allow crevice corrosion of some substrates. A potentiodynamic cyclic polarization technique was used to evaluate the corrosion resistance of 316L stainless steel and Inconel 625 (IN-625) coated with $Cr_2O_3$ by plasma spraying. The corrosion resistance of coated samples was found to be dominated by the characteristics of substrate alloys; coated 316L exhibited crevice corrosion while coated IN-625 exhibited no corrosion. The effectiveness of an epoxy sealant was also determined. The epoxy sealant used in this study was found to be very effective in enhancing the corrosion resistance of the thermally sprayed coated sample. Post-analyses of the corroded samples were carried out using optical and scanning electron microscopy. The results were compared with salt spray testing.

## 1. Introduction

Thermally-sprayed coatings are frequently used in corrosive environments, even when their major purpose is to provide wear resistance etc., rather than corrosion resistance. In these situations not only the corrosion resistance of the coating is important, but also the corrosion behavior of the substrate. This is because of the interconnected porosity which is almost always present in thermally sprayed coatings. This interconnected porosity may allow the corrosive media to reach the coating–substrate interface. Furthermore, galvanic effects between the coating and the substrate must be taken into account [1].

An example of the problem is the use of $Cr_2O_3$ coatings on stainless steels. Plasma sprayed $Cr_2O_3$ coatings have excellent wear resistance, particularly in self-mating wear and when mated against polymeric and carbon-graphite materials. They also have very good corrosion resistance in many environments. They are thus a good choice for a wide variety of seal surfaces, *e.g.* mechanical seals, O-ring seals, lip seals etc. Frequently these seals, or other components, operate in a corrosive environment and a stainless steel

0257-8972/89/$3.50 

such as 316L is chosen for the substrate alloy. Even though both the coating and the substrate, independently, are seemingly sufficiently corrosion resistant, the coatings occasionally blister as a result of crevice corrosion of the substrate even, on rare occasions, when sealed. It was with this problem in mind that the present study was initiated using UCAR* LC-4, a plasma sprayed $Cr_2O_3$, as the model coating.

While the general corrosion resistance of austenitic stainless steels is considered good, their crevice corrosion resistance is a concern. Crevice corrosion is a form of localized attack which usually occurs in chloride containing environments. It occurs due to the breakdown of passivity at narrow openings, cracks, and gaps where solution stagnation occurs due to restricted flow. The mechanism of crevice corrosion has been described in detail elsewhere [2]. Interconnected porosity in a thermally sprayed coating can act as a crevice, which may cause localized breakdown of the passivity of the substrate at the interface. Blistering of the coating occurs as a consequence of growth of corrosion products at the interface. In the advanced stages of corrosion or, in the case of poor general corrosion resistance of the substrate, a uniform layer of corrosion product may form at the interface and the coating may eventually spall.

Several different techniques have been used to protect the underlying substrate. These techniques include applying an impervious corrosion resistant undercoat on the substrate prior to thermal spray coating or sealing the porosity in the coating by a corrosion resistant sealant [3]. The effectiveness of these techniques, of course, depends upon the properties of the undercoat or sealant and their application technique. It should be noted that a sealant can only penetrate porosity that is open to the surface at the time of sealing. During finishing or wear, shorter, buried channels of porosity connected to the substrate may be opened. The sealant evaluated in this study was UCAR 100, an epoxy-based material that has been found to be particularly effective in many applications. The use of undercoats is not considered here, in part, because of their added cost.

An alternative approach to sealing the coating or using an impervious undercoat is to use a substrate which is highly resistant to crevice corrosion. To this end, Inconel 625 (IN-625) was evaluated in this study. The composition of IN-625 is nominally Ni–22Cr–9Mo–3Fe–3.5(Nb + Ta) while 316L stainless steel is Fe–17Cr–11Ni–2Mo–1.6Mn–0.02C. IN-625 was used in an annealed condition per AMS 5599C specification and 316L was used in a solution treated condition per AMS 5653B specification.

The most frequent means of evaluating the corrosion resistance of a coating–substrate system in an aqueous media has been by immersion or salt spray testing [4]. These techniques usually require long exposure times, hundreds or thousands of hours, and frequently yield limited information on the mechanism of corrosion. A few studies using electrochemical techniques have been reported [5 - 9]. These studies were not designed to elucidate some

---

*UCAR is a trademark of Union Carbide Corporation.

corrosion mechanisms, *e.g.* crevice corrosion of the substrate. In the Union Carbide Coatings Service studies, a potentiodynamic cyclic polarization technique is currently being used to investigate the corrosion characteristics of thermally sprayed coating systems. While this technique has been used on monolithic materials (bare alloys), it apparently has not been previously used on coatings. It appears that this technique requires only a very short exposure time and provides substantial insight on the mechanisms of corrosion.

The potentiodynamic cyclic polarization technique [10, 11] is an accelerated electrochemical corrosion test which is used to evaluate the localized corrosion resistance of alloys. In this test, specimens are anodically polarized in a corrosive medium and potential is increased in the positive direction up to a point where a certain value of resulting current is reached. At this point the potential is decreased. The applied potential and resulting current due to corrosion are plotted. The breakdown potential at which accelerated corrosion occurs and size of the hysteresis loop formed during the reverse scan are two important measures of pitting or crevice corrosion resistance of the sample used. A higher breakdown potential and a smaller hysteresis loop signify good corrosion resistance.

## 2. Experimental procedure

Flat panels of 316L and IN-625 approximately 6.35 mm (0.25 in) thick were used. When required, UCAR LC-4 $Cr_2O_3$ was applied with a standard Union Carbide plasma torch. The as-deposited thickness was 0.25 - 0.30 mm (0.010 - 0.012 in). When studying the effect of sealing on the crevice corrosion of the substrate, UCAR 100 sealant was applied to the as-deposited coating surface. The top 0.08 - 0.10 mm (0.003 - 0.004 in) of both sealed and unsealed coating was removed by grinding. This was particularly necessary for the sealed coatings not only to remove the top layer of epoxy, but also to simulate wear during service. Furthermore, for many applications, coatings are finish ground prior to being put in service. The remaining coating thickness was 0.18 - 0.20 mm (0.007 - 0.008 in). Samples approximately 20 mm (3/4 in) wide, 13 mm (1/2 in) long and 6 mm (1/4 in) thick were sectioned from the coated and bare alloys.

The electrochemical system used in the cyclic polarization studies consists of a corrosion cell, a potentiostat, personal computer and a printer. The corrosion cell was a five neck flask which included the following hardware: two carbon counter electrode rods, one reference saturated calomel electrode, one working electrode (the sample to be studied) and a gas sparging tube.

The working solution was a 3.56% NaCl solution as defined in ASTM G-61. Samples were immersed in the solution by a 3.2 mm (1/8 in) diameter stainless steel rod with one end screwed into the sample to be tested. Only a one centimeter square area of the sample was exposed to the solution. The rest of the sample and the portion of the stainless steel rod which was immersed in the solution were insulated by a thin coating of polyurethane varnish. This technique, rather than using a standard sample holder, was particularly useful since the gasket in a standard sample holder may be a source of crevice corrosion by itself.

Tests were conducted at room temperature (20 - 24 °C). The solution was de-aerated by purging with nitrogen gas for an hour prior to sample immersion. The corrosion potential of each sample was monitored for an hour prior to the polarization studies. This period was found to be enough to stabilize the corrosion potential of the samples used. Potential scans for polarization were initiated and ended at $-100$ *vs.* $E_{corr}$. A potential scan rate of 0.2 mV $s^{-1}$ was used. Scans were reversed when the current density reached about 2 mA.

## 3. Results and discussion

### *3.1. Bare alloys*

The potentiodynamic cyclic polarization curves for the bare alloys 316L stainless steel and IN-625 are given in Fig. 1. Current densities are plotted on the $x$-axis in microamperes per centimeter square and impressed potentials relative to the standard calomel electrode are plotted on the $y$-axis. These

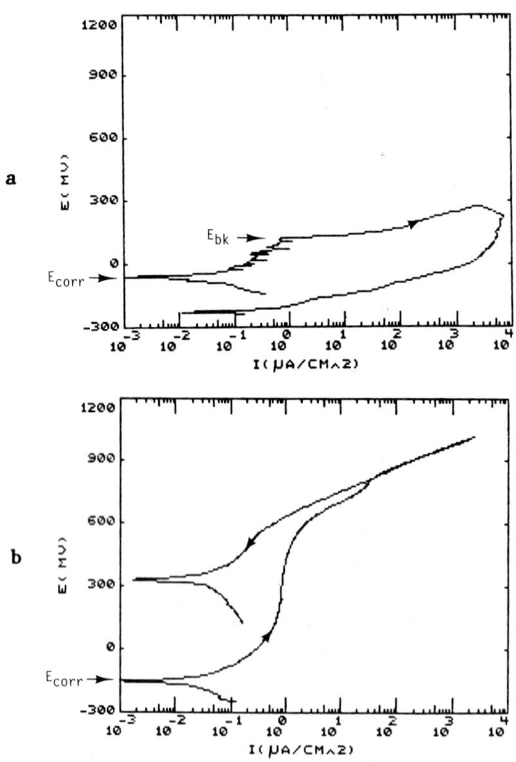


Fig. 1. Potentiodynamic cyclic polarization plots for (a) bare 316L and (b) bare IN-625.

alloys have very similar corrosion potentials ($E_{corr}$ values). The 316L SS sample shows a rapid increase in current at about 200 mV. The rapid increase in current is usually due to breakdown of passivity. In the absence of crevices or gaps rapid increase in current is due to breakdown of passivity at localized imperfections in the passive film on the alloy surface. This phenomenon is commonly known as pitting corrosion. The potential at which a rapid increase in current occurs is usually termed the 'pitting potential' or breakdown potential ($E_{pit}$ or $E_{bk}$). Pit growth occurs due to the autocatalytic conditions of increased acid and chloride ion concentration within pits. A propagating corrosion pit can be considered as a special form of crevice growing by the same mechanism.

In the case of IN-625 alloy, passivity is maintained until about 600 mV. The rapid increase in current at above 600 mV is not due to pitting, but it is due to a uniform corrosion in the transpassive or oxygen evolution region. In this region, the passive oxide layer starts to dissolve oxidatively, generally as a hydrolyzed cation in a higher oxidation state.

Potential scans were reversed in both cases when the current density reached about 2000 $\mu$A. The pit growth process did not slow down in 316L until a potential negative to the reversing potentials was reached. Furthermore, the current densities were higher during the reverse scan for a given potential. This causes a hysteresis in the curve. The reverse scan for IN-625 sample closely follows the forward scan. Since there were no pits, the corrosion of the alloy at a given potential remains the same in the reverse scan. The corrosion potentials of both alloys were modified at the end of the scan which may be due to an oxide film formation on the surfaces of the samples.

The better pitting corrosion resistance of IN-625 is well known and not surprising, since it is a nickel base alloy with a substantial amount of molybdenum (9 wt.%) while 316L is an iron base with much lower molybdenum (2 - 3 wt.%). Molybdenum is known to improve the pitting corrosion resistance of most alloys.

### *3.2. LC-4-coated samples*

Polarization curves for the unsealed LC-4-coated 316L and IN-625 samples are given in Fig. 2. The shapes of these curves are very similar to the corresponding curves for the bare alloys. Both coated and uncoated 316L samples show a hysteresis in the reverse scan, but none of the IN-625 samples show a hysteresis loop. The breakdown potentials for LC-4-coated and bare 316L are very similar, and the transpassivation potentials for the coated and uncoated IN-625 alloys are approximately the same. The similarities between the polarization curves for the LC-4-coated and -uncoated samples suggest that the corrosion behavior of the LC-4-coated samples is dominated by the corrosion behaviors of the substrates. This is due to interconnected porosity in the coating which provides a channel for the salt solution (electrolyte) through the coating to the substrate.

In addition to the above similarities there were some differences between the polarization curves for the coated and uncoated alloys. These a

b

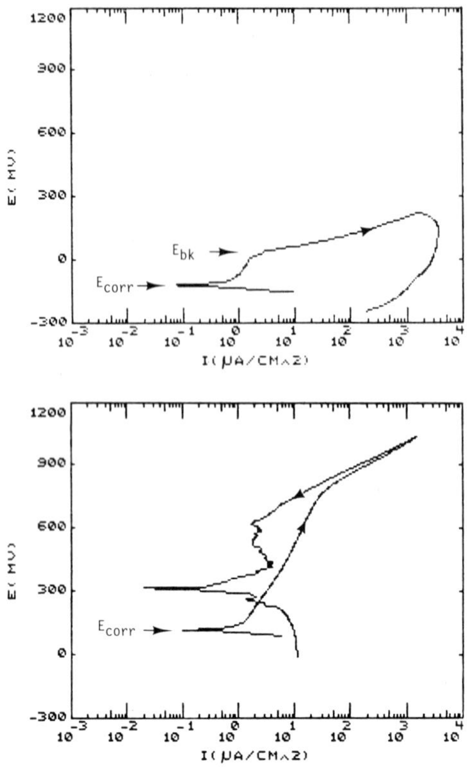


Fig. 2. Potentiodynamic cyclic polarization plots for (a) UCAR LC-4-coated 316L and (b) UCAR LC-4-coated IN-625 alloy.

differences may appear subtle, but important information can be inferred. It can be seen that the transition from passive to rapid corrosion at $E_{bk}$ is more abrupt for the bare 316L sample than that for the coated sample. A sharp transition usually indicates a pitting type of behavior while a gradual transition is a characteristic of crevice corrosion. The interconnected porosity in the LC-4 coating may be causing the crevice corrosion of the 316L substrate.

The range of potential over which passivity is maintained is a better measure of pitting corrosion than the breakdown potential *vs.* the reference electrode. The range over which passivity is maintained is greater for the bare 316L sample than this range is for the LC-4-coated sample. This indicates a lowering of the corrosion resistance of the 316L by the LC-4 coating. Furthermore, the hysteresis loop seems to be larger for the LC-4-coated sample than that for the bare sample, which is another indication of the poor crevice corrosion resistance of the LC-4-coated 316L sample.

The most significant difference in the polarization curves of coated and uncoated IN-625 samples is the continual increase in the corrosion current density in the passivation region for the coated sample. Higher current densities in the passivation region may be due to some corrosion of the LC-4 coating itself. These corrosion densities are, however, very low — of the order of a few microamperes per centimeter square. Consequently, corrosion of LC-4 coating would not be significant. This limited corrosion may be due to the presence of a small amount of oxide other than $Cr_2O_3$ which is soluble in a stronger, acidic solution. Such an acidic condition may be present in a crevice.

The irregularity near the end of the reverse potential scan of the LC-4-coated IN-625 sample is more difficult to explain, but this is also in a very low current density range which would have little significance to the crevice corrosion behavior of the sample.

*3.3. Sealed LC-4 coating*

Cyclic polarization curves for an LC-4-coated sample sealed with UCAR 100 on the 316L substrate are given in Fig. 3. Unlike the unsealed 316L sample, the sealed sample did not show a rapid increase in current during the forward potential scan. Current densities were very low throughout the potential scan and no hysteresis loop was observed during the reversed scan for the sealed sample. The sample thus demonstrated an excellent corrosion resistance, which is obviously due to complete sealing of pores in the coating and good corrosion resistance of the coating itself.

*3.4. Post-test analysis of samples*

All of the specimens were cross-sectioned after the electrochemical tests and examined optically and using a scanning electron microscope. The LC-4-coated IN-625 sample and the 316L sample sealed with UCAR 100 showed no signs of corrosion. The unsealed LC-4 316L sample, however, showed severe attack of the substrate. Energy dispersive X-ray (EDX) analysis of the corrosion product indicated it was mainly iron and chlorine, perhaps $FeCl_3$. This confirmed that the electrolyte penetrated the coating and attacked the 316L substrate.

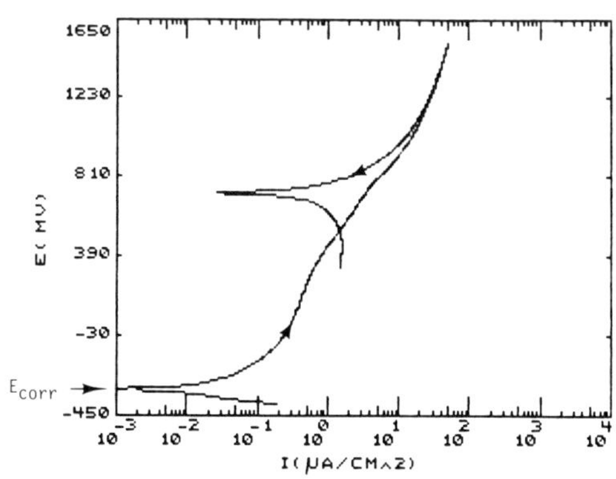


Fig. 3. Potentiodynamic cyclic polarization plot for UCAR LC-4-coated 316L sample sealed with UCAR 100 sealant.

### 3.5. *Long-term salt spray testing*

Duplicates of the samples used in the cyclic polarization studies were subjected to 500 h of salt spray testing using ASTM standard procedure B-117. Results of the salt spray testing were found to be consistent with the results described above. The unsealed LC-4 coating on 316L developed blisters. Corrosion product was detected at the coating/substrate interface and within the pores of the coating. The sealed LC-4 coating on 316L and unsealed LC-4 on IN-625 couples did not show any noticeable corrosion.

## 4. Conclusions

(1) This study clearly demonstrates the utility of the potentiodynamic cyclic polarization technique in the study of the corrosion characteristics of thermally sprayed coating–substrate systems as well as its ability to elucidate corrosion mechanisms.

(2) An unsealed plasma sprayed $Cr_2O_3$ 316L stainless steel couple is susceptible to crevice corrosion in an aqueous chloride solution. It is reasonable to expect similar behavior from the coupling of most thermally sprayed coatings on a substrate susceptible to crevice corrosion.

(3) An unsealed $Cr_2O_3$ coating on IN-625 is highly resistant to crevice corrosion as well as general corrosion in an aqueous chloride solution. Similar behavior could be expected with other substrates that are resistant to crevice corrosion.

(4) UCAR 100 is a very effective sealant. A note of caution is in order, however. Even though a quarter to a third of the coating thickness was removed by grinding after sealing in these tests without opening channels of porosity interconnected to the substrate, further reduction in thickness through wear may open unsealed channels through the coating and allow crevice corrosion to occur. Thus a preferred system is one with a crevice corrosion resistant substrate.

(5) No evidence of any significant corrosion of the $Cr_2O_3$ coating or any coating–substrate galvanic effect was noted in the systems studied.

## References

1 R. C. Tucker, Jr., in *Deposition Technology for Films and Coatings*, Noyes Publications, New Jersey, 1982, p. 454.
2 M. G. Fontana and N. D. Greene, *Corrosion Engineering*, McGraw-Hill, New York, 1967, p. 39.
3 R. H. Unger, in *Metals Handbook*, Vol. 13, ASM, Materials Park, OH, 9th edn., 1987, p. 459.
4 Standard Method of Salt Spray (Fog) Testing, *B117 (reapproved 1985)*, *Annual Book of ASTM Standards*, American Society for Testing and Materials, Philadelphia, PA.
5 J. D. Reardon *et al.*, *Thin Solid Films*, *108* (1983) 459.
6 B. Tsujino *et al.*, *Thin Solid Films*, *135* (1986) 51.

7 T. Kado and M. Araki, *Proc. Int. Symp. of Advanced Thermal Spraying Technology and Allied Coatings, Osaka*, 1988, p. 307.
8 B. Tsujino, *Proc. Int. Symp. Advanced Thermal Spraying Technology and Allied Coatings, Osaka*, 1988, p. 347.
9 Y. Arata *et al.*, *Thin Solid Films*, *156* (1988) 315.
10 Standard Practice for Conduction of Cyclic Potentiodynamic Polarization Measurements for Localized Corrosion, *G 61*, *Annual Book of ASTM Standards*, American Society for Testing and Materials, Philadelphia, PA.
11 J. W. Oldfield, *Int. Mat. Rev.*, *32* (1987) 153.

*Surface and Coatings Technology*, *39/40* (1989) 711 - 720


# EFFECT OF CONNECTED PORES ON THE CORROSION BEHAVIOR OF PLASMA SPRAYED ALUMINA COATINGS

M. FUKUMOTO, Y. WADA, M. UMEMOTO and I. OKANE

*Department of Production Systems Engineering, Toyohashi, University of Technology, Toyohashi 440 (Japan)*

(Received March 17, 1989)


## Summary

The role of connected pores in plasma sprayed alumina coatings on the corrosion of the substrate was investigated. Coatings of thickness ranging from 60 to 300 $\mu$m were prepared by plasma spraying using Fe–0.5 wt.% C commercial mild steel substrate. The density of connected pores was measured by the ferroxyl test and the anodic polarization test using 3% NaCl solution at 35 °C. Corrosion behavior of the substrate was investigated by means of the immersion test using 3% NaCl solution at 35 °C. The connected porosity was found to decrease with increasing the coating thickness. The connected porosity $P$ can be expressed using the coating thickness $t$ and a constant $\alpha$ as $P = \alpha t^{-0.82}$. From the immersion test, the corrosion resistance of plasma sprayed alumina coatings was found to be higher for the thicker coatings. However, in the more thickly coated specimens, the formation of the iron oxide layer at the coating–substrate interface was faster and the coating layer fell off earlier.


## 1. Introduction

Plasma sprayed ceramic coatings are known to be very effective in protecting metallic structures and equipment from severe thermal, abrasive and corrosive conditions. However, thermally sprayed ceramic coatings formed by the deposition of melted ceramic particles are inherently porous. Connected porosity strongly influences the corrosion resistance of plasma sprayed ceramic coated surfaces [1, 2]. Nevertheless, there exist almost no reports about the relationship between connected porosity in the coatings and the corrosion behavior of substrate materials.

Therefore, the present study aimed to investigate the influence of connected pores in the plasma sprayed alumina coatings on the corrosion behavior of substrate.

0257-8972/89/$3.50 

## 2. Experimental procedures

The ceramic powder for thermal spray coating used in the present study was 99.6% $Al_2O_3$ with a diameter of 10 - 44 $\mu$m. Alumina was selected because of its high corrosion resistance. Fe–0.5 wt.% C commercial mild steel (S50C) was used as a substrate material. The substrate was grit blasted just before being plasma sprayed. For plasma spraying, a mixture of $N_2$ (32 l min$^{-1}$)–$H_2$ (6.4 l min$^{-1}$) was used as a working gas. Arc current, arc voltage and spray distance selected were 500 A, 80 V and 100 mm respectively. Coatings with thickness ranging from 60 to 300 $\mu$m were produced by changing the spray period at constant powder feeding rate.

The ferroxyl test was conducted in accordance with the Japanese Industrial Standards [3] to measure the density of connected pores in the coating. In this test, the density of connected pores in the coating can be measured as the density of ferroxyl spots observed on the filter paper.

An anodic polarization test (Fig. 1) was carried out in order to investigate the connected porosity in the coatings. The electrolyte used for the measurement of the anodic polarization curves was 3% NaCl aqueous solution at 35 °C with initial pH 6.5. The anodic polarization curves were measured with a potentiostat using a potential control method. An Ag–AgCl electrode and a platinum plate were used as the reference electrode and counter-electrode respectively.

The immersion tests were also conducted using the same electrolyte. The volume of the electrolyte per unit area of the specimen surface was 50 ml cm$^{-2}$. For both anodic polarization and immersion tests, specimen surfaces were partially masked so that only the alumina coated surface was exposed. Surface observations and measurements of the weight change of the specimen were carried out after every 2 days immersion. The electrolyte was

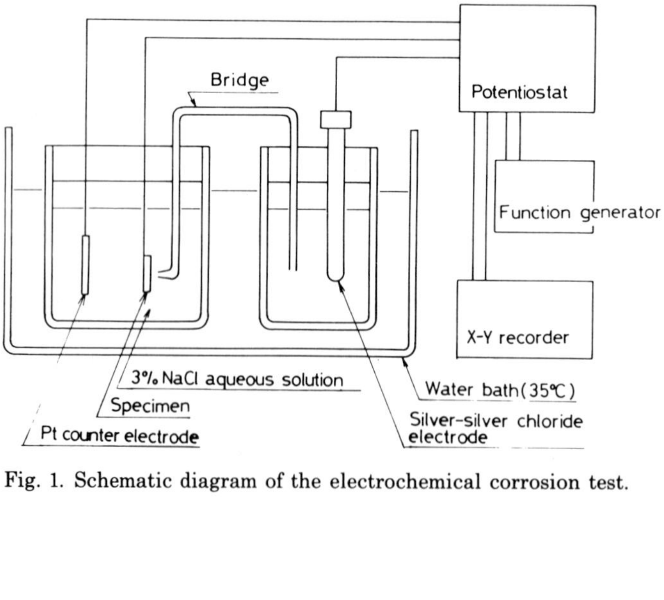


Fig. 1. Schematic diagram of the electrochemical corrosion test.

replaced with a new one at each measurement. The average weight loss of four specimens tested under the same conditions was obtained.

## 3. Results

The anodic polarization curves for specimens with various coating thickness are shown in Fig. 2. It is seen that before passivation the current density decreases with increasing coating thickness. This indicates that the total area of pores connected to the substrate decreases with increasing coating thickness. The passive current densities measured for specimens with various coating thickness (shown in region A of Fig. 2) are listed in Table 1.

Ferroxyl tests were performed for the specimens with various coating thickness. The measured ferroxyl spot densities are also listed in Table 1. The normalization of the passive current density and of ferroxyl spot density by those of 300 $\mu$m coated specimens respectively, was done in order to express values relative to a standard. These two normalized values are also listed in Table 1. Comparing these two normalized densities, it is noted that they coincide well with each other. In general, the current density corresponds to the total area of pores per unit area connected to the substrate and the

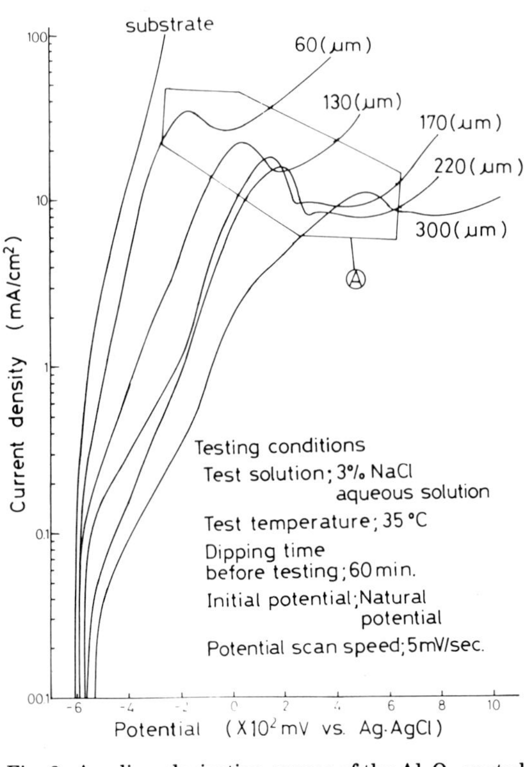


Fig. 2. Anodic polarization curves of the $Al_2O_3$-coated specimens with various coating thickness.

TABLE 1

The passive current densities, ferroxyl spot densities and those of normalized for various coating thickness specimens

| *Coating thickness* ($\mu$m) | *Passivate current density* (mA cm$^{-2}$) | *Normalized passive current density* | *Ferroxyl spot density* (spots cm$^{-2}$) | *Normalized ferroxyl spot density* |
|---|---|---|---|---|
| 60 | 26.1 | 3.4 | 6.9 | 3.7 |
| 130 | 14.5 | 1.9 | 3.9 | 2.1 |
| 170 | 8.9 | 1.1 | 2.7 | 1.5 |
| 220 | 7.8 | 1.0 | 2.3 | 1.3 |
| 300 | 7.8 | 1.0 | 1.9 | 1.0 |

ferroxyl spot density represents the number of the connected pores per unit area. Thus the coincidence of these two values suggests that the average cross-section area of a connected pore is independent of the coating thickness. Figure 3 shows the normalized passive current density and ferroxyl spot density as a function of coating thickness on logarithmic scales. A linear relationship can be seen. From Fig. 3 and Table 1, the passive current density $I$ (mA cm$^{-2}$) and ferroxyl spot density $S$ (cm$^{-2}$) of the present specimen can be given by

$$I = 735t^{-0.82} \tag{1}$$

and

$$S = 204t^{-0.83} \tag{2}$$

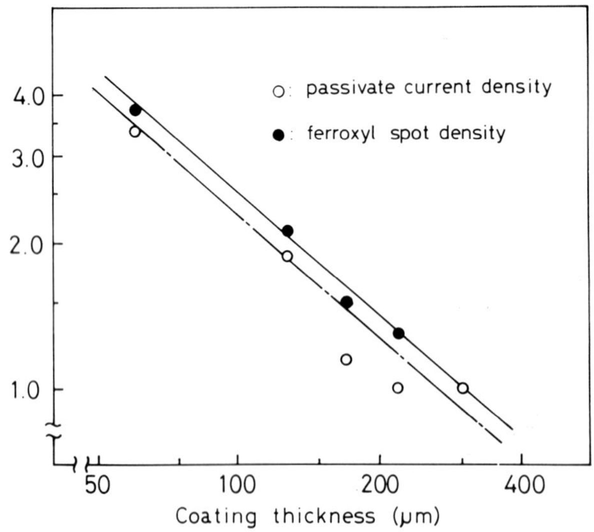


Fig. 3. The normalized passive current density and the ferroxyl spot density as a function of the coating thickness.

where $t$ ($\mu$m) is the coating thickness. The connected porosity $P$, which is proportional to $I$, can be expressed using a constant $\alpha$ as

$$P = \alpha t^{-0.82} \tag{3}$$

To investigate the corrosion behavior, the coating surfaces and specimen cross-sections were observed for various immersion periods. Figure 4 shows the optical micrographs of the surfaces of sprayed alumina coated specimens of three different coating thicknesses after various immersion periods. After 2 days immersion, spots of light-brown color appeared on the coating surface for specimens of every coating thickness. The density of these spots is smaller and the size is larger in the more thickly coated specimens. The light-brown color of the spots suggests that the spots are made of $Fe_2O_3$ [4]. After 10 days immersion, the iron oxide spot size and density are not altered but the spots are discolored. Comparing the spot density in Fig. 4 with those observed in the ferroxyl test, it was recognized that these two densities coincide well. This suggests that the spots shown in Fig. 4 are produced by the corrosion products exuded through the connected pores.

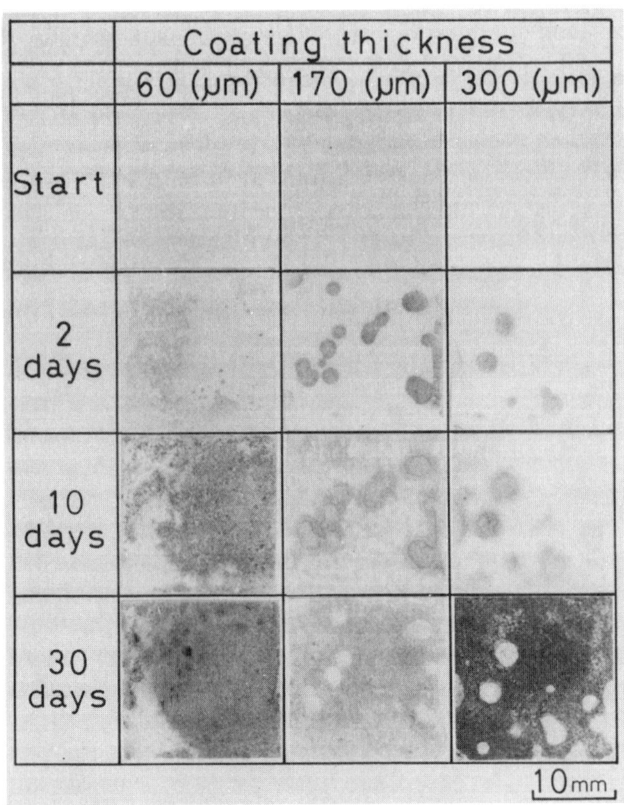


Fig. 4. Optical micrographs showing the change in the surface of the $Al_2O_3$ coatings with immersion period.

After 30 days immersion, in the specimen with a coating thickness of 60 μm the coating layer lost contact with substrate at most places. In the specimen with a coating thickness of 170 μm, the reversion of contrast occurred, that is, the color of the spots turned white and the matrix became brown. This coating layer did not fall off after 30 days immersion. In the specimen with a coating thickness of 300 μm, the coating layer completely fell off after 24 days immersion. After the coating layer fell off, small round shaped alumina with a diameter of about 1 mm were observed to be retained on the surface of the substrate exactly beneath the iron oxide spots observed on the coating surface. The picture of the coating of thickness 300 μm after 30 days (bottom right in Fig. 4) shows the substrate surface after the coating layer fell off. The black region is the corroded substrate surface and white round regions are alumina.

The progress of corrosion was investigated by scanning electron microscope (SEM) observations of specimen cross-section. The typical microstructure of the coating–substrate interfaces for the specimens with coating thicknesses of 60, 170 and 300 μm are shown in Fig. 5. It was noted that in the specimens with coating thicknesses of 170 and 300 μm, the corrosion behavior of the interfaces outside of the iron oxide spots on the coating surface and just beneath the iron oxide spots were different. Since the corrosion rate at the interface just beneath the iron oxide spots was very slow, the interfaces outside of the iron oxide spots were presented in Fig. 5 (except Fig. 5(j)). When the coating thickness was 60 μm, separation of substrate and coating was observed in some places even after 2 days immersion. After 30 days

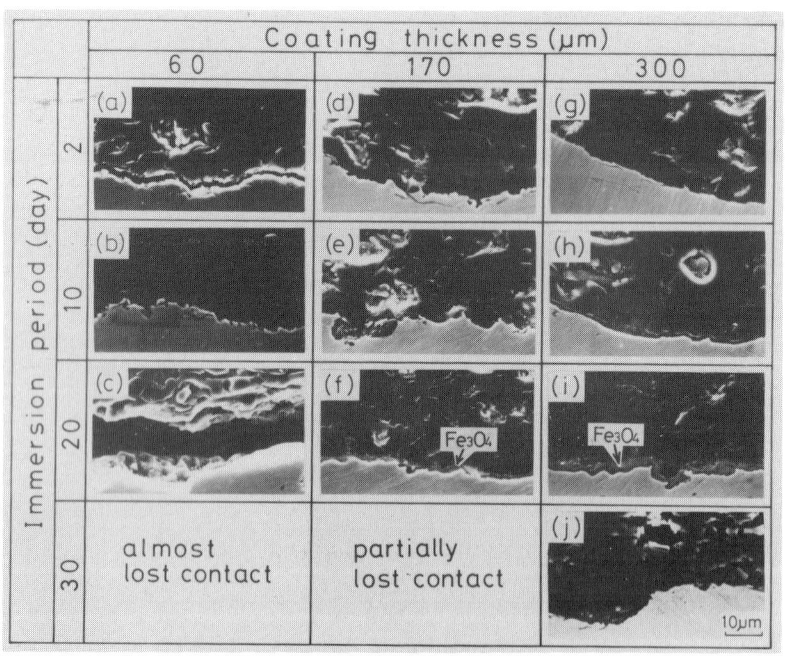


Fig. 5. SEM micrographs showing the coating–substrate interfaces.

immersion, the coating layer lost contact with the substrate at most of the places, but the coating layer did not fall off. In the specimen with a coating thickness of 170 $\mu$m, an oxide layer was formed at the coating–substrate interface after 10 days immersion. As the immersion period is increased, this oxide layer thickened. X-ray diffraction showed that this oxide layer was $Fe_3O_4$. After 30 days immersion, the coating layer lost contact with substrate at several places but did not fall off completely. In the specimen with a coating thickness of 300 $\mu$m, after 2 days immersion (Fig. 5(g)), the iron oxide layer had already formed at the coating–substrate interface. As the immersion period was increased, this oxide layer continuously thickened and after 20 days immersion (Fig. 5(i)) its thickness became about 3 $\mu$m. This oxide layer under the outside area of iron oxide spot seems to initiate the separation of the coating layer after immersion for long periods. Comparing the thickness of iron oxide layers at the interface in specimens with coating thicknesses of 170 and 300 $\mu$m after 20 days, the latter was observed to be thicker. After 30 days immersion, the coating layer completely fell off and only round shaped alumina remained. Figure 5(j) shows the round shaped retained alumina and substrate interface. At this interface, the oxide layer was scarcely observed. At the coating–substrate interface just beneath the iron oxide spot, strong adhesion between the coating and the substrate was maintained. After the separation of the coating layer, these regions remained as the small round shaped alumina.

As the immersion period increased the weight of the specimen increased. A larger weight gain was observed in the thicker coated specimen. This indicates that the formation of iron oxide increased with increasing the coating thickness. This leads to the earlier fall-off of the coating layer in the thicker coated specimen.

If we define the corrosion resistance of the coated specimen to be the length of immersion period before the fall-off of the coating, the corrosion resistance of the coated specimen is increased in accordance with the coating thickness. However, in the thicker coated specimen, the formation of the iron oxide layer at the coating–substrate interface is faster and the coating layer falls off earlier.

## 4. Discussion

From the results of the present study, the possible corrosion mechanism of plasma sprayed ceramic coated surfaces of the substrate can be considered as follows. The progress of corrosion can be divided into two stages. The first stage of corrosion is schematically illustrated in Fig. 6(a). Since the electrolyte initially penetrates through connected pores, the regions of substrate just beneath the connected pores are exposed to the electrolyte which contains dissolved oxygen. The exposed substrate has better access to oxygen than the substrate still covered with the alumina coating. According to the principle of differential aeration, oxygen-rich areas act as electron sources.

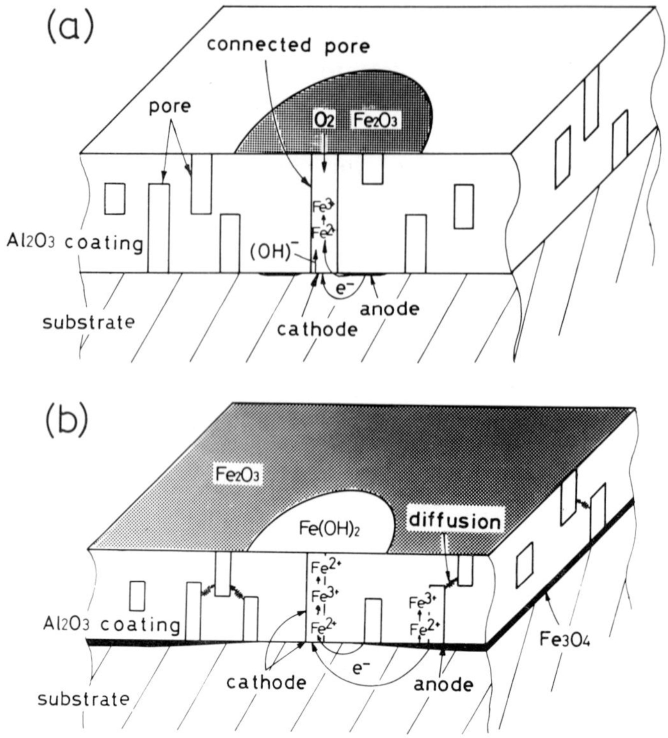


Fig. 6. A possible corrosion mechanism of the $Al_2O_3$-coated substrate during immersion test: (a) first stage, (b) second stage.

Thus the dissolution of substrate starts to occur at the regions adjoining the connected pores [5]. In neutral solutions, like the present electrolyte, iron dissolves into the solution as $Fe^{2+}$ [6] and simultaneously dissolved oxygen in solution reacts with water consuming electrons.
These reactions are

$$Fe \rightarrow Fe^{2+} + 2e^- \quad (4)$$

$$O_2 + 2H_2O + 4e^- \rightarrow 4(OH)^- \quad (5)$$

The dissolved $Fe^{2+}$ ions are unstable and will transform into $Fe^{3+}$ ions:

$$2Fe^{2+} + 4(OH)^- + H_2O + 1/2O_2 \rightarrow 2Fe^{3+} + 6(OH)^- \rightarrow 2Fe(OH)_3$$

$$(\rightarrow Fe_2O_3{\cdot}3H_2O) \quad (6)$$

$Fe^{3+}$ and/or hematite $Fe_2O_3$ diffuse through the connected pore and accumulate on the coating surface around the pores. The light-brown round spots are considered to be caused by such a process.

The second stage of corrosion, which begins after about 2 days immersion, can be schematically illustrated as in Fig. 6(b). As the immersion time progresses, the electrolyte diffusing through the alumina particles boundaries of coating begins to react with substrate. With this reaction, $Fe^{2+}$ ions are formed (anodic reaction) almost everywhere at the coating–substrate interface. Part of the $Fe^{2+}$ ions transform into $Fe^{3+}$ ions (eqn. 6) and diffuse out to the coating surface through the alumina particles boundaries. Consequently $Fe_2O_3$ compounds accumulate on the coating surface. Other $Fe^{2+}$ ions are considered to precipitate with $Fe^{3+}$ ions at the coating–substrate interface as $Fe_3O_4$ iron oxide. Since the oxygen concentration at the coating–substrate interface is low, the corrosion product is $Fe_3O_4$ instead of $Fe_2O_3$ [7]. Since the anodic reaction, which is the formation of $Fe^{2+}$ ions, occurs almost uniformly throughout the specimen, the area of anodic reaction becomes much larger than that of the cathodic reaction. The cathodic reaction rate should correspond to that of the anode. The ordinary cathodic reaction (eqn. 5) at the connected pores may not be fast enough. Then, the reduction reaction $Fe^{3+}$ to $Fe^{2+}$ may occur near the connected pore area to supply the ordinary cathodic reaction.

$$Fe^{3+} + e^{-} \rightarrow Fe^{2+} \tag{7}$$

Such $Fe^{2+}$ ions diffuse out through the connected pore, accumulate on the coating surface around the pore and cover the round $Fe_2O_3$ spots formed during the first stage of corrosion. $Fe^{2+}$ ions are nearly white, so the round spots originally light-brown gradually turn white. The areas around the connected pores are oxygen-rich and act as cathodes. Thus the dissolution rate of iron atoms should be smaller than that of the surrounding area. Good adhesion between substrate and coating will be maintained around the connected pores. This can explain the present observation that around the connected pores the round shaped alumina remained on the substrate after the coating fell off.

## 5. Conclusion

In the present study, the role of connected pores in plasma sprayed alumina coatings on the corrosion of the substrate was studied. Coatings with thickness ranging from 60 to 300 $\mu$m were prepared by plasma spraying using mild steel substrate. Connected porosity was measured by the ferroxyl test and the anodic polarization test using 3% NaCl solution at 35 °C. Corrosion behavior of substrate was investigated by means of the immersion test using 3% NaCl solution at 35 °C.

The main results obtained are summarized as follows.

(1) From the ferroxyl test, it was observed that the density of the connected pores decreases with increasing the coating thickness.

(2) The anodic polarization measurement indicated that the total area of connected porosity decreases with increasing the coating thickness. The connected porosity $P$ can be expressed using the coating thickness $t$ and a constant $\alpha$ as $P = \alpha t^{-0.82}$.

(3) The immersion test showed that in the first stage of corrosion, $Fe_2O_3$ spots were formed on the coating surface around the connected pores. In the second stage of corrosion, $Fe^{2+}$ ions were accumulated on the coating surface around the connected pores and $Fe_2O_3$ precipitated at other places. For the same immersion period, the thicker iron oxide layer was formed at the coating–substrate interface in the more thickly coated specimen.

(4) In the more thickly coated specimen, the formation of the iron oxide layer at the coating–substrate interface was faster and the coating layer fell off earlier.

## References

1 Y. Arata, A. Ohmori and C. J. Li, *Thin Solid Films*, *156* (1988) 315.
2 C. B. Andersen, *Proc. Int. Symp. on ATTAC'88*, 1988, p. 289.
3 Testing Methods for Thermal Sprayed Ceramic Coatings, *Japanese Industrial Standard*, *H8666* (1980) 4.
4 H. H. Uhlig, *Corrosion and Corrosion Control*, 2nd edn., C. E. Tuttle, Tokyo, 1974, p. 87.
5 J. O'M. Bockris and A. K. N. Reddy, *Modern Electrochemistry 2*, Plenum, New York, 1977, p. 1301.
6 M. G. Fontana, *Corrosion Engineering*, 3rd edn., McGraw-Hill, New York, 1986, p. 453.
7 G. Itoh, *Corrosion Science and Engineering*, Corona, Tokyo, 1969, p. 275.

*Surface and Coatings Technology*, *39/40* (1989) 721 - 731 

# ASSESSMENT OF CORROSION RESISTANT COATINGS FOR A DEPLETED U–0.75 Ti ALLOY

F. CHANG, M. LEVY and B. JACKMAN

*U.S. Army Materials Technology Laboratory, Watertown, MA 02172-0001 (U.S.A.)*

W. B. NOWAK

*Northeastern University, Boston, MA 02115 (U.S.A.)*

(Received March 22, 1989)

## Summary

Aluminum, zinc, magnesium, Al–Zn, Al–Mg, nickel, titanium, TiN and Al/TiN coatings were applied by the arc plasma physical vapor deposition (PVD) technique to a depleted uranium (DU) alloy for corrosion protection assessment. The as-deposited specimens were examined by scanning electron microscopy (SEM) for surface morphology and tested for adhesion. Electrochemical polarization tests and immersion tests were conducted in aerated 3.5 wt.% NaCl solution. Results of electrochemical polarization scans and observations after long-term exposure tests indicated that both Al–Zn and Al–Mg alloys appear to be the best sacrificial coating materials for improving the corrosion resistance of DU–0.75 Ti.

## 1. Introduction

High density depleted uranium (DU) alloys possess superior armor piercing capabilities but they are susceptible to corrosion when exposed to high humidity and salt laden environments. In order to mitigate corrosion, various metallic coatings such as aluminum and TiN have been applied by a cathodic arc plasma physical vapor deposition (PVD) process. The ions in the plasma have a high vapor energy level (50 - 150 eV) and the plasma has a high ratio of ions to neutral atoms (70 - 95%) [1]. This PVD system can operate over a wide range of pressures (high vacuum to $1 \times 10^{-1}$ Torr) and has the ability to deposit many materials. The high vapor energy level and the high ion ratio are the most attractive features among others which may be controlled to optimize the microstructure and composition of the deposited layer to produce coatings with better corrosion resistance.

0257-8972/89/$3.50 

In an earlier paper [2] on the corrosion behavior of an IVD aluminum coated* DU–0.75 Ti alloy, it was found that the coating provided general corrosion protection after 12 days of exposure to a 5% salt fog environment except for pits that initiated and grew at pin-hole defects in the coating. Defects such as cracks or porosity, either pre-existing or induced by the environment may be a controlling factor affecting corrosion behavior of coated specimens. For deposited metals which are anodic (sacrificial) with respect to DU–0.75 Ti, the presence of a small number of defects in the coated layer may be acceptable. For cathodic type (noble) coatings, the existence of a single pore or crack in the coated layer will be detrimental.

In this study, the following nine different metallic coatings were deposited on DU–0.75 Ti alloy specimens: zinc, magnesium, Al–Zn, Al–Mg (anodic coatings); aluminum, nickel, titanium, TiN (cathodic); and a dual layered coating of aluminum and TiN (Al/TiN). Reported here are SEM examinations of the deposit surface morphology, results of adhesion tests, and assessment of the corrosion resistance of the as-coated specimens by short-term electrochemical polarization tests and long-term immersion tests up to 120 days in 3.5 wt.% NaCl aqueous solution. Data obtained for the uncoated DU–0.75 Ti alloy are also included for comparison.

## 2. Experimental procedure

Specimens were fabricated from a long rod of DU–0.75 Ti alloy of diameter 25.40 mm, heat treated at 850 °C, water quenched at 0.46 m min$^{-1}$, and aged at 350 °C for 16 h to obtain the desired microstructure, which consisted of a major $\alpha$ phase and a minor $U_2Ti$ phase. This alloy had a martensitic microstructure and a grain size of about 400 $\mu$m. Two types of disk specimens were cut from the treated rods; larger ones with a diameter of 25.40 mm and thickness of 6.350 mm for immersion and adhesion tests, smaller ones with a diameter of 15.885 mm and thickness of 3.175 mm for electrochemical polarization tests. The disks were ground and polished through 600 grit silicon carbide paper to a surface roughness of about 0.25 $\mu$m and ultrasonically cleaned.

The coatings were applied by Nuclear Metals, Concord, MA, employing the cathodic arc plasma (PVD) process in a Multi-Arc Vacuum System (Multi-Arc Vacuum Systems, St. Paul, MN). Before placement into the deposition chamber, the specimens were degreased in ethyl alcohol, ultrasonically cleaned in detergent, rinsed and air dried. Electro-polishing and acid cleaning were also used as needed.

The specimens were loaded into the chamber and pumped down to a vacuum of $1.0 \times 10^{-5}$ Torr prior to starting the arc. This was followed by

*Applied by the McDonnell Douglas Ivadizer System; McDonnell Douglas Corp., P.O. Box 576, St. Louis, MO 63166, U.S.A.

nitrogen ion bombardment to the substrate for about 1 h to sputter off remaining contaminants on the surface and to heat the specimen. During deposition, the temperature of the specimen was controlled to remain below 300 °C and the vacuum was maintained at $20 \times 10^{-3}$ Torr. The thickness of the coatings ranged from 3 to 13 $\mu$m, deposition usually taking 1 - 2 h to complete, depending on the coating metals and the type of coatings: one or two layers, single metal or two-component alloy selected. The as-coated specimens were ultrasonically cleaned, then examined by SEM for surface morphology. It should be noted that coating parameters employed for deposition of these coatings were based on limited trials and do not represent optimum conditions.

Electrochemical corrosion testing of the coatings was conducted in aerated 3.5 wt.% NaCl aqueous solutions, pH 5.5, at room temperature (25 °C). A Princeton Applied Research potentiostat–galvanostat model 273 was used for potentiodynamic polarization of the specimens. An anodic scan rate of 20 mV min$^{-1}$ was used with a saturated calomel reference electrode (SCE), two high density, non-permeable graphite rod counter electrodes, and a PAR standard flat specimen holder model K105, which employed a sealing knife edge washer of Teflon to expose 1 cm$^2$ of specimen area to the test solution. Tests began after 1 h of immersion to achieve steady state. All electrochemical potentials are reported against SCE.

Adhesion tests were qualitative and consisted of placing transparent Scotch Magic Tape onto the coating, finger rubbing the tape many times until the air between the tape and coating seemed to be removed and then rapidly peeling the tape off the coating.

An immersion test was carried out to assess the relative corrosion resistance of the coatings that were tested electrochemically on other samples. The coated specimens used were the larger disks, containing an uncoated 1.600 mm diameter hole, drilled 9.525 mm deep in a direction normal to the through thickness. This hole represented a coating defect that exposed the substrate DU–0.75 Ti alloy to the NaCl solution and concomitant galvanic corrosion. Each disk was placed flat surfaces horizontal, separately in its own 250 ml Pyrex beaker filled to the 200 ml mark with a 3.5 wt.% NaCl aqueous solution. Each beaker was sealed by a parafilm sheet punctured with a pinhole, and left standing at ambient temperature (about 25 °C). Physical appearances were noted as a function of time and recorded.

## 3. Results and discussion

### *3.1. SEM examination of as-deposited coatings*

The surface morphology of representative coatings are shown in Fig. 1 (a - c). Generally the deposits contain microdroplets and defects such as pits or pores and cracks. Microdroplets are often contained in coatings deposited by the cathodic arc process [3]. The concentration of microdroplets can be controlled by optimization of process parameters such as cathode–substrate

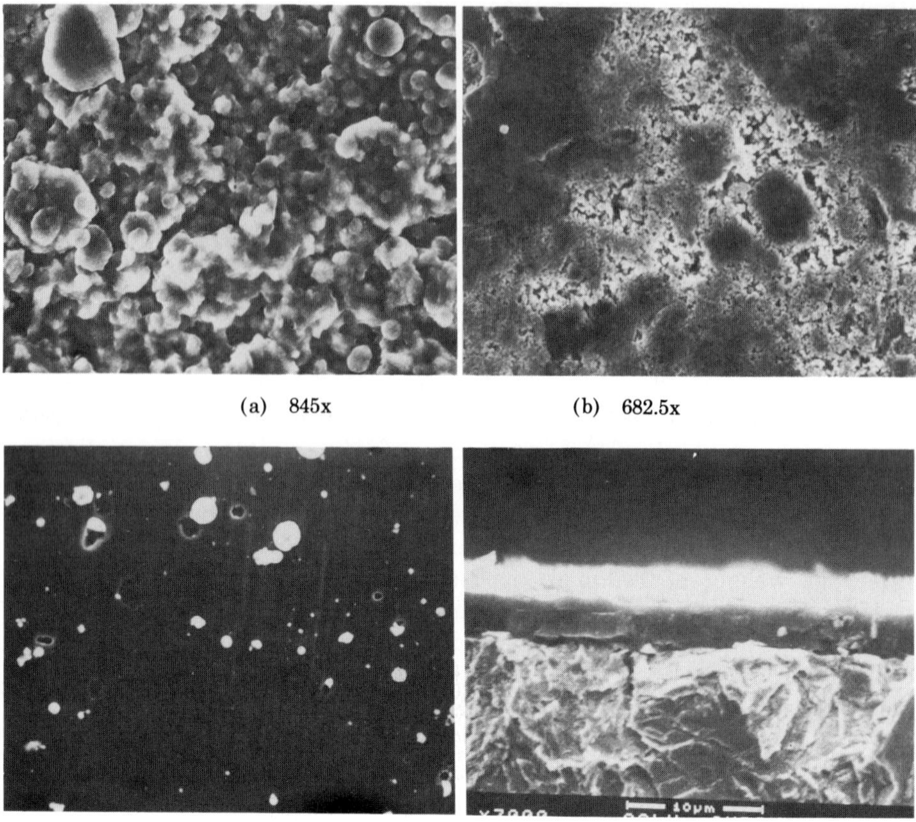

(a) 845x (b) 682.5x

(c) 910x (d) 1950x

Fig. 1. SEM photomicrographs of cathodic arc plasma PVD coated DU–0.75 Ti alloy specimens: (a) aluminum, (b) Al–Zn, (c) TiN, (d) cross-section of representative coating–substrate (TiN/DU–0.75 Ti).

geometry, arc current, cathode temperature, etc. The cracks are likely caused by the difference in thermal expansion coefficients of coating and substrate. The surface morphology of each coating is described in detail in Table 1. A cross-section of a representative coating/DU–0.75 Ti substrate is shown in Fig. 1(d). The coating is adherent, uniform in thickness and fine grained. A columnar structure was not observed.

### *3.2. Adhesion test*

The adhesion of all coatings was excellent as judged by the Scotch Tape test. No removal of coating from substrates was evident. Excellent adherence of the coating to its substrate is reported as one of the main advantages of the cathodic arc process due to the higher degree of ionization and energy level of the coating material leaving the arc source [1].

TABLE 1

Morphology of coatings (from SEM evaluation)

| *Coating* | *Morphology* |
|---|---|
| Aluminum (Fig. 1(a)) | Rough surface. Appears to be synthesis of a spray of droplets and an atomic vapor stream condensing on substrate. Particle size: 0.7 - 5.0 $\mu$m. Larger "droplets" are clusters. |
| Zinc | Fewer droplets than aluminum. Irregular pits. |
| Magnesium | Agglomeration of droplets and layering, with irregular islands of agglomerates with smooth surface. Deeper layers of agglomerates seen between islands. |
| Al–Zn (Fig. 1(b)) | Similar to zinc. No particles. Spongy structure with holes of diameter 1 $\mu$m and smaller. |
| Al–Mg | Similar to aluminum. Characteristic agglomerated particles. Porous, low density structure visible at surface. |
| Al/TiN | Identical to aluminum. Large agglomerated masses. Small particles of diameter 1.5 - 4 $\mu$m. Aluminum coating appears to have interconnected porosity. |
| Nickel | Some deep pits, large and small agglomerates, some with cracks surrounding them. Underlying smooth dense film. Particle size: 0.5 - 2.5 $\mu$m. |
| Titanium | Underlying dense, smooth film with numerous pits. Overlayer of particles 1.5 - 7 $\mu$m. Voids surrounding some agglomerates. |
| TiN (Fig. 1(c)) | Gold color. Scratches, droplets, some pits (may be sites where droplet broke away during handling and cleaning). |

### *3.3. Electrochemical polarization tests*

The anodic polarization curves for the coatings–alloy systems in 3.5 wt.% NaCl solution are shown in Figs. 2, 3 and 4. The important electrochemical parameters are corrosion rest potential ($E_C$), pitting potential ($E_P$) and protection range ($E_{PR}$); these have been extrapolated from these curves and are listed in Table 2. The protection range has been defined by Nowak *et al.* [4] as the difference in potential between the corrosion potential and the onset of pitting. The corrosion potential is the equilibrium or steady state potential of the alloy or coating.

If the corrosion potential of the coating is more active or electronegative than the substrate DU–0.75 Ti alloy, which is the case for zinc, magnesium, Al–Zn, and Al–Mg, then the coating is considered as sacrificial, that is, it provides galvanic protection to the DU–0.75 Ti. The direction of galvanic current through the NaCl solution is from coating to DU–0.75 Ti alloy and as a consequence the alloy is cathodically protected. McIntyre *et al.* [5] and

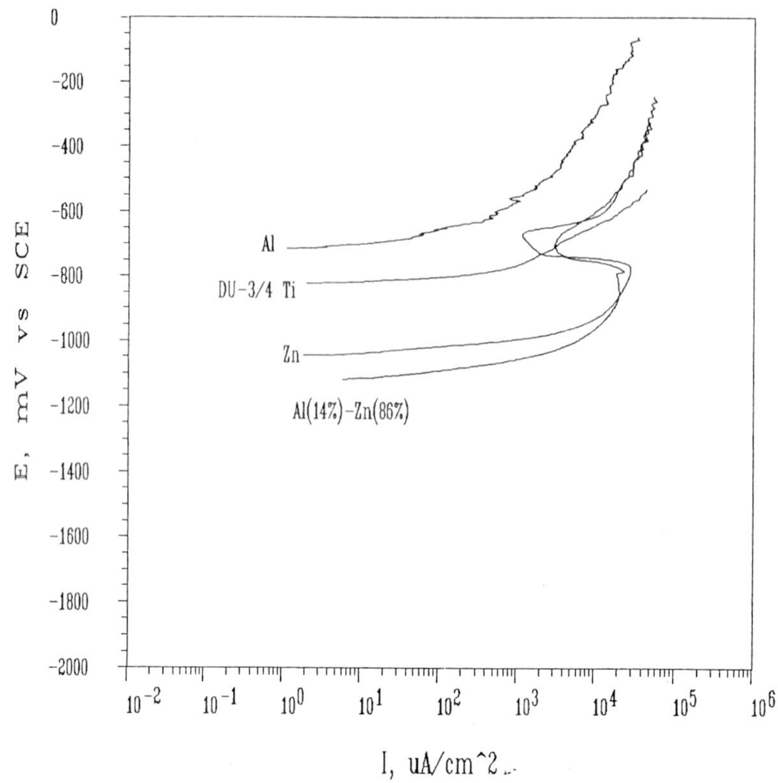


Fig. 2. Electrochemical anodic polarization curves of uncoated and coated (aluminum, zinc, Al–Zn) DU–0.75 Ti alloy in aerated 3.5 wt.% NaCl solution, 20 mV min$^{-1}$.

TABLE 2

Results from potentiodynamic polarization tests

| *Specimen* | $E_C^a$ ($V_{SCE}$) | $E_P^b$ ($V_{SCE}$) | $E_{PR}^c$ (V) |
|---|---|---|---|
| Titanium[d] | −0.27 C[g] | 0.20 | 0.47 |
| Nickel[d] | −0.39 C[g] | −0.08 | 0.31 |
| Al/TiN[e] | −0.71 C[g] | −0.60 | 0.11 |
| Aluminum[d] | −0.72 C[g] | −0.64 | 0.08 |
| TiN[f] | −0.78 C[g] | −0.78 | 0.00 |
| DU–0.75 Ti | −0.82 | −0.71 | 0.11 |
| Zinc[e] | −1.05 A[h] | −0.67 | 0.38 |
| Al–Zn[e] | −1.12 A[h] | −0.65 | 0.47 |
| Al–Mg[e] | −1.22 A[h] | −0.65 | 0.57 |
| Magnesium[e] | −1.45 A[h] | −0.75 | 0.70 |

[a]Corrosion (rest) potential at 55 min after specimen introduction to corrosion cell.
[b]Pitting potential. Designated as potential observed at onset of pitting.
[c]Protection range. Designated as the potential difference $E_P - E_C$.
[d]Film delaminated above $E_P$.
[e]Film broke and flaked off above $E_P$.
[f]Spontaneous pitting occurred upon specimen immersion in test solution; substrate visible through coating.
[g]$E_C$ is Anodic with respect to DU–0.75 Ti; coating is sacrificial.
[h]$E_C$ is Cathodic with respect to DU–0.75 Ti; coating is noble.

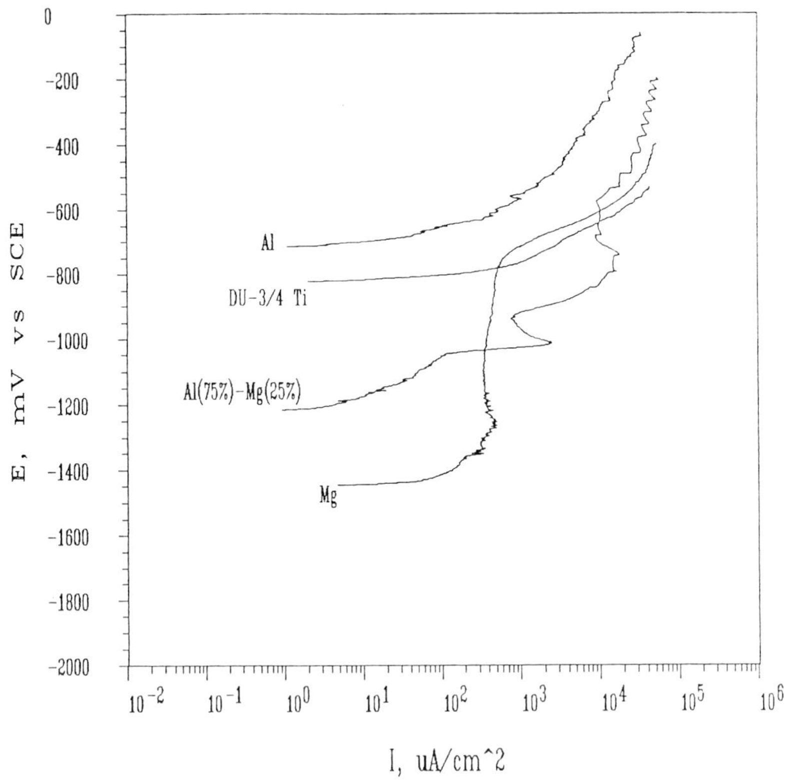


Fig. 3. Electrochemical anodic polarization curves of uncoated and coated (aluminum, magnesium, Al–Mg) DU–0.75 Ti alloy in aerated 3.5 wt.% NaCl solution, 20 mV $min^{-1}$.

Briggs [6] have also reported that zinc protective layers were active to DU–0.75 Ti and preferentially corroded at an accelerated rate. The degree of porosity of sacrificial coatings therefore is not of great significance but long term cathodic protection will depend on the thickness of the coating [7]. In comparison, the remaining coatings, i.e. aluminum, nickel, titanium and TiN, exhibit a more noble or electropositive potential with respect to the DU–0.75 Ti alloy. If any pores or cracks are present in the noble coating then the direction of galvanic current accelerates attack of the DU–0.75 Ti at these defects and eventually will undermine the coating. Weirick [8] also reported that aluminum and nickel coatings on DU–0.75 Ti alloy were noble and any flaws therein could result in an increased local corrosion rate.

The protection range of all the coatings listed in Table 2 indicates that passive behavior is observed here due to the formation of oxide films and in the case of the Al–Zn and Al–Mg coatings spinels may also form [4, 9]. Above this range pitting was observed. The sacrificial coatings Zn, Mg, Al–Zn, and Al–Mg exhibited the largest protection ranges with 14 at.% Al–86 at.% Zn offering the protection exceeding that observed by the individual films of aluminum and zinc. Nowak *et al.* [4] speculated that the excellent

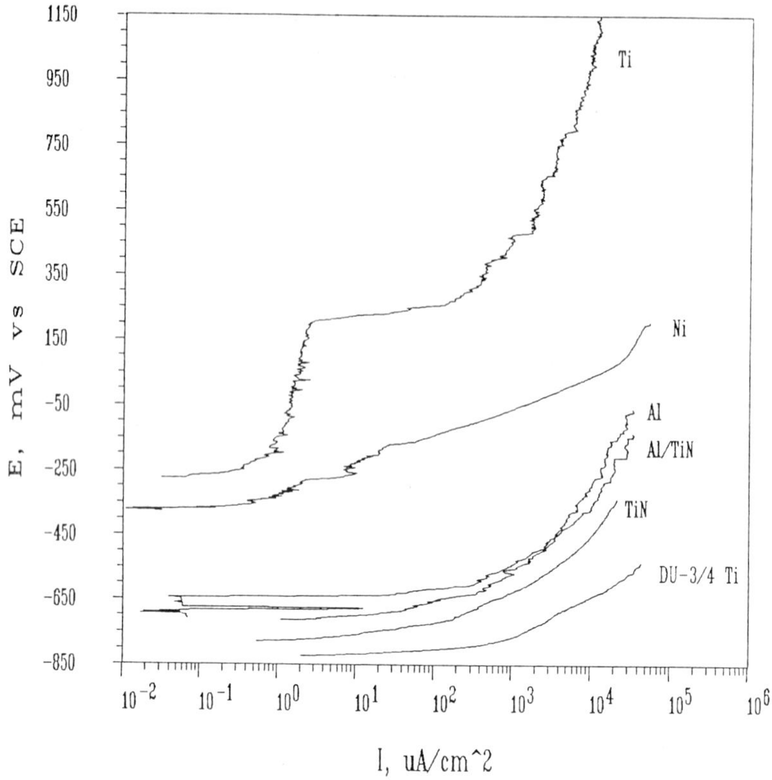


Fig. 4. Electrochemical anodic polarization curves of uncoated and coated (nickel, titanium, aluminum, TiN, Al/TiN) DU–0.75 Ti alloy in aerated 3.5 wt.% NaCl solution, 20 mV min$^{-1}$.

corrosion resistance of ion plated Al–Zn films was due to the Al–Zn alloys initially corroding preferentially at the zinc-rich sites followed by an aluminum corrosion character (formation of a hydrated $Al_2O_3$ passive layer) that delayed and was mediated by the presence of zinc. Also a very thin $Al_2ZnO_4$ spinel barrier layer may exist at the metal-oxide interface, providing enhanced passivity similar to the $Fe_3O_4$ spinel layer between iron and $Fe_2O_3$.

The 75 at.% Al–25 at.% Mg coating also exhibited a relatively large protection range. The surface protective layer could consist of a mixture of $Al_2O_3$ and MgO with different degrees of hydration or could be a spinel structure of $Al_2MgO_4$. The work of Nowak and Seyyedi [9] on ion plated Al–Mg alloy films with up to 52 at.% Mg showed that these alloys could develop the spinel structure which at higher magnesium content was an approximately stoichiometric spinel.

### *3.4. Immersion tests*

Photographs of specimens taken through the beaker walls and from above after 32 days of exposure are shown in Fig. 5(a - j). Visual observations conducted after 120 days of exposure are reported for each coating system in

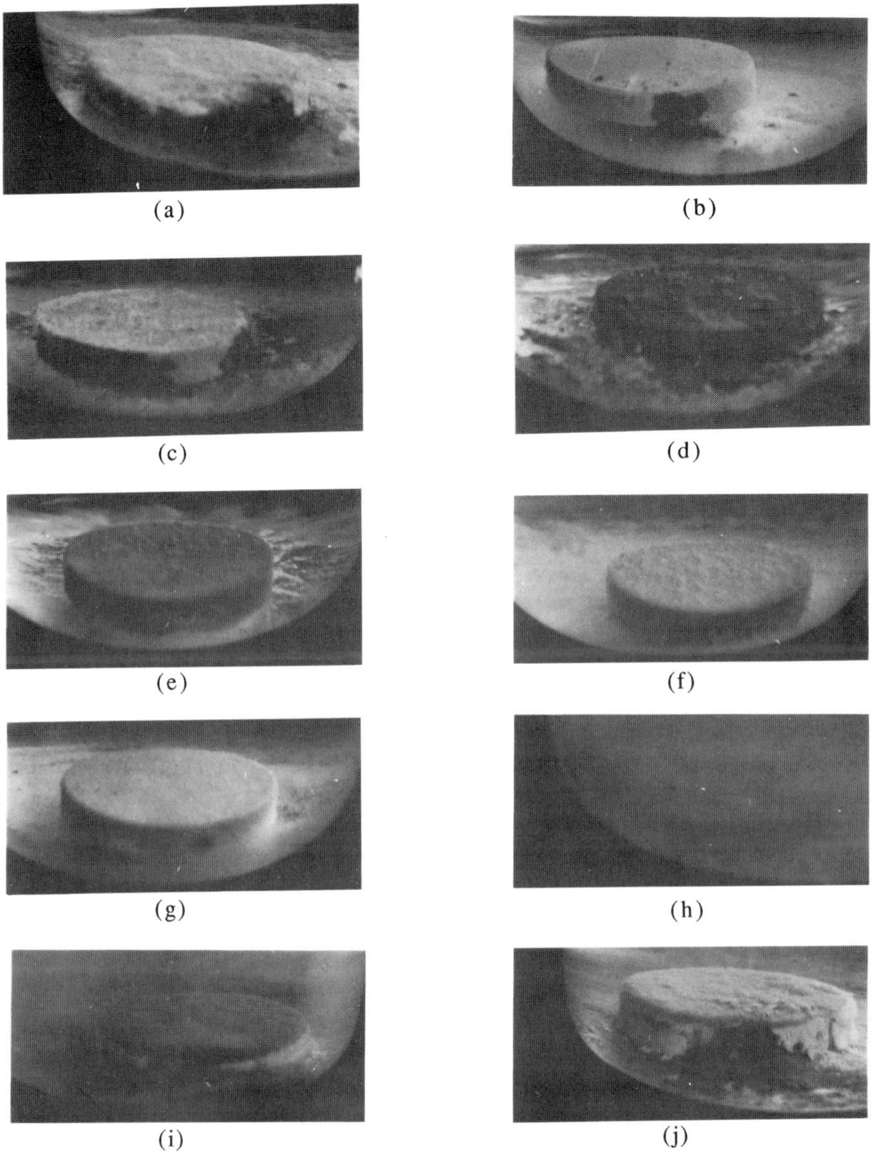


Fig. 5. Photographs of immersion test specimens 32 days after placement in aerated 3.5 wt.% NaCl solution at 25 °C: (a) DU–0.75 Ti, (b) aluminum coating, (c) zinc coating, (d) magnesium coating, (e) Al–Zn coating, (f) Al–Mg coating, (g) Al/TiN coating, (h) titanium coating, (i) nickel coating, (j) TiN coating.

Table 3. The degree of corrosion was rated on a scale of 0 (severe) to 10 (negligible). The DU–0.75 Ti alloy was considered baseline with a rating of 0.

A review of the observations in Table 3 showed that the sacrificial coatings such as Al–Zn and Al–Mg provided better corrosion protection to DU–0.75 Ti. The Al–Zn coating appeared to be the best of the sacrificial

TABLE 3

Observations after 120 days of immersion tests

| *Material* | *Rating* | *Observations* |
|---|---|---|
| DU–0.75 Ti | 0 | Black corrosion product over entire specimen ($UO_2$ or $UH_3$). |
| Aluminum | 6 | White corrosion product ($Al_2O_3$ hydrate). DU corrosion products: yellow ($UO_3$ hydrate), black ($UO_2$). No cathodic protection around defect. |
| Zinc | 4 | Black corrosion product over 75% of surface ($UO_2$), yellow deposits ($UO_3$ hydrate), some cathodic protection around defect. No evidence of white ZnO or $Zn_2OCl$. |
| Magnesium | 4 | Entire surface covered with black corrosion product ($UO_2$), some cathodic protection around defect. No evidence of white MgO hydrate. |
| Al–Zn | 10 | Some white corrosion product (mixed oxides or spinel), no evidence of DU corrosion. Cathodic protection around defect. |
| Al–Mg | 8 | Some white corrosion product (spinel), small area of black corrosion product ($UO_2$). Cathodic protection around defect. |
| Al/TiN | 6 | Same as aluminum. |
| Titanium | 2 | Black corrosion product ($UO_2$). Severe galvanic attack around defect. |
| Nickel | 2 | Black corrosion product ($UO_2$). Severe galvanic attack around defect. |
| TiN | 3 | Black corrosion product ($UO_2$), delamination of coating, galvanic attack around defect. |

coatings, forming little or no corrosion products with a clear solution. This coating afforded cathodic protection to the defect introduced into the specimen. There was no evidence of corrosion around the defect area. The Al–Mg coating provided the DU–0.75 Ti with similar cathodic protection. For the aluminum coating, the defect was corroded suggesting no cathodic protection was provided, thus its corrosion resistance should depend on the integrity of the deposited coating. This also will be true for the Al/TiN coating, but the Al/TiN coating system was more protective than the TiN coating, because of the presence of the aluminum top layer.

It is interesting to note that the zinc and magnesium coatings did not provide good corrosion protection to DU–0.75 Ti even though they are sacrificial to the substrate and exhibit a large protection range. The aluminum may extend the protective life of zinc and magnesium by moderating the rate at which each of these elements preferentially corrodes. Also, titanium performed poorly despite the relatively large protection range listed in Table 2.

## 4. Conclusions

It is known that most commerically prepared metal coatings contain porosity in some degree. The present work involving nine different coatings applied by the cathodic arc plasma PVD process also showed by SEM examination that fully dense coatings free of defects were not produced. From the corrosion standpoint, the galvanic action at the base of a defect such as a pit or crack which penetrates through the coating thickness will be an important factor in determining coating efficiency. Since the corrosion potential of the DU–0.75 Ti is relatively active, only materials such as zinc, magnesium, Al–Zn and Al–Mg may be used as sacrificial coatings. The other coatings, *i.e* nickel, titanium, aluminum and TiN, are shown to be more noble than DU–0.75 Ti. Therefore they should only be used if a defect free film can be obtained. Accordingly, the useful life of sacrificial coatings is governed by the thickness whereas that of noble coatings is controlled by the amount of defects.

Of the nine metallic coatings studied, it is apparent that the Al–Zn and Al–Mg alloy systems have the best potential for protecting DU–0.75 Ti. However, additional work is needed to optimize these coatings with respect to composition and producing a more desirable dense morphological structure which would exhibit an equiaxed grain structure.

## References


1 C. Bergman, in R. F. Hochman (ed.), *Ion Plating and Implantation Application to Materials*, American Society of Metals, Metals Park, OH, 1986, p. 115.

2 M. Levy and C. V. Zabielski, Corrosion of high density KE penetrator materials, *Proc. 1985 Tri-Service Corrosion Conf.*, pp. 137 - 164.

3 P. A. Linders, in R. F. Hochman (ed.), *Ion Plating and Implantation Application to Materials*, American Society for Metals, Metals Park, OH, 1986, p. 161.

4 W. B. Nowak, L. E. Burns and V. G. Harris, Corrosion-resistant ion-plated Al–Zn films, *J. Vac. Sci. Technol., A7* (1989) 2350 - 2354.

5 J. F. McIntyre, E. P. Lefeaue and K. A. Musselman, Galvanic corrosion behavior of depleted uranium in synthetic seawater coupled to aluminum, magnesium, and mild steel, *Corrosion, 44* (1988) 509.

6 J. L. Briggs, The corrosion resistance of zinc coatings on depleted uranium and uranium alloys, *Rockwell International RFP-3651*, January 27, 1985, p. 86.

7 H. H. Uhlig, *Corrosion and Corrosion Control*, John Wiley, New York, pp. 199 - 200.

8 L. J. Weirick, Evaluation of metallic coatings for the corrosion protection of a uranium–3/4 wt.% titanium alloy, *Sandia Laboratories Report SLL-73-5024*, February 1974.

9 W. B. Nowak and J. Seyyedi, Corrosion resistant ion-plated aluminum magnesium alloy films, *Corrosion Division Proceedings*, Vol. 84-3, Electrochemical Society, Pennington, NJ, 1984, pp. 89 - 96.

# AUTHOR INDEX

*Surface and Coatings Technology*, *39/40* (1989) 735 - 742 

# SUBJECT INDEX